Methods in Enzymology

Volume 162

IMMUNOCHEMICAL TECHNIQUES

Part L

Chemotaxis and Inflammation

METHODS IN ENZYMOLOGY

EDITORS-IN-CHIEF

John N. Abelson   Melvin I. Simon

Methods in Enzymology

Volume 162

Immunochemical Techniques

Part L

Chemotaxis and Inflammation

EDITED BY

Giovanni Di Sabato

DEPARTMENT OF MOLECULAR BIOLOGY
VANDERBILT UNIVERSITY
NASHVILLE, TENNESSEE

ACADEMIC PRESS, INC.
Harcourt Brace Jovanovich, Publishers
San Diego New York Berkeley Boston
London Sydney Tokyo Toronto

ACADEMIC PRESS, INC.
San Diego, California 92101

United Kingdom Edition published by
ACADEMIC PRESS INC, (LONDON) LTD.
24-28 Oval Road, London NW1 7DX

LIBRARY OF CONGRESS CATALOG CARD NUMBER: 54-9110

ISBN 0-12-182063-7 (alk. paper)

PRINTED IN THE UNITED STATES OF AMERICA
88 89 90 91 9 8 7 6 5 4 3 2 1

Table of Contents

Section I. Chemotaxis

A. Methods for the Study of Chemotaxis

B. Methods for the Study of Chemoattractants and Biochemistry of Chemotaxis

Secton II. Inflammation

A. Methods for the Study of the Cellular Phenomena of Inflammation and Experimental Models of Inflammation

B. Methods for the Study of Complement in Chemotaxis and Inflammation

Contributors to Volume 162

Article numbers are in parentheses following the names of contributors.
Affiliations listed are current.

GLENN C. ANDREWS (49), *Department of Molecular Genetics, Pfizer Central Research, Groton, Connecticut 06340*

DENNIS C. ARGENTIERI (31), *Division of Experimental Therapeutics, Ortho Pharmaceutical Corporation, Raritan, New Jersey 08869*

PHILIP J. BAILEY (29, 40), *Merck Sharp and Dohme Research Laboratories, Rahway, New Jersey 07065*

ASHWIN K. BHATT (18), *Division of Rheumatology, Washington University School of Medicine, St. Louis, Missouri 63110*

PIERLUIGI E. BIGAZZI (39), *Department of Pathology, University of Connecticut Health Center, Farmington, Connecticut 06032*

MARCIA L. BLIVEN (30), *Department of Immunology and Infectious Diseases, Central Research Division, Pfizer, Incorporated, Groton, Connecticut 06340*

ALAN N. BOHRER (8), *Department of Medicine, Veterans Administration Medical Center and Washington University School of Medicine, St. Louis, Missouri 63125*

MICHAEL D. P. BOYLE (10), *Department of Immunology and Medical Microbiology, College of Medicine, University of Florida, Gainesville, Florida 32610*

JAN R. BRENTJENS (41), *Department of Pathology, Microbiology, and Medicine, School of Medicine, State University of New York at Buffalo, Buffalo, New York 14214*

RICHARD A. F. CLARK (20), *Department of Medicine, National Jewish Center for Immunology and Respiratory Medicine, Denver, Colorado 80206*

ANDREY V. CYBULSKY (38), *Evans Memorial Department of Clinical Research and Department of Medicine, Boston University Medical Center, Boston, Massachusetts 02118*

SYAMAL K. DATTA (36), *Department of Medicine, New England Medical Center and Tufts University School of Medicine, Boston, Massachusetts 02111*

CARLOS C. DAUGHADAY (8), *Department of Medicine, Veterans Administration Medical Center and Washington University School of Medicine, St. Louis, Missouri 63125*

MARK J. DINUBILE (23), *Division of Infectious Diseases, Department of Medicine, Robert Wood Johnson School of Medicine, Camden, New Jersey*

ALFRED F. ESSER (45), *Department of Comparative and Experimental Pathology, University of Florida–Health Center, Gainesville, Florida 32610*

JILLIAN F. EVANS (7), *Merck Frosst Canada Incorporated, Pointe Claire-Dorval, Québec, Canada H9R 4P8*

DANIEL S. FLETCHER (40), *Merck Sharp and Dohme Research Laboratories, Rahway, New Jersey 07065*

ANTHONY W. FORD-HUTCHINSON (7), *Merck Frosst Canada Incorporated, Pointe Claire-Dorval, Québec, Canada H9R 4P8*

ARTHUR E. FRANKE (49), *Department of Molecular Genetics, Pfizer Central Research, Groton, Connecticut 06340*

RICHARD J. FREER (13), *Department of Pharmacology and Toxicology, Medical College of Virginia, Richmond, Virginia 23298*

JOHN I. GALLIN (5), *Intramural Research Program, National Institute of Allergy and Infectious Diseases, National Institutes of Health, Bethesda, Maryland 20892*

Adrian P. Gee (10), *Baxter Healthcare, Fenwal Division, Irvine, California*

Craig J. Gerard (49), *Department of Pediatrics, Harvard Medical School, Children's Hospital Medical Center, Boston, Massachusetts 02215*

Irma Gigli (47), *Division of Dermatology, University of California, San Diego, School of Medicine, San Diego, California 92103*

Frank M. Graziano (42, 43), *Department of Medicine, University of Wisconsin Hospital and Clinics, Madison, Wisconsin 53792*

Hans Gruler (11), *Department of Biophysics, University of Ulm, D7900 Ulm, Federal Republic of Germany*

Wendy S. Haston (1, 2), *Bacteriology Department, University of Aberdeen, Foresterhill, Aberdeen AB9 2ZD, United Kingdom*

Hideo Hayashi (14), *Department of Pathology, Kumamoto University Medical School, Kumamoto 862, Japan*

Michael J. Herron (4), *Department of Surgery, University of Minnesota, Minneapolis, Minnesota 55455*

Mitsuomi Hirashima (14), *Department of Pathology, Kumamoto University Medical School, Kumamoto 862, Japan*

Mitsuo Honda (14), *Department of Cellular Immunology, National Institute of Health of Japan, Tokyo 141, Japan*

Andrew C. Issekutz (27), *Departments of Pediatrics and Microbiology, Dalhousie University, Izaak Walton Killam Children's Hospital, Halifax, Nova Scotia B3J 3G9, Canada*

Thomas B. Issekutz (27), *Departments of Pediatrics and Microbiology, Dalhousie University, Izaak Walton Killam Children's Hospital, Halifax, Nova Scotia B3J 3G9, Canada*

Jarmila Janatova (46), *Department of Pathology, School of Medicine, University of Utah, Salt Lake City, Utah 84132*

Hugo E. Jasin (35), *Department of Internal Medicine, University of Texas Southwestern Medical Center at Dallas, Dallas, Texas 75235*

Suresh S. Kerwar (32, 34), *Inflammation and Immunology Research, American Cyanamid Company, Pearl River, New York 10965*

Douglas A. Lauffenburger (9), *Department of Chemical Engineering, University of Pennsylvania, Philadelphia, Pennsylvania 19104*

Michael J. P. Lawman (10), *Veterinary Infectious Disease Organization, University of Saskatchewan, Saskatoon, Canada*

Eufronio G. Maderazo (21), *Department of Medicine, Hartford Hospital, Hartford, Connecticut 06115, and Departments of Medicine and Pathology, University of Connecticut School of Medicine, Farmington, Connecticut 06032*

Wayne A. Marasco (19), *Department of Pathology, University of Michigan Medical School, Ann Arbor, Michigan 48109*

Richard A. McReynolds (31), *Department of Pathology, School of Medicine, East Carolina University, Greenville, North Carolina 27834*

Yasuhiro Mibu (14), *Department of Pathology, Kumamoto University Medical School, Kumamoto 862, Japan*

Peter F. Moore (28), *Department of Immunology and Infectious Diseases, Central Research Division, Pfizer, Incorporated, Groton, Connecticut 06340*

Natesa Muthukumaraswamy (13), *Department of Pharmacology and Toxicology, Medical College of Virginia, Richmond, Virginia 23298*

Paul H. Naccache (26), *Unité de Recherche "Inflammation et Immunologie-Rhumatologie," Centre Hospitalier de l'Université Laval, Sainte Foy, Québec G1V 4G2, Canada*

Yukifumi Nawa (16), *Department of Parasitology, Miyazaki Medical College, Miyazaki 889-16, Japan*

Robert D. Nelson (4), *Department of Sur-*

gery, University of Minnesota, Minneapolis, Minnesota 55455

BERNICE NOBLE (41), *Departments of Microbiology and Pathology, School of Medicine, State University of New York at Buffalo, Buffalo, New York 14214*

ARNOLD L. ORONSKY (32, 34), *Research and Development, American Cyanamid Company, Pearl River, New York 10965*

IVAN G. OTTERNESS (28, 30), *Department of Immunology and Infectious Diseases, Central Research Division, Pfizer, Incorporated, Groton, Connecticut 06340*

MAKOTO OWHASHI (16), *Department of Parasitology, Miyazaki Medical College, Miyazaki 889–16, Japan*

MICHAEL K. PANGBURN (48), *Department of Biochemistry, University of Texas Health Center at Tyler, Tyler, Texas 75710*

MARILYN C. PIKE (22), *Department of Internal Medicine, Massachusetts General Hospital, Boston, Massachusetts 02114*

SUSAN C. RIDGE (32, 34), *Inflammation and Immunology Research, American Cyanamid Company, Pearl River, New York 10965*

DAVID J. SALANT (38), *Evans Memorial Department of Clinical Research and Department of Medicine, Boston University Medical Center, Boston, Massachusetts 02118*

RAMADAN I. SHA'AFI (26), *Department of Physiology, University of Connecticut Health Center, Farmington, Connecticut 06032*

HENRY J. SHOWELL (49), *Department of Immunology and Infectious Diseases, Pfizer Central Research, Groton, Connecticut 06340*

CHARLES D. SMITH (24), *Department of Molecular Biology, The Burroughs-Wellcome Company, Research Triangle Park, North Carolina 27709*

JOHN A. SMITH (15), *Departments of Molecular Biology and Pathology, Massachusetts General Hospital, Boston, Massachusetts 02114*

RALPH SNYDERMAN (22, 24, 25), *Genentech, Incorporated, South San Francisco, California 94080*

JAMES M. SODETZ (45), *Department of Chemistry and School of Medicine, University of South Carolina, Columbia, South Carolina 29208*

FREDERICK S. SOUTHWICK (23), *Division of Infectious Diseases, Department of Medicine, University of Pennsylvania School of Medicine, Philadelphia, Pennsylvania 19104*

ISAIAS SPILBERG (8, 18), *Division of Rheumatology, Washington University School of Medicine, St. Louis, Missouri 63110*

NORMA P. STIMLER-GERARD (49), *Department of Medicine, Pulmonary Division, Harvard Medical School, Beth Israel Hospital, Boston, Massachusetts 02215*

ROBERT H. SWANBORG (37), *Department of Immunology and Microbiology, Wayne State University School of Medicine, Detroit, Michigan 48201*

JOEL D. TAUROG (31), *H. C. Simmons Arthritis Research Center, Department of Internal Medicine, University of Texas Southwestern Medical Center at Dallas, Dallas, Texas 75235*

FRANCISCO A. TAUSK (47), *Division of Dermatology, University of California, San Diego, School of Medicine, San Diego, California 92103*

ROBERT T. TRANQUILLO (9), *Department of Chemical Engineering and Materials Science, University of Minnesota, Minneapolis, Minnesota 55455*

MARGRITH W. VERGHESE (25), *Molecular Pharmacology, Glaxo Research Laboratories, Glaxo, Incorporated, Research Triangle Park, North Carolina 27709*

PETER A. WARD (19, 21), *Department of Pathology, University of Michigan Medical School, Ann Arbor, Michigan 48109*

NORMAN E. WIKNER (20), *Department of Medicine, National Jewish Center for Immunology and Respiratory Medicine, Denver, Colorado 80206*

PETER C. WILKINSON (1, 2, 3, 12, 17), *Department of Bacteriology and Immunology, University of Glasgow, Western Infirmary, Glasgow G11 6NT, Scotland*

PAUL H. WOOLEY (33), *Department of Medicine, Division of Rheumatology, UMDNJ-Robert Wood Johnson Medical School, New Brunswick, New Jersey 08903*

CHARLES L. WORONICK (21), *Department of Medicine, Hartford Hospital, Hartford, Connecticut 06115, and Department of Laboratory Medicine, University of Connecticut School of Medicine, Farmington, Connecticut 06032*

DANIEL G. WRIGHT (44), *Department of Hematology, Walter Reed Army Institute of Research, Walter Reed Army Medical Center, Washington, D.C. 20307*

SHUNSUKE YAMAMOTO (14), *Department of Pathology, Ohita Medical College, Ohita-ken 879-56, Japan*

MICHAEL YOUNG (10), *Department of Biochemistry and Molecular Biology, College of Medicine, University of Florida, Gainesville, Florida 32610*

JOHN B. ZABRISKIE (34), *The Rockefeller University, New York, New York 10021*

SALLY H. ZIGMOND (6, 9), *Biology Department, University of Pennsylvania, Philadelphia, Pennsylvania 19104*

Preface

Chemotaxis and inflammation, like many other biological processes, can be divided into humoral and cellular components. In this simplest sense soluble activators or mediators of host or external origin interact with cells that respond to signals received and transmitted via specific membrane receptors. The biological consequences are dramatic, and the biochemical mechanisms are complex and interrelated through a series of cascades that may involve several chemical messengers of different chemical classes. Volumes 162 and 163 of *Methods in Enzymology* cover *in vitro* and *in vivo* methodology that has been developed for the purpose of studying the biochemistry of these active humoral factors and the biology of the cells and their receptors that respond to the various signals.

Volume 162 consists of two sections. Section I is subdivided into two parts that cover techniques for studying chemotactic factors, including their isolation, characterization, synthesis in the case of active peptides, and the biochemical changes that take place in cells that respond to chemoattractants. Section II is also divided into two parts. The first several chapters deal with the various methods for studying cellular aspects of inflammation, with some emphasis given to discussions of experimental models of inflammatory disease. The last chapters cover the role in chemotaxis and inflammation of the classical and alternative complement pathways, including the individual complement components, their active fragments, and macromolecular complexes.

Volume 163 consists of three sections. Topics include a comprehensive coverage of the biochemistry and biology of individual mediators of inflammation and acute phase reactants, as well as methods for studying repair mechanisms in inflammation. In this volume special attention is given to the variety of enzymes involved in the inflammatory process, the use of specific inhibitors to study mechanisms at the molecular level, the role of oxidant-induced injury, and methods for studying growth factors that are involved in repairing damaged tissue.

The literature dealing with chemotaxis and inflammation is extensive and new techniques are constantly being developed. Therefore, some selection has been necessary to include the most commonly used and generally applicable techniques. Newer methods often involve significant modifications of established procedures, and we have tried to ensure that these innovations have been included.

The continued support of the Editors-in-Chief and our colleagues at Academic Press is gratefully acknowledged.

JOHN J. LANGONE

Giovanni Di Sabato
March 2, 1929–October 11, 1987

Giovanni Di Sabato

The scientific community lost a valued colleague and friend when Giovanni Di Sabato passed away on October 11, 1987, after an illness of many months. His enzymological studies at several institutions and his immunological work in the Department of Molecular Biology of Vanderbilt University brought him wide esteem.

Giovanni Di Sabato was born in Venice, Italy, on March 2, 1929. As a young student he worked at the Hospital of Venice, and in 1954 he was awarded an M.D. degree (magna cum laude) from the University of Padua. A series of research appointments in several countries and in several different specialty fields followed. From 1954 to 1958 he was a postdoctoral fellow with Dr. Enrico Ciaranfi at the Institute of General Pathology, University of Milan.

He joined Dr. William P. Jencks's laboratory at Brandeis University in 1959, just after the Graduate Department of Biochemistry was founded. Dr. Jencks recalls that Giovanni soon began studying the behavior of acetyl phosphate, the simplest "high-energy" compound in biology, and in 1960–1961 he reported one of the most extensive studies ever accomplished on the mechanisms of reactions of a high-energy phosphate compound. This work showed that transfer of the phosphate group is driven by a predominantly dissociative mechanism, rather than by interaction with the molecule that accepts the phosphate group. This work already reflected the excitement, thoroughness, and insight into scientific problems that characterized his later work in entirely different fields of science.

Later, as a senior research associate with Dr. Nathan O. Kaplan at Brandeis, he studied the role of sulfhydryl groups in lactic dehydrogenases. He continued his studies on enzymes during a two-year period with Dr. Martin Otteson at the Carlsberg Laboratories in Copenhagen, Denmark, and then, in 1965–1966, held the position of research assistant professor at Brandeis. Lactic dehydrogenases and complexes of NAD and pyruvate were his main research interests when he joined Vanderbilt's Department of Molecular Biology in 1966 as assistant professor. He was promoted to associate professor in 1968 and to professor of molecular biology in 1982.

In 1972, he decided to shift fields again, from the area of enzymology to the rapidly moving field of immunology. He took sabbatical leave to work as visiting professor of pathology in Professor Robert Good's outstanding immunological laboratory at the University of Minnesota School of Medicine. On his return to Vanderbilt in 1973 he and his graduate students began a series of studies which led in only two years to the discovery of a substance he named thymocyte-stimulating factor. They had initially observed the production by spleen cells of a factor that stimulated the responsiveness of thymus lymphocytes to phytohemagglutinin. After further characterization and partial purification of this stimulator, it became evident that this factor was the same as effectors simultaneously being discovered in other laboratories and assigned different names. Eventually this biological regulator was named interleukin 2. Interleukin 2 is now an important agent in the investigation of immunological phenomena and is being tested widely for the treatment of certain types of cancer. Dr. Di Sabato received wide recognition for his pioneering work that contributed to the present status of interleukin 2 as an important intermediary in biological processes. His research was supported by grants from the National Science Foundation and the National Institutes of Health.

Giovanni's training in medicine, biochemistry, and molecular biology put him in a unique position to assess and integrate the advances being made in the cellular aspects of immunology. In the Spring of 1982, Nathan O. Kaplan and Sidney P. Colowick invited him to help coedit a volume dealing with the study of lymphoid cells in the Immunochemical Technique Series of *Methods in Enzymology.* Largely because of the depth of knowledge that Giovanni brought to this topic and his desire to present a comprehensive work, coverage eventually required three volumes (Volumes 108 and 116 with coeditors and Volume 150 as sole editor). Giovanni was the driving force in organizing and editing this subseries. His coeditors and the staff of Academic Press were the beneficiaries of the energy, insight, and competence that he brought to these endeavors. He did more than his share, always with enthusiasm and good humor. This volume and Volume 163 were completed while he was terminally ill. He was justly proud of the four volumes that were published during his lifetime and would be equally proud of those published posthumously.

While Giovanni Di Sabato was especially dedicated to his research, his educational role was very important to him. He taught, with great effectiveness, Vanderbilt's courses in immunology. His graduate students responded to him with warmth and esteem. Initially he had a very European view of the role of professors, but as time passed he became more and more aware of, and involved with, the problems of his students. He also thought a great deal of his departmental colleagues. He was greatly interested in the progress of younger faculty and was concerned about the problems of his colleagues. His service as a freshman adviser and as a member of several college and university-wide committees reflected his deep interest in university affairs. He was a forthright person with standards and attitudes that he would vigorously defend. As chairman of the Department of Molecular Biology, I could always expect thoughtful and candid comments from him.

Giovanni Di Sabato was a vital, effective, and compassionate colleague for over twenty years. His associates at Vanderbilt and his many friends in the scientific community will miss him greatly. He is survived by his wife, Leida, to whom we express our sincere condolences.

OSCAR TOUSTER

METHODS IN ENZYMOLOGY

EDITED BY

Sidney P. Colowick and Nathan O. Kaplan

VANDERBILT UNIVERSITY
SCHOOL OF MEDICINE
NASHVILLE, TENNESSEE

DEPARTMENT OF CHEMISTRY
UNIVERSITY OF CALIFORNIA
AT SAN DIEGO
LA JOLLA, CALIFORNIA

METHODS IN ENZYMOLOGY

EDITORS-IN-CHIEF

Sidney P. Colowick and Nathan O. Kaplan

VOLUME VIII. Complex Carbohydrates
Edited by ELIZABETH F. NEUFELD AND VICTOR GINSBURG

VOLUME IX. Carbohydrate Metabolism
Edited by WILLIS A. WOOD

VOLUME X. Oxidation and Phosphorylation
Edited by RONALD W. ESTABROOK AND MAYNARD E. PULLMAN

VOLUME XI. Enzyme Structure
Edited by C. H. W. HIRS

VOLUME XII. Nucleic Acids (Parts A and B)
Edited by LAWRENCE GROSSMAN AND KIVIE MOLDAVE

VOLUME XIII. Citric Acid Cycle
Edited by J. M. LOWENSTEIN

VOLUME XIV. Lipids
Edited by J. M. LOWENSTEIN

VOLUME XV. Steroids and Terpenoids
Edited by RAYMOND B. CLAYTON

VOLUME XVI. Fast Reactions
Edited by KENNETH KUSTIN

VOLUME XVII. Metabolism of Amino Acids and Amines (Parts A and B)
Edited by HERBERT TABOR AND CELIA WHITE TABOR

VOLUME XVIII. Vitamins and Coenzymes (Parts A, B, and C)
Edited by DONALD B. MCCORMICK AND LEMUEL D. WRIGHT

VOLUME XIX. Proteolytic Enzymes
Edited by GERTRUDE E. PERLMANN AND LASZLO LORAND

VOLUME XX. Nucleic Acids and Protein Synthesis (Part C)
Edited by KIVIE MOLDAVE AND LAWRENCE GROSSMAN

VOLUME XXI. Nucleic Acids (Part D)
Edited by LAWRENCE GROSSMAN AND KIVIE MOLDAVE

VOLUME XXII. Enzyme Purification and Related Techniques
Edited by WILLIAM B. JAKOBY

VOLUME XXIII. Photosynthesis (Part A)
Edited by ANTHONY SAN PIETRO

VOLUME XXIV. Photosynthesis and Nitrogen Fixation (Part B)
Edited by ANTHONY SAN PIETRO

VOLUME XXV. Enzyme Structure (Part B)
Edited by C. H. W. HIRS AND SERGE N. TIMASHEFF

VOLUME XXVI. Enzyme Structure (Part C)
Edited by C. H. W. HIRS AND SERGE N. TIMASHEFF

VOLUME XXVII. Enzyme Structure (Part D)
Edited by C. H. W. HIRS AND SERGE N. TIMASHEFF

VOLUME XXVIII. Complex Carbohydrates (Part B)
Edited by VICTOR GINSBURG

VOLUME XXIX. Nucleic Acids and Protein Synthesis (Part E)
Edited by LAWRENCE GROSSMAN AND KIVIE MOLDAVE

VOLUME XXX. Nucleic Acids and Protein Synthesis (Part F)
Edited by KIVIE MOLDAVE AND LAWRENCE GROSSMAN

VOLUME XXXI. Biomembranes (Part A)
Edited by SIDNEY FLEISCHER AND LESTER PACKER

VOLUME XXXII. Biomembranes (Part B)
Edited by SIDNEY FLEISCHER AND LESTER PACKER

VOLUME XXXIII. Cumulative Subject Index Volumes I–XXX
Edited by MARTHA G. DENNIS AND EDWARD A. DENNIS

VOLUME XXXIV. Affinity Techniques (Enzyme Purification: Part B)
Edited by WILLIAM B. JAKOBY AND MEIR WILCHEK

VOLUME XXXV. Lipids (Part B)
Edited by JOHN M. LOWENSTEIN

VOLUME XXXVI. Hormone Action (Part A: Steroid Hormones)
Edited by BERT W. O'MALLEY AND JOEL G. HARDMAN

VOLUME XXXVII. Hormone Action (Part B: Peptide Hormones)
Edited by BERT W. O'MALLEY AND JOEL G. HARDMAN

VOLUME XXXVIII. Hormone Action (Part C: Cyclic Nucleotides)
Edited by JOEL G. HARDMAN AND BERT W. O'MALLEY

VOLUME XXXIX. Hormone Action (Part D: Isolated Cells, Tissues, and Organ Systems)
Edited by JOEL G. HARDMAN AND BERT W. O'MALLEY

VOLUME XL. Hormone Action (Part E: Nuclear Structure and Function)
Edited by BERT W. O'MALLEY AND JOEL G. HARDMAN

VOLUME XLI. Carbohydrate Metabolism (Part B)
Edited by W. A. WOOD

VOLUME XLII. Carbohydrate Metabolism (Part C)
Edited by W. A. WOOD

VOLUME XLIII. Antibiotics
Edited by JOHN H. HASH

VOLUME XLIV. Immobilized Enzymes
Edited by KLAUS MOSBACH

VOLUME XLV. Proteolytic Enzymes (Part B)
Edited by LASZLO LORAND

VOLUME XLVI. Affinity Labeling
Edited by WILLIAM B. JAKOBY AND MEIR WILCHEK

VOLUME XLVII. Enzyme Structure (Part E)
Edited by C. H. W. HIRS AND SERGE N. TIMASHEFF

VOLUME XLVIII. Enzyme Structure (Part F)
Edited by C. H. W. HIRS AND SERGE N. TIMASHEFF

VOLUME XLIX. Enzyme Structure (Part G)
Edited by C. H. W. HIRS AND SERGE N. TIMASHEFF

VOLUME L. Complex Carbohydrates (Part C)
Edited by VICTOR GINSBURG

VOLUME LI. Purine and Pyrimidine Nucleotide Metabolism
Edited by PATRICIA A. HOFFEE AND MARY ELLEN JONES

VOLUME LII. Biomembranes (Part C: Biological Oxidations)
Edited by SIDNEY FLEISCHER AND LESTER PACKER

VOLUME LIII. Biomembranes (Part D: Biological Oxidations)
Edited by SIDNEY FLEISCHER AND LESTER PACKER

VOLUME LIV. Biomembranes (Part E: Biological Oxidations)
Edited by SIDNEY FLEISCHER AND LESTER PACKER

VOLUME LV. Biomembranes (Part F: Bioenergetics)
Edited by SIDNEY FLEISCHER AND LESTER PACKER

VOLUME LVI. Biomembranes (Part G: Bioenergetics)
Edited by SIDNEY FLEISCHER AND LESTER PACKER

VOLUME LVII. Bioluminescence and Chemiluminescence
Edited by MARLENE A. DELUCA

VOLUME LVIII. Cell Culture
Edited by WILLIAM B. JAKOBY AND IRA PASTAN

VOLUME LIX. Nucleic Acids and Protein Synthesis (Part G)
Edited by KIVIE MOLDAVE AND LAWRENCE GROSSMAN

VOLUME LX. Nucleic Acids and Protein Synthesis (Part H)
Edited by KIVIE MOLDAVE AND LAWRENCE GROSSMAN

VOLUME 61. Enzyme Structure (Part H)
Edited by C. H. W. HIRS AND SERGE N. TIMASHEFF

VOLUME 62. Vitamins and Coenzymes (Part D)
Edited by DONALD B. MCCORMICK AND LEMUEL D. WRIGHT

VOLUME 63. Enzyme Kinetics and Mechanism (Part A: Initial Rate and Inhibitor Methods)
Edited by DANIEL L. PURICH

VOLUME 64. Enzyme Kinetics and Mechanism (Part B: Isotopic Probes and Complex Enzyme Systems)
Edited by DANIEL L. PURICH

VOLUME 65. Nucleic Acids (Part I)
Edited by LAWRENCE GROSSMAN AND KIVIE MOLDAVE

VOLUME 66. Vitamins and Coenzymes (Part E)
Edited by DONALD B. MCCORMICK AND LEMUEL D. WRIGHT

VOLUME 67. Vitamins and Coenzymes (Part F)
Edited by DONALD B. MCCORMICK AND LEMUEL D. WRIGHT

VOLUME 68. Recombinant DNA
Edited by RAY WU

VOLUME 69. Photosynthesis and Nitrogen Fixation (Part C)
Edited by ANTHONY SAN PIETRO

VOLUME 70. Immunochemical Techniques (Part A)
Edited by HELEN VAN VUNAKIS AND JOHN J. LANGONE

VOLUME 71. Lipids (Part C)
Edited by JOHN M. LOWENSTEIN

VOLUME 72. Lipids (Part D)
Edited by JOHN M. LOWENSTEIN

VOLUME 85. Structural and Contractile Proteins (Part B: The Contractile Apparatus and the Cytoskeleton)
Edited by DIXIE W. FREDERIKSEN AND LEON W. CUNNINGHAM

VOLUME 86. Prostaglandins and Arachidonate Metabolites
Edited by WILLIAM E. M. LANDS AND WILLIAM L. SMITH

VOLUME 87. Enzyme Kinetics and Mechanism (Part C: Intermediates, Stereochemistry, and Rate Studies)
Edited by DANIEL L. PURICH

VOLUME 88. Biomembranes (Part I: Visual Pigments and Purple Membranes, II)
Edited by LESTER PACKER

VOLUME 89. Carbohydrate Metabolism (Part D)
Edited by WILLIS A. WOOD

VOLUME 90. Carbohydrate Metabolism (Part E)
Edited by WILLIS A. WOOD

VOLUME 91. Enzyme Structure (Part I)
Edited by C. H. W. HIRS AND SERGE N. TIMASHEFF

VOLUME 92. Immunochemical Techniques (Part E: Monoclonal Antibodies and General Immunoassay Methods)
Edited by JOHN J. LANGONE AND HELEN VAN VUNAKIS

VOLUME 93. Immunochemical Techniques (Part F: Conventional Antibodies, Fc Receptors, and Cytotoxicity)
Edited by JOHN J. LANGONE AND HELEN VAN VUNAKIS

VOLUME 94. Polyamines
Edited by HERBERT TABOR AND CELIA WHITE TABOR

VOLUME 95. Cumulative Subject Index Volumes 61–74, 76–80
Edited by EDWARD A. DENNIS AND MARTHA G. DENNIS

VOLUME 96. Biomembranes [Part J: Membrane Biogenesis: Assembly and Targeting (General Methods; Eukaryotes)]
Edited by SIDNEY FLEISCHER AND BECCA FLEISCHER

VOLUME 97. Biomembranes [Part K: Membrane Biogenesis: Assembly and Targeting (Prokaryotes, Mitochondria, and Chloroplasts)]
Edited by SIDNEY FLEISCHER AND BECCA FLEISCHER

VOLUME 98. Biomembranes (Part L: Membrane Biogenesis: Processing and Recycling)
Edited by SIDNEY FLEISCHER AND BECCA FLEISCHER

VOLUME 99. Hormone Action (Part F: Protein Kinases)
Edited by JACKIE D. CORBIN AND JOEL G. HARDMAN

VOLUME 100. Recombinant DNA (Part B)
Edited by RAY WU, LAWRENCE GROSSMAN, AND KIVIE MOLDAVE

VOLUME 101. Recombinant DNA (Part C)
Edited by RAY WU, LAWRENCE GROSSMAN, AND KIVIE MOLDAVE

VOLUME 102. Hormone Action (Part G: Calmodulin and Calcium-Binding Proteins)
Edited by ANTHONY R. MEANS AND BERT W. O'MALLEY

VOLUME 103. Hormone Action (Part H: Neuroendocrine Peptides)
Edited by P. MICHAEL CONN

VOLUME 104. Enzyme Purification and Related Techniques (Part C)
Edited by WILLIAM B. JAKOBY

VOLUME 105. Oxygen Radicals in Biological Systems
Edited by LESTER PACKER

VOLUME 106. Posttranslational Modifications (Part A)
Edited by FINN WOLD AND KIVIE MOLDAVE

VOLUME 107. Posttranslational Modifications (Part B)
Edited by FINN WOLD AND KIVIE MOLDAVE

VOLUME 108. Immunochemical Techniques (Part G: Separation and Characterization of Lymphoid Cells)
Edited by GIOVANNI DI SABATO, JOHN J. LANGONE, AND HELEN VAN VUNAKIS

VOLUME 109. Hormone Action (Part I: Peptide Hormones)
Edited by LUTZ BIRNBAUMER AND BERT W. O'MALLEY

VOLUME 110. Steroids and Isoprenoids (Part A)
Edited by JOHN H. LAW AND HANS C. RILLING

VOLUME 111. Steroids and Isoprenoids (Part B)
Edited by JOHN H. LAW AND HANS C. RILLING

VOLUME 112. Drug and Enzyme Targeting (Part A)
Edited by KENNETH J. WIDDER AND RALPH GREEN

VOLUME 113. Glutamate, Glutamine, Glutathione, and Related Compounds
Edited by ALTON MEISTER

VOLUME 114. Diffraction Methods for Biological Macromolecules (Part A)
Edited by HAROLD W. WYCKOFF, C. H. W. HIRS, AND SERGE N. TIMASHEFF

VOLUME 115. Diffraction Methods for Biological Macromolecules (Part B)
Edited by HAROLD W. WYCKOFF, C. H. W. HIRS, AND SERGE N. TIMASHEFF

VOLUME 116. Immunochemical Techniques (Part H: Effectors and Mediators of Lymphoid Cell Functions)
Edited by GIOVANNI DI SABATO, JOHN J. LANGONE, AND HELEN VAN VUNAKIS

VOLUME 117. Enzyme Structure (Part J)
Edited by C. H. W. HIRS AND SERGE N. TIMASHEFF

VOLUME 118. Plant Molecular Biology
Edited by ARTHUR WEISSBACH AND HERBERT WEISSBACH

VOLUME 119. Interferons (Part C)
Edited by SIDNEY PESTKA

VOLUME 120. Cumulative Subject Index Volumes 81–94, 96–101

VOLUME 121. Immunochemical Techniques (Part I: Hybridoma Technology and Monoclonal Antibodies)
Edited by JOHN J. LANGONE AND HELEN VAN VUNAKIS

VOLUME 122. Vitamins and Coenzymes (Part G)
Edited by FRANK CHYTIL AND DONALD B. MCCORMICK

VOLUME 123. Vitamins and Coenzymes (Part H)
Edited by FRANK CHYTIL AND DONALD B. MCCORMICK

VOLUME 124. Hormone Action (Part J: Neuroendocrine Peptides)
Edited by P. MICHAEL CONN

VOLUME 125. Biomembranes (Part M: Transport in Bacteria, Mitochondria, and Chloroplasts: General Approaches and Transport Systems)
Edited by SIDNEY FLEISCHER AND BECCA FLEISCHER

VOLUME 126. Biomembranes (Part N: Transport in Bacteria, Mitochondria, and Chloroplasts: Protonmotive Force)
Edited by SIDNEY FLEISCHER AND BECCA FLEISCHER

VOLUME 127. Biomembranes (Part O: Protons and Water: Structure and Translocation)
Edited by LESTER PACKER

VOLUME 128. Plasma Lipoproteins (Part A: Preparation, Structure, and Molecular Biology)
Edited by JERE P. SEGREST AND JOHN J. ALBERS

VOLUME 129. Plasma Lipoproteins (Part B: Characterization, Cell Biology, and Metabolism)
Edited by JOHN J. ALBERS AND JERE P. SEGREST

VOLUME 130. Enzyme Structure (Part K)
Edited by C. H. W. HIRS AND SERGE N. TIMASHEFF

VOLUME 131. Enzyme Structure (Part L)
Edited by C. H. W. HIRS AND SERGE N. TIMASHEFF

VOLUME 132. Immunochemical Techniques (Part J: Phagocytosis and Cell-Mediated Cytotoxicity)
Edited by GIOVANNI DI SABATO AND JOHANNES EVERSE

VOLUME 133. Bioluminescence and Chemiluminescence (Part B)
Edited by MARLENE DELUCA AND WILLIAM D. MCELROY

VOLUME 147. Peptide Growth Factors (Part B)
Edited by DAVID BARNES AND DAVID A. SIRBASKU

VOLUME 148. Plant Cell Membranes
Edited by LESTER PACKER AND ROLAND DOUCE

VOLUME 149. Drug and Enzyme Targeting (Part B)
Edited by RALPH GREEN AND KENNETH J. WIDDER

VOLUME 150. Immunochemical Techniques (Part K: *In Vitro* Models of B and T Cell Functions and Lymphoid Cell Receptors)
Edited by GIOVANNI DI SABATO

VOLUME 151. Molecular Genetics of Mammalian Cells
Edited by MICHAEL M. GOTTESMAN

VOLUME 152. Guide to Molecular Cloning Techniques
Edited by SHELBY L. BERGER AND ALAN R. KIMMEL

VOLUME 153. Recombinant DNA (Part D)
Edited by RAY WU AND LAWRENCE GROSSMAN

VOLUME 154. Recombinant DNA (Part E)
Edited by RAY WU AND LAWRENCE GROSSMAN

VOLUME 155. Recombinant DNA (Part F)
Edited by RAY WU

VOLUME 156. Biomembranes (Part P: ATP-Driven Pumps and Related Transport: The Na,K-Pump)
Edited by SIDNEY FLEISCHER AND BECCA FLEISCHER

VOLUME 157. Biomembranes (Part Q: ATP-Driven Pumps and Related Transport: Calcium, Proton, and Potassium Pumps) (in preparation)
Edited by SIDNEY FLEISCHER AND BECCA FLEISCHER

VOLUME 158. Metalloproteins (Part A)
Edited by JAMES F. RIORDAN AND BERT L. VALLEE

VOLUME 159. Initiation and Termination of Cyclic Nucleotide Action
Edited by JACKIE D. CORBIN AND ROGER A. JOHNSON

VOLUME 160. Biomass (Part A: Cellulose and Hemicellulose)
Edited by WILLIS A. WOOD AND SCOTT T. KELLOGG

VOLUME 161. Biomass (Part B: Lignin, Pectin, and Chitin) (in preparation)
Edited by WILLIS A. WOOD AND SCOTT T. KELLOGG

VOLUME 162. Immunochemical Techniques (Part L: Chemotaxis and Inflammation)
Edited by GIOVANNI DI SABATO

VOLUME 163. Immunochemical Techniques (Part M: Chemotaxis and Inflammation) (in preparation)
Edited by GIOVANNI DI SABATO

VOLUME 164. Ribosomes (in preparation)
Edited by HARRY F. NOLLER, JR. AND KIVIE MOLDAVE

VOLUME 165. Microbial Toxins: Tools for Enzymology (in preparation)
Edited by SIDNEY HARSHMAN

VOLUME 166. Branched-Chain Amino Acids (in preparation)
Edited by ROBERT A. HARRIS AND JOHN R. SOKATCH

VOLUME 167. Cyanobacteria (in preparation)
Edited by LESTER PACKER AND ALEXANDER N. GLAZER

VOLUME 168. Hormone Action (Part K: Neuroendocrine Peptides) (in preparation)
Edited by P. MICHAEL CONN

VOLUME 169. Platelets: Receptors, Adhesion, Secretion (Part A) (in preparation)
Edited by JACEK HAWIGER

Section I

Chemotaxis

A. Methods for the Study of Chemotaxis
Articles 1 through 11

B. Methods for the Study of Chemoattractants and Biochemistry of Chemotaxis
Articles 12 through 26

[1] Chemotaxis: An Overview

By PETER C. WILKINSON and WENDY S. HASTON

Chemotaxis, the directional locomotion of organisms in relation to sources of chemical attractants or repellants (attractants usually in the best-studied examples), is a phenomenon of importance in widely diverse fields of biology. It was first described 100 years ago by Pfeffer[1] in bracken fern sperm, has been documented in many organisms since, and has recently come to be understood at the level of cellular biochemistry in three research fields. Probably the most advanced of these is the study of chemotaxis in bacteria, especially enterobacteria, in which the possibilities for genetic manipulation have allowed a detailed dissection of the chemoreceptors and their associated transduction apparatus. The two best-studied examples in eukaryotes are the slime molds and the leukocytes of mammalian blood, with which this book is concerned.

The great attraction of this area of research for many biologists is that it allows the possibility of understanding the sensory system of a single cell in great depth and therefore may have lessons for the much more daunting problems of sensory neurobiology in multicellular systems. An added incentive for studying leukocytes especially is that chemotaxis is a major mechanism (though maybe not the only one) by which the cells that defend the body against infection home in onto sites of microbial growth and of tissue injury. Research in this field is helped by the fact that the neutrophils (polymorphonuclear leukocytes, PMN), which show good chemotactic responses and are the most numerous of leukocytes in blood, are easy to obtain in high yield and purity. These cells appear to be homogeneous inasmuch as the majority of neutrophils behave alike when stimulated with chemotactic factors. Heterogeneity, if it exists, has not been a major problem. This is not the case for other leukocytes such as lymphocytes or mononuclear phagocytes, which also show chemotactic responses that are still poorly understood partly because the cell populations are heterogeneous.

Is All Leukocyte Locomotion Stimulated?

Chemotaxis is a particular form of locomotor behavior, and before discussing it we ought to ask a basic question about locomotion itself. Is all

[1] W. Pfeffer, *Unters. Bot. Inst. Tubingen* **1,** 363 (1884).

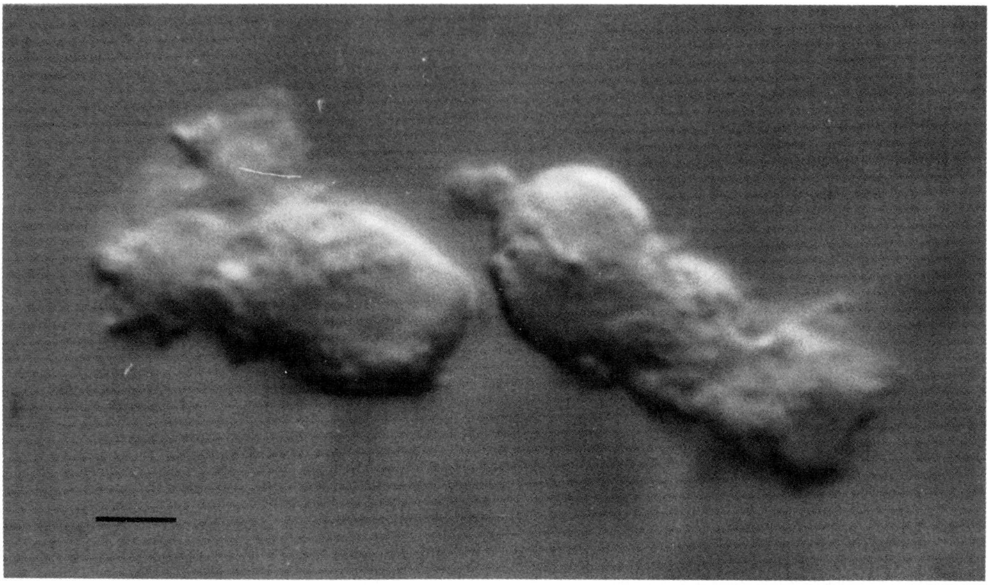

FIG. 1. Locomotor morphology in neutrophils. The flattened, ruffled, leading edge (lamellipodium) of each cell is facing away from the center of the picture. A constriction ring is present behind the lamellipodium of the cell to the left, and two constriction rings are seen in the cell to the right. These move from front to back as the cells move forward. The cell bodies are more compact and sharper in outline than the lamellipodia. A tail, seen in the cell to the right, may form when a contraction wave has moved all the way down the cell and is characteristic of but by no means universal in moving cells. Nomarski optics. Bar, 2 μm.

cell locomotion—and not just chemotaxis—signaled from outside the cell? Do leukocytes which are not receiving signals from the environment move around at random, possibly in response to internal signals, or do such unstimulated leukocytes remain completely immobile? Studies from this laboratory suggest that the latter is more likely to be correct.[2,3] When neutrophils are prepared carefully from blood, centrifuged at low speed, not allowed to rest as pellets for any length of time, and not cooled to 4°, and the cells are examined in suspension in the absence of an attractant and without being allowed to adhere to a surface, almost the entire population remains spherical and immotile. On addition of any of a range of chemotactic factors (those studied include formylmethionylleucylphenylalanine, C5a, leukotriene B_4, and denatured albumin), the cells, still in

[2] J. M. Shields and W. S. Haston, *J. Cell Sci.* **74,** 75 (1985).

[3] W. S. Haston and J. M. Shields, *J. Immunol. Methods* **81,** 229 (1985).

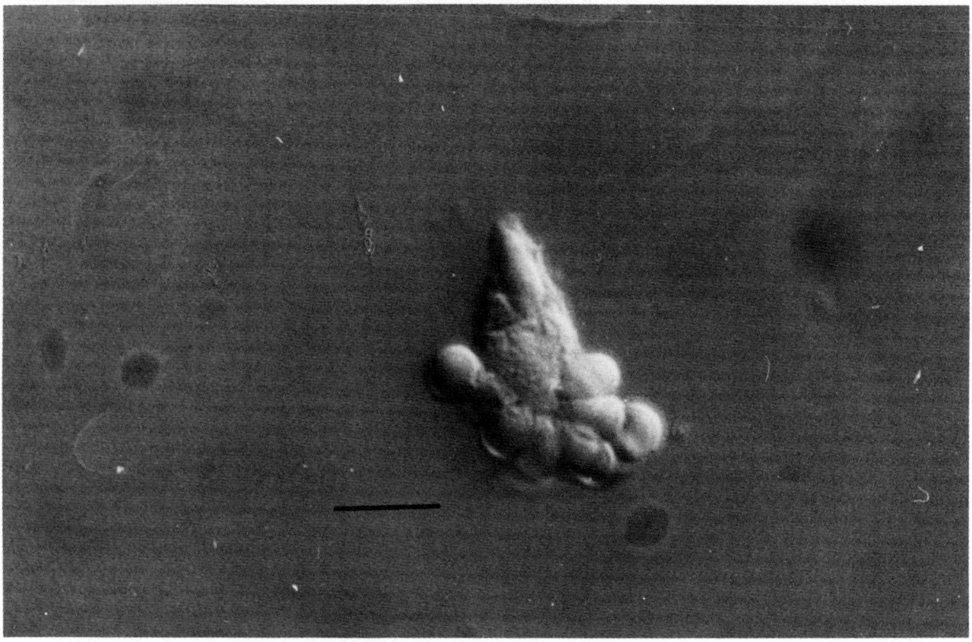

FIG. 2. Forward redistribution of Fc receptors in a polarized neutrophil. Polarization was induced by incubation with a chemotactic factor (no gradient) at 37°. The cells were then chilled to 4° and incubated with sheep erythrocytes coated with IgG–anti-sheep antibody. The Fc rosette has formed at the leading edge. Nomarski optics. Bar, 10 μm.

suspension, begin to change shape within 30 sec. Within 2–3 min, polarization of the cells into an anterior lamellipodium and posterior tail is evident. Polarization is accompanied by the appearance of contraction waves which move down the cell from front to back. These take about 30 sec to traverse the cell. Often they are easily visible as constriction rings running anteroposteriorly[4-6] (Fig. 1). All these events are evident in cells in suspension. If the cell is allowed to adhere to, or to gain traction from, a surface these events are translated into locomotion. Then the contraction waves can be seen in time-lapse films to remain stationary relative to the substratum, and the cell moves forward, almost as if it were moving through the contraction wave.

[4] W. H. Lewis, *Bull. Johns Hopkins Hosp.* **49,** 29 (1931).

[5] H. Senda, H. Tamura, N. Shibata, J. Yoshitake, K. Kondo, and K. Tanaka, *Exp. Cell Res.* **91,** 393 (1975).

[6] W. S. Haston and J. M. Shields, *J. Cell Sci.* **68,** 227 (1984).

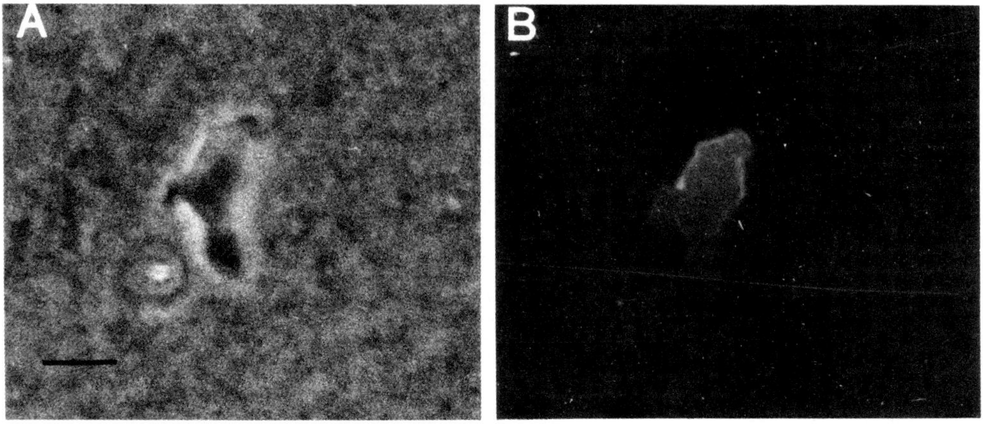

FIG. 3. Forward distribution of the Thy-1 antigen in a mouse lymphocyte migrating within a collagen gel. (A) Phase constrast; (B) fluorescence microscopy of the same cell. The gel with its contained cells was fixed with paraformaldehyde, and FITC–anti-Thy-1 monoclonal antibody was added. The cell is moving toward the top of the picture, and Thy-1 is present on the front but not the back of the cell. Bar, 4 μm. From Haston and Shields.[6]

Polarization is accompanied by receptor redistribution. This was first studied looking at Fc or C3b receptor distribution in neutrophils in fMet-Leu-Phe (fMLP)[6a] gradients. These receptors move to the front end of the cell[7,8] (Figs. 2 and 3). More interestingly the presence of a gradient is unnecessary for this. Receptors also redistribute to the front end of neutrophils in suspension in isotropic concentrations of chemotactic factors,[2] and the formyl peptide receptor itself has been shown to redistribute toward the front end of neutrophils during polarization.[9] Note that this receptor redistribution is *not* ligand specific. Stimulation with fMLP causes forward movement of Fc receptors as well as its own receptors,[6–8] and stimulation with leukotriene B_4 (LTB_4) causes forward distribution of fMLP receptors.[9] Polarization and receptor redistribution are the earliest visible responses of a leukocyte to signals from ligands in its environment, and these are entirely independent of the presence of a *gradient* of the ligand. Thus *polarization and locomotion, rather than chemotaxis or chemokinesis, are the primary responses of leukocytes to external signals.* The form the locomotion takes, e.g., chemokinesis or chemotaxis, depends on the dispo-

[6a] Abbreviations: fMLP, formyl-Met-Leu-Phe; LTB_4, leukotriene B_4; BSA, bovine serum albumin; HEVs, high endothelial venules.

[7] P. C. Wilkinson, J. Michl, and S. C. Silverstein, *Cell Biol. Int. Rep.* **4,** 736 (1980).

[8] R. J. Walter, R. D. Berlin, and J. M. Oliver, *Nature (London)* **286,** 724 (1980).

[9] S. J. Sullivan, G. Daukas, and S. H. Zigmond, *J. Cell Biol.* **99,** 1461 (1984).

sition of the attractant. If this is the case, then the classification of locomotor attractants as chemotactic factors, chemokinetic factors, and so on may not be of primary importance, since the primary effect of these factors is to stimulate locomotion. Inherent in this may be the fact that they induce receptor redistribution to the front of the cell, a phenomenon which is probably of considerable importance for directional locomotion in gradients.

A question to which there is not a detailed answer is how, when an attractant is added to leukocytes in isotropic concentration, so that attractant molecules are binding to the cell surface on all parts of the cell, does the cell manage to polarize in a particular direction. Possibly the cell responds to the first hit (signaled event) by redistribution of receptors to the point of hit and by polarization with this point as the front of the cell. This would be a stochastic process which could only work if the attractant concentration were low enough for the first hit to be distinguished from later hits. Dose–response studies suggest that this is indeed so. Studies of the efficiency of polarization of neutrophils in different concentrations of fMLP[2] suggest that at low concentrations ($<10^{-9}$ M) only a minority of cells polarize, but these polarize well. As the concentration is raised to an optimum (fMLP 10^{-8} M), the proportion of responding cells rises to almost 100% of well-polarized cells. At supraoptimal concentrations (fMLP 10^{-6} M) the cells protrude more than one pseudopod and therefore polarize poorly. These findings are compatible with the idea that at suboptimal and optimal concentrations of ligand the cells can distinguish a first hit from subsequent hits, but that once the ligand concentration is raised above optimum receptors become occupied simultaneously at many points on the cell surface and the cell is unable to redistribute receptors and to show a clear anteroposterior polarization.

Locomotor Reactions of Leukocytes

As discussed above, chemotaxis and chemokinesis may be regarded as locomotor reactions that cells show to particular environments rather than as inherent properties of attractant molecules. It is clear that a single molecular attractant can cause leukocytes to show different locomotor reactions depending on whether the attractant is present in isotropic or anisotropic, fixed or changing concentration. The locomotor reactions (Fig. 4) can be subdivided into those in which random locomotion is altered, but in a way that still allows it to be described as random (these are the *kineses*), and those in which the form of locomotion is changed from random to directional (these include the *taxes*). Random locomotion is actually not a simple concept. Since this volume is about locomotor re-

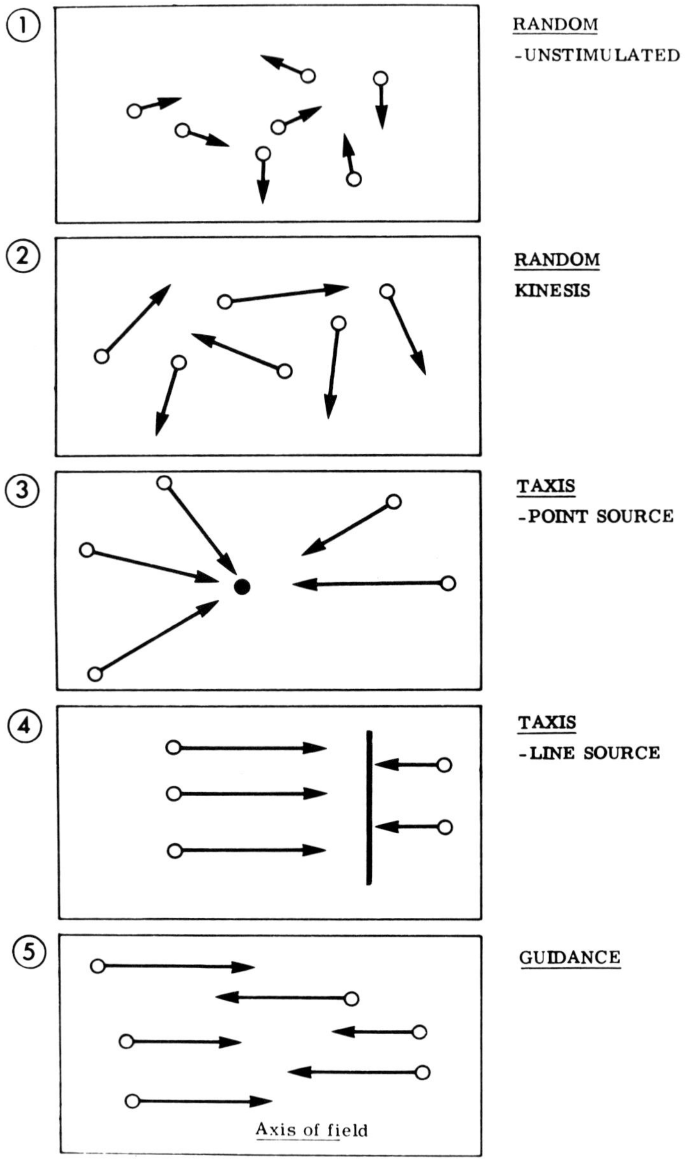

FIG. 4. Schematic diagrams to show locomotor responses of cells in different environments. The arrows indicate cell displacements (velocity and direction), not detailed cell tracks. Diagrams 1 and 2 show random locomotion, which in 2 has been accelerated by positive orthokinesis; 3 and 4 show directional locomotion toward different types of chemotactic gradient source; and 5 shows the influence of contact guidance in an aligned field. See text for details. From P. C. Wilkinson, J. M. Lackie, and R. B. Allan, *in* "Cell Analysis" (N. Catsimpoolas, ed.), p. 145. Plenum, New York, 1982.

sponses rather than about locomotor theory, I shall not discuss it here. It has been defined elsewhere,[10] and there are several useful discussions of how cells move at random.[11-14]

Kineses

The term *chemokinesis* refers to a kinesis in which the cell responds to a chemical agent (cf. photokinesis, etc.). Two forms of chemokinesis can be distinguished as follows:

1. *Orthokinesis,* in which the rate of locomotion is altered by a chemical substance, e.g., stationary cells are induced to start moving; cells moving slowly in random paths are induced to move faster but still in random paths; or, conversely, moving cells may be induced to stop or slow down, a process that has been described as "negative orthokinesis." [10]
2. *Klinokinesis,* in which the frequency or rate with which the cell turns is altered (cf. tumbling in bacteria).

Obviously, other factors being equal, a population of cells moving randomly in straight runs will show greater displacement than a population which turns frequently. Klinokinesis has not been easy to demonstrate in leukocytes. "Turning" in these cells is complex, and they often change direction by taking serpentine, curved paths rather than by making discrete, measurable turns. Also the effect of chemical agents on turning behavior has not been defined clearly yet.[15]

Orthokinesis is easy to measure using time-lapse cinematography, and it is not difficult to show in appropriate experimental conditions, using protein-coated substrata which permit cell movement, that factors such as fMLP or C5a in uniform concentration can mobilize neutrophils and cause them to move rapidly in random paths. This process on its own would be insufficient to allow accumulation of cells, since there is no directional cue. However, orthokinesis alone *may* lead to cell accumulation, but only providing that the orthokinetic agent is anisotropically disposed. We made a study of neutrophils moving randomly in a microscope field on one side of which they were moving on a coat of bovine

[10] H. U. Keller, P. C. Wilkinson, M. Abercrombie, E. L. Becker, J. G. Hirsch, M. E. Miller, W. S. Ramsey, and S. H. Zigmond, *Clin. Exp. Immunol.* **27,** 377 (1977).

[11] M. H. Gail and C. W. Boone, *Biophys. J.* **10,** 980 (1970).

[12] G. A. Dunn, *in* "Biology of the Chemotactic Response" (J. M. Lackie and P. C. Wilkinson, eds.), p. 1. Cambridge Univ. Press, Cambridge, England, 1981.

[13] G. A. Dunn, *Agents Actions Suppl.* **12,** 14 (1983).

[14] P. C. Wilkinson, J. M. Lackie, J. V. Forrester, and G. A. Dunn, *J. Cell Biol.* **99,** 1761 (1984).

[15] H. U. Keller, G. Meier, and A. Zimmermann, *Blood Cells* **10,** 505 (1984).

serum albumin (BSA) alone, on the other side of which they were moving on BSA–anti-BSA immune complexes.[14] The cells moved rapidly on BSA but become slowed on the immune complexes, by binding of substratum-linked IgG to their Fc receptors. The result was a gradual accumulation of randomly moving neutrophils on the immune complexes by adhesive trapping. The point is worth making that chemotaxis is not the only mechanism for cell accumulation, nor are gradients always necessary. For example, cells moving at random may meet, and adhere preferentially to, other cells, and it may not be necessary to invoke chemotactic mechanisms for all forms of cell clustering, e.g., in immune responses. The mechanisms for such phenomena remain, in most cases, unexplored.

Chemotaxis

Chemotaxis is a reaction in which the direction of locomotion of cells is determined by a chemical substance. A substance which, if present in isotropic, unchanging concentration, stimulates an increase in the rate of random locomotion can also, if present in anisotropic or changing concentration, stimulate a change from random to directional locomotion. In time-lapse films, this is seen as an effect on the direction of turning. Cells which, when moving at random, move in sinuous paths with frequent changes of direction, will, when exposed to a chemotactic gradient, move in straight paths toward the gradient source with few changes of direction, these being corrected if away from the gradient source.[16] Chemotactic direction-finding in neutrophils is extremely efficient, and cells in good gradients move in near straight-line paths to the gradient source.[16,17]

The method for gradient perception and for this very efficient directional response has interested many workers, and evidence for various proposals for spatial and temporal mechanisms for gradient perception by neutrophils has been put forward.[16,18,19] On the basis that cell polarization and receptor redistribution occur in stimulated cells quite independently of whether or not a gradient of the stimulant is present, it may not be necessary to postulate that neutrophils detect gradients by either mechanism.[2] If polarization takes place in the direction of the first hit, then in a uniform attractant concentration, the direction of polarization of the different cells in a population will be random. In a gradient, however, the chance that the first hit will be on the side of the cell facing the gradient

[16] S. H. Zigmond, *Nature (London)* **249,** 450 (1974).
[17] R. B. Allan and P. C. Wilkinson, *Exp. Cell Res.* **111,** 191 (1978).
[18] S. H. Zigmond, H. I. Levitsky, and B. J. Kreel, *J. Cell Biol.* **89,** 585 (1981).
[19] G. Gerisch and H. U. Keller, *J. Cell Sci.* **52,** 1 (1981).

source will be high, and the steeper the gradient and the greater the affinity of the attractant for its receptor, the higher the chance will be that cells will polarize toward, and move toward, the gradient source. Since the majority of receptors are now at the front of the cell, this will increase the chance that further signals will be recognized there and that the cell will continue to move forward toward the source.

The evidence about receptor redistribution in polarized leukocytes is still very scanty, but it is probable that not all receptors redistribute to the anterior pole and that some remain on other parts of the cell.[9] This may be a dynamic process with receptors in continued movement but with some mechanism favoring distribution to the front of the cell (e.g., attachment to the cytoskeleton or membrane recycling—but there is no good evidence for either). The presence of some receptors at the rear would seem necessary to explain the responses of cells already polarized by gradients in one direction, when exposed to new gradients in which the attractant concentration is highest at the side or tail of the cell as in the studies with micropipets by Gerisch and Keller.[19] These cells sometimes make U turns toward the new gradient as might be expected of a cell whose sensory apparatus is concentrated at its front end, but they also sometimes put out a new pseudopod from the side or tail toward the new source. This would certainly require the presence of stimulable receptors at sites other than the front of the cell. The idea that neutrophils respond to the first received signal by receptor redistribution, polarization, and subsequent movement in the direction of that pole of the cell at which the signal was received in some ways makes movement up gradients easier to understand, since, instead of complex computation of differences in receptor occupancy at different sites, required for spatial gradient sensing, or a cellular memory required for temporal sensing, it requires simply that the cell polarizes and moves in the direction in which probability dictates that it is most likely to be stimulated, and so long as the stimulus continues to be maximal at the front the cell goes on moving toward it.

Another aspect of gradient detection that requires exploration is how a cell detects a gradient in the face of random fluctuations in ligand concentration and in receptor binding. The idea that, in the absence of a mean gradient, fluctuations in *perceived* gradients (i.e., noise) are detected and account for chemokinetic motility, whereas chemotactic motility results from perceived fluctuations around a mean gradient, has recently been proposed.[19a,19b]

[19a] R. T. Tranquillo and D. A. Lauffenburger, *J. Math. Biol.* **25,** 229 (1987).

[19b] R. T. Tranquillo, B. E. Farrell, E. S. Fisher, and D. A. Lauffenburger, *Math. Biosci.* in press (1987).

Mechanisms for Cell Accumulation

The leukocytes are the major cellular defense mechanism following infection or tissue injury, and their locomotion serves to allow them to leave the blood and to accumulate in the lesions of infection or tissue injury in large numbers. Obviously chemotaxis is an excellent mechanism for directing cells into such lesions, but even if we understood leukocyte chemotaxis fully, there would still be many other questions to ask about how cells accumulate *in vivo.* The neutrophils in blood spend part of the time circulating (a passive process) and part of the time as marginated cells, attached to vascular endothelium, usually in sites of slow blood flow. Under physiological conditions, margination is transient, and the stationary cells rejoin the circulating pool. However, in postcapillary venules near inflamed sites, margination is increased, and the marginated cells leave the vessel by crawling through gaps between endothelial cells and across the vascular basement membrane into the tissue. There has been considerable recent progress in the study of leukocyte–endothelial interactions. It is clear that molecules of the LFA-1, CR3 (Mac-1), gp150:95 series are important leukocyte-surface adhesive proteins with a role in binding blood leukocytes to endothelium.[20,21] Leukocytes from patients with a deficiency of these proteins adhere poorly to endothelium.[20] Leukocyte-surface LFA-1 class molecules are believed to bind specifically to ICAM-1, a molecule found on endothelial and other cells[22] which is inducible by treatment with lipopolysaccharide, tumor necrosis factor, γ-interferon, or interleukin 1. This binding is an important determinant of leukocyte–endothelial adhesion. In addition, *in vitro,* neutrophils can be induced to cross layers of vascular endothelium in response to chemotactic factors placed on the other side.[23,24]

The influence of the patterning and architecture of tissues on the migration of leukocytes has been little studied. Three-dimensional fibrous gels of collagen or fibrin can be used to provide a simple model to study how the alignment of connective tissues influences leukocyte locomotion.

[20] J. M. Harlan, P. D. Killen, F. M. Senecal, B. R. Schwartz, E. K. Yee, R. F. Taylor, P. G. Beatty, T. H. Price, and H. D. Ochs, *Blood* **66,** 167 (1985).

[21] D. Haskard, D. Cavender, P. Beatty, T. Springer, and M. Ziff, *J. Immunol.* **137,** 2901 (1986).

[22] M. L. Dustin, R. Rothlein, A. K. Bhan, C. A. Dinarello, and T. A. Springer, *J. Immunol.* **137,** 245 (1986).

[23] E. B. Cramer, L. C. Milks, and G. K. Ojakian, *Proc. Natl. Acad. Sci. U.S.A.* **77,** 4069 (1980).

[24] R. G. Russo, L. A. Liotta, U. Thorgeirsson, R. Brundage, and E. Schiffmann, *J. Cell Biol.* **91,** 459 (1981).

When such gels are stretched, the protein fibrils line up parallel to one another. Studies of neutrophils migrating through aligned 3-dimensional collagen or fibrin gels show that locomotion in the axis of alignment is preferred to locomotion in other axes.[25] This is known as *contact guidance,* a response that has been described in a number of cell types. Various mechanisms, discussed by Dunn,[26] have been proposed to explain it, but it is still not well understood. The term contact guidance covers locomotor responses in which the direction of locomotion is determined by the shape or curvature of the substratum through or on which the cell is moving. Contact guidance is thus a form of directional locomotion, but not one which, on its own, causes cells to accumulate at a single locus. This is because, though the cells are constrained to move in one axis, they are free to move in *both* directions in that axis. The ability of cells moving through aligned gels to show chemotactic responses to gradients is influenced by the alignment of the gel. Much better chemotactic responses are seen when the axis of alignment of the fibers (the guidance field) is parallel to the axis of the chemotactic field than when the two fields are at right angles to one another, in which case the cells are receiving two conflicting directional cues.[27] This is a fairly simple example of an effect of tissue patterning on the direction of cell locomotion. It is very likely that in tissues which are arranged in more complex patterns, such as exist *in vivo,* the success or failure of cell accumulation may depend in part on the constraints imposed on cell locomotion by tissue architecture.

Another factor which may influence cell accumulation is cell adhesion. Neutrophils and mononuclear phagocytes adhere to substrata and gain traction for locomotion from these adhesive contacts. The LFA-1, Mac-1, gp150:95 family of molecules is certainly important for these contacts, but other adhesive mechanisms probably also contribute. There is presumably an optimum interaction with the substratum at which forces of attachment and detachment are balanced so as to allow optimal locomotion, and above or below this optimum locomotion is less efficient. However, this is a very difficult field in which to get precise information. Chemotactic factors produce transient changes in cell-to-cell adhesion over periods of a few minutes[28,29] and may produce complex changes in attachment of

[25] P. C. Wilkinson, J. M. Shields, and W. S. Haston, *Exp. Cell Res.* **140,** 55 (1982).

[26] G. A. Dunn, *in* "Cell Behaviour" (R. Bellairs, A. S. G. Curtis, and G. A. Dunn, eds.), p. 247. Cambridge Univ. Press, Cambridge, England, 1982.

[27] P. C. Wilkinson, and J. M. Lackie, *Exp. Cell Res.* **145,** 255 (1983).

[28] P. R. Craddock, D. Hammerschmidt, J. G. White, A. P. Dalmasso, and H. S. Jacob, *J. Clin. Invest.* **60,** 260 (1977).

[29] J. T. O'Flaherty, D. L. Kreutzer, and P. A. Ward, *J. Immunol.* **119,** 232 (1977).

leukocytes to substrata over longer periods of up to 1 hr.[30] As mentioned earlier, changes in adhesion, e.g., adhesive trapping, can cause leukocytes to accumulate even in the absence of chemotactic gradients.

Mononuclear Phagocytes and Lymphocytes

Studies of locomotion and chemotaxis in mononuclear phagocytes and lymphocytes have advanced much more slowly than those in neutrophils, though the importance of these cells in chronic inflammation and in immune reactions demands that these studies be done. Heterogeneity is a major problem with both cell types. It is clear that the blood monocytes, the most readily accessible of mononuclear phagocytes for study, may undergo differentiation along a number of pathways to become macrophages with different functional properties and membrane markers. The locomotor capacity of monocytes has probably been studied in more detail than that of the different types of macrophage. Monocytes have receptors for and respond to well-defined neutrophil chemotactic factors such as fMLP, C5a, or LTB_4, though only about 60% of human blood monocytes possess receptors for such factors.[31,32] Receptors may be slowly inducible if the cells are stimulated to differentiate, as suggested by studies of the human monocyte line U937 cultured in the presence of a lymphokine.[33] Macrophages taken from the peritoneal cavity of the mouse 4 days after an eliciting inflammatory stimulus are more motile than resident peritoneal macrophages, a higher proportion ($>70\%$ as opposed to $<20\%$) can be stimulated into locomotion, and the inflammatory macrophages respond more vigorously to a chemotactic gradient from *Candida albicans* spores in serum than the resident macrophages.[34] Possibly macrophages newly derived from monocytes retain locomotor capacity, but prolonged residence in the unstimulated peritoneal cavity causes the cells to lose locomotor capacity. The differences in locomotor behavior among different inflammatory macrophages, immunologically activated macrophages, etc. remain to be defined.

When lymphocytes are considered, there would appear to be at least two important migration pathways *in vivo* to which active locomotion contributes, the first being the recirculation of lymphocytes and their active

[30] R. P. C. Smith, J. M. Lackie, and P. C. Wilkinson, *Exp. Cell Res.* **122,** 169 (1979).

[31] G. J. Cianciolo and R. Snyderman, *J. Clin. Invest.* **67,** 60 (1981).

[32] W. Falk and E. J. Leonard, *Infect. Immun.* **29,** 953 (1980).

[33] M. C. Pike, D. Fischer, H. S. Koren, and R. Snyderman, *J. Exp. Med.* **152,** 31 (1980).

[34] P. C. Wilkinson, *Immunobiology* **161,** 376 (1982).

emigration from blood across the high endothelial venules (HEVs) into lymphoid tissue, which is a property of small, unprimed lymphocytes, and the second being the recruitment of lymphocytes into sites of immune inflammation. A number of studies (reviewed in Ref. 35) suggest that the population of lymphocytes which leaves the circulation to migrate into inflammatory sites is not the recirculating population and that the lymphocytes which migrate best into inflammatory sites are those which have been activated by priming with antigen or by other means. Possibly the ligands involved in recirculating across HEVs are different from those responsible for attracting lymphocytes into inflammatory sites. Using monoclonal antibodies, separate lymphocyte surface molecules can be defined that mediate adhesion to peripheral lymph node HEVs on the one hand and to Peyer's patch HEVs on the other in rodents.[36–38] LFA-1 mediates binding of lymphocytes to nonspecialized endothelium[21] and possibly to HEVs as well. Many authors have reported that only a minority of lymphocytes taken directly from peripheral blood has locomotor capacity, but that this proportion can be increased by culture *in vitro* especially in the presence of mitogens.[35] Lymphoblasts are highly motile. Studies with activators such as phytohemagglutinin (PHA), purified protein derivative (PPD), or anti-CD3 antibody suggest that lymphocyte locomotor capacity is enhanced by cell growth and that large lymphocytes which are active in RNA and protein synthesis are more motile than small cells showing lower rates of synthesis.[39,40]

What does all this mean for the understanding of chemotaxis in lymphocytes? Lymphocytes can certainly be stimulated to change from a round to a polarized form by materials such as fetal calf serum or cell culture supernates, but the factors that induce this change are not defined biochemically. The ligands that stimulate locomotion in lymphocytes may be different from those in neutrophils, and classic neutrophil chemotactic factors such as fMLP, C5a, and LTB_4 have little effect on blood lymphocytes. It is possible that the requirement for cell growth is actually a requirement for receptor synthesis, or that lymphocytes respond best to

[35] D. M. V. Parrott and P. C. Wilkinson, *Prog. Allergy* **28,** 193 (1981).

[36] Y.-H. Chin, R. Rasmussen, A. G. Cakiroglu, and J. J. Woodruff, *J. Immunol.* **133,** 2961 (1984).

[37] Y.-H. Chin, R. A. Rasmussen, J. J. Woodruff, and T. G. Easton, *J. Immunol.* **136,** 2556 (1986).

[38] S. Jalkanen, R. A. Reichert, W. M. Gallatin, R. F. Bargatze, I. L. Weissman, and E. C. Butcher, *Immunol Rev.* **91,** 39 (1986).

[39] P. C. Wilkinson, *Immunology* **57,** 281 (1986).

[40] P. C. Wilkinson and A. Higgins, *Immunology* **60,** 445 (1987).

factors such as products released following cell activation or growth,[41-43] or that they are stimulated by cell–cell contact. The heterogeneity of lymphocytes also raises the possibility that different subsets respond to different ligands. It is tempting to believe that the morphological events that occur during immune induction, such as lymphocyte clustering around accessory or antigen-presenting cells, may result from chemotactic signaling by the accessory cells, but there is no good evidence for this and, as discussed above, alternative explanations are possible. Studies of these and similar events have certainly been hampered by the difficulty of finding suitable assay systems for studying lymphocyte behavior. Lymphocytes are too poorly adhesive for their locomotion to be studied on the plane substrata that are used for other cells. However, in 3-dimensional environments such as collagen gels lymphocytes show vigorous locomotion by a mechanism that is independent of adhesion to a substratum,[44,45] and it may be possible to use collagen gels to study how lymphocytes respond to contact with, or to factors released from, accessory cells or other lymphocytes under different conditions of activation.

In this chapter, the idea has been emphasized that the basic response of leukocytes to stimulation is locomotion itself rather than any particular form of locomotor behavior (chemotaxis, chemokinesis, etc.) and that these behavioral responses reflect the environmental disposition of the ligand (uniform or changing concentration). Assays for studying leukocyte locomotion must be adequate to measure the effects of ligands on locomotion and on locomotor behavioral responses accurately and quantitatively and to distinguish the behavioral responses from one another. Locomotion is inherently difficult to measure, and there is no one assay which gives full information about all its different aspects. Another difficulty is that not only are chemotactic factors stimulants of locomotion but they also stimulate a whole range of nonlocomotor cellular functions, such as secretion and oxidative metabolism. These may not be linked to the locomotor response either operationally or in the biochemical events involved. If binding of a chemotactic factor to its receptor activates several responses, the mechanisms for stimulus–response coupling may be different in each case. At present the coupling mechanism for locomotion is not clearly distinguished from the others, nor is there much information to suggest how locomotion is sustained and controlled over long time periods.

[41] D. E. Van Epps, J. W. Potter, and D. A. Durant, *J. Immunol.* **130,** 2727 (1983).
[42] D. M. Center and W. Cruikshank, *J. Immunol.* **128,** 2563 (1982).
[43] J. W. Potter and D. E. Van Epps, *Cell. Immunol.* **105,** 9 (1987).
[44] W. S. Haston, J. M. Shields, and P. C. Wilkinson, *J. Cell Biol.* **92,** 747 (1982).
[45] S. L. Schor, T. D. Allen, and B. Winn, *J. Cell Biol.* **96,** 1089 (1983).

[2] Visual Methods for Measuring Leukocyte Locomotion

By WENDY S. HASTON and PETER C. WILKINSON

Introduction

Chemotaxis can be measured using a number of techniques based on different principles. For example, micropore filter assays and under-agarose assays measure the net population movements resulting from a chemotactic response but give little information about the detailed behavior of the individual cells that make up those populations. To study this behavior, methods are needed which allow small populations or single cells to be followed by direct observation. These are the visual assays which are the subject of this chapter.

We begin by describing a useful, simple, and accurate assay of the initial locomotor events that are activated by adding chemotactic factors to leukocytes. These events result in a change in shape from a spherical to a polarized, locomotor, morphology. The proportion of polarized cells, the degree to which they become polarized, and the time course of polarization can all be measured. The polarization assay is useful, but it does not give any information about the locomotor events which follow polarization. To measure chemotaxis we need to know about the direction of cell movement, the speed of cells, the way they turn, and so on. In this chapter, we describe visual assays of locomotion, using 3-dimensional physiological matrices such as collagen or fibrin or studying the detailed tracks made by leukocytes migrating on 2-dimensional substrata toward sources of chemotactic gradients.

Polarization Assay

All assays currently in use for chemotaxis measure the response of cells to gradients of factors. This is not very surprising as chemotaxis is defined as the directional locomotion of a cell toward (or away from in negative chemotaxis) a gradient source. Several groups have reported that neutrophil leukocytes respond to uniform concentrations of chemotactic factors by taking up a polarized morphology,[1,2] and it has also been demonstrated

[1] S. H. Zigmond, H. I. Levitsky, and B. J. Kreel, *J. Cell Biol.* **89**, 585 (1981).
[2] C. W. Smith, J. C. Hollers, R. A. Patrick, and C. Hassett, *J. Clin. Invest.* **63**, 221 (1979).

that neutrophils become equally well polarized when floating in suspension.[3,4]

These observations suggest that neither the presence of a gradient nor adhesion to a substratum plays any part in the initial response of neutrophils to chemotactic factors. Neutrophil leukocytes purified from peripheral blood using a standardized procedure have no intrinsic polarity or motility, and 95% remain spherical even after incubation for 2 hr or more at 37° in Hanks' balanced salt solution buffered with morpholinopropanesulfonic acid (HBSS/MOPS). This is in contrast to peritoneal neutrophils (e.g., from the rabbit) which are generally elicited by the administration of thioglycolate or glycogen intraperitoneally 4 hr before collection of the peritoneal exudate. Neutrophils obtained by this method can be 100% polarized, presumably because they have already responded *in vivo* to a chemotactic factor. Storage of the cells at 4° reduces the number of polarized cells, but on rewarming to 37° they once more become polarized. Incubation of neutrophils (from blood or peritoneal exudate) at 4° disassembles microtubules which can lead to morphological changes resembling polarization.

We have proposed that, in leukocytes, polarity is determined by the direction of waves of contraction which are sometimes visible as refractile bands in the cytoplasm of moving leukocytes or as constriction rings in cells fixed while moving or floating in suspension.[4] Contraction waves determine the direction of polarity, i.e., they move from the front to the back of the cell in the direction opposite to that of locomotion, and they provide the locomotive force for translocation when the cell is moving over a substratum.[5,6] Contraction waves, resulting in morphological polarization, are stimulated by chemotactic factors, and therefore very reproducible polarizing responses to known chemotactic factors can be obtained in nongradient conditions. As the assay can be performed equally well on cells in suspension[7] it is possible to separate direct polarizing responses, to known or supposed chemotactic factors, from indirect effects which can result from interactions among the factor, the substratum, and the cell.

Procedure

The assay depends on obtaining a starting population of round unstimulated cells. In the case of human blood neutrophils it is possible to obtain,

[3] H. U. Keller and H. Cottier, *Cell Biol. Int. Rep.* **5**, 3 (1981).

[4] J. M. Shields and W. S. Haston, *J. Cell Sci.* **74**, 75 (1985).

[5] H. Senda, H. Tamura, N. Shibata, J. Yoshitake, K. Kondo, and K. Tanaka, *Exp. Cell Res.* **70**, 129 (1975).

[6] W. S. Haston and J. M. Shields, *J. Cell Sci.* **68**, 227 (1984).

[7] W. S. Haston and J. M. Shields, *J. Immunol. Methods* **81**, 229 (1985).

routinely, a cell suspension of >95% round cells using a standard separation procedure as described by Haston and Shields.[7] Two volumes of heparinized blood are mixed with 1 volume of 6% dextran and left to sediment for 40 min. The supernatant is removed, layered onto Ficoll/Hypaque lymphocyte separation mixture (Pharmacia; specific gravity = 1.078, ratio 7 ml plasma/3 ml separating mixture), and centrifuged at 400 *g* for 30 min. Conical bottomed tissue culture plastic tubes are used throughout as prolonged contact with glass causes (1) adhesion and loss of cells and (2) release of endogenous chemotactic factors leading to a high background of polarized cells.

After centrifugation, the cell pellet containing neutrophils and red cells is removed to a fresh tube and washed once with HBSS/MOPS. [The composition of HBSS/MOPS is as follows. Hanks' balanced salt solution: $CaCl_2$, 185.5 mg/liter; KCl, 400 mg/liter; KH_2PO_4, 60 mg/liter; $Mg_2SO_4 \cdot 7H_2O$, 200 mg/liter; NaCl, 8000 mg/liter; NaH_2PO_4, 47.5 mg/liter; D-glucose, 1000 mg/liter; phenol red sodium salt, 17 mg/liter. This can be obtained from Flow Laboratories (McLean, VA; Rickmansworth, UK) as either a 10× or a 1× solution, to which a buffer must be added. MOPS (10 m*M*) or HEPES (10 m*M*) may be used. The pH is adjusted to 7.2 with NaOH.] The red cells are lysed by resuspending the pellet in 1 ml distilled water and mixing for 45 sec. After lysis 9 ml HBSS/MOPS is added to the tube, and the neutrophils are washed 2 times more in HBSS/MOPS. If, after this processing, more than 10% of the neutrophils are polarized then further washes should reduce the numbers of polarized cells. Final cell preparations should be >95% spherical. Keeping the neutrophils pelleted or cooling to 4° always increases the background numbers of polarized cells and reduces the sensitivity of the assay.

For assays, the neutrophils are adjusted to 1×10^6 cells/ml HBSS/MOPS to which the factor to be tested is added. The tubes are incubated for 30 min at 37°, and then equal volumes of 2.5% glutaraldehyde in HBSS/MOPS are added to each tube to fix the cells. After 5 min the fixed cells are washed 2 times in HBSS/MOPS and resuspended in the remaining drop (about 200 μl) of medium. The fixed cells can be stored for several days at 4° if necessary. The cells are examined by phase microscopy (×40), and any cell which is longer in one axis than the other and with an identifiable front and back is counted as polarized. Figure 1b shows the results of incubating neutrophils in an optimal dose of the peptide formylmethionylleucylphenylalanine (fMLP) for 30 min. Clearly the cells are polarized compared with Fig. 1a (no fMLP). Different degrees of polarity can be assessed by simple measurements of cell length either directly or by drawing outlines using a camera lucida attachment for the microscope followed by manual measurement or by computer-assisted image analysis.

a

b

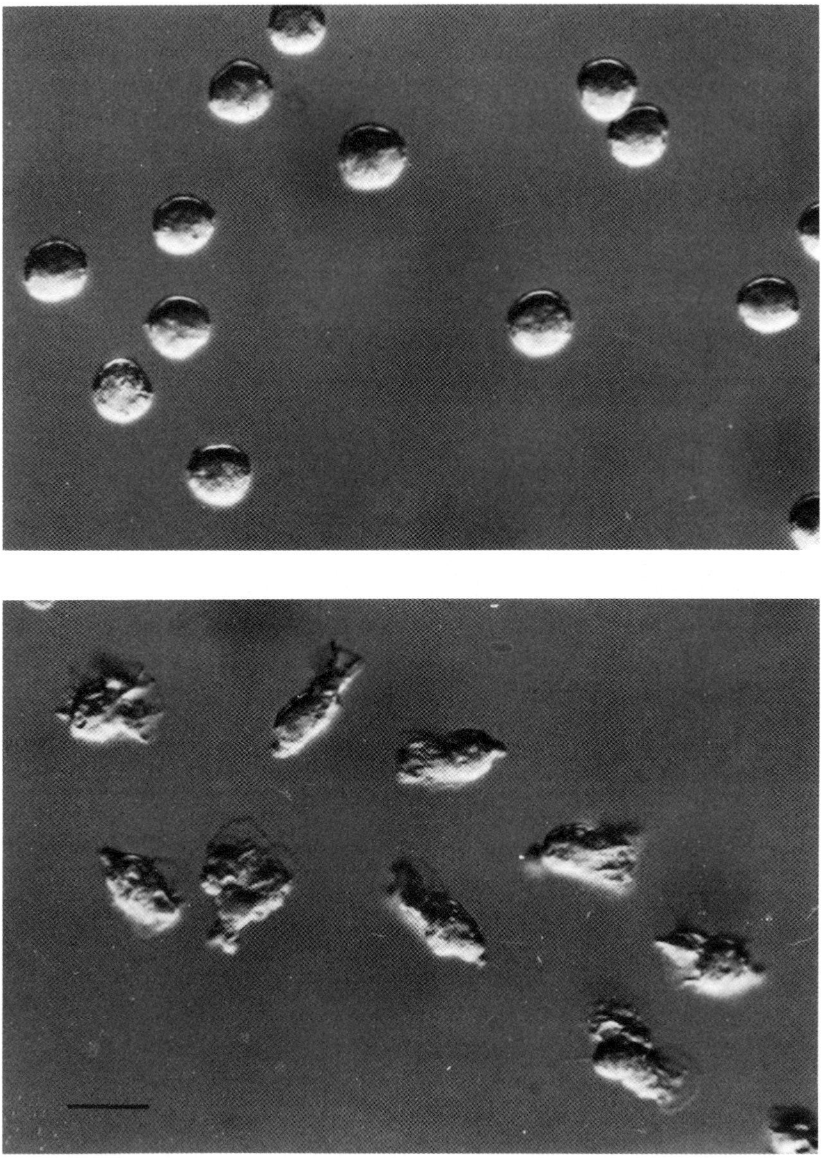

FIG. 1. (a) Neutrophils incubated in suspension for 30 min in HBSS/MOPS and then fixed with 2.5% glutaraldehyde. (b) Neutrophils incubated for 30 min in HBSS/MOPS plus 1×10^{-8} *M* fMLP and then fixed with 2.5% glutaraldehyde. Bar, 10 μm.

Dunn and Brown[7a] have recently described a sophisticated analysis of cell shape, to which the interested reader is referred, that has proved useful in our own studies of cell polarization. A time course of the response of neutrophils to fMLP is shown in Fig. 2 which illustrates the development of polarity.

Advantages of the Polarization Assay

Advantages of the polarization assay are that large numbers of potential factors can be screened for stimulating activity, which can then be followed up to see if the factor induces directional locomotion in gradient conditions. It also provides a rapid assessment of the immediate locomotor response to factors which can then be manipulated to determine the effects of various inhibitors, channel blockers, agents which disrupt the cytoskeleton, ion fluxes, etc. The effects of this type of manipulation on the locomotor component of the response to chemotactic factors have been neglected because other responses, e.g., the respiratory burst or degranulation, can be measured immediately. Previously used assay systems for the locomotor response to chemotactic factors involve long incubation periods, and the substratum dependence of the assays introduces many other variables which cannot be controlled easily.

One disadvantage of the polarization assay is that certain compounds induce polarization but do not cause chemoattraction. For example, microtubule-disassociating agents such as colchicine induce morphological changes which resemble polarization (although the time course is different),[8] but other assay systems indicate that they are not chemoattractants. High extracellular potassium concentrations (125 m*M*) which, like fMLP, cause membrane depolarization,[9] also cause morphological polarization, but K^+ ions cannot be regarded as chemoattractants. Certain other agents have a marked effect on neutrophil morphology in that they cause significant membrane ruffling, e.g., the calcium ionophore A23187 and the tumor-promoting phorbol esters,[9a] but they fail to induce polarization.

In general, we have found that factors which induce a chemotactic response in micropore filters induce significant polarization in suspension. The assay is, therefore, very useful for rapid screening of factors and cells. The polarization assay may be equally useful for investigating the immediate responsiveness of lymphocytes to potential stimulating factors. However, few well-defined chemotactic factors are generally available for lym-

[7a] G. A. Dunn and A. F. Brown, *J. Cell Sci.* **83,** 313 (1986).

[8] H. U. Keller, A. Naef, and A. Zimmermann, *Exp. Cell Res.* **153,** 173 (1984).

[9] R. L. Roberts, N. L. Mounessa, and J. I. Gallin, *J. Immunol.* **132,** 2000 (1984).

[9a] F. J. Roos, A. Zimmermann, and H. U. Keller, *J. Cell Sci.* **88,** 399 (1987).

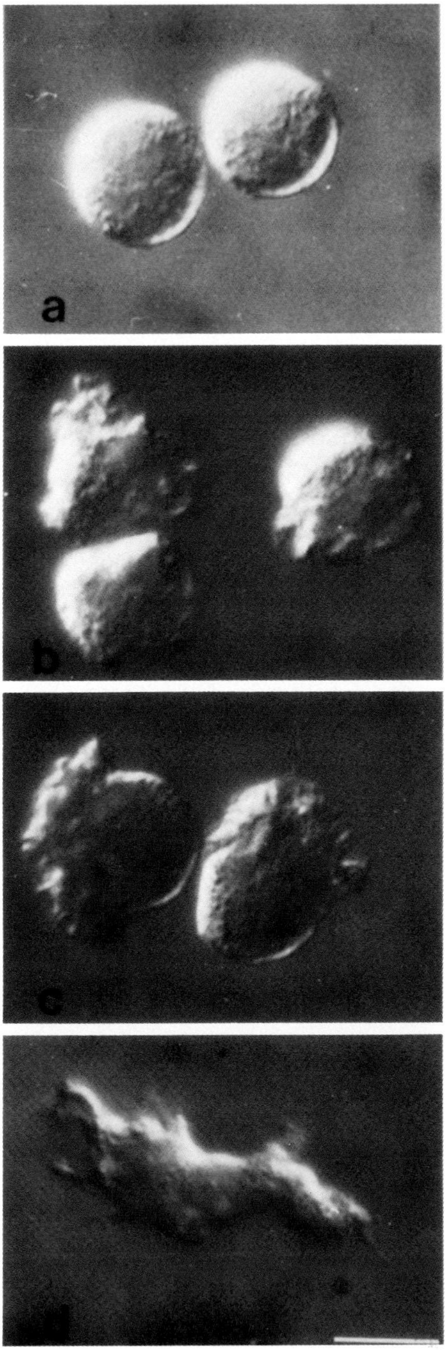

FIG. 2. Micrographs a, b, c, and d show neutrophils fixed after incubation in suspension at 37° in 1×10^{-8} M fMLP for 0, 30 sec, 2 min, and 30 min, respectively. Bar, 10 μm. From W. S. Haston and J. M. Shields, *J. Cell Sci.* **82,** 249 (1986).

phocytes. In spite of this, the assay has been used successfully to quantify the relationship between the acquisition of locomotor morphology and activation of growth *in vitro*.[10]

Collagen and Fibrin Gel Assays

By observing the behavior of individual cells using time-lapse cinematography the effects of defined compounds on cell speed, turning behavior, and on general morphology can be analyzed in considerable detail. The disadvantage is that only small numbers of cells in a population can be analyzed. Methods which use transparent, 3-dimensional gels of collagen or fibrin allow for the analysis of individual cell responses but can also be adapted for population measurements and therefore provide the maximum information.

The collagen or fibrin gel assay is similar to the micropore filter assay in that leukocytes are allowed to migrate into the 3-dimensional fibrous matrix of the gel in response to a diffusing gradient of attractant. The advantages are that the gel matrix is transparent so the path of individual cells can be followed, but after an appropriate time interval the gel can be fixed with glutaraldehyde and the net population movement assessed either by using the leading-front measurement (see chapter [3]) or by counting the number of cells at intervals through the gel. Migration into the gel by leukocytes is independent of direct adhesive interaction with the collagen fibers (Ref. 10 and Fig. 3) so that no protein need be present in the media to reduce the adhesive interaction between the cells and the matrix. Figure 4 shows a lymphocyte moving through a collagen gel with both a lateral bleb and a constriction ring, which are consistent features of lymphocyte locomotion.

Preparation of Collagen and Fibrin Gels

Stock solutions of type I collagen can be prepared from rat tail tendons using a method described by Elsdale and Bard.[11] These tendons run anteroposteriorly down the tail and are readily identifiable once the skin is removed. The tail tendons from three rats are removed, washed in phosphate-buffered saline and dissolved in 400 ml 3% acetic acid. The resultant solution is centrifuged to remove gross debris, and the collagen is precipitated with 2 volumes of 20% NaCl, redissolved in 3% acetic acid and centrifuged once more, precipitated in NaCl, and dissolved finally in 3%

[10] P. C. Wilkinson, *Immunology* **57**, 281, (1986).

[11] T. Elsdale and J. Bard, *J. Cell Biol.* **54**, 626 (1972).

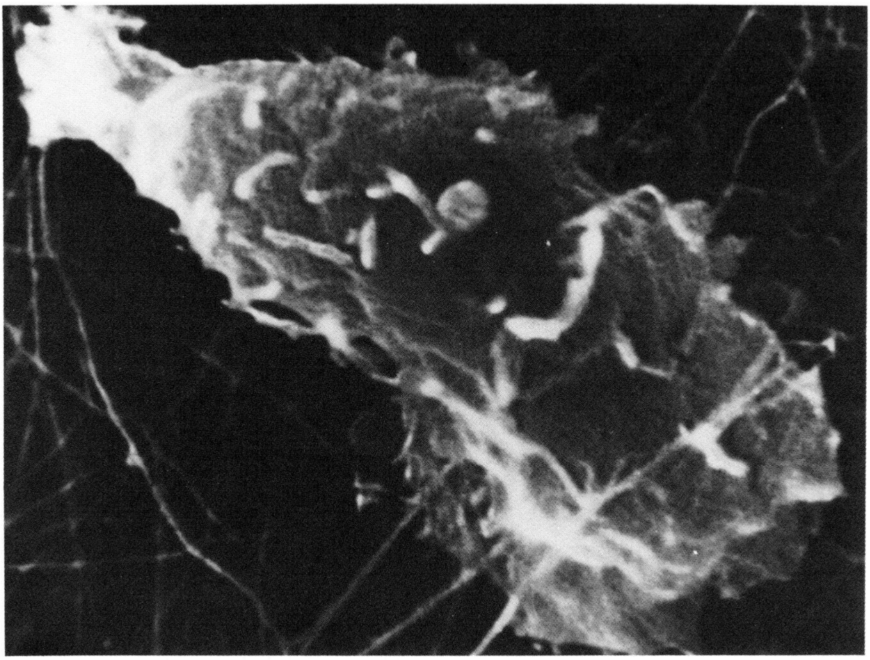

FIG. 3. Scanning electron micrograph of a mouse lymphocyte within a collagen gel. Lymphocytes do not adhere to 2-dimensional coats of collagen, but they are able to gain traction in 3-dimensional matrices of collagen by extending processes among the network of fibers which can be seen in this micrograph. From Haston and Shields.[6]

acetic acid. The solution is dialyzed against water at pH 4.0. The resultant aqueous solution of collagen can be stored at $-20°$ for several months. A standard curve can be prepared by freeze-drying a sample of the solution and measuring the optical density of known concentrations at 230 nm. Collagen gels are prepared from the stock solution (which is adjusted to 1.5 mg/ml) by bringing both the pH and osmolarity to physiological levels. For example, appropriate volumes of $10\times$ HBSS/MOPS buffer and NaOH are added to the aqueous collagen solution, which is then rapidly mixed and aliquoted into suitable dishes, e.g. 2 ml multiwell tissue culture plates with 0.5 ml solution in each. Gelling of the collagen is temperature dependent and takes 5–10 min at room temperature. This can be delayed by cooling.

Chemoattractants can be incorporated into the liquid collagen, but as the addition of attractants may change the gelling characteristics and therefore the topography of the final matrix it is better to incorporate attractants into a layer of collagen or agarose which can then be overlayed, after it has

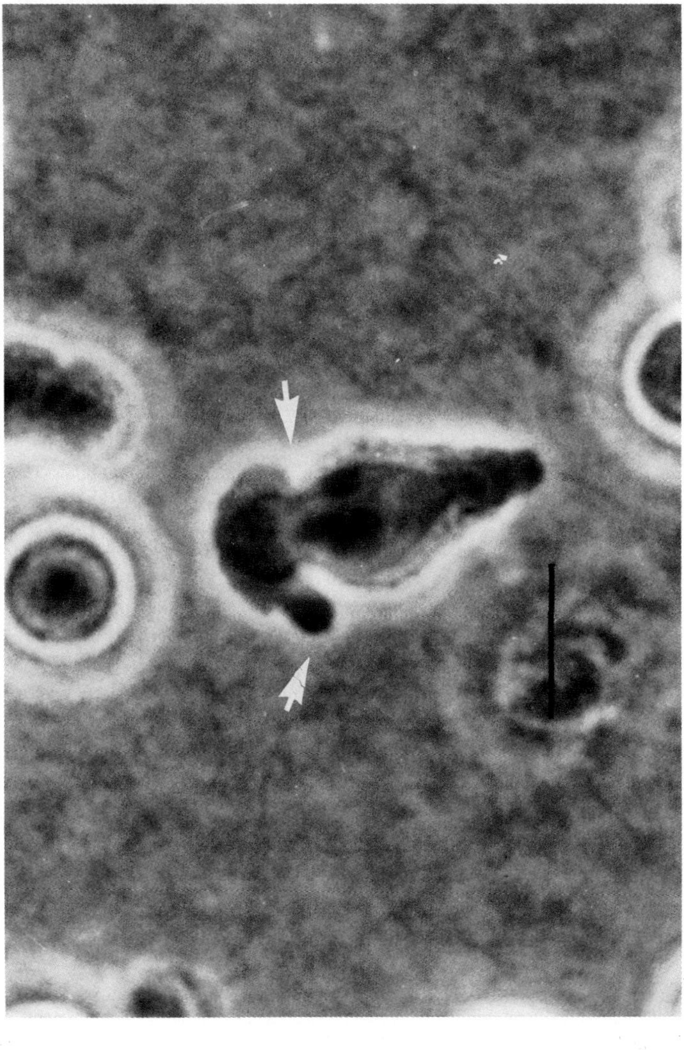

FIG. 4. Lymphocyte moving through a collagen gel, the fibers of which are just visible as a gray flocculent background. A lateral bleb and constriction ring are arrowed. Bar, 5 μm. From Haston and Shields.[6]

gelled, with unaltered collagen. Alternatively, when no gradient is required attractants can be allowed to diffuse into the gel from a liquid overlay. Cells can also be incorporated into the collagen before it gels, but incorporated cells may act as nucleation centers for the polymerizing collagen fibers which could restrict movement. Gels are allowed to set for up to 1 hr, covered closely to prevent surface dehydration which results in shrinkage of the gel and close packing of collagen fibers to form an inpenetrable surface skin. It is better to overlay the set gel with medium as quickly as possible after setting to minimize dehydration. Such gels can be kept for several days at 4° before using.

Figure 5 shows various assay systems that can be used to measure leukocyte locomotion in gels under different conditions. In Fig. 5a, the concentration of fMLP is uniform throughout, and locomotion measured in this system is therefore not directed to an fMLP source. In Fig. 5b, an

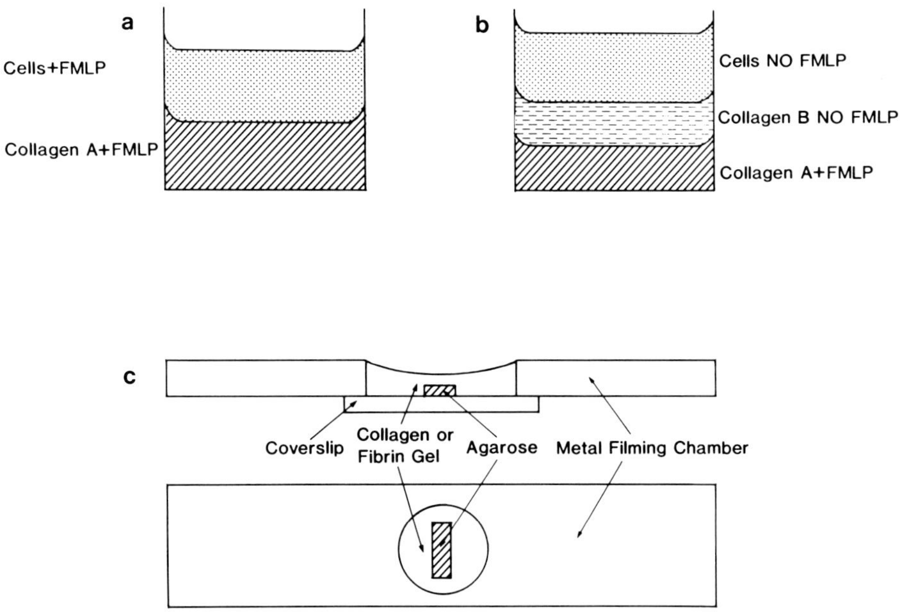

FIG. 5. Schematic diagrams of assays used for studying leukocyte locomotion and chemotaxis in collagen gels. (a) Assays for leukocyte invasion of collagen in an isotropic concentration of fMLP. (b) Assay for leukocyte invasion of collagen in a gradient of fMLP. (c) Visual assay of chemotaxis using a metal filming chamber. The fMLP source is incorporated in a block of agarose which is overlaid with collagen or fibrin. In all diagrams the hatched areas indicate solid media containing fMLP. The dotted areas indicate fluid media containing cells with or without fMLP. These cells settle on the surface of the collagen which they then invade. The area with broken hatching indicates collagen without fMLP. From Islam *et al.*[12]

fMLP-containing gel (A) is overlaid with an fMLP-free gel (B), so that a gradient forms through gel B. After 20 min, to allow gel B to set and the gradient to establish, gel B is overlaid with the cell suspension. These cells should migrate directionally into the gel toward the gradient source (i.e., gel A). These assays thus compare cell distributions in the *vertical* plane in isotropic (Fig. 5a) and anisotropic (Fig. 5b) attractant concentrations as described below. The methods for evaluation are similar to those used for the micropore filter assay (see this volume [3]). In Fig. 5c, the assay is designed to study directional locomotion of leukocytes in the *horizontal* plane toward a gradient source (the agarose block) by time-lapse filming (see below). In practice, it gives better evidence for chemotaxis than the assay in Fig. 5b.[12]

Preparation of Fibrin Gels

Fibrin gels are prepared by adding 1 part of thrombin (at 1.0 NIH units/ml) to 9 parts of fibrinogen (final concentration of 1.0 mg/ml) in HBSS/MOPS. Fibrin gels set within 30 sec so the mixed solution should be dispensed rapidly into dishes. After setting the fibrin should be overlayed with medium. The advantages of fibrin are (1) pure preparations are commercially available (e.g., Sigma) and (2) gelling is rapid and predictable. However, the structure of the gel alters with time, and it is therefore more suitable for short-term assays of 1–4 hr.

Assessment of Cell Distribution in Gels

Cells added, in suspension at 2×10^6/ml, to the top surface of collagen or fibrin gels are allowed to invade the gel matrix for an appropriate time (1–3 hr at 37°). The gel is then fixed by adding an equal volume of 2.5% glutaraldehyde in HBSS/MOPS for 30 min. The fixed gel is washed with three 10-min changes of medium after which the following measurements can be made using an inverted microscope. The first is measurement of the leading front, i.e., the distance that the leading two cells in the same focal plane in one microscopic field (×200) have migrated from the gel surface, as measured using a micrometer-calibrated fine adjustment focusing on the microscope.[13] At least four randomly chosen fields in a minimum of three dishes should be counted. As a second measurement, the vertical distribution of the cell population throughout the gel can be determined by counting the number of cells in each of a series of planes at 40 μm intervals. A

[12] L. N. Islam, I. C. McKay, and P. C. Wilkinson, *J. Immunol. Methods* **85,** 137 (1985).

[13] S. H. Zigmond and J. G. Hirsch, *J. Exp. Med.* **137,** 387 (1973).

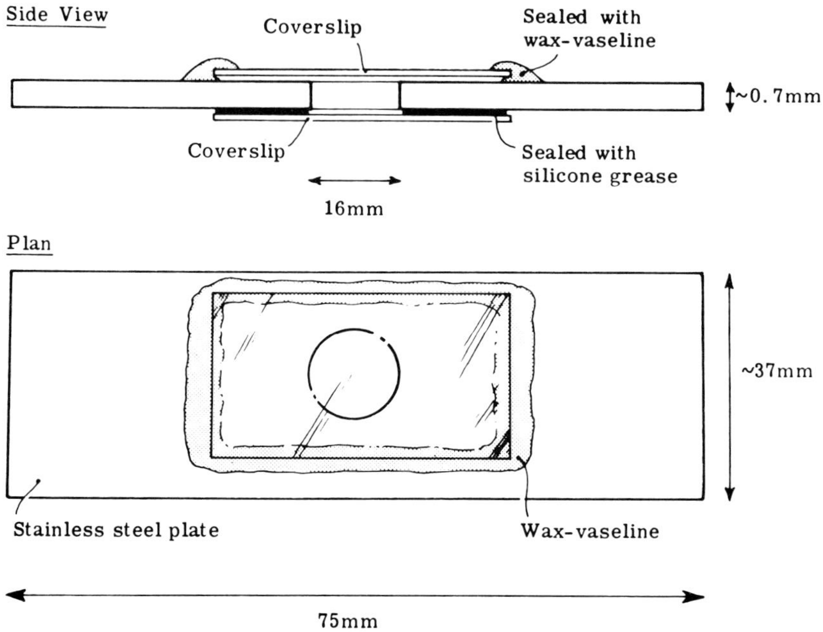

FIG. 6. Method for constructing a filming chamber from a metal slide with a central hole to the underside of which a glass or plastic coverslip can be attached to act as a substratum. From P. C. Wilkinson, J. M. Lackie, and R. B. Allan, *in* "Cell Analysis" (N. Catsimpoolas, ed.), p. 153. Plenum, New York, 1982.

mathematical analysis of the migration of neutrophils in collagen and fibrin gels is described by Islam *et al.*[12]

Direct Observation of Chemotaxis in 3-Dimensional Gels

The transparent nature of the gels described above allows for direct microscopic observation and filming of cells migrating toward a chemotactic source. This method has been used successfully for both neutrophil leukocytes and lymphocytes.[13,14] A source of attractant can be made by embedding a small agarose block, containing the factor, within a collagen or fibrin gel (Fig. 5c). The agarose block is about 1 mm long by 4 mm wide, and the chamber used is shown in Fig. 6. The cells can then be incorporated into the collagen or fibrin which is poured over the agarose block containing the attractant and allowed to set. For filming purposes, the gels can be prepared in chambers made from stainless steel slides in which a central circular hole (1.6 cm) is cut and to the underside of which a

[14] P. C. Wilkinson, *J. Immunol. Methods* **76,** 105 (1985).

coverslip can be attached with silicone grease, thus forming a shallow well (Fig. 6). This can be filled with collagen or medium and then sealed with another coverslip. These chambers can easily be machined in any workshop. The dimensions are given in Fig. 6.

Time-Lapse Cinematography

To analyze the movement of cells observed microscopically, a filmed record is essential. It is possible to follow one or two cells by eye, but not more. Also leukocytes usually move at speeds of slower than 20 μm/min, and it is difficult to concentrate on watching cells moving at such speeds for very long. Therefore it is helpful to have a speeded-up record of events, i.e., to use time-lapse cinematography. Time-lapse filming also has the great merit that the details of an experiment remain available permanently and can be reanalyzed years later to test new ideas which were not in the mind of the investigator when the film was made. The two recording media available at present are cinefilm and videotape. While videotape has the advantage that it can be played back immediately, it still gives poorer images than cinefilm at the time of writing, and analysis of cell tracks is subject to image distortion. This situation may improve as image enhancement techniques become more generally available. Cinefilm is certainly still superior for high-power filming of changes in details of cell shape during movement.

Equipment

The basic requirements for time-lapse cinematography are a microscope with adequate optics for the task at hand, and a heated stage, together with a cinecamera. For analysis of behavior of a cell population, phase-contrast optics and low magnifications (e.g., 10× or 20× objectives) are adequate, because what is needed is a sufficiently large field to contain enough cells for statistical analysis without overcrowding the cells so that they continually collide. For studying detailed shape changes in single cells, higher magnifications are needed as well as differential interference–contrast (Nomarski) optics, which gives sharply focused images free of phase halos. The microscope stage can be heated accurately to 37° using an air curtain. We use a fan heater to blow air across the stage, the temperature being controlled by a miniature-bead thermistor. This gives quite precise control.

The cinecamera used in our laboratory during the last several years has been a Bolex H16 reflex with automatic exposure and an automatic timer (Nikon) which allows selection of a range of desired frame intervals from 4 frames/sec to 1 frame per 30 min. An important requirement is for a beam-splitter so that the field seen by the camera can, at the same time, be monitored by the eye. This greatly facilitates field selection and focusing. For filming, we use a slow reversal film (Kodak Plus-X reversal) for most purposes. For filming transient shape changes, a faster film (Kodak Tri-X reversal) is preferred.

Analysis of Cell Tracks

Time-lapse cinematography is an exact and quantitative technique, though it has often not been used as such. To use it as a quantitative tool, the film, once developed, must be analyzed. This is done most easily using an analytical projector. The model in our laboratory is from the Lafayette Instrument Company (Lafayette, IN). It is a stop-action projector which allows the film to be moved forward or backward, either continuously at 16 or 24 frames/min or step-by-step manually or automatically at set intervals (1, 2, 4, 6, 8, or 12 frames/sec). The film is projected onto drawing paper, and cell tracks can be recorded, e.g., by marking the cell position at intervals of 10 frames (40 sec) or any other suitable interval, then joining the dots with straight lines to give a track represented as a series of steps. This can also be done using a digitizer tablet attached to a desk-top computer. Alternatively, planimetry can be used to give a direct record of the cell track. The time base chosen for the step analysis should be short enough to allow the detailed cell track to be reproduced but not so short that minor nontranslocational movements are recorded. An interval of between 40 sec and 1 min is ideal for neutrophils, which are about 10 μm in diameter and frequently move at 10–20 μm/min, so that each step is about a cell diameter in length. Provided the time base of the film is recorded, parameters such as cell speed and direction can be calculated from analysis of the tracks.

Recently a method has been developed, based on image analysis, by which simultaneous tracking and track analysis of many cells can be achieved using a microcomputer.[15] This is rapid, and the necessary apparatus is not expensive. It is suitable for analysis of cell tracks on plane surfaces, though no apparatus exists that can fully automate the analysis of tracks of cells moving through 3-dimensional matrices.

Figure 7 gives an outline of the procedures that can be adopted using the step analysis. It shows an example of a brief track of three steps (S_1, S_2, and S_3) and shows how speed and turning behavior can be analyzed. Figure 8 shows examples of tracks of cells from a real experiment in which

[15] J. A. T. Dow, J. M. Lackie, and K. V. Crocket, *J. Cell Sci.* **87,** 171 (1987).

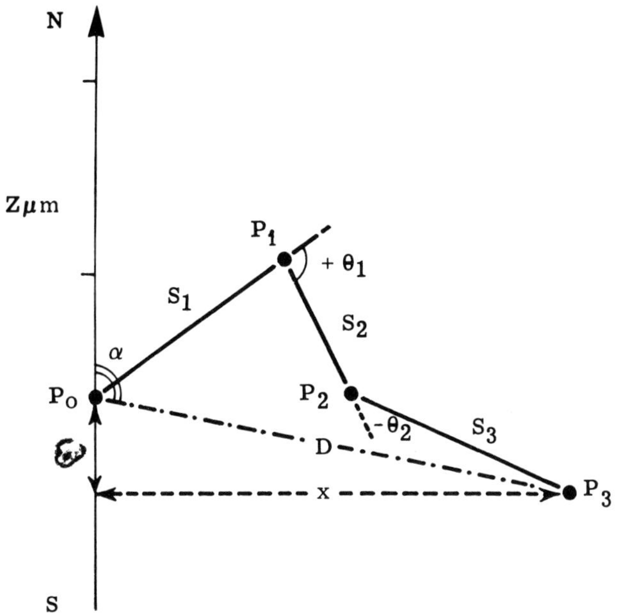

FIG. 7. Diagram illustrating information that can be obtained from an analysis of cell tracks based on steps, i.e., by marking the positions of the cell at equal time intervals and considering the track as a series of straight-line segments. Analysis is carried out either manually or using a bit pad/computer system. An axis $N-S$ is defined with reference to the film frame (arbitrary unless an orientation, e.g., chemotactic gradient or contact guidance field, can be defined, in which case $N-S$ should be the axis of the field). The axis $N-S$ is also used as the scale bar. Positions (P_0, P_1, P_2, . . . P_n) of the cell center are marked at equal time intervals (e.g., every tenth frame), and each step (S_1, S_2, etc.) is measured; the total step length ($\Sigma_i^n S$) is divided by n, the number of steps, to give the mean step length, an index of *speed. Displacement* (D) is computed from the coordinates of P_0 and P_n, and the ratio $\Sigma_i^n S/D$ is the persistence index. Angles of turn (θ) are defined as positive when clockwise and negative when anticlockwise. The rectangular coordinates of P_n (x,y) can be computed easily from the polar coordinates (D,α). This has been done in Fig. 9 (see below).

Note that values for speed and turning behavior obtained in this way are not only determined by the cell behavior but are also biased by the sampling time interval. (For example, obviously, if an irregular cell track is sampled only at long intervals, it will look straighter than if the short-interval details are included, i.e., the long-interval sampling will give a lower speed (d/t) than the short-interval sampling.) Unbiased estimates of speed and turning behavior can be obtained by fitting a mathematical model of cell dispersion to the experimental data, in which one parameter is the root mean square speed of the cells and the other is the persistence in direction time, which is a reciprocal measure of the rate of change in direction of the cells with time. This model is described by G. A. Dunn [*Agents Actions Suppl.* **12,** 1 (1983)], and its use in an experimental analysis is documented by P. C. Wilkinson *et al.* [*J. Cell Biol.* **99,** 1761 (1984)]. From P. C. Wilkinson, J. M. Lackie, and R. B. Allan, *in* "Cell Analysis" (N. Catsimpoolas, ed.), p. 153. Plenum, New York, 1982.

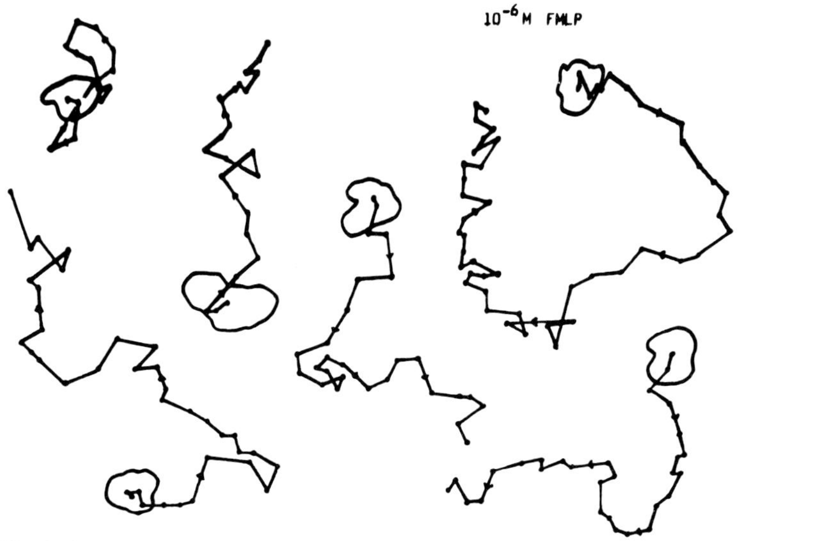

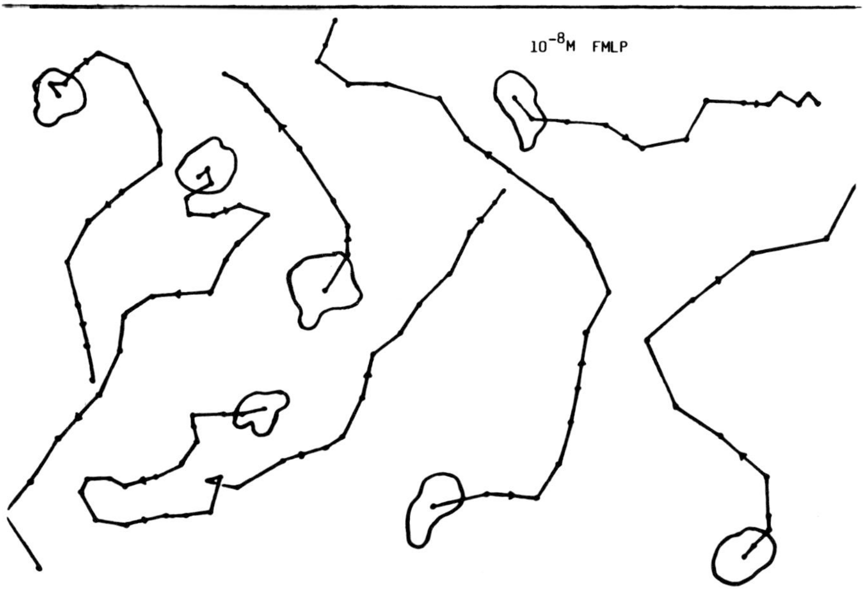

FIG. 8. Two sets of tracks made by neutrophils moving on a 2-dimensional surface in two different concentrations of the chemotactic peptide fMLP. The obvious differences in the straightness or persistence of direction shown by these two sets of tracks can be quantified as described in the text. The mean speed of cells in 10^{-8} *M* fMLP was 11 μm/min, and their mean angle of turn was 62°; mean speed in 10^{-6} *M* fMLP was 6 μm/min and mean angle of turn 88° (calculated as outlined in Fig. 7). From Shields and Haston.[4]

neutrophil locomotion was analyzed in two different, isotropic, concentrations of fMLP. Even cursory inspection of these suggests that the cells in 10^{-8} *M* fMLP move faster, and in straighter paths, than those in 10^{-6} *M* fMLP. This impression can be put on a firm quantitative basis by adopting the procedures outlined in the legend of Fig. 7. Obviously this can be done manually, but the job is made much easier using a bit pad linked to a computer or even more so using the method described in Ref. 15. Probably the most soundly based analysis is that described by Dunn[16] in which the true speed (S) is derived mathematically as also the persistence time (P), a measure of the straightness of cell path. The diffusion coefficient (R) is easily derived from these ($R = S^2P$). See Refs. 16 and 17 for details. Possible reasons for the different responses of neutrophils to optimal (10^{-8} *M*) and supraoptimal (10^{-6} *M*) concentrations of fMLP are discussed in this volume [1]. The theoretical and practical difficulties that may be encountered in this analysis lead into a complex discussion which is beyond the scope of this chapter, and the reader is referred to a number of discussions of this subject, the best introduction to which is probably that by Dunn.[18] See also Refs. 17, 19, and 20.

Assessment of Directional Responses Using Time-Lapse Cinematography

A useful method of measuring the response of cells to gradients is described by Wilkinson.[14] The velocity of each moving cell is averaged over the whole length of its track. The displacement for each cell is then plotted on a vector scatter diagram, and a comparison can be made between the mean displacements of cell populations in gradient or nongradient conditions. An example of such a vector scatter diagram (Fig. 9) shows the displacements of two populations of cells (a, 17 cells; b, 21 cells) relative to a chemotactic gradient source. The diagram is oriented so that the source lies to the north (i.e., $y+$). The origin of the displacement of each cell is taken to be the zero point (at the crossing of x and y axes). For each cell, rectangular coordinates (see Fig. 7) are determined and displacements plotted (as crosses in Fig. 9a; as dots in Fig. 9b). In Fig. 9a, visual inspection shows that most of the cells have displaced to the north (i.e., toward the gradient source). In Fig. 9b, the cells seem to have displaced randomly from the point of origin. The mean displacement and the SEM in the x and

[16] G. A. Dunn, *Agents Actions Suppl.* **12,** 14 (1983).

[17] P. C. Wilkinson, J. M. Lackie, J. V. Forrester, and G. A. Dunn, *J. Cell Biol.* **99,** 1761 (1984).

[18] G. A. Dunn, *Soc. Exp. Biol. Semin. Ser.* **12,** 1 (1981).

[19] S. H. Zigmond, *Nature (London)* **249,** 450 (1974).

[20] M. H. Gail and C. W. Boone, *Biophys. J.* **10,** 980 (1970).

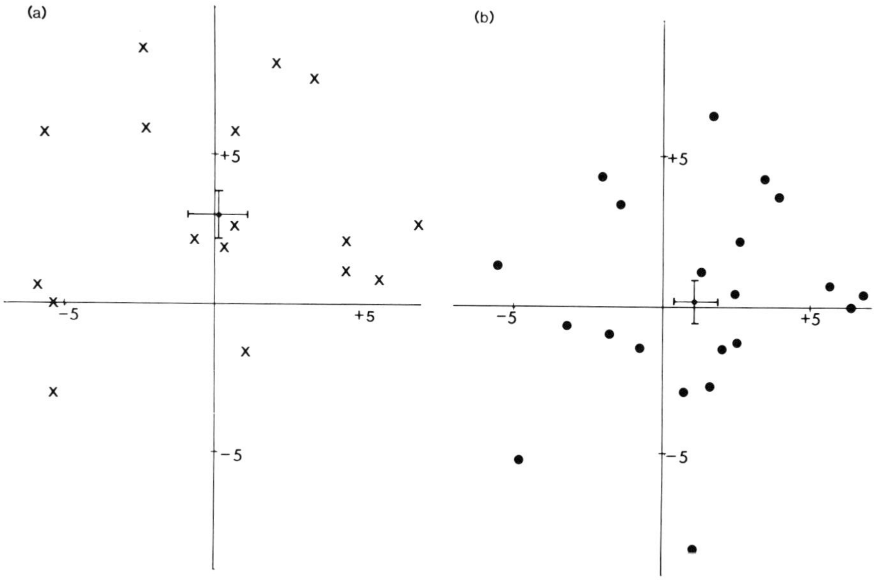

FIG. 9. (a) Displacement of individual cells moving in a gradient, with the population mean marked at the large cross. The bias in the distribution indicates that there is a directional component to the population movement. (b) The symmetrical distribution of the displacements indicates that the direction of locomotion is random. See text for explanation. From Wilkinson.[14]

y axes are then calculated. These are represented by the large crosses (the arms are standard error bars for the two axes; the point of transection is the mean). In Fig. 9a, mean displacement in the y axis is $+3.0 \pm 0.8$ μm/min ($p < 0.01$), and mean displacement in the x axis is $+0.1 \pm 1.0$ μm/min (not significant): i.e., there is significant displacement of the cell population to the gradient source, thus there is evidence of chemotaxis. In Fig. 9b, mean displacement in the y axis is $+0.2 \pm 0.7$ μm/min (not significant) and in the x axis $+1.1 \pm 0.7$ μm/min (not significant); thus these cells show no evidence of chemotaxis.

This method of distinguishing directional locomotion can be used for studying leukocyte movement either on 2-dimensional substrata (see below) or in 3-dimensional gels. Note that, when filming cells in 3-dimensional gels, one is in fact considering only two of the three dimensions. Cell movement at right angles to the focal plane cannot be analyzed. For this reason, chemotactic gradients should be set up in the horizontal plane for analysis of cell locomotion in gels (as in Fig. 5c).

Visual Assays on Plane 2-Dimensional Substrata

It is possible to gain valuable information on the behavioral responses of leukocytes to chemoattractants by filming and tracing tracks made by cells in response to chemoattractants on 2-dimensional substrata such as serum- or albumin-coated glass or tissue culture plastic. Lymphocytes cannot easily be investigated using these assays because of their inability to gain traction on protein-coated 2-dimensional substrata.[21] In order for neutrophils to move on 2-dimensional glass or plastic, these highly adhesive substrata must be coated with a protein which will reduce the adhesiveness enough to allow locomotion but which still allows the cells to gain traction. Evidence suggests that leukocytes obtain traction by adhering to the substratum through gaps in the protein coat and that the protein coat does not contribute directly to the adhesion of the cells. The protein used should not stimulate the cells. For example, purified human or bovine serum albumin is not chemotactic, but at concentrations between 5 and 20 mg/ml it provides a coating which allows neutrophils to move when stimulated with an additional factor. Glass coverslips are soaked in the albumin solution for a few minutes, or the solution is added to the wells of tissue culture plastic dishes as appropriate. The glass or plastic is then washed free of unattached protein. Many other proteins can be used to coat plane surfaces in the same way. Serum-coated surfaces are frequently used. Note that serum contains fibronectin, which may act as an attachment protein, e.g., for monocytes. Fibronectin has no effect on attachment of neutrophils or lymphocytes to artificial substrata. The efficiency of coating of substrata with protein can easily be monitored using protein labeled with ^{125}I or any other suitable isotope.

Visual Assay of Chemotaxis Using Candida albicans

Chemotactic responses can be investigated by adding factors in anisotropic concentrations. Data such as turning angles, speed, and persistence can be obtained from the resulting cell tracks (see Fig. 8). It is considerably more difficult to set up stable gradients using this type of assay. Allan and Wilkinson[22] described a method based on the generation of chemotactic factors in plasma by spores of *Candida albicans.* Each spore, when suspended in plasma, is opsonized by antibodies or by alternative pathway complement activation and acts as a point source of chemoattractants generated from the consequent activation of complement in the plasma

[21] W. S. Haston, J. M. Shields, and P. C. Wilkinson, *J. Cell Biol.* **92,** 474 (1982).

[22] R. B. Allan and P. C. Wilkinson, *Exp. Cell Res.* **111,** 191 (1978).

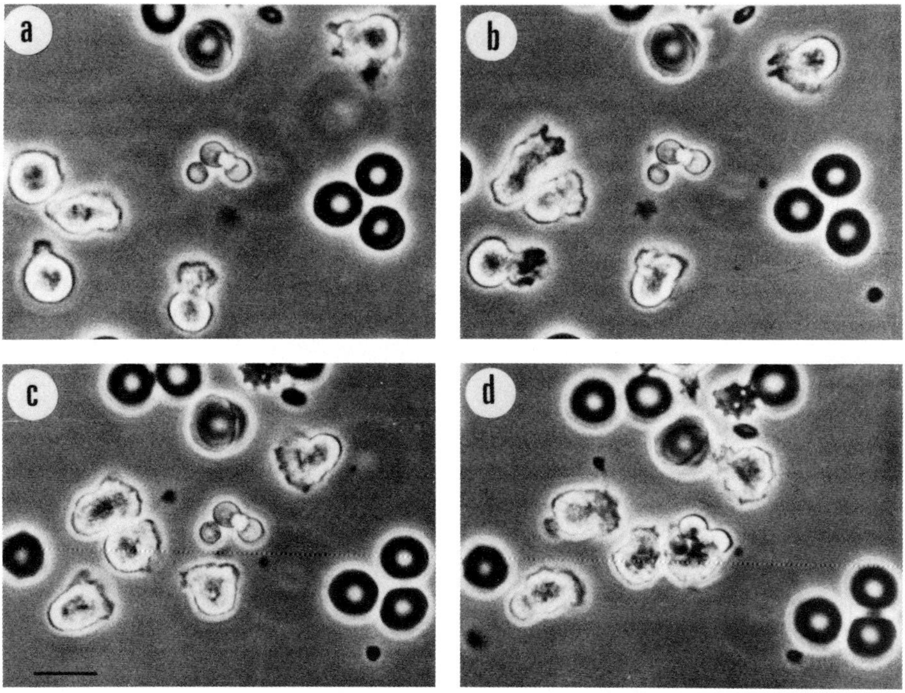

FIG. 10. Sequential micrographs of chemotaxis of neutrophils toward and subsequent phagocytosis of blastospores of *Candida albicans.* (a) A group of blastospores is present in the center of the micrograph, and two of the neutrophils are oriented toward them. (b) All of the neutrophils are oriented toward the blastospores. (c) The neutrophils have moved toward the blastospores, which are engulfed (d). Bar, 10 μm. From P. C. Wilkinson, R. B. Allan, and R. J. Russell, *Agents Actions Suppl.* **3,** 61 (1977).

(see Fig. 10). This type of assay is particularly useful for obtaining data on the effects of agents which alter the locomotor behavior of cells. For example, Allan and Wilkinson described the effects of microtubule-disrupting agents on the chemotaxis of neutrophils toward *C. albicans.*[22] Many microorganisms other than candida could be used for this assay. However, candida spores are conveniently large, and it is easy to track leukocytes moving chemotactically toward them.

The assay using *C. albicans* is performed using the chamber shown in Fig. 6. Candida may be derived from clinical isolates and is cultured on dextrose–peptone–agar (Sabouraud's medium) derived from any commercial supplier of bacteriological media (e.g., Oxoid, UK). This medium encourages the growth of blastospores about 3 μm in diameter but not the growth of pseudomycelial forms. Blastospores, being round, provide suitable point gradient sources; obviously, analysis of the response to a gradient from a sticklike mycelium is more difficult. The blastospores are easy to see under the microscope, and chemotaxis and phagocytosis of neutrophils can be analyzed accurately. The neutrophils (10^6 cells/ml) in HBSS/MOPS plus 20% fresh human serum are allowed to settle on the glass coverslip on the lower surface of the filming chamber (Fig. 6). After 5–10 min unattached cells are washed off, and the chamber is filled with a suspension of candida spores in HBSS/MOPS, pH 7.2, to which 20% fresh human serum has been added immediately before. Normal serum contains anti-candida antibody and is also a complement source, so that the candida spores become opsonized and also act as sources from which chemotactic gradients (e.g., of C5a) are generated. The candida are suspended in this mixture at 2×10^5 spores per ml. The well is then sealed by gently lowering a coverslip (ringed with silicone grease) over its upper surface. The candida spores settle on the lower coverslip. The filming chamber is placed on the warmed microscope stage and filming is begun.

We film at 15 frames/min using a magnification of $\times 100$ which gives a field with 50–60 cells, so that many chemotactic events can be recorded. Higher magnifications are used for recording detailed behavior of single cells. Figure 10 is a series of still photographs illustrating the sequence of events during 1–2 min of filming. Once the film has been made and developed, it has to be analyzed. Having recorded all the events on the film sequence, parameters such as direction of movement relative to the nearest candida spore and speed of locomotion can be calculated using the procedure outlined in Fig. 7, either with a ruler and protractor or, more frequently nowadays, by tracing the cell paths onto a digitizer tablet linked to a microcomputer which has been programmed suitably. The design and use of such programs is beyond the scope of this chapter. The authors' colleague, Dr. J. M. Lackie (Department of Cell Biology, University of Glasgow, UK), has written a number of programs for analysis of leukocyte locomotion and chemotaxis and would be ready to advise.

The response of neutrophils to gradients derived from candida spores or from other microorganisms is rapid. Cells within 50 μm of the nearest spore move in nearly straight-line paths toward the spore which they then ingest within 5 min. A simple and long-established method of comparing responses in different experiments is to use the chemotactic ratio or chemotropism index described by McCutcheon.[23] The total length of the path taken by the cell is estimated, and the straight-line distance from the cell's starting point to the source of the gradient is measured. The McCutcheon index is the distance from start to source/the length of cell path from start

[23] M. McCutcheon, *Physiol. Rev.* **26**, 319 (1944).

to source or, in other words, cell displacement/distance covered by the cell. In Fig. 7 this ratio would be $D/(S_1 + S_2 + S_3)$ if the cell track were from P_o to a spore at P_3. For a perfect chemotactic response, the McCutcheon index is 1.0, i.e., the cell moves in a straight line to the source. The lower the index, the less efficient the response. This can be used to evaluate the effects of drugs on chemotaxis. For example, colchicine-treated neutrophils give lower McCutcheon indices than untreated neutrophils.[22] The effect of the environment on the efficiency of chemotaxis can be studied in other ways. For example, neutrophils moving in an aligned collagen gel (a contact guidance field; see this volume [1]) show higher McCutcheon indices when moving toward spores in an axis parallel to the alignment of the collagen fibers than when moving at right angles to that axis.[24] Moving across aligned collagen fibers hinders the ability of the cells to respond by directional migration in the gradient.

[24] P. C. Wilkinson and J. M. Lackie, *Exp. Cell Res.* **145,** 255 (1983).

[3] Micropore Filter Methods for Leukocyte Chemotaxis

By PETER C. WILKINSON

Introduction

The filter assay has, since its introduction by Boyden,[1] been the most popular assay for leukocyte chemotaxis. For a considerable time it was the only assay in use, though there are now a number of alternatives. The principle and the apparatus needed are simple. A filter with pores of a size sufficient to allow cells to crawl through actively, but not to drop through passively, is used to separate two compartments. Cells are placed in the upper compartment and an attractant in the lower compartment. The attractant diffuses up through the filter to form a gradient, to which the cells respond by migrating, through the pores of the filter, toward the lower surface. After an appropriate time the filters are fixed and locomotion is evaluated by one of several possible methods. The simplicity of the apparatus required probably accounts for the popularity of the method, especially in clinical laboratories. However, it also accounts for some of its limitations as a cell biological tool, which have led to the development of other assays described in the accompanying chapters of this book. An

[1] S. Boyden, *J. Exp. Med.* **115,** 453 (1962).

outline of one version of the method, together with some discussion of problems that arise and the limitations of the technique, is presented in this chapter. Further discussion of these problems will be found in Ref. 2.

Methods

Chemotaxis Chambers

There are numerous variations on the basic chemotaxis chamber which fulfill the requirements of the technique. Some of these are blind-well chambers[3] (e.g., from Neuro Probe, Cabin John, MD, or Cellophex, Basel, Switzerland); in others an upper chamber is suspended in a lower. In our laboratory we use an apparatus of the latter type, machined locally, in which the lower chamber is a machined multiwell plastic compartment which is placed in a plastic box with holes bored in its lid. The upper chambers are sawed-off tuberculin syringe barrels to the lower end of which a filter has been glued. This apparatus is shown in Fig. 1, and the technical procedure outlined below is based on it. With this apparatus, it is important to use an alcohol-soluble glue to attach the filters to the upper chambers. This dissolves during alcohol fixation after the experiment and the filters drop off. In Europe, Uhu glue is suitable for this purpose and can be bought in any hardware shop. Another useful and commercially available (Neuro Probe) multiwell filter apparatus is the one described by Falk *et al.*[4]

Filters

The filters of choice are composed of cellulose esters and are available from several commercial sources including Sartorius (Göttingen, Germany) or Millipore (Bedford, MA). These filters have a thickness of 150 μm, and cells must penetrate this distance to traverse them which allows adequate assessment of the response to the factor. The pores of these filters are in the form of a meshwork or matrix. Filters come in various pore sizes, and the basic requirement is that the pore size be less than that of the cell population under test, but not so narrow that the cells cannot penetrate the pores. For neutrophils filters of 3 or 5 μm are commonly used. Filters of 3-μm pore size exclude other types of blood leukocyte, e.g.,

[2] P. C. Wilkinson, "Chemotaxis and Inflammation." Churchill-Livingstone, Edinburgh, Scotland, 1982.

[3] R. Snyderman, *in* "Methods for Studying Mononuclear Phagocytes" (D. O. Adams, P. J. Edelson, and H. Koren, eds.), p. 535. Academic Press, New York, 1981.

[4] W. Falk, R. H. Goodwin, and E. J. Leonard, *J. Immunol. Methods* **33,** 239 (1980).

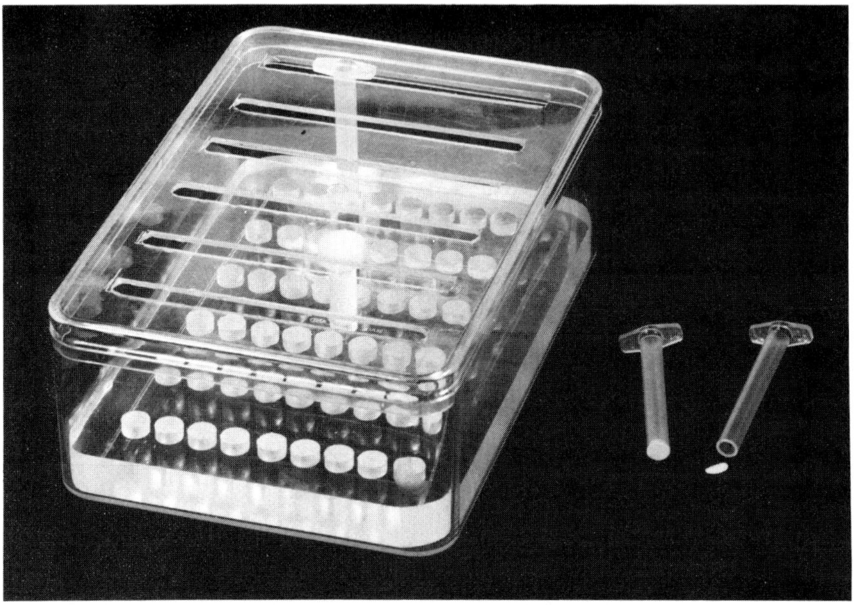

FIG. 1. An apparatus for micropore filter assays. A Perspex block is machined to contain a series of wells to which chemotactic factor is added. The block fits into a plastic box, and the upper compartments, which contain the cells, are suspended into the wells through slits in the lid of the box. These upper compartments are made from sawed-off tuberculin syringes to the lower end of which a filter is glued (right). From P. C. Wilkinson, *in* "Handbook of Experimental Immunology" (D. M. Weir and L. Herzenberg, eds.), 4th ed. Blackwell, Oxford, England, 1986.

monocytes and lymphocytes. For monocytes 8-μm-pore size filters are satisfactory. Blood lymphocytes also migrate readily into 8-μm-pore size filters, though, in studies of the large blast cells from established human B cell lines, we found that a 12-μm-pore size filter was best. Note that the polarity of certain filters (e.g., from Sartorius) is important. Cells migrate much better into these filters from one side than from the other.[5] This needs to be tested for with each new batch.

There is another type of filter in wide use, especially for studies of monocytes. This is the polycarbonate filter[6,7] (Nuclepore, Neuro Probe, or Bio-Rad, Richmond, CA), which is not a matrix, like the cellulose ester filters, but a solid phase punctured by holes of an exact pore size. The main

[5] A. F. Nind, *J. Immunol. Methods* **40,** 39 (1981).

[6] D. A. Horwitz and M. A. Garrett, *J. Immunol.* **106,** 649 (1971).

[7] R. Snyderman, L. C. Altman, M. S. Hausman, and S. E. Mergenhagen, *J. Immunol.* **108** 857 (1972).

drawback of these filters is their thickness (12 μm, about the diameter of a leukocyte), which does not allow any assessment of effects of absolute concentration compared to gradients, though they have been used on many occasions. Their advantage, in my opinion not a sufficient one, is that the morphology of the migrated cells is more easily studied microscopically than the morphology of cells in cellulose ester filters.

Media

Most balanced salt solutions (e.g., Hanks', Gey's) are suitable. They should be well buffered, e.g., with HEPES or MOPS (morpholinopropane sulfonate), to pH 7.2. An essential requirement is that the medium should contain protein. Most leukocytes bind strongly to protein-free filters and do not move even in the presence of powerful stimulants. Thus for studies of formyl peptides, LTB_4, etc., the medium should contain a protein such as human serum albumin (HSA, 10 mg/ml).

Cells

Cells are prepared by any of the well-established techniques, and there is an extensive literature describing these (see various articles in Vol. 108 of this series). They should not be refrigerated. Cooling to 4° inhibits the functions of ion pumps and causes microtubule depolymerization. Thus, cells may take some time to recover after refrigeration, and, in our hands, cells warmed from 4° show a higher background locomotor activity than cells maintained at room temperature.

Chemotactic Factors

Obviously choice of chemotactic factor will depend on the purpose of the assay. A negative control (medium plus HSA) and a positive control should be included. Formylmethionylleucylphenylalanine (fMLP, 10^{-8} to 10^{-9} *M*) can be used as a positive control, but it tends to decay with time due to oxidation of the methionine sulfur. Casein (e.g., from Merck, Darmstadt, Germany) is an inexpensive and very stable positive control, though very impure.

Procedures

1. Cells are prepared at $1-2 \times 10^6$/ml in buffered balanced salt solution (pH 7.2) plus HSA (10 mg/ml).
2. The lower compartments of the chemotaxis chamber are filled with appropriate reagents also at pH 7.2. In the apparatus shown in Fig. 1, the lower compartments are filled with 170 μl of these solutions.

3. The upper compartments are then filled with the cell suspension. For the apparatus shown in Fig. 1, 100 μl is needed. No chemotactic factor is added to the upper compartments (except in the special case of the checkerboard assay, see below). The filters are allowed to wet from the top and then placed in the lower compartments. This ensures that the filter is presoaked with nonchemotactic medium. Thus, when it is placed in the lower compartment, a diffusion gradient of attractant forms from the *bottom* of the filter. If dry filters were placed in the lower compartment, the filters would wet by filling with attractant from below giving a uniform attractant concentration throughout the filter and no gradient for the cells to detect. The volumes given (100 μl above, 170 μl below) are determined to ensure that the fluid levels in lower and upper compartments are the same; obviously, this needs to be determined in advance for the apparatus in use in any particular laboratory.
4. The tests are incubated at 37° to allow cell locomotion and are not disturbed during incubation. The time of incubation should be the minimum which ensures that stimulated cells migrate well into the depths of the filter. There are some published calculations of the time needed to set up a gradient, and of the time after which a gradient approaches close to equilibrium on both sides of a filter.[8] These times vary depending on the diffusion coefficient of the compound under study. A gradient of small molecules such as formyl peptides or LTB_4 is likely to form within a few seconds but to have become very shallow in amplitude across a filter after about 45 min so gradient effects are seen best early in the assay. On the other hand, if the experiment is stopped too early, evaluation is difficult because the cells have not penetrated very far into the filters. For neutrophils 60 min is an adequate incubation time, but monocytes and lymphocytes need longer (2 hr). These times are for leading-front assays (see below). For assays which rely on the cells penetrating the whole depth of the filter, substantially longer times are needed. This is one of several disadvantages of such assays.
5. After the appropriate incubation time, the upper compartments are removed and the fluid in them is inverted. They are dipped in 70% ethanol to fix the cells present on and in the filters. After a few minutes the glue melts and the filters become detached from the tuberculin syringes. Note that 70% ethanol is preferred to absolute ethanol. Filters from some sources become very soft and unmanageable in absolute ethanol.

[8] D. A. Lauffenburger and S. H. Zigmond, *J. Immunol. Methods* **40,** 45 (1981).

6. The filters are then taken through a staining procedure. It is important to use a stain that marks only the cells and not the filters. Romanowsky stains (e.g., Giemsa) may not be suitable. A suggested staining procedure is as follows:

Fix 70% ethanol	5 min
Distilled water	1 min
Hematoxylin	30 sec to 1 min
Distilled water	1 min
Tap water	2–3 min
70% ethanol	2 min
95% ethanol	2 min
A mixture of 80% ethanol and 20% butanol	5 min
Xylene	Enough time to allow filters to become transparent

Note that xylene dissolves laboratory plastic and that this step must be carried out in glass. The xylene clarifies the filters, which is essential for counting cells in the depths of the filter. If the filters are not clear after a few minutes in xylene, they are not properly dehydrated; they must be taken back through the alcohols to remove residual water. After clarification, the filters are mounted in a slide and coverslip preparation using any suitable mountant.

Counting Procedures. There is no ideal method for quantifying filter assays. It can be done in one of three possible ways (see Fig. 2): (a) to determine the distance migrated by the leading front of cells; (b) to determine the number of cells which have reached or migrated past a given point; or (c) to determine the total number of cells per field entering the filter and their distribution in the filter (both are required). Procedure (c) is very time consuming and only practicable with automated counting equipment. There are a number of references in the literature describing such equipment,[9–12] which is expensive. This procedure gives information about all the cells which have entered the filter in a field, and how they are distributed. The other two procedures are methods of sampling this population.

a. The leading-front assay. The filters are mounted topside up, and the micrometer fine adjustment is racked down past the leading front of

[9] N. H. Valerius, *Acta Pathol. Microbiol. Scand., Sect.* **85,** 289 (1977).

[10] T. E. Van Dyke, A. A. Reilly, H. Horoszewicz, N. Gagliardi, and R. J. Gencio, *J. Immunol. Methods* **31,** 271 (1979).

[11] V. A. Moss, H. K. L. Simpson, and J. A. Roberts, *J. Immunol. Methods* **27,** 293 (1979).

[12] S. R. Turner, *J. Immunol Methods* **28,** 355 (1979).

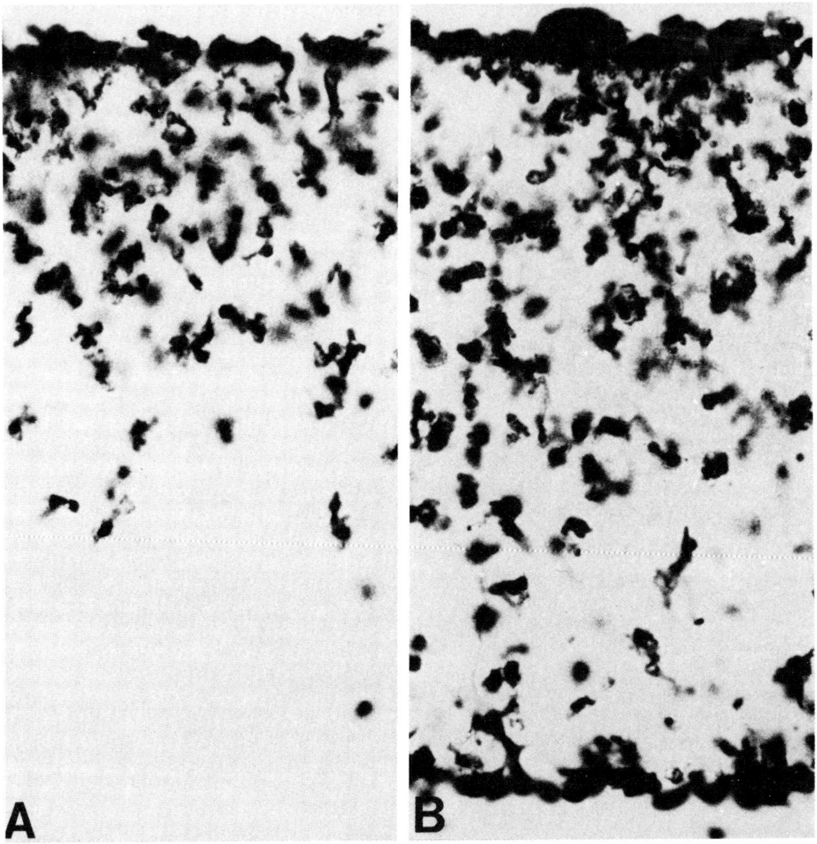

FIG. 2. Cross section through a cellulose ester filter to show the migration of neutrophils. The filter is about 150 μm thick, and cells have migrated from the top surface into its depths. (A) Cells have migrated about halfway; (B) cells have migrated through the whole depth of the filter to its lower surface. From R. Snyderman, H. Gewurz, and S. E. Mergenhagen, *J. Exp. Med.* **128,** 259 (1968).

cells. It is then racked back up to the leading plane in which the nuclei of two or more cells are in focus. A micrometer reading is taken. The fine adjustment is then racked up to the top of the filter and a second reading is taken. The difference between the two is the distance migrated by the leading front (though note that mounting media do not have refractive indices of 1.0; nobody to my knowledge in the extensive literature on this assay has made a correction for this). This procedure is repeated in several fields and a mean taken.

b. Counting cells at or beyond a chosen level. The oldest version of this method is to count cells on the bottom surface of the filter (still used by some). It is inaccurate because cells may drop off once they reach the bottom surface.[13] These nonadherent cells can be trapped by a second nonpermeable filter,[13] or, as in the chromium labeling assay,[14] the radiolabeled cells are allowed to penetrate the whole thickness of the upper filter, then allowed to penetrate a second, cell-permeable, filter. Radioactivity of cells penetrating the latter filter gives a measure of the number of cells which have migrated farthest. Another method is to choose a plane *within* the filter (say 40 or 60 μm below the top) and to count all the cells at and beyond this plane. Counting the cells *at* any given plane does not give much useful information, but counting cells beyond any plane is useful and correlates very well with the leading-front method. The reason for this is made clear by Fig. 3. I have discussed these different counting methods in more detail elsewhere[2] with some experimental data comparing them.

Advantages and Disadvantages of Counting Methods

The leading-front method is quite accurate and reproducible because it is less dependent on the number of cells placed on the filter at the beginning of the experiment than counting methods are. The number of cells in the filter will be directly proportional to the number in the original population, whereas the relation for the leading front is that $(\text{distance})^2$ is proportional to $\log_{10}$ cell number, so that a 10-fold increase in cells used results in only an approximately 1.4-fold increase in the distance migrated. However, though this is an advantage in some respects it is a disadvantage in others. If the objective of a particular experiment is to determine what proportion of a population of cells is responding, the leading-front method will be less sensitive than counting cells in the filter. For example, aggregated IgG has a chemokinetic and chemotactic effect on neutrophils,[15] probably because it is a denatured protein (see this volume [17]). However, since it binds to filters, the neutrophils attach to the filter-bound IgG, and most of them become immobilized. In consequence, a small proportion of the neutrophils are motile, but these few cells penetrate quite deep into the filter. In clinical studies, where it might be suspected that a proportion of neutrophils were immobilized, a counting assay would be more sensitive

[13] H. U. Keller, J. F. Borel, P. C. Wilkinson, M. Hess, and H. Cottier, *J. Immunol. Methods* **1**, 165 (1972).

[14] J. I. Gallin, R. A. Clark, and H. R. Kimball, *J. Immunol.* **110**, 233 (1973).

[15] P. C. Wilkinson, *Immunology* **41**, 457 (1980).

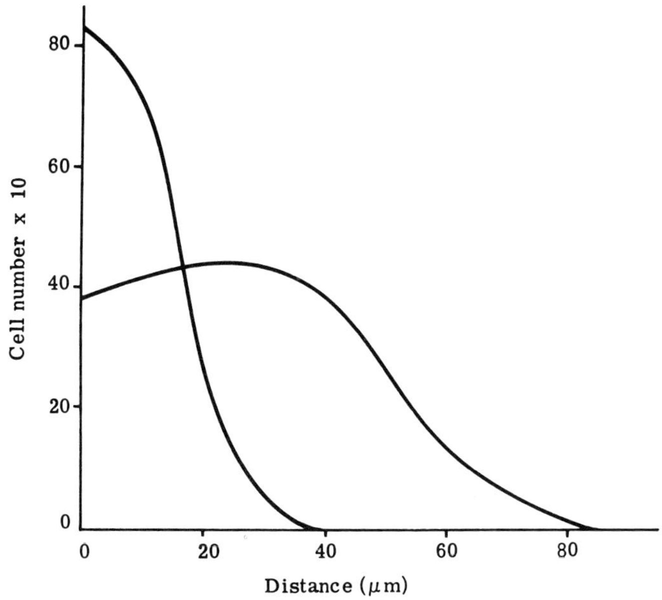

FIG. 3. Curves to show theoretical distribution within a micropore filter of cells responding by random (left-hand curve) or by directional locomotion (right-hand curve). The random locomotion curve shows a normal distribution. Log cell number is linearly proportional to d_2 (square of distance). On the other hand, cells moving directionally may show a flux such that the peak cell number is not at the origin but at some point in the filter. This affects the counting method that should be chosen. Counting cells *at* a given level in the filter is not useful. In the example shown, a count at 20 μm would give the same result for both cell distributions. However, counting all the cells *beyond* 20 μm would show the difference between the two distributions very clearly. From P. C. Wilkinson, J. M. Lackie, and R. B. Allan, *in* "Cell Analysis" (N. Catsimpoolas, ed.), p. 177. Plenum, New York, 1982.

than a leading-front assay. Another possible disadvantage of the leading-front assay is that the leading front may be a small and unrepresentative sample of the cell population, though Zigmond and Hirsch[16] provided evidence that, in their assays, it was representative. This can be checked by comparing the leading-front with an assay of cell distribution within the filter. One of the chief advantages of the leading-front assay is that it can be adapted for checkerboard studies. The principle of these is outlined below.

The Checkerboard Assay

One of the major drawbacks of the micropore filter assay is that, since it is impossible to make direct observations of cells migrating into the filters,

[16] S. H. Zigmond and J. G. Hirsch, *J. Exp. Med.* **137,** 387 (1973).

it is impossible to track them and thus to determine whether their locomotion is random or directed. It is therefore impossible to distinguish whether any factor is chemokinetic, chemotactic, or a combination of the two. Thus, evidence obtained using this assay that any factor is chemotactic is, at best, inferential. In many cases, the simple fact that cells migrate into a filter in the presence of a substance underneath has led workers to conclude that this substance is a chemotactic factor. This is clearly not adequate, and the checkerboard assay was introduced[16] as a best attempt to provide evidence whether or not a directional response had been shown by the cell population. In this assay, a series of chambers are set up using the procedures described above, in which the cells are exposed to (1) a range of attractant concentrations but no gradient (this provides a dose–response curve for nonchemotactic effects of the attractant on locomotion); (2) a range of positive gradients; and (3) a range of negative gradients. The setup is shown in Table I. In Table I, each of the figures represent a different chamber. In some of these (Table IA, those down the diagonal from upper left to lower right) the fMLP concentration is uniform throughout [condition (1) above]; above the diagonal are the results for positive gradients (concentration below filter > above); and below it are those for negative gradients (concentration above filter > below). If the factor has a chemotactic effect (Table IA), cells would be expected to penetrate deeper into the filter in (2) than in (1), and less far into the filter in (3) than in (1). However, this is inexact, and anomalous results may be obtained, e.g., if high concentrations of the attractant were inhibitory to locomotion.

Zigmond and Hirsch[16] gave a method of calculation which gives better evidence for chemotaxis as follows. From the dose–response curve under condition (1), the velocity of the leading front cells in different attractant concentrations is known, and the acceleration of the cells between any two concentrations can easily be calculated (assuming a linear gradient between the two concentrations). The calculations give a series of figures which tell what the effect of the gradient would be, on the assumption that the cells responded by accelerating or decelerating only in response to the change in absolute concentration and not by directional locomotion in the gradient. The difference between the experimental value obtained for migration in any gradient and the calculated migration provides evidence whether or not a chemotactic response is present (see Table I). This works out fairly well in practice and provides indirect evidence for chemotaxis using factors such as fMet-Leu-Phe or C5a which can be shown directly to be chemotactic by visual assays. The checkerboard assay is dependent on homogeneous behavior by the cells under study and may not be suitable for heterogeneous populations like blood lymphocytes. The assumption that a linear gradient is formed is also probably not justified, though use of the assay suggests that, on the ascending slope of the dose–response curve

TABLE I
CHEMOTACTIC AND NONCHEMOTACTIC CHECKERBOARD ASSAYS

A. Chemotaxis: Neutrophils responding to different concentrations and gradients of fMet-Leu-Phe in the presence of HSA (1 mg/ml)[a,b]

fMLP (M) above filter	fMLP (M) below filter: 0	10^{-10}	10^{-9}	10^{-8}	10^{-7}
0	18				
10^{-10}		26	72 (40)	81 (52)	82 (58)
10^{-9}		29 (65)	79	84 (85)	90 (82)
10^{-8}		62 (80)	67 (82)	87	93 (82)
10^{-7}		53 (82)	76 (84)	75 (80)	75

B. No chemotaxis: Neutrophils responding to different concentrations and gradients of BSA[c]

BSA (μm/ml) above filter	BSA (μg/ml) below filter: 0	50	200	350	500
0	19				
50		20	20 (20)	21 (21)	23 (21)
200		23 (24)	24	27 (25)	27 (26)
350		33 (31)	33 (32)	33	35 (34)
500		41 (37)	38 (37)	38 (38)	39

[a] Figures without parentheses represent experimental observations. Those inside parentheses represent a calculation of migration expected if, based on the figures obtained for migration observed in the absence of a gradient (down the diagonal), the cells were responding to the absolute concentration and not to the gradient. The calculation is explained in the text. Note that cells move farther in positive gradients and less far in negative gradients than calculated.

[b] Leading front measure (μm).

[c] Explanation of results as for A. Note that the migration distances obtained experimentally are very close to those obtained by calculation of migration expected in response to the absolute concentration of BSA alone. There is no evidence for chemotaxis.

at least, it is a reasonable approximation. The checkerboard assay was first developed for use with a leading-front assay, but, more recently, the calculations have been adapted to cell-counting assays.[17]

[17] D. Lauffenburger, C. Rothman, and S. H. Zigmond, *J. Immunol.* **131,** 940 (1983).

Calculation for the Checkerboard Assay Using the Leading-Front Method

An example of the calculation for the checkerboard assay is as follows. Suppose we have a chemotactic factor, present at a concentration of 10 μg/ml below a filter of 120-μm thickness, and a linear gradient exists so that the concentration of factor at the top of the filter is 0 μg/ml, at a 60-μm depth it is 5 μg/ml, and at the bottom, 10 μg/ml. How do we find out how far cells would move into the filter in 1 hr if they were showing no chemotactic response, and if their locomotion rate were determined only by the absolute concentration of the attractant? To determine the velocity of cells in each of the three concentrations, we could set up three chambers in each of which the absolute concentration was uniform throughout. In the first chamber there is no attractant: we find that the cells move 30 μm into the filter in 1 hr (leading front) (V_0). In the second, a uniform attractant concentration of 5 μg/ml is present: the cells move 100 μm/hr (V_1). In the third, containing 10 μg/ml of attractant, the cells move 110 μm/hr (V_2). We now know the velocities of the cells at three attractant concentrations and can calculate the acceleration between any two of these known velocities. We also know that, in our gradient, the distance between the sites of 0 and 5 μg/ml concentrations is 60 μm (d_1) and the distance between the sites of 5 and 10 μg/ml concentrations is 60 μm (d_2). The procedure for calculating where the leading front would be in our test gradient is outlined by Zigmond and Hirsch[16] as follows:

$\overline{V}_1$ (mean velocity over distance d_1) $= (V_0 + V_1)/2 = 65$ μm/hr
T_1 (time to move from top of filter to d_1) $= d_1/\overline{V}_1 = 0.92$ hr
$\overline{V}_2$ (mean velocity over d_2) $= (V_1 + V_2)/2 = 105$ μm/hr
T_2 (time to move from d_1 to d_2) $= d_2/\overline{V}_2 = 0.57$ hr

Since $T_1 + T_2$ is greater than 1 hr, the cells will not get all the way across d_2. How far they get across d_2 is calculated as follows:

a_2 (acceleration across d_2) $= (V_2 - V_1)/T_2 = 17.5$ μm/(hr)^2
T_f (time remaining after cells have crossed d_1) $= 1 - T_1 = 0.08$ hr
Distance into $d_2 = V_1 T_f + (a_2/2)(T_f)^2 = 8 + 0.056$ μm

The total distance the leading front would move would be

$$d_t = \overline{V}_1 T_1 + V_1 T_f + (a_2/2)(T_f)^2$$
$$\overline{V}_1 T_1 = d_1$$
$$\therefore d_t = 60 + 8 + 0.056 = 68.056 \quad \mu\text{m}$$
$$d_t = 68 \quad \mu\text{m (to the nearest round figure)}$$

This is the figure we would expect on the basis of random migration alone, and the figures in parentheses in Table I are based on similar calculations (but with more points, i.e., V_0, V_1, V_2, V_3, etc. and d_1, d_2, d_3, etc. (based on the figures on the diagonals). If, in an actual experiment, the cells had moved 95 μm/hr instead of 68 μm in a linear gradient from 0 to 10 μg/ml, this would be much higher than would be expected from random locomotion alone, and would indicate a directional component to the locomotion. In Table IA there is strong evidence for a directional component (the disparity between the experimental and the calculated figures for each test chamber); e.g., for the test at upper right, calculated migration was 58 μm, but the cells actually moved 82 μm. In Table IB, the figures are close to each other, and the locomotion observed is consistent with random locomotion. This simple calculation has been widely, and unjustifiably, ignored by people reporting results of checkerboard filter assays.

[4] Agarose Method for Human Neutrophil Chemotaxis

By ROBERT D. NELSON and MICHAEL J. HERRON

Introduction

Chemotaxis of leukocytes under agarose was first described for guinea pig peritoneal neutrophils by Cutler in 1974.[1] Soon after, Nelson *et al.*,[2] Johns and Sieber,[3] and Repo[4] described measurement of human neutrophil chemotaxis under agarose. Over the following decade the agarose method, in its original or modified forms, has been used by numerous investigators for measurement of random and/or directed movement of leukocytes as well as cells of other types.

In the basic form of the assay, three wells are cut in line in a gel in a tissue culture dish, and cells are placed in the central well with chemoattractant and control medium on opposite sides. Cells migrate radially from the central well by crawling in the space between the agarose and the surface of the dish. After a period of incubation, a normal distribution pattern of migrating cells will be egg shaped with the pointed end of the pattern facing the well containing chemoattractant.

[1] J. E. Cutler, *Proc. Soc. Exp. Biol. Med.* **147,** 471 (1974).

[2] R. D. Nelson, P. G. Quie, and R. L. Simmons, *J. Immunol.* **115,** 1650 (1975).

[3] T. J. Johns and O. F. Sieber, Jr., *Life Sci.* **18,** 177 (1976).

[4] H. Repo, *Scand J. Immunol.* **6,** 203 (1977).

In this chapter we describe in detail the agarose method we use routinely to measure migratory functions of human neutrophils. Many other versions of the agarose method have been described, however to accommodate special needs. Because these have become too numerous to detail in this chapter, references to some of these methods will be provided together with a discussion of considerations which apply to selection of assay conditions.

Methods

Preparation of Agarose Gel

Agarose gel is prepared by mixing two solutions prepared separately, one consisting of agarose dissolved in water and the other consisting of double-strength HEPES-buffered tissue culture medium supplemented with gelatin. To prepare agarose medium suffient for four tissue dishes:

1. Dissolve 0.25 g agarose in 12.5 ml of distilled water by heating in a boiling water bath for 10–15 min. Transfer to a 56° water bath.
2. Dissolve 0.125 g of gelatin powder (Difco) in 12.5 ml of double-strength HEPES-buffered minimal essential medium (MEM)[4a] by warming in the 56° water bath with occasional swirling. This will require approximately 10 min. [To prepare stock double-stength HEPES-buffered MEM, mix 10 ml of 10× MEM, 1.0 ml stock HEPES buffer (1 *M* HEPES solution in distilled water adjusted to pH 8.1 by addition of 10 *M* NaOH), and 39 ml distilled water. Sterilized by filtration and store at 4°.]
3. Add the gelatin-supplemented MEM to the agarose solution, mix by swirling, and pipet 6-ml volumes to each of four 60 × 15 ml tissue culture dishes. The final agarose medium will be 1% with regard to agarose, 0.5% gelatin, and 1× MEM.

The agarose used in our laboratory is Litex agarose, type HSA, obtained from Accurate Chemical and Scientific Corp., Westbury, NY. Both the source and concentration of agarose are somewhat arbitrary. Other investigators have used agarose from a variety of sources (i.e., Seakem ME agarose from Marine Colloids, Rockland, ME, and agarose of unspecified types of Calbiochem, San Diego, CA, and Sigma Chemical Co., St. Louis, MO) with apparently comparible success. A choice of agarose may be

[4a] Abbreviations: MEM, minimal essential medium; AAS, agarose-activated human serum; ZAS, zymosan-activated human serum; fMLP, *N*-formylmethionylleucylphenylalanine.

made in special cases, however, based on a need to eliminate protein[5] or to control the deformability required for cells to enter and migrate in the subagarose space.[2,5]

A number of choices are also possible for the tissue culture medium. In addition to MEM, other investigators have used RPMI 1640, Medium 199, or Krebs–Ringer phosphate buffer. No information is available to direct one's choice of medium. The buffer may be either HEPES or bicarbonate, depending on the availability of an incubator with a CO_2 atmosphere and the need for a protein-free system.[5]

Although protein may not be an absolute requirement for all agarose–buffer systems,[5] protein supplementation of the gel medium has several advantages. It can provide for easy removal of the gel after fixation and appears to provide for improved cell motility. It also optimizes formation of gradients of purified C5-derived chemoattractants by eliminating nonspecific adsorption of chemoattractant to the substratum.[6] Complement-sufficient human serum is not an acceptable protein source for studies of human leukocyte migratory functions because of the ability of agarose to activate the alternate complement pathway. Products of activation of the fifth component of complement in the gel will stimulate random motility and inhibit chemotaxis to C5-derived chemoattractants. Even heat-decomplemented serum is not recommended for this purpose because it will also contain products of complement activation. Any other protein used for this purpose (i.e., ovalbumin, human or bovine serum albumin, or fetal calf serum) should be tested for such influences before its use is accepted.

We have not specified a particular kind of tissue culture dish. This is because we have found that dishes from various sources, either treated or untreated for tissue culture, seem to work equally well for assays of human neutrophil migratory functions. For cells of other types, however, this may be an important consideration. The use of glass microscope slides has also been described,[6] providing the advantages of easier viewing and storage. Disadvantages of slides include the requirements for cleaning, pretreatment with gelatin when using certain chemoattractants,[6] and the need to provide a humid atmosphere during setup and incubation.

Cutting Wells

Cutting the wells is best done with the aid of a punch and template. Our punch is round with an o.d. of 3.1 mm (11-gauge hypodermic stock) and an inside bevel. Wells of this size easily accept a 10-μl volume of fluid. Our

[5] K. W. Mollison, G. W. Carter, and R. A. Krause, *Proc. Soc. Exp. Biol. Med.* **167,** 419 (1981).

[6] D. E. Chenoweth, J. G. Rowe, and T. E. Hugli, *J. Immunol. Methods* **25,** 337 (1979).

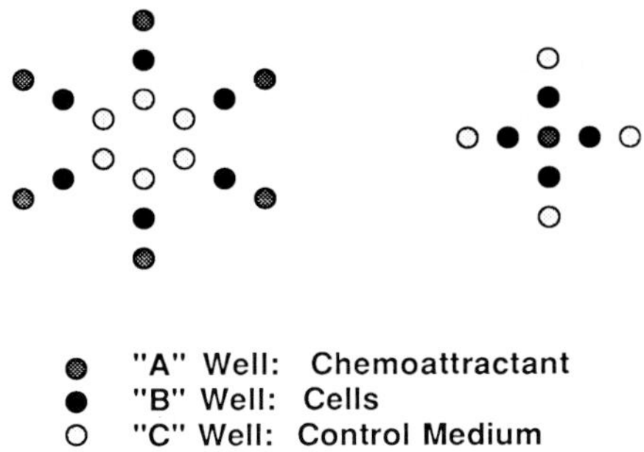

FIG. 1. Two patterns for cutting wells in agarose gels in tissue culture dishes. Chemoattractant, cells, and control medium are placed in the wells according to the legend. When tissue culture dishes are prepared with 6 ml of agarose gel, wells of 3.1-mm diameter will accept a 10-μl volume. The C well containing control medium is optional. We recommend that measurements of random motility be derived from either single wells or wells cut in pairs to accept cells and control medium (see text).

template provides for separation of wells in each group by a distance, edge-to-edge, approximating the well diameter. A punch and template can be purchased from the authors, if not otherwise available. Care must be taken not to scratch the surface of the tissue culture dish by twisting the punch or by applying too much pressure. Scratches will provide a barrier for migrating cells. Cores can be extracted using either a hypodermic needle (with the bevel bent to form a hook) or a pipet (with a diameter smaller than the core) attached to a vacuum. Care must also be taken not to lift the surrounding gel from the substrate. Lifting the gel changes the gel–dish interface, allowing easier movement of cells into such areas. To aid in removal of the cores, the dishes can be refrigerated for a short time. When wells are cut and cores removed in advance, fluid will bleed into the prepared wells. This fluid can be removed using either a pipet or a twisted strip of tissue paper as a wick. In some situations, it may be appropriate to increase or decrease the well diameters or the distance between wells. In other situations, rectangular wells or troughs may be preferred to the round wells.[7]

Alternative well configurations are diagrammed in Fig. 1. Using three-well arrangements and placing chemoattractant in the well designated A, cells in well B, and control medium in well C provides for measurement of both random and chemotactic migratory functions from the same cell population. Alternative placement of the well containing chemoattractant at the center of the dish will reduce the number of wells needed for the

[7] D. A. Lauffenburger, C. Rothman, and S. H. Zigmond, *J. Immunol.* **131,** 940 (1983).

chemoattractant. Two-well arrangements of wells A and B can also be used when the control medium does not differ from the medium in the agarose gel and when it may be more appropriate to measure independently the random migratory function of the cells, i.e., when the incubation time and/or chemoattractant concentration allow the gradient to reach the well containing cells.

Applying Cells and Chemoattractants

The number of cells added to well B influences their rate of migration and determines the density of the migration pattern. We use 5×10^5 human neutrophils in 10 μl, requiring a stock concentration of 5×10^7 cells/ml. A reduced density of the migration pattern would be an advantage when migratory functions are quantified by enumeration methods[8] or when cellular morphology and orientation are evaluated.[9]

Chemoattractants used routinely include agarose- or zymosan-activated human serum (AAS or ZAS) as a source of chemotactic product of the fifth component of complement, C5adesArg, and *N*-formylmethionylleucylphenylalanine (fMLP). Using AAS requires only that complement-sufficient serum be added to the appropriate wells and the dishes incubated for 1 hr at 37° before addition of the cells. We use undiluted serum for this purpose and suggest that sera from different individuals may vary in terms of chemotactic potential. Increased chemotactic activity of serum may also be achieved by preventing conversion of C5a to C5adesArg through addition of a serum carboxypeptidase inhibitor. DL-2-mercaptomethyl-3-guanidoethylthiopropanoic acid (Calbiochem),[10] or the chemoattractant can be concentrated in the form of ZAS–HCl.[11] We use fMLP dissolved in at a concentration of 1.1×10^{-7} *M* (5×10^{-5} mg/ml). The arachidonic acid metabolite, leukotriene B_4 (LTB_4),[12] and cell-free culture supernatants from a variety of microbial species have also been used as chemoattractants. For any chemoattractant, the concentration used will of necessity be greater than that required to produce a chemotactic response across a membrane filter. This difference is related to the greater distance over which the gradient must be established and the small volume of chemoattractant used with the agarose method.

[8] W. Orr and P. A. Ward, *J. Immunol. Methods* **20,** 95 (1978).

[9] J. Palmblad, A.-M. Uden, and N. Venizelos, *J. Immunol. Methods* **44,** 37 (1981).

[10] T. H. Plummer, Jr. and T. J. Ryan, *Biochem. Biophys. Res. Commun.* **98,** 448 (1981).

[11] C. Gerard and T. E. Hugli, *Biol. Chem.* **254,** 6346 (1979).

[12] J. Palmblad, C. L. Malmsten, A.-M. Uden, O. Radmark, L. Engstedt, and B. Samuelsson, *Blood* **58,** 658 (1981).

Incubation and Termination

The incubation period for routine assays of human neutrophil migratory functions is 2 hr at 37° in a humidified atmosphere. For special applications this period has been extended to as long as 8 days.[13] Longer incubation periods demand use of sterile technique and incorporation of antibiotics in the agarose medium. The incubation period will be determined in part by the time over which an effective reagent gradient can be maintained[13,14] and the time over which cells remain viable and retain their motility. At any time during the migration period it is possible to monitor cellular migration microscopically. This can be done by examining the inverted plate using an ordinary microscope, focusing initially on the bottom edge of a well. We advise that this be done routinely before addition of fixative to confirm that cellular migration has occurred.

We accomplish termination and fixation by flooding the dishes with 2.5% glutaraldehyde. The fixative is prepared by dilution of a stock 25% aqueous solution (Sigma) in Dulbecco's phosphate-buffered saline. Migration patterns are not preserved if the agarose is removed without prior fixation. Use of formaldehyde, as buffered formalin, and methanol for this purpose is also acceptable but does not provide for optimal preservation of cellular morphology, if such detail is needed. Treatment for either 1 hr at ambient temperature or overnight at 4° will fix the cells adequately to the dish to allow removal of the agarose. Removal of the agarose is best done in a hood or ventilated space, holding the dish facing downward at a 45° angle, and catching an edge of the gel with a narrow spatula. Do not allow the gel to rotate in the dish during this step or the migration patterns may be smeared. Following removal of the agarose, the dishes are rinsed in tap water and stained with Wright stain or air dried.

Migration Patterns

Migration patterns illustrating chemotactic migration patterns of human neutrophils are presented in Fig. 2. Figure 2A represents a typical migration pattern obtained with either AAS or fMLP as the chemoattractant. Figure 2B represents a migration pattern reflecting the blunting which can occur when the concentration of chemoattractant is too high. Grossly high concentrations of a chemoattractant can inhibit directed cellular motility to abrogate a differential response.

[13] S. U. Orredson, D. R. Knighton, H. Scheuenstuhl, and T. K. Hunt, *J. Surg. Res.* **35,** 249 (1983).

[14] D. A. Lauffenburger and S. H. Zigmond, *J. Immunol. Methods* **40,** 45 (1981).

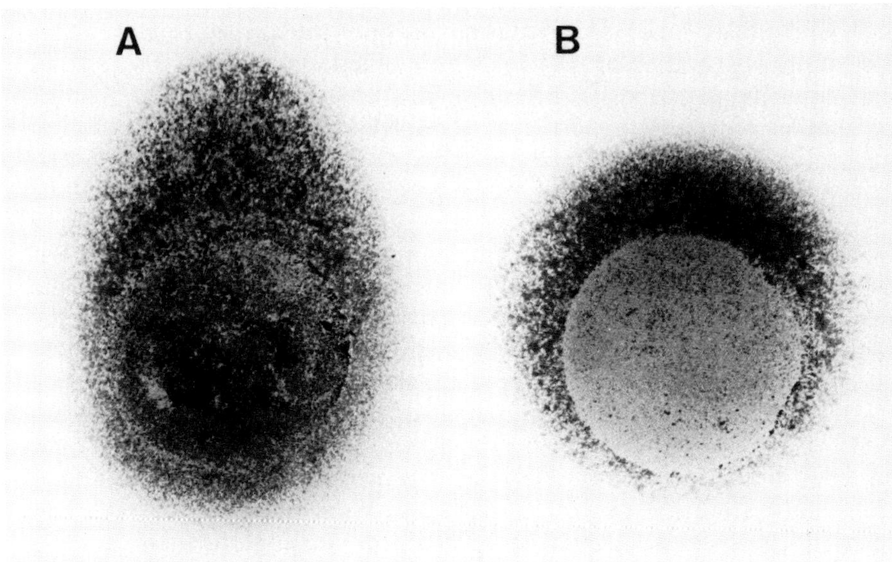

FIG. 2. Two patterns of migration of human neutrophils. (A) Egg-shaped migration pattern typical of a chemotactic response to either AAS or fMLP. The leading edge of the migration pattern toward the source of the chemoattractant may be more or less "rocket shaped," depending in part on the chemoattractant. (B) Blunted migration pattern observed when the concentration of chemoattractant applied is too high. Grossly high concentrations of chemoattractant can produce a migration pattern with a constant radius, falsely indicative of defective chemotaxis.

Quantification of Migratory Functions

Quantification of migratory functions can be accomplished by either of two general methods, one involving counting of migrated cells and another involving measurement of the distance of migration, or leading front. Each method has its counterpart in the membrane filter assay. For two methods of measurement involving counting of cells the reader is referred to the report by Orr and Ward.[8]

For measurement of migration distance we project migration patterns onto a piece of graph paper using a Tri-Simplex microprojector (Bausch and Lomb) fitted with a 12× objective, giving a final magnification of approximately ×40. This information can alternatively be obtained using a microscope fitted with an appropriate ocular micrometer disk.[8] In either

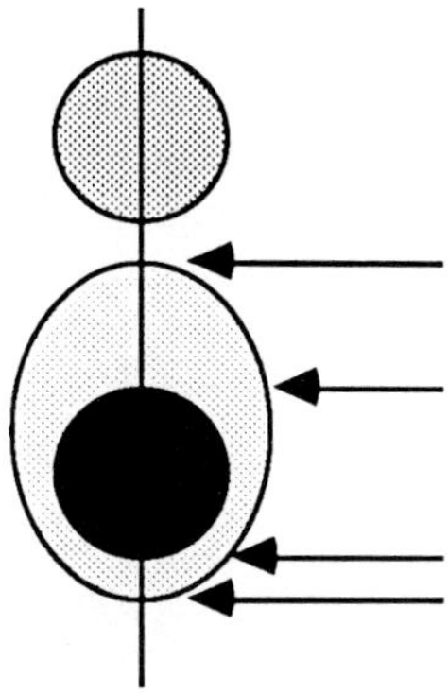

FIG. 3. Quantification of migratory functions by the leading-front method. The distance of cellular migration from the edge of the central well outward toward the well containing chemoattractant, distance A, represents chemotaxis. The distance of cellular migration from the edge of the well outward toward the well containing control medium, distance B, represents random motility. These distances can be measured macroscopically with the aid of a projecting microscope or microscopically using a microscope with grid disk inserted in the ocular lens. The text describes methods of manipulation of these distance measurements to provide a chemotactic differential or chemotactic index.

case, the distance of cellular migration outward from the well margin is recorded as illustrated in Fig. 3. The distal limit of the migration pattern can be identified subjectively by ignoring cells that have "beaten the pack." It can be identified more objectively by excluding some specified number of front-runner cells[5] or by defining the limit as the farthest point at which some number of cells (i.e., three) are aligned in the same plan and parallel to the margin of the central well.[6] For measurements of chemotactic responses, the migration pattern should be positioned on the graph paper so that the measurement is taken on the axis joining the well centers. Our projected (×40) and actual (×1) values for random and chemotactic migration distances fall within the ranges described in Table I.

TABLE I
RANDOM AND CHEMOTACTIC MIGRATION DISTANCES

Function	Well	Distance ×40 (cm)	Distance ×1 (mm)
Random motility	B	2.5 to 4.5	0.6 to 1.1
Chemotaxis, ZAS	A	3.5 to 5.5	0.9 to 1.4
Chemotaxis, fMLP	A	6.5 to 9.0	1.6 to 2.3

In some situations further manipulation of migration distance data may be attempted. For example, patient neutrophils or neutrophils receiving *in vitro* pretreatments often exhibit reduced distances of migration for all migratory functions tested. In this situation differentiation between loss of chemotactic function related nonspecifically to loss of random motility or specifically to loss of responsiveness to an individual chemoattractant is not possible by consideration of the separate migration distance values. However, by calculation of a "chemotactic differential" or a "chemotactic index," such differentiation may be attempted. These calculations are illustrated below.

$$\text{Chemotactic differential} = \text{distance chemotaxis} - \text{distance random motility}$$

$$\text{Chemotactic index (alternative 1)} = \frac{\text{distance chemotaxis}}{\text{distance random motility}}$$

$$\text{Chemotactic index (alternative 2)} = \frac{\text{distance chemotaxis} - \text{distance random motility}}{\text{distance random motility}}$$

Of these, we prefer chemotactic index (alternative 2) for this purpose because the chemotaxis value is corrected for its random motility component before the two functions are related.

Troubleshooting

There will be sporadic occasions when cell migration will not occur and no explanation can be found for the result. This is not atypical of bioassays. Repeated failure to observe cell migration, however, seems to occur for one of two identifiable reasons. One is improper preparation of the tissue culture medium to be mixed with the agarose. This must be double-strength medium so that dilution with the agarose yields an isotonic solution. The other involves handling of the neutrophils and seems most often to involve either the isolation protocol or hypotonic lysis of erythrocytes. To test these possibilities, we recommend use of neutrophils as buffy coat leukocytes or purified neutrophils without elimination of erythrocytes.

Special Applications

The agarose method provides the investigator with opportunities for analysis of cellular migration which are either not available or at best difficult with the membrane filter assay. These include opportunities for examining migratory functions by microcinematography and cellular features by both scanning[15] and transmission electron microscopy (unpublished observations). To facilitate sectioning for TEM analysis, cells are allowed to migrate on a Thermanox coverslip (Miles Scientific).

The agarose method can also be used for analyses of the random and directional component of leukocyte chemotaxis. Two reports describe measurement of cell orientation toward the chemoattractant. Palmblad *et al.*[9] have applied the visual chemotaxis assay of Zigmond and Hirsch[16] to obtain this information, and MacFarlane *et al.*[17] have applied computer-interfaced video microscopy to analysis of cell shape and orientation. The latter authors descibe a "chemotactic behavior index" which compares abnormal with control neutrophil chemotactic function using both cell orientation and migration distance data together. Reports by Lauffenburger and colleagues[7,18] describe a more sophisticated mathematical approach to analysis of the chemosensory movement of cells. A formula incorporating random motility and chemotaxis coefficients together with attractant concentration provides for precise quantitative characterization of the relative contributions of chemokinesis and chemotaxis to population migration as well as for future investigation of the consequences of changes in speed, persistence time, and orientation bias on this phenomenon.

Space does not allow for a review of conditions which have been applied to studies of migratory functions of cells of other types from both human and other sources. For information on such applications of the agarose method, the reader is invited to contact the authors.

[15] R. D. Nelson, V. D. Fiegel, and R. L. Simmons, *J. Immunol.* **117,** 1676 (1976).

[16] S. H. Zigmond and J. G. Hirsch, *J. Exp. Med.* **137,** 387 (1973).

[17] G. D. MacFarlane, M. C. Herzberg, and R. D. Nelson, *J. Leukocyte Biol.* **41,** 307 (1987).

[18] R. T. Tranquillo, S. H. Zigmond, and D. A. Lauffenburger, "Cell Motility and the Cytoskeleton." In Press, 1988.

[5] Chromium-51 Radioimmunoassay for Chemotaxis

By JOHN I. GALLIN

Introduction

The ^{51}Cr radioassay for chemotaxis was introduced over 15 years ago for the study of neutrophil locomotion.[1-3] Subsequently, it has been modi-

[1] E. J. Goetzl and K. F. Austen, *Immunol. Commun.* **1,** 421 (1972).

[2] J. I. Gallin, R. A. Clark, and H. R. Kimball, *J. Immunol.* **110,** 233 (1973).

[3] J. I. Gallin, *Antibiot. Chemother. (Washington, D.C.)* **19,** 146 (1974).

fied for use with monocytes and other leukocytes,[4,5] and some laboratories have used microassays to minimize the amount of material employed.[5] In this assay leukocytes are labeled with chromium-51 and then placed in the upper compartment of a chemotactic chamber which is separated from a lower compartment (containing a chemoattractant) by two filters. The type of filter used varies depending on the type of leukocyte being tested. The chambers are then incubated for 3 hr at 37° The amount of radioactivity in the lower filter, which reflects the number of cells that migrated through the upper and into the lower filter, is determined. A detailed description of the original assay used for human leukocytes follows. Various modifications of the assay that are now available are described at the end of this chapter.

Method

Equipment

1. 37° incubator (100% humidity with 5% CO_2)
2. Single-well chemotactic chambers (Neuroprobe, Cabin John, MD, or Ahlco Scientific, Granby, CT)
3. 3-μm-pore, 13-mm-diameter cellulose nitrate filters (Sartorius, Hayward, CA) for neutrophils or 5-μm polycarbonate and 8-μm cellulose nitrate filters (Neuroprobe) to monocytes
4. Gamma counter

Reagents

1. Chromium-51 (e.g., New England Nuclear, NEZ-030)
2. Gey's balanced salts solution with 2% bovine serum albumin, penicillin, and streptomycin (e.g., M. A. Bioproducts, Walkersville, MD)
3. Hanks' balanced salt solution with calcium and magnesium
4. Chemotactic factors [for standard assays we use formylmethionylleucylphenylalanine (fMet-Leu-Phe, 10^{-8} *M*, Peninsula Labs, Inc., San Carlos, CA), *Escherichia coli* endotoxin-activated serum (5% v/v), and casein (5 mg/ml)]; see Ref. 6 for preparation of chemoattractants
5. 0.85% NaCl

Cell Preparation

1. Peripheral blood anticoagulated with heparin (20 units/ml) is subjected to Hypaque–Ficoll gradient centrifugation followed by dex-

[4] J. I. Gallin, J. A. Klimerman, G. A. Padgett, and S. M. Wolff, *Blood* **45,** 863 (1975).

[5] R. H. Rubin, A. B. Cosimi, and E. J. Goetzl, *Clin. Immunol. Immunopathol.* **6,** 376 (1976).

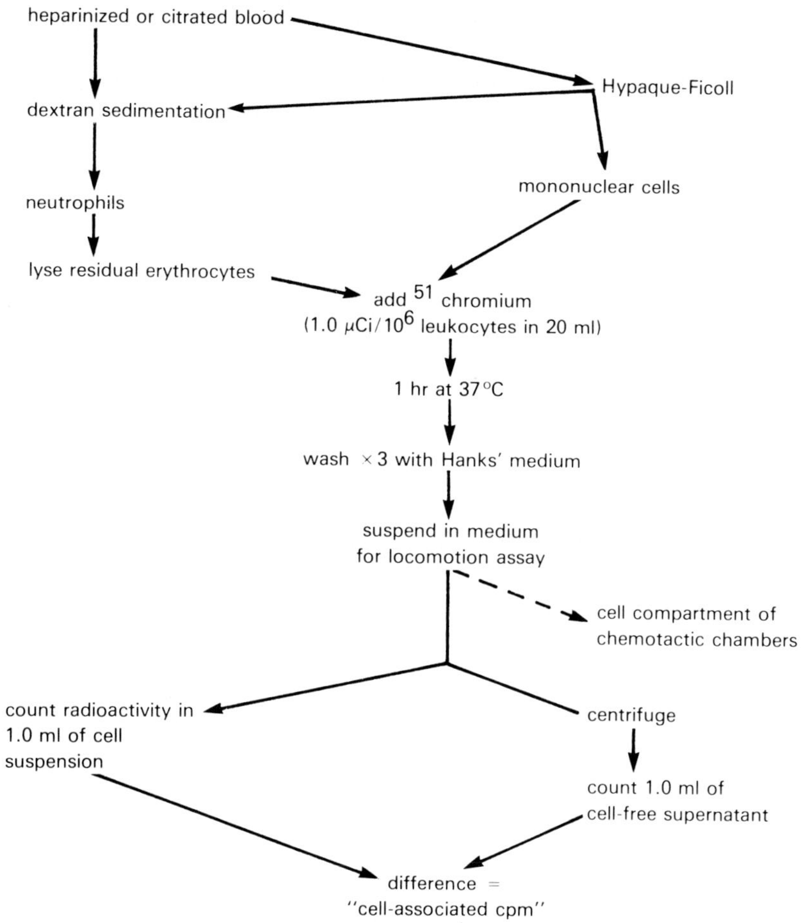

FIG. 1. Leukocyte preparation for ^{51}Cr radioassay of chemotaxis.

tran sedimentation[6] (see this series, Vol. 108 [9]). The purified neutrophils or monocytes are then suspended in Hanks' balanced salts solution at a density of 30×10^6 cells/ml for labeling with ^{51}Cr (Fig. 1).

2. Add 30 μCi ^{51}Cr and incubate 60 min at 37°.
3. Wash 2 times and resuspend in Gey's balanced salts solution to 2.5×10^6 cells/ml.
4. Save 2 ml of cells for determining cell-associated cpm.
   a. Count 1.0 ml cell suspension for total radioactivity.

[6] J. A. Metcalf, J. I. Gallin, W. M. Nauseef, and R. K. Root, "Laboratory Manual of Neutrophil Function." Raven, New York, 1986.

b. Centrifuge the other 1.0 ml and determine radioactivity in the supernatant.
c. Subtract b from a to obtain cell-associated cpm.
d. Divide cell-associated cpm by 2.5 to obtain cpm/10^6 cells.

Procedure

1. Insert two filters (3-μm cellulose nitrate filters for neutrophils and one 5-μm polycarbonate on top of an 8-μm cellulose nitrate filter for monocytes) in the upper compartment of a chemotactic chamber and hold in place by a threaded Teflon cylinder to form side walls for the leukocytes.
2. Load chemotactic factor (usually 1.6 ml) in the bottom chamber. The chamber may have to be tilted to allow air bubbles to escape. Make sure there are no air bubbles.
3. Load cells (usually 0.3–0.8 ml, depending on the chemotactic chamber used) into the upper chamber.
4. Incubate for 3 hr at 37° in a humidified incubator.
5. Dismantle chamber and discard upper filter.
6. Rinse lower filter 3 times in 0.85% NaCl.
7. Place filter in counting tubes and determine radioactivity.

Calculation

The number of cells migrating into the lower filter is expressed as counts per minute (cpm) of cell-associated ^{51}Cr, where the number of cpm is adjusted each day for variability in ^{51}Cr uptake by the cells. To do this, isotope incorporation into the leukocytes is adjusted to an arbitrary value of 10,000 cpm/10^6 cells. Cell migration is then expressed as corrected counts per minute in the lower filter (cor cpm LF) according to the formula[2]:

$$\text{cor cpm LF} = \frac{\text{observed cpm LF} \times 10{,}000}{\text{cpm}/10^6 \text{ cells}}$$

Typical Data for Neutrophils

Using the above calculation, typical data obtained with normal human neutrophils are shown in Table I.

Limitation and Advantages of the Radioassay

With the double-filter radioassay leukocytes migrate through the upper and into the lower filter with little or no barrier to migration offered by the interface between the filters. With the cellulose nitrate filter system, mix-

TABLE I
TYPICAL DATA FOR NEUTROPHILS

Stimulus	Cells in lower filter (corrected cpm)[a]
Buffer	250 ± 15
E. coli endotoxin-activated serum (5% v/v)	2800 ± 250
fMet-Leu-Phe (10^{-8} *M*)	3100 ± 320
Casein (5 mg/ml)	4200 ± 380

[a] Mean ± SEM of four replicates, one experiment.

tures of neutrophils and mononuclear cells can be used to assess neutrophil migration since the slower migrating mononuclear cells do not cross the top filter during the incubation period.

With the chromium-51 technique, individual observer variability and subjectivity and observer-to-observer variability, which are considerable in the morphologic assay, are eliminated. The sources of variability in the assay include inherent differences among cells from different individuals; this variability exceeds day-to-day variability using a single cell source and is comparable to that noted with morphologic techniques.[2]

The radioassay of leukocyte locomotion has proved a reliable functional screening test for clinical studies, which for some diseases is more reliable than many of the other available assays.[7] The method is easy to learn and eliminates the time-consuming task of microscopic enumeration of leukocytes in stained filters. The double-filter radioassay is a leading-front technique that measures the response of the leading population of cells which enter the lower filter. The migrating cells comprise 15–20% of the cell population originally placed in the upper compartment. This provides a good sampling of the migrating cell population and therefore renders the method preferable to the morphologic leading-front technique, which measures only a small sample of the cell population. The radioassay does not, however, provide as much information about the entire population of responding cells as does the morphologic technique of counting cells at different depths in a single cellulose nitrate filter.[6]

Other limitations of the technique need to be emphasized. The radioassay is unable to distinguish cell types (i.e., eosinophils from neutrophils). When such morphologic distinction is required either purified cells should

[7] J. I. Gallin, R. A. Clark, and E. J. Goetzl, *in* "Leukocyte Chemotaxis: Methods, Physiology, and Clinical Implications" (J. I. Gallin and P. G. Quie, eds.), p. 79. Raven, New York, 1978.

be employed or the under agarose technique, which enables distinction of cell types, should be used. Although the radioassay is suitable for distinguishing chemokinesis from chemotaxis by using different concentrations of chemotactic stimuli in both the cell and stimulus compartments of the chemotactic chambers, the technique does not provide information on cell orientation. Cell orientation is best measured with either the orientation assay of Zigmond, employing cells stuck to a glass slide[8] or the assay of Malech *et al.*, using cells adherent to narrow pore cellulose nitrate filters.[9] With the radioassay of chemotaxis leukocytes must be labeled with ^{51}Cr, which adds 1–2 hr to cell preparation. Standard controls must also be included with any unknown or new chemotactic stimulus to assess loss of cellular ^{51}Cr, which is usually very small. The potential hazards of using chromium-51 are minimal and easily controlled.

Although chromium-51 has no detectable effect on leukocyte migration and over 95% of the isotope remains within the cells, several laboratories have suggested using other isotopes. Technetium-99m radiocolloid-labeled granulocytes have been reported to reduce the time of cell preparation,[10] and diisopropylfluoro[^{32}P]phosphate and indium-111 oxine have been reported to be good isotopes for labeling the leukocytes.[11]

A major limitation of the radioassay of chemotaxis is that relatively large numbers of cells are used. This renders the technique impractical for study of neutrophils from neonates and small children. However, recently the radioassay method has been modified for use with multiwell microchambers (Neuroprobe).[12] The modified microassay has been used with neutrophils and uses two 3-μm polycarbonate filters on top of one 3-μm cellulose nitrate filter. The optimal incubation period is reduced from 3 hr to 1 hr, and the assay requires about one-fifth as many cells as the standard assay described above. Although the microassay is appealing and may be useful in settings of limited chemoattractant or cells, it has not yet been subjected to sufficient evaluation to know its reliability.

Conclusion

After recognition of the limitations, it is concluded that the chromium-51 radioassay is useful in laboratories studying leukocyte migration. In particular, it has application in basic and clinical research programs where automated quantitation of the data is desired.

[8] S. Zigmond, *J. Cell Biol.* **77,** 269 (1978).

[9] H. L. Malech, R. K. Root, and J. I. Gallin, *J. Cell Biol.* **75,** 666 (1977).

[10] D. English and T. A. McPherson, *Am. J. Hematol.* **3,** 245 (1977).

[11] D. English and J. A. Clanton, *J. Nucl. Med.* **8,** 913 (1984).

[12] T. J. Sims, F. T. Geissler, and R. C. Page, *J. Immunol. Methods* **78,** 279 (1985).

[6] Orientation Chamber in Chemotaxis

By SALLY H. ZIGMOND

Introduction

The aim of this chapter is to describe an assay which determines if cells can orient their locomotion toward a given agent. The chamber described allows one to observe the behavior of individual cells under conditions where both the concentration and the steepness of the gradient of a chemotactic factor are controlled. The ability of the cells to detect a chemotactic gradient can be investigated by evaluating the cell orientation, a process necessitating only the cell movement required for shape change. The speed of cell movement can be separately scored. Thus, one can directly differentiate the chemotactic and the chemokinetic effects of a given chemoattractant. This differentiation is difficult in the micropore filter or under agarose assays where only the distribution of a population of cells at a given time is evaluated. Furthermore, observations of individual living cells can provide clues about the basis for any abnormal responses.

Methods

The visual assay chamber shown in Fig. 1 is designed to allow observation of living cells in a relatively stable linear gradient.[1] By varying the concentration of chemotactic factor placed in each of the two wells, both the steepness of the gradient and the mean concentration of the attractant can be controlled. The chamber is a 1 × 3 inch piece of 3-mm-thick Plexiglas acrylic plastic into which two wells 1 mm apart have been cut. Each well is a 1 mm deep and about 4.5 mm across. The two wells are separated by 1 mm, the "bridge." The gradient is established across the 1-mm bridge after a coverslip, to which the cells are attached, is laid over the chamber and held in place by spring clips. The distance between the top of the bridge and the coverslip is sufficiently small (optimally about 5 μm, see below) that it provides resistance to fluid flow over the bridge. Thus fluid, from a pipet placed at the edge of one well, flows into that well and is drawn onto the bridge by capillary action. However, the fluid does not flow into the second well. The second well can then be filled with a second solution. The gradient over the 1-mm bridge is established by diffusion. The concentration of the chemotactic factor at the midpoint of

[1] S. H. Zigmond, *J. Cell Biol.* **75**, 606 (1971).

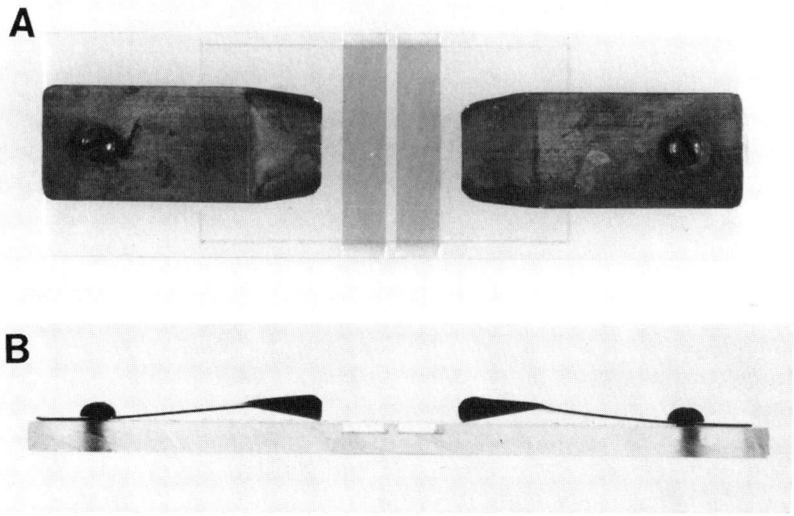

FIG. 1. The visual assay chamber. (A) Top view: the wells are filled with a stain solution to give them contrast; the coverslip is outlined. (B) Side view: the wells are empty.

the bridge is the mean of the concentrations of the solutions in the two wells.

Chamber Design

The width of the bridge affects (1) the steepness of the gradient. Since the gradient between the two wells is formed across the bridge, the steepness of the gradient is inversely proportional to the width of the bridge, i.e., the wider the bridge, the shallower the gradient. (2) It also affects the time required to establish a linear gradient. The gradient is established by diffusion of the chemotactic factor. Thus, the time required for a linear gradient of a factor to become established is inversely proportional to the square of the width of the bridge. Over a 1-mm bridge, a linear gradient of a tripeptide (with a diffusion coefficient of 1×10^{-5} cm^2/sec) is established to 99% completion with 10 min.[1,2] A 2-mm bridge requires 40 min. A larger chemotactic factor (larger diffusion coefficient) requires a longer time.[2,3] (3) The optical properties of the field being scored are also affected by the bridge width. For best optics when using phase contrast, the ring of light from the condenser needs to be entirely on the bridge. Bridges less

[2] D. A. Lauffenburger and S. H. Zigmond, *J. Immunol. Methods* **40,** 45 (1981).

[3] D. A. Lauffenburger, R. T. Tranquillo, and S. H. Zigmond, this volume [9].

than 1 mm may not be sufficient. (4) Finally, the resistance of fluid flow and thus the stability of the gradient are affected. It appears that the resistance to fluid flow across the bridge is critical for the development of a reproducible gradient. In addition, resistance to fluid flow may prevent convection from disturbing the gradient once it is established. The narrower the bridge, the greater the convection.

The size of the wells affects the stability of the gradient. If the wells were of infinite volume and constantly mixed, a stable linear gradient would be formed across the bridge. In fact, the wells in the chamber described hold about 0.1 ml of medium and are not stirred. Thus the gradients established (1) are not limited to the 1 mm of the bridge but extend to some extent into the well and (2) are not stable but decay with time. The rate of decay of the gradient is a function of the diffusion coefficient of the chemotactic factor.[2,3] With small peptides, we find the degree of orientation is fairly constant between 20 and 60 min. It is worth noting that although the steepness of the gradient decays with time, the mean concentration of the midpoint of the bridge remains constant.

I have attempted to stir the solutions in the two wells by perfusing media through the wells. However, it is difficult to balance the perfusion pressure in the two wells exactly. Uneven pressure results in fluid flow across the bridge, disrupting the gradient.

The material of the chamber can be either acrylic plastic or glass. The acrylic chambers give surprisingly good images with phase-contrast microscopy but cannot be used for Nomarski optics. Plastic has the additional advantage of being nonbreakable. Most acrylic chambers are 3 mm thick and thus require a microscope with a condenser able to focus light through this distance. The glass chambers, although more difficult to make, have better optics. They are also more resistant to cleaning agents and to becoming scratched with normal use. Both acrylic and glass chambers are available through Neuroprobe (7621 Cabin Rd., Cabin John, MD 20818).

Procedures

Cleaning the Chambers. It is essential to have clean chambers. Chambers sitting in a room adsorb material from the air. We routinely wash the chambers with a tissue culture detergent such as 7× just before use. The bridge is wiped with a tissue wetted with 70 or 95% ethanol and rinsed well with distilled water. The chambers are blotted dry with a paper towel. Occasionally we observe that cells lying over the bridge die while cells lying over the wells are alive. Such chambers require further washing. Sometimes such cell toxicity can be eliminated by incubating the chambers in a 1% bovine serum albumin (BSA)[3a] solution or by increasing the protein concentration of the medium.

Cells. Human neutrophils can be obtained from a finger prick. About 50 μl of blood is placed across the center of a 22 × 40 mm coverslip. The coverslip is then placed in a moist chamber (wet filter paper in the bottom of a petri dish works well) at 37° until the blood has clotted and the clot retracts. Clots are normally incubated between 45 and 120 min. When the clot is ready to use, fluid can be seen around the edge. At this point the clot and residual red cells can be gently rinsed off with 0.9% saline. A monolayer of cells, containing granulocytes and monocytes, remains on the coverslip. It is important to put enough blood on the coverslip so that the blood clots and retracts without drying. The blood is prevented from spreading out on the coverslip by the hydrophobic nature of the glass; the drop should be a couple of millimeters thick.

Isolated cells of any species can be similarly settled from suspension onto a coverslip. To obtain a cell monolayer of appropriate density, approximately 50 μl of a cell suspension (10^5 to 10^6 cells/ml in Hanks' balanced salt solution containing 1% BSA) is placed across the coverslip, and the cells are allowed to settle and attach. Isolated cells in contact with glass in the absence of protein die. Thus, some protein such as BSA must be present while the cells settle. Gelatin can not be used for this purpose because the cells are not sufficiently adhesive in solutions containing gelatin and they wash off the coverslip.

Medium. The wells are filled with a Hanks' balanced salt (HBSS) solution that has been modified by replacing the bicarbonate buffer with HEPES (2.4 g/liter). The pH of this medium (7.2) is stable in room air. The movement of cells is improved if the medium contains 1% gelatin (Knox Gelatin, Inc., Englewood Cliffs, NJ). Gelatin does not itself stimulate the cells, but it improves their motility by reducing their adhesion to the substrate. We prepare stocks of 10% gelatin by dissolving the gelatin in boiling water. The stocks can be stored in aliquots in the refrigerator and then warmed to redissolve the gelatin before addition to the HBSS. It is important that the gelatin remains in solution (particularly if the medium is cooled) since particles of gelatin can interfere with the formation of a thin fluid layer over the bridge. The medium described here is slightly acidic and slightly hypotonic. These conditions contribute to good cell locomotion. Alkalinity and hypertonicity inhibit cell locomotion.

Chamber Assembly. The cell layers on the coverslip, after rinsing with saline, is rinsed with a few drops of the incubation medium containing the lower concentration of chemotactic factor. Most of the fluid on the cover-

[3a] Abbreviations: BSA, bovine serum albumin; HBSS, Hanks' balanced salt solution.

slip is drained off onto a paper wick. The fluid on the sides of the cell layer is also wiped off with a Kimwipe or tissue. The coverslip is then inverted onto the chamber such that the cells lie over the bridge. In order to avoid formation of air bubbles on the bridge during assembly, the coverslip is lowered at an angle so that the fluid on one side makes contact with the bridge. The coverslip is then gently lowered onto the other side such that the fluid flows across the bridge. Sometimes it helps to put a small drop of medium on one edge of the bridge. This fluid can then be used to make contact with the coverslip before it is fully lowered onto the bridge. The clips to hold the coverslip in place are then placed on each side without moving the coverslip. Any movement of the coverslip at this time will shear the cells lying on the bridge.

The two wells are then filled by bringing a Pasteur pipet containing the appropriate solution to the well at the edge of the coverslip. The fluid flows into the well aided by surface tension. The resistance to fluid flow over the bridge prevents the fluid from the first well from flowing into the second well. The second well can then be filled. It is advisable to fill the well containing buffer or the lower concentration of chemotactic factor first to prevent desensitization of the cells to the higher concentration. The chambers are then placed in a moist environment at 37° for about 20 min. During this time the gradient will be established and the cells will begin to respond to the gradient.

Thickness of Fluid Layer over the Bridge. We find that the most common problem preventing optimal orientation of the cells is that the fluid layer between the bridge and the coverslip is too thick. For reasons that are not entirely understood, the optimal orientation of the cells requires the final distance between the top of the bridge and the bottom of the coverslip to be about 5 μm. In a 5-μm space, the cells are slightly squeezed and appear gray rather than yellow when viewed by phase-contrast microscopy. The thickness can be measured by using the micrometer on the fine-focus knob of most microscopes to determine the difference in focal plane between a scratch on the bridge and a fine process of a cell attached to the coverslip. The ability to obtain this thin fluid layer appears to depend primarily on having only a small amount of fluid present on the coverslip as it is lowered onto the bridge. The depth of the fluid layer can also be affected by the pressure exerted by the clips. Factors that can prevent formation of a thin layer of fluid include particulate substances in the medium, burrs on the edges of the wells, or a concave curvature of the chamber itself.

Scoring of the Cell Response. The cells on the bridge are observed under the phase-contrast microscope. The cell orientation can be evaluated on the basis of the morphology of a locomoting polymorphonuclear leuko-

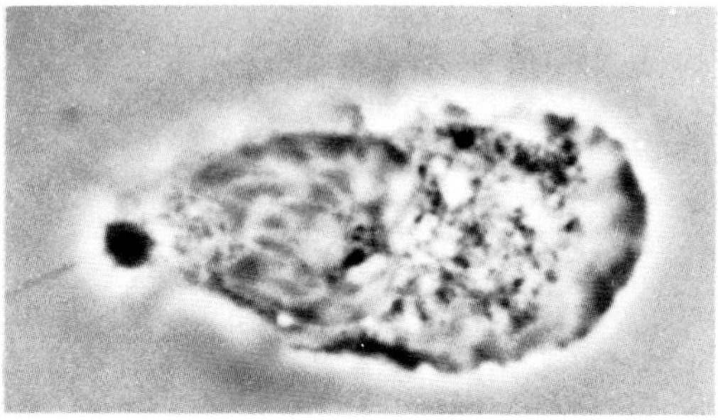

FIG. 2. Phase-contrast micrograph of a locomoting neutrophil. The granule-free lamellipodium (the cell front) extends to the right, and the knoblike tail and a retraction fiber are to the left. ×4000.

cyte (PMN) which has a broad lamellipodium at the front and a knoblike or drawn-out tail at the rear (Fig. 2). The rear can also be marked by retraction fibers. Cells with a polarized morphology can be scored according to the direction of their movement. It is also useful to determine what percentage of the cells have a polarized morphology. We routinely score the orientation of 100 cells along the center of the bridge. The center of the bridge is chosen since the concentration of the chemotactic factor present there does not vary with time once the gradient is established. Usually chambers are scored between 20 and 60 min after they are assembled. It is also possible to observe the bridge continuously. After long incubations, the cells often accumulate along the side of the bridge with the higher concentration of chemotactic factor. It appears that cells do not easily move onto or off of the bridge once the chamber is assembled.

The movement of cells can also be recorded by film or videotape.[4] In order to prevent evaporation of the fluid in the wells, the ends of the wells should be sealed with petroleum jelly or a nontoxic oil. The rate as well as the direction of movement can be evaluated from films, and the net flux of cells in the gradient can be determined.

Several factors are known to interfere with (or enhance) the orientation of cells in the chamber.[5] Cells are known to digest chemotactic factors. The amount of fluid actually over the bridge is only about 1 μl, resulting in an effective cell concentration on the bridge as high as 10^8 cells/ml. Therefore, the cells on the bridge can alter dramatically the gradient by locally de-

[4] W. A. Marasco, E. L. Becker, and J. M. Oliver, *Am. J. Pathol.* **98,** 749 (1980).

[5] S. H. Zigmond, R. D. Klausner, R. Tranquillo, and D. A. Lauffenburger, *in* "Membrane Receptors and Cellular Regulation"(M. P. Czech and C. R. Kahn, eds.), p. 347. Liss, New York, 1984.

grading the chemotactic factor. Often the best orientations are observed when there is a high cell concentration on the bridge. Presumably this is because the cells digest the peptide on the bridge and thereby steepen the gradient. A further consequence of digestion is the orientation of cells along each edge of the bridge toward the nearest well, even the well that contains the lower concentration of a chemotactic factor. We have referred to this as an edge effect.[1] Presumably the cells have digested chemotactic peptide causing a depletion of chemotactic factor at the center of the bridge and a gradient toward both sides. Decreasing the cell concentration can lessen these artifacts.

Cells are also known to release their own chemotactic factors. It is often possible to see cells accumulate in large clumps on the bridge apparently being attracted by some factors being released *in situ.* This is a frequent occurrence along the edge of an air bubble. The air bubble per se does not appear to attract cells, but once a cell is at the edge of an air bubble it seems to attract other cells. Air bubbles can also alter the gradient of chemotactic factors across the bridge. Excellent orientation of cells can be seen where two large air bubbles approach one another along the length of the bridge. The concentration gradient which normally forms over the 1-mm bridge appears to be focused across this small gap thus forming a steep gradient locally.

In practice we achieve the best orientation when one well is filled with a concentration of a chemotactic factor that is about 2 times the concentration of the dissociation constant, K_D, of that factor for the receptor and the second well is filled with buffer. Under these conditions (1) the concentration at the midpoint of the bridge is fixed at the concentration of the K_D, (2) the steepness of the gradient is at its maximum for the particular chamber and this concentration of the chemotactic factor, and (3) by having buffer in the other well there is no edge orientation toward this well caused by digestion of peptide on the bridge.[1]

The actual level of orientation achieved in several repetitions of an experiment using one cell population and the same preparation of buffer and chemotactic factor is unfortunately still variable. If the mean orientation is 80%, the standard deviation is likely to be 10%. The reasons for this variation are not clear. Certainly the distance between the bridge and the coverslip is an important variable. The cell concentration on the bridge can also affect the response as described above. Other factors include the amount of cell-derived factor that competes with the peptide in directing cell orientation and the fluid convection present in a given chamber. This variation in the response makes the assay of limited use for detecting fine differences between populations of cells. The assay is most appropriate for definitely identifying a chemotactic, i.e., directional response, in individual cells and for characterizing dramatic differences in the responses to different chemotactic agents or between different cell populations.

Acknowledgments

This work was supported by National Science Foundation Grant PCM-8303017. I would like to thank Richard Goodwin, Elizabeth Shell-Frederick, and Mary Wilde for helpful comments on the manuscript.

[7] Neutrophil Aggregation and Chemokinesis Assays

By ANTHONY W. FORD-HUTCHINSON and JILLIAN F. EVANS

Introduction

Interaction of chemotactic factors, such as leukotriene B_4, C5a, and formylmethionylleucylphenylalanine, with receptor sites on neutrophil cell membranes initiates a series of biochemical and cellular events.[1] These include alterations in ion fluxes and transmembrane potential, shape changes, secretion of lysosomal enzymes, production of superoxide anions, and enhanced locomotion (chemotaxis and chemokinesis). Thus a number of functional readouts may be utilized to study these receptor interactions, and two such assays which are ameanable to rapid analysis of large numbers of samples are described below, namely, the aggregation of rat neutrophils and the chemokinesis of human neutrophils. These assays are illustrated through the use of leukotriene B_4, a potent chemotactic, chemokinetic, and aggregating substance for polymorphonuclear leukocytes.[2,3] These functional assays for leukotriene B_4 have been correlated with binding to high-affinity receptor sites.[4]

Aggregation of Rat Neutrophils[5]

Interaction of chemotactic factors with neutrophils results in immediate shifts in ion fluxes and volume and shape changes including formation

[1] R. Snyderman and M. C. Pike, *Annu. Rev. Immunol.* **2,** 257 (1984).

[2] A. W. Ford-Hutchinson, M. A. Bray, M. V. Doig, M. E. Shipley, and M. J. H. Smith, *Nature (London)* **286,** 264 (1980).

[3] A. W. Ford-Hutchinson, *Agents Actions Suppl.* **12,** 154 (1983).

[4] A. W. Ford-Hutchinson, S. Charleson, and J. F. Evans, this series, Vol. 163 [30].

[5] F. M. Cunningham, M. E. Shipley, and M. J. H. Smith, *J. Pharm. Pharmacol.* **32,** 377 (1980).

of pseudopods.[6,7] If during this initial interaction the cell populations are stirred, then cell–cell interactions will occur with formation of neutrophil aggregates which can be monitored by nephelometry.[5] This process may be conveniently studied using rat peritoneal neutrophils, and the resultant experiment is a sensitive and reproducible assay for chemotactic factors such as leukotriene B_4,[2] C5a, formylmethionylphenylalanine,[5] and platelet activating factor.[8] The assay uses a cell population readily obtainable, and results are obtained quickly. Neutrophil aggregation may also be performed on peripheral human neutrophils purified by dextran sedimentation, Ficoll–Hypaque centrifugation, and hypertonic lysis.[2] Using this technique the yields of neutrophils are low, and spontaneous aggregation of the cells may occur. This latter problem may be partially overcome by isolating and storing the cells in Ca^{2+}-free medium until immediately prior to initiation of the aggregation response. An *in vivo* correlate of this *in vitro* aggregation response is the transient neutropenia observed following systemic infusion of chemotactic factors.[9]

Preparation of Rat Neutrophils (see also this series, Vol. 108 [9] and [25])

Elicited rat peritoneal polymorphonuclear leukocytes are obtained from male Sprague–Dawley rats (250–400 g) injected i.p. with 8 ml of 12% (w/v) sodium caseinate in saline and sacrificed 16–18 hr later.[10] Most commercial preparations of sodium caseinate are of a fine grade and give poor yields of elicited neutrophils. In order to obtain good yields a coarse-grade sodium caseinate must be used. Difco Grade 0187-17 (coarse grade) is suitable for these purposes and is injected through a 16-gauge needle to lightly anesthetized (ether) rats. Animals are sacrificed by CO_2 asphyxiation and injected i.p. with 20 ml of Eagle's minimum essential medium (Flow Laboratories) buffered with 30 m*M* HEPES [4-(2-hydroxyethyl)-1-piperazineethanesulfonic acid], pH 7.4. The peritoneums are massaged, the skin cut and carefully pulled aside to avoid hairs entering the cell suspension, and a midline incision made through which the cells are removed using a disposable plastic pipet. Care must be taken to avoid red cell contamination through any bleeding points at the sites of incision. The

[6] J. T. O'Flaherty, D. L. Kreutzer, and P. A. Ward, *Am. J. Pathol.* **90,** 537 (1978).

[7] J. T. O'Flaherty, H. J. Showell, E. L. Becker, and P. A. Ward, *Am. J. Pathol.* **92,** 155 (1978).

[8] A. W. Ford-Hutchinson, *Int. J. Immunopharmacol.* **5,** 17 (1983).

[9] J. T. O'Flaherty, H. J. Showell, and P. A. Ward, *J. Immunol.* **118,** 1586 (1977).

[10] F. M. Cunningham, M. J. H. Smith, A. W. Ford-Hutchinson, and J. R. Walker, *J. Pathol.* **128,** 15 (1979).

cells are centrifuged for 5 min at 200 *g* at room temperature, the supernatant is discarded, and the cells are resuspended in medium, filtered through coarse lens paper (e.g., Fisher lens paper, Fisher Scientific, Montréal, Que., Cat. No. 11-995), and centrifuged as before. Supernatants are again discarded, and the cells are resuspended in medium at a concentration of 10^7 cells/ml. This simple cell preparation gives a very pure population of polymorphonuclear leukocytes (>90%) with a yeild of $2-4 \times 10^8$ cells per rat.

Aggregation Assay

Rat neutrophil aggregation is carried out in a Payton dual channel aggregation module (Payton Scientific Ltd., Scarborough, ON) using siliconized cuvettes and stirrer bars. The cells are stirred at 800 rpm, and the following instrument settings used: zero 0–30, output 50–60, range 5, and level 3. The output from the aggregation module is connected to a pen recorder (1 mV), and the chart is run at a speed of 1 cm/min. The changes in light scattering observed during neutrophil aggregation are much smaller than those observed during conventional platelet aggregation. Thus, instead of setting up the aggregometer with neutrophil-rich and neutrophil-poor medium, the limits of the aggregation response may be set between two cuvettes, the first containing 10^7 cells/ml, the second containing 0.8×10^6 cells/ml.

In order to study the aggregation response of arachidonic acid metabolites, rat neutrophils (0.5 ml, 10^7 cells/ml) are preincubated in the aggregometer for 1 min at 37° prior to addition of the putative agonist. Aggregation responses are measured arbitrarily in millimeters from the recorder tracing (Fig. 1). Rat neutrophils respond less to leukotriene B_4 after stand-

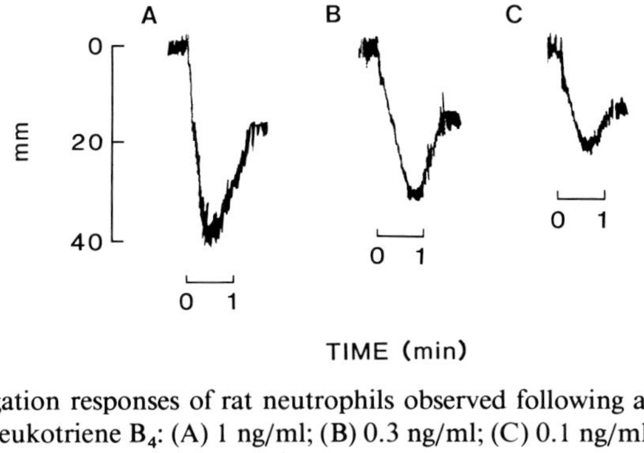

FIG. 1. Aggregation responses of rat neutrophils observed following addition of various concentrations of leukotriene B_4: (A) 1 ng/ml; (B) 0.3 ng/ml; (C) 0.1 ng/ml. The *y* axis shows the height of the aggregation response measured arbitrarily on the chart recorder in millimeters.

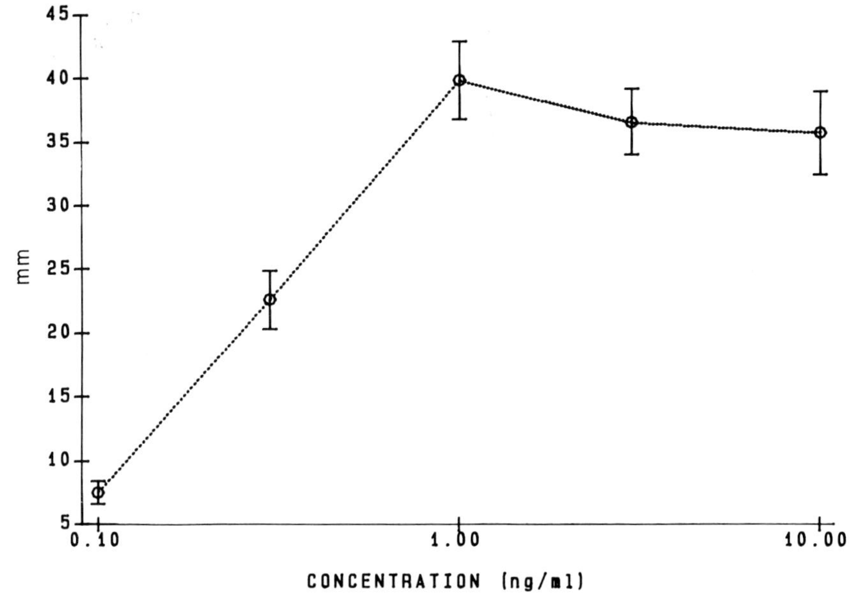

FIG. 2. Dose–response curve for the aggregation of rat neutrophils by leukotriene B_4. Results are expressed as the height of the aggregation response measured arbitrarily on the chart recorder in millimeters and shown as means ± SEM ($n = 5-10$).

ing at room temperature for 5–6 hr, and for optimal responses the cells should be used within 4 hr of isolation. Loss of activity may be minimized if the bulk preparations of cells are not allowed to sediment to the bottom of the tube during storage, and occasional filtering of the cells helps to maintain a steady response during the course of a series of experiments. Reproducible dose–response curves to leukotriene B_4 and related agonists may be obtained both within experiments and from day to day (Fig. 2). Results using this technique correlate closely with receptor-binding studies on rat neutrophil membrane preparations.[4]

Agarose Microdroplet Chemokinesis Assay[11]

The effects of arachidonic acid metabolites on the chemokinesis (stimulated random movement) of human polymorphonuclear leukocytes may be readily determined using an agarose microdroplet assay. A large number of samples may be assayed at one time using this technique, unlike more conventional assays for stimulated cell movement such as the Boyden

[11] M. J. H. Smith and J. R. Walker, *Br. J. Pharmacol.* **69,** 473 (1980).

chamber. The assay is useful for comparing the relevant potencies of various chemokinetic agents including arachidonic acid metabolites, and differences may be observed both in the relative potencies and in the maximum degree of stimulation observed. The assay is also an extremely sensitive bioassay for leukotriene B_4 in biological fluids when coupled with purification by high-performance liquid chromatography and, for example, has been used in the analysis of leukotriene B_4 in samples from psoriatic skin.[12]

Reagents

Agarose: Indubiose agarose A37 (Industrie Biologique Française, 16 Boulevard de General Ledere, 92115 Clichy, France) is made up as a 0.8% (w/v) solution (80 mg/10 ml of distilled water). Prior to use, the solution is boiled and cooled to 37–40°. The solution may be stored at room temperature and used up to 10 times (i.e., boiling the solution on 10 separate occasions).

Double-strength medium: The medium is made by dissolving 100 mg of Eagle's minimal essential medium powder (Flow) in 3.65 ml of distilled water. To this is added 0.05 ml of gentamicin (5 mg/ml stock solution), 0.3 ml of 60 m*M* HEPES, pH 7.4, and 1 ml of heat-inactivated (56° for 30 min) normal human serum. The solution may be stored in 1-ml aliquots at −20°. Prior to use the solution is warmed to 37° and filtered through a 0.22-μm Millex filter (Millipore).

Final agarose solution: Equal volumes of agarose and double-strength medium are added to give a 0.4% (w/v) solution of agarose in single-strength medium containing 10% human serum buffered with 30 m*M* HEPES, pH 7.4 (e.g., 0.2 ml + 0.2 ml).

Single-strength medium: Eagle's minimum essential medium (Flow) is made up freshly at the standard concentration, containing 30 m*M* HEPES buffered to pH 7.4.

Dextran: Dextran 75-S is obtained as a 6% solution from Abbott Laboratories and has an average molecular weight of 75,000 (range 20,000–200,000).

Preparation of Cells

Twenty milliliters of heperanized (10 units/ml) venous blood is sedimented with dextran (4 ml, to give a 1% final concentration) at room temperature for approximately 45 min. The leukocyte-rich cell layer is

[12] S. Brain, R. Camp, P. Dowd, A. Kobza Black, and M. Greaves, *J. Invest. Dermatol.* **83,** 70 (1984).

removed and centrifuged at 200 *g* for 5 min at room temperature. The remaining erythrocytes are lysed by addition of 0.2% (w/v) NaCl (contact time 20–30 sec) and then returned to isotonicity with an equal volume of 1.6% (w/v) NaCl. The cells are recentrifuged at 200 *g* for 5 min prior to resuspension.

Chemokinesis Assay

After the final centrifugation the cells are resuspended in a minimum volume of single-strength Eagle's minimum essential medium (0.15 ml). Cells are warmed for 6–10 min at 37°. Then an equal volume (0.15 ml) of the 0.4% final agarose solution is added to the cells, the solution vortexed, mixed, and immediately 2-μl volumes placed onto the bottom of microtiter plate wells. The microtiter plate has previously been placed on an ice bath, and the cooling of the plates allows the droplets to solidify and stick. The flat-well microtiter plates (e.g., Sterilin microtiter system, Sterilin, Middlesex, UK) must not be of tissue culture grade otherwise difficulties in the adhesion of the droplets to the bottom of the plates will be encountered. The droplets are then gently covered with 0.1-ml volumes of either test or control fluids. For example, arachidonic acid metabolites may be dissolved in single-strength Eagle's minimum essential medium at various concentrations, or biological fluids may be assayed although the final serum concentration may not exceed 5% (v/v). Great care must be taken not to displace the microdroplets from the plastic surfaces during the pipetting procedure. The microtiter plates are then incubated at 37° in a humid atmosphere for 2 hr. If other cell types are used in this assay

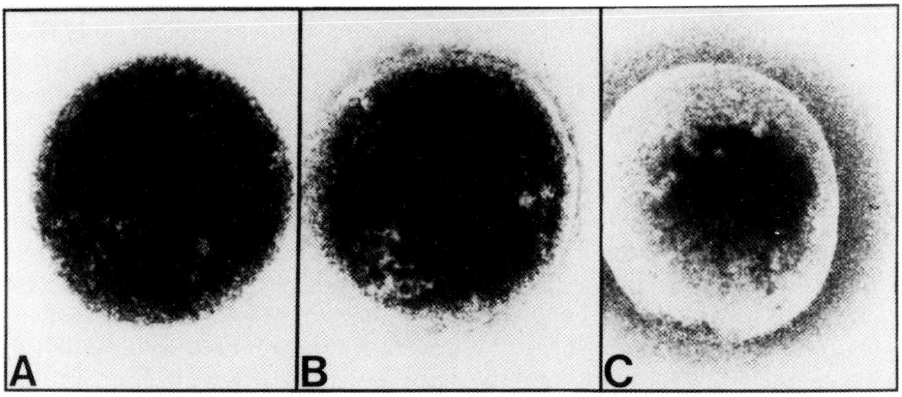

FIG. 3. Migration of human neutrophils away from an agarose microdroplet in the presence of buffer alone (control) (A), 0.1 n*M* leukotriene B_4 (B), and 10 n*M* leukotriene B_4 (maximal response) (C). Micrographs taken with a Leitz microscope. ×17.

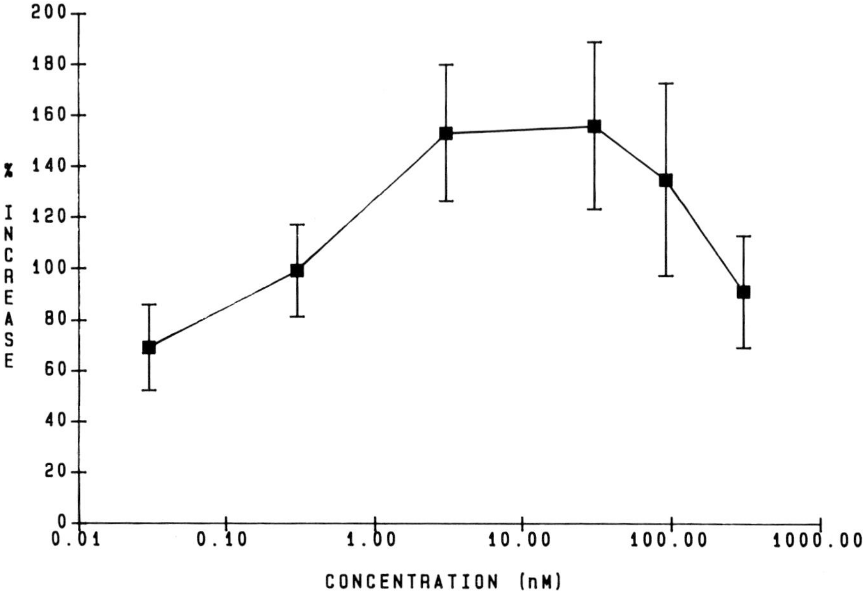

FIG. 4. Dose–response curve for induction of chemokinesis of human polymorphonuclear leukocytes by leukotriene B_4 as measured by the agarose microdroplet assay. Results are expressed as the percentage increase in the area of migration in the presence of leukotriene B_4 and shown as the means ± SEM ($n = 12$) derived from three separate experiments from three separate donors.

different incubation times are required (e.g., 1 hr for eosinophils, 6 hr for human peripheral monocytes, and 24 hr for rat peritoneal macrophages).

After the incubation period the area of migration is then measured using a projecting microscope (e.g., Projectina or Leitz). Figure 3 shows typical migration responses observed in medium alone (control) and in the presence of 0.1 and 10 n*M* leukotriene B_4. The area of migration and the area of the initial microdroplet may then either be drawn on a piece of paper and the area calculated by planimetry or by cutting out and weighing the paper or be directly entered onto a *x,y*-digitizer and the areas of migration fed directly to a microcomputer. The effects of arachidonic acid metabolites are expressed as a percentage increase in migration as compared to migration in medium alone. Bell-shaped dose–response curves are obtained as shown in Fig. 4. The decrease in migration at the higher concentrations of the agonists may reflect either immobilization induced by interaction with a low-affinity receptor site or desensitization of the high-affinity receptor sites. Using this technique, some variation in neutrophil responsiveness will be obtained from blood donor to donor although pooled results may be meaned together.

[8] Semiautomated Measurement of Neutrophil Chemotaxis with an Image Analyzer

By CARLOS C. DAUGHADAY, ALAN N. BOHRER, and ISAIAS SPILBERG

Introduction

Neutrophil chemotaxis can be evaluated quantitatively *in vitro* by several methods including surface counting, leading-front measurements, and collecting ^{51}Cr-labeled cells in a second filter. The first two techniques are often tedious, and observer fatigue and/or bias could easily influence results. These shortcomings were reduced by the introduction of the ^{51}Cr technique,[1] but its application has been limited to neutrophils; investigators still rely on the older methods when studying monocytes or other slowly moving cells. To reduce tediousness and to improve objectivity, a commercially available automatic image analyzer has been adapted to measure leading fronts in micropore filters.[2,3] In this chapter we describe the features of an image analyzer combined with an automatic data acquisition system and compare it with other methods on the basis of sensitivity, reproducibility, and technician time.

Materials and Methods

Preparation of Cells

Polymorphonuclear neutrophils (PMN) are prepared by sequential dextran sedimentation and Ficoll–Hypaque gradient centrifugation[4] (see also this series, Vol. 108 [9]). The final cell suspension should be greater than 95% pure PMN with 100% viability as assessed by Trypan blue dye exclusion.

Chemotaxis Assay

Assays are performed in Lucite chambers (Bio-Rad Laboratories, Richmond, CA), using methylcellulose filters with 5-μm pores (Millipore). Cells

[1] J. I. Gallin, R. A. Clark, and H. R. Kimball, *J. Immunol.* **110,** 233 (1973).

[2] T. E. Van Dyke, A. A. Reilly, H. Horoszewicz, N. Gagliardi, and R. J. Gencio, *J. Immunol. Methods* **31,** 271 (1979).

[3] B. L. Shore, C. C. Daughaday, and I. Spilberg, *Am. Rev. Respir. Dis.* **128,** 481 (1983).

[4] A. Bøyum, *Scand. J. Clin. Lab. Invest.* **21** (Suppl. 97), 77 (1968).

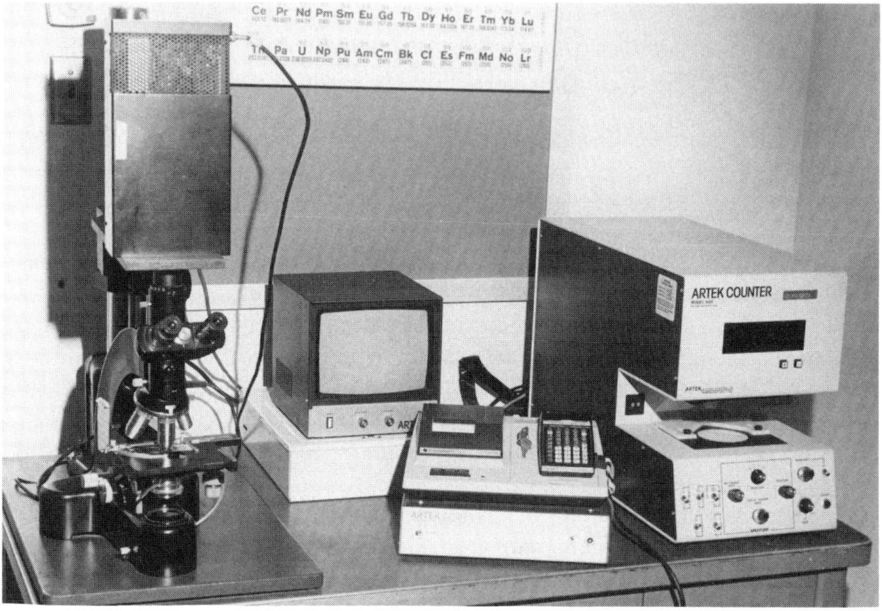

FIG. 1. The system for measurement of neutrophil migration by image analysis consists of a video camera mounted on a microscope, an automatic particle counting unit, a television monitor, and a programmable calculator mounted on a Compu-print.

are suspended at 4×10^6 PMN/ml in minimum essential medium (MEM) with Earle's salts (Gibco) containing 2% bovine serum albumin (Gibco) (MEM-BSA). *N*-Formylmethionylleucylphenylalanine (fMLP) (Sigma Chemical Co.) is prepared as a stock solution of 10^{-3} *M* in dimethyl sulfoxide and diluted in MEM-BSA to final concentrations of 10^{-8} to 10^{-12} *M*. After incubation for 30 min at 37° in 5% CO_2, the filters are fixed in buffered formalin, stained with Giemsa, washed, dried, cleared with xylene, and mounted with Permount (Fisher). Leading fronts are measured visually as described by Zigmond and Hirsch.[5] For comparison, chemotaxis with ^{51}Cr-labeled PMN may be performed as previously reported.[6]

Image Analyzer and Data Acquisition System

The system is illustrated in Fig. 1 and consists of a video camera mounted on a microscope, a television monitor, and the automatic particle counting unit (Artek Counter, Model 880, Artek Systems Corp., Farming-

[5] S. H. Zigmond and J. G. Hirsch, *J. Exp. Med.* **137,** 387 (1973).

[6] I. Spilberg, A. Gallacher, J. M. Mehta, and B. Mandell, *J. Clin. Invest.* **58,** 815 (1976).

dale, NY) interfaced with a programmable calculator and printer (Texas Instruments TI 59 and PC 100C) through an Artek Compu-print 700. Details of the design and performance of the automatic particle counter have been published elsewhere.[7] The filters are examined with a Leitz microscope equipped with a 40× Leitz PL objective, N.A. 0.65. Using the monitor screen, the operator focuses on the cell monolayer on top of the filter and adjusts the illumination system for optimal resolution by the Köhler principle as recommended by the microscope manufacturer. The size and level discriminators are adjusted so that an illuminated dot appears over each cell on the monitor screen and areas of the filter containing no cells show no dots. After the initial settings are selected, little or no change is required in subsequent experiments as long as illumination conditions remain constant. When the operator presses the count button, the number of cells for the monolayer are automatically printed and stored by the calculator. The operator manually advances the fine-focus knob in 10-μm increments, pressing the count button at each level until reaching a level with 0 counts. User-defined keys reset the calculator and allow the operator to accumulate data for three to five random fields in each filter before calculating and printing the average chemotactic parameters for each duplicate pair of filters.

Data Analysis

The chemotactic response is expressed in three different ways. The number of cells counted in the plane of focus 30 μm below the surface, as described by Swanson and Becker,[8] is taken directly from the raw data. The statistical equivalent of the leading front is calculated as the depth of penetration reached by the 99th percentile of the cell population. The central tendency of the population is represented by the weighted arithmetic mean, corresponding to the leukocyte chemotaxis index described by Maderazo and Woronick.[9] The latter two indices are calculated using standard equations for grouped univariate data[10] and incorporated into a program for the programmable calculator.

The number of cells recorded at each depth in the filter actually represents the number of cells in a horizontal slice whose thickness is determined by the depth of focus of the microscope objective. The depth of focus of the 40× objective used in these studies was found to be approxi-

[7] P. A. Bunn, S. S. Ford, and S. E. Shackney, *J. Natl. Cancer Inst.* **59,** 285 (1977).

[8] M. J. Swanson and E. L. Becker, *J. Immunol. Methods* **13,** 191 (1976).

[9] E. G. Maderazo and C. L. Woronick, *in* "Leukocyte Chemotaxis" (J. I. Gallin and P. G. Quie, eds.), p. 43. Raven, New York, 1978.

[10] R. P. Runyon and A. Haber, "Fundamentals of Behavioral Statistics." Addison-Wesley, Reading, Massachusetts, 1980.

mately 10 μm. Therefore it has been assumed that the count at 10 μm represents the frequency of cells falling into the interval between 5 and 15 μm. The frequency of each slice and the sum of the frequencies of the slices from the top of the filter through each slice (cumulative frequency) are used to calculate chemotactic parameters. The programmable calculator finds the depth penetrated by the 99th percentile (D_{99}) by linear interpolation using the following equation:

$$D_{99} = X_{\mathrm{LL}} + \frac{cum\ freq_{99} - cum\ freq_{\mathrm{LL}}}{f_i} \times i$$

where X_{LL} is the lower limit of the interval containing the 99th percentile (midpoint − 5 μm); *cum freq*$_{99}$, the cumulative frequency of 99th percentile or 0.99 × cumulative frequency at the deepest penetration; *cum freq*$_{\mathrm{LL}}$, the cumulative frequency at the lower limit of the interval containing D_{99} (*cum freq* at the previous interval); f_i, the frequency within the interval containing D_{99}; and i, the thickness of interval = 10 μm.

The weighted arithmetic mean is determined by the equation:

$$D = \frac{1}{N} \sum X_i f_i$$

where N is the total number of cells and $X_i f_i$ the product of the depth times the number of cells for each interval. A sample calculation is given in Table I. The programmable calculator will calculate and print these results using the raw data supplied to it by the Artek Counter.

Reproducibility and Sensitivity

The depth of penetration of the 99th percentile was calculated for five experiments. As shown in Table II, the standard deviations were approximately 15–20% for both random migration and the chemotactic response to 10^{-9} M fMLP, but the individual variations closely followed the results of visually measured leading fronts, indicating that the source of variability was the chemotaxis assay itself and not the method of data collection. The D_{99} was highly correlated with the visual leading fronts ($R = 0.99$).

The weighted arithmetic mean migration was calculated from the same data as a measure of the central tendency of the cell population. Since only a portion of the cell population responds to fMLP and the incubation time was only 30 min, nearly half the cells remained on the upper surface, making the population median an insensitive index. The weighted mean reflects both depth of penetration and density of cells. The mean random migration (±SD) was 6 (±1.7) μm, and the mean chemotactic migration (±SD) was 11 (±2.0) μm for 10^{-9} M fMLP.

TABLE I
SAMPLE CALCULATION[a]

Depth (μm)	Number of cells	Cumulative frequency	Frequency × depth
0	165	165	0
10	99	264	990
20	52	316	1040
30	35	351	1050
40	12	363	480
50	6	369	300
60	4	373	240
70	1	374	70
			4170

[a] The cumulative frequency for the 99th percentile is

$$cum\ freq_{99} = 0.99 \times 374 = 370.25$$

Since this falls in the interval whose midpoint is 60, it is partway between 55 and 65. The distance is calculated by linear interpolation over the interval.

$$D_{99} = 55 + \frac{(370.25 - 369)}{4} \times 10 = 58 \quad \mu\text{m}$$

The weighted mean is $4170/374 = 11.1\ \mu$m.

TABLE II
LEADING FRONTS MEASURED BY VISUAL AND IMAGE ANALYZER TECHNIQUES[a]

	Random migration		10^{-9} *M* fMLP	
Experiment	Visual	Image analyzer	Visual	Image analyzer
1	29	25	37	34
2	31	28	67	53
3	27	25	66	58
4	39	37	46	40
5	34	32	46	38
Mean	32	29	52	45
SD	4.7	5.1	13.4	10.3

[a] Correlation coefficient, $R = 0.99$.

TABLE III
COMPARISON OF DOSE RESPONSES TO FORMYLMETHIONYLLEUCYLPHENYLALANINE[a]

fMLP (M)	Visual leading front	Image analyzer leading front	Mean migration (μm)	Number of cells at 30 μm	^{51}Cr
10^{-8}	58[b]	54[b]	17.5[b]	390[b]	16.2[b]
10^{-9}	47[b]	40[b]	9.5[b]	226[b]	8.6[b]
10^{-10}	36	33[b]	7.0[b]	116[b]	1.2
10^{-11}	33	33[b]	6.5[b]	111[b]	1.2
10^{-12}	28	31	6.1	110	0.8
Random	32	30	5.8	69	0.8

[a] $n = 7$.
[b] Significantly different from random, $p < 0.05$ by Student's t test.

The sensitivity of the image analyzer system for detecting responses to low concentrations of chemotactic factors was investigated in dose–response studies. The results are shown in Table III for image analyzer-derived leading front (D_{99}), mean migration, and number of cells per high-power field in the 30-μm plane (N_{30}). The sensitivity of the D_{99} measure was comparable to the visual leading-front method. Significant differences from random were observed at a concentration of 10^{-10} M fMLP, and small effects were statistically detectable at concentrations as low as 10^{-11} M fMLP. The results were similar for mean migration. The N_{30}, however, showed greater sensitivity with definite increases observed at 10^{-11} M. For comparison, results from a similar experiment using the ^{51}Cr method are given in the last column of Table III. Using the same fMLP preparation, the ^{51}Cr method could not detect neutrophil chemotaxis at concentrations of 10^{-10} M fMLP or less.

Technician Time and Fatigue

The time required for preparation of cells and loading chambers is identical for all techniques except the ^{51}Cr method which has an additional hour-long incubation and wash. Staining and mounting of filters require time depending on the staining method used. Obtaining the data from the image analyzer requires 2 or 3 min per filter, but no additional time for data analysis is needed when an interfaced programmable calculator is used. The time is comparable to that required for visual leading-front measurements, but operator fatigue is greatly reduced. In comparison with the ^{51}Cr assay, the leading front techniques use less time with an overall savings of about 2 hr. In addition, the technique is easy to learn, and there is little or no variation between operators with different levels of experience once the proper illumination conditions are set.

The principal advantages of the image analyzer system over visual measurement of leading fronts are improved objectivity and reduced operator fatigue. A further advantage is that three indices may be obtained simultaneously from the same filter. The leading front as a measure of the best response, the mean migration as the central tendency of the whole population, and cell counts in a single plane as the most sensitive index represent a more complete profile of chemotaxis than a single parameter alone.

Although no direct comparisons were made to surface counting techniques, several potential difficulties with the latter technique have been pointed out by other authors.[8,9] Variability in the thickness of micropore filters, cell detachment from the lower surface, and variations in the number of cells applied to the upper surface are among the pitfalls that are avoided by leading-front measurements. Semiautomation of surface counting using a system similar to that described in this chapter has been reported by other investigators.[11] That method overcomes the drudgery of hand counting but retains all the disadvantages of surface counting inherent in the technique.

The image analyzer system shares the advantages of objectivity with the ^{51}Cr assay but has superior sensitivity and avoids the need for using a radioisotope. The most promising application of the image analyzer chemotaxis system is likely to be in investigations of low concentrations of chemotactic factors where increased sensitivity is required.

[11] N. H. Valerius, *Acta Pathol. Microbiol. Scand., Sect. B* **86,** 149 (1978).

[9] Concentration Gradients of Chemotactic Factors in Chemotaxis Assays

By DOUGLAS A. LAUFFENBURGER, ROBERT T. TRANQUILLO, and SALLY H. ZIGMOND

Introduction

The response of cells to a chemotactic factor depends on both the concentration and the concentration gradient of the factor. Therefore, in assays designed to experimentally investigate the chemotactic behavior of cell populations, it is extremely important to know what attractant concentrations and concentration gradients are present in the assay system. In current assays, this means that the attractant concentration profiles must be described as a function of time and spatial position throughout the field of cell movement.

When quantitative measurements of chemotactic responses are desired, either for cell populations or for individual cells, full description of the attractant concentration profiles present during the experiment is imperative. Cell population behavior can be quantitatively characterized in a meaningful way only by the use of phenomenological parameters involved in a mathematical model of cell population movement.[1] These parameters can be determined by application of the model to experimental data on cell population migration obtained from an assay experiment,[2] provided that the attractant concentration profile present during the experiment can be described. In studies of individual cell behavior, quantities such as the chemotactic index are commonly used to characterize cell responses. Such quantities are of little value, however, unless the attractant concentrations and gradients to which the observed cells are exposed are known. Thus, useful quantitative information concerning cell chemotactic responses can be obtained only when the relevant attractant concentration profiles can be measured or predicted.

Despite the central importance of attractant concentration profiles in meaningful interpretation of chemotaxis assays, there are very few reports of experimental measurements of these in commonly used assay systems.[3] Given this situation, profile behavior must be predicted theoretically. Since in most assays the attractant gradient is formed by diffusion of the attractant, the profiles should be governed by the diffusion equation applied to the particular geometry appropriate to a specific assay (see, for example, Crank[4]).

The attractant concentration profiles will depend on a number of considerations. Foremost among these is the attractant diffusion coefficient. This coefficient varies inversely with attractant molecular weight, generally to the $\frac{1}{3}$ to $\frac{1}{2}$ power.[5] The smaller the diffusion coefficient, the longer it will take for attractant concentration profiles to change with time. This means that it will take longer to establish a stable concentration gradient for a higher molecular weight attractant but that such a gradient

[1] D. A. Lauffenburger, *Agents Actions Suppl.* **12,** 34 (1983).

[2] D. A. Lauffenburger, C. Rothman, and S. H. Zigmond, *J. Immunol.* **131,** 940 (1983).

[3] D. F. Stickle, D. A. Lauffenburger, and S. H. Zigmond, *J. Immunol. Methods* **70,** 65 (1984).

[4] J. Crank, "The Mathematics of Diffusion." Oxford Univ. Press, London, 1970.

[5] C. Tanford, "Physical Chemistry of Molecules." Wiley, New York, 1961.

will also take longer to decay. The commonly used peptide attractants have diffusion coefficients of the order of 1×10^{-5} cm^2/sec at 37°,[3] so that the larger C5a molecule should have a diffusion coefficient of about 3×10^{-6} cm^2/sec. One should expect, therefore, that the time scales for gradient formation and decay should be longer by a factor of 3 for C5a than for peptide attractants. The concentration profile will also depend on geometrical aspects of the assay, such as the spatial distance over which a gradient occurs as well as the configuration of the system (i.e., linear versus cylindrical geometry). In addition, attractant concentration can be affected by chemical processes, such as degradation of the attractant by the cells. This can have a profound effect on the attractant gradient, depending on the cell density. Finally, release of cell-derived chemotactic factors, such as leukotriene B_4 (LTB_4),can cause additional, unintended gradients to be present. The magnitudes and net results of all these effects can be determined only by actually solving of the diffusion equations for each particular assay system of interest.

The purpose of this chapter is to present quantitative information concerning attractant concentrations and gradients present in commonly used assays, obtained by solution of the diffusion equation for a number of assays of interest. We will consider the bridge visual assay,[6] the Millipore assay,[7] and the under agarose assay in both cylindrical[8,9] and linear[2] geometries. We will analyze the effect of cell degradation of attractant in the latter system, for we have experimentally found it to have a potentially significant influence on assay behavior.[10]

Bridge Visual Assay

The bridge visual assay was designed for purposes of observation of individual cells within specified attractant concentration gradients.[6] A thin channel is formed over a bridge between two large reservoirs of solution, with the cells placed over the bridge. If the reservoirs contain different concentrations of attractant, a concentration gradient will form by diffusion of the attractant across the channel. The resulting concentration profile varies with both time and spatial position across the bridge; proper interpretation of cell behavior observations thus requires knowledge concerning this variation. In the absence of experimental determination of the

[6] S. H. Zigmond, *J. Cell Biol.* **75,** 606 (1977).
[7] S. Boyden, *J. Exp. Med.* **115,** 453 (1962).
[8] J. E. Cutler, *Proc. Soc. Exp. Biol. Med.* **147,** 471 (1974).
[9] R. D. Nelson, P. G. Quie, and R. L. Simmons, *J. Immunol.* **115,** 1650 (1975).
[10] R. T. Tranquillo, Ph.D. thesis. University of Pennsylvania, Philadelphia, 1986.

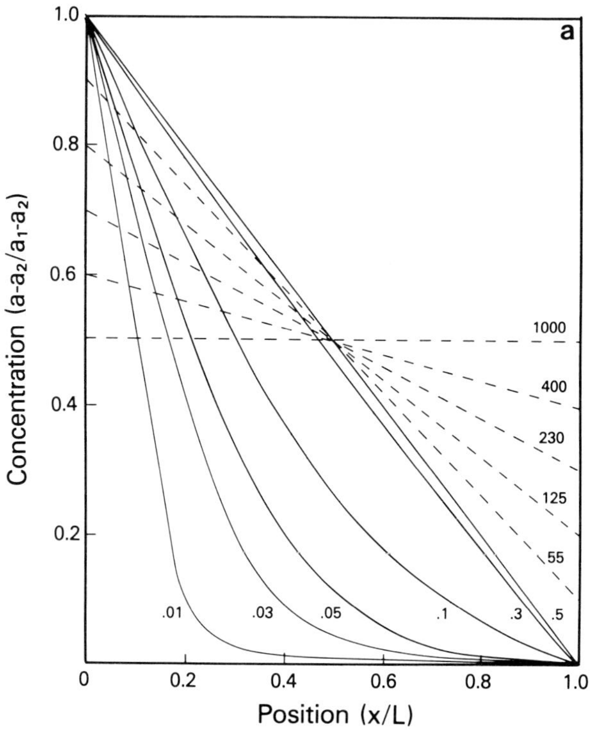

FIG. 1. (a) Attractant concentration as a function of spatial position across the bridge in the bridge visual assay. The concentration is expressed as the fraction of the original concentration difference between the two reservoirs. The spatial position is expressed as the fraction of the distance L between the two reservoirs. The numbers near each curve correspond to a sequence of times, expressed in terms of the quantity $\tau = Dt/L^2$. At time $\tau = 0$ the concentration of the bridge is assumed to be a_2. (b) Specific gradient as a function of spatial position across the bridge in the bridge visual assay. The specific gradient is expressed in units of d/L, where d is a chosen gradient detection distance (see text). The numbers near each curve again correspond to values of the dimensionless time quantity $\tau = Dt/L^2$.

concentration profiles, such knowledge must be obtained theoretically from solution of the diffusion equation applied to this system.

We have previously reported approximate calculations for this solution, assuming the reservoir volumes to be well mixed and large compared to the channel volume.[11] For a typical ratio of reservoir volume to channel volume (roughly 500), our results are shown in Fig. 1a and 1b. In Fig. 1a the concentration profile over the bridge is plotted as a function of spatial position. L is the distance across the bridge between the two reservoirs,

[11] D. A. Lauffenburger and S. H. Zigmond, *J. Immunol. Methods* **40,** 45 (1981).

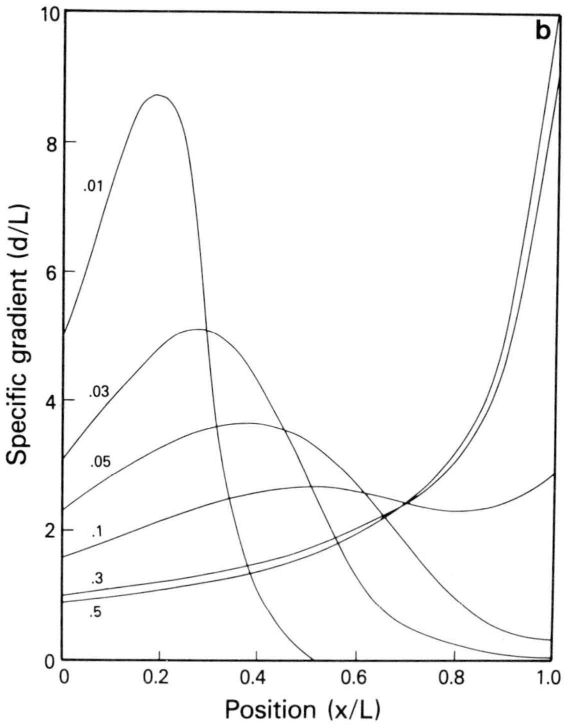

FIG. 1. *(continued)*

usually about 1 mm. a_1 and a_2 are the initial reservoir concentrations. The profiles are given for a series of values of time, expressed in units of the quantity $\tau = Dt/L^2$. This quantity is useful for generalizing the computational results, because it allows the behavior for different systems to be shown on the same plot. The quantity L^2/D has units of time; this roughly represents the period of time necessary for a significant number of molecules to diffuse a distance L. The solid curves show the development of a quasi-steady linear concentration profile across the bridge; this profile is essentially attained when $\tau = 0.3$. From this time on, the attractant concentration at the midpoint of the bridge is $(a_1 + a_2)/2$. The dashed curves represent the decay of this profile with time until the concentrations are equalized between the reservoirs. Significant deviation from the quasi-steady profile begins to appear at about $\tau = 55$. For a peptide attractant, then, a quasi-steady linear profile should exist between about 5 min and 20 hr, while for C5a it should exist between about 15 min and 60 hr. The actual gradients and decay times will be smaller, however, because in practice the reservoirs are usually not mixed intentionally. An estimate of this effect will be given in the section on the Millipore filter assay.

Figure 1b illustrates another important aspect of the concentration profile: the specific gradient. This quantity is defined as the ratio of the absolute concentration gradient at any point to the absolute concentration at that point. It is a useful quantity because it provides a crude criterion for assay construction. Zigmond[6] has found that the limit of gradient detection by polymorphonuclear leukocytes is approximately a specific gradient of 0.5% over a distance of 10 μm, at an absolute concentration near K_d, the equilibrium dissociation constant for binding of attractant to cell surface receptors. In order to be capable of allowing a noticeable chemotactic response, then, an assay must generate specific gradients at least this great at concentrations near K_d for significant periods of time. Figure 1b shows the specific gradients generated within the bridge visual assay during the formation of the quasi-steady linear gradient in units of d/L, where d is any specified gradient detection distance, for the particular case $a_1/a_2 = 10$. For instance, choosing $d = 10$ μm, the specific gradient at the center of the bridge during the quasi-steady period ($\tau \geq 0.3$) is roughly 1.5%. Since the absolute concentration at the center of the bridge can be maintained constant at any desired concentration [equal to $(a_1 + a_2)/2$] during the quasi-steady period, the desired gradient can be generated easily in this assay. Notice that the specific gradient varies significantly as a function of position on the bridge. Cells in different places should thus be expected to show different orientation responses.

Millipore Filter Assay

The Millipore filter assay was designed for quantitative assessment of cell population migration in response to chemotactic attractants.[7] A filter is placed between two reservoirs of solution, with cell migration occurring into the filter from one of its surfaces. If the reservoirs contain different concentrations of attractant, a concentration gradient will form by diffusion of the attractant through the filter. As with the bridge visual assay, the resulting concentration profile varies with time and spatial position within the filter, and determination of useful parameters quantifying the cell response depends on describing the concentration profile behavior.

We have previously used a solution of the diffusion equation to predict the concentration profiles present in this assay, assuming that the reservoirs are of equal volume and that there is no mixing of the reservoir volumes.[11] If the volumes were, in fact, well mixed the results discussed in the previous section would be applicable here. Conversely, if there is no mixing of the reservoirs in the bridge visual assay, the concentration

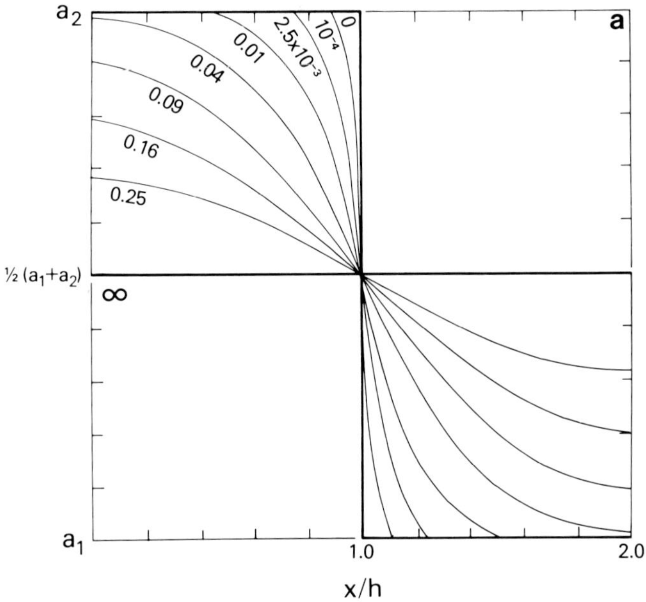

FIG. 2. (a) Attractant concentration in the Millipore filter assay as a function of distance from the top of the assay chamber. The filter is located at the position $x = h$, where h is the depth of a reservoir. The numbers near each curve correspond to a sequence of times, expressed in terms of the quantity $\tau = Dt/h^2$. (b) Specific gradient across the filter in the filter assay as function of time. The specific gradient is expressed in units of d/h, where d is a chosen gradient detection distance (see test). Time is expressed in units of h^2/D.

profiles in that system would be more accurately described by the results to follow in this section.

Figure 2a shows the profile of absolute concentration as a function of spatial position within the entire assay system for a series of values of time, now parameterized by the quantity Dt/h^2 (h is the depth of each of the assay reservoirs). Since the filter thickness is generally very small compared to the reservoir depth, there is relatively little variation in concentration across the filter in this system because of the assumption of no mixing of the reservoir contents. (In contrast, if the reservoir volumes were well mixed, as in the bridge visual assay, the entire gradient would occur across the filter itself. Thus, mixing of the reservoirs provides a sharper gradient across the cell movement field; it correspondingly enhances the rate of decay of the gradient.) Hence, the filter concentration and gradient can be considered to be simply that occurring at the center of the assay system; on Fig. 2a this is the spatial position $x/h = 1$. The concentration within the filter remains constant at the value $(a_1 + a_2)/2$, the average of the initial

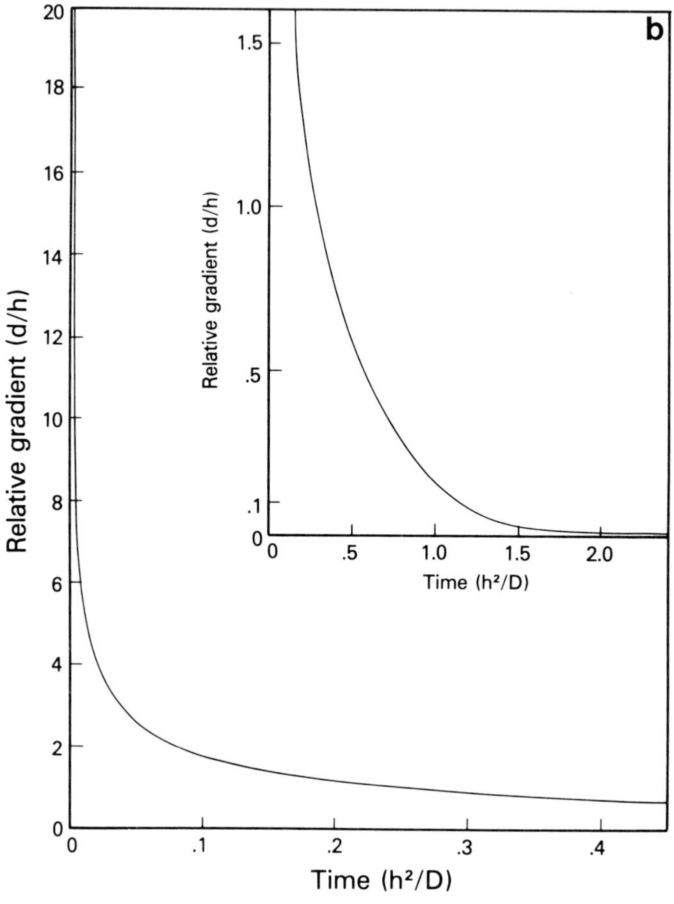

FIG. 2. *(continued)*

reservoir concentrations. Figure 2b shows the specific gradient present within the filter as a function of time, with time given in units of h^2/D. The specific gradient is given in units of d/h, where d is a specified gradient detection distance as before. Notice that the gradient decays dramatically with time. For a typical value of $h = 3$ mm, and $d = 10\ \mu$m, this figure shows that the specific gradient within the filter will be greater than 0.5% over 10 μm for roughly 30 min for a peptide attractant, and for about 90 min for C5a. This result is independent of the filter thickness, because this thickness is negligible compared to the length of the assay system.

If the reservoir contents in the filter assay were well mixed, Fig. 1a and 1b would apply, but with L equal to the filter thickness. The appropriate volume ratio would now be h/L; for typical Millipore filters $L = 150\ \mu m$ so that h/L would be roughly 20. Surprisingly, it turns out that the desired specific gradients of at least 0.5% over 10 μm would be present during approximately the same time period as for the case with unmixed reservoirs. Apparently the state of mixing of the reservoir contents in this assay does not affect this criterion greatly. In the well-mixed situation the gradients will be sharper, but will dissipate faster, so that the opposing effects largely cancel. (However, the observed cell response might be affected significantly because the gradients actually present during this time period will thus be different.) The net result will depend on the filter thickness. For the thinner nitrocellulose-type filters, with L equal to about 10 μm, h/L is roughly 300. For these filters the desired specific gradient will be present for a much shorter period of time: about 5 min for a peptide attractant and about 15 min for C5a. It can be more generally shown[11] that, with all other aspects unchanged, the ratio of specific gradient duration times, t_1/t_2, for filters of lengths L_1 and L_2 will be equal to $L_1/L_2 \ln(L_2/L_1)$. Thus, it can be expected in general that for thinner filters the gradients will be somewhat sharper but for a significantly shorter period of time when the reservoirs are well mixed.

Figures 2a and 2b are relevant in an approximate way to the bridge visual assay, as well, if its reservoir contents are not well mixed. In this situation $2h$ would be the distance across the entire assay system, roughly 5 mm. Given this value the desired specific gradient would exist for a peptide attractant for about 3.5 hr, and for C5a for about 10 hr. Notice that these values are substantially smaller than predicted by the earlier solution assuming well-mixed reservoirs for this assay. However, these values *underestimate* the gradient decay time because the solution on which they are based neglects the fact that the channel across the bridge is much thinner than the reservoirs.

Under Agarose Assay

The under agarose assay possesses the advantage of potentially allowing individual cell observations along with population migration measurements, but it has the most complicated concentration profile behavior. This problem is alleviated somewhat by the modification in which a 1-dimensional linear configuration is used[2] instead of the original 2-dimensional cylindrical geometry.[8,9] Both systems will be discussed here.

Cylindrical Assay

In the cylindrical assay system the attractant is initially placed in a cylindrical well cut into a layer of agarose on a surface. It diffuses radially through the surrounding agarose, creating concentration profiles that depend on time and radial distance from the well. The cells are initially placed in another cylindrical well at a distance from the attractant well and migrate out of the well under the agarose layer. One problem with this system is that cells migrating from different areas of the cell well experience different attractant concentrations and gradients.

We have previously presented calculations of the attractant concentra-

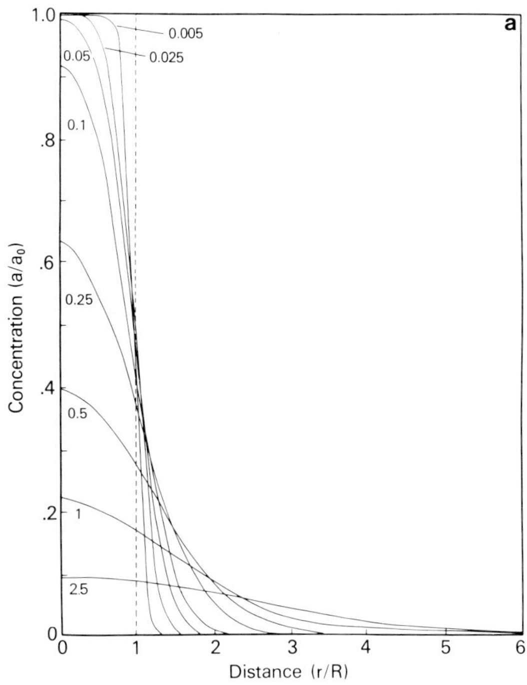

FIG. 3. (a) Attractant concentration in the cylindrical under agarose assay as a function of radial distance from the center of the attractant well. The concentration is expressed as the fraction of the initial well concentration. Distance is expressed in terms of the ratio of the distance to the well radius R. The numbers near each curve correspond to a sequence of times, expressed in terms of the quantity $\tau = Dt/R^2$. (b) Specific gradient in the cylindrical under agarose assay as a function of radial distance from the center of the attractant well. The specific gradient is expressed in units of d/R, where d is chosen gradient detection distance (see text). The numbers near each curve again correspond to the dimensionless time quantity $\tau = Dt/R^2$.

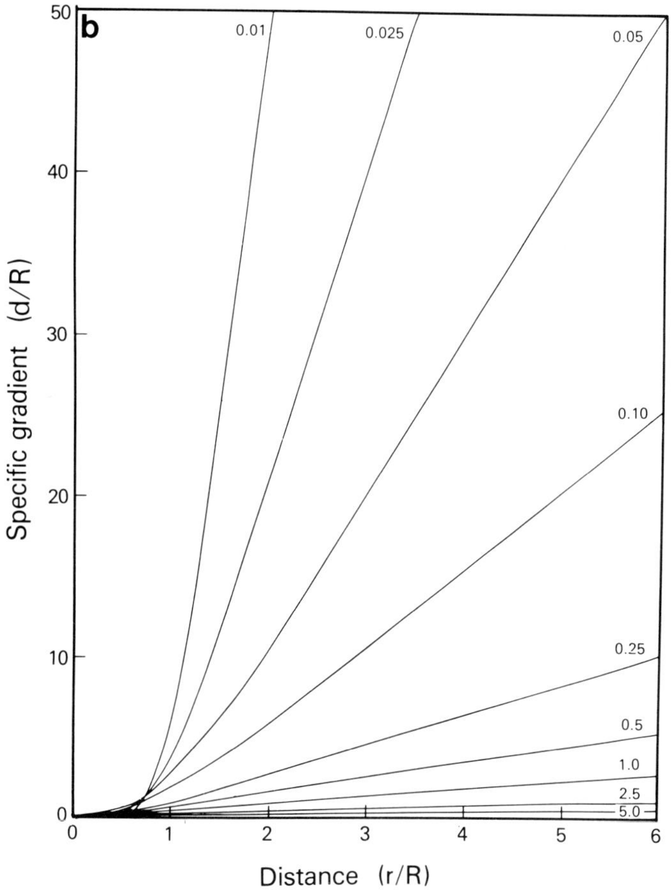

FIG. 3. *(continued)*

tion profile as a function of time and radial distance from the attractant well.[11] The results are shown in Fig. 3a and 3b; the curves are given for a sequence of values of time, in units of the quantity Dt/R^2. R is the attractant well radius, and a_0 is the initial attractant concentration in that well. The cell well is typically placed such that its edge nearest the attractant well is a distance of about $2R$ away from the edge of that well. Notice from Fig. 3a that the attractant concentration at the edge of the cell well never exceeds $0.1a_0$. This means that the initial attractant well concentration must be at least $10K_d$ in order for the concentration near the cell well to be about K_d. The specific gradient near the cell well can be quite large, on the other hand, as can be seen from Fig. 3b. For a typical value of $R = 0.3$ cm,

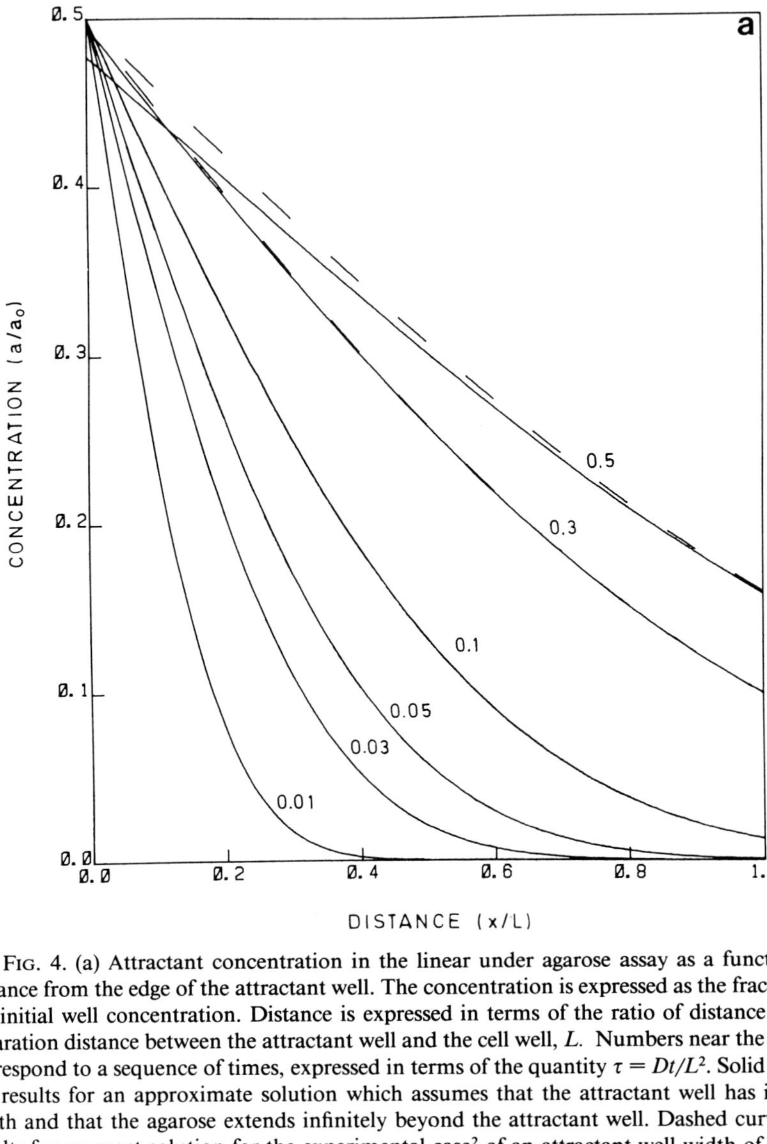

FIG. 4. (a) Attractant concentration in the linear under agarose assay as a function of distance from the edge of the attractant well. The concentration is expressed as the fraction of the initial well concentration. Distance is expressed in terms of the ratio of distance to the separation distance between the attractant well and the cell well, L. Numbers near the curves correspond to a sequence of times, expressed in terms of the quantity $\tau = Dt/L^2$. Solid curves are results for an approximate solution which assumes that the attractant well has infinite width and that the agarose extends infinitely beyond the attractant well. Dashed curves are results for an exact solution for the experimental case[2] of an attractant well width of 5 mm, agarose extending 25 mm beyond the attractant well, and $L = 5$ mm. For a typical 3-hr experiment using a peptide attractant, $\tau = 0.45$, the approximate solution is in excellent agreement with the exact solution over the normal range of cell population migration ($x/L = 0.5 - 1.0$). Increasing deviations occur for times greater than 0.5 for these dimensions and would occur at earlier times for smaller dimensions. (b) Specific gradient in the linear under agarose assay as a function of distance from the edge of the attractant well. The specific gradient is expressed in units of fractional concentration difference over the distance L. The solid curves coincide exactly with the exact curves, indicating that the approximate solution is in excellent agreement with the exact solution. Numbers near the curves again correspond to the dimensionless time quantity $\tau = Dt/L^2$.

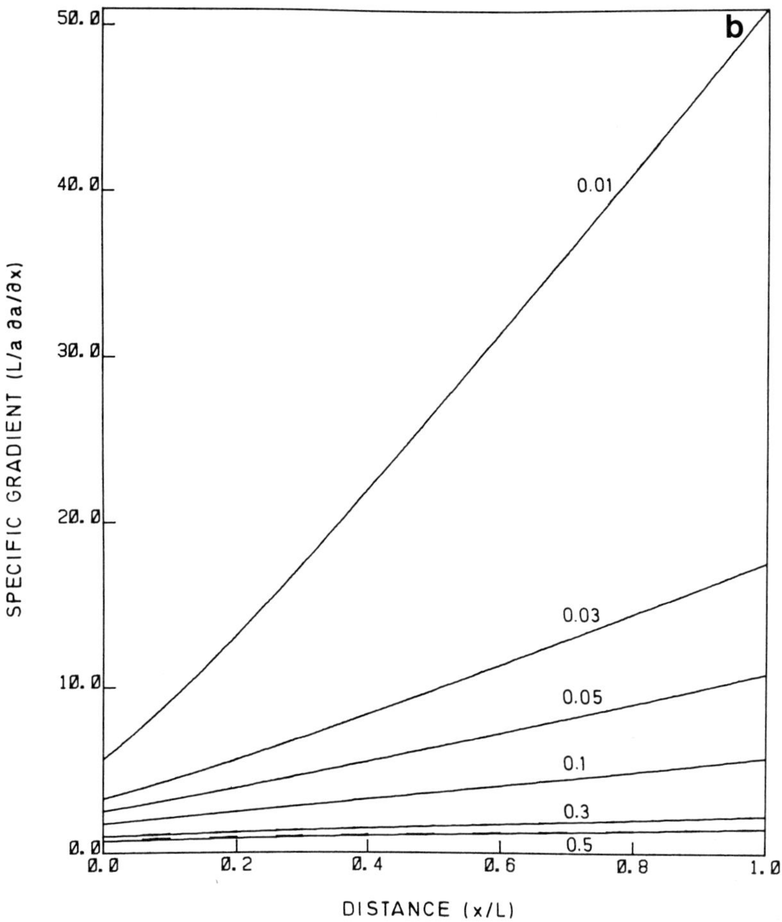

FIG. 4. *(continued)*

the specific gradient at the cell well will be greater than 0.5% over 10 μm for about 1.5 hr for a peptide attractant, and for about 4.5 hr for C5a.

Linear Assay

Interpretation and analysis of the under agarose assay is greatly aided by a recent modification: instead of cylindrical wells, parallel linear wells are used for the attractant and cells.[2] Since attractant diffusion is now 1-dimensional, the concentration and gradient are uniform along the cell well. An approximate solution to the diffusion equation for this system was provided by Rothman and Lauffenburger,[12] and an exact solution more recently by Tranquillo.[10] Both of these results are shown in Fig. 4a and 4b, which give the concentration and specific gradient as functions of distance from the attractant well for a sequence of times. a_0 is again the initial concentration in the attractant well, and L is the distance between wells. Thus, the cell well is located at position $x/L = 1$ on these plots. The values of time for these curves are given in terms of the quantity Dt/L^2. Notice that in this linear geometry, the attractant concentration at the cell well is a greater fraction of a_0 than in cylindrical geometry. To achieve a given concentration at the cell well, then, a_0 can be lower in the linear assay. The specific gradient is still quite large. In Fig. 4b the specific gradient is given in terms of the relative concentration change over the distance L. For a typical value of $L = 3$ mm, a specific gradient of 0.5% over 10 μm will be present for just about the same period of time as in the cylindrical under agarose assay. Thus, the change in geometry does not seem to affect the rate of gradient decay noticeably.

Attractant Degradation

One complication that has not yet been considered in any of the assays is the degradation of attractant by the cells. The effects of cell processing of the attractant can be significant when high cell densities are used. Tranquillo[10] has recently obtained an approximate solution for the diffusion equation which includes degradation of attractant by the cells in the well, since that is where the cells will be present at highest density. Degradation of attractant by the smaller number of migrating cells is neglected. Results of these calculations are shown in Fig. 5a and 5b, where the concentration and specific gradient are plotted as a function of distance from the attractant well for a sequence of times, as before. Additionally, the effect of a parameter characterizing the rate of attractant degradation by the cells is illustrated. This parameter is $\beta = kL^2N_0/Da_0$, where k is the degradation rate constant per cell and N_0 the cell density in the cell well. k is assumed to be independent of attractant concentration for this approximate calculation. Figure 5a shows that increasing β decreases the attractant concentration, but *increases* the specific gradient—especially near the cell well. This suggests that at high cell densities attractant degradation might cause a substantial enhancement of cell migration. We have recently obtained experimental evidence that supports this suggestion. Figure 6 shows experimental cell migration results in the linear under agarose assay for a range of cell densities, for the case in which attractant is initially present at a

[12] C. Rothman and D. A. Lauffenburger, *Ann. Biomed. Eng.* **11**, 451 (1984).

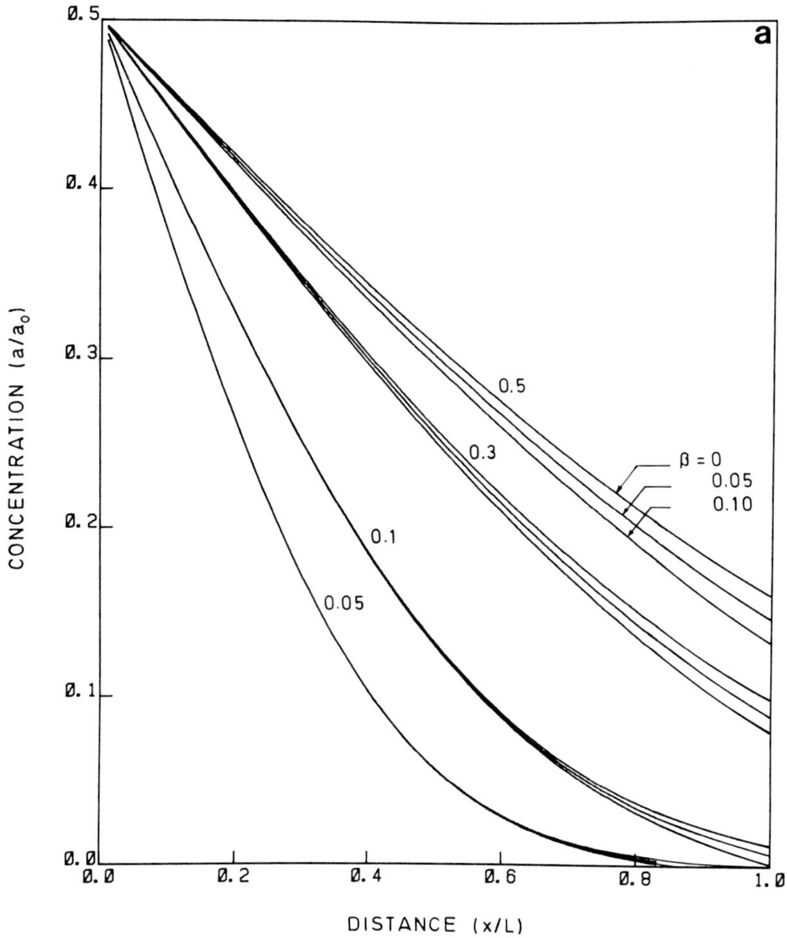

FIG. 5. (a) Attractant concentration in the linear under agarose assay as function of distance from the edge of the attractant well, including the effect of cell degradation of attractant. The concentration is expressed as the fraction of the initial concentration in the attractant well. Results are given for four values of the dimensionless time quantity $\tau = Dt/L^2$. For each of these times, results are given for three values of the dimensionless degradation parameter $\beta = kL^2N_0/Da_0$, where k is the degradation rate constant per cell and N_0 the cell well density. The results for no degradation, $\beta = 0$, correspond to the results for the approximate solution in Fig. 4a. Increasing degradation decreases the attractant concentration, especially near the cell well and for longer times. (b) Specific gradient in the linear under agarose assay as a function of distance from the edge of the attractant well, including the effect of cell degradation of attractant. The specific gradient is given in terms of the fractional concentration change over the distance L. Results are shown for three values of the dimensionless degradation parameter β, for each of three values of the dimensionless time parameter $\tau = Dt/L^2$. The results for no degradation, $\beta = 0$, correspond to the solutions shown in Fig. 4b. Increasing levels of degradation increase the specific gradient, especially for short times and near the cell well.

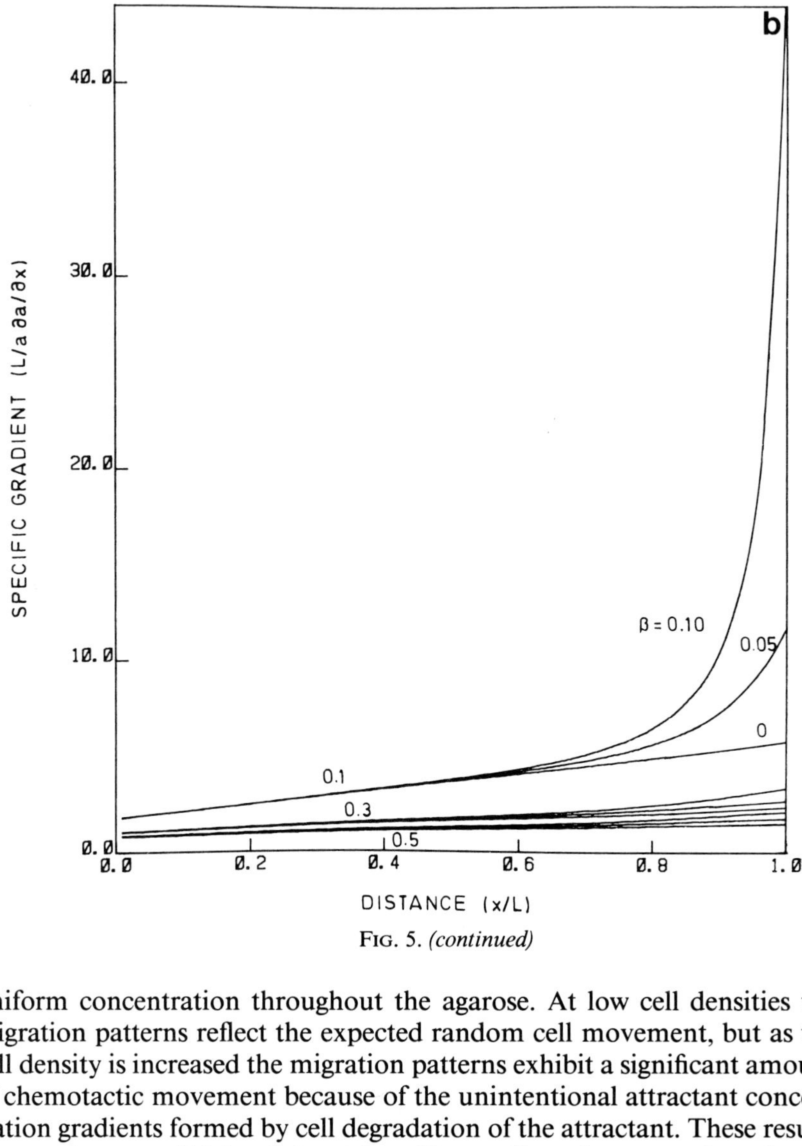

FIG. 5. *(continued)*

uniform concentration throughout the agarose. At low cell densities the migration patterns reflect the expected random cell movement, but as the cell density is increased the migration patterns exhibit a significant amount of chemotactic movement because of the unintentional attractant concentration gradients formed by cell degradation of the attractant. These results suggest that use of high cell densities will complicate interpretation of assay measurements. If the degradation rate constant can be determined independently,[13] however, Fig. 5a and 5b can be used quantitatively to analyze the measurements even when degradation is significant.

[13] S. H. Zigmond, S. J. Sullivan, and D. A. Lauffenburger, *J. Cell Biol.* **92,** 34 (1982).

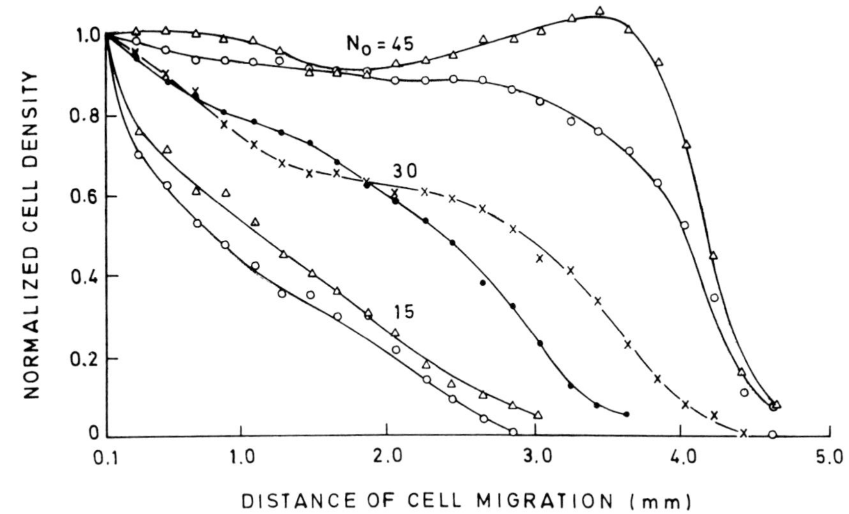

FIG. 6. Experimental density profiles of migrating cells in the linear under agarose assay.[10] Cell density is expressed as the fraction of the density present at the edge of the cell well, after an experimental time period of 3 hr. These results are for human polymorphonuclear leukocytes migrating in response to a chemotactic peptide initially present at *uniform* concentration everywhere throughout the assay chamber. Duplicate results are shown for each of three initial cell well densities, N_0 (cells/100 μm^2). A transition from the expected random (chemokinetic) migration profiles to a chemotactic response is observed as the cell density is increased, presumably due to generation of an attractant concentration gradient arising from cell degradation of the attractant.

Acknowledgments

This work has been supported by National Institutes of Health Grants DE-00143-02 (Research Career Development Award) and AI-21538-02 to D. A. Lauffenburger, and by a grant from the Monsanto Company in conjunction with a National Science Foundation Presidential Young Investigator Award to D. A. Lauffenburger.

[10] Measurement of Leukocyte Chemotaxis *in Vivo*

By MICHAEL D. P. BOYLE, MICHAEL J. P. LAWMAN, ADRIAN P. GEE, and MICHAEL YOUNG

Measurement of Chemotaxis *in Vivo*

Several methods have been developed to study the vectorial chemotactic migration of purified cell populations *in vitro*. These methods (see this volume [2]–[6]) (1) have established the existence of a number of different chemotactic agents, (2) have demonstrated the presence of specific receptors for these agents on the surface of migrating cells, and (3) have been used to characterize the responding cell population.[1-3] Since all of these methods use isolated chemotactic agents, and pure or highly enriched sources of migrating cells, directed migration can be readily demonstrated. By contrast, measurements of chemotactic responses under physiological conditions are more complex.

The response of an animal to a chemotactic agent represents the net effect of (1) the direct chemotactic activity of the agent itself and (2) any physiological chemotactic factors that may be generated by this agent. In addition, the *in vivo* response will also be influenced by various local inhibitory or amplifying systems and the architecture (e.g., lymphatic drainage, capillary access) at the injection site. Thus, it is difficult to prove that accumulation of cells at any site is a consequence solely of chemotaxis and that enhanced random motion of cells does not contribute to the response.

A variety of animal models have been described to examine cellular localization *in vivo* and to evaluate the role of chemotaxins in inflammatory reactions. These methods can be grouped into four major categories: (1) histopathological approaches, (2) use of radioactively labeled leukocytes to monitor cellular localization, (3) the skin window technique, and various modifications of this procedure, including implantations of chambers and sponges, and (4) the air sac procedure. Each of these procedures has limitations and advantages. The purpose of this chapter is to compare the various methods and to describe, in detail, the air sac technique which we have used in our laboratory to study early inflammatory responses to various chemotactic stimuli in mice. In examining these methods, a variety of considerations will be addressed: (1) simplicity and versatility of the method, (2) Ability to measure responses, both qualitatively and quantitatively, and (3) Ability to distinguish responses mediated by noncomplement-dependent reactions from those involving activation of the complement system and generation of the chemotactic split product C5a.

Histopathological Approaches

Much of the early information on inflammatory responses to chemotactic agents has been derived from analysis of the cells that accumulate

[1] A. P. Gee, *Mol. Cell. Biochem.* **62,** 5 (1984).

[2] W. A. Marasco, K. M. Becker, D. E. Feltner, C. S. Brown, P. A. Ward, and R. Nairn, *Biochemistry* **24,** 2227 (1985).

[3] I. Olsson and P. Venge, *Allergy* **35,** 1 (1980).

and participate in granuloma formation.[4–6] Frequently these studies involved gross measurement of the extent of reaction by determining the degree of foot pad swelling, skin induration, etc.[7,8] Samples from around the site of injection of the eliciting stimulus are surgically removed, fixed in an appropriate fixative [e.g., 10% formalin in 75 m*M* phosphate-buffered saline (PBS), pH 7.0], and decalcified by treatment in 5% formic acid. Samples are then embedded in paraffin wax, sectioned, and stained by standard procedures.[9,10] This approach has been used to study inflammatory responses in many anatomical sites. In general, the pattern of cellular infiltration in all sites is similar, with polymorphonuclear leukocytes being the earliest cell type to respond.[3]

The role of the complement system in this model can be determined by carrying out the experiments in mice genetically devoid of C5, or in other animals in which the terminal complement components have been depleted by treatment with cobra venom factor.[11,12] This latter approach has limitations, since cobra venom factor depletes terminal complement components but does not inhibit their synthesis. Furthermore, depletion of the terminal complement components occurs by a process that involves activation of C5, and so, in these experiments, the animals are exposed to very high concentrations of C5a, which may affect their normal inflammatory response patterns.

In general, the histopathological techniques have been of value in establishing the morphological nature of the cells involved in inflammatory reactions. By using a variety of histological stains, it is possible to study the distribution of different activation states of cells, and electron microscopy can be used to examine, in detail, the fine structure of cells present in granulomatous tissue. Although this approach allows identification of the cellular composition within the inflammatory site, the number of sections used and the time and skill required to prepare good samples

[4] R. Ross, *Biol. Rev.* **43,** 51 (1968).

[5] J. J. Oppenheim, D. L. Rosenstreich, and M. Potter (eds.), "Cellular Functions in Immunity and Inflammation." Elsevier, New York, 1981.

[6] S. Shoshan, *Int. Rev. Connect. Tissue Res.* **9,** 1 (1981).

[7] G. L. Asherson and C. G. Allwood, *Immunology* **22,** 493 (1972).

[8] D. F. Gray and P. A. Jennings, *Am. Rev. Tuberc.* **72,** 171 (1953).

[9] S. W. Thompson, "Selected Histochemical and Histopathological Methods." Thomas, Springfield, Illinois, 1966.

[10] L. G. Luna, "Manual of Histological Staining Methods of the Armed Forces Institute of Pathology." McGraw-Hill, New York, 1960.

[11] J. T. Hicks, F. A. Ennis, E. Kim, and M. Verbonitz, *J. Immunol.* **121,** 1437 (1978).

[12] R. L. Hirsch, D. E. Griffin, and J. A. Winkelstein, *J. Immunol.* **121,** 1276 (1978).

make this method inefficient for routine use in studying chemotactic responses *in vivo.*

Use of Radiolabeled Leukocytes

A number of methods have been described that use radiolabeled cells to study the directed movement of cells to a site of inflammation or to a chemotactic agent.[13-19] A variety of *in vivo* and *in vitro* methods for labeling cells have been used. *In vivo* labeling can be achieved by injection of a radioisotope, e.g., [^{3}H]thymidine, into the animal, allowing the isotope to be incorporated into cells during cell division.[13] Cells labeled in this way can then be transferred to an experimental animal, either by isolation and injection or by surgically joining the circulations of the "labeled" and experimental animals.[13] The fate of the labeled cells in the experimental animal is then determined by autoradiographic examination of tissue sections at the site of inflammation where the labeled cells will localize.[13] This approach is extremely complex and is no longer widely used.

A second approach is to label cells *in vitro.* In this approach, blood is removed either from the experimental animal or from a syngeneic or allogeneic donor, and the leukocyte population is isolated and labeled with a suitable radioisotope. In most cases, the isotopes of choice are gamma emitters such as ^{51}Cr, ^{111}In, ^{133}Xe, or ^{99}Tc.[14-19] These isotopes have the advantages that they are readily incorporated into the cytoplasm of leukocytes and that the distribution in the experimental animal can be readily monitored by counting blood or tissue samples, or, in some instances, can be continuously monitored, without sacrificing the animal, by use of sophisticated counting instruments, e.g., the gamma camera.[19] This approach enables the distribution of injected cells to be followed efficiently, and it permits accurate quantitation of accumulation of cells at different sites in the animal. The validity of extrapolating the results of these studies to normal physiological responses assumes that the isotope remains associated with the labeled cells following injection, and that the cell type which was originally labeled has not been modified in any way during the isolation and labeling procedures. Furthermore, since only the cells from

[13] W. G. Spector, A. W. J. Lykke, and D. A. Willoughby, *J. Pathol. Bacteriol.* **93,** 101 (1967).
[14] M. F. Franco and J. Morley, *J. Immunol. Methods* **11,** 7 (1976).
[15] T. J. Williams, *Br. J. Pharmacol.* **65,** 517 (1979).
[16] I. G. Colditz and H. Z. Movat, *J. Immunol.* **133,** 2163 (1984).
[17] I. G. Colditz and H. Z. Movat, *J. Immunol.* **133,** 2169 (1984).
[18] D. M. Bamberger, D. N. Gerding, K. M. Bettin, M. K. Elson, and L. A. Forstrom, *J. Infect. Dis.* **152,** 903 (1985).
[19] A. W. Segal, M. L. Thakur, R. N. Arnot, and J. P. Lavender, *Lancet* **2,** 1056 (1976).

the peripheral blood that are labeled will be quantified by this approach, the contribution of cells from other sources, e.g., bone marrow, would not be assessed.

These *in vitro* techniques have proved successful for locating abscesses and other sites of inflammation in man[19] and have confirmed in animal models that the neutrophil is the first cell type to respond to inflammatory stimuli.[14-18] The approach, however, is of limited value for long-term studies, and it can be difficult to obtain large numbers of counts at an inflamed site versus a control site. In addition, problems associated with complying with the appropriate regulations for handling animals injected with radioactive materials limit the number of animals that can be conveniently studied.

The importance of the complement chemotactic products in studies using radioactively labeled cells can be determined by using C5-deficient mice or by depletion of complement by treating the animals with cobra venom factor, as described above. The major disadvantage of methods which use labeled cells is that the responses measured are those mediated by a select population of cells which have been extensively manipulated *in vitro*. The advantages of this method are its ability to obtain quantitative results and the potential for continually monitoring cell migration without having to remove samples or sacrifice the animal.

Implantation of Chambers, Sponges, and Other Modifications of the Skin Window Technique

One of the most extensively used techniques for studying chemotaxis *in vivo* is the skin window technique and its various modifications. In these approaches, a coverslip, or some form of permeable plastic, Teflon, or stainless steel chamber containing a chemotactic factor, is directly attached to a suitable site on the skin.[20-28] The cells that migrate into the chamber, or become attached to the coverglass, can then be counted and their morphological type determined. Methods that use chambers containing a fixed volume of liquid are more versatile, since all the migrating cells can

[20] J. W. Rebuck and J. H. Crowley, *Ann N.Y. Acad. Sci.* **59,** 757 (1955).

[21] H. S. Goldsmith, A. G. Levin, and C. M. Southam, *Surg. Forum* **16,** 102 (1965).

[22] R. Samak, R. Edelstein, D. Bogucki, M. Samak, and L. Israel, *Biomedicine* **32,** 165 (1980).

[23] P. M. Bedard, B. Zweiman, and P. C. Atkins, *J. Clin. Immunol.* **3,** 84 (1983).

[24] M. L. Goodman, B. A. Way, and J. W. Irwin, *J. Pathol.* **128,** 7 (1979).

[25] J. F. Borel and C. Feurer, *J. Pathol.* **124,** 85 (1978).

[26] K. Sveen and T. Hofstad, *Acta Pathol. Microbiol. Scand., Sect. B* **84,** 252 (1976).

[27] M. B. Sporn, A. B. Roberts, J. H. Shull, J. M. Smith, and J. M. Ward, *Science* **219,** 1329 (1983).

[28] C. Feurer and J. F. Borel, *Antibiot. Chemother.* **19,** 161 (1974).

be sampled, whereas, when a glass coverslip is used, the pattern of cellular buildup is frequently uneven and aggregation of cells makes quantitation extremely difficult.

The basic technique of using implanted chambers, cover glasses, or sponges to measure chemotaxis *in vivo* has been applied to studies in humans,[20–23] rabbits,[24–26] dogs,[29] and mice.[30] The method for applying or implanting the chambers is similar in all cases. Skin at a suitable site, e.g., the human forearm or the rabbit ear, is shaved, and an area corresponding to the size of the chamber to be implanted is stripped of the epidermis to reveal the underlying corium.

Three basic approaches to preparing the skin site have been described. These are skin abrasion using (1) surgical tape or a scalpel,[31,32] (2) a rotary wheel,[33] or (3) a suction blister device.[34–36] It can be difficult to prepare skin lesions that are identical in area and demonstrate no disruption of blood vessels, without causing direct tissue damage and inflammation. Bleeding is frequently observed when surgical or mechanical abrasion techniques are employed,[32,33] and although tape stripping techniques do not usually result in bleeding considerable variation in inflammatory responses accompanying the skin site by this approach has been reported.[35] A variety of suction devices have been described for preparing suitable skin sites for implantation of chambers.[30,34] The blisters produced by these devices are uniform and reproducible. They can be formed with minimal inflammation at the site, and bleeding or infection has not been a major problem.[34–36] It is possible to prepare multiple suction blisters on the forearm simultaneously, enabling parallel studies on the same individual to be performed.[35]

A sterile cover glass or chamber is attached to the prepared site by surgical tape or surgical cement. A variety of cell collection chambers have been developed, including some manufactured from materials commonly available in the laboratory. Senn *et al.*[33] have used a chamber consisting of 20-mm polypropylene bottle caps into which two disposable 18-gauge

[29] J. M. Riddle and M. I. Barnhart, *Blood* **25,** 776 (1965).

[30] J. W. Osebold, P. M. Outteridge, L. D. Pearson, and R. A. Capua, *Infect. Immun.* **2,** 127 (1970).

[31] B. S. Goldberg, W. L. Weston, P. F. Kohler, M. B. Harris, and J. R. Humbert, *J. Invest. Dermatol.* **72,** 248 (1979).

[32] D. C. Dale and S. M. Wolfe, *Blood* **38,** 138 (1981).

[33] H. Senn, J. F. Holland, and T. Banerjee, *J. Lab. Clin. Med.* **74,** 742 (1969).

[34] K. B. Hellum and C. O. Solberg, *Acta Pathol. Microbiol. Scand. Sect. C* **85,** 413 (1977).

[35] P. Koivuranta-Vaara, *J. Immunol. Methods* **79,** 71 (1985).

[36] U. Kiistala, *J. Invest. Dermatol.* **50,** 129 (1968).

hypodermic needles have been heat sealed. The needles are inserted, one near the tip of the cap and the other near the base, to enable the contents of the chamber to be readily aspirated and refilled.[33] Hellum and Solberg[34] used a siliconized hollow rubber stopper as a chamber. Liquids may be inserted or removed from these chambers without leakage by use of a sterile needle attached to a syringe. Other groups have described chambers made of Pyrex,[37] stainless steel,[27] plastic,[25,26,28,31,33,35,38] and glass.[39,40] One form of the plastic chamber is commercially available from Sandoz (Basle). In most cases, the chambers have been designed to enable fluid to be withdrawn and replaced, thereby enabling responses to be monitored over time. This has led to the development of methods to measure not only cellular responses but also deposition of extracellular matrix as a function of time.[27,28] While there has been no definitive comparison of different chambers, Mass *et al.*[38] have reviewed the data from a number of studies using different chamber designs.

The implantation technique can also be used to quantify responses either by microscopic cell counting or by chemical methods which quantitate cellular products.[23,27] These approaches are particularly valuable for experiments carried out over prolonged periods, i.e., 2–3 weeks,[27,28] but they are technically difficult and not practical for experiments on large numbers of animals. Implantation methods, in particular the skin window technique, however, have been used extensively in humans for short-term studies of inflammation.[20–23] In animals, these approaches are of particular value for longer term studies of wound healing.

Air Sac Procedure

The "granuloma pouch" technique was first developed as a method for studying inflammatory responses in rats.[41] Air is injected subdermally to generate a symmetrical air pouch on the back of the animal, resulting in the disruption of the normal architecture of the skin and creating a space into which a chemotactic factor, or putative inflammatory stimulus, can be introduced. On its ventral surface, the air sac is composed of a thin membranous layer of fascia which can be removed, stained, and microscopically examined. The infiltrating cells can be counted, and the morphological type can also be readily determined. This technique has been

[37] P. E. Perillie and S. C. Finch, *J. Clin. Invest.* **43,** 425 (1964).

[38] M. F. Mass, P. B. Dean, W. L. Weston, and J. R. Humbert, **86,** 1040 (1975).

[39] R. G. Grayton, P. E. Stokes, M. S. Schwartz, and D. B. Louria, *N. Engl. J. Med.* **270,** 123 (1970).

[40] H. J. Senn and J. F. Holland, *Blood* ***30,*** 888 (1967).

[41] H. Selye, *Proc. Soc. Exp. Biol. Med.* **82,** 328 (1953).

successfully used to measure inflammatory reactions in mice,[42–45] rats,[41,46–49] and rabbits.[50] Difficulties have been observed in creating an appropriate air sac in the skin of guinea pigs.[51]

The creation of a connective tissue air sac is simple. Extensive surgical manipulation is not required, and it is easy to generate sterile air blebs on a large number of experimental animals. The simplicity of this technique and its ability to produce semiquantitative estimates of cellular infiltrates makes it particularly valuable for measuring chemotaxis *in vivo*. Consequently, we describe this technique in detail and demonstrate its versatility.

Outline of the Air Sac Assay. Several authors have described methods for generating air sacs on different experimental animals.[41–49] The procedure that we describe is one we have adapted from earlier studies and used to investigate chemotactic agents in mice.[42,43] The procedure is carried out in four stages, as follows.

In the first stage, a connective tissue air bleb is generated by injecting 0.9 ml of air subdermally on the back of the mouse. The air is injected slowly via a 25-gauge needle from a 1-ml tuberculin syringe (Fig. 1A). Once formed, 0.1 ml of the chemoattractant solution, present in the same syringe, is introduced into the air bleb. We have found that the ideal position for the bleb is to one side of the midline and away from the neck region and that it is technically difficult to have more than one bleb per mouse.

At the appropriate time after injection of the chemoattractant, mice are sacrificed and the air bleb and surrounding skin are surgically excised (stage 2). The intact air bleb is then reflected onto a microscope slide and the surrounding tissue carefully removed (stage 3). The thin membranous lining of the bleb is gently stretched onto a microscope slide and fixed and stained with Camco Quick Stain II (Cambridge Chemical Products, Inc., Ft. Lauderdale, FL) (stage 4).

The air bleb forms a discreet bubble in the skin that can be easily

[42] M. D. P. Boyle, M. J. P. Lawman, A. P. Gee, and M. Young, *J. Immunol.* **134,** 564 (1985).

[43] M. J. P. Lawman, M. D. P. Boyle, A. P. Gee, and M. Young, *J. Immunol. Methods* **69,** 197 (1984).

[44] R. D. Higginbotham, *J. Immunol.* **95,** 867 (1965).

[45] J. M. Clark, H. Menduke, and E. F. Wheelock, *J. Reticuloendothel. Soc.* **25,** 255 (1979).

[46] S. Konno and S. Tsurufuji, *Br. J. Pharmacol.* **80,** 269 (1983).

[47] S. Tsurufuji, S. Yoshino, and K. Ohuchi, *Int. Arch. Allergy Appl. Immunol.* **69,** 189 (1982).

[48] M. Fukuhara and S. Tsurufuji, *Biochem. Pharmacol.* **18,** 475 (1969).

[49] K. Ohuchi, S. Yoshino, K. Kanaoka, S. Tsurufuji, and L. Levine, *Int. Arch. Allergy Appl. Immunol.* **68,** 326 (1982).

[50] M. D. P. Boyle, unpublished observations.

[51] M. J. P. Lawman, M. D. P. Boyle, A. P. Gee, and M. Young, *Exp. Mol. Pathol.* **43,** 274 (1985).

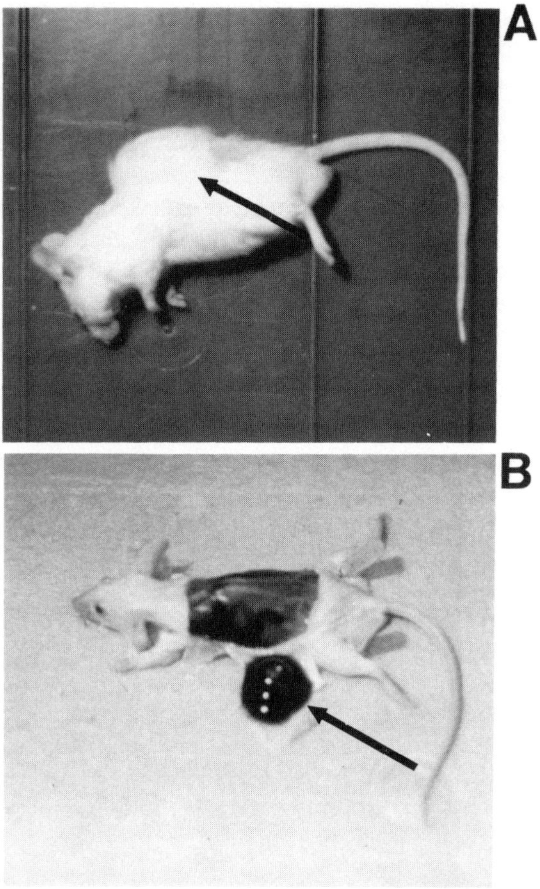

FIG. 1. (A) Appearance of an air sac on a mouse. (B) Appearance of an air sac in the skin following surgical resection. The sac has been filled with Trypan blue for easy visualization.

excised without bursting (see Fig. 1). If the air bleb bursts during removal of the skin, it is difficult to recover the appropriate membranous fragment. This particular part of the procedure is critical and technically demanding; however, with some practice, it is possible even for an inexperienced animal surgeon to recover the appropriate sample successfully as a single thickness membrane.

By virtue of the simplicity of the method, it is possible to examine the response of mice to a wide variety of chemotactic agents at different concentrations and at various times. A demonstration of the response of mice injected with the chemotactic peptide fMet-Leu-Phe is presented in

Ref. 43. For these studies, a stock solution of 10^{-2} M fMet-Leu-Phe dissolved in methanol is prepared. Various concentrations of fMet-Leu-Phe are injected into the air sac and the infiltrating cell number and morphological type scored by examining photomicrographs of the stained membranes at different magnifications. Figure 2 shows the response of mice to injection of 0.1 ml of a 10^{-10} M solution of fMet-Leu-Phe or the appropriate vehicle control. Mice were sacrificed at varying times, the membranes removed, fixed, stained, and photographed. The results must be considered to be semiquantitative because of the subjective nature by which the photographed fields are selected. This is not a major problem

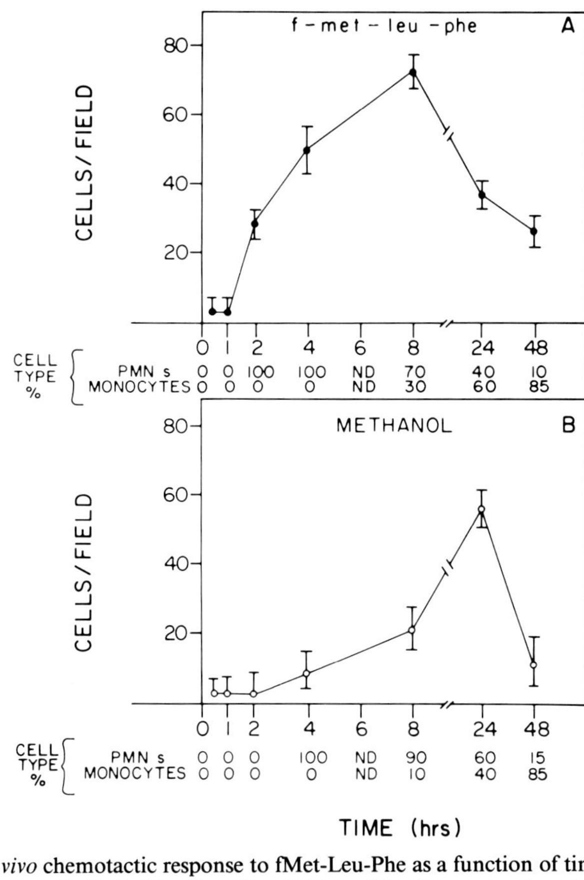

FIG. 2. *In vivo* chemotactic response to fMet-Leu-Phe as a function of time. (A) Kinetics of the response to 10^{-10} M fMet-Leu-Phe. Differential cell counts at each time point are shown. (B) Kinetics of the response to an equivalent concentration of methanol in PBS (10^{-6}%). Differential cell counts are shown for each time point. ND, Not determined.

when a strong response is observed. Under such circumstances, the pattern of infiltrating cells is remarkably uniform. Difficulties arise when the concentration of the chemotactic agent is decreased. In such cases, one frequently sees a localized infiltrate of cells around a blood vessel. These patchy infiltrates of cells represent a consistent response to a chemotactic stimulus, and they are not observed in mice injected with buffer alone.

In determining the chemotactic potential of an unknown agent, it is important to be able to distinguish between a response due to the selective accumulation of a given cell population and nonspecific leakage of cells as a result of vascular disruption. In this model, these two forms of cellular accumulation can readily be distinguished. Injection of buffer containing 1% methanol into an air sac results in an accumulation of cells containing a large number of erythrocytes. This is in marked contrast to the infiltrates seen in mice injected with buffer alone, and to mice injected with 10^{-10} *M* fMet-Leu-Phe, which contain a large number of infiltrating polymorphonuclear leukocytes but no detectable erythrocytes.[43]

Reproducibility of the in Vivo Method for Measuring Chemotaxis. The response of mice injected with fMet-Leu-Phe suggest that the air bleb technique is a rapid and useful assay for chemoattractants *in vivo.* The reproducibility of this method, both from mouse to mouse and from day to day, has been established.[43] In Table I, the response in four BALB/c mice on the same day is shown, as well as the response observed when similar

TABLE I
RESPONSE OF MICE TO INJECTION OF 10^{-10} *M* fMet-Leu-Phe

Variation	Number of cells/field[a]	
	Range	Mean ± SD
Mouse-to-mouse variation		
A	19–31	24.6 ± 5.8
B	19–38	28.0 ± 7.6
C	20–29	25.6 ± 3.4
D	24–28	22.4 ± 4.1
Experiment-to-experiment variation		
1	24–29	26.8 ± 2.2
2	19–28	24.5 ± 4.0
3	27–31	29.0 ± 1.8
4	23–27	26.3 ± 3.0

[a] The number of polymorphonuclear leukocytes in five randomly selected fields was scored 2 hr after injection with 10^{-10} *M* fMet-Leu-Phe (10^{-6}% methanol). Injection of 10^{-6}% methanol alone did not result in any infiltration of cells within the 2-hr period.

experiments were carried out on four different days. These results indicate that the response to fMet-Leu-Phe is remarkably consistent from mouse to mouse and from day to day. With the endotoxin-nonresponsive C3H/HeJ strain and the C5-deficient DBA-2J strain, among others, we have found the response to fMet-Leu-Phe to be comparable from strain to strain and from mouse to mouse within a given strain.

Role of C5a Generation in the Response to Various Chemotaxins. Using the *in vivo* air bleb system, we have demonstrated that, in addition to fMet-Leu-Phe, staphylococcal protein A and the γ subunit of nerve growth factor (NGF) can mediate the accumulation of polymorphonuclear leukocytes in the air sac.[42,43] In order to determine the role C5a plays in mediating the responses of these different agents, the air sac technique has been carried out on DBA-1J (C5-sufficient) and DBA-2J (C5-deficient) mice. The results of experiments using NGF as the chemotactic agent demonstrate that its activity is the same with both C5-deficient and C5-sufficient mice. In contrast, C5-sufficient mice are required to produce a cellular infiltrate when protein A is used (see Refs. 42 and 43). The availability of inbred strains of mice deficient in C5 enables ready identification of chemotactic factors that mediate their activity through the generation of the chemotactic complement split fragment, C5a.

Measurement of Cellular Infiltrates as a Function of Time. In the preceding sections, we have demonstrated that the air sac method reliably measures *in vivo* chemotaxis. The approaches described can be applied to monitoring the kinetics of cellular infiltration.[51] The results of a typical time course following injection with buffer or 150 n*M* NGF are presented in Figs. 3 and 4. The normal response to injection of buffer is shown in Fig. 3 (low magnification) and Fig. 4 (high magnification). Figure 3 also presents a scoring evaluation (− to ++++) to indicate the semiquantitative degree of cellular infiltration. Within 4–8 hr, a low number of infiltrated polymorphonuclear leukocytes is observed. Over the next 24 hr, the membranes contain predominantly polymorphonuclear leukocytes with a few highly vacuolated monocytes (Fig. 4). Between 24 and 72 hr, the percentage of polymorphonuclear leukocytes declines, and there is an increase in vacuolated mononuclear type cells and the appearance of a small number of fibroblastlike cells. In contrast, mice injected with 150 n*M* NGF demonstrated a markedly different response. Within 2–8 hr, large numbers of polymorphonuclear leukocytes are present in the air sac of the NGF-injected mice (Figs. 3 and 4). The response of this cell type peaks by 8 hr (Fig. 4). At 24 hr, the polymorphonuclear leukocytes are replaced by a highly vacuolated mononuclear cell population (Fig. 4). By 72 hr, the predominant cell type is fibroblastic. When compared with the control, the total number of cells infiltrating the air sac at each time point

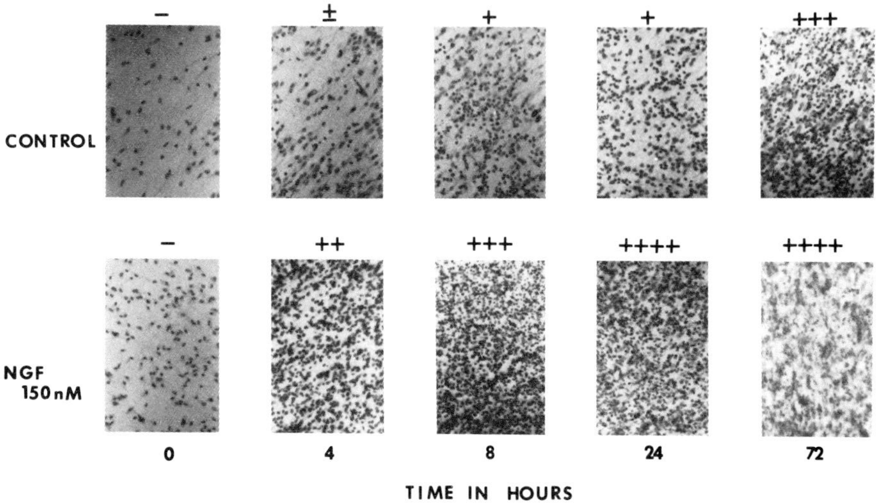

FIG. 3. Kinetics of cellular infiltration into a connective tissue air sac in response to injection of 150 n*M* nerve growth factor (NGF) or buffer. The cellular infiltrate was graded from − to ++++ to indicate the degree of infiltration. Magnification: ×60.

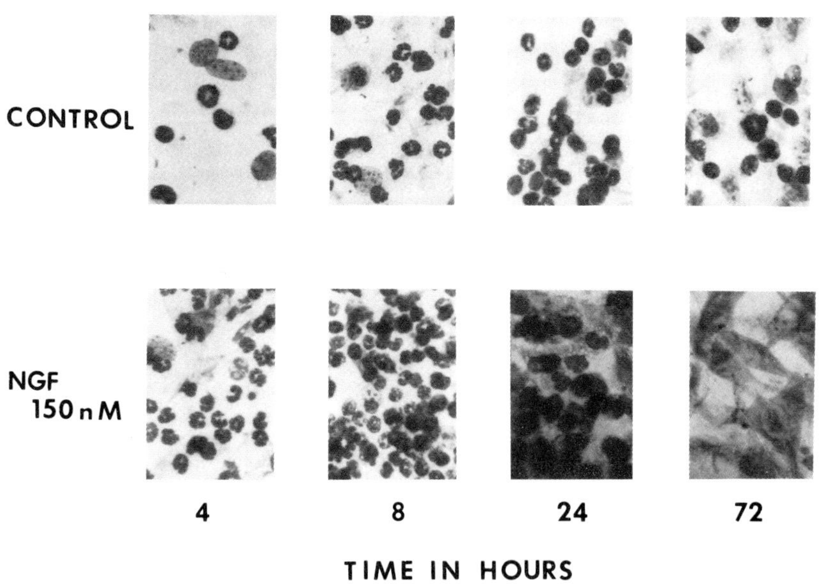

FIG. 4. Morphological nature of cells infiltrating into the connective tissue air sac at various times following injection of 150 n*M* nerve growth factor (NGF) or buffer. Magnification: ×340.

is greater in the NGF-treated mice. In addition, the morphology of the cellular infiltrate is more uniform in the mice injected with NGF. That is, at 8 hr the cell populations are >90% polymorphonuclear leukocytes, at 24 hr >90% vacuolated monocytes, and at 72 hr >90% fibroblastlike cells. In contrast, mixtures of cell types are present in the mice injected with buffer, and a heterogeneous population of cells is present at all of the time points studied (Fig. 4).

The limit for time-course experiments in the air sac model is around 5 days. Attempts to carry out experiments over longer time periods are complicated by collapse of the air sac and increasing cellular response in the control animals. It is possible to maintain an air sac for more than 7 days by "reinflating" the sac by injection of more air at approximately 3-day intervals. This technique is not particularly valuable because the response of control mice to air alone at 7 days is strong.

Advantages and Limitations of the Air Sac Technique. The results presented above demonstrate that the air sac technique is a reliable, reproducible method for studying chemotaxis *in vivo.* By comparison with skin window and chamber implantation techniques, the air sac is generated with minimal disruption of the surrounding tissue and thus can be achieved reproducibly without infection. The major limitation of the air sac technique is in the subjective nature of the quantitation. Unlike approaches using labeled leukocytes, it is not possible to obtain a totally objective measure of cell migration. As noted above, it is difficult to score accurately responses that are patchy in nature. This problem can usually be minimized by increasing the concentration of the chemotactic stimulus tested, in order to increase the degree of response to obtain a uniform infiltrate. The major advantage of the air sac technique is its simplicity and its accommodation of large numbers of samples. This enables studies to be carried out at multiple time points, with different concentrations of a chemotactic stimulus, and facilitates the use of multiple animals for each experimental condition.

Problems Common to All in Vivo Measurements of Chemotactic Agents

All of the techniques for measuring *in vivo* chemotaxis can be influenced by the presence of endotoxin. Previous studies have demonstrated that, depending on concentration and source of endotoxin, the inflammatory response can be either enhanced or inhibited.[16,17,23,52] It is therefore of importance in any of the *in vivo* systems for measuring chemotaxis that appropriate controls be included to assess the response of the buffer used as diluent. A positive chemotactic agent should also be included in each experiment.

[11] Necrotaxis and Galvanotaxis

By HANS GRULER

Introduction

Polymorphonuclear leukocytes (PMNs) are sent to sites of infection to combat invading microorganisms. There must exist some mechanism to account for the fact that these cells gravitate toward and remain in the vicinity of sites of infection. The direction-finding mechanism(s) involved must be a function of some information transmitted to the leukocytes by the infected cells. The directed locomotion of leukocytes has been investigated using *in vitro* experiments with clusters of bacteria or concentration gradients of bacteria-derived factors.

There are various ways in which tissues can be damaged. One is by bacteria, in which case it could be assumed that the resultant chemotaxis is due to the bacteria. However, directed locomotion of leukocytes toward damaged tissue can also be induced without bacterial infection, suggesting that the chemotactic factors originate in the damaged cells. Buckley[1] investigated directed locomotion *in vivo* by making an aseptic, localized heat injury in the "ear chamber" of a rabbit. Leukocytes appeared in the injured area within 4–24 hr. This experiment clearly demonstrates that leukocytes are attracted by factors unrelated to microorganisms.

Bessis and Nomarski[2] were the first to notice the attraction of phagocytes to dying cells in an *in vitro* experiment (Fig. 1). Bessis[3] termed this type of directed locomotion necrotaxis.[4] The necrotactic response of leukocytes is not due to bacteria-related factors, nor is it a function of the nature of the target because red cells from various animals, leukocytes from mammals, and cells from different tissue cultures can be used as targets. Even hemoglobin or albumin stained with Janus green can be coagulated with a microbeam and used as a nectrotactic stimulus. However, cells which have been dead for a long time do not act as necrotactic stimuli. The actual nature of the necrotactic stimulus is, therefore, not clear.[5]

The signals transmitted by infected or dying cells may be of an electrical nature. For instance, when a cell is lysed, ions inside and outside the cell

[1] I. K. Buckley, *Exp. Mol. Pathol.* **2,** 402 (1963).
[2] M. Bessis and G. Nomarski, *J. Biophys. Biochem. Cytol.* **8,** 777 (1960).
[3] M. Bessis, *Cell. Inj., Ciba Found. Symp., 1963,* p. 287 (1964).
[4] A film is available from Sandoz, Motion Picture Department, CH-4000 Basel, Switzerland.
[5] M. Bessis and A. de Boisfleury-Chevance, *Blood Cells* **10,** 5 (1984).

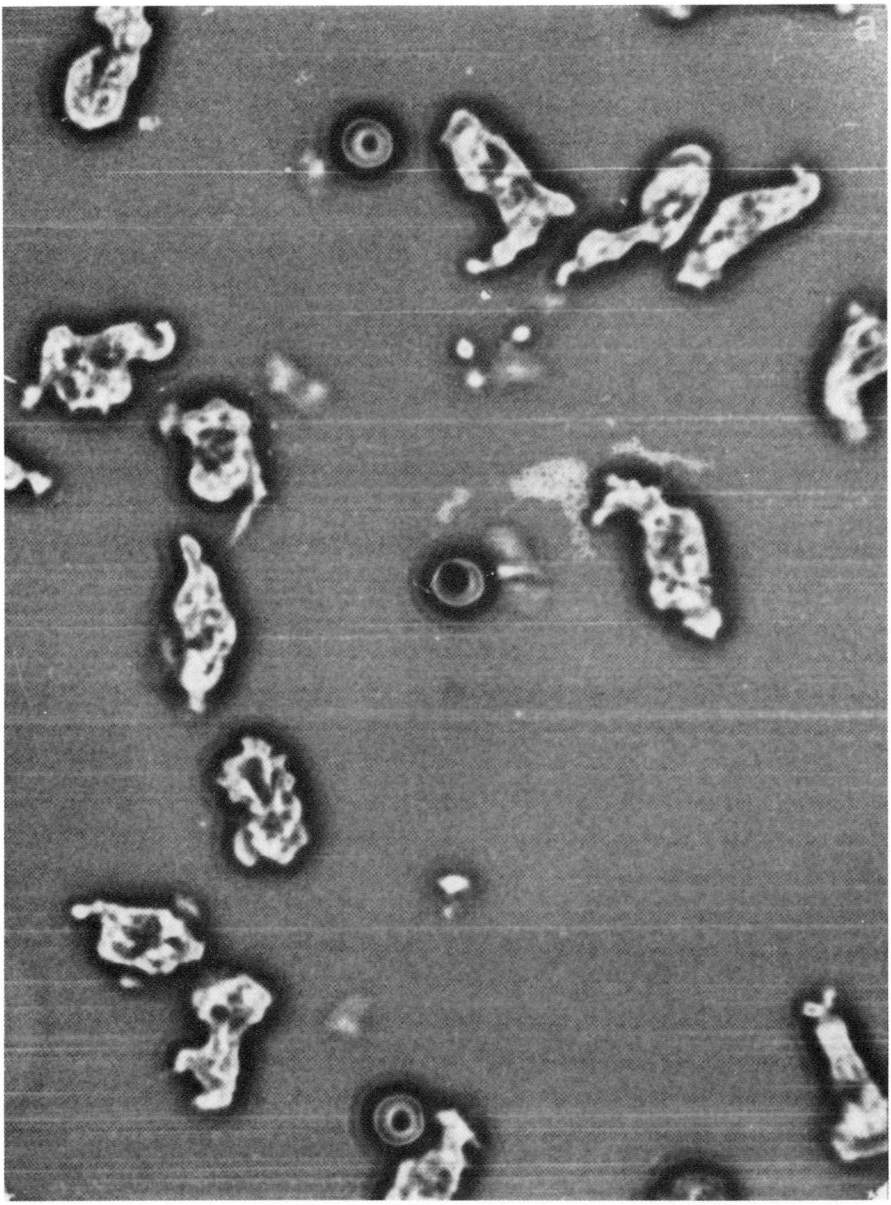

FIG. 1. Necrotaxis of human granulocytes in plasma containing 100 μM colchicine (phase contrast). (a) Random locomotion immediately prior to lysis of the target cell (erythrocyte at center). (b) Locomotion immediately after lysis of the target cell. (c) Strong directed locomotion toward the lysed cell 2 min after lysis of the target cell.

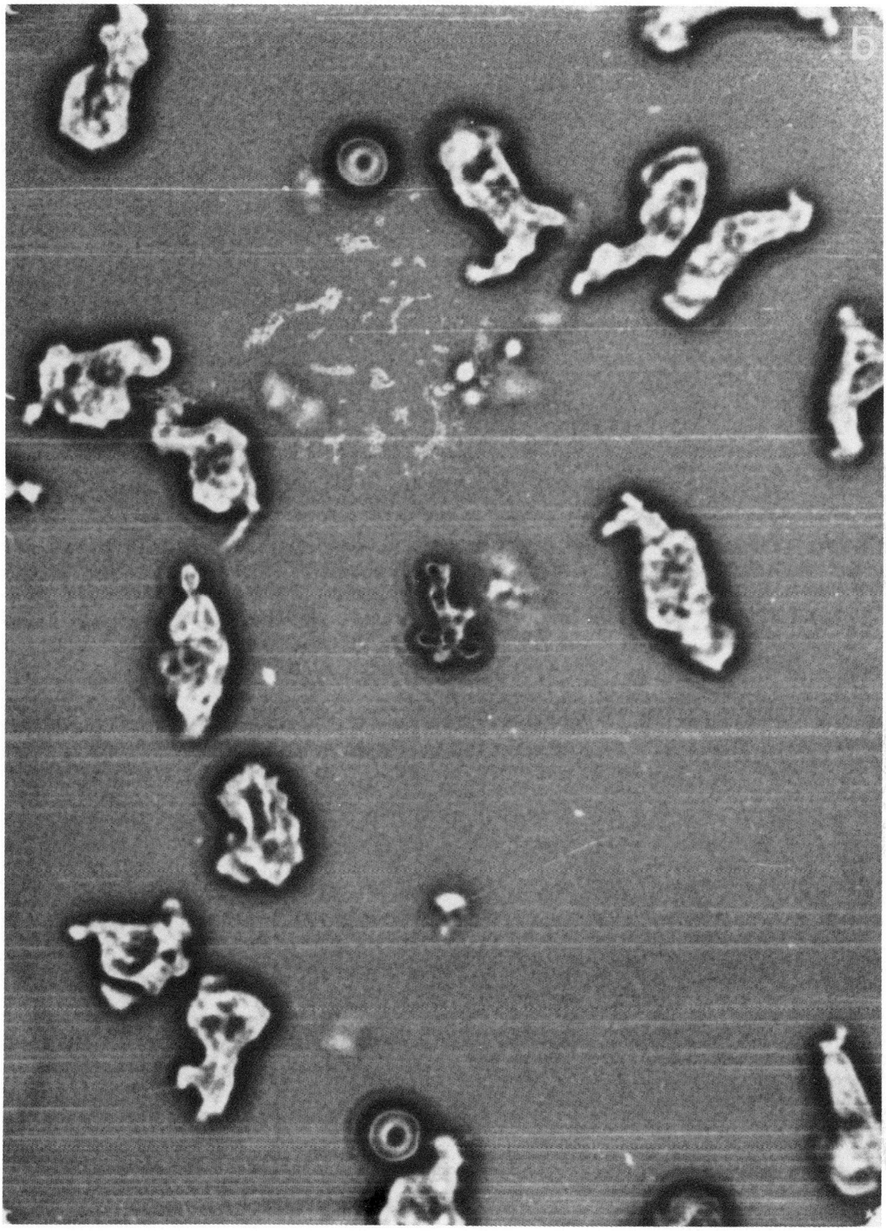

FIG. 1. *(continued)*

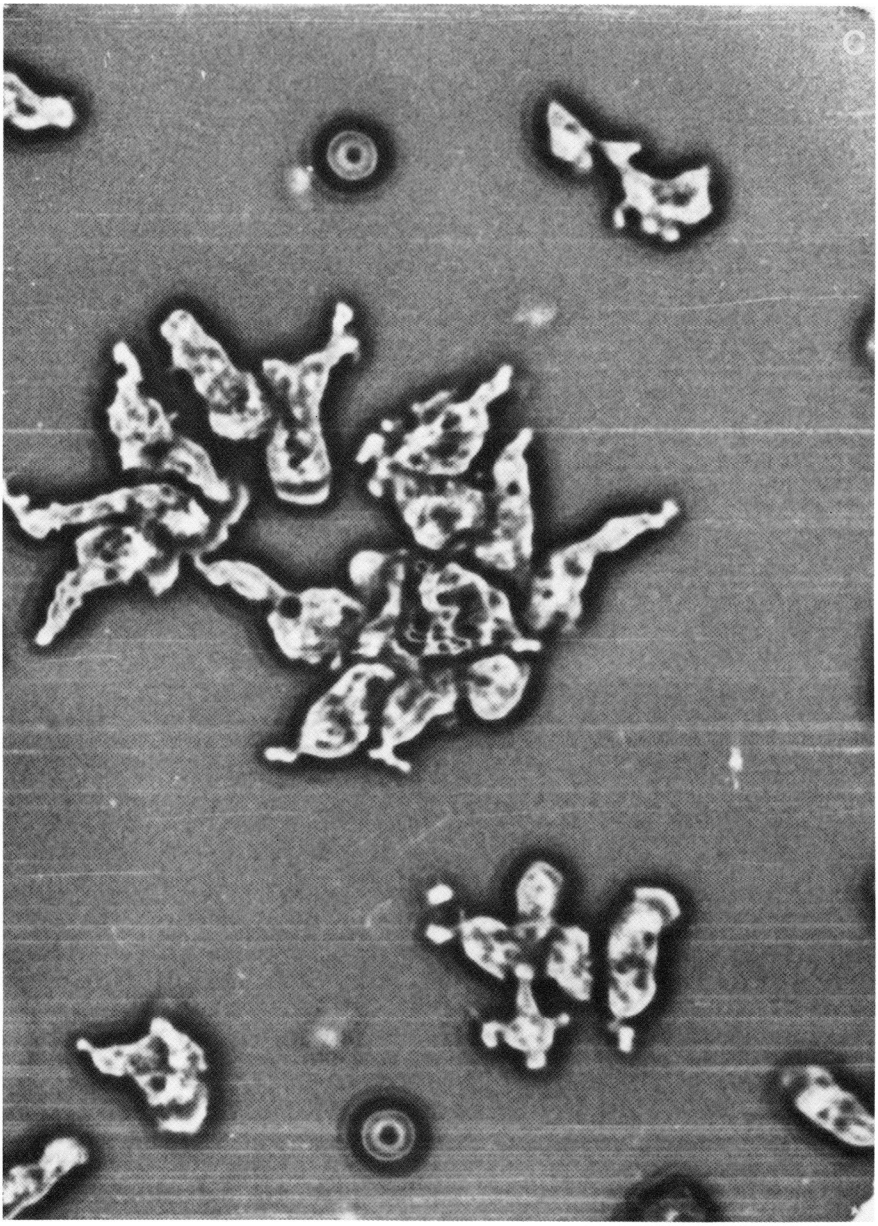

FIG. 1. *(continued)*

diffuse to reestablish a concentration equilibrium. This diffusion process is a function of the size of the ions, i.e., smaller ions diffuse faster than larger ones. The different diffusion constants of the ions involved result in the separation of small and large ions, and a diffusion potential is created. In intact cells, the plasma membrane keeps different types of ions separate from one another and thus maintains this difference of potential. When the plasma membrane is disrupted, the electric field is not localized at the membrane but rather spread over a relatively large area (up to 500 μm). The strength of the electric field in the vicinity of wounds has been measured and found to be approximately 0.2 V/mm.[6] Leukocytes can orient themselves in such electric fields. *In vitro* experiments with leukocytes and macrophages show that these cells have the ability to recognize electric fields.[7] Leukocytes and macrophages migrate toward the anode, whereas other cell types, e.g., neural crest and epithelial cells, fibroblasts, and developing neurons, move toward the cathode.[8] The response of leukocytes to damaged cells may therefore involve these electric fields (galvanotaxis).

In vitro experiments can be used to investigate the responses involved in chemotaxis, galvanotaxis, and necrotaxis. Only a few experiments involving the galvanotaxis of leukocytes[7,9–11] are reported in the literature. Most of this chapter is devoted to the necrotaxis of leukocytes. For experimental purposes, necrotaxis and galvanotaxis present several advantages: (1) measurements are very reproducible; (2) directed and nondirected locomotion can be investigated before and after lysis of the target cell in a necrotactic assay and before and after applying the electric field in a galvanotactic assay; and (3) the response of cells to external factors can be determined as a function of time. On the other hand, there are also disadvantages associated with necrotaxis: (1) the external factors are unknown, and (2) the magnitude of the external factors is a function of both time and space.

The necrotactic or galvanotactic response of locomoting cells can be described mathematically to obtain information about the cellular response.[12–14] It is thus possible (1) to determine whether the equipment used functions properly; (2) to quantify chemokinetic and necrotactic

[6] A. T. Baker, L. F. Jaffe, and J. W. Vanable, *Am. J. Physiol.* **242,** R358 (1982).

[7] K. Fukushima, N. Senda, H. Innui, H. Mirua, Y. Tamai, and Y. Murakami, *J. Osaka Univ.* **4,** 195 (1953).

[8] C. A. Erickson and R. Nuccitelli, *J. Cell Biol.* **98,** 296 (1984).

[9] E. Dineur, *Bull. Seances Soc. Belg. Microsc. (Brussels)* **18,** 113 (1891).

[10] J. Monguio, *Z. Biol. (Munich)* **93,** 553 (1933).

[11] N. Orida and J. D. Feldmann, *Cell Motil.* **2,** 243 (1982).

[12] H. Gruler and B. D. Bültmann, *Blood Cells* **10,** 61 (1984).

[13] H. Gruler, *Blood Cells* **10,** 107 (1984).

[14] H. Gruler and R. Nuccitelli, *Clin. Biol. Res.* **210,** 337 (1986).

activity; and (3) to determine properties inherent in locomoting cells, including, for example, the internal clock of cells, which governs the state of locomotion, the period of refraction, during which a cell is insensitive to new external signals, the directional memory of cells, and the amount of information cells obtain from their environment.

Methods

Cell Preparation (see also this series, Vol. 108 [9])

1. Granulocytes are prepared by drawing venous blood into an anticoagulant solution (sodium citrate or heparin) and allowing it to sediment at room temperature for about 2 hr in tubes tilted at about 45°. The buffy layer containing leukocytes and few erythrocytes is used for further preparations.

2. Leukocytes are separated from heparinized blood by dextran sedimentation on a Hypaque–Ficoll density gradient. Samples (20 ml) of heparinized blood and 2 ml of 4.5% dextran solution are incubated at 37°. After half of the erythrocytes have sedimented, the supernatant containing the white blood cells is collected. Seven milliliters of the supernatant is carefully placed on top of 3 ml of Hypaque–Ficoll in a siliconized tube and centrifuged at 600 *g* for 20 min. The remaining erythrocytes sediment to the bottom of the tube, and the Hypaque–Ficoll contains the granulocytes. The supernatant is plasma. The cells are washed 3 times in Hanks' solution.

A drop of the cell suspension from the buffy coat or the Hypaque–Ficoll preparation is transferred to a coverslip. After 5–10 min in a humid incubator, the granulocytes adhere to the surface of the glass. The glass is then rinsed with the medium used in the necrotactic or galvanotactic experiments. The control medium can be plasma, a medium consisting of 80% minimum essential medium (MEM) and 20% fetal calf serum, or Hanks' solution enriched with a chemokinetic factor such as the peptide fMet-Leu-Phe.

Galvanotaxis Chamber

An electric field is created in the galvanotaxis chamber by sending an electric current through the medium (see above). However, there are two factors that interfere with the experiments. First, there is a voltage drop at the power electrodes, which makes it impossible to measure the electric field even if the chamber has a well-defined geometry. Second, products of electrolysis generated at the power electrodes diffuse through the medium and affect the movement of the cells.

These problems can be eliminated by using an apparatus with four electrodes, two of which are used to apply the current (power electrodes) and two to measure the electric field (measuring electrodes). The power electrodes are configured in such a way that the products of electrolysis cannot reach the cells. The two measuring electrodes do not carry current. With this apparatus, there is no voltage drop (measuring electrodes) and electrolysis is eliminated.

The galvanotaxis experiment can be performed by transferring a coverslip (2 × 1 cm) with a drop of medium to a 3-mm-thick glass plate (2.5 × 2.5 cm) with four parallel grooves on it (see Fig. 2). The grooves are cut with a diamond saw. (Hellma GmbH, Postfach 69, D-7840 Müllheim, West Germany, supplies such plates with precut grooves.) The same base plate is also used to study chemotaxis. A glass chamber is more stable than a plastic chamber, making the results more reproducible. The grooves on the base plate have a cross section of 1 × 1 mm and are separated by 1-mm-wide barriers. The distance between the coverslip and the base plate at the barrier should be about 20 μm. The coverslip is then glued onto the base plate with a few drops of wax.

The two measuring electrodes (made of 0.5-mm platinum wire) lie in the two inner grooves of the base plate (Fig. 2). Since platinum is brittle, silver wire (also 0.5 mm in diameter) is often used instead. However, the surface of the wires must be chlorinated before every experiment. This is done by first cleaning them with concentrated nitric acid for about 2 sec and then washing them with water passed through a Millipore filter. They are then chlorinated by keeping them in a bath containing 25 vol HCl

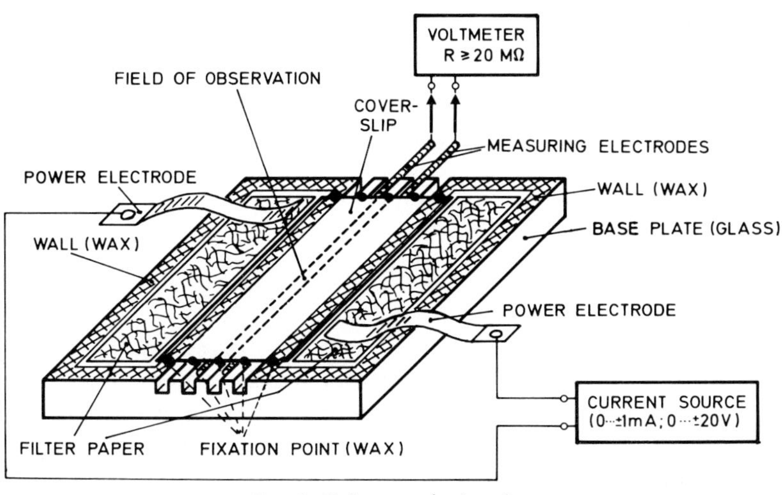

FIG. 2. Galvanotaxis chamber.

(32%), 25 vol saturated $Fe^{III}Cl$ solution, and 50 vol water for up to 5 min. These two (Ag–AgCl) electrodes are then inserted into the two inner grooves of the base plate and connected to a voltmeter with a high resistance ($\geq$20 MΩ) so that the measurement is practically without current. With properly functioning measuring electrodes, the voltage is 0 or less than 2 mV when no electric current is being sent through the galvanotaxis chamber.

The power electrodes are then introduced into the galvanotaxis chamber. In order to minimize the consequences of the products of electrolysis generated at the power electrodes, several layers of filter paper (0.5 $\times$ 1.5 cm) are placed in the two grooves (Fig. 2). An appropriate amount of Hanks' solution is added to the filter paper, and the power electrodes are pressed onto the wet filter paper (Fig. 2) and connected to the source of current. After assembly, the chamber is sealed with wax to prevent evaporation.

With this apparatus, the current is less than 1 mA for a field of 1 V/mm in the chamber. The strength of the electric field to which the cells are exposed is expressed as the voltage measured at the measuring electrodes divided by the width of the barrier. The disadvantage of this apparatus is that products of electrolysis are not prevented from entering the chamber. As a result, the measuring time is restricted. For example, the measuring time is about 10 min for 1 V/mm. With weaker electric fields, the measuring time is correspondingly longer. It is necessary to keep in mind that a field strength of 1 V/mm is already high because it usually damages cells. Directed locomotion can be investigated by applying different voltages, and the locomoting cells are observed with a phase-contrast microscope.

Necrotaxis Chamber

The coverslip with a drop of medium is transferred to a microscope slide to build up a sandwich cell. The edges of the coverslip are sealed with paraffin to prevent evaporation. The thickness of the aqueous phase should be about 20 μm. The necrotactic response is obtained when a target cell is lysed. The "knife" used in this operation is a strong light beam focused onto the target cell.

Instrumentation for Necrotaxis

Cellular necrotaxis can be observed with any microscope equipped with a microbeam device and a time-lapse or video camera. For example, microscopes developed to perform radiation microsurgery on living cells can be used. A microscope originally designed for photobleaching experiments can also be used. These microscopes are all based on the same principles[15–17]: (1) the light source is focused on a pinhole, and (2) the pinhole is projected onto the object by the objective of the microscope. Thus, the greater the magnification the smaller the focused microbeam, and beams as small as 0.5 μm in diameter can be obtained.

The light can have an effect on a cell only if all or part of it is absorbed by components of the cell. The following may be involved in lysis of the target cell in the light-excited state: (1) the light energy may be transformed into thermal energy, in which case extremely localized thermal energy may lyse the target cell; or (2) components of the cell may undergo reversible or irreversible changes and give rise to states with new physical and chemical properties (e.g., different surface tension or molecular structure), which may cause lysis of the target cell. Many components of cells, e.g., proteins, absorb wavelengths less than 400 nm (UV light). Various cells are therefore lysed with UV light. Bessis and Nomarski[2] used the 370-nm line of a cadmium lamp. A nitrogen "laser" with a line at 355 nm can also be used.

The quality and intensity of the microbeam can be increased if the light source is replaced by a laser with a high coherence length and small beam divergence. Bessis and Burté[15] used a ruby laser flash (694 nm, 0.5 μsec). Using an erythrocyte as the target cell, they achieved the destruction of an entire cell. This situation is shown in Fig. 1. The photographs were taken using the microscope described by Bessis and Burté.[15] If a leukocyte is used as the target cell, vital staining is necessary (Janus green and methylene blue, Bessis and Storb[18]). The microbeam consists of monochromatic light from an argon laser (488 nm), which produces a small hole in the target cell. Roch-Arveiller and Giroud[19] obtained such holes in leukocytes, erythrocytes, and mast cells.

Microscope manufacturers now offer a device to connect a laser to a microscope to produce a microbeam. For example, a laser adapter is available for Zeiss IM microscopes (Zeiss, Department of Microscopy, D-7082 Oberkochen, West Germany).

Analysis of Data

The kinetics of locomoting cells can be determined if the center of gravity of a cell is found on the basis of consecutive frames. The microscopic image contains a large amount of information which is also a

[15] M. Bessis and B. Burté, *C. R. Seances Soc. Biol. Ses Fil* **11,** 139 (1964).

[16] M. Bessis, F. Gires, G. Mayer, and G. Nomarski, *C. R. Hebd. Seances Acad. Sci.* **255,** 1010 (1962).

[17] M. Bessis and M. M. Ter-Pogossian, *Ann. N.Y. Acad. Sci.* **122,** 689 (1965).

[18] M. Bessis and R. Storb, *Nouv. Rev. Fr. Hematol.* **5,** 459 (1965).

[19] M. Roch-Arveiller and J. P. Giroud, *Blood Cells* **10,** 135 (1984).

function of the light microscopy technique used. The center of gravity can be approximated by taking the center of the area defined by the contour line of the cell.

Data reduction is carried out when the contour line is determined by computerized image analysis (e.g., IBAS Kontron, Munich). Semimanual analysis of the images is also possible by (1) projecting an image onto a graphic tablet (e.g., Apple II); (2) determining the contour line of the cell visually and feeding it into a computer; and (3) determining the center of the area defined by the contour line by computer.

The interval between consecutive images is chosen as a function of the rate of locomotion of the cell. For example, a 4-sec interval provides a good time-lapse film for an average track velocity of 24 μm/min. Good results are obtained by analyzing every other frame. If the shape of the cells changes significantly with time (e.g., fibroblast locomotion), it is necessary to take the average center of gravity from two or more frames.

Results

A typical example of random movement is shown in Fig. 1a. The cells are elongated along the path of the directed locomotion. The orientation of the cells is random. Random movement is observed if the path of one cell is followed over a long period of time or if the paths of many cells are followed. The random locomotion of the cells then becomes directed when an erythrocyte is lysed by a laser flash (Fig. 1b and 1c). Typical paths obtained by time-lapse photomicrography are shown in Fig. 3. The cells orient themselves along the gradient of the necrotactic field (dashed lines in Fig. 3). The paths appear snakelike and lead to the lysed cell. The cell density near the lysed cell increases dramatically with time, and a closely packed monolayer of cells ultimately results. Since the granulocytes still possess their chemokinetic activity, the whole cluster of cells, which is referred to as a rosette, may exhibit a cooperative clockwise or counterclockwise movement around the lysed cell.

Random locomotion of cells also becomes directed when an electric field is applied. The cells orient themselves along the gradient of the electric field. The paths appear snakelike and lead to the anode. The advantage of necrotaxis and galvanotaxis is that both the random locomotion and tactic movement of the cells can be viewed.

One function of granulocytes is to travel to sites of infection or inflammation. In general, locomoting cells in a polar field (electric field, concentration gradient of chemotactic molecules, necrotactic gradient) drift parallel to the gradient of the polar field. The mean displacement of cells $\bar{x}$ exposed to a polar field for the time t is the average drift velocity $\bar{v}_{\parallel}$ times

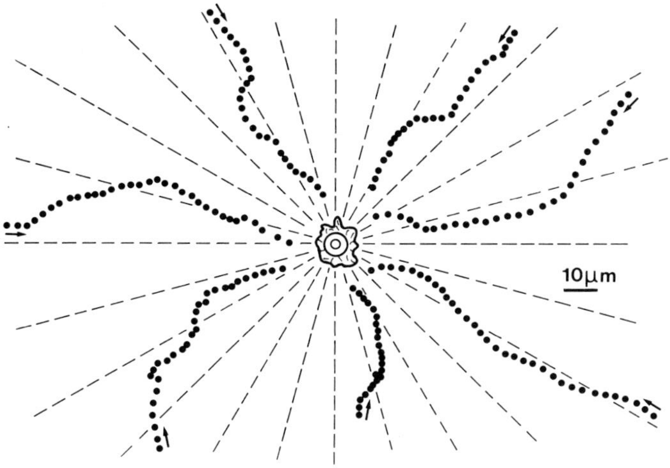

FIG. 3. Necrotactic response of seven granulocytes to a lysed cell. The points represent the movement of the center of the area occupied by a locomoting cell. The time difference Δt between two points is 8 sec.

the time t. This type of movement can be characterized by (1) the average drift velocity $\bar{v}_{\parallel}$; (2) the chemotropism index $\bar{v}_{\parallel}/\bar{v}_c$, where $\bar{v}_c$ is the mean track velocity; (3) the McCutcheon index (ratio between travel parallel to the gradient of the polar field and the actual distance traveled); (4) the polar order parameter $\bar{P}_1$ (the average of cos ϕ, where the angle ϕ describes the actual direction of movement with respect to the gradient of the polar field); and (5) the necrotaxis index $N(t_k)/N(t_0)$, where $N(t_0)$ is the number of cells in the viewing field before the target cell is lysed and $N(t_k)$ the number of cells in the viewing field at the time t_k with $t_k - t_0$ being the time after the target is lysed.

The chemotropism index, the McCutcheon index, and the polar order parameter equal 0 for random movement and 1 if the cells move in a straight line parallel to the gradient of the polar field. Values between 0 and 1 therefore indicate the extent to which cells move in the direction of a target. The chemotropism index, the McCutcheon index, and the polar order parameter for necrotaxis are 0.82 ± 0.05, 0.79 ± 0.07, and 0.85 ± 0.05, respectively.[13] These three independent measurements should agree if the chemokinetic activity of a cell as expressed in terms of the mean track velocity and the decision-making mechanism which transforms information obtained from the polar field into tactic movement are independent of one other. In this case, the mean drift velocity is the product of the mean track velocity times the polar order parameter.

Directed locomotion can also be quantified by measuring the increase in the number of cells in the vicinity of the lysed cell. The necrotaxis index is easy to obtain but presents the disadvantage of being a function of the elapsed time $t_k - t_0$ and the size of the viewing field. The question also arises as to whether the polar order parameter, the McCutcheon index, and the necrotaxis index are somehow interrelated. For a viewing field with a radius R_0 and a mean cell density n_0, the total number of cells at the time $t < t_0$ is

$$N(t_0) = n_0 \pi {R_0}^2 \tag{1}$$

The necrotactic stimulus is created at the time t_0, and $N(t_0)$ cells then locomote toward the lysed target cell. In addition, new cells enter the viewing field. The directed locomotion of these cells is ΔR for the time $t_k - t_0$:

$$\Delta R = \bar{v}_c \, \overline{P}_1 \, (t_k - t_0) \tag{2}$$

The cells in a viewing field with the radius R_0 come from a circular area with the radius $R_0 + \Delta R$:

$$N(t_k) = n_0 \, \pi \, (R_0 + \Delta R)^2 \tag{3}$$

The necrotaxis index is thus

$$\frac{N(t_k)}{N(t_0)} = (1 + \Delta R/R_0)^2 = \left(1 + \bar{v}_c \overline{P}_1 \frac{t_k - t_0}{R_0}\right)^2 \tag{4}$$

Equation (4) describes the relationship between the chemotaxis index and the polar order parameter. The polar order parameter is calculated to be between 0.5 and 0.6 for a viewing field with a diameter of 200 μm and a time of 3 min. However, the experimental results for the McCutcheon index and the polar order parameter are found to be 0.8 ± 0.07 and 0.85 ± 0.05, respectively. These results are surprising because they differ from those obtained by the calculation shown above.

The reason for this discrepancy is that the necrotaxis index is determined by counting the cells entering a circle with the radius R_0, which means directed locomotion outside the circle $R > R_0$ is quantified. The McCutcheon index and the polar-order parameter $\overline{P}_1$, on the other hand, are determined in the vicinity of the target cell for $R < R_0$. It is thus necessary to determine the number of cells for a very small viewing field. However, since the laser could harm locomoting cells, there should be no locomoting cells within this field. It is therefore necessary to wait until there are no granulocytes in close proximity to the target. Spontaneous variations in cell density are not taken into account when the average number of cells within the viewing field is determined on the basis of the average cell density.

The polar order parameter is found to be between 0.8 and 0.85 for a viewing field with a diameter of 30 μm and a period of 3 min. The polar order parameter determined on the basis of the necrotaxis index is thus in agreement with the measured polar order parameter, the McCutcheon index, and the chemotropism index.

It might initially seem that necrotactic assays are disadvantageous because they are both time and space dependent. However, since the dependence of the polar order parameter on the strength of the polar field can be seen in a single experiment, this also represents an advantage. The polar order parameter is large in the vicinity of the lysed cell and decreases as the distance from the cell increases.

The McCutcheon index and the polar order parameter for human granulocytes are 0.85 ± 0.07 in a necrotactic assay ($R_0 = 30\ \mu$m) and the polar order parameter for human granulocytes 0.65 ± 0.1 in a galvanotactic assay (electric field strength of 0.3 V/mm).[7] The polar order parameter determined on the basis of experiments in a Zigmond chamber with 10 nM/mm fMet-Leu-Phe is 0.8 ± 0.05.[12]

Acknowledgment

I would like to thank Marcel Bessis for helpful discussions. I am also particularly indebted to Anne de Boisfleury-Chevance for excellent technical assistance. This work was supported by a NATO travel grant and a grant from the Fond der chemischen Industrie.

[12] Chemotactic Factors: An Overview

By PETER C. WILKINSON

What Is Meant by a Chemotactic Factor?

It is becoming apparent that the term chemotactic factor does not describe the effects of the so-named factors on leukocytes at all adequately. Strictly speaking, a chemotactic factor is a ligand which induces a chemotactic response, i.e., directional locomotion in a gradient, as defined in this volume [1]. However, the best characterized of these factors act not only on chemotactic responses but on metabolic activity, oxygen metabolite generation, secretion, adhesion, and other functions of leukocytes. Moreover, these functions are not all part and parcel of an identical response. Low concentrations of ligands such as formyl peptides have marked effects on locomotor responses and little effect on metabolism, secretion, and microbicidal activity, whereas at higher concentrations of the ligands the latter functions are activated strongly but locomotor responses are poor.[1] Thus a major problem in studying signal–response coupling using such ligands is unravelling how each of the different cellular biochemical changes observed contributes to each of the different functions that are stimulated.

Another inadequacy of the term chemotactic factor, as discussed elsewhere in this volume [1], is that even when we consider locomotion in isolation from all the other functions of cells it is probable that the primary effect of these factors is not to stimulate chemotaxis *sensu stricto,* but to stimulate locomotion. These factors cause polarization and translocation of leukocytes whether there is a concentration gradient or not.[2–7] If such a gradient is present, as is likely but not inevitable at inflammatory sites, then the response will take the form of morphological orientation and locomotion toward the gradient source, i.e., chemotaxis. If this view, elaborated at greater length elsewhere in this volume [1], is correct, then a molecule which is a chemotactic factor in one set of circumstances is a chemokinetic factor in another. It is also probable that the ability of leukocytes to respond chemotactically to a gradient of ligand is dependent on the ligand concentration being low enough for an unpolarized cell to discriminate between a first signal received at one point on the cell surface and subsequent signals received elsewhere, and so to polarize in the direction of the first stimulus. It is unclear how a cell which has polarized toward a first stimulus no longer responds immediately to subsequent stimuli at other points on the cell, though receptor redistribution to the anterior pole is probably important.[6,8] It can be concluded from experimental observations[6] that if the ligand concentration is too high the cell will be stimulated at many sites simultaneously, and, although the machinery for stimulus–response coupling is activated, it is activated globally and the cell is unable to polarize or translocate effectively. Hence there is a decline in locomotor response as ligand concentration is raised above an optimum.

In the remainder of this chapter, I shall use the term chemotactic factor in the inexact sense which it usually has, i.e., as a synonym for locomotor

[1] R. Snyderman (ed.), *Contemp. Top. Immunobiol.* **14** (1984) (contains several relevant articles).

[2] G. F. Schreiner, J. Braun, and E. R. Unanue, *J. Exp. Med.* **144,** 1683 (1976).

[3] G. J. Cianciolo and R. Snyderman, *J. Clin. Invest.* **67,** 60 (1981).

[4] C. W. Smith and J. C. Hollers, *J. Clin. Invest.* **67,** 60 (1981).

[5] H. U. Keller, A. Zimmermann, and H. Cottier, *J. Cell Sci.* **64,** 89 (1983).

[6] J. M. Shields and W. S. Haston, *J. Cell Sci.* **74,** 75 (1985).

[7] W. S. Haston and J. M. Shields, *J. Immunol. Methods* **81,** 229 (1985).

[8] S. J. Sullivan, G. Daukas, and S. H. Zigmond, *J. Cell Biol.* **99,** 1461 (1984).

stimulant, bearing in mind the caveats above and also that in many cases the evidence that such factors actually stimulate chemotaxis *sensu stricto* is incomplete.

Chemotaxis, Chemokinesis, and Microtubules

Since chemotactic factors are locomotor stimulants that also act as chemokinetic factors, the question should be asked whether the reverse is also true, namely, whether all factors which affect chemokinesis (changes in cell speed or turning frequency) also of necessity act as chemotactic factors. The answer is probably "no" because chemokinesis is likely to be affected by a wider range of circumstances than chemotaxis. To respond by chemotaxis, a cell must detect an anisotropically disposed ligand. The cell can respond chemokinetically to the same group of ligands, but it is evident that the speed at which a cell moves is also determined by all sorts of other factors unrelated to ligand–receptor binding. The simplest effect of any agent on chemokinesis is to cause immotile cells to start moving or, vice versa, to cause cells to stop. To take a trivial example, metabolic inhibitors cause cells to stop moving, i.e., they have negative effects on chemokinesis.

A more interesting example is that of agents such as colchicine, which prevent microtubule polymerization. These drugs induce polarization and translocation of previously immotile neutrophils and lymphocytes in the absence from the medium of any chemotactic factor[9–11]; thus, by definition, they have a positive effect on chemokinesis (orthokinesis). The effect of colchicine on polarization and locomotion is slow (15–30 min) probably because it is mediated from inside the cytoplasm. It does not involve a ligand–receptor interaction, the cell cannot detect a stimulus from a particular direction, and a chemotactic response is therefore precluded. In contrast, formyl peptides bind to receptors, cause polarization and locomotion within 5 min, and, since the occupied receptors have a precise geographical location, the cell can respond directionally. Drugs like deuterium oxide which stabilize microtubule polymers have an effect opposite to that of colchicine inasmuch as they cause rounding-up of polarized, motile cells.[9,11]

This interesting regulatory effect of microtubules on the locomotor capacity of leukocytes is still not understood either at the ultrastructural or at the molecular level. There is a large and conflicting literature on the

[9] H. U. Keller, A. Naef, and A. Zimmermann, *Exp. Cell Res.* **153,** 173 (1984).

[10] S. C. Jarvis, R. Snyderman, and H. J. Cohen, *Blood* **48,** 717 (1976).

[11] P. C. Wilkinson, *Immunology* **57,** 281 (1986).

effects of colchicine and similar drugs on stimulated leukocyte locomotion, some of which suggests that these drugs inhibit chemotaxis. It is clear from visual studies of colchicine-treated leukocytes migrating in gradients that colchicine has no such inhibitory effect.[12,13] Colchicine-treated neutrophils detect gradients efficiently and move faster toward their sources than untreated cells, though the paths described by colchicine-treated cells may be more irregular than those of control cells,[12,13] as has also been shown in patients with the immotile cilia syndrome[14] in whom a microtubular abnormality is postulated.

General Problem of Recognition of Chemotactic Factors

The evidence is now beyond dispute that the effects of the best studied chemotactic factors are mediated by binding to cell membrane receptors and that this binding is stereospecific as judged by the effect of structural modifications to the ligand on binding affinity and on efficacy in producing a cellular response. Most such studies have been done using formyl peptides,[15,16] and similar studies have been done using leukotriene B_4 (LTB_4).[17] However, a very large number of chemotactic factors has been described, documented in varying detail, and the information is lacking, in most cases, to demonstrate that these too bind to specific receptors. The question has to be asked, although it is a question which is exceedingly difficult to answer experimentally, whether all ligands which activate leukocyte locomotion do so through conventional receptors (excluding for present purposes agents such as colchicine whose action is intracellular). I have argued elsewhere[18,19] that some forms of recognition may be relatively nonspecific. An example is discussed below in reference to denatured proteins. A similar problem exists in phagocytic recognition, which may be very specific in the case of phagocytosis mediated by receptors for IgG Fc

[12] R. B. Allan and P. C. Wilkinson, *Exp. Cell Res.* **111,** 191 (1978).

[13] S. Malawista and A. de Boisfleury-Chevance, *J. Cell Biol.* **95,** 960 (1982).

[14] L. L. Englander and H. L. Malech, *Exp. Cell Res.* **135,** 468 (1981).

[15] H. J. Showell, R. J. Freer, S. H. Zigmond, E. Schiffmann, S. Aswanikumar, B. A. Corcoran, and E. L. Becker, *J. Exp. Med.* **143,** 1154 (1976).

[16] R. J. Freer, A. R. Day, J. A. Radding, E. Schiffmann, S. Aswanikumar, H. J. Showell, and E. L. Becker, *Biochemistry* **19,** 2040 (1980).

[17] R. A. Lewis, E. J. Goetzl, J. M. Drazen, N. A. Soter, K. F. Austen, and E. J. Corey, *J. Exp. Med.* **154,** 1243 (1981).

[18] P. C. Wilkinson, "Chemotaxis and Inflammation," 2nd Ed. Churchill-Livingstone, Edinburgh, Scotland, 1982.

[19] P. C. Wilkinson, *in* "Biology of the Chemotactic Response" (J. M. Lackie and P. C. Wilkinson, eds.), p. 53. Cambridge Univ. Press, Cambridge, England, 1981.

or for C3b or C3bi but which may also be nonspecific, e.g., in the case of phagocytosis of latex beads or glutaraldehyde-treated red cells. It is difficult to envision receptors for latex beads, nor is it obvious how a cell would have evolved a specific mechanism for ingesting such artificial (and newly manufactured) objects. In this sort of discussion, much rests on the definition of "receptor" and the degree of specificity required to couple a signal to a response. At present there is no information to suggest whether or not denatured proteins or latex beads, for example, bind to a discrete class of cell membrane molecules.

Chemotactic Factors

The chemotactic factors of major interest at the time of writing are formyl peptides, C5a, and LTB_4, the structure of each of which has been determined. Of these, the formyl peptides are the most widely used, since they have the advantage of being easy to synthesize in large quantity and have powerful and immediate actions on leukocytes which can be assayed accurately. There is now an extensive literature on these peptides, and they, C5a, and LTB_4, are discussed in other chapters ([7] and [46]) of this volume. The activity of these factors is not neutrophil-specific, and each has been shown to have activity for mononuclear phagocytes.

Table I shows the effect of these and other factors on the different leukocytes of human blood, the data being obtained for the greater part from polarization assays (see this volume [2]). It will be noticed that, while fMet-Leu-Phe (fMLP), C5a, and LTB_4 all have powerful activity for neu-

TABLE I
ACTIVITIES OF SOME CHEMOTACTIC FACTORS IN THE POLARIZATION ASSAY[a] USING DIFFERENT HUMAN BLOOD LEUKOCYTE TYPES

	Maximal effective concentration (M) for cell type		
Factor	Neutrophil	Monocyte	Lymphocyte
C5a	10^{-8}	10^{-8}	Inactive
$C5a_{desArg}$	10^{-7}	10^{-7}	Inactive
fMLP	10^{-9}	5×10^{-9}	Inactive
LTB_4	10^{-8} to 10^{-9}	10^{-9}	Inactive
Platelet activating factor	10^{-5} to 10^{-6}	10^{-7}	Inactive
Alkali-denatured human serum albumin	5×10^{-5}	5×10^{-5}	5×10^{-5}
TNF-α	10^{-9}	10^{-9}	Inactive

[a] See this volume [2].

trophils and monocytes, they have no effect on lymphocytes. Denatured proteins (discussed in [17]) are more weakly active but do stimulate locomotion of a minority population of blood lymphocytes,[11] as does casein as reported in several studies.[20-22] All the leukocyte types from blood are stimulated by cell-released products, lymphokines, and other cytokines, and, although much more work is needed to purify these factors and to define their activity, it is likely that there are cell-specific lymphokines, e.g., the mononuclear phagocyte-specific lymphocyte-derived chemotactic factor[23,24] and the T-lymphocyte chemotactic factor (LCF) defined in a preliminary way in more than one laboratory.[25,26] Tumor necrosis factor (TNF-α) has recently been described as a chemotactic factor for neutrophils and monocytes.[27]

[20] P. C. Wilkinson, J. A. Roberts, R. J. Russell, and M. McLoughlin, *Clin. Exp. Immunol.* **25,** 280 (1976).
[21] A. L. El-Naggar, D. E. Van Epps, and R. C. Williams, *Cell. Immunol.* **56,** 365 (1980).
[22] A. L. El-Naggar, D. E. Van Epps, and R. C. Williams, *Cell. Immunol.* **60,** 43 (1981).
[23] L. C. Altman, *in* "Leukocyte Chemotaxis" (J. I. Gallin and P. G. Quie, eds.), p. 267. Raven, New York, 1978.
[24] L. C. Altman, R. Snyderman, J. J. Oppenheim, and S. E. Mergenhagen, *J. Immunol.* **110,** 801 (1973).
[25] D. E. Van Epps, J. W. Potter, and D. A. Durant, *J. Immunol.* **130,** 2727 (1983).
[26] D. M. Center and W. Cruikshank, *J. Immunol.* **128,** 2563 (1982).
[27] W. J. Ming, L. Bersani, and A. Mantovani, *J. Immunol.* **138,** 1469 (1987).

[13] Synthesis of Chemotactic Peptides

By NATESA MUTHUKUMARASWAMY and RICHARD J. FREER

Introduction

Since the initial report from Schiffmann's group[1] that N^{α}-formyl peptides exhibit chemotactic activity for neutrophil leukocytes, there has followed an extensive evaluation of the structural requirements for interaction of these small peptides with their specific receptors. The prototype peptide, N^{α}-formyl-Met-Leu-Phe-OH (fMLP), was first described by Sho-

[1] E. Schiffmann, B. A. Corcoran, and S. Wahl, *Proc. Natl. Acad. Sci. U.S.A.* **72,** 1059 (1975).

well *et al.*[2] and has served as the model for subsequent evaluation.[3-7] In addition, specific antagonists have been described.[8,9]

Exposure of neutrophil leukocytes to fMLP initiates a host of biochemical, physiological, and functional responses, and this peptide and related analogs have proven invaluable in elucidating basic cellular mechanisms in these cells.[10-12] Based on the extensive structure–activity data[2-7] as well as other studies,[13,14] a model of the ligand–receptor interaction has also been proposed.[5]

While the specific compounds described herein are available from commercial sources, the general synthetic strategies have been applied in our laboratory to scores of analogs with equivalent success. Thus, the procedures will not only serve as a guide to the specific compounds but should also be applicable to preparation of new analogs.

Synthesis of Agonists

A general synthetic scheme using solution methods is given in Fig. 1. For the purposes of this chapter we have outlined the synthesis of the prototype peptide fMLP. The steps are a series of repetitive deprotections of *t*-Boc-amino acids or peptides using trifluoroacetic acid (TFA), subsequent neutralization with *N*-methylmorpholine (NMM), followed by coupling of the next *t*-Boc-amino acid using a rapid mixed anhydride method.[15] Although several methods are available for formylation of

[2] H. J. Showell, R. J. Freer, S. H. Zigmond, E. Schiffmann, S. Aswanikumar, B. A. Corcoran, and E. L. Becker, *J. Exp. Med.* **143,** 1154 (1976).

[3] R. J. Freer and A. R. Day, *in* "Peptides: Structure and Biological Function" (E. Gross and J. Meienhofer, eds.), p. 749. Pierce Chemical Co., Rockford, Illinois, 1979.

[4] R. J. Freer, A. R. Day, J. A. Radding, E. Schiffmann, S. Aswanikumar, H. J. Showell, and E. L. Becker, *Biochemistry* **19,** 2404 (1980).

[5] R. J. Freer, A. R. Day, N. Muthukumaraswamy, D. Pinon, A. Wu, H. J. Showell, and E. L. Becker, *Biochemistry* **21,** 257 (1982).

[6] C. Toniolo, P. M. Bonora, H. J. Showell, R. J. Freer, and E. L. Becker, *Biochemistry* **23,** 698 (1984).

[7] P. M. Bonora, C. Toniolo, R. J. Freer, and E. L. Becker, *in* "Peptides" (U. Ragnarsson, ed.), p. 529. Almquist & Wiskell, Stockholm, Sweden, 1984.

[8] S. Aswanikumar, B. Corcoran, E. Schiffmann, C. B. Pert, J. L. Morell, and E. Gross, *in* "Peptides" (M. Goodman and J. Meienhofer, eds.), p. 141. Wiley, New York, 1977.

[9] A. R. Day, D. Pinon, N. Muthukumaraswamy, and R. J. Freer, *Peptides* **1,** 289 (1980).

[10] E. L. Becker, *J. Reticuloendothel. Soc.* **26** (Suppl.), 701 (1979).

[11] E. Schiffmann, *Annu. Rev. Physiol.* **44,** 553 (1982).

[12] R. Snyderman and M. C. Pike, *Contemp. Top. Immunobiol.* **14,** 1 (1984).

[13] E. L. Becker, H. E. Bleich, A. R. Day, J. A. Glasel, M. Latina, and J. Visintainer, *Biochemistry* **18,** 4656 (1979).

[14] W. A. Marasco, H. J. Showell, R. J. Freer, and E. L. Becker, *J. Immunol.* **128,** 956 (1982).

[15] M. A. Tilak, *Tetrahedron Lett.* **11,** 849 (1970).

TABLE I
PROTOCOL FOR SOLID-PHASE SYNTHESIS OF CHO-Met-Leu-Phe-OH[a]

Step	Reagent (times × volume)[b]	Time (min)[c]
1	CH_2Cl_2 (3 × 20 ml)	2
2	25% TFA/CH_2Cl_2 (1 × 20 ml)	2
3	25% TFA/CH_2Cl_2 (1 × 20 ml)	30
4	CH_2Cl_2 (6 × 20 ml)	2
5	10% DIEA/CH_2Cl_2 (1 × 20 ml)	2
6	10% DIEA/CH_2Cl_2 (1 × 20 ml)	10
7	CH_2Cl_2 (6 × 20 ml)	2
8	*t*-Boc-amino acid (1.0 mmol) in CH_2Cl_2 (5–10 ml)	5
9	DCC (4 ml of 0.25 *M*) in CH_2Cl_2	120
10	CH_2Cl_2 (3 × 20 ml)	2
11	Ethanol (3 × 20 ml)	2
12	Repeat Step 1	

[a] Volumes and molar amounts are based on approximately 1 g of Phe—resin with a substitution of 0.4 mmol/g.

[b] CH_2Cl_2, Dichloromethane; TFA, trifluoroacetic acid; DIEA, *N,N*-diisopropylethylamine; DCC, dicyclohexylcarbodiimide, *t*-Boc, *tert*-butyloxycarbonyl.

[c] Where more than 1 wash is indicated the time noted applies to each wash.

peptides,[16-21] we have had consistently excellent results using the anhydride of isovaleryl chloride and formic acid.[16] Final deprotection is usually carried out with anhydrous hydrogen fluoride (HF) (45 min at 4°) in the presence of 50 Eq of anisole. All reactions are monitored by TLC and in general are complete in 60 min. These analogs are also readily prepared by solid-phase synthesis (protocol given in Table I).

Synthesis of Antagonists

The same general protocol is used for the synthesis of the antagonists (Fig. 2). The prototype analog here is the pentapeptide, *t*-Boc-Phe-Leu-Phe-Leu-Phe-OH. The final step, however, is a catalytic hydrogenolysis to yield the *t*-Boc-free acid form of the desired peptide. It should be noted that although we have presented the synthesis for the pentapeptide, the tripep-

[16] A. R. Day, N. Muthukumaraswamy, and R. J. Freer, *Peptides* **1,** 187 (1980).

[17] J. C. Sheehan and D. D. H. Yang, *J. Am. Chem. Soc.* **80,** 1154 (1958).

[18] M. Waki and J. Meienhofer, *J. Org. Chem.* **42,** 2019 (1977).

[19] J. O. Thomas, *Tetrahedron Lett.* **4,** 335 (1967).

[20] G. Lajoie and J. L. Kraus, *Peptides* **5,** 653 (1984).

[21] J. Martinez and J. Lour, *Synthesis* **11,** 979 (1982).

tide *t*-Boc-Phe-Leu-Phe-OH is also fully active and only slightly less potent.[9] No solid-phase synthesis for the antagonists have been reported.

Materials and Methods

Protected amino acids may be purchased from Bachem, Inc. (Torrance, CA) or Peninsula Laboratories (Belmont, CA). All solvents and chemicals are reagent grade.

Thin-layer chromatography is carried out on Merck silica gel F254 precoated glass plates. Plates are developed using ninhydrin and/or chlorine/*o*-tolidine spray reagents. Solvent systems are (A) benzene: acetic acid: water (9:9:1), (B) methanol: chloroform (1:1), (C) chloroform: methanol: acetic acid: water (60:30:4:1), and (D) *n*-butanol: acetic acid: water (4:1:1).

Melting points are determined on a Thomas Hoover apparatus (Unimelt) and are uncorrected. Optical rotations are measured at the sodium D line using a Perkin-Elmer 141 polarimeter.

Experimental

Synthesis of CHO-Met-Leu-Phe-OH (Fig. 1)

Boc-Leu-Phe-OBzl. Boc-Ley-OH (4.62 g, 20 mmol) is dissolved in ethyl acetate (50 ml) and cooled to −15°. NMM (2.54 ml, 20 mmol) and isobutyl chloroformate (2.55 ml, 19 mmol) are added successively such that the temperature remains at −15°. After 2 min a solution of H-Phe-OBzl · TsOH (8.11 g, 19 mmol) in dimethylformamide (DMF) (20 ml) containing NMM (2.54 ml, 20 mmol) is added. The acetone/dry ice bath is removed, and the reaction mixture is stirred until the reaction is complete (generally 1 hr). The solution is cooled to 4° and stirred with *N,N*-dimethylamine-3-aminopropane (1.2 ml, 10 mmol) for 20 min. The reaction mixture is evaporated to dryness *in vacuo,* and the oily residue is dissolved in ethyl acetate (EtOAc). The organic phase is washed with cold 1 *M* HCl (3 × 50 ml), saturated $NaHCO_3$ solution (3 × 50 ml) and saturated NaCl solution (3 × 50 ml). After drying the organic phase (anhydrous Na_2SO_4), it is filtered and evaporated *in vacuo.* The crude residue is crystallized from EtOAc/petroleum ether. Yield 8.3 g (98%); mp 100°; R_f^A 0.87; R_f^C 0.91; R_f^D 0.82; $[\alpha]_D^{22}$ −36.2° (*c* 1, MeOH); Anal. Calc. for $C_{27}H_{36}N_2O_5$: C, 69.20; H, 7.74; N, 5.98; found: C, 69.48, H, 7.73; N, 6.02.

TFA · Leu-Phe-OBzl. The dipeptide is deprotected in anhydrous TFA (3 ml/mmol of peptide), and when complete (usually 60 min) the acid is

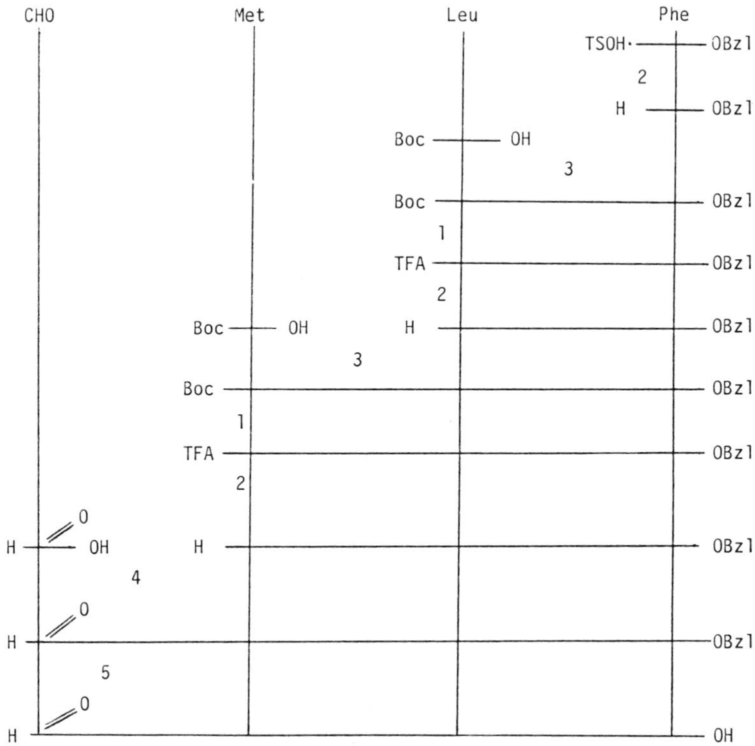

FIG. 1. Synthesis of CHO-Met-Leu-Phe-OH. Steps are as follows: 1, deprotection with trifluoroacetic acid (TFA); 2, neutralization with *N*-methylmorpholine (NMM); 3, mixed anhydride coupling with isobutyl chloroformate; 4, formylation with anhydride of isovaleryl chloride and formic acid or diimide coupling of formic acid; 5, final deprotection with anhydrous hydrogen fluoride (HF). Details are given in the text.

removed *in vacuo.* The TFA salt is precipitated with ether, filtered, dried, and used without further purification.

Boc-Met-Leu-Phe-OBzl. The tripeptide prepared on a 12.5 m*M* scale as detailed above using Boc-Met-OH and TFA·Leu-Phe-OBzl. The crude product, obtained following evaporation of the reaction mixture, is dissolved in EtOAc and washed in sequence with cold 1*N* HCl, saturated $NaHCO_3$, and saturated NaCl solutions. The organic phase is dried (Na_2SO_4), and the pure product is obtained by crystallization from EtOAc/petroleum ether. Yield 6.7 g (90%); mp 144°; R_f^A 0.94; R_f^C 0.91; Anal. Calc. for $C_{32}H_{45}N_3SO_6$: C, 64.02; H, 7.56; N, 7.00; found: C, 63.32; H, 7.68; N, 6.86.

Formyl-Met-Leu-Phe-OBzl. The formyl group is most effectively introduced by a conventional anhydride coupling using formic acid and isovaleryl chloride. This gives an excellent product, but the offensive odor of the resulting isovaleric acid is problematical. A water-soluble diimide 1-ethyl-3-(3-dimethylaminopropyl)carbodiimide (EDC) coupling with formic acid is also effective, and a dicyclohexylcarbodiimide (DCC) coupling has been used for both solution and solid-phase protocols. The odor problem is not present with diimide couplings, but these are somewhat slower, and, in the case of DCC, dicyclohexylurea is a difficult to remove by-product. Specific reaction conditions are given below.

1. *Anhydride coupling.* Ninety-nine percent formic acid (4 Eq) is dissolved in tetrahydrofuran (1.25 ml/mEq) and cooled to $-15°$. NMM (4 Eq) is added and followed by isovaleryl chloride (3.6 Eq). The reaction mixture is stirred for 1 min, and a cold solution of TFA·Met-Leu-Phe-OBzl (1 Eq) (prepared in the same way as for TFA·Leu-Phe-OBzl) in DMF (2 ml.mEq) containing NMM (1 Eq) is added. The reaction mixture is allowed to come to room temperature and stirred until the reaction is complete (usually 60 min). The reaction mixture is then cooled in an ice bath and *N,N*-dimethylamine-3-aminopropane (4 Eq) added. After 20 min the solvent is evaporated *in vacuo* and the residue dissolved in EtOAc. The organic phase is then washed successively with cold 1 *N* HCl, saturated $NaHCO_3$, and saturated NaCl and dried over Na_2SO_4. The solvent is evaporated, and, if necessary, the product is purified by crystalization. Systems such as EtOAc/hexane, chloroform/hexane, and EtOAc/petroleum ether have been generally useful. Yields above 90% are usually obtained by this method. *Note:* All of these steps should be carried out in a good hood as the isovaleryl chloride and isovaleric acid have a strong and offensive odor.

2. *Diimide coupling.* Ninety nine percent formic acid (1.1 Eq) is dissolved in EtOAc and cooled on an ice bath. Hydroxybenzotriazole (1.1 Eq) and the diimide (1.1 Eq) (either DCC or EDC can be used) are added, and the reaction mixture is stirred for 10 min. A cold solution of TFA·Met-Leu-Phe-OBzl (1 Eq) in DMF (1 ml/mEq) containing NMM (1 Eq) is added, and the reaction mixture is stirred for 2 hr at 0°. If necessary the reaction can be stirred at room temperature overnight. If DCC is used there will be a precipitate of dicyclohexylurea present which should be removed by filtration. The solvent is removed by rotary evaporation and the residue redissolved in EtOAc (an additional filtration may be necessary to remove a further amount of urea). The crude product is then treated as described for the anhydride coupling. The overall yield from this procedure is generally 70–80%. The product has chromatographic mobilities of R_f^A 0.97, R_f^B 0.9, and R_f^D 0.75.

CHO-Met-Leu-Phe-OH. CHO-Met-Leu-Phe-OBzl (1.3 g; 2.5 mmol) is treated with anhydrous HF (15 ml) in the presence of anisole (2 ml) for 45 min at 4°. The HF is removed *in vacuo,* and the residue is dissolved in MeOH (10 ml). The product is precipitated with ether (200 ml), filtered, washed with ether, and dried. The final product is crystallized from acetone/water. Yield 0.6 g (55%); mp 215°; R_f^A, 0.9; R_f^B, 0.78; R_f^D, 0.67; Anal. Calc. for $C_{21}H_{31}N_3SO_5$; C, 57.64; H, 7.14; N, 9.60; found; C, 57.34; H, 6.91; N, 9.38.

Solid-Phase Synthesis of CHO-Met-Leu-Phe-OH

The synthesis of CHO-Met-Leu-Phe-OH and analogs proceeds readily using standard solid-phase synthesis protocols (Table I and Ref. 22). The formyl group is introduced into the resin-bound peptide by using 99% formic acid (2.5 Eq) and DCC (2.5 Eq). Final deprotection and removal from the resin is with anhydrous HF essentially as noted above.

Synthesis of Boc-Phe-Leu-Phe-Leu-Phe-OH (Fig. 2)

The dipeptide Boc-Leu-Phe-OBzl is prepared as detailed above for CHO-Met-Leu-Phe-OH. Specific intermediates and analytical data are given below. The deprotected peptide, TFA·Leu-Phe-OBzl, is also prepared as outlined previously, and all TFA deprotections are carried out in the same manner. Specific intermediates and analytical data are given below.

Boc-Phe-Leu-Phe-OBzl. Yield 83%; mp 159–160°; R_f^A 0.99; R_f^C 0.95; R_f^D 0.82; $[\alpha]_D^{22}$ −19.0° (*c* 1, DMF); Anal. Calc. for $C_{36}H_{45}N_3O_6$: C, 70.22; H, 7.37; N, 6.82; found: C, 69.43; H, 7.11; N, 6.74.

Boc-Leu-Phe-Leu-Phe-OBzl. Yield 73%; mp 170–172°; R_f^A 0.98; R_f^C 0.97; R_f^D 0.83; $[\alpha]_D^{22}$ −50.7° (*c* 1, MeOH); Anal. Calc. for $C_{42}H_{56}N_4O_7$: C, 69.21; H, 7.74; N, 7.68; found: C, 69.42; H, 7.86; N, 7.58.

Boc-Phe-Leu-Phe-Leu-Phe-OBzl. Yield 83%; mp 214–216°; R_f^A 0.97; R_f^C 0.96; R_f^D 0.80; $[\alpha]_D^{22}$ −26.9° (*c* 1, DMF); Anal. Calc. for $C_{51}H_{65}N_5O_8$: C, 69.92; H, 7.48; N, 7.99; found: C, 70.02; H, 7.69; N, 7.85.

Boc-Phe-Leu-Phe-Leu-Phe-OH. Hydrogenolysis is carried out by addition of 1 mmol of Boc-Phe-Leu-Phe-Leu-Phe-OBzl to a suspension of 10% palladium/charcoal (50 mg) in 12 ml of methanol : water : acetic acid (10 : 1 : 1). The reaction proceeds for 16 hr at room temperature at 18 psi. The reaction mixture is filtered and evaporated to dryness, and the product is collected from petroleum ether. Yield 95%; mp 210°; R_f^A 0.9; R_f^C 0.84;

[22] J. M. Stewart and J. D. Young, "Solid-Phase Peptide Synthesis." Pierce Chemical Co., Rockford, Illinois, 1985.

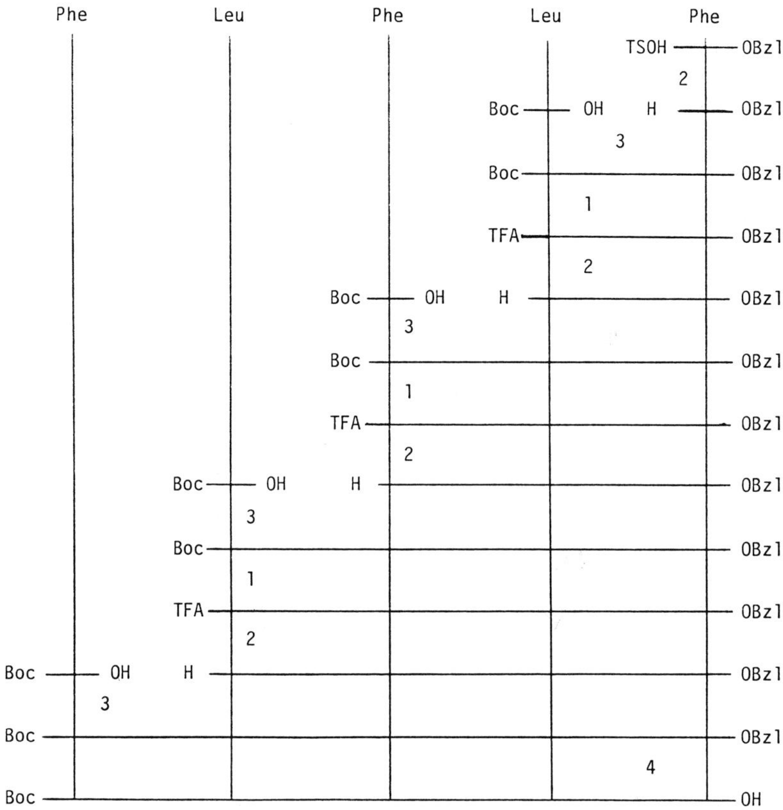

FIG. 2. Synthesis of Boc-Phe-Leu-Phe-Leu-Phe-OH. Steps are as follows: 1, deprotection with trifluoroacetic acid (TFA); 2, neutralization with *N*-methylmorpholine (NMM); 3, mixed anhydride coupling; 4, catalytic hydrogenolysis. Details are given in the text.

R_f^D 0.79; $[\alpha]_D^{22}$ $-21°$ (*c* 1, DMF); Anal. Calc. for $C_{44}H_{59}N_5O_8 \cdot 1H_2O$: C, 65.75; H, 7.64; N, 8.71; H_2O, 2.2; found: C, 65.53; H, 7.56; N, 8.91; H_2O, 1.9.

Other compounds prepared by this procedure include the following:

Boc-Phe-Leu-Phe-OH. Yield 61% mp 151–153°; R_f^A 0.84; R_f^C 0.83; R_f^D 0.79; $[\alpha]_D^{22}$ $-16.2°$ (*c* 1, MeOH); Anal. Calc. for $C_{29}H_{39}N_3O_6$: C, 66.26; H, 7.48; N, 7.99; found: C, 66.45; H, 7.65; N, 7.86.

Boc-Leu-Phe-Leu-Phe-OH. Yield 87%; mp 185–191°; R_f^A 0.76; R_f^C 0.84; R_f^D 0.87; $[\alpha]_D^{22}$ $-29.4°$ (*c* 1, DMF); Anal. Calc. for $C_{35}H_{50}N_4O_7$: C, 65.81; H, 7.89; N, 8.77; found: C, 65.51; H, 7.72; N, 8.57.

[14] Natural Mediators of Leukocyte Chemotaxis

By HIDEO HAYASHI, MITSUO HONDA, YASUHIRO MIBU, SHUNSUKE YAMAMOTO, and MITSUOMI HIRASHIMA

Introduction

Putative chemotactic factors must fulfill some if not all of the conditions.[1-3] (1) They should be present at the inflammation site to induce leukotaxis. (2) The amount (or activity) of the chemotactic factor(s) should be proportional to the extent of leukotaxis. (3) They should induce morphologic changes (detectable by light and electron microscopy) similar to those observed in the original area of inflammation, when injected in concentrations comparable to those detected at the inflammation site. (4) Their action should be specific for the associated leukocytes and not induce vascular permeability change and hemorrhage. (5) They should be inhibited by specific antagonist(s) present *in situ.* (6) Removal of the chemotactic factor or its precursor(s) should cause a decrease in the leukotaxis. (7) The precursor(s) of the chemotactic factor or the enzyme(s) involved in their production should be present at the site of inflammation.

Natural Mediators of Macrophage Chemotaxis

Induction of Cutaneous Allergic Inflammation and Quantification of Tissue Leukocytosis

Guinea pigs are commonly used for study of the skin hypersensitivity reaction. The purified protein derivative of tuberculin (PPD)[3a] and DNP-*Ascaris* extract (DNP-As) induce different types of allergic skin lesions in male Hartley guinea pigs weighing 300–500 g.

[1] H. Hayashi, M. Yoshinaga, and S. Yamamoto, *in* "Chemotaxis: Its Biology and Biochemistry" (E. Sorkin, ed.), p. 296. Karger, Basel, Switzerland, 1974.

[2] H. Hayashi, *Int. Rev. Cytol.* **40,** 101 (1975).

[3] H. Hayashi, M. Honda, Y. Shimokawa, and M. Hirashima, *Int. Rev. Cytol.* **89,** 179 (1984).

[3a] Abbreviations: PPD, purified protein derivative of tuberculin; DNP-As, DNP-Ascaris; BCG, Bacillus Calmette–Guérin; PBS, phosphate-buffered saline; CFA, Freund's complete adjuvant; DHR, delayed hypersensitivity reaction; MCF, macrophage chemotactic factor; PB, phosphate buffer; MCA, macrophage chemotactic activity; LCF, lymphocyte chemotactic factor; PCA, passive cutaneous anaphyloxis; PEC, peritoneal exudate cells; ECF, eosinophil chemotactic factor; BGG; bovine gamma globulin; Con A, concanavalin A; NCF, neutrophil chemotactic factor.

Immunization Protocols for Induction of PPD Lesions.[4] At day 0, inject 1.6 mg of Bacillus Calmette–Guérin (BCG; Trudeau Research Institute, Saranac Lake, NY) i.d. using a 26-gauge, 1/2-inch needle, in a total volume of 1.0 ml of phosphate-buffered saline (PBS), pH 7.2, per guinea pig; 0.1 ml of the solution is given at each of 10 sites, i.e., 8 sites on the footpad and thigh and 2 sites on the neck. At days 42–45, inject i.d. 10 μg of PPD (Connaught Laboratories, Ontario, Canada) in a total volume of 1.0 ml of PBS per site at 10 sites on the shaved back of the guinea pigs. The recommended procedure is to first dialyze the PPD solution against PBS and then dilute the stock solution of antigen to 100 μg/ml in PBS. Inflammation is observed for 72 hr after challenge.

Immunization Protocols for Induction of DNP-As Lesions.[5] At day 0, inject each guinea pig i.d. with 1.5 mg of DNP-As (for preparation, see Ref. 6) in 0.3 ml of PBS emulsified in 0.2 ml of Freund's complete adjuvant (CFA, Difco) with *Mycobacterium tuberculosis* (H37 Ra-Difco) using a 23-gauge, 5/8-inch needle. Simultaneously inject i.p. 1×10^{10} cells of killed *Bordetella pertussis* (Institute de Microbiologie et d'Hygiene de l'Université de Montréal, Quebec), in 1 ml of PBS.

At day 7, repeat the immunization procedure with the same dose of antigen emulsified in CFA and *B. pertussis.*

At day 12–15, inject i.d. 0.25 mg of DNP-As in a total volume of 0.6 ml in PBS per site at 12 sites on the shaved back of each animal. Inflammation is observed for 72 hr after challenge.

Quantification of Tissue Leukocytosis.[7,8] At various intervals following challenge with antigen, the animals are sacrificed and the injection sites are immediately excised using a metallic punch (1×1 cm). The necropsies include all layers of the skin. Among the mononuclear cells accumulated, macrophages can be differentiated from lymphocytes by the number and fragility of the lysosomes which are identified as cytoplasmic granules and stained with acid phosphatase[7] (see also this series, Vol. 132 [3]).

1. The skin tissue is placed with the fatty surface directly in contact with the inner wall of a glass vial in liquid nitrogen.
2. Sections, 5 μm thick, of each specimen are made with a cryostat at $-30°$.

[4] M. Honda and H. Hayashi, *Am. J. Pathol.* **108,** 171 (1982).

[5] M. Honda, M. Hirashima, M. Nishiura, and H. Hayashi, *Virchows Arch. B* **27,** 317 (1978).

[6] G. Strejan and D. H. Campbell, *J. Immunol.* **98,** 893 (1967).

[7] J. J. Turk, C. J. Heather, and J. V. Diengdoh, *Int. Arch. Allergy Appl. Immunol.* **29,** 278 (1966).

[8] A. B. Kay, *Clin Exp. Immunol.* **6,** 75 (1970).

3. Sections are attached to slides and then fixed in precooled 10% formalin at 4° for 12 hr.
4. After washing for 15 min, the sections are incubated at 37° with Gomori–Holt acid phosphatase substrate mixture for 10, 20, 30, and 60 min, respectively.
5. Sections are rinsed in distilled water and transferred to H_2S-saturated water for 1 min.
6. After washing, the sections are finally stained with hematoxylin and mounted in PVP.
7. Using stained 5-μm-thick sections, a total of 15 random strips 0.3 mm wide between the junction of the epidermis and dermis and the lower limit of the panniculus carnous muscle are examined using a calibrated graticule[8] (Fig. 1A and 1B).

Method for Quantifying Macrophage Chemotaxis[4,5]

The modified Boyden's method using polycarbonate filters is best suited for the screening study of macrophage chemotaxis. This method as well as the checkerboad assay to differentiate chemotaxis from chemokinesis have been described elsewhere in this volume ([3] and [6]).

Preparation of Skin Extracts[9,10]

Procedure

1. Excise the inflamed skin site, cut it with scissors into several pieces, and freeze with dry ice immediately.
2. Cut the pieces of frozen skin into approximately 50-μm-thick slices with a freezing microtome.
3. Dehydrate the slices with 10 vol of cold double-distilled acetone at 4° for 30 min.
4. Repeat dehydration procedure 3 times.
5. Air-dry the dehydrated skin in the cold room for 48 hr.
6. Extract the skin with 10 vol cold 67 m*M* sodium phosphate buffer (PB), pH 7.2, per gram of skin (approximately 2.5 g/animal) in the cold for 4 hr, with continuous stirring.
7. Centrifuge the extract for 30 min at 10,000 *g*, collect the supernatant, and then filter through membrane filter.
8. Ultrafiltrate the supernatant through a UM2 filter (Amicon) to give an absorbancy between 35 and 40 at 280 nm and then sterilize.

[9] K. Udaka, *Kumamoto Med. J.* **16,** 55 (1963).
[10] H. Hayashi, K. Udaka, H. Miyoshi, and S. Kudo, *Lab. Invest.* **14,** 665 (1965).

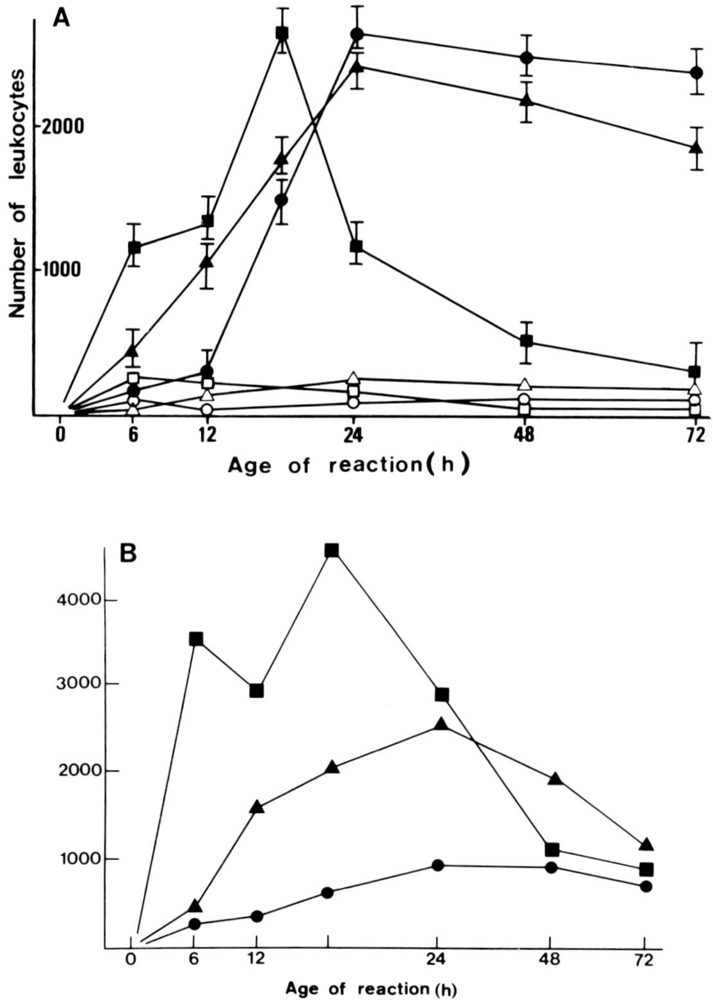

FIG. 1. Time course of tissue leukocytosis in (A) PPD (10 μg)-induced and (B) DNP-*Ascaris* extract (250 μg)-induced allergic skin lesions in guinea pigs. ▲, Macrophage reaction in sensitized animals; △, macrophage reaction in nonsensitized; ●, lymphocyte reaction in sensitized; ○, lymphocyte reaction in nonsensitized; ■, neutrophil reaction in sensitized; □, neutrophil reaction in nonsensitized.

Proteinase inhibitors (typically phenylmethylsulfonyl fluoride, $1 \times 10^{-5} M$) are used in the extraction. At various stages of PPD- or DNP-As-induced skin lesions, chemotactic assays may be performed on extracts from the lesions. The extracts from skin lesions of sensitized animals at

24 hr after challenge represent the best samples for separating macrophage chemotactic factors (MCFs).

Purification and Characterization of MCFs[4,5]

Isolation of MCFs from PPD-Induced Delayed Hypersensitivity Reaction Sites. Three types of MCF (MCF-a, -b, and -c) can be isolated from the inflamed skin sites of guinea pigs. A summary of MCF-c purification is shown in Fig. 2. Recovery of activity and each step are given in Table I.

Procedure[4]

Step 1. Apply 5 ml of skin extract to a Sephadex G-100 column (3 × 50 cm) equilibrated and eluted with 67 m*M* PB, pH 7.4. Three peaks are detected at 280 nm. MCF-a and -c are present in the first, and MCF-b in the second (Fig. 2A).

Step 2. Pool fractions containing MCF-a and -c, concentrate by ultrafiltration on a UM2 filter, and dialyze versus 20 m*M* PB, pH 7.4. Apply the dialysate to a DEAE-Sephadex column (1 × 12 cm) equilibrated and eluted with 20 m*M* PB, pH 7.4. Figure 2B shows that the first peak has marked macrophage chemotactic activity (MCA). MCA is also found in the eluate with a salt gradient between 0 and 1 *M* NaCl in 20 m*M* PB, pH 7.4.

Step 3. Dialyze MCF-c-containing fractions (hatched area in Fig. 2B) against 20 m*M* acetate buffer, pH 5.5, at 4° for 16 hr. Then apply the material to a CM-Sephadex column (1 × 10 cm) equilibrated with 20 m*M* PB, pH 7.4. As illustrated in Fig. 2C, a linear gradient elution between 0 and 1 *M* NaCl in 20 m*M* acetate buffer, pH 5.5 (10 ml/hr), yields at least three peaks of optical density at 280 nm, the second of which exhibits strong MCA. No chemotactic activity is detected in the other two.

Step 4. Dialyze the MCA-containing fractions against 20 m*M* acetate buffer, pH 5.5, and rechromatograph on a CM-Sephadex column under the same conditions as specified above. As shown in Fig. 2D, the active peak is quite symmetric and the MCA parallels absorbancy at 280 nm.

Step 5. Concentrate the pooled MCF-c fractions (stippled area in Fig. 2D), dialyze versus 25 m*M* Tris–HCl, 23 m*M* ϵ-aminocaproic acid pH 8.5, and add ampholine (pH 5–7, 6–8, and 7–9, LKB). Apply the preparation to an LKB 7960 isotachophoresis column containing a 3.3% polyacrylamide gel. The run is carried out for 24 hr at 4°, at 7mA, and at a flow rate of 10 ml/hr. The cathode buffer is 25 m*M* Tris–HCl, ϵ-aminocaproic acid, pH 8.5; the anode buffer is 66 m*M* Tris–sulfate buffer, pH 7.0. Dialyze exhaustively the effluent fraction against 67 m*M* PB, pH 7.2, for 72 hr. Four peaks of optical density at 280 nm are obtained (Fig. 2E). MCA is present in the second peak. Lymphocyte chemotactic activity is

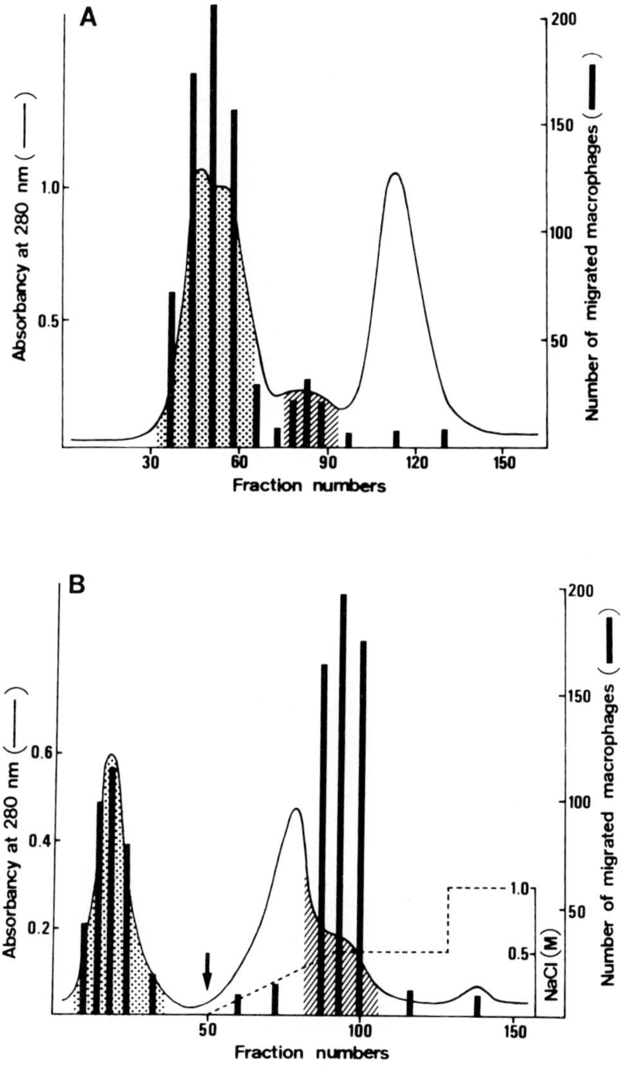

FIG. 2. Purification of MCF-c in the isolation of MCF-a, -b, and -c from PPD-induced skin lesions in guinea pigs. The sequence of purification is as follows: (A) Sephadex G-100 chromatography, (B) DEAE-Sephadex A-50, (C) CM-Sephadex C-50, (D) CM-Sephadex, (E) isotachophoresis. The chemotactic activities were assayed on 2-fold diluted fractions of C, D, and E.

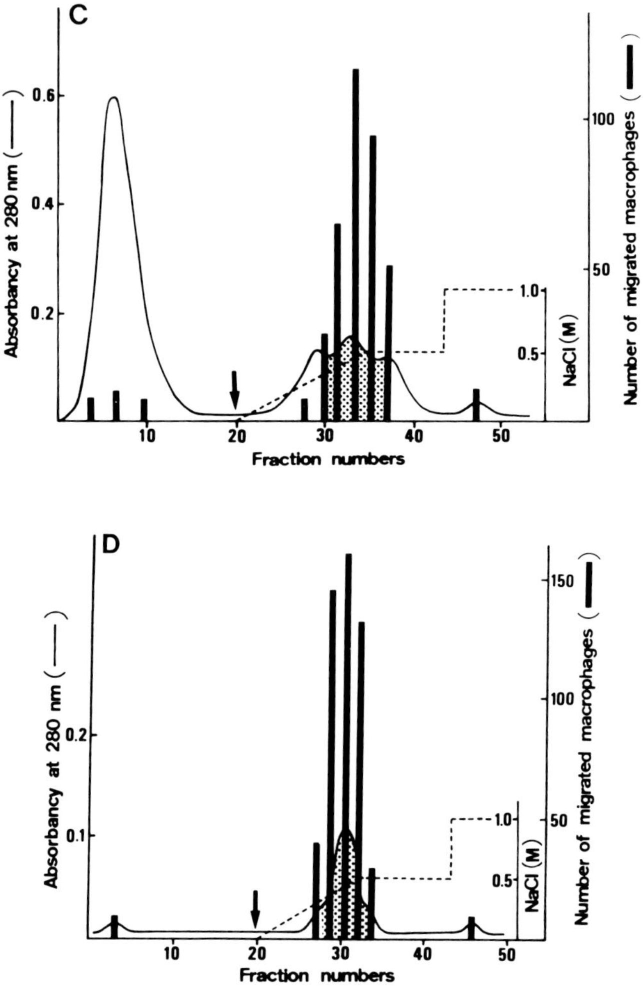

FIG. 2. (C) and (D).

detected in the fourth peak (see below). No MCA is detected in the other two peaks.

Analysis of purified MCF-c by polyacrylamide disk gel electrophoresis shows a single band. Approximately 0.5 mg of MCF-c is obtained from 2.5 g of dried skin; the amount of MCF-c in a 24-hr skin site is calculated to be approximately 30 μg. The final yield of MCF-c is about 15–20%, and the increase in specific activity is about 700-fold (Table I). In PPD-induced

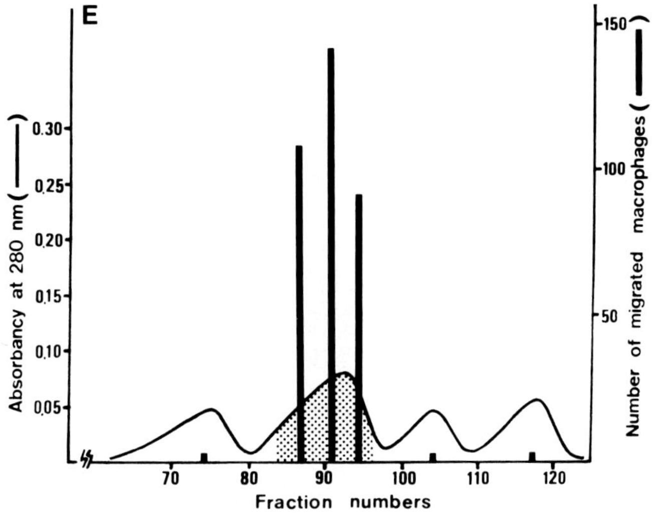

FIG. 2. (E).

TABLE I
MCF-c PURIFICATION

Sample tested[a]	Volume (ml)	Protein (mg)	Total activity (macrophage count)	Recovery (%)	Specific activity[b]
Original skin extract[c]	5.0	183	20,100	100	101
MCFs on Sephadex G-100					
a and c[d]	33.9	76.3	14,400	71.6	189
b	16.0	10.2	1010	5.02	99
MCFs on DEAE-Sephadex A-50					
a	25.0	22.3	3990	19.9	179
c	6.22	12.3	6560	32.6	540
MCF-c on CM-Sephadex C-50	2.19	3.01	6060	30.1	2000
MCF-c on CM-Sephadex C-50 rechromatography	3.49	2.0	5250	26.1	2645
MCF-c after isotachophoresis	10.0	0.50	3635	18.1	7270

[a] Mean of 5–7 experiments.

[b] Macrophage count/$E_{280\,nm}$.

[c] Extracted from 2.5 ± 0.4 g powderized 24-hr skin lesions from BCG-sensitized, PPD-challenged guinea pigs.

[d] MCF-a and -c coexist in the same fraction.

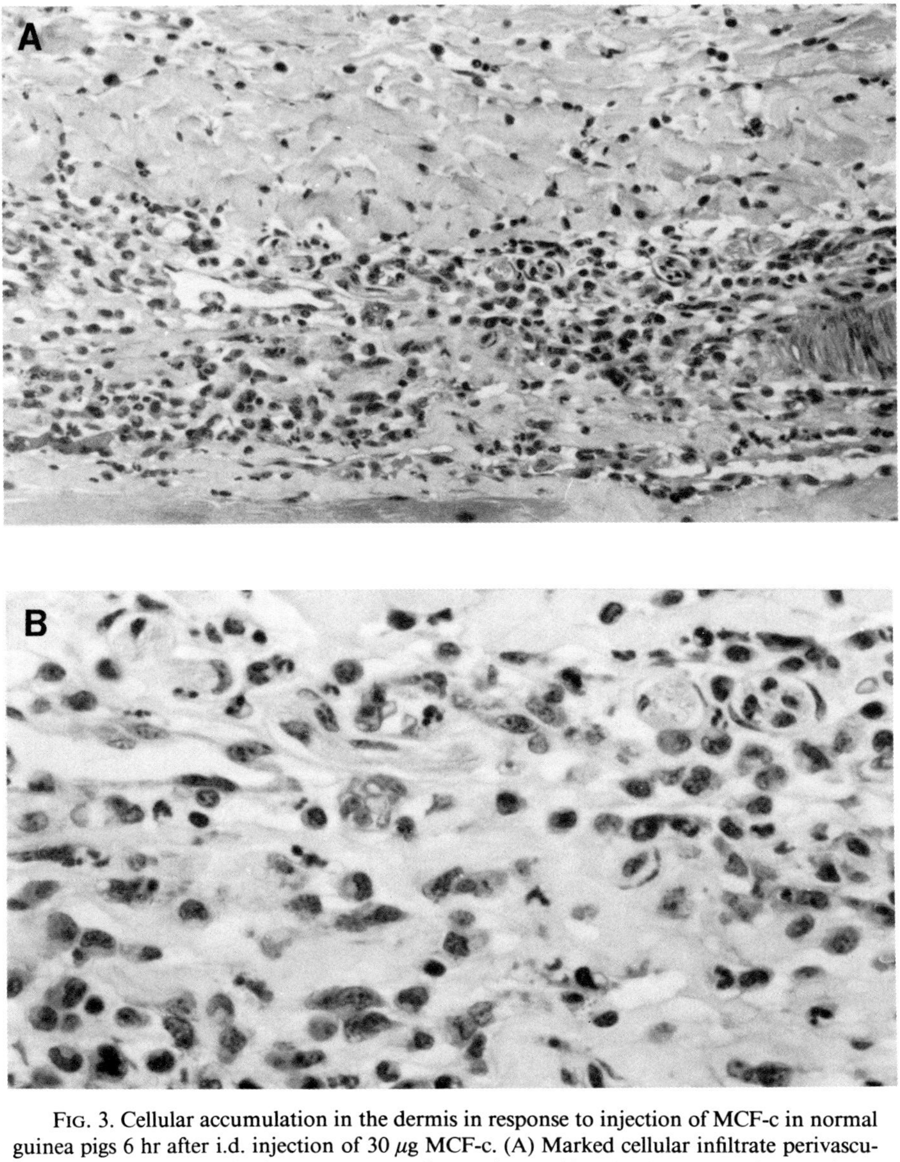

FIG. 3. Cellular accumulation in the dermis in response to injection of MCF-c in normal guinea pigs 6 hr after i.d. injection of 30 μg MCF-c. (A) Marked cellular infiltrate perivascularly throughout the skin tissue. Hematoxylin–eosin. $\times$188. (B) The same lesion at higher magnification demonstrating that most of the cells are macrophages. $\times$423.

lesions, MCF-c is produced in greater amount than MCF-a, while MCF-b is produced in negligible amounts.[4]

Intradermal injection of purified MCF-c (30 μg in 0.1 ml PBS) into normal guinea pigs induces intense mononuclear cell migration through the venular walls into the adjacent tissues in the deep dermis with no vascular permeability change and hemorrhage. The majority of the cells observed are macrophages (Fig. 3).

Isolation of MCFs from DNP-As Induced DHR Sites.[5] MCF-a, -b, and -c can be isolated by similar procedures from skin extracts of DNP-As-induced skin lesions. In the skin lesions, MCF-a is obtained in greater amount than MCF-b. MCF-c is obtained in negligible amounts.[5] A summary of MCF-a purification is shown in Fig. 4.

Procedure

Step 1. Apply 5 ml of skin extract from DNP-As lesions to a Sephadex G-100 column (3 × 50 cm) eluted with 67 m*M* PB, pH 7.2. Three peaks of optical density at 280 nm are obtained. MCF-a and -c are in the first peak; MCF-b is in the second. The yield of the total activity is greater than that of PPD lesions (Fig. 4A).

Step 2. Pool the first peak, concentrate, and dialyze against 20 m*M* PB, pH 7.4. Apply it to a DEAE-Sephadex column (1 × 12 cm) equilibrated and eluted as specified for MCF. As shown in Fig. 4B, MCA in the first peak (MCF-a) is higher than in the third (MCF-c).

Step 3. Dialyze the MCF-a-containing fractions (stippled area in Fig. 4B), apply the dialysate to a CM-Sephadex column (1 × 15 cm) equilibrated with 20 m*M* PB, pH 6.8, and carry out the elution with a linear gradient between 0 and 0.3 *M* NaCl in 20 m*M* PB, pH 6.8 (Fig. 4C). The second peak exhibits strong MCA.

Step 4. Pool the MCF-a-containing fraction, dialyze against the equilibration buffer, and rechromatograph on a CM-Sephadex column under the same condition as Step 3. Pool the active fractions (stippled area in Fig. 4D).

Step 5. Dialyze against 67 m*M* PB, pH 7.4, apply to an anti-IgG affinity column prepared by coupling anti-guinea pig IgG rabbit antibody (Cappel) to Sepharose 4B according to the manufacturer's specifications. The column (1 × 3.5 cm) is eluted first with 67 m*M* PB, pH 7.4, at a flow rate of 1.0 ml/hr at 4°. MCF is eluted with 1 *M* acetic acid. The samples are neutralized immediately with 1.0 *M* Tris and dialyzed versus PBS, pH 7.2.

Characterization of MCF-c.[3] MCF-c, like the chemotactic factors obtained *in vitro* by the action of PPD on T-lymphocytes,[11] binds to various

[11] M. Honda, K. Miura, J. Kuratsu, and H. Hayashi, *Cell. Immunol.* **67**, 212 (1982).

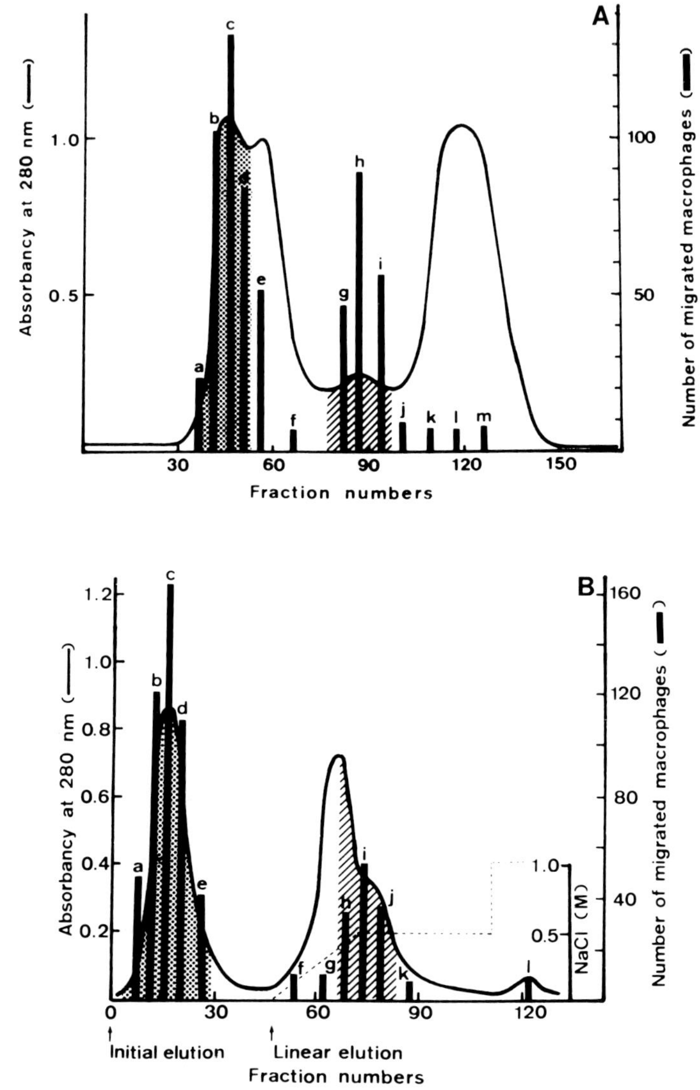

FIG. 4. Purification of MCF-a from DNP-As-induced skin lesions in guinea pigs. (A) Sephadex G-100. (B) DEAE-Sephadex A-50. (C) CM-Sephadex C-50. (D) CM-Sephadex.

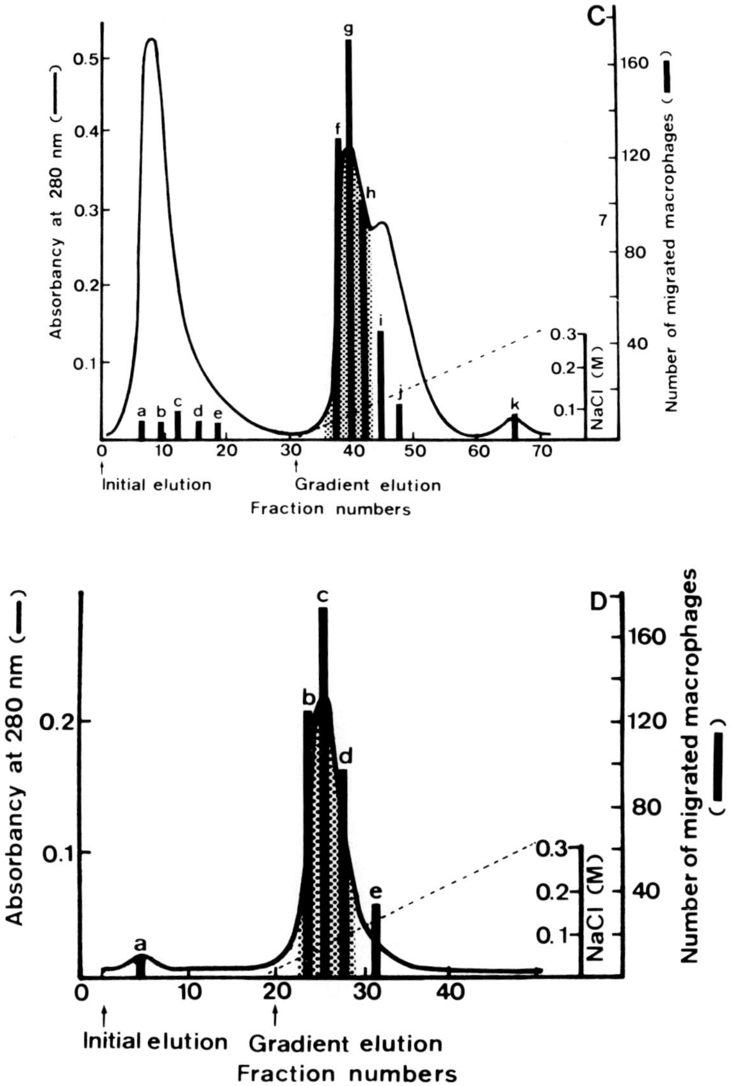

FIG. 4. (C) and (D).

proteins and particularly to α-globulin. These complexes are formed at the inflammatory site, which has a pH near neutrality, and are dissociated at pH 3–4. Dissociated MFC-c has a molecular weight of about 12,500.

Purified MCF-c is specific for chemotaxis of macrophages and blood monocytes of guinea pigs, mice, and humans.[3] MCF-c selectively attracts

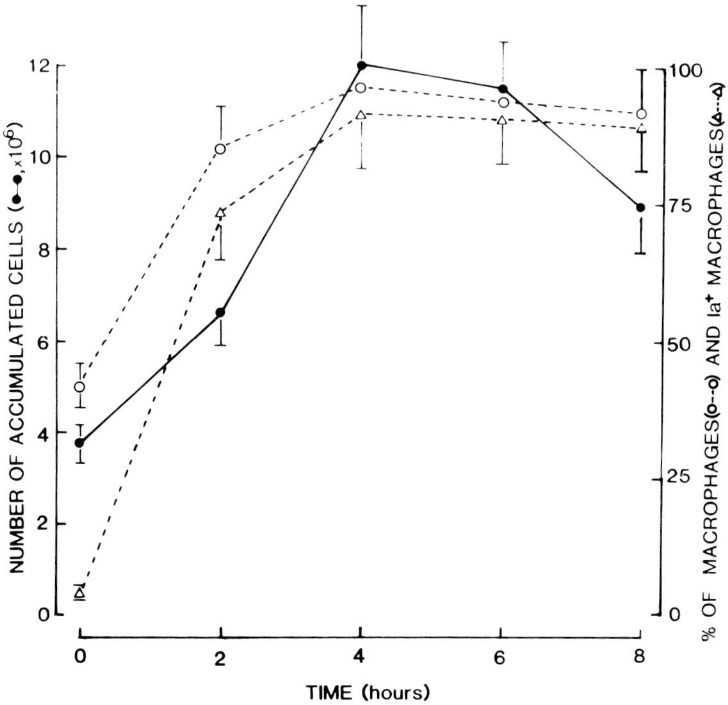

FIG. 5. Kinetics of Ia-positive macrophage accumulation after injection of MCF-c into normal mice. Peritoneal cells are harvested at 2-hr intervals after i.p. injection of 1 μg MCF-c. Ia positivity is confirmed by immunofluorescence using monoclonal antibody.

Ia-positive macrophages and monocytes *in vitro* and *in vivo*.[12-15] MCF-c does not convert Ia-negative macrophages to Ia positive.[12] MCF-c (1 μg), when injected i.p. into normal mice, induces selective accumulation of Ia-positive macrophages (Fig. 5). Such an experimental approach is useful for the analysis of macrophage accumulation in inflammation sites, particularly delayed hypersensitivity reaction (DHR) sites.[16,17]

Characterization of MCF-a.[3] MCF-a (monoegresin) is a heat-labile (56° for 30 min) protein and has an approximate MW of 150,000. Intra-

[12] M. Honda, T. Yoshimura, T. Watanabe, and H. Hayashi, *J. Immunol.* **131,** 2989 (1983).

[13] M. Honda, T. Masuda, T. Yoshimura, and H. Hayashi, *Cell. Immunol.* **86,** 1 (1984).

[14] K. Miura, M. Honda, T. Yoshimura, and H. Hayashi, *Immunology* **52,** 641 (1984).

[15] T. Yoshimura, M. Honda, K. Miura, and H. Hayashi, *Immunology* **52,** 269 (1984).

[16] H. Hayashi, M. Honda, and T. Yoshimura, *in* "Macrophage Biology" (S. Reichard and M. Kojima, eds.), p. 231. Liss, New York, 1985.

[17] H. Hayashi, T. Yoshimura, and J. Chen, *Comp. Immun. Microbiol. Infect. Dis.* **8,** 73 (1985).

dermal injection of MCF-a (30 μg) into guinea pigs provokes marked macrophage accumulation without changes in vascular permeability or hemorrhage.

MCF-a shares common antigenicity with guinea pig IgG; its chemotactic activity is completely absorbed by anti-IgG and anti-light chain antibodies but not by anti-C_5 and anti-MCF-c antibodies. MCF-a is generated *in vitro* from serum IgG by neutrophil seryl proteinase (elastaselike).[3,18] Its generation is maximal at 6 hr digestion with the enzyme, accompanied by release of dialyzable peptide(s) from the IgG molecule. The enzyme digests the Fc fragment but not the Fab fragment.[3] The enzyme (MW 27,000) is separated from rabbit peritoneal exudate neutrophil granules and purified by elution on DEAE-Sephadex, CM-Sephadex, and Sephadex G-75 columns.[18,19] It is optimally active at pH 7.0 for hemoglobin.

In contrast to MCF-c, MCF-a selectively attracts Ia-negative macrophages and monocytes.[12-15] I.p. injection of MCF-a (1 μg) into normal mice induces selective accumulation of Ia-negative macrophages.

Characterization of MCF-b[3]. MCF-b has not been purified. It is heat stable and has an approximate MW of 14,000. Its chemotactic activity is completely absorbed by anti-C_5 antibody but not by anti-C_3, anti-IgG, or anti-MCF-c antibody.[3,5] A macrophage chemotactic factor with similar molecular weight is generated *in vitro* from C_5 when digested with the neutral chymotrypsin-like enzyme (MW 35,000) from neutrophil granules.[20]

Natural Mediators of Lymphocyte Chemotaxis

Induction of Cutaneous Allergic Inflammation and Quantification of Tissue Leukocytosis

DHR, as described earlier, is induced by i.d. challenge of PPD to BCG-sensitized guinea pigs. The tissue leukocytosis induced is shown in Fig. 1A.

Method for Quantifying Lymphocyte Chemotaxis

Quantification of lymphocyte chemotaxis has been described in detail elsewhere in this volume ([3] and [6]) and will not be duplicated here.

[18] M. Ishida, M. Honda, and H. Hayashi, *Immunology* **35,** 167 (1978).
[19] M. Ishida and H. Hayashi, *Kagaku (Tokyo) Suppl.* **93,** 47 (1981).
[20] P. A. Ward and J. H. Hill, *J. Immunol.* **104,** 535 (1970).

Preparation of Skin Extracts

The preparation of skin extracts was described earlier. The supernatant is diluted to give an absorbancy between 25 and 30 at 280 nm and then sterilized by filtration. Chemotactic assays for lymphocytes are performed on the extracts at various stages of PPD lesions (Fig. 1A). The extract from 24-hr skin lesions is the most active for lymphocyte,[21] which corresponds to the peak in lymphocyte infiltration.

Purification and Characterization of LCFs

Step 1.[21] Apply 20 ml of a 24-hr skin extract to a Sephadex G-100 column (8 × 80 cm) equilibrated and eluted (50 ml/hr, 12-ml fractions) with 67 m*M* PB, pH 7.4. Three peaks of optical density at 280 nm are obtained in which chemotactic activity for lymphocytes is detected (Fig. 6A). LCF-a is found in the third, LCF-b in the second, and LCF-c and -d in the first peak.

Step 2. Purification of LCF-a.[22] Pool the fractions containing LCF-a (stippled area in Fig. 6A) and concentrate by rotary evaporation (absorbancy between 4 and 6 at 280 nm). Apply the fractions (3 ml) to a Sephadex G-15 column (1.5 × 60 cm) equilibrated and eluted with 67 m*M* ammonium bicarbonate, pH 8.0. The first peak, corresponding to about 2–4% of the applied sample, exhibits chemotactic activity for lymphocytes (Fig. 6B).

Lyophilize the LCF-a-containing fraction, dissolve it in distilled water, and apply it to a sheet of Whatman 3 MM paper. Descending chromatography is carried out for 24 hr at room temperature with the organic layer of a mixture of *n*-butanol : acetic acid : water (4 : 1 : 5). After that, high-voltage paper electrophoresis on the perpendicular direction is carried out in an electrophoresis apparatus equipped with a cooling system and operated at 40–50 V/cm at 10° using pyridine–acetate buffer, pH 3.7. A fluorescamine solution (1 mg/100 ml in acetone) is sprayed on the paper, and the fluorescent spots are detected under a long-wave (336-nm) UV lamp. Strips (3 × 5 cm) of paper are eluted with 30% acetic acid. The sample is relyophilized and dissolved in PBS for the chemotactic assay.

At least three fluorescent spots are demonstrated (Fig. 7). Chemotactic activity for lymphocytes is detected only in strip a (Fig. 7) moving 2.4–3.5 cm toward the cathode and descending 2.4–4.8 cm from the origin. No chemotactic activity is detected in other strips. LCF-a is heat stable, and its molecular weight is less than 5,000.

[21] Y. Shimokawa, S. Harita, Y. Higuchi, and H. Hayashi, *Br. J. Exp. Pathol.* **63,** 355 (1982).
[22] Y. Shimokawa, S. Harita, Y. Higuchi, and H. Hayashi, *Br. J. Exp. Pathol.* **63,** 362 (1982).

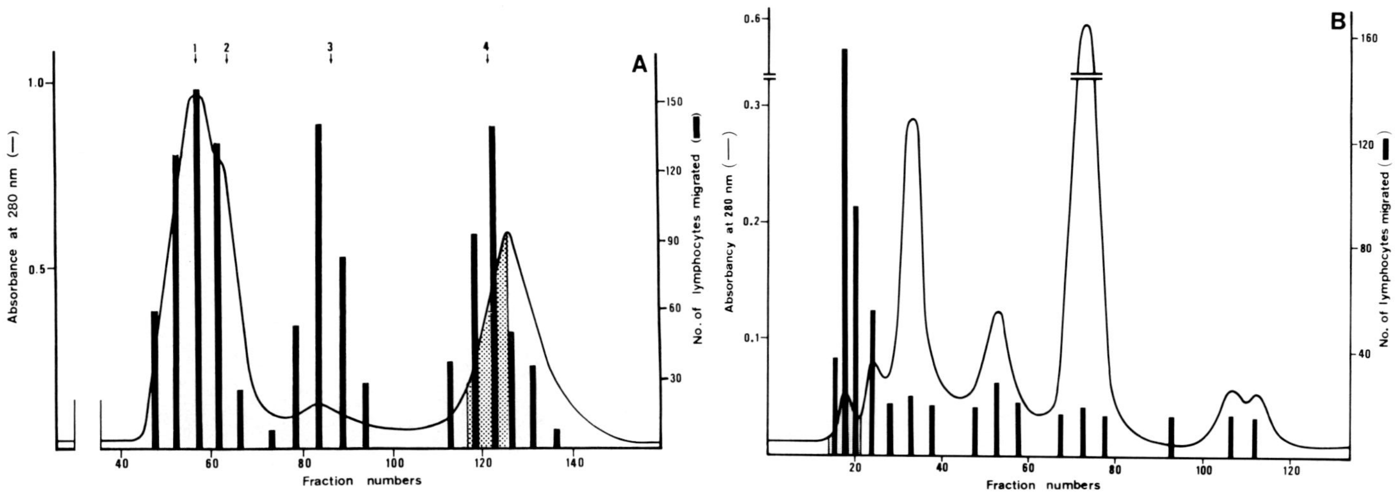

FIG. 6. Sephadex G-100 chromatography of an extract from PPD-induced 24-hr skin lesions. The chemotactic assay is carried out on undiluted fractions. LCF-a is in the third, LCF-b in the second, and LCF-c and -d in the first peak. Molecular weight standards: bovine serum albumin dimer; 2, bovine serum albumin monomer; 3, cytochrome *c*; 4, bacitracin. (B) Sephadex G-15 chromatography of LCF-a (stippled area in A). The chemotactic assay is carried out on undiluted fractions.

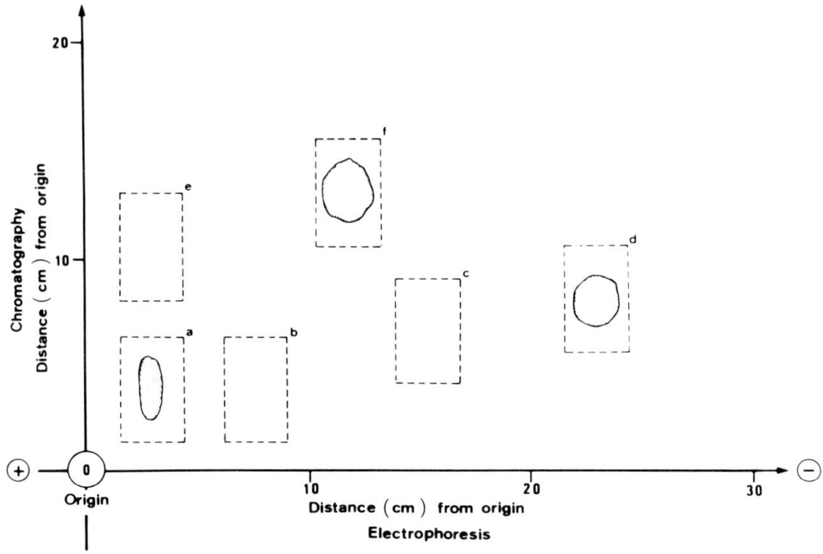

FIG. 7. Two-dimensional paper chromatography and electrophoresis of LCF-a (stippled area in Fig. 6A). Each spot was demonstrated with fluorescamine, and each strip of paper (a, b, c, d, e, and f; each 3 × 5 cm in area) was eluted with 30% acetic acid.

Step 3. Separation of LCF-c and -d.[23,24] Pool the fractions containing LCF-c and -d (Fig. 6A), concentrate by ultrafiltration on a YM10 membrane (Amicon) (absorbancy between 2 and 3 at 280 nm), and dialyze against 20 m*M* PB, pH 7.4. Apply the material (30 ml) to a DEAE-Sephadex A-50 column (1.5 × 20 cm) equilibrated with 20 m*M* PB, pH 7.4, and elute with a linear gradient elution between 0 and 1 *M* NaCl in 20 m*M* PB, pH 7.4. The flow rate is 20 ml/hr, and 5-ml effluent fractions are collected. Fig. 8A shows that the first (LCF-c) and third (LCF-d) peaks obtained exhibit marked chemotactic activity for lymphocytes.

Step 4. Purification of LCF-c.[24,25] *CM-Sephadex chromatography.* Pool LCF-c-containing fractions, concentrate, and dialyze against 20 m*M* PB, pH 6.8. Apply the sample (10 ml; absorbancy 2 at 280 nm) onto a CM-Sephadex C-50 column (1.5 × 20 cm). Fig. 8B shows that a linear gradient elution between 0 and 1 *M* NaCl in 20 m*M* PB, pH 6.8, yields three peaks of optical density at 280 nm. Chemotactic activity is present predominantly in the second peak.

[23] Y. Shimokawa, S. Harita, and H. Hayashi, *Imminology* **51,** 275 (1984).
[24] Y. Shimokawa, S. Harita, Y. Mibu, and H. Hayashi, *Immunology* **51,** 287 (1984).
[25] S. Harita, Y. Shimokawa, Y. Mibu, and H. Hayashi, *Immunology* **51,** 295 (1984).

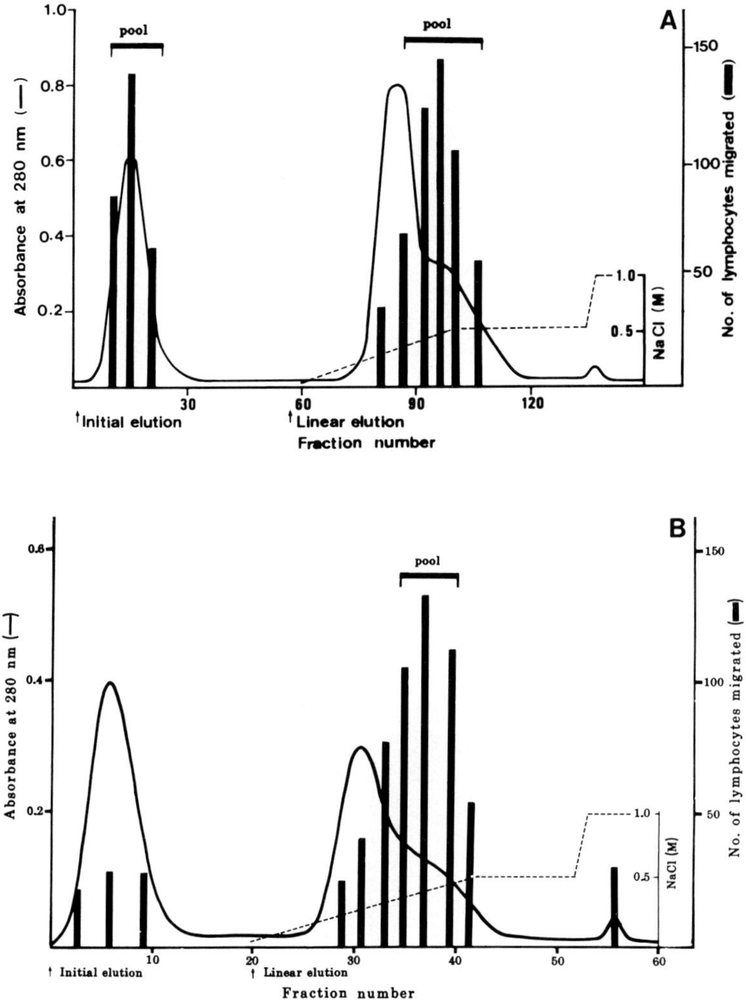

FIG. 8. (A) Separation of LCF-c and -d by DEAE-Sephadex chromatography. LCF-c is present in initial eluates, and LCF-d is found in the eluates obtained at increased salt concentration. (B) CM-Sephadex chromatography of LCF-c (Fig. 8A). (C) Chromatofocusing analysis of LCF-c. LCF-c (Fig. 8B), after elution from an anti-IgG affinity column, is applied to the chromatofocusing column. The activity is focused at pH 8.1 ± 0.2. (D) CM-Sephadex chromatography of LCF-d (Fig. 8A). (E) Isotachophoresis of LCF-d (Fig. 8D). LCF-d is present in the fourth and MCF-c in the second peak (see Fig. 2E).

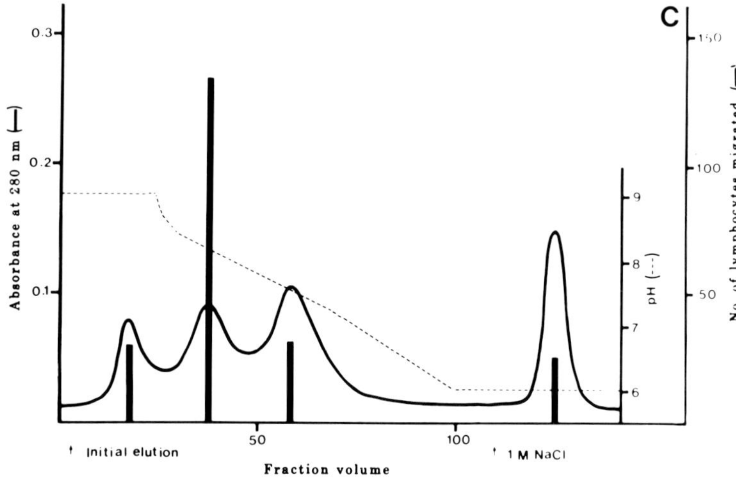

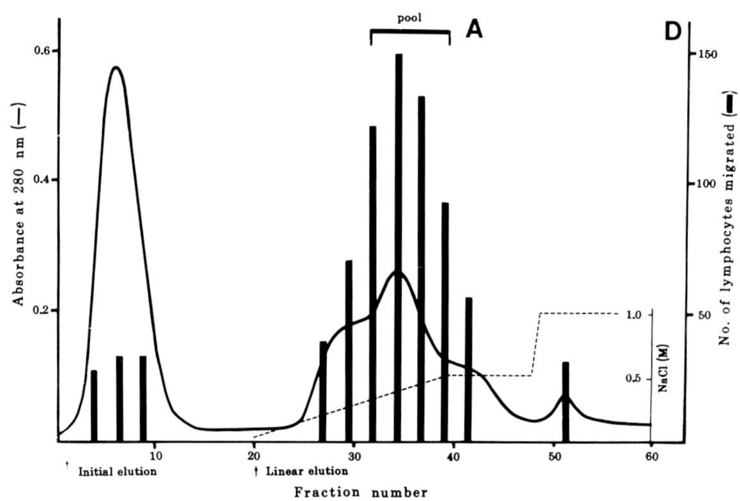

FIG. 8. (C) and (D).

Immunoadsorbent chromatography. IgG or MCF-a, if present in the fraction, must be separated from LCF-c. Apply the fraction (10 ml; absorbancy 2 at 280 nm) to a column (1.5 × 10 cm) of activated Sepharose 4B coupled with anti-guinea pig IgG rabbit antibody and equilibrated with 67 m*M* PB. The unabsorbed fractions containing LCF-c are pooled, concentrated, and dialyzed against 25 m*M* ethanolamine acetate buffer, pH 9.4.

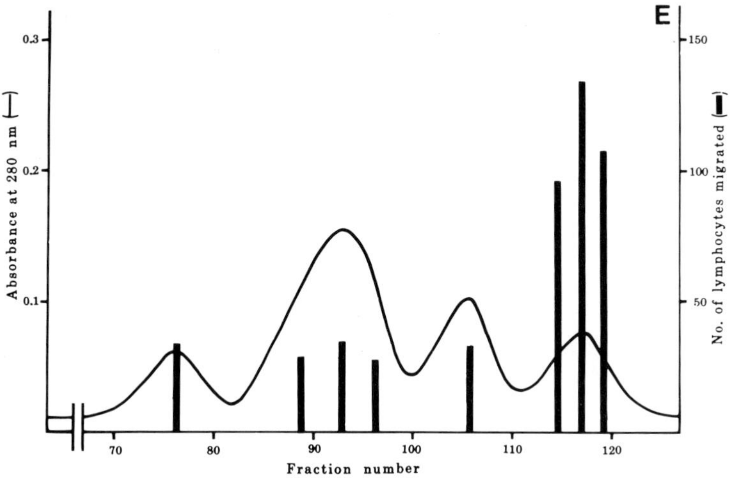

FIG. 8. (E).

Chromatofocusing. Apply LCF-c-containing fractions (5 ml; absorbancy 0.3–0.6 at 280 nm) to a chromatofocusing column PBE 94 (Pharmacia; 1.0 × 15 cm; pH 9 to 6). The sample is eluted with 200 ml of 10% Polybuffer 96 (Pharmacia) adjusted to pH 6.0 with sodium acetate buffer at a constant flow rate of 25 ml/hr. As shown in Fig. 8C, the chemotactic activity for lymphocytes is focused at pH 8.1 ± 0.2. LCF-c is a heat-labile and acid-stable protein with a MW of about 160,000. Analysis of LCF-c by polyacrylamide disk gel electrophoresis shows a single protein band.

Step 5. Purification of LCF-d.[24,25] *CM-Sephadex chromatography.* Apply LCF-d-containing fractions (Fig. 8A) (10 ml; absorbancy 1.6–2.0 at 280 nm), dialyzed against 20 m*M* sodium acetate buffer, pH 5.5, to a CM-Sephadex C-50 column (1.5 × 20 cm). Fig. 8D shows that a linear gradient elution between 0 and 1 *M* NaCl in 20 m*M* sodium acetate buffer, pH 5.5, yields three peaks of optical density at 280 nm. The second shows marked chemotactic activity.

Preparative isotachophoresis. This procedure was described earlier. Pool the LCF-d-containing fractions, concentrate (absorbancy 0.5–0.8 at 280 nm), and dialyze against the cathode buffer to which 0.3 ml ampholine is added. The chemotactic activity of LCF-d is predominantly concentrated in the fourth peak (Fig. 8E). Analysis of purified LCF-d by polyacrylamide disk gel electrophoresis shows a single protein band. LCF-d is a heat- and acid-labile protein and has an approximate MW of 300,000 and p*I* of 6.2.

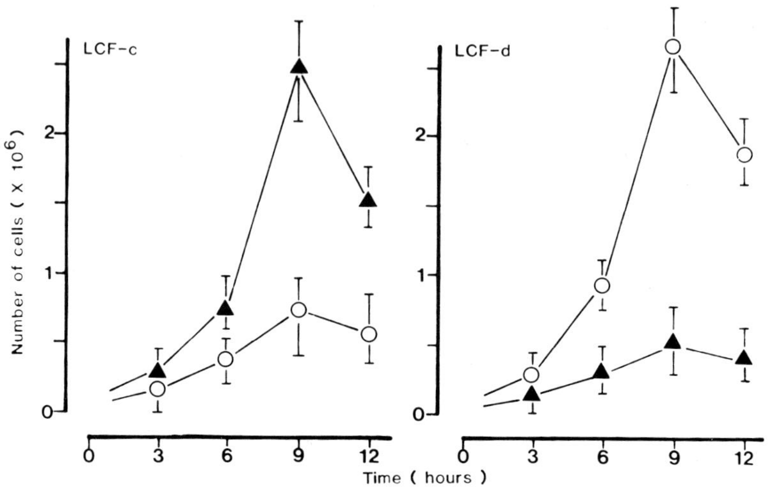

FIG. 9. Kinetics of Lyt-1 and Lyt-2 T-lymphocyte accumulation in the peritoneal cavity of normal mice in response to injection of LCF-c and LCF-d. Lyt-1 and Lyt-2 T subsets are assessed by immunofluorescent staining with anti-Lyt-1.1 and anti-Lyt-2.1 antibodies, respectively. ○, Lyt-1 T subset; ▲, Lyt-2 T subset. Peritoneal cells are harvested at 3-hr intervals after i.p. injection of 1 μg LCF-c and -d.

Characterization of LCFs[3]

LCF-a may be generated *in vitro* by neutrophil neutral thiol proteinase,[2] accompanied by release of dialyzable peptide(s) from IgG[3]; the chemotactic activity may be generated from the Fc fragment but not from the Fab fragment.[26] The Fc and Fab fragments themselves exhibit no chemotactic activity for lymphocytes. Importantly, LCF-a attracts B lymphocytes (in resting and blast form) of guinea pigs, mice, and humans.[22,27]

LCF-b has not been purified, and its precursor is unknown. It has an approximate MW of 14,000. LCF-b, like LCF-d, attracts T lymphocytes.

LCF-c is a T-lymphocyte-derived lymphokine with an approximate MW of 160,000.[3,28] LCF-c attracts cells with both suppressor/cytotoxic and helper/inducer phenotypes *in vitro*.[27,29] Interestingly, LCF-c (1 μg), when injected i.p. into normal mice, induces selective migration of cells with the suppressor/cytotoxic phenotype (Fig. 9). Intradermal injection of LCF-c

[26] Y. Higuchi, M, Ishida, and H. Hayashi, *Cell. Immunol.* **46,** 297 (1979).

[27] Y. Mibu, Y. Shimokawa, and H. Hayashi, *Immunology* **55,** 473 (1985).

[28] S. Harita, Y. Shimokawa, and H. Hayashi, *Int. Arch. Allergy Appl. Immunol.* **70,** 118 (1983).

[29] Y. Shimokawa, K. Miura, M. Hifumi, and H. Hayashi, *Immunology* **49,** 95 (1983).

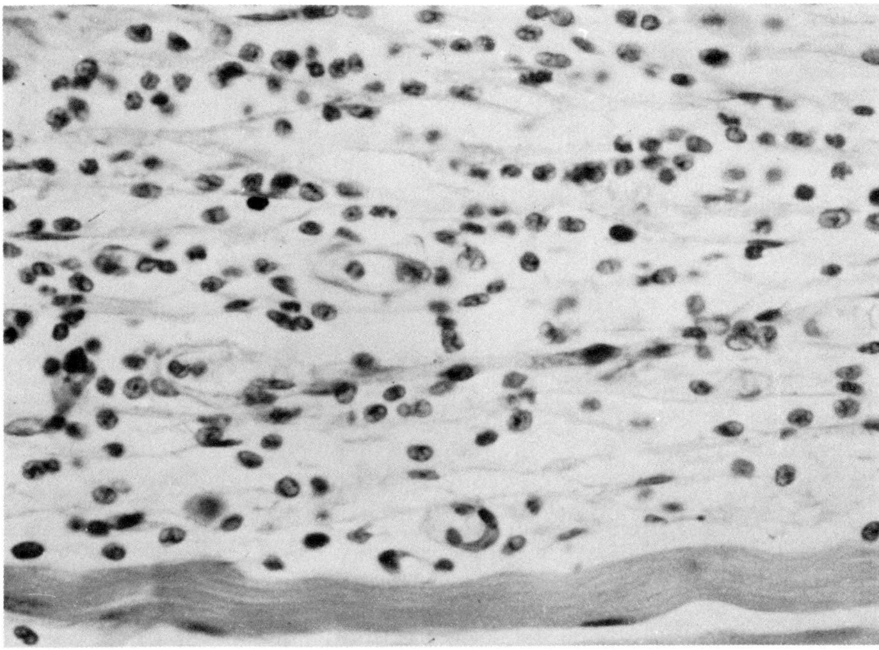

FIG. 10. Selective accumulation of lymphocytes in the skin of normal guinea pigs in response to i.d. injection of 10 μg LCF-d. Hematoxylin–eosin. ×225.

(10 μg) into normal guinea pigs induces marked lymphocyte migration without vascular permeability changes or hemorrhage.

LCF-d is a T-lymphocyte-derived lymphokine with a MW about 27,000, though it exists in the form of complex (MW 300,000) with serum protein present in the exudate. *In vitro* LCF-d attracts cells with the helper/inducer phenotype[27,29] and induces selective migration of cells with the same phenotype when injected i.p. (Fig. 9). Intradermal injection of LCF-d (10 μg) into normal guinea pigs induces intense lymphocyte accumulation (Fig. 10) without vascular permeability changes or hemorrhage.

Natural Mediators of Neutrophil Chemotaxis

Natural mediators of neutrophil chemotaxis are found primarily in Arthus lesions. The immunization protocol for induction of the Arthus reaction has been described elsewhere in this volume [40]. Methods for the quantification of tissue leukocytosis have been described above.

Preparation of Skin Extracts

The procedure for preparation of skin extracts was described above. Increased vascular permeability reaches its peak at 4.5 hr, while neutrophil accumulation reaches a peak at 12 hr.[2,30] The skin having undergone the Arthus reaction is excised at 4.5, 12, 24, and 48 hr after challenge with antigen (bovine serum albumin).

Purification and Characterization of NCF (Leucoegresin)

Purification of NCF[31,32]

Step 1. The extract of Arthus skin lesions is fractionated with ammonium sulfate (33%, 50%, and 70%) into three fractions, containing chiefly euglobulin, pseudoglobulin, and albumin. Protein fractions are desalted on a column of Sephadex G-50 equilibrated with 67 m*M* PB, pH 7.4. Chemotactic activity is predominantly concentrated in the pseudoglobulin fraction. The mean specific activity (neutrophil count/$E_{280\text{ nm}}$) of a fraction from an extract of 12-hr Arthus skin lesions showing maximal neutrophil infiltration is 24.

Step 2. DEAE-Sephadex chromatography. Dialyze the fractions against 67 m*M* PB, pH 7.4, and apply 100-ml aliquots of the active fractions (absorbancy 18–22 at 280 nm) to a DEAE-Sephadex A-50 column (3.5 × 40 cm) equilibrated and eluted with 20 m*M* PB, pH 7.4, followed by a linear gradient between 0 and 0.3 *M* NaCl in the same buffer (Fig. 11A). Chemotactic activity is present in the first fraction. Its mean specific activity is about 200.

Step 3. Concentration of active fractions on CM-Sephadex. Collect the active fractions (Fig. 11A) and dialyze them against 20 m*M* PB, pH 6.8, for 16 hr. Apply 700-ml aliquots of the active fractions to a CM-Sephadex column (2 × 5 cm) equilibrated and eluted with the same buffer followed by elution with a linear gradient between 0 and 0.3 *M* NaCl in the same buffer. Chemotactic activity is eluted with 0.2 *M* NaCl at a flow rate of 24 ml/hr. By this procedure, 700 ml is concentrated to about 50 ml.

Step 4. CM-Sephadex chromatography. Dialyze the concentrated fractions against 20 m*M* PB, pH 6.8, for 16 hr. Apply 50-ml aliquots of the active fractions (absorbancy 0.5–0.8 at 280 nm) to a CM-Sephadex C-50 column (1.5 × 40 cm) as described above (Fig. 11B). A linear gradient elution between 0 and 0.2 *M* NaCl in 20 m*M* PB, pH 6.8, yields two peaks.

[30] H. Hayashi, M. Yoshinaga, M. Koono, H. Miyoshi, and M. Matsumura, *Br. J Exp. Pathol.* **45,** 419 (1964).

[31] K. Yoshida, M. Yoshinaga, and H. Hayashi, *Nature (London)* **218,** 977 (1968).

[32] M. Yoshinaga, K, Yoshida, A. Tashiro, and H. Hayashi, *Immunology* **21,** 281 (1971).

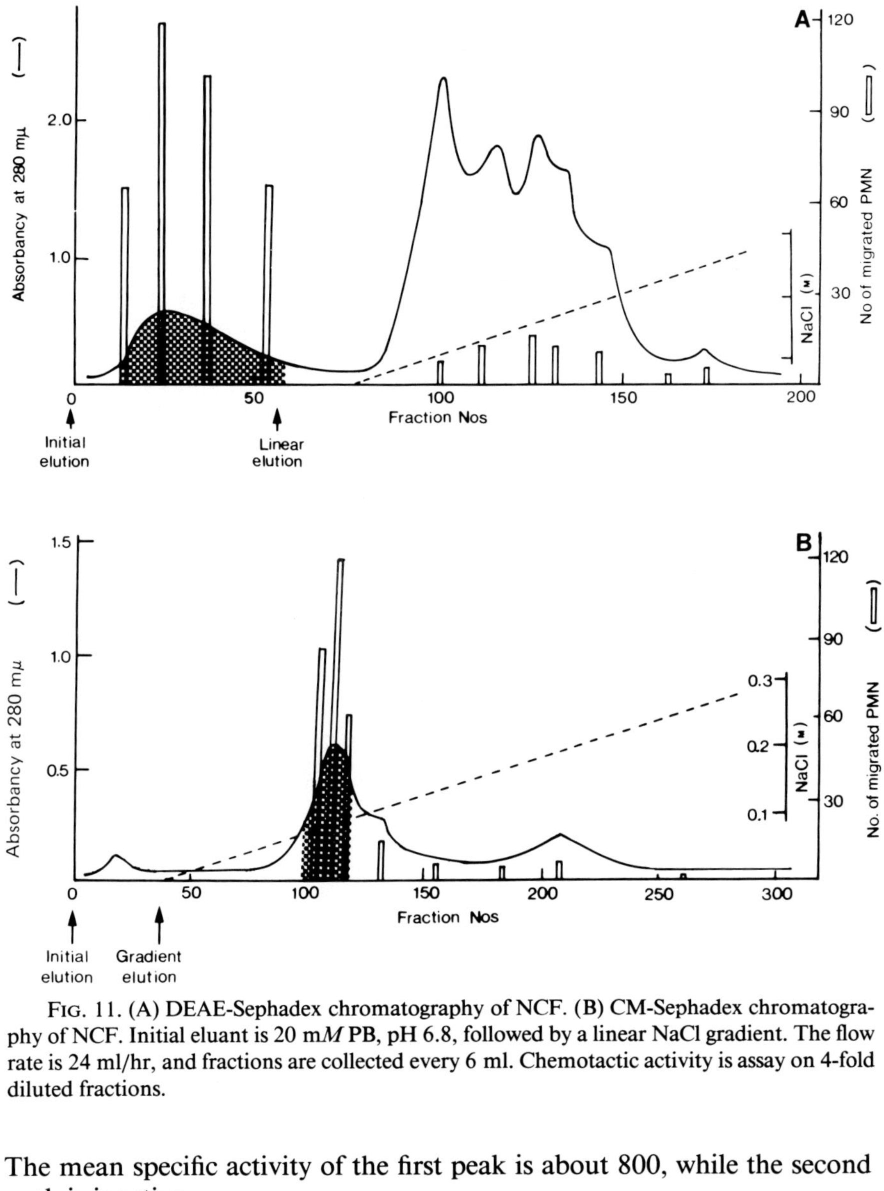

FIG. 11. (A) DEAE-Sephadex chromatography of NCF. (B) CM-Sephadex chromatography of NCF. Initial eluant is 20 m*M* PB, pH 6.8, followed by a linear NaCl gradient. The flow rate is 24 ml/hr, and fractions are collected every 6 ml. Chemotactic activity is assay on 4-fold diluted fractions.

The mean specific activity of the first peak is about 800, while the second peak is inactive.

Step 5. Rechromatography on CM-Sephadex. Collect the active fractions (Fig. 11B), lyophilize, and dialyze them against 20 m*M* PB, pH 6.8. Rechromatograph the active fractions on CM-Sephadex as for Step 4. The active fractions have a specific activity of about 1450. Approximately

30 mg of chemotactic factor is obtained from 114 g of dried skin. Thus, the amount of the chemotactic factor per Arthus skin lesions (at 12 hr) is about 250 μg. The active fractions may be applied to an anti-rabbit IgG affinity column and eluted in 0.1 *M* glycine–HCl buffer, pH 3.0, without loss of activity. The fractions have a specific activity of about 1500.[33]

Characterization of NCF[1,3]

Pure NCF behaves as a homogeneous substance on ultracentrifugation and on moving boundary electrophoresis. The factor is a protein free of nucleic acid and is relatively heat stable. It has an approximate MW of 140,000, an isoelectric point p*I* 5.0, and a sediment coefficient of 6.58 S.[32] A similar factor is isolated from burned skin sites in rabbits; its amount is approximately 50 μg per 4-hr skin site showing maximal neutrophil infiltration.[34] Differences in the NCF amount are to be expected because neutrophil accumulation in Arthus sites is more intense than that in burned sites.

NCF shares common antigenicity with serum IgG.[35,36] NCF is produced from serum IgG by a neutral thiol proteinase,[1,37,38] accompanied by release of dialyzable peptide(s) from the Fc portion of the IgG molecule. The enzyme is found in the euglobulin fraction (prepared at 33% ammonium sulfate saturation) from extracts of Arthus skin sites[9,10] and purified by chromatography with DEAE-Sephadex A-50 then GE-cellulose.[2,39] The enzyme is released by antigen-stimulated sensitized resident macrophages[40] and by immune complexes phagocytosed by neutrophils.[2] The enzyme may satisfy many of the criteria necessary for an inflammatory proteinase.[2]

NCF is clearly selective in causing neutrophil chemotaxis *in vitro* and *in vivo*.[1,2] Intradermal injection (20 μg) into normal rabbits induces marked neutrophil migration (Fig. 12) without change in vascular permeability or hemorrhage. It is active on neutrophils of rabbits, guinea pigs, rats, and mice.

[33] M. Yoshinaga, S. Yamamoto, and H. Hayashi, *Immunology (Tokyo)* **7,** 31 (1972).

[34] S. Maeda, M. Yoshinaga, and H. Hayashi, *Virchows Arch. B* **19,** 37 (1975).

[35] S. Yamamoto, M. Yoshinaga, and H. Hayashi, *Immunology* **20,** 803 (1971).

[36] M. Yoshinaga, M. Mayumi, S. Yamamoto, and H. Hayashi, *Nature (London)* **225,** 1138 (1970).

[37] M. Yoshinaga, S. Yamamoto, S. Maeda, and H. Hayashi, *Immunology* **20,** 809 (1971).

[38] H. Hayashi, M. Koono, M. Yoshinaga, and M. Muto, *in* "Inflammation Biochemistry and Drug Interaction" (A. Bertellin and J. C. Houck, eds.), p. 34. Excerpta Medica, Amsterdam, 1974.

[39] M. Koono, M. Muto, and H. Hayashi, *Tohoku J. Exp. Med.* **94,** 231 (1968).

[40] H. Hayashi, *in* "Biochemistry of Acute Allergic Reactions" (K. F. Austen and E. L. Becker, eds.), p. 141. Blackwell, Oxford, England, 1968.

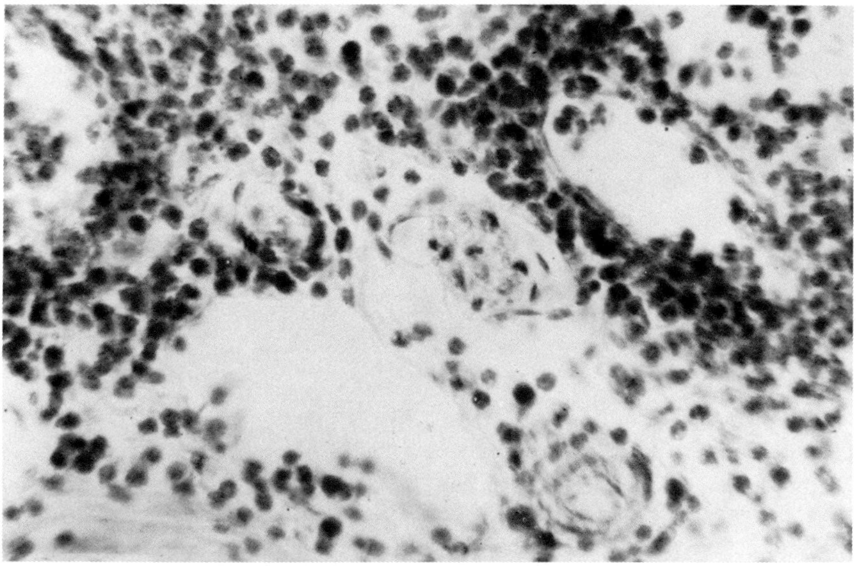

FIG. 12. Selective neutrophil migration in the skin of normal rabbits after i.d. injection of 20 μg NCF. No cells other than neutrophils are seen. Micrograph taken 3 hr after injection. Hematoxylin–eosin. ×288.

Other neutrophil chemotactic factors such as C5a and diffusible peptide are also separated from Arthus skin sites or DNP-As-induced skin lesions, but they have not been purified. Their role in inflammatory leukotaxis seems less significant.[1,3]

Natural Mediators of Eosinophil Chemotaxis

Eosinophils accumulate in inflamed sites (e.g., in certain allergic states such as bronchial asthma or hay fever and metazoan parasite infestation). Mediators of eosinophil chemotaxis have been isolated from DNP-As-induced eosinophil-rich allergic skin lesions in guinea pigs.

Induction of Cutaneous Allergic Tissue Eosinophilia and Its Quantification

Immunization Protocols for Induction of DNP-As Lesions.[41] Induction of DNP-As lesions is carried out following the method described above. The antigen (0.1 mg/site) is injected i.d. at 8 sites on the shaved back of guinea pigs. After Susa's fixation and subsequent Luna's staining for eosin-

[41] M. Hirashima and H. Hayashi, *Immunology* **30,** 203 (1976).

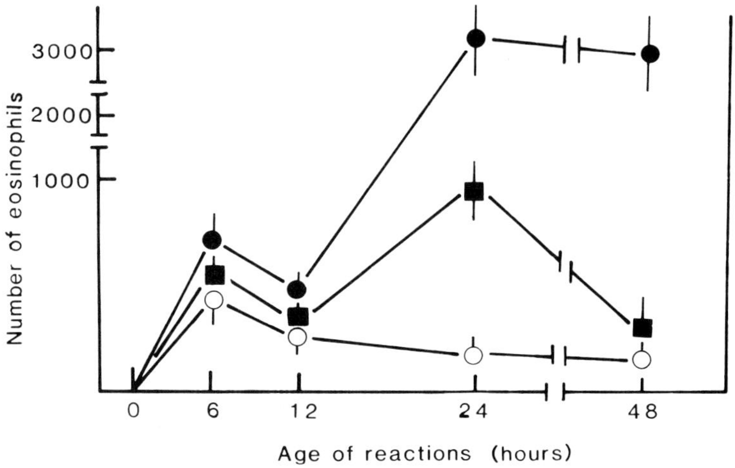

FIG. 13. Time course of tissue eosinophilia following i.d. injection of DNP-As in guinea pigs. ●, Eosinophils in animals sensitized with antigen with *B. pertussis* and challenged; ■, eosinophils in animals immunized with antigen with CFA plus *B. pertussis* and challenged; ○, eosinophils in animals immunized by the antigen with CFA and challenged.

ophils,[42] the accumulated eosinophils are counted by the method described above. Tissue eosinophilia appears in two phases: an early phase is weak and short-lasting, peaking at about 6 hr; a delayed phase, more intense and long-lasting, peaks at about 24 hr (Fig. 13). The extract from a 6-hr skin site (showing maximal early phase) is used for obtaining an early eosinophil chemotactic factor (ECF), and that of a 24-hr skin site (showing maximal delayed phase) for obtaining a delayed ECF.

The injection of adjuvant with the antigen affects tissue eosinophilia. Treatment with *B. pertussis* or alum hydroxide gel prepared from aluminum potassium sulfate[43] (Wako Pure Chemical Industries, Ltd., Osaka, Japan) alone strongly enhances the eosinophilia,[44] whereas CFA alone suppresses it almost completely[45] (Fig. 13).

Method for Quantifying Eosinophil Chemotaxis[41]

Quantification of eosinophil chemotaxis is performed following the method described elsewhere in this volume ([3] and [6]), using cellulose

[42] L. G. Luna, "Manual of Histologic Staining Methods of the Armed Forces Institute of Pathology," 3rd Ed., p. 111. McGraw-Hill, New York, 1968.

[43] J. S. Garvery, N. E. Cremer, and D. H. Sussdorb, "Methods in Immunology," 3rd Ed., p. 185. Benjamin, Reading, Massachusetts, 1977.

[44] K. Sakata, K. Tashiro, M. Hirashima, and H. Hayashi, *J. Immunol.* **135,** 3463 (1985).

[45] M. Hirashima, K. Tashiro, and H. Hayashi, *Immunology* **51,** 441 (1984).

nitrate filters (8-μm pore size) and a chemotactic chamber containing two 200-μl compartments. The chambers are incubated for 3 hr at 37° in a 5% CO_2 atmosphere. The filters are fixed with a mixture of 1 part of saturated mercuric chloride and 1 part of 100% ethanol for 1–18 hr at room temperature, rinsed in water for 10 min, stained with Carazzi's hematoxylin for 10 min, dipped in 1% HCl–alcohol (10 ml concentrated HCl: 1000 ml 70% alcohol) for a few seconds, rinsed in water for 10–20 min, stained with 0.1% Chromotrope 2R for 2 min, rinsed in water 1 min, dipped in 100% ethanol 3 times, rinsed in propanol for 1 min, rinsed in a mixture of propanol and xylol (1 : 1) for 1 min, rinsed in 100% xylol for 10 min, and mounted on slides.[7] After staining, the cells that had migrated through the filter to the lower surface of it are counted. The chemotactic activity is expressed as the mean count of migrated eosinophils in 20 high-power fields randomly selected.

Eosinophils are obtained by peritoneal lavage from guinea pigs which received weekly multiple (8–12) injections of 2 ml horse serum. Peritoneal cells, containing 40–60% eosinophils at 4–5 days after the last injection, are centrifuged. To enrich eosinophils, contaminating macrophages are removed by passage through a glass wool-packed column (0.4 g/10 ml) after incubation for 30 min at 37° (see this series, Vol. 108 [28]). Eosinophil-rich (70–80%) cell suspensions (1×10^6 cells/ml) in Hank's balanced salt solution containing 5% fetal calf serum are used in chemotaxis measurements.

Purification and Characterization of ECFs

Early ECF.[46] Apply the skin extract (prepared as detailed above) (7 ml; absorbancy 40 at 280 nm) of a 6-hr skin site to a Sephadex G-100 column (3 × 50 cm) equilibrated with 67 m*M* PB, pH 7.4. Two chromatographic peaks are obtained. Eosinophil chemotactic activity is present in the second (Fig. 14A). The chemotactic fraction (20 ml; absorbancy 40 at 280 nm) is applied to a Sephadex G-25 column (5 × 100 cm) equilibrated and eluted with the same buffer. The eluted factor (early ECF) has a MW of approximately 300. It is heat stable at 56° for 30 min or at 80° for 15 min and is resistant to freeze–thawing and lyophilization. It may also be derived from antigen-stimulated mast cells.[41,46]

Delayed ECF-a.[46,47] Apply the extract (7 ml; absorbancy 40 at 280 nm) of a 24-hr skin site to a Sephadex G-100 column (3 × 50 cm) equilibrated with 67 m*M* PB, pH 7.4 (Fig. 14B). Two peaks of optical density at 280 nm are obtained. Chemotactic activity is present in the first peak (early

[46] M. Hirashima, Y. Hirotsu, and H. Hayashi, *in* "Immunobiology of the Eosinophil" (T. Yoshida and M. Torisu, eds.), p. 213. Elsevier New York, 1983.

[47] Y. Hirotsu, M. Hirashima, and H. Hayashi, *Immunology* **48,** 59 (1983).

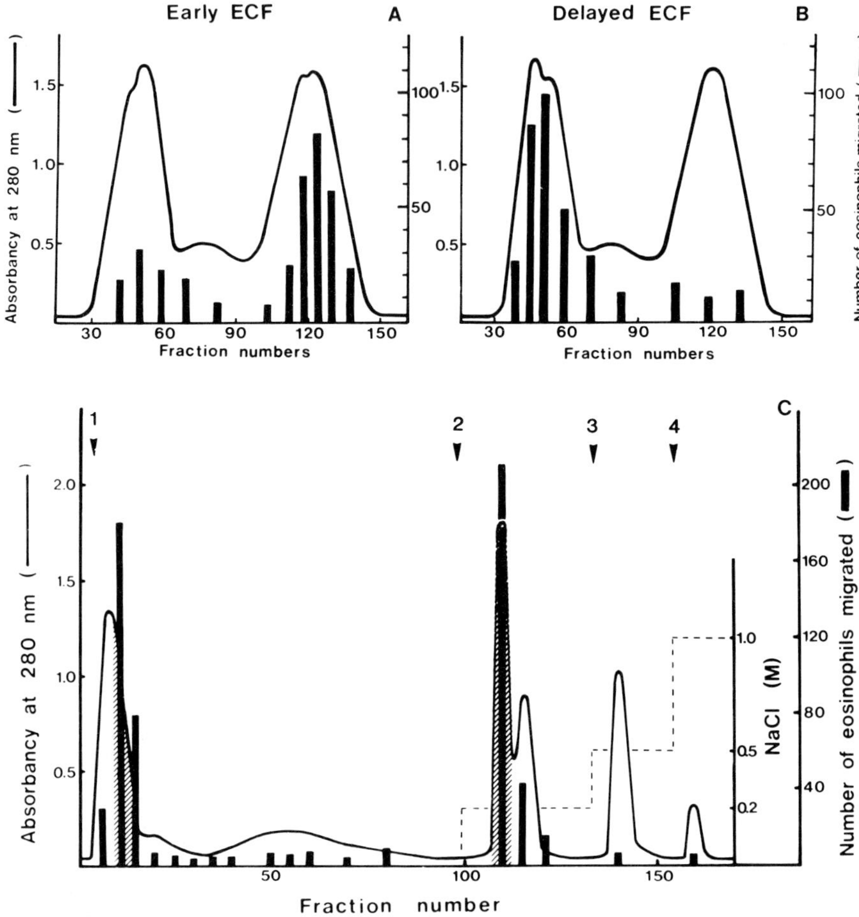

FIG. 14. Demonstration of eosinophil chemotactic activity of guinea pig skin extracts after gel filtration on Sephadex G-100. Extracts were prepared from 6-hr (A) and 24-hr skin lesions (B). The chemotactic activity of each fraction is assayed without dilution. (C) DEAE-Sephadex chromatography of delayed ECF-containing fractions (first peak in B). Changes in the ionic strength of eluants at pH 7.2 are indicated by a discontinuous line. Chemotactic activities are assayed without dilution. Delayed ECF-a activity is found in the first and delayed ECF-b in the third peak.

ECF is not found in the second peak because its chemotactic activity has disappeared by 24 hr). Apply the chemotactic fraction (10 ml; absorbancy 3 at 280 nm) to a DEAE-Sephadex A-50 column (2 × 30 cm) equilibrated with 67 m*M* PB, pH 7.2, and eluted (12 ml/hr) with a linear gradient between 0 and 1 *M* NaCl in 67 m*M* PB, pH 7.2 (Fig. 14C). Delayed ECF is a mixture of at least two chemotactic factors; delayed ECF-a is present in the first peak and ECF-b in the third.[47] They have similar molecular weights (about 70,000) but differ in antigenicity; the former is absorbed by anti-delayed ECF-a antibody but not by anti-delayed ECF-b antibody, and the latter is absorbed by anti-delayed ECF-b antibody but not by anti-delayed ECF-a antibody.

Delayed ECF-a can also be obtained from inflammation sites with eosinophilia mediated by peritoneal exudate cells (PECs) from DNP-As-sensitized guinea pigs.[41] The PECs (2×10^8 in RPMI 1640), collected by rinsing the peritoneal cavity of sensitized animals with 50 ml PBS containing 10 units/ml heparin 5 days after injection of 10 ml Bay oil (Sigma), are transferred i.p. into normal guinea pigs, and 2 days later DNP-As (0.1 mg/site) is injected i.d. Delayed tissue eosinophilia peaking at 24–30 hr is obtained.[46]

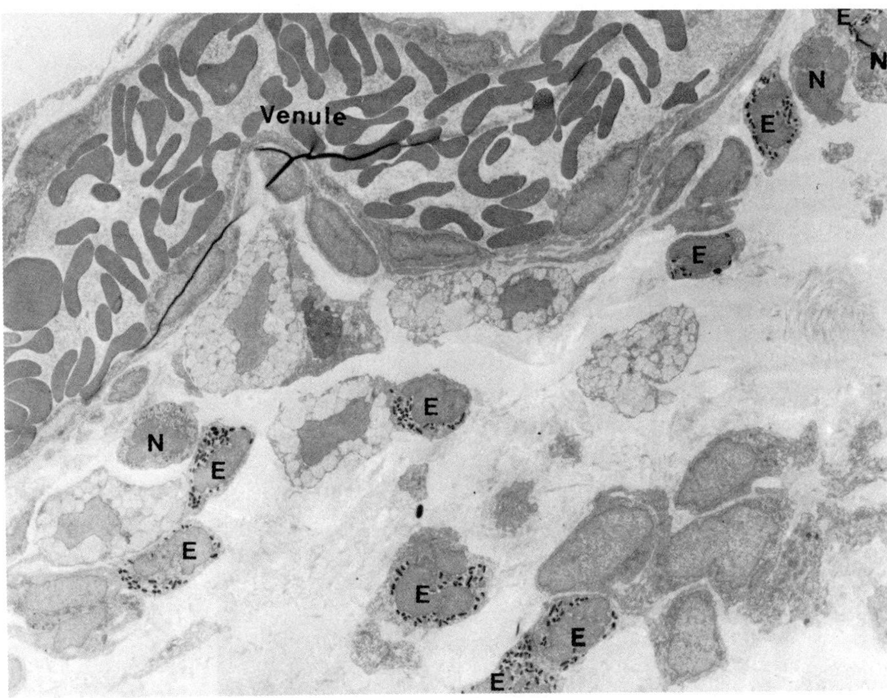

FIG. 15. Electron micrograph of eosinophil migration induced by injection of delayed ECF into the cremaster muscle of a rat. A number of eosinophils with characteristic highly electron-dense granules can be seen in the injected site, indicating the selective *in vivo* effect of the factor. Only a few neutrophils are observed. E, Eosinophil; N, neutrophil. ×1440. Courtesy of Y. Ishimaru.

A delayed ECF-a-like factor may be produced *in vitro* by antigenic or concanavalin A (Con A) stimulation.[45] PECs (3×10^6/ml) are resuspended in RPMI 1640 supplemented with 20 m*M* HEPES, 100 units penicillin, 10 μg streptomycin, 2×10^{-5} *M* 2-mercaptoethanol, and 2 m*M* L-glutamine, free of serum, and stimulated with 5 μg/ml DNP-As at 37° for 48 hr in an atmosphere of 5% CO_2. The culture supernatant contains the delayed ECF-a-like factor, produced by peritoneal exudate T lymphocytes. For the production of Con A-induced ECF,[46] lymphoid cells (3×10^6/ml), collected from mesenteric lymph nodes and spleens of normal guinea pigs, are washed and suspended in RPMI 1640 and cultured with 5 μg/ml Con A for 24 hr at 37°. The source of the factor is T lymphocytes. Delayed ECF-a induces *in vivo* eosinophil migration (Fig. 15).

Delayed ECF-b.[46,47] Delayed ECF-b is not well defined. Its activity is absorbed by anti-guinea pig serum rabbit antibody.

[15] Eosinophilic Chemotactic Factors of Anaphylaxis

By JOHN A. SMITH

Introduction

Eosinophilic chemotactic factors of anaphylaxis (ECF-A) are preformed, granule-associated peptides and polypeptides, which are released during immediate hypersensitivity reactions. Although initially described as mediators released by IgG-dependent mechanisms from guinea pig[1] and human[2] lung preparations, ECF-A is also found in rat pleural and peritoneal mast cells,[3] human leukemic basophils,[4] and in human nasal polyps[5] which contain increased numbers of mast cells. The biological activities attributed to ECF-A include selective activation[1,2] and deactivation[6] of human eosinophil chemotaxis, release of human eosinophil, granule-associated arylsulfatase,[7] activation of the human eosinophil hexose mono-

[1] A. B. Kay, D. J. Stechschulte, and K. F. Austen *J. Exp. Med.* **133,** 602 (1971).
[2] A. B. Kay and K. F. Austen, *J. Immunol.* **107,** 399 (1971).
[3] S. I. Wasserman, E. J. Goetzl, and K. F. Austen, *J. Immunol.* **112,** 351 (1974).
[4] R. A. Lewis, E. J. Goetzl, S. I. Wasserman, F. H. Valone, R. H. Rubin, and K. F. Austen, *J. Immunol.* **114,** 87 (1975).
[5] M. A. Kaliner, S. I. Wasserman, and K. F. Austen, *N. Engl. J. Med.* **289,** 277 (1973).
[6] S. I. Wasserman, D. Whitmer, E. J. Goetzl, and K. F. Austen, *Proc. Soc. Exp. Biol. Med.* **148,** 301 (1975).
[7] S. I. Wasserman, E. J. Goetzl, and K. F. Austen, *J. Immunol.* **114,** 645 (1975).

phosphate shunt pathway,[6] enhancement of human and/or rat eosinophil C3b complement,[8] IgG Fc,[9] and IgE Fc[10] receptors, and augmentation of human eosinophil superoxide production.[11]

Both low-molecular-weight (300–1,000) and intermediate-molecular-weight (1,500–2,500) forms of ECF-A have been isolated from human lung[12] and rat peritoneal mast cells.[13] However, only the structures of the low-molecular-weight ECF-A peptides from human lung have been determined, and the low-molecular-weight ECF-A is known to consist of two tetrapeptides: H-Ala-Gly-Ser-Glu-OH and H-Val-Gly-Ser-Glu-OH.[14] This chapter deals with the established procedures employed for the isolation of the low-molecular-weight ECF-A tetrapeptides from human lung, as described by Goetzl and Austen.[14]

Procedures

Source of Material

As described by Goetzl and Austen[14]:

1. Fresh human lung surgical specimens are divided into 100-g portions.

2. The minced tissue is extracted twice with 1-butanol: glacial acetic acid (10:1, v/v) (100 ml each) or alkaline Tyrode's buffer–50 mM Tris-HCl (pH 8.2) (100 ml each) by homogenization using a tissue grinder (Polytron, Brinkman, Lucerne, Switzerland) followed by sonication for 15 min at 4° (Ultrasonics Instruments International, Farmingdale, NY). Immunologically released ECF-A can also be generated from minced human lung that is passively sensitized with serum obtained from an human donor sensitive to ragweed and then challenged with ragweed antigen E.[14] As described by Orange *et al.*,[15] the lung fragments are suspended in human ragweed-sensitized serum diluted 1:3 to 1:5 in Tyrode's buffer (1 g of fragmented lung/5 ml of the diluted serum), incubated for 16–22 hr at 22°, washed 3 times with Tyrode's solution, resuspended in Tyrode's

[8] A. R. E. Anwar and A. B. Kay, *Nature (London)* **269,** 522 (1977).
[9] M. Capron, A. Capron, E. J. Goetzl, and K. F. Austen, *Nature (London)* **289,** 71 (1977).
[10] M. Capron, A. Capron, J. P. Dessaint, G. Torpier, S. G. O. Johansson, and L. Prin, *J. Immunol.* **126,** 2087 (1981).
[11] P. H. Beswick and A. B. Kay, *Clin. Exp. Immunol.* **43,** 399 (1981).
[12] E. J. Goetzl and K. F. Austen, *Prog. Immunol.* **3,** 439 (1977).
[13] R. N. Boswell, K. F. Austen, and E. J. Goetzl, *J. Immunol.* **120,** 15 (1978).
[14] E. J. Goetzl and K. F. Austen, *Proc. Natl. Acad. Sci. U.S.A.* **72,** 4123 (1975).
[15] R. P. Orange, W. G. Austen, and K. F. Austen, *J. Exp. Med.* **134,** 136s (1971).

solution (1 g of fragmented lung/15 ml of Tyrode's solution) containing ragweed antigen E (1–5 protein nitrogen units per milliliter of Tyrode's solution), and incubated at 37° for 15 min. The diffusate contains the ECF-A peptides.

Alkaline Tyrode's Buffer–50 mM Tris (pH 8.2)

NaCl	8 g
Glucose	1 g
$NaHCO_3$	1 g
KCl	0.2 g
$NaH_2PO_4 \cdot H_2O$	0.05 g
$CaCl_2$	0.2 g
$MgCl_2$	0.1 g
Tris–HCl	7.88 g
Water	Sufficient quantity to bring to 1 liter, after pH is adjusted to 8.2

3. The extracts or diffusates are clarified by centrifugation, and the supernatants are removed.

Assay of Biological Activity (see also this volume [3])

The selective chemotaxis of human eosinophils is assessed by a Boyden micropore filter technique, using mixed leukocyte suspensions in Hanks' balanced salt solution (Microbiological Associates, Inc., Bethesda, MD) containing 0.4% ovalbumin and 5 m*M* Tris–HCl, pH 7.4. Each modified Boyden chamber contains approximately 1×10^6 eosinophils/chamber.

Purification

Gel Filtration on Sephadex G-25. A Sephadex G-25 (fine) (Pharmacia, Inc., Piscataway, NJ) column (e.g., 1.5 × 85 cm) is equilibrated with either 0.1 *M* glacial acetic acid for extracted ECF-A or Hanks' balanced salt solution for immunologically released ECF-A. For example, fractions of 1.5 ml are collected at a flow rate of 4.5–6 ml/hr at 4°. The low-molecular-weight ECF-A elutes in a broad range at approximately 80% total bed volume. The active fractions, assessed by selective eosinophil chemotaxis in a Boyden chamber assay, are pooled and lyophilized.

Ion-Exchange Chromatography. Dowex AG1-X8 (formate form) (Bio-Rad Laboratories, Richmond, CA) is prepared by washing the resin with 1 *N* NaOH (1 liter), water (2–3 liters), 1 *N* HCl (1 liter), water (2–3 liters), 1 *N* formic acid (1 liter), and water (2–3 liters). Initially, the column (e.g., 1.5 × 25 cm) of the Dowex AG1-X8 is washed with 0.1 *M* pyridine (200 ml). After dissolving the ECF-A isolated from the Sephadex G-25 column in 0.1 *M* pyridine (1–2 ml), it is applied to the column. The column is eluted with 0.1 *M* pyridine, followed by a 2-liter, linear gradient between 0.1 *M* pyridine (8 ml/liter) and 0.5 *M* formic acid (25 ml/liter). For example, fractions of 16 ml are collected at a flow rate of 25 ml/hr. A pool of the late-eluting fractions (pH 2.2–2.6) is desalted on a Sephadex G-10 column and lyophilized.

Descending Paper Chromatography. The lyophilized sample is dissolved in 0.1 *M* NH_4OH (e.g., 1 ml) and is applied in a thin band at the origin marked on a sheet of prewashed, Chromatocarb A-125 paper (Scientific Manufacturing Industries, Emeryville, CA) and separated by descending paper chromatography with a mixture of water:1-butanol:acetic acid (5:4:1, v/v/v). Two distinct bands of selective eosinophil chemotactic activity are detected with R_f ranges of 0.5–0.6 and 0.6–0.7, relative to 3,5-dinitrophenylglutamic acid (an external standard). The individual bands are eluted from paper strips, encompassing these R_f values, with 10 m*M* NH_4OH (3–4 ml) and lyophilized twice from water. The recovery of the lower R_f ECF-A is 3–10 nmol, while the recovery of the higher R_f ECF-A is 7–8 nmol, as determined by amino acid analysis.[16]

Sequence Analysis

The sequence of the peptides was determined by a combination of methods, including the dansyl–Edman method,[17] carboxypeptidase A digestion,[18] hydrazinolysis,[19] selective tritiation, hydrolysis, and amino acid analysis,[20] and amino acid analysis after acid hydrolysis or enzymatic digestion. The primary structure of the lower R_f ECF-A is determined to be H-Ala-Gly-Ser-Glu-OH, while that of the higher R_f peptide is H-Val-Gly-Ser-Glu-OH.

Peptide Synthesis

The solid-phase peptide synthesis method of Merrifield[21] is used to synthesize the two ECF-A peptides isolated from human lung. These peptides are also available from commercial sources (e.g., Peninsula Laboratories, Inc., Belmont, CA). As shown by Goetzl and Austen,[14] the chemical and biological properties of these synthetic peptides mimic those of the

[16] D. H. Spackman, W. H. Stein, and S. Moore, *Anal. Chem.* **30,** 1190 (1958).
[17] W. R. Gray, this series, Vol. 11 [52].
[18] R. P. Ambler, this series, Vol. 11 [49].
[19] H. Fraenkel-Conrat and C. M. Tsung, this series, Vol. 11 [13].
[20] H. Matsuo, Y. Fujimoto, and T. Tatsuno, *Biochem. Biophys. Res. Commun.* **22,** 69 (1966).
[21] R. B. Merrifield, *J. Am. Chem. Soc.* **85,** 2149 (1963).

low-molecular-weight ECF-A isolated by sequential purification using Sephadex G-25, Dowex 1, and descending paper chromatography. The maximal chemotactic activity of the synthetic peptides, used separately or together in chemotactic assays, was achieved between 3×10^{-8} and 3×10^{-7} *M*. Additional peptide analogs have been synthesized by Goetzl and Austen[22] and Smith *et al.*[23] and have been used to clarify which functional groups of the parent tetrapeptides are required for biological activity. Recently, proton nuclear magnetic resonance spectroscopic measurements, including temperature coefficients of the amide proton chemical shifts, of H-Ala-Gly-Ser-Glu-OH and the D-Val^1, Ala^2, and Ala^3 analogs indicate that the solution conformation of the ECF-A peptides is a reverse turn (i.e., a type I β turn).[24] However, the conformation of the pharmacologically active peptide remains uncertain.

Acknowledgments

This work was supported by a grant from Hoechst Aktiengesellschaft (Federal Republic of Germany). The author thanks Professor K. Frank Austen for his continued encouragement.

[22] E. J. Goetzl and K. F. Austen, *J. Exp. Med.* **144,** 1424 (1976).

[23] J. A. Smith, E. J. Goetzl, and K. F. Austen, *in* "Peptides: Structure and Biological Functions" (E. Gross and J. Meienhofer, eds.), p. 753. Pierce Chemical Co. Rockford, Illinois, 1979.

[24] G. R. Beilharz, J. A. Smith, K. F. Austen, and P. E. Wright, *Int. J. Pept. Protein Res.* **25,** 337 (1985).

[16] High-Molecular-Weight Eosinophil Chemotactic Factor from *Schistosoma japonicum* Eggs

By MAKOTO OWHASHI and YUKIFUMI NAWA

Introduction

Schistosoma japonicum eggs contain two eosinophil chemotactic factors (ECF)[1,1a]: one a high-molecular-weight (>900,000) glycoprotein

[1] M. Owhashi and A. Ishii, *J. Immunol.* **129,** 2226 (1982).

[1a] Abbreviations: ECF, eosinophil chemotactic factor; ECF-SJE, high-molecular-weight ECF from *S. japonicum* eggs; SEA, soluble egg antigen; PBS, phosphate-buffered saline; Con A, concanavalin A.

(ECF-SJE) and the other of low molecular weight.[2] ECF-SJE can be isolated from crude soluble egg antigen (SEA) in milligram-quantities by rather simple biochemical techniques so that it can be used as a standard parasite-derived ECF.

Assay Methods

Preparation of Indicator Cells

To obtain eosinophil-rich cell suspensions, C57BL/6 female mice weighing about 20 g are given an i.p. injection of 200 mg/kg cyclophosphamide, and, 2 days later, they are infected by oral administration of 1,000 embryonated eggs of *Toxocara canis* by means of a stomach tube.[3] Eggs of *T. canis* are collected from the uterus of adult female worms obtained from mongrel puppies spontaneously infected. The eggs are decorticated in a 2% sodium hypochlorite solution. Embryonation of eggs is carried out at 26° in 0.1 *N* sulfuric acid with gentle shaking and aeration for 2 weeks. The number of embryonated eggs is determined under a microscope before use. Embryonated eggs can be stored at 4° in 0.1 *N* sulfuric acid up to 1 year without decrease of infectivity. Eosinophil-rich peritoneal exudate is induced in infected mice by i.p. injection of 1.5 ml of 10% proteose–peptone on day 12 of infection. Peritoneal lavage is collected 48 hr later. By this method, $1.5-2.0 \times 10^7$ nucleated cells can be recovered from one mouse and 55–70% are eosinophils. The major contaminants are macrophages and neutrophils. The cells are washed 3 times with phosphate-buffered saline (PBS) and resuspended in RPMI 1640 culture medium (Gibco) supplemented with 10% heat-inactivated fetal bovine serum at the cell density of 1×10^6 eosinophils/ml.

In Vitro Chemotactic Assay

Blind-well chambers (Bio-Rad) are equipped with Millipore membrane filters having a pore size of 3 μm. The chambers are filled with 0.2 ml of the eosinophil-rich cell suspension in the upper compartment and 0.2 ml of each sample in the lower compartment. The chambers are incubated for 2 hr at 37° in 5% CO_2 95% air. The staining procedures of the membrane are those described by Litt.[4] The Millipore membranes are fixed in Bouin–Hollande solution [picric acid 5.0 g, copper(II) acetate 12.5 g, 40% forma-

[2] M. Owhashi, Y. Horii, and A. Ishii, *Am. J. Trop. Med Hyg.* **32,** 359 (1983).

[3] K. Sugane and T. Oshima, *Parasite Immunol.* **7,** 255 (1985).

[4] M. Litt, *Am. J. Pathol.* **42,** 529 (1963).

lin 50 ml, glacial acetic acid 7.5 ml, made up to 500 ml with distilled water] for 5 min. After 3 washings with distilled water, the membranes are stained in a mixture of 0.5% orange G and 0.5% eosine Y in 0.1 *M* phosphate buffer, pH 6.2, for 10 min. They are quickly rinsed with the phosphate buffer and are counterstained with 0.3% toluidine blue O for 30 sec. The membranes are then washed 3 times with distilled water, dehydrated in 3 changes of absolute ethanol (1 min each) and once in 2-propanol (1 min), cleared by xylene, and mounted on a glass slide. Eosinophils that have migrated from the upper surface to the lower surface of the membrane are counted at 10 high-power (×400) fields.

Purification Procedure

Starting Material

Schistosoma japonicum eggs are obtained from the intestines of infected mice by the enzymatic digestion method. Mice from the BALB/c or any other susceptible strain (e.g., C57BL/6, C3H/He), 4 to 5 weeks old, are infected by intraperitoneal injection of 30–60 cercariae of *S. japonicum* (Kofu strain) (obtainable from the University of Lowell Research Foundation, 450 Aiken St., Lowell, MA 01854). To obtain milligram quantities of ECF, at least 100 mice should be used. Eight weeks after infection, mice are starved for 48 hr to minimize contamination by the intestinal contents. The intestines are finely minced, homogenized in a Waring Blendor, and digested with 0.01% Pronase (Kaken, Tokyo, Japan) in 50 m*M* phosphate-buffered saline (PBS), pH 7.0, at 37° for 1 hr. The digested material is centrifuged at 2,000 *g* for 5 min, and the sediment containing eggs is digested again with 0.05% collagenase (Sigma C-0330, St. Louis, MO) and 0.01% Pronase in 50 m*M* PBS, pH 7.0, at 37° for 2 hr. The digested material is passed through a 150-mesh stainless steel sieve and washed with 1.0% NaCl solution (to avoid hatching of eggs). With this method, approximately 2×10^5 eggs are obtained from one mouse. The eggs are harvested and suspended in PBS to a final concentration of 5×10^5 eggs/ml. The suspension is homogenized with a Teflon homogenizer on ice. The homogenate is stirred at 4° overnight and then centrifuged at 60,000 *g* for 1 hr. The supernatant is used as crude SEA. When 1×10^7 eggs are used as starting material, approximately 100 mg SEA can be obtained.

Step 1. Anion-Exchange Chromatography on DE-52

Crude SEA is dialyzed against 17 m*M* Tris–HCl buffer, pH 7.5. A DE-52 (Whatman, Springfield, England) column (1.6 × 34.5 cm) is equili-

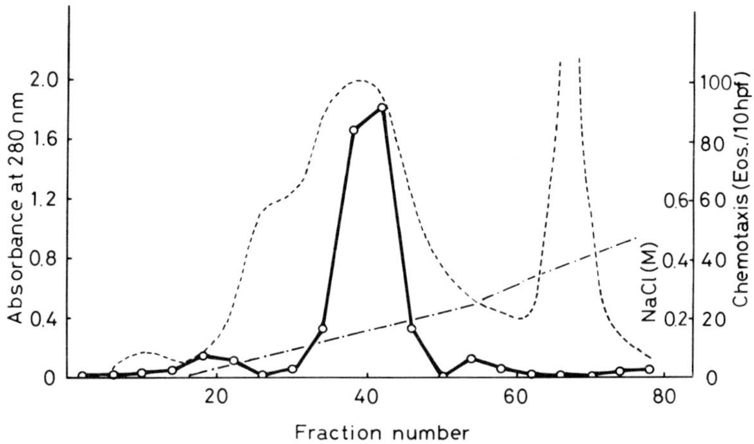

FIG. 1. Chromatography of crude soluble egg antigen (SEA) on DEAE-cellulose. A 600-mg quantity of SEA (6×10^7 eggs) was applied. The flow rate was 32 ml/hr; 11-ml fractions were collected.

brated with the same buffer. After application of the sample, elution is carried out at 4° with a linear gradient of NaCl (0–0.5 *M*) in Tris buffer. A typical elution profile is shown in Fig. 1. The protein recovery is about 17%.

Step 2. Gel Filtration on Sephacryl S-300

ECF-containing fractions from the previous step are concentrated by ultrafiltration using UM10 membrane filter (Amicon, Lexington, MA). A Sephacryl S-300 (Pharmacia Fine Chemicals, Uppsala, Sweden) column (2.6×70 cm) is equilibrated with PBS. A typical elution profile is shown in Fig. 2. The protein recovery is 35%.

Step 3. Affinity Chromatography on Con A–Sepharose 4B

ECF-containing fractions from the previous step are pooled and concentrated by ultrafiltration using a DM5 membrane filter (Amicon). The concentrated material is dialyzed against concanavalin A (Con A) buffer (0.1 *M* acetate buffer, pH 6.0, containing 0.5 *M* NaCl, 1 m*M* $CaCl_2$, 1 m*M* $MnCl_2$, and 1 m*M* $MgCl_2$). A Con A–Sepharose 4B (Pharmacia) column (1.2×6.5 cm) is equilibrated with Con A buffer. Elution is carried out at 4°. The bound material is subsequently eluted with Con A buffer containing 0.2 *M* methyl-α-D-glucoside (Fig. 3). Protein recovery is about 11%. This material can be used as purified ECF-SJE.

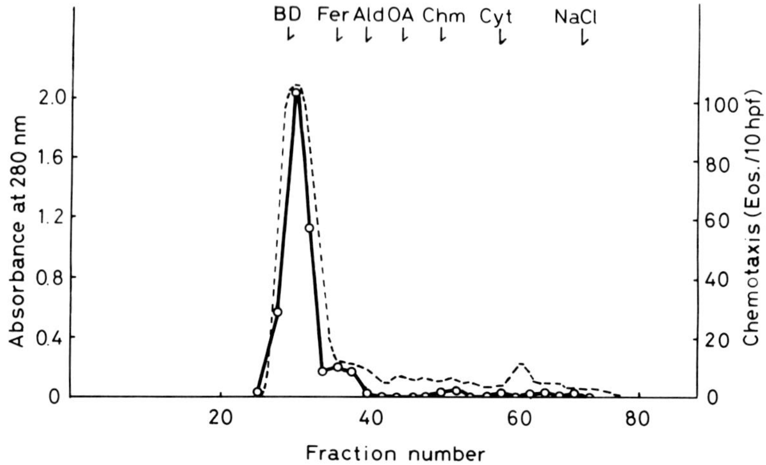

FIG. 2. Elution profile of ECF on Sephacryl S-300 gel chromatography. ECF-positive fractions from the DEAE-cellulose column (Step 1) were pooled, concentrated, and 5 ml of sample (20 mg protein/ml) was loaded onto a Sephacryl S-300 column. Blue dextran (BD), ferritin (Fer), aldolase (Ald), ovalbumin (OA), chymotrypsinogen (Chm), cytochrome *c* (Cyt), and NaCl were used as molecular weight markers. The flow rate was 21 ml/hr; 5-ml fractions were collected.

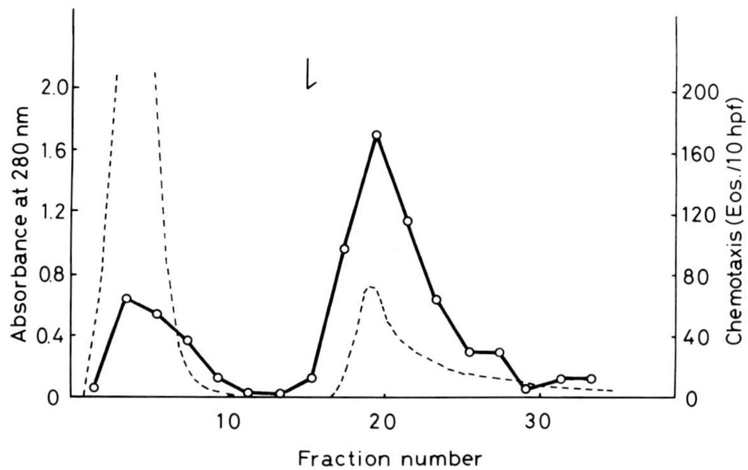

FIG. 3. Affinity chromatography on Con A–Sepharose 4B. ECF-positive fractions from the Sephacryl S-300 column (Step 2) were pooled, concentrated, and dialyzed against Con A buffer. After loading the sample onto a Con A–Sepharose 4B column, unbound material was washed with Con A buffer. Bound material was eluted with 0.2 *M* methyl-α-D-glucoside in Con A buffer (arrow). The flow rate was 3.2 ml/hr; 2-ml fractions were collected.

Properties of Purified ECF-SJE

Yield and Specificity

With this purification scheme, 4 mg ECF-SJE can be isolated from 6×10^7 eggs. Thus, the ECF-SJE content per egg is 6×10^{-8} mg. When the ECF activity of purified ECF-SJE is compared to that of crude SEA by dose–response curves (Fig. 4), the protein concentration required to obtain half-maximal activity is 1.9 μg/ml for ECF-SJE and 36 μg/ml for crude SEA. Thus, the increase in specific activity after purification is about 19.

Properties of ECF-SJE in Comparison with Low-Molecular-Weight ECF

ECF-SJE is a high-molecular-weight (> 900,000) glycoprotein containing about 40% carbohydrate.[1] ECF-SJE is antigenic as shown by the fact that serum obtained 7 or more weeks after infection of mice with *S. japonicum* gives a clear precipitin band with purified ECF-SJE in double diffusion in agar. It is stable to heating (100°, 60 min) or Pronase treatment but is sensitive to $NaIO_4$, indicating that the active site resides in the carbohydrate moiety. When Pronase-treated ECF-SJE is chromatographed on molecular sieve HPLC (SW-3000, Toyo Soda, Tokyo), ECF activity is found at a molecular weight of about 30,000. In contrast, when crude SEA

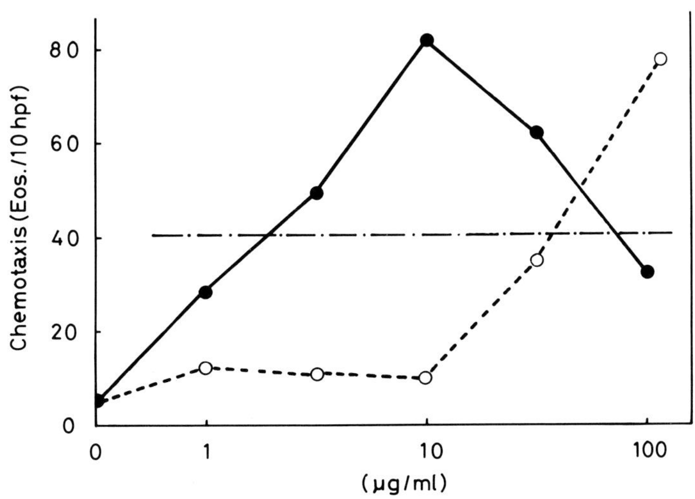

FIG. 4. Dose–response curves of eosinophil chemotactic activity for purified ECF-SJE (●) and crude SEA (○). The increase in the specific activity after purification was determined at the half-maximal eosinophil chemotactic activity of ECF-SJE (indicated by the dot–dash line).

is chromatographed on a Sephadex G-150 column, the low-molecular-weight ECF elutes with an apparent molecular weight below 1,000. The low-molecular-weight ECF is extracted with toluene and is stable to heating (100°, 1 hr), Pronase, or carboxypeptidase A treatment.[2]

[17] Denatured Proteins as Chemotactic Agents: Mitogens as Lymphocyte Locomotion Activators

By PETER C. WILKINSON

Introduction

There are two groups of agents with actions on leukocyte locomotion for which the molecular mechanisms are not clarified. The first consists of denatured proteins which may, as discussed in this chapter, activate chemotaxis and other leukocyte functions by recognition mechanisms of low specificity. Methods for preparing these proteins in a form in which they can be used to stimulate leukocyte locomotion *in vitro* are described here. The locomotion assays that can be used are the same as for other chemotactic factors and are described elsewhere in this volume ([2] and [3]). The second group of agents consists of mitogens which activate lymphocyte locomotion. These are not chemotactic factors, and their effect is to confer locomotor capacity on previously immotile lymphocytes. This, unlike chemotaxis, is a delayed effect requiring hours rather than minutes of culture, and it requires protein synthesis by, and probably growth of, the cells. Lymphocytes which have acquired locomotor capacity in this way probably also acquire the ability to respond to lymphocyte chemoattractants, though the latter molecules are not discussed here.

Preparation of Conformationally Altered Proteins for Chemotaxis Studies

There is much evidence that denatured proteins are removed rapidly from the circulation and from tissues by phagocytic cells. Studies from our laboratory suggest that similar proteins can induce chemotactic responses in leukocytes *in vitro*, and it seems likely that these responses assist in clearance of damaged protein molecules *in vivo*. Here I describe methods used for the study of three of these, namely, serum albumin, hemoglobin, and α_2-macroglobulin (α_2M).[1] I have preferred the term "conformationally altered" to "denatured" in the title above, because, in the case of α_2M, it is clear that the conformational alteration is part of a physiological process rather than a disordered state. One of the difficulties in work with such proteins is that conformational alterations are unpredictable and are usually achieved by crude means which result in a heterogeneous product. Thus different preparations vary considerably in activity. Consequently, it is difficult to get precise information about the interactions of these proteins with cell surfaces.

Serum Albumin

Conformational change in serum albumin has been well reviewed by Foster.[1a] Many of the best studied methods of denaturing serum albumin, e.g., at pH 2 or with urea, yield products which return to the native state at physiological pH and ionic strength. When tested with cells in physiological media, they have little or no activity. Treatment with alkali or reduction–alkylation, on the other hand, yield irreversibly denatured products. Serum albumin is rich in disulfide bonds which are broken by such treatments. Heat denaturation (e.g., at 70° for 30 min) is also irreversible, but heat-denatured albumin is poorly soluble and therefore does not form good gradients in chemotaxis assays. The treatment we have used most often is alkali denaturation which is described below.

Source of Protein. Serum albumin from commercial sources may be used. It is preferable to use a source which does not contain traces of denatured protein. Albumin prepared by salt precipitation is preferable to albumin prepared by cold ethanol fractionation. Human serum albumin (HSA) from Behringwerke (Marlburg, West Germany) is excellent. Some commercial preparations of bovine serum albumin (BSA) have chemotactic activity at high concentrations, probably because of traces of denatured BSA.

Method for Denaturation. HSA (200 mg) is dissolved in 100 ml of sterile distilled water. The pH is raised to 12.0 by dropwise addition of concentrated NaOH with stirring. The protein is left at 20° for 24 hr. The pH is then returned to 7.4 at 0° with stirring by slow dropwise addition of

[1] Abbreviations: α_2M, α_2 macroglobulin; HSA, human serum albumin; BSA, bovine serum albumin; ΔE, difference in extinction, E; HEPES, *N*-2-hydroxyethylpiperazine-*N*′-2-ethanesulfonic acid; MOPS, 3-(*N*-morpholino)propanesulfonic acid; Hb, hemoglobin; ANS, 8-anilinonaphthalene sulfonate; HEV, high endothelial venules; PHA, phytohemagglutinin; PPD, purified protein derivative; FCS, fetal calf serum.

[1a] J. F. Foster, *in* "Albumin Structure, Function and Uses" (V. M. Rosenoer, M. Oratz, and M. A. Rothschild, eds.), pp. 53–84. Pergamon, Oxford, England, 1977.

concentrated HCl to bring the pH to about 9.0 and then using NaH_2PO_4 to lower it more slowly. The protein is then dialyzed at 4° against a large volume of distilled water (several changes).

Measurement of Denaturation. Denaturation is measured by difference spectroscopy against native HSA in a double-beam spectrophotometer. Native HSA is placed in the sample beam and the denatured protein at the same concentration in the reference beam. Characteristically 2 peaks are seen, at 280 and 287 nm, the latter being the larger. The amplitude of the peak at 287 nm gives a measure of protein unfolding.[2,3] Typically $\Delta E_{287}^{2\ \mathrm{mg/ml}}$ is between 0.02 and 0.06. The higher this peak (ΔE), in general, the more likely is the sample to have activity in chemotaxis assays.

Dialysis. The protein is dialyzed overnight against phosphate-buffered saline, pH 7.2. The buffer is changed 2–3 times. Finally the protein is dialyzed against Hanks' balanced salt solution buffered with HEPES (*N*-2-hydroxyethylpiperazine-*N'*-2-ethanesulfonic acid) or MOPS [3-(*N*-morpholino)propanesulfonic acid] (10 m*M*). It is important to use sterile procedures throughout these dialyses. Even small numbers of bacteria may introduce extraneous chemotactic activity.

Concentration. The protein is concentrated approximately 10-fold (to about 20 mg/ml) using an Amicon PM10 ultrafilter (Amicon Corp., Danvers, MA). It is then sterile-filtered and is ready for use. It may be stored at −20° but with repeated freezing and thawing tends to aggregate and to lose activity.

Use of Denatured HSA. The protein is used in a concentration range between 250 μg/ml and 15 mg/ml and is active within this range. Table I shows a dose–response in a polarization assay. In molar terms this concentration range is high (approximately 3×10^{-6} to $10^{-4}M$) compared to low-molecular-weight factors. However, the range of protein concentrations may not reflect accurately the range of ligand concentration binding to the leukocyte surface. Activity has been seen using the following assays:

1. The micropore filter assay[4] (see this volume [3]) using the checkerboard method.
2. The orientation chamber (see this volume [6]). Little activity is seen using this assay at protein concentrations below about 8–10 mg/ml, possibly because it is not a very sensitive way of detecting weak chemotactic responses.

[2] E. J. Williams and J. F. Foster, *J. Am. Chem. Soc.* **81,** 865 (1959).

[3] D. B. Wetlaufer, *Adv. Protein Chem.* **17,** 303 (1962).

[4] P. C. Wilkinson, *in* "Leukocyte Chemotaxis" (J. I. Gallin and P. G. Quie, eds.). Raven, New York, 1978.

TABLE I
DOSE–RESPONSE OF HUMAN BLOOD NEUTROPHILS TO ALKALI-DENATURED HSA IN A POLARIZATION ASSAY

Concentration of alkali-denatured HSA (mg/ml)	Percent of neutrophils[b]
0[a]	8
0.8	28
1.6	57.3
3.2	73.3
6.5	78
13	94.3

[a] Cells in 10 mg/ml native HSA.
[b] In polarized morphology after 30 min at 37°.

3. The polarization assay[5] (see this volume [2]). Polarization is seen within 5 min and is fully developed in 15 min.
4. Denatured HSA also has activity in inducing enzyme release from human neutrophil leukocytes, e.g, neutral proteases, measured by azocasein digestion,[6] and lysozyme.

Samples vary considerably in activity. The reasons for this are not clear, but it is usually best to prepare several samples and select the most active.

Hemoglobin

Hemoglobin unfolds (loses α-helical structure) when the heme group is dissociated from the globin molecule. After removal of heme, the tetrameric protein dissociates into $\alpha\beta$-chain dimers. The α-helical content of native ferrihemoglobin decreases from 65 to 52% on removal of heme.[7] The unfolding can be reversed completely by adding hemin back to the heme-free globin solution.[7] These events can be monitored accurately by loss of absorption intensity of the spectrophotometric peak in the Soret region (406 nm for hemoglobin) on removal of heme and recovery of absorption intensity at the same wavelength on readdition of hemin. Thus, this provides a system in which biological activity can be related accurately to chemical change. Methods for preparation of, and modification of,

[5] W. S. Haston and J. M. Shields, *J. Immunol. Methods* **81**, 229 (1985).
[6] P. C. Wilkinson and G. R. Bradley, *Immunology* **42**, 637 (1981).
[7] K. Javaherian and S. Beychok, *J. Mol. Biol.* **37**, 1 (1968).

hemoglobin and myoglobin are treated in detail in a monograph by Antonini and Brunori.[8]

Preparation of Hemoglobin. Commercial preparations were not found satisfactory for these experiments. Human hemoglobin is prepared freshly from erythrocytes from blood-bank blood by the method of Rossi-Fanelli *et al.*[9,10] One pint of blood is centrifuged and the supernatant removed. The cells are washed 4 times with 0.15 *M* NaCl (4 parts to 1 part cells) spinning at 1200 *g* for 10 min between each wash. The packed red cells are then lysed with 4 parts of distilled water to 1 part cells at 4°. The lysate is precipitated by addition of 20 parts saturated $(NH_4)_2SO_4$ to 80 parts lysate for 30 min at 4°. Hemoglobin remains soluble at this salt concentration. The preparation is centrifuged at 25,000 *g* for 30 min at 4° to remove precipitate and dialyzed extensively against distilled water. During dialysis a further precipitate may form, in which case the preparation is recentrifuged. The supernatant is retained. This should contain approximately 30 mg/ml hemoglobin. The product is checked and its concentration determined by spectroscopy. Hemoglobin gives several peaks in the UV and visible regions, the λ_{max} in the latter regions being slightly different for oxyhemoglobin, carboxyhemoglobin, etc. Details of these are found in Ref. 8. The latter variants, which show minor ligand-induced conformational changes, are all inactive in chemotaxis assays. Likewise, change from the ferrous to the ferric state has no effect on chemotactic activity. The preparation is over 95% pure, and the hemoglobin is in the native form.

Preparation of Globin from Hemoglobin. A satisfactory method is that of Rossi-Fanelli *et al.*[10] To 30 volumes of cold acetone, acidified by addition of 2 *N* HCl (3 ml/liter of acetone), 1 volume of hemoglobin solution at approximately 30 mg/ml (prepared as above) in H_2O is added slowly with stirring. This is maintained at −20° for 30 min in a well-sealed vessel. Heme dissociates from the globin molecule. The former remains in solution, but the latter forms a precipitate. The precipitate is centrifuged at 0°, then redissolved in H_2O, readded to acid–acetone as before, and recentrifuged. The precipitate is redissolved in H_2O then dialyzed against a solution of $NaHCO_3$ (100 mg/ml) at 4°, and finally against balanced salt solution (e.g., Hanks'–HEPES). The globin remains in solution at 4° but comes partially out of solution at 37°. Concentrations of globin and hemoglobin (Hb) are determined spectrophotometrically. The $E_{280}^{1\,mg/ml}$ for globin is 0.85; $E_{274}^{1\,mg/ml}$ for Hb is 1.8; $E_{276}^{1\,mg/ml}$ for HbO_2 is 2.1.[10] In each case, the

[8] E. Antonini and M. Brunori, "Hemoglobin and Myoglobin in Their Reactions with Ligands." North-Holland, Amsterdam, 1971.

[9] A. Rossi-Fanelli, E. Antonini, and A. Caputo, *J. Biol. Chem.* **236,** 391 (1961).

[10] A. Rossi-Fanelli, E. Antonini, and A. Caputo, *Biochim. Biophys. Acta* **30,** 608 (1958).

wavelength given is the UV λ_{max} for the protein. The removal of heme from globin causes unfolding of the globin molecule.[7] Heme-free globin (apohemoglobin) is chemotactically active as judged by checkerboard micropore filter assays at concentrations between 50 μg/ml and 2 mg/ml. Native hemoglobin tested at the same concentrations is inactive.

Renaturation of Globin by Addition of Hemin. The method is adapted from Ref. 7. Hemin chloride (5 mg) is dissolved in 3 ml of 0.1 *N* KOH and diluted to 15 ml with H_2O. Measured amounts are added to globin and neutralized. In our experiments quantities of hemin from 0.04 to 0.72 mg are added to 18 mg globin, giving molar ratios in the reaction mixture from 1 hemin : 10 globin to 2 hemin : 1 globin. These are dialyzed against indicator-free balanced salt solution, pH 7.2. The amount of hemin bound to globin is determined by measuring absorption at 405 nm. At this wavelength, heme-free globin does not absorb. Absorption increases with increasing hemin binding up to the point where the globin molecule is completely refolded at 1 mol of hemin per mole globin dimer ($E_{405}^{1\ \mathrm{mg/ml}} =$ 5.6). In practice, the latter ratio is usually achieved by adding 2 parts hemin to 1 part globin. The products are tested in chemotaxis assays for activity with neutrophils. Using 24 such samples, a very good negative correlation was found between the heme:globin ratio and the distance migrated by neutrophils into filters ($r = -0.82$, $p < 0.0001$), suggesting that unfolding of the globin is recognized by neutrophils and that, if allowed to refold, the globin is no longer recognized.[11] A disadvantage of the use of globin in such assays is its poor solubility at 37°.

Myoglobin can be denatured and renatured using similar procedures to those described above[8] and with identical effects on chemotactic activity.[11]

α_2-Macroglobulin and α_2M–Trypsin Complexes

A further example of a relationship between conformational change and stimulation of leukocyte locomotion is furnished by α_2-macroglobulin. This protein acts as an esterase inhibitor by covalently binding to any of a wide range of esterases. On binding esterase, the α_2M, which is a large molecule, undergoes a conformational change such that the esterase molecule is enclosed in the interior of the α_2M molecule, thus sterically inhibiting the action of the esterase. Conformational changes in α_2M on binding esterase can be monitored by measuring increases in fluorescence of 8-anilinonaphthalene sulfonate (ANS) added to the preparation. If trypsin is added to α_2M in increasing concentrations, a dose-dependent increase in ANS fluorescence is seen and, in parallel, a dose-dependent increase in

[11] P. C. Wilkinson, *Nature (London)* **244,** 512 (1973).

chemoattractant activity for neutrophils and monocytes as well.[12] α_2M is a large molecule which probably forms gradients slowly, and it is much easier to demonstrate a chemokinetic than a chemotactic effect of α_2M–trypsin complexes. It has also been established that mononuclear phagocytes possess a receptor which binds α_2M–trypsin complexes but not native α_2M.[13] Such complexes are cleared rapidly by monocytes both *in vitro* and *in vivo*,[14] and it seems likely that this is a mechanism for clearing enzymes in the healing of inflammatory lesions.

Preparation of Human α_2M. The method is modified from that in Ref. 15. Blood is collected into acid–citrate–dextrose containing 0.5 mg/ml soybean trypsin inhibitor, and the plasma (25 ml) is separated by centrifugation. The plasma is mixed with 0.28 volumes of aqueous 25% (w/v) polyethylene glycol 6000, pH 6.5, and left at 20° for 30 min. The precipitate is removed by centrifugation, and a further 0.72 volumes of polyethylene glycol is added to the supernatant. The second α_2M-rich precipitate is centrifuged and redissolved in 50 m*M* Tris–HCl, pH 8, and the 25-ml sample applied to a Cibacron blue–Sepharose CL-6B column (4 × 40 cm) (Pharmacia, Uppsala, Sweden). The sample is eluted with 50 m*M* Tris buffer, pH 8.0. Fractions containing α_2M are eluted early in the separation procedure. Their purity is checked by SDS-PAGE.

The α_2M-rich fractions (> 95% pure) are pooled and concentrated using an Amicon P30 ultrafilter and dialyzed against 1% glycine for 24 h. The α_2M is freed of haptoglobin, fibrinogen, and immunoglobulins by affinity chromatography. IgG antibodies to human fibrinogen, total immunoglobulins, and haptoglobin (Scottish Antibody Production Unit, Carluke, UK) are linked to a cyanogen bromide-activated Sepharose 4B gel (Pharmacia) as follows. The antibodies are each diluted to 5 mg/ml in coupling buffer (0.1 *M* bicarbonate in 0.5 *M* NaCl, pH 8.3) and mixed. The mixture is added to the Sepharose and mixed by shaking for 2 hr. The gel is then transferred to 0.2 *M* glycine buffer (pH 8) for 16 hr at 4°. Excess absorbed protein is washed away with 0.1 *M* acetate in 0.5 *M* NaCl (pH 4.0). The gel with bound antibody is poured to form a column (2.5 × 10 cm). The α_2M-rich fraction prepared as above is gravity-fed through the column (20 mg of α_2M in 18 ml 1% glycine). Unabsorbed protein is eluted with 100 m*M* phosphate buffer, pH 7.0, and concentrated to 1 mg per ml. Purity is checked by SDS–PAGE and by immunoelectrophoresis against anti-whole human serum. Both should show pure α_2M free of fibrinogen, haptoglobin, and immunoglobulin.

[12] J. V. Forrester, P. C. Wilkinson, and J. M. Lackie, *Immunology* **50,** 251 (1983).
[13] J. Kaplan and M. L. Nielsen, *J. Biol. Chem.* **254,** 7329 (1979).
[14] M. T. Debanne, R. Bell, and J. Dolovich, *Biochim. Biophys. Acts* **411,** 295 (1975).
[15] G. D. Virca, J. Travis, P. K. Hall, and R. C. Roberts, *Anal. Biochem.* **89,** 274 (1978).

The proteinase inhibitory activity of the preparation may be assayed by the azure hide method.[16] Hide powder azure (Sigma), a particulate substrate, produces a color reaction on digestion. Proteinase activity was assayed using 0.2 μg trypsin, 1–50 μg α_2M, and 10 mg hide powder azure as detailed in Ref. 16 and estimating E_{595} after 20 min at 37°. The inhibition achieved by adding α_2M was measured against a standard curve obtained with trypsin plus hide powder azure without α_2M.

Structural Modification of α_2M: α_2M–Trypsin Complexes. The α_2M sample is incubated with trypsin for 5 min at 37° at various ratios, and the conformational change is monitored by addition of the fluorescent probe, ANS.[17] We prepared α_2M–trypsin complexes at ratios from 1:0 to 1:6 of α_2M to trypsin and found that complexes at 1:3 α_2M:trypsin were fully active. ANS (BDH, Poole, UK) at pH 6.0 is added to the α_2M and α_2M–trypsin complexes (at 1 mg/ml α_2M) to give a final ANS concentration of 10^{-6} *M*. Fluorescence is monitored in a fluorimeter (excitation 366 nm, emission 475 nm) using 10^{-6} *M* ANS alone as a negative control. The results are expressed in arbitrary units using the fluorescence emitted by a solution of ANS–HSA (1 mg/ml) as 100 units, since serum albumin has a high affinity for ANS and fluoresces strongly in its presence. As the α_2M–trypsin ratio is increased from 1:0 to 1:3, fluorescence increases, but thereafter the curve flattens out,[12] suggesting that sufficient trypsin has been added to saturate α_2M–enzyme binding sites.

Assay of Activity of α_2M–Trypsin Complexes in Neutrophils. α_2M–trypsin complexes have been studied by the micropore filter assay (see this volume [3]). α_2M is a very large molecule (M_r 725,000) and would be expected to diffuse slowly. The complexes, but not uncomplexed α_2M, stimulate locomotion of both neutrophils and monocytes, but, as mentioned above, a chemotactic checkerboard may be seen only with monocytes. The activity is optimally present at an α_2M–trypsin ratio of 1:3. Conformational change is maximal at this ratio. No chemotactic effect of α_2M–trypsin complexes is seen using an orientation chamber. However, gradient formation of these large complexes may be very poor in this short-term assay.

Use of Mitogens as Activators of Lymphocyte Locomotion

Many studies both *in vivo* and *in vitro* suggest that the locomotor capacity of small lymphocytes, e.g., from blood or from unprimed lymph nodes, is not fully expressed (reviewed in Ref. 18). Many of these cells are

[16] A. J. Barrett, M. A. Brown, and C. A. Sayers, *Biochem. J.* **181,** 401 (1979).
[17] Y. Jacquot-Armand and G. Krebs, *Biochim. Biophys. Acta* **303,** 138 (1973).
[18] D. M. V. Parrott and P. C. Wilkinson, *Prog. Allergy* **28,** 193 (1981).

nonmotile. On the other hand, activated lymphocytes, e.g., following challenge with antigen *in vivo* or with mitogens *in vitro*, are much more motile. These differences may reflect differences in the way these cells behave in the body. The recirculating pool of lymphocytes which circulates from blood across high endothelial venules (HEV) into lymph nodes then back via lymphatics to the blood consists mainly of small nonactivated T lymphocytes, and a specialized recognition event may be necessary for these cells to traverse HEV.[19,20] These recirculating lymphocytes do not cross nonspecialized endothelia or migrate into inflammatory sites. After activation, blood lymphocytes appear to leave the recirculating pool and to acquire the ability to cross endothelia at many sites and to enter inflammatory foci. This activation of locomotion can be studied *in vitro* by culturing human blood lymphocytes in the presence of mitogens and observing the effect on locomotor capacity. During the first 24–48 hr of culture in the presence of mitogens, many of the cells become motile.[21] Note that this is *not* due to a chemotactic effect of the mitogen, and binding of mitogen is not followed by an immediate locomotor response. There is a considerable delay (> 8 hr), probably because it is necessary for the lymphocytes to enter the G_1 phase of the cell cycle and begin protein synthesis before acquiring mobility.

Source of Lymphocytes

Lymphocytes are obtained from heparinized human blood by centrifugation of the blood on Ficoll–Hypaque, SpG 1078 (Flow, Rickmansworth, UK) (see Vol. 108 of this series [9]). The mononuclear cell fraction which forms a cushion on top of the separating mixture is removed and washed 3 times with Hanks' solution containing 10 m*M* MOPS, pH 7.4 (Hanks'–MOPS; composition detailed in chapter [2] in this volume). This fraction contains about 80% lymphocytes and 20% monocytes.

Mitogens

Many mitogens can be used as activators of lymphocyte locomotion. Those we have studied include phytohemagglutinin (PHA, purified grade; Wellcome, Dartford, Kent, UK); purified protein derivative (PPD, tuberculin; Evans Medical, Greenford, Middlesex, UK) which is mitogenic in tuberculin-positive subjects; staphylococcal protein A (Pharmacia), a spe-

[19] W. L. Ford, M. E. Smith, and P. Andrews, *in* "Cell–Cell Recognition" (A.S.G. Curtis, ed.), pp. 359–392. Cambridge Univ. Press, Cambridge, England, 1978.

[20] W. M. Gallatin, I. L. Weissman, and E. C. Butcher, *Nature (London)* **304,** 30 (1983).

[21] P. C. Wilkinson, *Immunology* **57,** 281 (1986).

cific mitogen for B lymphocytes; and the monoclonal antibody, OKT3 (Ortho Diagnostics, Raritan, NJ), a mouse IgG_{2a} antibody which binds to the CD3 receptor on T lymphocytes. Since the CD3 receptor is complexed to the antigen receptor (Ti), it is possible that OKT3 serves as a model for antigen-induced lymphocyte activation.[22] Most of our studies have been with either the OKT3 antibody[23,24] or with PHA, and the culture conditions using these two mitogens are detailed below.

Lymphocyte Culture

Lymphocytes are cultured in tissue culture plastic dishes whose size is chosen depending on the quantity of test reagents available. We routinely use 17 × 16 mm multiwell dishes (Flow) and 1 ml of cell suspension, but, for smaller volumes, flat-bottomed multiwell plates with covers (0.6 × 1 cm; well capacity 0.35 ml) are more suitable. The lymphocytes are cultured at 37° in RPMI 1640 medium buffered to pH 7.2 with 20 m*M* HEPES or 20 m*M* MOPS. The RPMI medium is available as a 10× stock solution. This is diluted appropriately, and glutamine (final concentration 2 m*M*), penicillin (final concentration 100 U/ml), and streptomycin (100 μg/ml) are added. The medium is buffered as above, and fetal calf serum (FCS) is added. The lymphocytes are usually cultured at $2-4 \times 10^6$ cells/ml. Lymphocyte polarization by mitogens is cell density dependent, so, in any test, the cell concentration should be constant in all the wells.

In this medium without added mitogen, about 20–30% of lymphocytes at the densities stated above assume a polarized morphology in response to unidentified factors in FCS. Addition of a mitogen increases this proportion over a period of 24–48 hr. If PHA (1 μg/ml) is added, 60–70% of the cells become polarized after 24 hr in culture, increasing to 80–90% after 48 hr. These cells also increase in size and in synthesis of RNA and protein during this time as measured by uptake of [^{3}H]uridine and [^{3}H]leucine, respectively.[21] There is no DNA synthesis (measured by [^{3}H]thymidine uptake), before 48 hr. Background polarization due to FCS can be reduced by culturing the lymphocytes in a simpler medium, e.g., RPMI–MOPS plus 20 mg/ml HSA (Behringwerke). The number of lymphocytes in locomotor morphology in this medium without a mitogen is usually around 10%. The percentage of polarized cells in the presence of a mitogen in HSA-containing medium is usually 20–30% lower than in FCS-containing medium.

PHA is not a selective ligand for specific surface glycoproteins. The

[22] J. P. van Vauwe, J. R. de May, and J. G. Goossens, *J. Immunol.* **124,** 2708 (1980).

[23] P. C. Wilkinson and A. Higgins, *Immunology* **60,** 445 (1987).

[24] P. C. Wilkinson and A. Higgins, *Immunology* **61,** 311 (1987).

OKT3 monoclonal antibody has a more precise target, namely, the CD3 marker of T lymphocytes. OKT3 is fully active at concentrations as low as 250 pg/ml. We use it routinely at 1 ng/ml. At 48 hr, 60–80% of lymphocytes acquire locomotor capacity in OKT3. The OKT3 as supplied by the manufacturer contains azide, and this must be dialyzed out before adding the antibody to cell cultures. We do not attempt to separate the monocytes from lymphocytes for culture of the mononuclear cell fraction in OKT3. When the cells are cultured, monocytes and many of the B lymphocytes become adherent to the dish, whereas T lymphocytes remain nonadherent.[25] The adherent cell population contains the majority of IgG Fc receptor-bearing cells. These cells are probably necessary for the locomotor response in OKT3. If they are removed by Fc rosetting, the response to OKT3 is lost.[23] Furthermore, the $F(ab')_2$ fragment of OKT3 has no activity, suggesting the OKT3 cross-links T cells to Fc receptor-bearing cells.[23]

After appropriate times in culture, the culture supernatants, including the nonadherent cells, are decanted. These are assayed for locomotor activity as detailed below.

Assays for Locomotor Activity

The assay in routine use is a modification of the polarization assay described in this volume [2]. This is a rapid and reproducible assay: however, since it is not a direct measure of locomotion it may be supplemented by assay of lymphocyte invasion of collagen gels (also described in this volume [2]).

Polarization Assay. Lymphocytes decanted as above from culture media at appropriate times are immediately fixed in glutaraldehyde (final concentration 1.25%) before being allowed to cool. They are fixed for approximately 10 min, then washed twice with buffered saline. The pellet (in ~100 μl) remaining after decanting the second wash solution is used to make slide-and-coverslip preparations. The percentage of polarized cells is determined by counting 300 cells using a 40× phase-contrast objective (see Fig. 1). At the same time, a morphological assessment is made of cell viability. This is more reliable than dye exclusion. Usually the number of damaged or dead cells is low unless the cultures have been allowed to lose fluid by evaporation. Polarization is expressed as a percentage of viable cells.

Collagen Gel Assay. The procedure for this is described in this volume [2]. A collagen gel (1.5 mg collagen/ml in RPMI–MOPS) is cast and

[25] W. C. van Voorhis, L. S. Hair, R. M. Steinman, and G. Kaplan, *J. Exp. Med.* **155,** 1172 (1982).

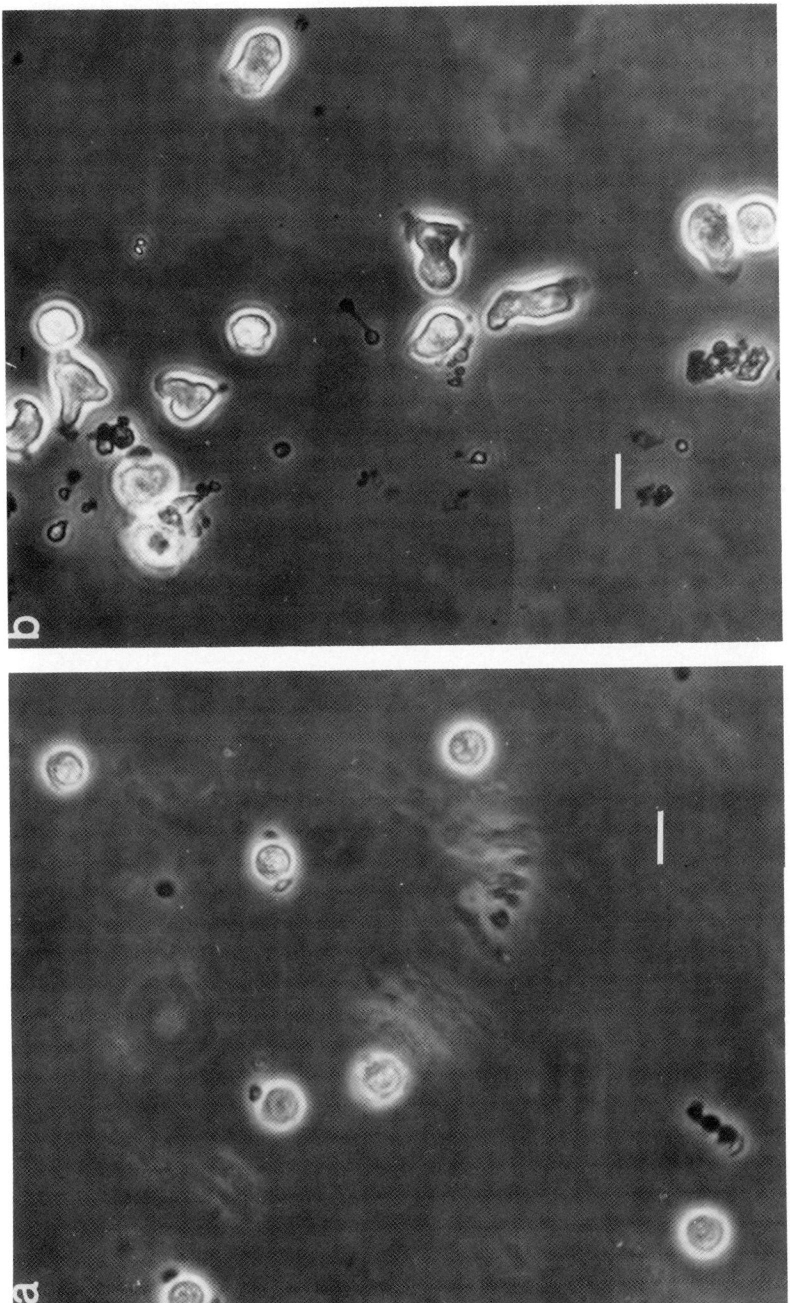

FIG. 1. Effect of a mitogen on the polarization of blood lymphocytes. In (a) cells are shown direct from blood in Hanks'–HSA in spherical morphology. In (b) the cells have been cultured in the presence of a mitogen for 24 hr, and most of them have assumed locomotor morphology. Scoring of the assay is done on the basis of this shape change. The mitogen used here was PPD, but exactly similar changes are seen using the others described in the chapter. Bars, 10 μm.

allowed to set. Lymphocytes (2×10^6 cells/ml in RPMI–MOPS) are allowed to invade the gel for 2 hr at 37°. The gel is then fixed by adding an equal volume of glutaraldehyde (2.5%) and washed. The proportion of lymphocytes which have penetrated the gel from its upper surface is determined. This proportion is usually close to the proportion of polarized cells.[21,23]

Cell Surface Markers

To determine the proportions of T and B lymphocytes in polarized and nonpolarized populations, an aliquot is removed from the culture and fixed by addition of an equal volume of 4% paraformaldehyde. After washing and preparation of a cell pellet, the cells are treated as follows:

1. Estimation of B-cell numbers is made by counting surface-Ig$^+$ cells. A mixture of FITC-labeled anti-κ light chain and anti-λ light chain antibodies (Cappel, West Chester, PA) is added for 20 min. The optimum dilution of antibody to use must be predetermined for each batch by testing a range of dilutions. The cells are washed twice to remove free FITC and examined by fluorescence microscopy using a 40× oil-immersion objective (see also Vol. 108 of this series [41]).

2. Estimation of T-cell numbers is made using the OKT3 antibody, or, in the case where lymphocytes had been cultured in OKT3, the OKT11 antibody (which binds to another T-cell marker, the CD2 molecule) can be used. The antibody (10 μl of a solution prepared as recommended by the manufacturer, i.e., at 10–100 μg/ml depending on which antibody is used) is added to the lymphocyte pellet for 20 min. The cells are then washed twice, and a second layer of FITC goat anti-mouse (Becton Dickinson, Mountain View, CA) or FITC rabbit anti-mouse IgG (Miles, Stoke Poges, UK) is added at a dilution predetermined as above. After 20 min, the cells are again washed twice and examined by fluorescence microscopy (see also Vol. 108 of this series [41]). T-cell subsets can be evaluated by using appropriate monoclonal antibodies against CD4, CD8, or other appropriate markers as described above.

[18] Crystal-Induced Chemotactic Factor

By ISAIAS SPILBERG and ASHWIN K. BHATT

Introduction

The accumulation of neutrophils at the site of injury is the histopathological hallmark of acute inflammation. Data from several laboratories support the concept that local generation of chemotactic factors plays an important role in the specific attraction of neutrophils to the site of tissue injury. Several chemically diverse substances stimulate polymorphonuclear leukocyte (PMN)[1] random motility directly (chemokinesis) and, in the appropriate experimental format, can be shown to specifically stimulate oriented motility of these cells (chemotaxis) toward areas of increased mediator concentration. Because of the ubiquitous and often labile nature of many of these substances, it has been difficult to delineate a primary role for any one factor in many of the models of inflammation that mimic human pathology.

The initial pathogenic feature of human acute gout and of pseudogout is the presence of urate or calcium pyrophosphate crystals, respectively, in the articular space. Although various humoral mediator systems may play a modulating role in the inflammation induced by the crystals, an extensive array of experimental evidence suggests that the acute phlogistic event can occur independently of Hageman factor, kinins, or complement. PMN depletion, however, markedly abrogates the inflammatory reaction. Phelps in 1969[2] published the observation that PMNs that had phagocytosed sodium urate crystals *in vitro* released a substance that was chemotactic for other PMNs and mononuclear cells. Work since then has extended this initial observation and showed crystal-induced chemotatic factor (CCF) to be a peptide of MW ~ 15,000 and to bind a specific receptor in human neutrophil membranes. The production of the peptide is exquisitely sensitive to suppression by pharmacologic doses of colchicine. The intraarticular injection of CCF into the joints of rabbits induces an acute synovitis without a significant increase in vascular permeability.[3-5] The peptide

[1] Abbreviations: PMN, polymorphonuclear leukocyte; CCF, crystal-induced chemotactic factor; MSU, monosodium urate; DDT, dithiothreitol; SDS, sodium dodecyl sulfate; BSA, bovine serum albumin.

[2] P. Phelps, *Arthritis Rheum.* **12,** 197 (1969).

[3] I. Spilberg, B. Mandell, J. Mehta, L. Simchowitz, and D. Rosenberg, *J. Clin. Invest.* **64,** 775 (1979).

[4] I. Spilberg and J. Mehta, *J. Clin. Invest.* **63,** 85 (1979).

[5] I. Spilberg, J. Mehta, and L. Simchowitz, *J. Lab. Clin. Med.* **100,** 399 (1982).

exists in the cell in a glycosylated and a nonglycosylated form. The glycosylated CCF contains mannose, galactose, *N*-acetyl-D-glucosamine, and fucose.[6] Summarized below is the methodology for the isolation and biological assay of CCF.

Methods

Induction and Isolation of CCF

Microcrystalline monosodium urate (MSU) is prepared by the method described by Seegmiller *et al.*[7] Eight grams of uric acid (Sigma Chemical Co.) is dissolved, with stirring, in 1600 ml of boiling deionized water containing 49 ml of 1 *N* NaOH. The pH of the solution is adjusted to 7.2 by the addition of HCl, and the solution is slowly cooled with stirring at room temperature, then stored overnight at 4°. The crystals are sterilized by heating at 200° for 1 hr and stored for future use in aseptic vials. Before use, the crystals are suspended in water and sonicated for 1 min to obtain crystals of uniform size 5–8 μm in length.

The MSU crystals (50 mg) are added to 5×10^8 human PMNs which had been isolated by dextran sedimentation and Ficoll–Hypaque (Pharmacia–Winthrop) centrifugation[8] (see Vol. 108 of this series [9]) in 25 ml of RPMI medium (Gibco), buffered to pH 7.4, shaken gently, and incubated in a Corning 25-cm^2 tissue culture flask at 37° in 5% CO_2 for 1 hr (Fig. 1). Phagocytosis of the crystals by the cells is monitored visually with the aid of dark-field and polarized-compensated microscopy. At the end of 1 hr at least 50% of the cells should contain one or more intracellular crystals. The yield of CCF correlates directly to the degree of phagocytosis. The cells are centrifuged and washed 3 times to remove extracellular particles, transferred to plastic centrifuge tubes (50 ml), resuspended in 3 ml complete Hanks' medium buffered to pH 7.4, and the cell membranes are disrupted mechanically[9] by aspirating and expelling the suspension 20 times through a Millipore grid attached to a 20-ml syringe. The now viscous, white cell-free suspension is centrifuged at 400 *g* for 10 min at 4°. The supernatant is saved and centrifuged at 8,000 *g* at 4° for 15 min; the resultant pellet represents the granular fraction of the PMNs and contains most of the chemotactic activity. To the granular fraction 200 μl

[6] A. R. Bhatt and I. Spilberg, *Agents Actions* (in press) (1988).

[7] J. E. Seegmiller, R. R. Howell, and S. Malawista, *JAMA, J. Am. Med. Assoc.* **180,** 468 (1962).

[8] A. Bøyum, *Scand. J. Clin. Lab. Invest.* **21** (Suppl. 97), 77 (1968).

[9] Z. A. Cohn and J. G. Hirsch, *J. Exp. Med.* **76,** 403 (1970).

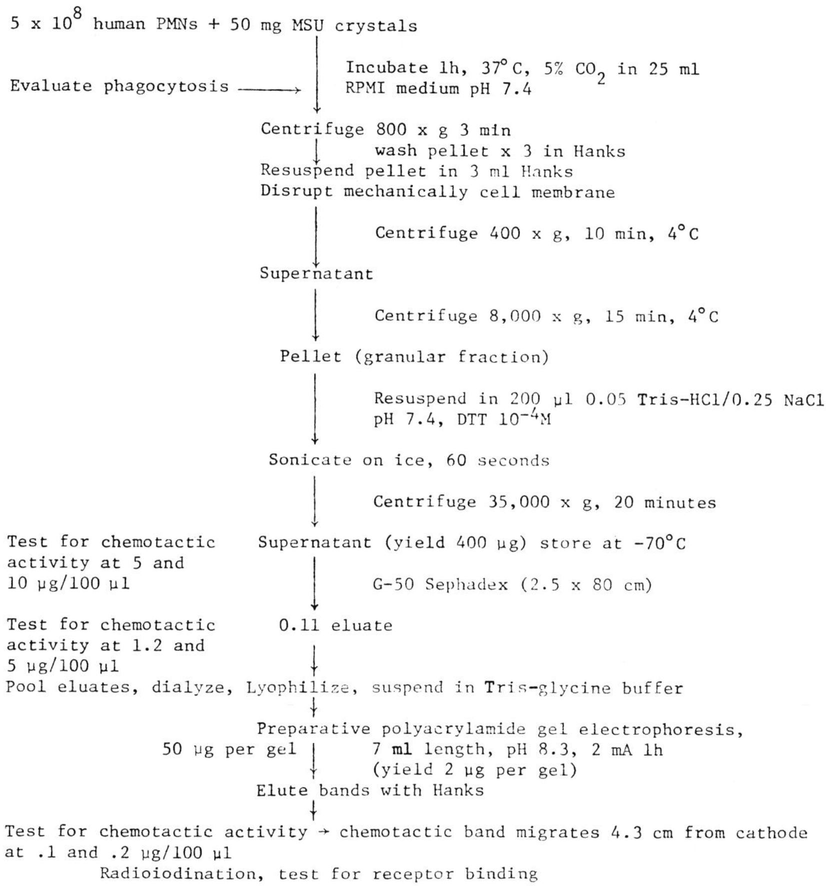

FIG. 1. Isolation of crystal-induced chemotactic factor.

of 50 m*M* Tris–HCl/25 m*M* NaCl/0.1 m*M* dithiothreitol (DTT, Sigma), pH 7.4, is added, and the suspension is sonicated on ice for 60 sec to disrupt granular membranes and centrifuged at 35,000 *g* for 20 min. The supernatant is saved, tested for chemotactic activity, and stored at −70°. The chemotactic activity of the 35,000-*g* supernatant contains about two-thirds of the total chemotactic activity of the original cell–crystal incubation mixture.[10]

Five batches of 35,000 *g* supernatants are pooled and passed consecutively over a calibrated Sephadex G-50 column (2.5 × 80 cm, Pharmacia)

[10] I. Spilberg, B. Mandell, and R. D. Wochner, *J. Lab. Clin. Med.* **83**, 56 (1974).

and eluted with 0.85% sodium chloride solution. The bulk of the chemotactic activity appears at 11% of the effluent volume.[11] The next step in the purification procedure is electrophoresis, and for this the chemotactically active fractions from five experiments are pooled, dialyzed, lyophilized, and suspended in Tris–glycine buffer (380 m*M* glycine, 5 m*M* Tris), pH 8.3, with 10% glycerol, 5 m*M* 2-mercaptoethanol, and bromphenol blue as the leading dye. The fractions are electrophoresed (50 μg protein/gel) in a 7.5% preparative polyacrylamide gel using a Bio-Rad electrophoresis system[6,12] in the same glycine buffer pH 8.3 at 2 mA per tube for 1 hr.[6] The gels are 7 cm in length. The major peaks of protein are localized at 4.3, 3.3, and 1.6 cm from the cathode when stained with Coomassie blue.[12] Only the fastest moving protein band stains for carbohydrate.[13] The unstained gels are cut in areas corresponding to the stained gel. The gel segments are macerated to elute the proteins into Hanks' medium for testing of chemotactic activity. Chemotactic activity is restricted to the fastest moving band. When the chemotactically active band is run on sodium dodecyl sulfate (SDS) gels for molecular weight estimation, it is first placed in SDS buffer (0.2% SDS, 10% sucrose, 5 μM 2-mercaptoethanol, and bromphenol blue) and electrophoresed again in 10% and 15% SDS–acrylamide gel (gel length 21 cm).[6,12] On SDS, the chemotactic protein migrates as a single band 7.8 cm from the cathode with a relative mobility of 0.805 (MW ~ 15,000). Protein is assayed by the fluorescamine method.[14]

Assay of CCF

Two methods were used: (1) a chemotactic assay which is a semiautomatic adaptation of the Boyden chamber technique using an image analyzer to measure the leading front of migrating cells into Millipore filters (see this volume [8]) and (2) a receptor binding assay.

Radioiodination of CCF

CCF is iodinated[15,16] utilizing the Bio-Rad radioiodination reagent (Enzymobeads), which contains lactoperoxidase and glucose oxidase (Bio-

[11] I. Spilberg, A. Gallacher, J. M. Mehta, and B. Mandell, *J. Clin. Invest.* **58,** 815 (1976).

[12] J. V. Maizel, *Methods Virol.* **5,** 179 (1971).

[13] R. M. Zacharius, T. E. Zell, J. H. Morrison, and J. J. Woodlock, *Anal. Biochem.* **30,** 148 (1969).

[14] N. Nakai, C. Y. Lai, and B. L. Horecker, *Anal. Biochem.* **58,** 563 (1974).

[15] I. Spilberg and J. Mehta, *J. Clin. Invest.* **63,** 85 (1979).

[16] I. Spilberg and J. Mehta, *J. Lab. Clin. Med.* **104,** 939 (1984).

Rad, Inc., Richmond, CA). CCF (50 μg protein) in 50 μl of 0.2 *M* sodium phosphate-buffered saline, pH 7.4, is mixed with 50 μl of the Bio-Rad reagent; 0.25–0.5 mCi $Na^{125}I$ (Amersham-Searle Corp., Arlington Heights, IL) is then added. After gentle mixing at room temperature, the iodination reaction is initiated by the addition of 25 μl 2% D-glucose solution and continued for 45 min. After centrifugation at 400 *g* for 10 min at 15°, the iodinated CCF in the supernatant is isolated from unreactive iodine by passing the mixture through a Sephadex G-50 Pharmacia column (1.2 × 25 cm) eluted with 50 m*M* sodium phosphate-buffered saline, pH 7.4, containing 0.1% bovine serum albumin (BSA) and stored at 4° until used. The specific activity of the iodinated CCF obtained is about 0.1–0.5 μCi/μg protein. The iodinated CCF retains its chemotactic activity for neutrophils as measured by the Boyden chamber technique and migrated identically as the unlabeled material on thin-layer chromatography on precoated silica gel 60 plates (American Scientific Products, McGaw Park, IL) in 1-propanol/water (7:5, R_f 0.65) and 1-propanol/34% ammonium hydroxide (7:3, R_f 0.14). Binding of labeled CCF to whole cells averaged 70% of the total count added to the cell–^{125}I-CCF mixture.[15,16]

Binding Assay

Binding experiments[15,16] are performed at 4° using plasticware. Briefly, to determine total binding, 10^7 PMNs are incubated with ^{125}I-CCF (final concentration 14.8 n*M*) in a microfuge tube in a total volume of 0.5 ml Hanks' solution containing 0.1% BSA, pH 7.4. Nonspecific binding is determined by incubating the cells under similar conditions with ^{125}I-CCF (final concentration 14.8 n*M*) along with unlabeled CCF (final concentration 10 μM). After 15 min of incubation, 0.5 ml of ice-cold Hanks' solution–0.1% BSA is added, followed by centrifugation at 8000 *g* for 1 min in a Microfuge (Beckman Instrument Co., Palo Alto, CA). The pellet is resuspended in Hanks' solution–0.1% BSA, transferred to another tube, and pelleted again. The cell-associated ^{125}I-CCF is determined in a gamma counter with 60% efficiency. The specific binding is calculated by the difference between the total and nonspecific binding. The nonspecific binding averages approximately 15% of the total binding.

[19] Chemotactic Factors of Bacterial Origin

By WAYNE A. MARASCO and PETER A. WARD

Background

Most bacteria elaborate chemotactic factors in tissue culture filtrates.[1,2] During the inflammatory response associated with most bacterial infections, neutrophils migrate into tissues containing viable bacteria and participate in their elimination from the tissue. These events are believed to be initiated by the release from bacteria of chemotactic factors as well as by the local activation of the complement cascade.[3] In addition, this process may be further amplified by the release from damaged host cell membranes of lipid and protein constituents that have chemotactic activity.[4,5]

Historically, however, detailed structural analysis of these bacterial chemotactic substances has been hampered by their occurrence in very low concentrations. Although several preliminary reports described the presence in *Escherichia coli* culture filtrates of chemotactic activity soluble in chloroform–methanol with properties of a fatty acid,[3,6,7] structural analysis of this material has not been pursued. In contrast, partial characterization by Ward *et al.*[1] of chemotactic factors produced by various strains of streptococci, staphylococci, and pneumococci demonstrated that these substances were low-molecular-weight peptides. Other investigators had established that the active components in culture filtrates from both *E. coli*[3] and *Streptococcus sanguis*[8] were small heterogeneous peptides with blocked amino groups.

These studies, as well as the recognition that a number of bacteria elaborated similar chemotactic substances, suggested that a specific class of compounds characteristic of these cells could be involved. Schiffman *et al.*[9]

[1] P. A. Ward, I. H. Lepow, and L. J. Newman, *Am. J. Pathol.* **52,** 725 (1968).
[2] H. U. Keller and E. Sorkin, *Int. Arch. Allergy Appl. Immunol.* **31,** 505 (1967).
[3] E. Schiffmann, H. J. Showell, B. A. Corcoran, P. A. Ward, E. Smith, and E. L. Becker, *J. Immunol.* **114,** 1831 (1975).
[4] S. R. Turner, J. A. Campbell, and W. S. Lynn, *J. Exp. Med.* **141,** 1437 (1975).
[5] H. Carp, *J. Exp. Med.* **155,** 264 (1982).
[6] J. A. Tainer, S. R. Turner, and W. S. Lynn, *Am. J. Pathol.* **81,** 401 (1975).
[7] M. Solymossy, Z. Nagy, F. Antoni, and L. Szallar, *Acta Microbiol. Acad. Sci. Hung.* **31,** 345 (1984).
[8] Y. Miyake, T. Yasuhara, K. Fukui, H. Suginaka, T. Nakajima, and T. Moriyama, *Biochim. Biophys. Acta* **758,** 181 (1983).
[9] E. Schiffmann, B. A. Corcoran, and S. M. Wahl, *Proc. Natl. Acad. Sci. U.S.A.* **72,** 1059 (1975).

hypothesized that such prokaryotic cell products could consist of formylmethionyl peptides derived from the NH_2-terminal regions of newly synthesized proteins. With some proteins, including those found in the cytoplasm, the formylmethionyl peptides are cleaved posttranslationally and are not found in the mature protein. In the case of membrane and secretory proteins which possess NH_2-terminal signal peptides, these NH_2-terminal extensions are cleaved by a signal peptidase following appropriate polypeptide transport[10] and, therefore, could be released into the extracellular fluid surrounding the cell membrane.[11] The signal peptide is encoded in the protein in the form of a short-lived sequence extension and is believed to direct the transport of newly synthesized polypeptides across the appropriate membrane following their synthesis. This hypothesis has been supported by the findings that simply synthetic formylmethionyl peptides are potent chemoattractants for leukocytes.[9,12] Other indirect evidence that these attractants are formylmethionyl peptides has come from both receptor–binding[13] and immunochemical[14,15] data. The chemical properties of the bacterial factors and the synthetic peptides are also similar.[3,8]

This chapter reviews the methodology that has been used to obtain partial or complete purification of bacterial peptide chemoattractants including the chemoattractant fMet-Leu-Phe (fMLP)[15a] from *E. coli* culture filtrates.[16] For the latter, definitive demonstration of the formylmethionyl group has been obtained following high-pressure liquid chromatography (HPLC), using dipeptidyl carboxypeptidase/gas chromatography–mass spectroscopy (DCP/GC–MS) to determine the amino acid sequences.

[10] G. Blobel, P. Walter, C. N. Chang, B. M. Goldman, A. H. Erickson, and V. R. Lingappa, *Symp. Soc. Exp. Biol.* **33,** 9 (1979).

[11] J. P. Bennett, K. P. Hirth, E. Fuchs, M. Sarvas, and G. B. Warren, *FEBS Lett.* **116,** 57 (1980).

[12] H. J. Showell, R. J. Freer, S. H. Zigmond, E. Schiffmann, S. Aswanikumar, B. A. Corcoran, and E. L. Becker, *J. Exp. Med.* **143,** 1154 (1976).

[13] S. Aswanikumar, B. Corcoran, E. Schiffmann, A. R. Day, R. J. Freer, H. J. Showell, E. L. Becker, and C. B. Pert, *Biochem. Biophys. Res. Commun.* **74,** 810 (1977).

[14] W. A. Marasco, H. J. Showell, R. J. Freer, and E. L. Becker, *J. Immunol.* **128,** 956 (1982).

[15] W. A. Marasco and E. L. Becker, *J. Immunol.* **128,** 963 (1982).

[15a] Abbreviations: fMLP, fMet-Leu-Phe, formylmethionylleucylphenylalanine; HPLC, high-pressure liquid chromatography; DCP/GC–MS, dipeptidyl carboxypeptidase/gas chromatography–mass spectrometry; SBF, stock bacterial chemotactic factor; DMSO, dimethyl sulfoxide; RIA, radioimmunoassay; TCA, trichloroacetic acid; BSA, bovine serum albumin; PBS, phosphate-buffered saline; ACE, angiotensin-converting enzyme; TMS, trimethylsilylated.

[16] W. A. Marasco, S. H. Phan, K. Krutzsch, H. J. Showell, D. E. Feltner, R. Nairn, E. L. Becker, and P. A. Ward, *J. Biol. Chem.* **259,** 5430 (1984).

Isolation of Purified Fractions Containing Chemotactic Activity

Figure 1 outlines the general procedures for the isolation and purification of the *E. coli* chemotactic materials. Procedure I[3] was performed approximately 10 years prior to procedure II[16]; however, in both cases, the same crude butanol extracts were used as starting material. In procedure I, the active material eluted from an AG 50-X8 column is subjected to gel filtration, gel electrophoresis, and repeat gel filtration to produce material suitable for analysis. In procedure II, the crude butanol extract is dried in a

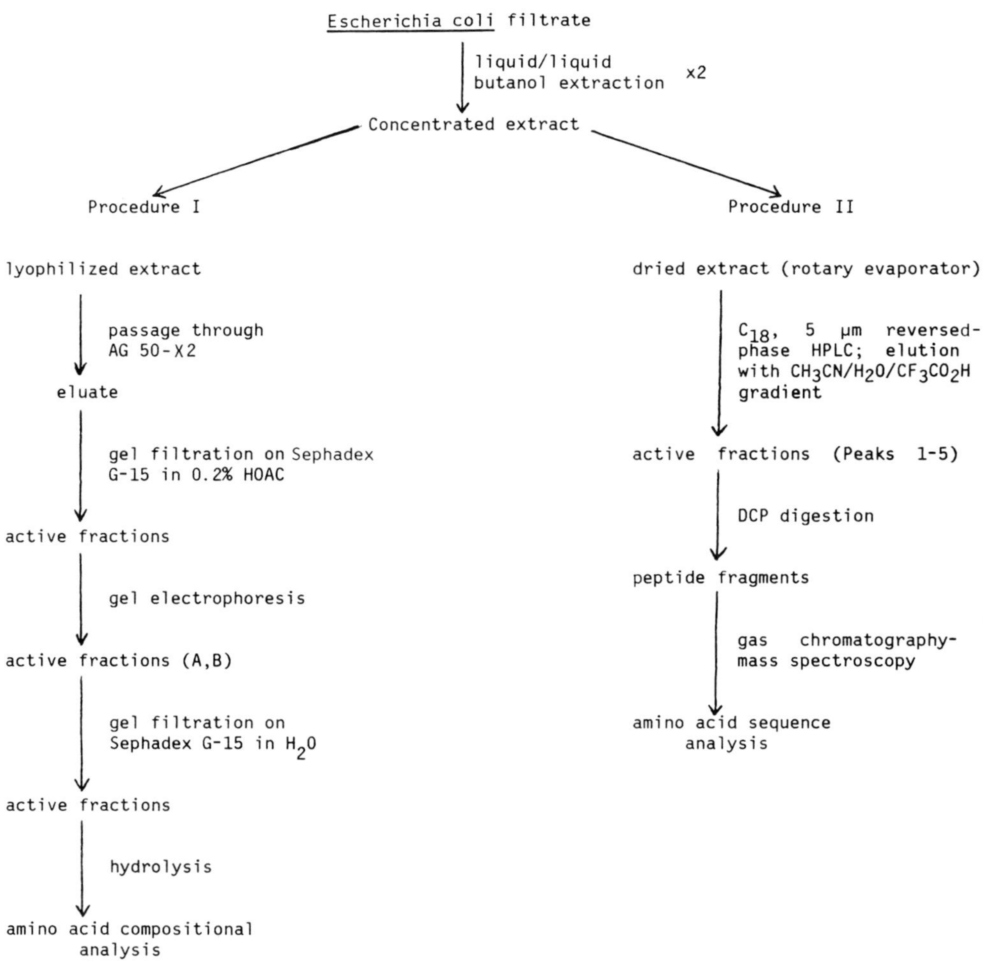

FIG. 1. Flow chart of purification procedures for *Escherichia coli* filtrates. Procedure I from Schiffmann *et al.*[3] Procedure II from Marasco *et al.*[16]

rotary evaporator, resuspended in HPLC buffer (0.1% CF_3CO_2H in H_2O) and applied to a reversed-phase C_{18} 5-μm particle size column. After gradient elution in the same buffer containing between 20 and 45% CH_3CN, individual fractions are collected, dried by rotary evaporation, and either resuspended in water for analysis of biological and immunogenic activity or subjected to DCP/GC–MS methodology for peptide sequence analysis.

Preparation of Butanol Extracts from Bacterial Culture Filtrates

As a general rule, butanol extracts of bacterial culture filtrates are prepared from either aqueous preparations or lyophilized material[3,7]; however, other investigators report the use of butanol–glacial acetic acid (10:1 v/v) mixed 1:2 with filtrate,[3] 80% ethanol extracts,[7] and filtrates that have not been extracted[8] as starting material. For our studies, 24-hr culture filtrates (250 liters) of *E. coli* WRAIR 1485 (obtained from Walter Reed Army Institute of Research) grown in minimal media (Difco Laboratories, Detroit, MI) are used as starting material. To optimize the recovery of chemotactic material, culture filtrates are harvested 24–48 hr after incubation at 37°[1] when the cells approach the end of their log growth phase, with maximal release of chemotactic activity into the growth medium. After centrifugation at 5000 *g* for 15 min at 4° to remove the cell pellet and filtration through a Millipore filter (0.45 μm, Millipore Corp.), the resulting supernatants are extracted twice with water-saturated butanol in a liquid extractor. The extracts (49 liters of the first extract and 39 liters of the second extract are combined and concentrated to 1 liter by lyophilization or rotary evaporation.

Procedure I: Ion-Exchange and Molecular Sieve Chromatography

The butanol is removed from the concentrated extract on a rotary evaporator. The resulting residue is dried *in vacuo* over P_2O_5, extracted with distilled water in the proportion of 2 ml of H_2O to 10 ml of the concentrated butanol extract. The water extract is centrifuged at 45,000 *g* to remove the insoluble residue prior to ion-exchange chromatography. The water-soluble supernatant is next applied to a column (0.9 × 120 cm) of AG 50-X8 (Bio-Rad, Richmond, CA), 100–200 mesh, H^+ form, exchange resin that is equilibrated in H_2O. The eluate is monitored at 260 nm and assayed for chemotactic activity using the modified Boyden chamber technique.[17] The chemotactic activity which in our studies elutes between 45 and 95 ml of H_2O is lyophilized to dryness, reconstituted in

[17] M. J. Swanson, *J. Immunol. Methods* **116,** 385 (1977).

1–2 ml of 0.2% acetic acid, applied to a column of Sephadex G-15 (1.5 × 180 cm), and eluted with 0.2% acetic acid. Two peaks of chemotactic activity elute between 180–220 ml and 276–342 ml. These peaks are kept separate and lyophilized.

Purification of Sephadex Fractions. Polyacrylamide gel electrophoresis. Polyacrylamide electrophoresis is performed on chemotactically active fractions isolated from Sephadex G-15 in acetic acid. The 7% polyacrylamide gels are polymerized in tubes (1.25 × 14 cm) with no stacking gel. A continuous buffer system (8.3 m*M* Tris, 30 m*M* barbital at pH 7.5) is employed. The sample is added in 0.5 ml containing 0.1 ml of 40% sucrose, and electrophoresis performed at 10–12 mA/gel for 4 hr at 4°. The gels are removed and cut into 3-mm sections which are next extracted with 1 ml of 0.15 *M* NaCl overnight at 4° and then assayed for chemotactic activity (as above). Samples with activity are pooled and lyophilized.

The salts and impurities accompanying the electrophoretic fractions are removed by gel filtration in water on a column of Sephadex G-15 (1.5 × 90 cm). Fractions containing chemotactic activity but no detectable absorption at 210 nm elute before the salt front as determined by measuring conductivity. These components are pooled and lyophilized. For amino acid sequence analysis, the dry active fractions are hydrolyzed in a sealed tube under nitrogen with distilled 6 *N* HCl for 36 hr. After removal of the acid, the samples are applied to a Beckman analyzer.[3]

To our knowledge, only two other groups have been involved in the purification of chemotactically active peptides from bacteria, the flow charts of the procedures used being shown in Fig. 2. The first group, headed by Dr. Elliott Schiffmann (NIH), obtained partial purification of *E. coli* (ATCC-0127) chemotactic peptides (Fig. 2, left-hand column). A highly purified, but not completely characterized, fraction was found to contain aspartic acid, serine, glutamic acid, alanine, and glycine. This compositional analysis was the same as we found for material purified by procedure I (Fig. 1). Figure 2 (right-hand column) also shows the procedure used by Miyake *et al.*[8] to obtain structural analysis of two NH_2-terminally blocked peptides from *Streptococcus sanguis* (ATCC-10556). These investigators were able to obtain compositional analysis of both peptides (one peptide contained equimolar amounts of methionine, isoleucine, leucine, and phenylalanine while the other peptide contained proline, valine, methionine, isoleucine, and leucine); however, sequence analysis was not performed.

Procedure II: HPLC of Partially Purified Escherichia coli Culture Filtrates

We established a liquid chromatography system (0.1% CF_3CO_2H in H_2O with gradient elution using CH_3CN in the same buffer) which results

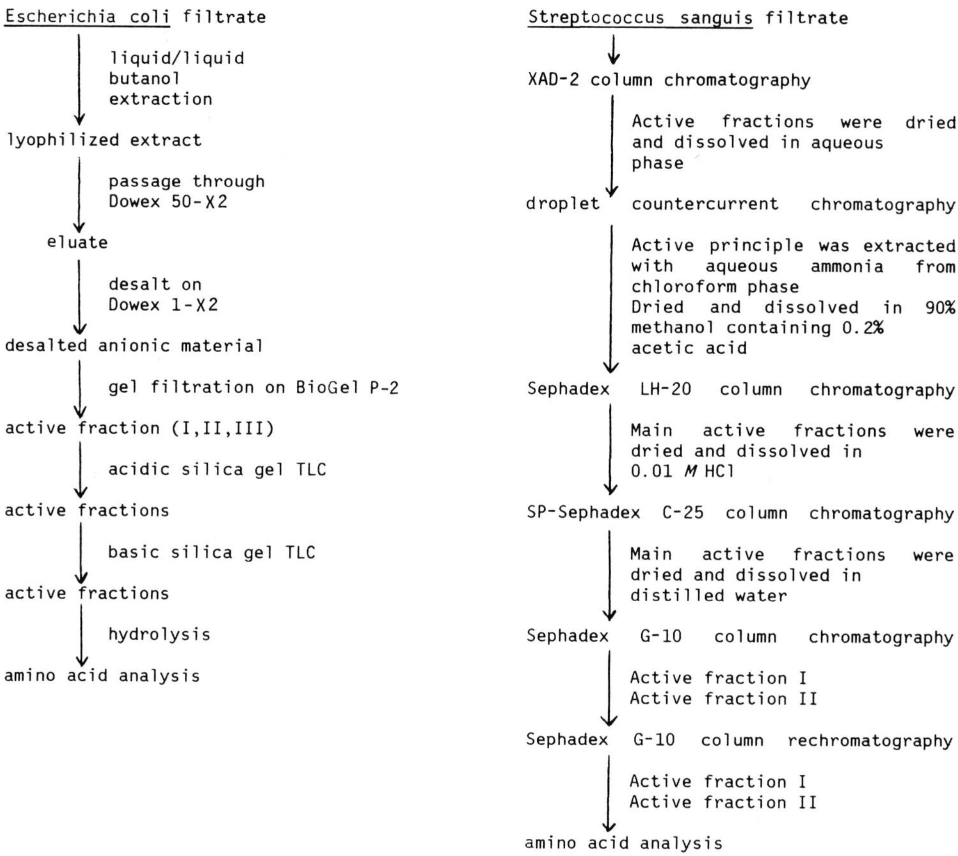

FIG. 2. Flow chart of purification procedures for Sephadex fractions of *Escherichia coli*[3] and *Streptococcus sanguis*[8] filtrates.

in marked differences in retention times for several synthetic formylated tri- and tetrapeptides, all of which elute at CH_3CN concentrations higher than 25%. One to three milliliters of partially purified bacterial chemotactic factor (crude butanol extracts) is dried in a Speed Vac (Savant Instruments) rotary evaporator and resuspended in the HPLC buffer (0.1% CF_3CO_2H in H_2O).[16] The liquid chromatography system consists of a Varian LC-5020 and Varichrom variable wavelength detector. The sample is run on a reversed-phase, C_{18}, 10-μm (MCH 10) or 5-μm (MCH5) particle size column (Varian Instruments). Chromatography is performed under the following conditions: flow rate 1.0 ml/min, temperature 30°, gradient elution from 20 to 45% CH_3CN in H_2O containing 0.1% CF_3CO_2H throughout, over 60 min. Fractions are collected every 1 or 2 min, thoroughly dried in a rotary evaporator, resuspended to 0.5 ml in

H_2O, and assayed for biological activity. Spectrophotometric detection is at 210 nm. Under these conditions, column pressure should never exceed 250 atm. Figure 3 shows a typical ultraviolet absorption profile of the partially purified bacterial chemotactic factor demonstrating numerous peaks throughout the entire 20–45% CH_3CN elution gradient. Most of the material that elutes early gives the greatest absorbance at 210 nm. However, several small peaks are seen at CH_3CN concentrations higher than 25%, which is the region where the more hydrophobic and biologically active formyl peptides elute (see below and Fig. 4).

Analysis of Recovery of Biological Activity from Individual HPLC Fractions. For assessing biological activity, aliquots from each fraction recovered from the HPLC step are tested for their ability to induce lysosomal enzyme secretion from rabbit neutrophils. Rabbit neutrophils are harvested 18 hr after intraperitoneal injection of 400 ml 0.1% oyster glycogen (Sigma Chemical Co., St. Louis, MO) in saline.[14] This is accomplished using a 16-gauge needle inserted into the peritoneal cavity while the rabbits are strapped to a restrainer board and inverted. Lysosomal enzyme release from rabbit peritoneal neutrophils is performed with 5×10^6 cells/ml in

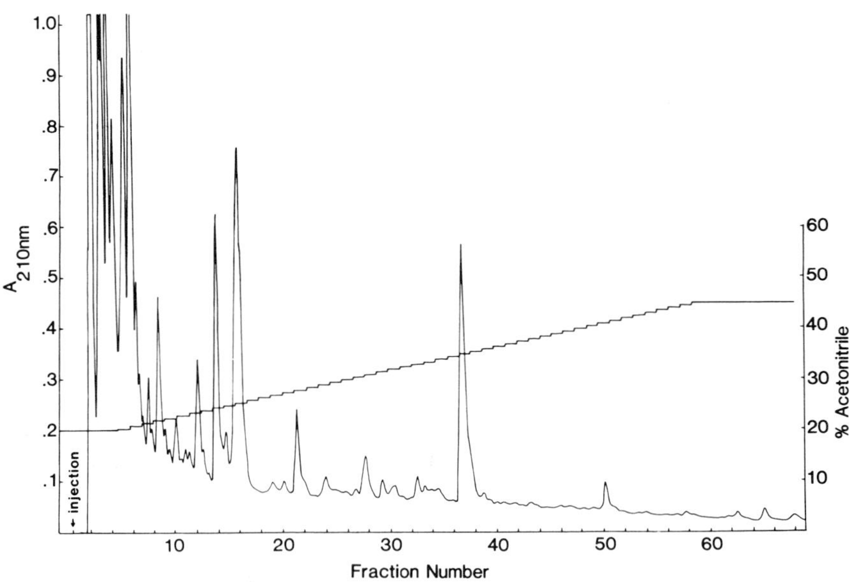

FIG. 3. HPLC profile of partially purified bacterial chemotactic factor. Five hundred microliters of 3-fold concentrated material (1.5 ml original) was applied to the column (MCH5; 5-μm particle size) under the buffer and gradient conditions described. Chart speed, 0.25 cm/min. The run was monitored at 210 nm and at an attenuation of 1028. Fractions were collected every 2 min. Percent acetonitrile (⌐).

Hanks' buffer plus 1.6 mM $CaCl_2$ and 1 mg/ml dextrose in the presence of 5 μg/ml cytochalasin B. The reaction is terminated after 5 min at 37° by immersing the tubes in an ice bath followed by centrifugation.

N-Acetyl-β-D-glucosaminidase is assayed in the supernatant by measuring the release of *p*-nitrophenol from its substrate *p*-nitrophenyl-*N*-acetyl-β-D-glucosaminide.[18] For this assay 200 μl of supernatant is added to 400 μl of a 1.4 mg/ml solution of *p*-nitrophenyl-*N*-acetyl-D-glucosaminidase in 50 mM sodium citrate buffer, pH 4.5. Supernatants of cells treated with 0.1% Triton X-100 are used to determine total enzyme release. After incubation for 15–30 min, the reactions are quenched by the addition of 500 μl 0.4 M glycine and analyzed colorimetrically for the release of *p*-nitrophenol with a spectrophotometer at 410 nm. Various dilutions (1 : 100 to 1 : 10,000 final) of each HPLC fraction are assayed so as to quantitate completely the amount of biological activity in each fraction. Dilutions of stock bacterial chemotactic factor (SBF, i.e., crude butanol extract), concentrated bacterial chemotactic factor (dried and resuspended in HPLC buffer), and fMet-Leu-Phe (Sigma) [1×10^{-8} to $1 \times 10^{-11} M$, prepared from a frozen stock at $1 \times 10^{-2} M$ in dry dimethyl sulfoxide (DMSO)] are assayed along with the HPLC fractions to allow the generation of standard curves for quantification of recovery of biological activity, and for conversion of recovery into fMLP mole equivalents.

To quantitate recovery, an exponential curve is fitted to the SBF data using a Sharp PC-1500 microcomputer with a modified, exponential curve-fitting program.[19] Lysosomal enzyme release data (absorbance units or percent specific release) are then computed to estimate SBF values (microliters) using the formula $y = ab^x$, where y is microliters of SBF; x, absorbance units or percent specific enzyme release; and a and b are constants for a given experiment. The estimated microliters of SBF equivalents recovered in each HPLC fraction are summed for all fractions recovered, corrected for deviation of the fitted curve from (0,0), and expressed as percent SBF recovered. In our studies, five peaks of activity are obtained using the 5-μm particle size column. Percent of recovered SBF and of original SBF is calculated for each peak.

In order to determine the fMLP mole equivalents of the HPLC fractionated material, the fMLP dose–response data (percent specific enzyme release versus molarity) is fitted to an exponential curve using the same modified Sharp program. This program is used to generate fMLP mole equivalents from the percent specific enzyme release data for the fractions of recovered bacterial chemotactic factor. These values from each HPLC

[18] W. A. Marasco, J. C. Fantone, R. J. Freer, and P. A. Ward, *Am. J. Pathol.* **111,** 273 (1983).

[19] Sharp Electronics Corp., Pocket Computer Applications Manual, PC-1500.

fraction are then summed and multiplied by the percent recovered SBF for each of the peaks.

Figure 4 demonstrates the results we have obtained using the HPLC-separated butanol extracts. Analysis of activity in individual HPLC fractions (at a 1 : 100 final dilution) for their ability to induce lysosomal enzyme release from rabbit neutrophils reveals an average percent recovery (of applied sample) of 50.18 ± 4.55% (SEM), with a range of 37.7–66.3% ($n = 7$). Gradient elution from the column resolves the starting material into five major peaks of activity with retention times of 27, 33, 45, 55, and 69 min, respectively (Fig. 3). The majority of activity appears in peak 3 (Fig. 4, fractions 21–26); this is the only sample in which activity is still present at a 1 : 10,000 dilution. At 1 : 100 and 1 : 1000 dilutions, peak 3 contains 55 and 69% of the total recovered activity, respectively, whereas peak 4 (fractions 27–31) contains 32 and 17%, respectively. Peak 1 (fractions 13–15), peak 2 (fractions 16–18), and peak 5 (fractions 33–37) contain activity only at a 1 : 100 dilution, with each peak representing

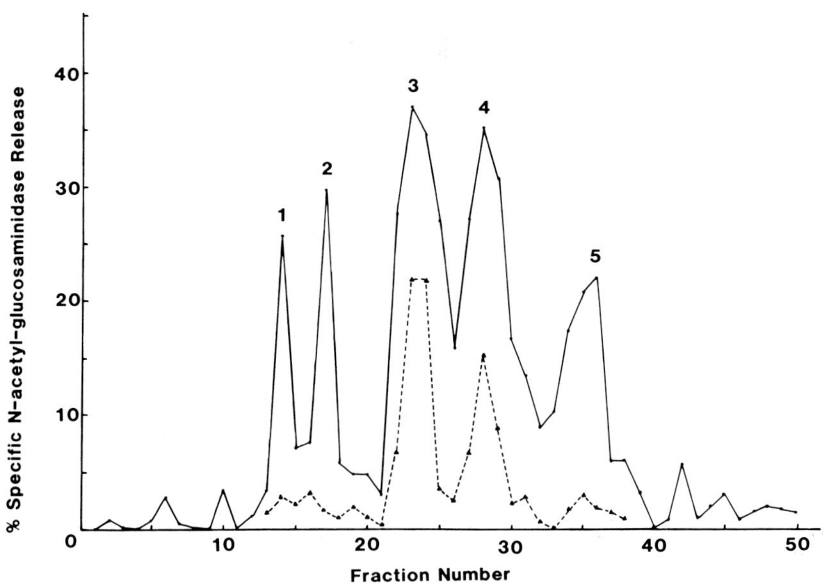

FIG. 4. Biological activity of individual HPLC fractions from the purification shown in Fig. 3. Fractions were collected every 2 min, thoroughly dried in a rotary evaporator, and resuspended to 0.5 ml in H_2O. The release of *N*-acetyl-β-D-glucosaminidase from rabbit neutrophils is used to assay for biological activity. Each fraction was run at two dilutions (1 : 100, solid curve, and 1 : 1,000, dashed curve, final) to ensure that the activity was representing a portion of the dose–response curve for the unknown peptide.

5–7% of the total recovered activity. The broad nature of these peaks is due to the extremely high specific activity of this material.[12,13] Control experiments in which the activity profile of fMet-Leu-Phe has been assessed in individual fractions after being applied to the column has provided a similarly broad peak. Authentic fMet-Leu-Phe, when subsequently chromatographed on the same column, appears as a peak at 45 min, corresponding exactly to the retention time of 45 min for peak 3.

RIA Analysis of Recovery of Antigenic Activity from Individual HPLC Fractions. For analysis of the antigenic character of the chemotactic peptides, a radioimmunoassay (RIA) is used. The peptide–protein conjugate used for the production of this antisera is prepared as follows[14]: fMet-Leu-Phe (1.04 mg) dissolved in 50 μl DMSO is added in 10-μl aliquots with immediate vortexing as a 20-molar excess to 4 ml solution of 2 mg/ml BSA (essentially fatty acid free, Sigma) in H_2O to which was added 50–150 pmol of fMet-Leu-[^{3}H]Phe (47.6 Ci/mol, New England Nuclear). One milliliter of a 500 mg/ml solution of the coupling reagent 1-ethyl-3-(3-dimethylaminopropyl)carbodiimide—HCl (Sigma) is added next. The reaction mixture is maintained at 24° and at neutral pH during the first 2 hr of the 18–24 hr by addition of 0.1 *N* NaOH as required. The reaction is then dialyzed for 24 hr against 0.8% NaCl, and covalent coupling is determined by 10% trichloroacetic acid (TCA) precipitability. Typically 8–10 mol fMet-Leu-Phe are covalently bound per mole of bovine serum albumin (BSA) with >95% TCA precipitability. For the production of the antisera, adult New Zealand rabbits are immunized with approximately 20 intradermal injections of antigen (1 ml of 500 μg total protein) along the back emulsified in an equal volume of Freund's complete adjuvant (Difco). The animals are injected in the same way with antigen in Freund's incomplete adjuvant twice biweekly. Starting 1 week after the last injection, rabbits are bled weekly for 4 weeks.

The partially purified butanol extract from *E. coli* can compete with fMet-Leu-[^{3}H]Phe for binding to anti-fMet-Leu-Phe produced in a rabbit against $fMLP_{10}$–BSA. In addition, anti-fMet-Leu-Phe bound to protein A–Sepharose can remove the biologically active components of the chemotactic factor-enriched crude butanol extract.[14] Furthermore, since the *N*-formyl group is mandatory for maximal antibody binding activity[14] (i.e., equivalent nonformylated parent peptides are more than 10,000-fold less active), this reagent is highly specific for formylmethionyl peptides. In addition, structurally related formylmethionyl peptides, with subtle changes in their amino acid composition and relative position, bind to the antibody with varying degrees of affinity. We therefore use this radioimmunoassay to monitor the separation and to quantitate the antigenic recovery of formylmethionyl peptides in the individual HPLC fractions.

To analyze by RIA the ability of the individual HPLC fractions to compete with fMet-Leu-[^{3}H]Phe (47.6 Ci/mmol) for binding to rabbit anti-fMet-Leu-Phe, 50-μl aliquots of each HPLC fraction (dried and resuspended in H_2O), 50 μl containing 88 fmol fMet-Leu-[^{3}H]Phe in phosphate-buffered saline (PBS), and 50 μl of an appropriate dilution of rabbit anti-fMet-Leu-Phe (which bound >95% of the radiolabeled peptide at its greatest dilution) are incubated at 24° for 30 min in 12 × 75 mm polystyrene tubes and then analyzed for competitive binding.[9] After this incubation, 100 μl of saturated ammonium sulfate is added to each tube and the mixtures are incubated for an additional 1 hr at 4°. The precipitates are pelleted by centrifugation, 25 μl of the resulting supernatant is removed, and the radioactivity is determined in a liquid scintillation counter. To quantify recovery of SBF using the RIA data, inhibition of fMet-Leu-[^{3}H]Phe binding to rabbit anti-fMet-Leu-Phe by SBF is plotted versus microliters SBF in a given dilution. Linear regression analysis of the data yields a standard curve. Using this best-fit line, percent inhibition is converted to microliters SBF for individual fractions. These data are corrected for deviation of the best-fit line from (0,0), summed, and converted to percent recovery of original material. The percentage of recovered material in each peak is then calculated.

The fMLP mole equivalents are calculated by fitting percent inhibition of fMet-Leu-[^{3}H]Phe binding to rabbit anti-fMet-Leu-Phe versus fMet-Leu-Phe concentration (molar) data to a curve generated by the modified, exponential curve-fitting program in the Sharp PC-1500 microcomputer. The computer estimates fMLP equivalents (molar) from percent inhibition data from the individual fractions, and these molar quantities are then converted to moles.

The analysis of antigenic activity in the same individual HPLC fractions (above) provides a pattern of reactivity shown in Fig. 5 with an average percent recovery of original applied material of 69.2 ± 25.1% (SEM), with a range 38–100% ($n = 3$). This is in good agreement with the recovery of enzyme releasing activity (Fig. 4). Several distinct peaks of activity appear which overlap with the peaks of biological activity seen in Fig. 4. The largest antigenic peak (fraction 23), containing 66% of the recovered antigenic activity, corresponds to bioassay peak 3 which is also where the majority of the biological activity is found. A summary of the overall recovery of activity in both assays expressed as fMLP mole equivalents is shown in Table I. Good agreement in the overall recovered activity units by both assays is evident with less than 15% difference between the two methods of measuring recovery. For peaks 1, 3, and 5, good overall agreement is seen, with peaks 3 and 5 showing less than 15% difference and peak 1 showing a 3-fold difference in activity. The major differences,

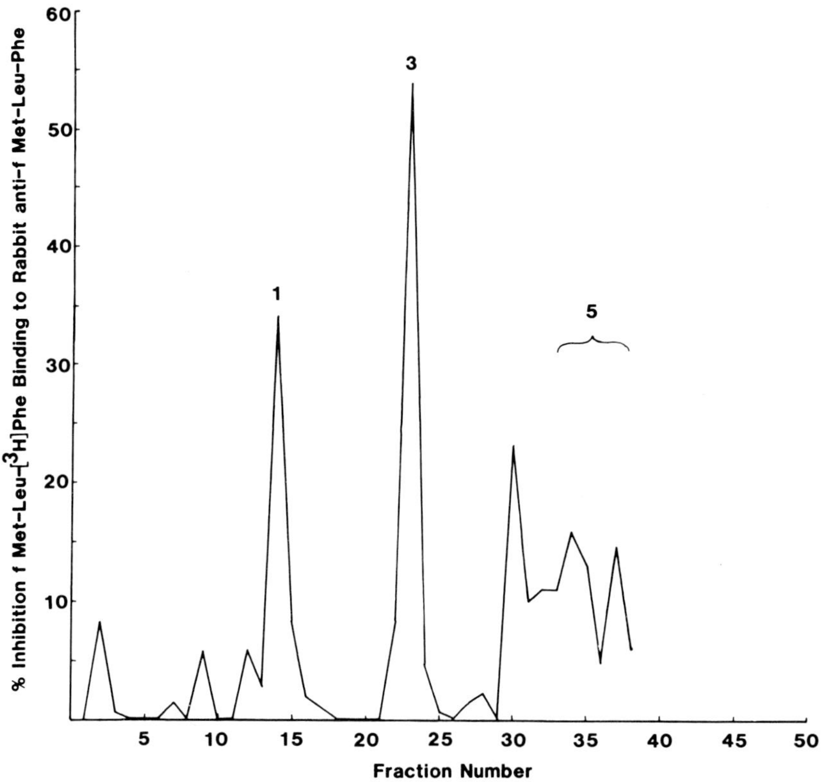

FIG. 5. Antigenic activity of individual HPLC fractions from the purification shown in Fig. 3. To assay for antigenic activity, 50 μl from each HPLC fraction is analyzed for its ability to compete with 88 fmol of fMet-Leu-[^{3}H]Phe for binding to anit-fMet-Leu-Phe. Peaks 1, 3, and 5 correspond to the same peaks seen in Fig. 4.

however, are seen in peaks 2 and 4 where a respective 45- and 8-fold greater enzyme releasing activity versus antigenic activity is seen (discussed below). Based on these data and correcting for an overall 50% recovery of original activity, the fMLP concentration of the partially purified filtrate (starting material) is approximately 2.76×10^{-7} *M*. This value is in good agreement with the ED_{50} value (the dose of agonist which gives 50% of the maximal specific releasable enzyme) in the enzyme release assay which is around 1 : 2,500 to 1 : 5000. Furthermore, since the starting material is the extract of 250 liters of culture filtrate concentrated to 1 liter, the concentration of the formyl peptide in the original culture filtrate is approximately 1.1×10^{-9} *M*. Approximately 60% of this starting material (peak 3) appears to be naturally occurring fMet-Leu-Phe (see below).

TABLE I

FRACTIONATION OF PARTIALLY PURIFIED *E. coli* CULTURE FILTRATE BY HPLC[a]

Peak	Fraction numbers	Calculated recovery of fMLP (mol equiv)[b]		Activity ratio,[e] RIA/ER
		Enzyme release (ER) assay[c] (mol)	Radioimmunoassay[d] (mol)	
1	13–15	7.61×10^{-12}	2.39×10^{-11}	3.12
2	16–18	1.28×10^{-11}	2.82×10^{-13}	0.022
3	21–26	1.18×10^{-10}	1.47×10^{-10}	1.25
4	27–31	7.58×10^{-11}	9.7×10^{-12}	0.128
5	33–37	1.05×10^{-11}	1.16×10^{-11}	1.11
Total	—	2.24×10^{-10}	1.92×10^{-10}	0.86

[a] From Marasco *et al.*[16]

[b] For quantification of recovery of biological and antigenic activity and for conversion of quantified recovery into fMLP mol equiv, see appropriate methods section.

[c] Analysis of lysosomal enzyme releasing activity in individual HPLC fractions was determined at a 1 : 100 final dilution.

[d] Analysis of competition with fMet-Leu-[^{3}H]Phe by individual HPLC fractions for binding to rabbit anti-fMet-Leu-Phe. Individual fractions were assayed at a 1 : 3 final dilution.

[e] Activity ratio is the calculated ratio of recovered fMLP mol equiv determined by radioimmunoassay divided by fMLP mol equiv obtained by the enzyme release assay (RIA/ER).

DCP/GC–MS Analysis for fMLP. Analysis of HPLC fractions for the presence or absence of fMLP is carried out using the DCP/GC–MS methodology for peptide sequence analysis.[16,20] The DCP digestion step is required because fMLP itself is too polar to be rendered into a volatile form that would be suitable for GC–MS analysis. Conditions are used that give the best possible conversion of fMLP into fMet and Leu-Phe with authentic fMLP (Sigma) as the test substrate. Angiotensin-converting enzyme (ACE) (EC 3.4.15.1, dipeptidyl carboxypeptidase I) (Sigma), in acetone is used as the DCP enzyme.[21] The following procedure minimizes losses of Leu-Phe from residual dipeptidase activity, for which Leu-Phe is an excellent substrate. The digestion time is shortened to 2 hr, the ACE used is rechromatographed on an Affi-Gel 202 (Bio-Rad) column eluted with 5 m*M* *N*-methylmorpholine acetate buffer, pH 7.0, to remove as much residual dipeptidase activity as possible, and 0.005 units of this ACE

[20] H. C. Krutzsch, this series, Vol. 91, p. 511.

[21] M. L. Yarmush, H. C. Krutzsch, and J. J. Kindt, *Mol. Immunol.* **17,** 319 (1980).

preparation is used. One unit of ACE activity is defined as that amount of enzyme which digests 1 μmol of substrate/min. The substrate used to define units of activity is 10^{-4} *M* furanacryloyl-L-phenylalanylglycylglycine (Vega Biochemical), and activity is measured by observing the loss of absorbance at 328 nm as a function of time.[20]

These experiments have the added benefit of providing authentic mass spectra and gas chromatographic elution temperatures for fMet and Leu-Phe by comparison to results from analyses carried out with unknown samples. A similar set of analyses has been previously carried out on another peptide using the complementary DCP/GS–MS approach.[21] Once digestion of fMLP is accomplished, the products are identified by HPLC. These procedures have been described in detail in this series (Vol. 91 [46] and Vol. 47 [39]) and will not be duplicated here.

The HPLC fractions are analyzed as follows. To ensure removal of residual volatile HPLC solvents, the dried HPLC fractions (in a 100-μl Reacti-Vial, Pierce Chemical) are first extracted with 100 μl of methanol (Burdick and Jackson), which is then evaporated under a stream of nitrogen. The dried fractions are then extracted with 50 μl of distilled water, which is added to the same Reacti-Vial as before and freeze-dried. The residue in the Reacti-Vial is then incubated for 2 hr at 37° with an appropriate aliquot of the purified ACE (see above) and then freeze-dried. The freeze-dried digests are then analyzed for the presence of fMet and Leu-Phe by GC–MS, following trimethylsilylation, as previously described.[20] Briefly, 10 μl each of acetonitrile and bis(trimethylsilyl)trifluoroacetamide (Supelco) are added to the dried digestion mixture, and the vials are sealed with a Teflon-lined cap and heated at 140° for 10 min. After cooling, 5 μl of this solution is injected into a 25-m SE-54 fused silica capillary column (Hewlett-Packard) using splitless injection. The flash heater is set at 280°, and the column temperature is programmed at 10°/min from 110° to 270°. The GC–MS used in our experiments is an LKB Model 2091 with the trap current set at 50 μA.

The fraction with the most activity within each of the bioactive HPLC peaks is chosen to be further tested for the presence of fMLP using DCP/GC–MS analysis. Individual peak fractions are used since detection at 210 nm usually fails to demonstrate a peak of activity, suggesting that the peaks contain less than 1 nmol of fMet-Leu-Phe (as compared to calibration using pure standards). The GC–MS data obtained from each fraction are then compared to authentic fMLP. The total ion chromatograms for the various samples and authentic fMLP are then compared. The total ion chromatograms also furnish GC elution temperatures for fMet and Leu-Phe, from DCP-digested fMLP, and thus allow comparison to the elution temperatures for putative fMet and Leu-Phe from the DCP-treated fractions. GC elution temperatures also confirm the identity of the compounds as obtained from their mass spectra. Furthermore, the identity of the labeled peaks in the chromatograms is subsequently determined from their mass spectra.

The computer-generated total ion and selected ion plots from two GC–MS runs (authentic fMLP and a peak fraction) are first compared. The selected ion plot allows one to select materials exhibiting a particular ion in their mass spectra, thus simplifying the search of the computer stored data for a given possible mass spectrum. The total ion plots allow reference to the original total ion chromatograms for such purposes as determining the GC elution temperature of a particular peak. The selected ion mass of 247 is used to search for TMS-fMet, and the ion mass of 158 is used to search for TMS-Leu-Phe. The ion mass of 247 results from loss of $CH_3SCH{=}CH_2$ from the molecular ion of TMS-fMet, and the ion mass of 158 corresponds to the ion $[TMS\text{-}NH{=}CHCH_2CH(CH_3)_2]^+$. This technique is also employed to search for other dipeptides. One peak from each of the two selected ion traces from the GC–MS analysis of DCP treated fractions is compared with the correct GC retention temperatures for fMet and Leu-Phe.

The mass spectra of TMS-fMet and TMS-Leu-Phe as obtained from GC–MS analysis of either DCP-treated fMLP or the HPLC peak fractions are next analyzed. Because of sample contamination by siloxanes from HPLC column bleed, the mass spectra of TMS-derivatized, DCP-treated HPLC fractions may have some extraneous peaks. However, by reference through the computer-generated selected ion plot and total ion plot to the total ion chromatogram, it can be determined if the materials exhibiting these correct mass spectra also have GC elution temperatures identical to those of authentic fMet and Leu-Phe, respectively. In our published studies, this occurred only for fraction 23 of peak 3.[16]

Concluding Remarks

The simultaneous analysis of both biological and antigenic activities in the bacterial culture filtrates are invaluable steps in the purification of the chemotactic factors, especially the formylmethionyl peptides. Since multiple structurally diverse biological substances are known to act as chemotactic factors for leukocytes (reviewed in Ref. 22), analysis of antigenic activity quickly allows the establishment of the class of chemoattractant.

The DCP/GC–MS methodology combined with separation and purifi-

[22] E. L. Becker and W. A. Marasco, *In* "Symposium on Chemical Mediators of Inflammation" (S. Cohen, H. Hayashi, K. Saito, and A. Takada, eds.), p. 220. Academic Press, Orlando, Florida, 1985).

cation of the bacterial chemotactic peptides by HPLC is thus far the only technique which has unequivocally resulted in determination of both the complete amino acid sequence of a bacterial chemotactic peptide as well as the presence of the formyl group. Because of the presence of artifactual contaminants that accumulate during isolation of fMLP such as siloxanes from HPLC column bleed, other mass spectral methodologies that use direct rather than GC introduction into the mass spectrometer to analyze for this compound, such as fast-atom bombardment MS, are not as usable, because the ions identifying fMLP would be obscured by ions arising from these contaminants. Also, as we can deduce from the peak heights in the total ion chromatogram in our studies, the amount of fMLP present was in the range of only about 5 pmol, an amount that is too low to be analyzed by other MS approaches. Unfortunately, application of greater amounts of starting material (crude butanol extract) to the HPLC system is not possible due to insolubility of these hydrophobic peptides when resuspended in HPLC buffer.

Despite the success of this approach to identify one of the bacterial peptides, equal success in obtaining amino acid sequence analysis of components of four of the five peaks of activity using the DCP/GC–MS methodology has not been forthcoming and may represent the limitations of the DCP/GC–MS technique. For example, the lack of a detectable formylmethionyl peptide in two HPLC peaks using the DCP/GC–MS methodology, despite the presence of substantial antigenic activity and UV-detectable material, could arise from several possibilities. One is that these materials are peptides with an even number of amino acids that upon DCP digestion would yield an fMet-X dipeptide, which, depending on X, might not elute from the GC. A second possibility is that these materials cannot be degraded by DCP because of a blocked carboxy terminus or an unfavorable confirmation. Another possibility is that these materials may be present in amounts below the limits of detection using this methodology.

The inability to detect antigenic activity in two other HPLC peaks despite strong enzyme releasing activity in these same fractions is most likely explained with the hypothesis that these activities may represent very active formyl peptides containing at least four amino acids. This hypothesis is based on the fact that addition of a hydrophobic fourth amino acid to the tripeptide fMet-Leu-Phe (the most active tripeptide synthesized thus far[12,23,24]) will increase biological activity 20- to 100-fold. Thus, substantial

[23] R. J. Freer, A. R. Day, E. Schiffmann, S. Aswanikumar, H. J. Showell, and E. L. Becker, *Biochemistry* **19,** 2404 (1980).

[24] R. J. Freer, A. R. Day, N. Muthukumaraswamy, D. Pinon, A. Wu, H. J. Showell, and E. L. Becker, *Biochemistry* **21,** 257 (1982).

biological activity may be detected in spite of a negative RIA (i.e., anti-fMet-Leu-Phe has minimal recognition of any amino acid beyond position 3[14]). This same hypothesis would explain the inability to detect any formylmethionyl peptides by DCP/GC–MS. It is also possible that these activities are associated with peptides structurally unrelated to the formyl peptides.

Acknowledgments

Along with those cited in the reference list, we would specifically like to acknowledge the contributions of the following people to the development of the methods we have outlined here: Henry Showell, Henry Krutzsch, Elliott Schiffmann, and Elmer Becker. We would also like to acknowledge the expert secretarial and word processing capabilities of Jeny Brown who prepared the manuscript. Financial support of W.A.M. and P.A.W. by National Institutes of Health Grant HL28422 and of W.A.M. by the Biomedical Research Support Grant of University of Michigan Medical School are gratefully acknowledged.

[20] Chemotactic Fragments of Fibronectin

By NORMAN E. WIKNER and RICHARD A. F. CLARK

Introduction

In previous studies performed in this laboratory it has been shown that intact purified fibronectin (Fn)[1] has no chemotactic activity for peripheral blood mononuclear cells (PBMC), while preparations containing proteolytic fragments of Fn are chemotactic for PBMC.[1a] Recently we have identified this activity as residing with a 120-kDa fragment produced during digestion of Fn by thermolysin.[2] Limited thermolysin digestion cleaves Fn into several major fragments, including 30-kDa fibrin, 45-kDa collagen, and 150- and 160-kDa fragments. The latter two contain cell- and heparin-binding domains. Further digestion of this mixture, or digestion of purified 150- and 160-kDa fragments, yields 30- and 40-kDa heparin-binding fragments and a non-heparin-binding 120-kDa fragment. This frag-

[1] Abbreviations: Fn, fibronectin; PBMC, peripheral blood mononuclear cells; PMSF, phenylmethylsulfonyl fluoride; CPD, citrate–phosphate–dextrose buffer; PBS, phosphate-buffered saline; EACA, ϵ-aminocaproic acid; SBTI, soybean trypsin inhibitor.

[1a] D. A. Norris, R. A. F. Clark, L. M. Swigart, J. C. Huff, W. L. Weston, and S. E. Howell, *J. Immunol.* **129,** 1612 (1982).

[2] R. A. F. Clark, N. E. Wikner, D. A. Norris, and S. E. Howell, *J. Cell Biol.* **101,** 217a (1985).

ment contains both the fibroblastic cell-binding site and chemotactic activity for monocytes. During the course of these studies we have developed a more rapid method for the isolation and purification of this 120-kDa fragment.

Fibronectin Preparation

The preparation of purified Fn fragments is facilitated by starting with a pure, undegraded Fn preparation. We use the method of Furie and Rifkin[3] modified to include the presence of phenylmethylsulfonyl fluoride (PMSF) in all buffers (with exceptions noted below).

Columns

Lysine–Sepharose: diameter 2.7 cm, length 12 cm; available from Pharmacia

Gelatin–Sepharose: diameter 2.7 cm, length 12 cm; prepared from CNBr-activated Sepharose (Pharmacia) according to manufacturer's instructions

DE-52: diameter 1.5 cm, length 17.5 cm (Whatman)

Reagents

Citrate–phosphate–dextrose (CPD) buffer: 89.6 m*M* sodium citrate, 142 m*M* dextrose, 11.1 m*M* citric acid, 14.2 m*M* monosodium phosphate

0.1 *M* EDTA, pH 7.4

PMSF 100×: 0.1 *M* PMSF in ethanol

Phosphate-Buffered Saline (PBS): 50 m*M* P_i, pH 7.4, 160, m*M* NaCl

PBS+: PBS with 0.02% sodium azide

PBS++: PBS+ with 1 m*M* PMSF

GS wash 1: 0.1 *M* Tris–HCl, pH 8.5, 0.5 *M* NaCl

GS wash 2: 0.1 *M* sodium acetate, pH 4.5, 0.5 *M* NaCl

GS eluent: 4 *M* urea in water (no PMSF)

LS wash: 50 m*M* P_i, pH 7.5, 0.5 *M* NaCl, 0.02% sodium azide, 1 m*M* PMSF

LS eluent: 0.2 *M* ϵ-aminocaproic acid (EACA), 1 m*M* PMSF in water

LS stripping buffer: 0.2 *M* EACA, 50 m*M* P_i, pH 7.5, 1 *M* NaCl

Fn DE-52 running buffer: 10 m*M* Tris–HCl, pH 8.1, 50 m*M* NaCl, 1 m*M* PMSF

Fn DE-52 elution buffer: 10 m*M* Tris–HCl, pH 8.1, 0.3 *M* NaCl, 1 m*M* PMSF

Fn DE-52 wash buffer: 10 m*M* Tris–HCl, pH 8.1, 0.5 *M* NaCl

[3] M. B. Furie and D. B. Rifkin, *J. Biol. Chem.* **255,** 3134 (1980).

Solids

Soybean trypsin inhibitor (SBTI)
Barium chloride
Ammonium sulfate

Miscellaneous

Single unit (~600 ml) blood bag
Large dialysis tubing (2 ml/cm, MW cutoff 10,000)
Polycarbonate fraction tubes (Fisher)

Procedure

Except where indicated, all procedures are performed at room temperature. To avoid losses of Fn due to precipitation and adherence to containers, the following measures should be taken: (1) The steps from the drawing of blood to loading and washing the gelatin–Sepharose column must be done in 1 day. (2) Turbulence during transfers must be carefully avoided. Where elution from a column is followed by dialysis, it is better to elute the peak directly into dialysis tubing suspended in the dialysis buffer than to collect the peak into fraction tubes. If this is not practical, polycarbonate fraction tubes should be used, and the distance drops will fall should be minimized. (3) All steps should be performed in plastic containers at room temperature, unless otherwise indicated.

The lysine–Sepharose column is stacked on top of the gelatin–Sepharose column. The two are washed with PBS++, 200 ml.

Sixty milliliters each of CPD buffer and 0.1 *M* EDTA, pH 7.4, are introduced into a standard 1 unit blood bag. Blood is then allowed to flow directly from the donor's arm into the bag until the bag is full. Blood thus obtained is decanted into centrifuge tubes and spun at 2000 *g* for 15 min (preferably in a swinging-bucket rotor). The resulting plasma supernatant is pooled in a 500-ml graduated beaker and the volume noted. SBTI, 0.1 mg/ml, is added and dissolved with gentle stirring. The beaker is placed in an ice bath on a magnetic stirrer. Vitamin K-dependent clotting factors are removed by adsorbtion onto barium citrate.[4] Solid barium chloride, 15 mg/ml, is added, and the mixture is stirred gently for 30 min. This is then centrifuged at 5000 *g* at 4° for 20 min. The supernatant is decanted into a 500-ml beaker. Removal of excess barium chloride is achieved by adding ammonium sulfate, 9.5 mg/ml. The mixture stirred gently for 30 min on ice and then again centrifuged at 5000 *g* at 4° for 20 min. The supernatant is decanted into a graduated plastic beaker and a one-1/100th volume of

[4] K. G. Mann, this series, Vol. 45, [13].

PMSF 100× is added. The resulting mixture is loaded onto the stacked lysine–Sepharose and gelatin–Sepharose columns. The lysine–Sepharose column absorbs plasminogen.[5] The gelatin–Sepharose column absorbs fibronectin.[6] When all of the plasma is loaded on, the columns are washed with PBS++. When all of the plasma has passed through the lysine–Sepharose column (as indicated by the absence of a yellow color) the columns are separated. The gelatin–Sepharose column is washed overnight with PBS++ (500 ml).

The lysine–Sepharose column is washed with LS wash (500 ml) then eluted with LS eluent (500 ml). The resulting peak is plasminogen. The lysine–Sepharose column is washed with PBS+ (500 ml) and stored in the cold. Every five uses the lysine–Sepharose column is stripped with LS stripping buffer and then PBS+ (500 ml each). The gelatin–Sepharose column is eluted with GS eluent (500 ml). If possible, the peak is collected directly into dialysis tubing (about 40 cm) suspended in Fn DE-52 running buffer. The gelatin–Sepharose column is washed with 500 ml each of GS wash 1, GS wash 2, and sterile PBS+ and stored in a cold room.

The Fn, eluted from the gelatin–Sepharose, is dialyzed against Fn DE-52 running buffer (1 liter times 4 changes, 2–4 hr per change). Care is taken to prevent the dialysis tubing from moving within the dialysis buffer. Movement of the tubing can result in precipitation of the Fn along the walls of the dialysis tubing. During dialysis the DE-52 column is washed with Fn DE-52 running buffer (200 ml). Then the dialyzed Fn is loaded onto the DE-52 column. The column is washed with another 200 ml of Fn DE-52 running buffer.

In order to separate the Fn from any gelatin-binding contaminants, the DE-52 column is eluted with a 50 m*M* to 0.3 *M* salt gradient. This is formed by adding 250 ml each of Fn DE-52 running buffer and Fn DE-52 elution buffer to opposite chambers of a double-cylinder gradient maker such that the lower salt concentration comes off first. The resulting gradient is used to elute the DE-52 column. Minor peaks are variably seen in the effluent and in the early part of the eluate. The pooled fractions from the major peak in the eluate contain pure Fn as judged by a single (doublet) band on silver-stained polyacrylamide gels. Following elution the DE-52 column is washed with 500 ml of Fn DE-52 wash buffer followed by 500 ml of Fn DE-52 running buffer and stored in a cold room.

The fibronectin concentration of the pooled peaks is about 0.2–0.5 mg/ml as judged by the optical density at 280 n*M* (extinction coefficient 1.26 ml/mg). The yield of Fn ranges from 0.15 to 0.3 mg/ml of starting

[5] D. G. Deutsch and E. T. Mertz, *Science* **170,** 1095 (1970).

[6] E. Engvall and E. Ruoslahti, *Int. J. Cancer* **20,** 1 (1977).

plasma, depending primarily on the donor. Fn thus obtained may be stored at −70°. For best results freeze no more than 50-ml aliquots and freeze quickly, preferably in a dry ice–alcohol bath. Fn should be used immediately or kept frozen until use. Fn stored at room temperature for 2 months in the absence of PMSF is degraded completely; the principal degradation product is a 120-kDa fragment.

Vacuum Dialysis

The vacuum dialysis method of concentrating protein solutions minimizes losses in our hands. Between 20 and 30 cm of narrow dialysis tubing (MW cutoff 10,000, 0.3 ml/cm) is tied off at one end. The other end is fixed with a rubber band to the distal 5 cm of a 5-ml disposable plastic pipet which has been inserted through a #8 rubber stopper. The latter is then placed in the neck of a 1-liter side-arm vacuum Erlenmeyer flask. The end of the plastic pipet protruding from the stopper is attached with a short piece of plastic tubing to a reservoir of suitable volume (usually a 60-ml plastic syringe with the plunger removed). Bench-top vacuum is applied to the side arm until the desired reduction in volume has been achieved. If the initial volume of solution is greater than the reservoir will hold, more solution can be added when the reservoir is half empty.

The disadvantage of this method is the large amount of time required (up to several days, if large volume reduction is needed). Thus all Fn solutions are concentrated in the presence of PMSF. The process can be speeded by using a longer length of dialysis tubing at the expense of greater losses due to deposition of protein on the sides of the tubing. If a large volume of solution is being concentrated, protein concentrations in the tubing can become much higher than those in the reservoir; this can cause precipitation of proteins within the tubing. Therefore it is advisable, every 8–12 hr, to release the vacuum and dilute the tubing contents back up into the reservoir by rolling the tubing on a plastic stirring rod then allowing it to refill.

Fibronectin Digestion

The following procedure was adapted from the method of Sekiguchi *et al.*[7]

Reagents

Digest buffer: 25 m*M* Tris–HCl, pH 7.6, 0.5 m*M* EDTA, 50 m*M* NaCl. Note: as large volumes of this buffer are needed for this and

[7] K. Sekiguchi, M. Fukuda, and S. Hakomori, *J. Biol. Chem.* **256,** 6452 (1981).

subsequent procedures, this solution is most conveniently prepared as a 10× solution.

0.1 *M* EDTA, pH 7.6

Calcium chloride, 1.0 *M*

Thermolysin (Sigma), reconstituted with water to a protein concentration of 1 mg/ml (approximately 80 units/ml)

A fibronectin solution comprising 50–90 mg of total fibronectin is vacuum dialyzed (at room temperature) to a concentration of 0.5–1 mg/ml. This is then dialyzed against digest buffer (1 liter times 4 changes, 2–4 hr per change at room temperature). The dialyzed Fn solution is transferred to a plastic screw-cap bottle. Digestion is begun by adding 1.0 *M* calcium chloride to a final concentration of 2.5 m*M* and thermolysin (1 mg/ml) to a final concentration of 2.5 μg/ml (approximately 0.2 units/ml). The digestion is allowed to proceed at room temperature for 8–30 hr, depending on which fragment(s) is of interest.

Fn is degraded very quickly to high-molecular-weight fragments of 150K and greater. The 120K fragment is produced far more slowly. Thus, to maximize the yield of this fragment, long digestion times (approximately 24 hr) should be used. We have found it useful to perform a trial digestion on a 1 mg sample of Fn, especially when a new lot of thermolysin is used. Aliquots for polyacrylamide gel electrophoresis are taken every 2–4 hr, and a silver-stained polyacrylamide gel of the aliquots examined before a full-scale digestion is performed.

Digestion is terminated by adding 0.1 *M* EDTA, pH 7.6, to the reaction mixture to a final concentration of EDTA of 5.0 m*M*. This reaction mixture may be used directly in the isolation of the individual fragments as described below.

Isolation and Purification of Fragments

An outline of this procedure is presented in Fig 1.

Columns

Gelatin–Sepharose: length 17 cm, diameter 1.5 cm; prepared as the gelatin–Sepharose column for Fn isolation

Heparin–Sepharose: length 12.5 cm, diameter 1.0 cm; prepared according to manufacturer's instructions (Pharmacia)

DE-52: length 9.5 cm, diameter 1.5 cm; prepared according to manufacturer's instructions

Reagents

PBS, digest buffer, 4 *M* urea, GS wash 1, and GS wash 2 are as above

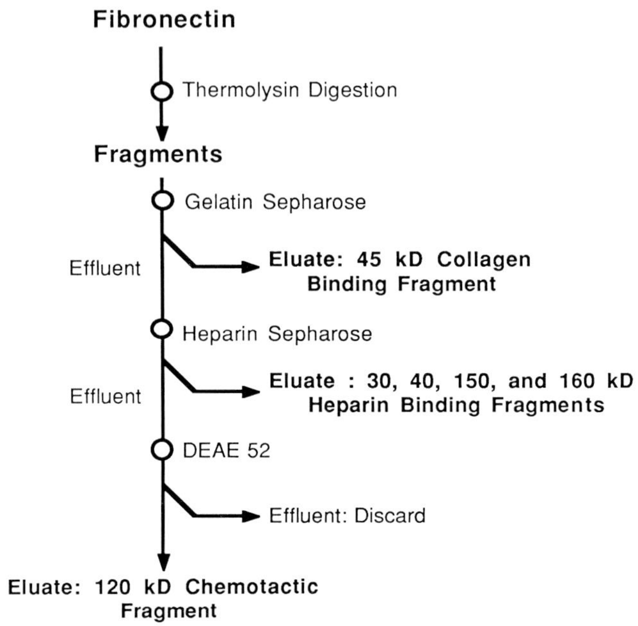

FIG. 1. Schematic representation of the method for separating thermolytic fragments of fibronectin.

Elution buffer: 25 m*M* Tris–HCl, pH 7.6, 0.5 m*M* EDTA, 500 m*M* NaCl (this is digest buffer with 500 m*M* NaCl instead of 50 m*M* NaCl and is most easily made up from 10× digest buffer)

Stripping buffer: 25 m*M* Tris–HCl, pH 7.6, 0.5 m*M* EDTA, 1 *M* NaCl

Procedure

The three columns are stacked in the following order: (1) gelatin–Sepharose, (2) heparin–Sepharose, (3) DE-52. This assembly is washed with 400 ml of digest buffer. The Fn digest mixture is then loaded on the columns, and, subsequently, the columns are washed with another 400 ml of digest buffer. The effluent from this assembly contains a small amount of a large number of low-molecular-weight fragments that we have not characterized. After the columns have been washed, they are separated.

The DE-52 column is eluted with a 50–500 m*M* NaCl gradient by placing 250 ml each of digest buffer and elution buffer into opposite chambers of a double-cylinder gradient maker in a manner similar to the elution of the Fn preparation DE-52 column described above. An early, very low, broad peak consists of several low-molecular-weight (< 40K) proteins which have not been characterized. The large peak, which elutes at a salt concentration of just under 300 m*M*, is the 120-kDa chemotactic fragment. The pooled fractions comprising this peak (usually 40–60 ml) are concentrated by vacuum dialysis down to a volume of 3–4 ml, a volume entirely contained within the dialysis tubing used for vacuum dialysis. The upper end of this dialysis tubing is clamped off, and the solution is dialyzed against PBS for subsequent biological assays. The concentration is measured (the extinction coefficient at 280 nm is 1.12 ml/mg), and 0.1- to 0.5-ml aliquots are frozen at −70°. As shown in Fig. 2, the resulting protein migrates as a single band on sodium dodecyl sulfate–

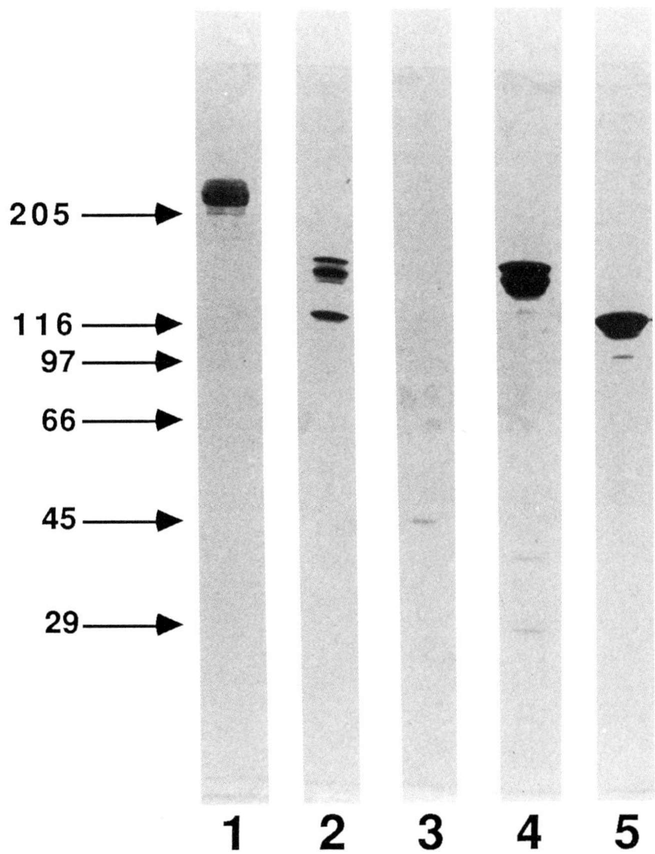

FIG. 2. Silver-stained 5–15% gradient polyacrylamide gel of fibronectin and its fragments. Samples were run under reducing conditions. Lanes contain samples as follows: 1, intact Fn monomer (220 kDa); 2, thermolytic digest at 8 hr; 3, gelatin-binding fragment (45 kDa); 4, heparin-binding fragments (160, 150, 40, and 30 kDa); 5, purified 120-kDa fragment. Migration positions of standards (kDa) are shown by arrows.

polyacrylamide gels. The DE-52 column is washed with 500 ml each of stripping buffer and digest buffer and stored in a cold room.

The heparin–Sepharose column is eluted with elution buffer (100–200 ml). A single peak elutes immediately. SDS–PAGE shows this peak to consist of the 160-, 150-, 40-, and 30-kDa heparin-binding fragments. If desired these fragments can be further purified by dialyzing them against digest buffer, loading them onto the DE-52 column, and eluting the column with a 50–500 m*M* gradient of NaCl in digest buffer as described in the preceding paragraph. The early peak is a dilute solution of the 30- and 40-kDa fragments. The second peak contains the 150- and 160-kDa fragments. Both of these fractions may be concentrated, dialyzed against PBS, and stored as was done for the 120-kDa fragment. Alternatively, these fragments can be separated by molecular sieve chromatography. The heparin–Sepharose column is washed with 300 ml each of stripping buffer and digest buffer and stored in a cold room.

The 45-kDa collagen-binding fragment is eluted from the gelatin–Sepharose column with 300 ml of 4 *M* Urea. It is then concentrated, dialyzed, and stored as above. The column is washed sequentially with 300 ml each of GS wash 1, GS wash 2, and digest buffer and stored in a cold room.

Assay for Chemotaxis

A detailed discussion of the methods of chemotaxis assays is presented elsewhere in this volume ([2]–[6]). Our method is a modification of the method of Campbell.[8] Mononuclear cells are separated from citrated blood by Ficoll–Hypaque centrifugation and used as a suspension of 500,000 cells/ml in RPMI 1640 containing 1% human serum albumin. Assays are performed in blind-end well chambers. A filter sandwich, consisting of a 5-μm pore polycarbonate filter overlying a 0.45-μm pore cellulose nitrate filter, is used to separate the upper and lower compartments. An aliquot of 225 μl of the sample to be tested for chemotactic activity is placed in the lower chamber, and 600 μl of the mononuclear cell suspension is placed in the upper chamber. Migrating cells are trapped in the lower (cellulose nitrate) filter. The chambers are incubated for 90 min at 37° in a 100% humidity chamber containing 95% air and 5% CO_2. Then the lower filters are removed, fixed in methanol, and stained with Gill hematoxylin dye. The cells in 20 random high-power fields are counted. Four replicate filters are used per assay. As measured by this assay, the 120-kDa fragment shows maximal chemotactic activity at concentrations between 80 and 300 nmol/liter.

[8] P. B. Campbell, *J. Lab. Clin. Med.* **90,** 381 (1977).

[21] Inhibitors of Chemotaxis

By EUFRONIO G. MADERAZO, CHARLES L. WORONICK, and PETER A. WARD

Introduction

As with all cell responses, neutrophil (PMN)[1] chemotaxis requires regulation in order to participate in an effective and controlled host response. In the absence of strictly regulated chemotactic responses, uncontrolled inflammation would result. Natural inhibitors of chemotaxis are, therefore, to be expected and are indeed present.

Since there are two participants in the chemotactic response, the chemotactic signal and the responding cell, inhibitors may be best classified according to their specific target, i.e., inhibitors of chemotactic factors and cell-directed inhibitors. There are, at present, three relatively well-defined inhibitors, two of which have chemotactic factors for their target and one of which acts directly to inhibit the responding cells. This discussion will detail these three groups of inhibitors.

Inactivators of Chemotactic Factors

There are two relatively well-defined chemotactic factor-directed inhibitors. Both are irreversible inhibitors of their target chemotactic attractant and thus are called inactivators: the anaphylatoxin inactivator (or carboxypeptidase N, EC 3.4.17.3, arginine carboxypeptidase) and the chemotactic factor inactivator. The principal properties of these serum factors are summarized in Table I.

Anaphylatoxin Inactivator (Carboxypeptidase N)

Anaphylatoxin inactivator (AI) was first described, characterized, and partially isolated by Bokisch and Muller-Eberhard in 1970.[1a] It is present in the pseudoglobulin fraction of serum which may be prepared by diluting human serum with approximately 4 volumes of distilled water, adjusting the pH to 7.0 with 1 *N* HCl, letting the mixture stand at 4° overnight, and removing the precipitate by centrifugation. AI activity is found in the

[1] Abbreviations: PMN, polymorphonuclear leukocytes; AI, anaphylatoxin inactivator; CFI, chemotactic factor inactivator; EACA, ϵ-aminocaproic acid; CDI, Cell-directed inhibitor.

[1a] V. A. Bokisch and H. J. Muller-Eberhard, *J. Clin. Invest.* **49,** 2427 (1970).

TABLE I
ANAPHYLATOXIN INACTIVATOR AND CHEMOTACTIC FACTOR INACTIVATOR: SUMMARY OF CHARACTERISTICS

Characteristic	AI	CFI
Solubility in 45% saturated $(NH_4)_2SO_4$	Precipitates (pseudoglobulin)	Soluble
Electrophoretic mobility	α-Globulin	α- and β-globulin
Sedimentation velocity	9.5 S	4 and 7 S
Molecular weight	270,000–280,000	—
Activity		
Heat sensitivity	Heat labile	Heat labile
pH range	pH 5–10, destroyed at pH 4 in 1 hr at 20°	pH 7–7.6
Inhibited by		
10 m*M* phenanthroline	Yes	No
10 m*M* EDTA	Yes	No
ε-Aminocaproic acid	Yes	No
Speed of inhibitory action on its own substrate	Within seconds	Within minutes, 75% in 20 min
Molecules substrate inactivated/molecule inhibitor	300–400 molecules C3a in 2 min at 20°	—

supernate of this preparation. Pseudoglobulin can also be obtained by precipitation of the euglobulin component with ammonium sulfate at 34–46% of saturation.

The purification of AI and its assay were described by Plummer and Erdos in another volume of this series, to which the reader is referred for details.[2] Briefly, plasma is dialyzed against 5 m*M* Tris–HCl, pH 7.0, containing diisopropylphosphorofluoridate and phenylmethylsulfonyl fluoride. It is then applied to a DE-52 cellulose (Whatman) column and eluted with a linear NaCl gradient. It is adsorbed to a *p*-aminobenzoyl-L-arginine–Sepharose 6B column and eluted with the competitive inhibitor guanidinoethylmercaptosuccinic acid. The yield is 30% at a 2608-fold purification.

The enzyme is assayed either by its esterase activity, using benzylglycylargininic acid, or by its peptidase activity, using Bz-Ala-Lys or furylacryloylalanyllysine. The latter compound is preferred for nonpurified solutions such as plasma. The concentration of carboxypeptidase N in normal serum is 30–40 μg/ml.[1a]

[2] T. H. Plummer, Jr., and E. G. Erdos, this series, Vol. 80, [36].

Since AI is derived from the liver, serum activity is diminished in cirrhosis.[3] Low levels are observed in the presence of intravascular complement activation such as in cardiopulmonary bypass,[4] hemodialysis,[5] dextran and Dengue fever shock,[6] and in blunt and postsurgical trauma.[7] A familial carboxypeptidase N deficiency has been described in a patient with an 11-year history of angioedema.[8] Increased levels occur in patients with neoplasms such as Hodgkin's disease,[3] in pregnancy,[3] and in cystic fibrosis and chronic lung disease.[9]

Chemotactic Factor Inactivator

Shortly after the description of AI, Berenberg and Ward[10] observed the presence in fractionated normal human serum of an inactivator of a broad spectrum of chemotactic factors that include complement-derived chemotactic factors and bacterial chemotactic factor from culture filtrates of *Escherichia coli*. They called this activity the chemotactic factor inactivator (CFI). Subsequent studies characterized CFI activity into two distinct factors: an α-globulin with sedimentation velocity of approximately 4 S and a β-globulin with sedimentation velocity of approximately 7 S. The characteristics of CFI are summarized and compared with AI in Table I.

Isolation. The initial procedure follows that described by Till and Ward.[11] The terminal procedure (repeated hydrophobic chromatography) follows the procedure of Kreutzer *et al.*[12] CFI is precipitated from 1 liter of serum by 45–65% saturated ammonium sulfate. The precipitate at 65% saturation is redissolved in distilled water and dialyzed against 8 m*M* phosphate buffer (ionic strength 0.02 μ, pH 7.3). This is then applied to a Sephadex G-150 column (7 × 80 cm) equilibrated with 0.1 *M* phosphate buffer, pH 7.2 (10 mmho at 4°) at a flow rate of 2 ml/min. The active fractions coinciding with the first protein peak (approximately the IgG

[3] E. G. Erdos, I. M. Wohler, M. I. Levine, and M. P. Westerman, *Clin. Chim. Acta* **11**, 39 (1965).
[4] D. L. Kreutzer, J. R. McCormick, A. Despins, M. Moore, J. H. Rousou, W. Dobbs, and R. N. Engelman, *J. Exp. Pathol.* **1**, 183 (1984).
[5] J. R. McCormick, D. L. Kreutzer, H. J. Keating, J. Hupp, A. Despins, and M. Moore, *Am. J. Pathol.* **109**, 283 (1982).
[6] N. C. Corbin, T. E. Hugli, and H. J. Muller-Eberhard, *Anal. Biochem.* **73**, 41 (1976).
[7] E. G. Maderazo, M. Moore, and C. L. Woronick, *J. Infect. Dis.* **157**, 364 (1988).
[8] K. P. Mathews, P. M. Pan, N. J. Gardner, and T. E. Hugli, *Ann. Intern. Med.* **93**, 443 (1980).
[9] J. Lieberman, *Am. Rev. Respir. Dis.* **111**, 100 (1975).
[10] J. L. Berenberg and P. A. Ward, *J. Clin. Invest.* **52**, 1200 (1973).
[11] G. Till and P. A. Ward, *J. Immunol.* **114**, 843 (1975).
[12] D. L. Kreutzer, W. D. Claypool, M. L. Jones, and P. A. Ward, *Clin. Immunol. Immunopathol.* **12**, 162 (1979).

position) and a second zone of activity on the ascending shoulder of the major protein (albumin) peak are pooled, concentrated, and dialyzed against 20 mM phosphate buffer (pH 7.3) at 4° (Fig. 1). Two to three liters of human serum may be prepared in this fashion and the CFI-rich fractions pooled for the next step. This material is then applied to a DE-52 cellulose column (5 × 50 cm) equilibrated with 20 mM phosphate buffer (pH 7.3), and a salt gradient, consisting of 1000 ml of 20 mM phosphate buffer (pH 7.3) as the starting buffer and 1000 ml of 0.5 M NaCl in the same buffer as the terminal solution, is applied. The active fractions, coinciding with the first protein peak, are saved as pool A, and the fractions coinciding with the major protein peak are saved as pool B (Fig. 2). Pool A contains the β-globulin CFI and pool B the α-globulin CFI. Pool B is concentrated

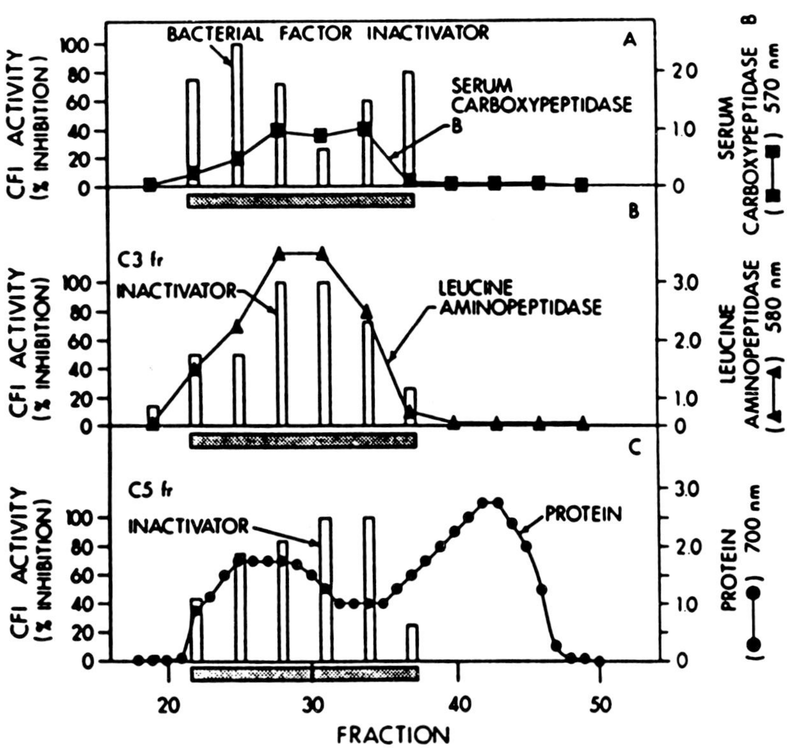

FIG. 1. Elution pattern of CFI activity from Sephadex G-150. (A) Fractions containing activity against chemotactic factor derived from bacteria (bacterial factor). (B) Recovery of activity against C3 complement-derived chemotactic fragments (C3fr). (C) Recovery of activity against C5 complement-derived chemotactic fragments (C5fr). The shaded bars indicate the fractions which were pooled for further purification. From Kreutzer *et al.*,[12] with permission.

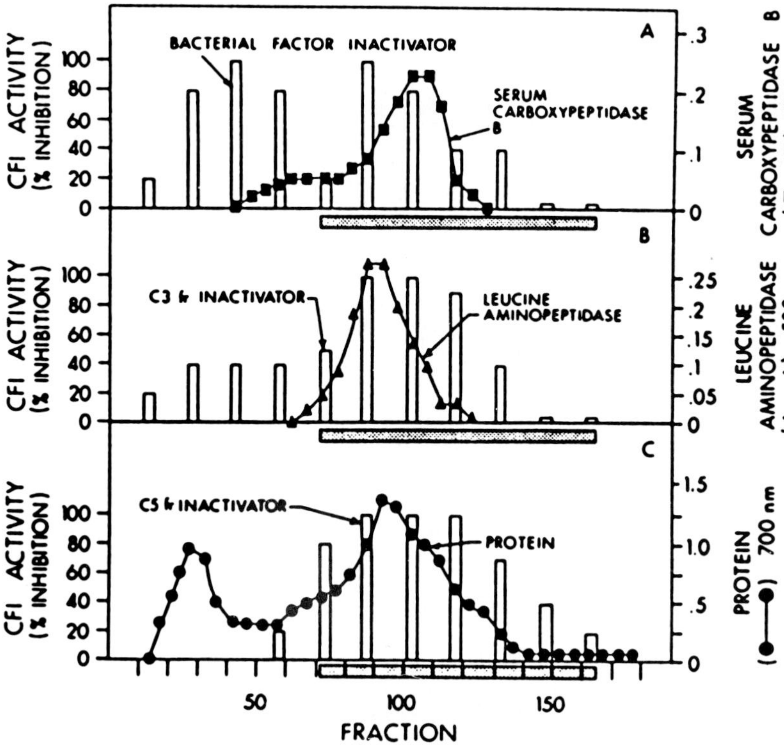

FIG. 2. Elution pattern of CFI activity on a DE-52 anion-exchange column eluted with a linear NaCl gradient. The CFI pool obtained from the Sephadex G-150 column was applied to this column. The CFI activity against bacterial factor (fractions 10–51) was pooled separately from the CFI activity against C3fr and C5fr (fractions 72–170, shaded bars). From Kreutzer *et al.*,[12] with permission.

and chromatographed on a Sephadex G-200 column (5 × 80 cm) equilibrated with 0.1 *M* phosphate buffer, pH 7.2 (10 mmho at 4°), at a flow rate of 2 ml/min. The CFI activity is found in the first protein peak and in the ascending limb of the second, major protein peak (Fig. 3).

The CFI-active fractions are pooled, concentrated by ultrafiltration using the Amicon Concentrator (PM10 membrane, Amicon Corp., Lexington, MA), and the concentrate is diluted with 5 m*M* phosphate buffer (pH 7.2) to a conductance of 1.5 mmho at 4°. Further purification of CFI is accomplished by hydrophobic chromatography using a ω-aminohexylagarose column (5 × 10 cm) (Miles Laboratories, Elkhart, IN). CFI is eluted using a linear NaCl–sodium phosphate gradient made with 250 ml of 10 m*M* sodium phosphate (pH 7.2, 1.5 mmho conductance at 4°) as the starting buffer and 250 ml of 0.65 *M* NaCl in 0.1 *M* sodium phosphate

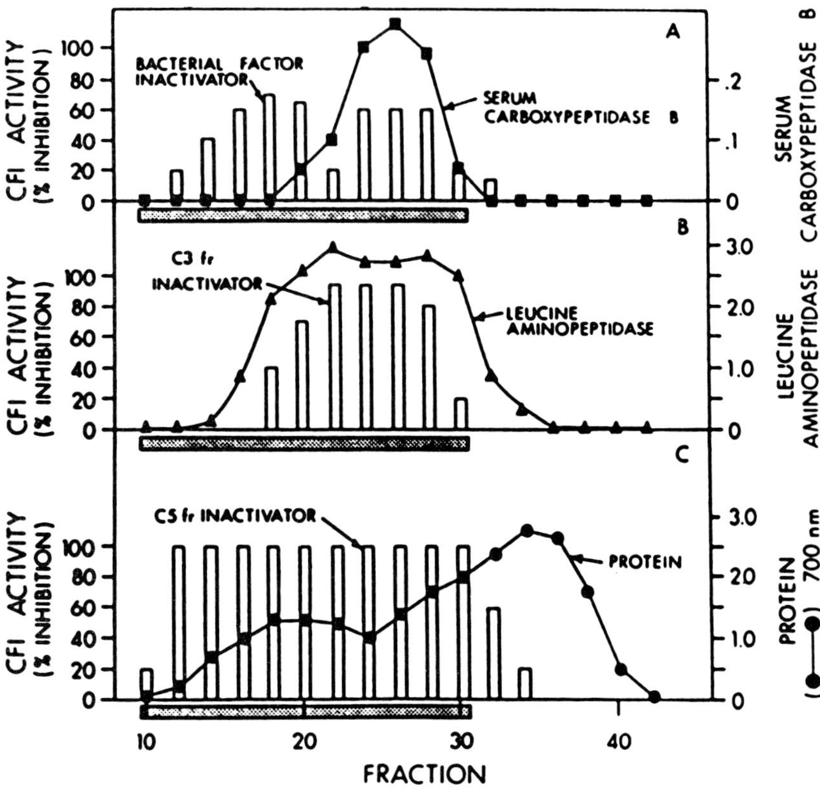

FIG. 3. Elution pattern of CFI activity against complement-derived chemotactic fragments (C3fr and C5fr) separated on a Sephadex G-200 column. The C3fr–C5fr pool from the DE-52 column was applied to this column. Fractions containing CFI activity against C3fr and/or C5fr were pooled (shaded bars) for further purification. From Kreutzer *et al.*,[12] with permission.

buffer (pH 7.2, 35 mmho conductance at 4°) as the terminal buffer. Fractions of 3–4 ml are collected. The bacterial factor-active CFI, carboxypeptidase B-like activity, and leucine aminopeptidase activity coincide with the protein peak (fractions 10–14), while the C5a-active CFI is found in the trailing edge of the protein peak and in subsequent fractions which contain almost no protein (Fig. 4). Fractions 15–35 are pooled, concentrated, diluted, and rechromatographed 5–6 times on the hydrophobic column using an identical linear gradient. The original 3 liters of human serum will provide a highly purified C5a-active CFI (α-globulin CFI) containing less than 1 mg of protein and devoid of AI, aminopeptidase, and bacterial factor inactivator activities. Pool A is chromatographed on a

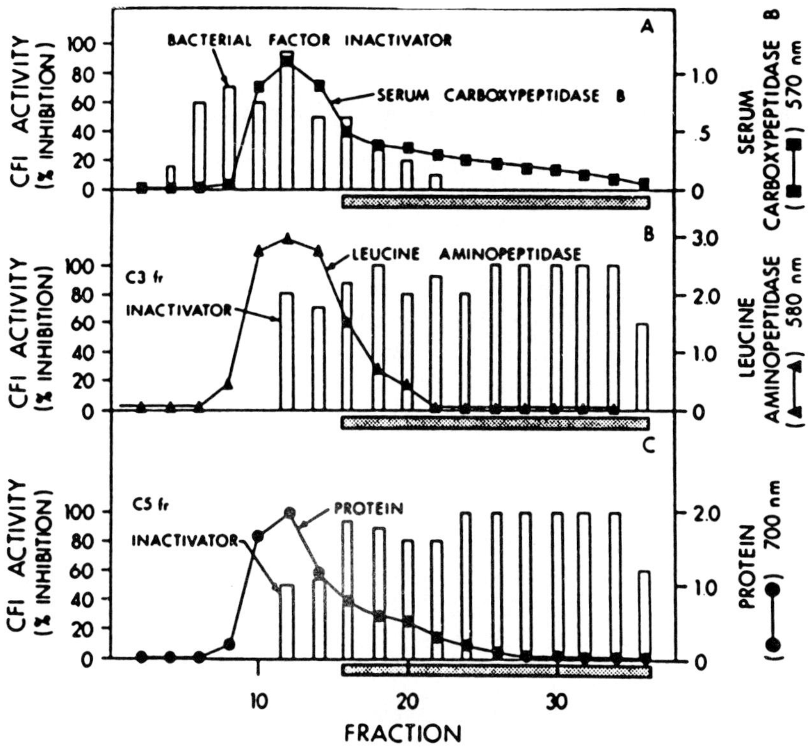

FIG. 4. Elution pattern of C3fr–C5fr CFI activity on an ω-aminohexylagarose column. The C3fr–C5fr pool obtained from the Sephadex G-200 column was applied to this column. The hydrophobic column was eluted with a linear NaCl–sodium phosphate gradient. The fractions containing CFI activity against complement-derived factors were pooled (shaded bars). The pool was concentrated and rechromatographed 5–6 times on the same column using an identical gradient. From Kreutzer *et al.*,[12] with permission.

Sephadex G-200 column, as described above, to yield a bacterial factor-active CFI (β-globulin CFI) which is devoid of activity against C5a.

Testing of CFI Activity (see also this volume [2]–[7]). The tests are based on the ability of CFI to reduce chemotactic responses of neutrophils.[13,14] Chemotactic responses are measured by assaying locomotory response by the standard micropore filter assay or the under agarose assay, or by quantitating the amount of lysosomal enzymes released from cytochalasin B-treated neutrophils on exposure to chemotactic factor. Both

[13] J. C. Fantone, R. M. Senior, D. L. Kreutzer, M. Jones, and P. A. Ward, *J. Lab. Clin. Med.* **93**, 17 (1979).

[14] R. Tannous, N. Cavender-Zylich, and B. H. Ginsberg, *J. Lab. Clin. Med.* **98**, 238 (1981).

chemotactic end points will give results that are almost superimposable. However, the enzyme assay is less tedious, more reliable, and more reproducible than the biological assay.

For assay of CFI the following materials are employed:

1. Human blood neutrophils or rabbit peritoneal leukocytes. Normal human PMN are prepared from blood to which is added 50 units of preservative-free sodium heparin per milliliter of whole blood. Six parts of blood are mixed with 1 part of 6% hetastarch in saline (McGaw Laboratories, Irvine, CA), and allowed to stand for 30 min to allow the erythrocytes to settle. The leukocyte-rich plasma is centrifuged at 500 *g* for 5 min, and the cell pellet is washed once with Medium 199 (pH 7.4) and resuspended in the same medium to a final concentration of 5×10^6 PMN/ml. For each chamber 5×10^5 cells are used.

2. Reagents: Cytochalasin B, yeast *(Saccharomyces cerevisiae)*, ϵ-aminocaproic acid (EACA), and *p*-nitrophenyl-*N*-acetyl-*β*-D-glucosaminide (Sigma Chemical Co., St. Louis, MO).

3. C5a fragment preparation (see also this volume [46]): This is prepared from 1 unit of fresh frozen plasma clotted by the addition of 2 ml of 1.1 *M* $CaCl_2$. The clotted plasma is centrifuged to yield 180 ml of serum. Zymosan is prepared by suspending 3 g yeast in 10 ml distilled water, heating in a boiling water bath for 30 min, centrifuging at 2000 *g* for 10 min, washing with 10 ml distilled water, and centrifuging again. The serum is activated in the presence of 1 *M* EACA by the addition of the prepared zymosan and incubation for 1 hr at 37°. After incubation the zymosan is removed by centrifugation and the supernate is heat-inactivated at 56° for 1–2 hr. The C5a fragment is then purified from this material by cation-exchange chromatography using a procedure developed by Kreutzer *et al.*[15] which follows. Activated serum is adjusted to pH 5 using ice-cold 1 *N* HCl and pumped through a SP 250 cartridge (ZetaPrep, Pharmacia LKB Biotechnology Inc., Piscataway, NJ) that is equilibrated with 0.1 *M* ammonium formate buffer, pH 5.0. The cartridge is sequentially eluted with pH 5.0 ammonium formate buffers of increasing molarities at a flow rate of 2 ml/min. The cartridge is first eluted overnight with 0.1 *M* buffer, followed by 0.25 *M* buffer until the absorbance at 280 nm is less than 0.05 (1–1.5 liters). These fractions are discarded. The C5a activity is eluted with 0.5 *M* buffer, and the activity is found in the early fractions. The cartridge is cleaned with 1 *M* buffer and reequilibrated with 0.1 *M* buffer before reuse. Active fractions are pooled and dialyzed against phosphate-buffered

[15] D. L. Kreutzer, R. Nelson, N. Herkert, R. Mandero, and K. Hou, manuscript in preparation.

saline, pH 7.4. The C5a activity of the fractions is determined by lysozomal release of glucosaminidase (discussed in the next section), and the pool usually has a K_m of 20–30 μl. The C5a concentration of the pool is determined using a C5a desArg radioimmunoassay kit (Upjohn Diagnostics, Kalamazoo, MI). The active pool usually has a C5a fragment concentration of about 900 ng/ml.

4. The reaction mixtures for the lysosomal enzyme release assay are made up as follows. Serum or plasma in diluted 1:200 with Hanks' solution, from which four serial 1 to 1 dilutions are made. A series of tubes is prepared by adding 30 μl of Hanks' solution to the first tube and 30 μl of each serum or plasma dilution to subsequent tubes. To each tube is then added 30 μl ($3 \times K_m$) of the C5a preparation, and the tubes are incubated at 37° for 30 min. To each tube is added 30 μl of human or rabbit neutrophils (2×10^7 cells/ml) which are suspended in Hanks' solution containing 5 μg/ml cytochalasin B. After incubation for 5 min at 37°, the mixtures are centrifuged (200 *g*) for 5 min at 4°, and 50 μl of the supernatants is removed for assay of glucosaminidase activity. Experiments are run in triplicate.

5. Glucosaminidase assay. To 50 μl of the supernatants is added 400 μl of 4 m*M* *p*-nitrophenyl-*N*-acetyl-β-D-glucosaminide in 50 m*M* sodium citrate buffer (pH 4.5). The mixture is incubated at 37° for 30 min, and the reaction is terminated by adding 0.5 ml of 0.4 *M* glycine buffer, pH 10.5. The absorbance of the mixture is determined at 410 nm.

6. Calculation of CFI values. Using the mean value from the tubes with no added serum or plasma as the control, the percent inhibition is calculated for each sample concentration as follows:

$$\%I = 1 - \text{Sample/control} \times 100$$

The $\%I$ values are plotted against the logarithm of the volume of sample on the x axis. The data follow a sigmoidal curve, from which the value of K_i, the volume of sample giving 50% inhibition, is estimated by performing linear regression analysis on the approximately linear portion of the curve.[13]

High serum CFI levels are found in neoplastic states such as Hodgkin's disease[16] and "hairy cell" leukemia,[17] liver cirrhosis,[18] sarcoidosis,[19] and

[16] P. A. Ward and J. L. Berenberg, *N. Engl. J. Med.* **290,** 76 (1974).

[17] C. R. Kjeldsberg, *Ann. Intern. Med.* **88,** 268 (1978).

[18] E. G. Maderazo, P. A. Ward, C. L. Woronick, and R. Quintiliani, *J. Lab. Clin. Med.* **89,** 190 (1977).

[19] E. G. Maderazo, P. A. Ward, C. L. Woronick, J. Kubick, and A. C. DeGraff, *Ann. Intern. Med.* **84,** 414 (1976).

lepromatous leprosy,[20] as well as in neonates.[21] Another substance found in sera from patients with systemic lupus erythematosus has CFI-like activity[22]; however, it differs from AI and CFI because it is heat stable and its inhibitory activity is reversible. The nature of this factor is still not well defined.

Cell-Directed Inhibitor (CDI)

The presence of CDI in fractionated normal serum was demonstrated by Maderazo *et al.*,[23] following observation of increased serum CDI activity and its characterization in a patient who was found to have defective neutrophil locomotory responses. Prior to this case, three others had been described where the neutrophil chemotactic defect could be ascribed to a serum factor that inhibits neutrophil chemotactic response when placed with the cells in the upper compartment of the chemotactic chamber.[24–26]

Both CDI isolated from the serum of the patient described above as well as CDI isolated from normal human serum directly inhibit neutrophil locomotory (chemokinetic) responses, including unstimulated migration (in the absence of chemoattractant).[23] Inhibitory activity persists even after exposed cells are washed but disappears when pretreated and washed cells are reincubated with normal serum and subsequently either washed or left unwashed.[23] This suggests that this type of CDI is either displaced from the cell by a substance in normal serum or that its inhibitory activity is reversed by a stimulatory factor in the serum.

Major CDI activity in serum is precipitated between 20 and 35% saturated ammonium sulfate. It is nondialyzable and is not destroyed by incubation at 56° for 2 hr. By ultracentrifugal analysis in a linear sucrose density gradient, CDI activity resolves into two activities with estimated sedimentation coefficients of 7 and 10 S. Further studies consistently identified IgG as the major component in both of these fractions (the 10 S fraction is probably polymerized or other forms of IgG), and, indeed, IgG was subsequently identified as a CDI in human serum.[27,28] Among the

[20] P. W. Ward, S. Goralnick, and W. E. Bullock, *J. Lab. Clin. Med.* **87,** 1025 (1976).

[21] R. Tannous, R. E. Spitzer, W. R. Clarke, C. P. Goplerud, and N. Cavendar-Zylich, *J. Lab. Clin. Med.* **99,** 331 (1982).

[22] H. D. Perez, M. Lipton, and I. M. Goldstein, *J. Clin. Invest.* **62,** 29 (1978).

[23] E. G. Maderazo, P. A. Ward, C. L. Woronick, and R. Quintiliani, *J. Lab. Clin. Med.* **89,** 190 (1977).

[24] P. A. Ward and R. V. Schlegel, *Lancet* **2,** 344 (1969).

[25] C. W. Smith, J. C. Hollers, E. Dupree, A. S. Goldman, and R. A. Lord, *J. Lab. Clin. Med.* **79,** 878 (1972).

[26] R. B. Soriano, M. A. South, A. A. Goldman, and C. W. Smith, *J. Pediatr.* **83,** 951 (1973).

[27] N. Kramer, H. D. Perez, and I. M. Goldstein, *N. Engl. J. Med.* **303,** 1253 (1980).

[28] C. L. Woronick, J. Malnick, and E. G. Maderazo, *J. Lab. Clin. Med.* **98,** 58 (1981).

TABLE II
CDI ACTIVITY OF HUMAN IMMUNOGLOBULINS AND FRAGMENTS[a]

Factor	From average distance migrated		From cell number	
	K_i (μM)	$\%I_{max}$	K_i (μM)	$\%I_{max}$
PMN values				
IgG	0.083	40	0.051	50
Fc from IgG	2.4	46	4.2	88
Fab from IgG	No effect		No effect	
λ-IgM	No effect		No effect	
κ-IgA_1	0.042	40	0.12	72
Monocyte values				
IgG	0.034	69	0.0037	69

[a] K_i is the concentration of factor that produces 50% of maximum inhibition. $\%I_{max}$ is the maximum inhibition of locomotion produced when the concentration of the factor is extrapolated to infinity.[26]

purified immunoglobulins studied, only IgG and IgA have CDI activities, whereas IgM has no inhibitory activity. In the case of IgG, the inhibitory activity is associated with the Fc fragment, the Fab fragment being entirely inactive. Thus, CDI from normal serum cannot be related to an autoantibody. This suggests that the Fc receptor sites are involved in CDI action on cells.

Quantitative characteristics of CDI from normal serum are summarized in Table II. In addition to its inhibitory effects on neutrophil locomotion, CDI also inhibits monocyte migration (Table II) and neutrophil phagocytosis.[23,27]

Testing of CDI Activity. The concentration of CDI activity in serum is quantitated by measuring its effect on leukocyte locomotion using the standard micropore filter assay that uses 13-mm-diameter, 5-μm-pore size cellulose nitrate filters (Sartorius, Hayward, CA) and clear acrylic modified Boyden chambers[28,29] (see also this volume [3]). Table III shows a routine protocol to screen for cell- and serum-associated neutrophil locomotory abnormalities. A comparison of tests 7 or 8 versus tests 7 or 10 will identify serum with suspected CDI activity.

To further estimate the quantity of CDI activity in serum, 0.1 ml of the PMN suspension (5×10^5 PMN) is placed in a tube along with 50 μl of normal serum and various volumes (from 0 to 100 μl) of patient serum. Each tube is brought to a final volume of 250 μl by adding the appropriate volume of Medium 199. After 30 min of incubation in a 37° shaking water

[29] E. G. Maderazo and C. L. Woronick, *Clin. Immunol. Immunopathol.* **11,** 196 (1978).

TABLE III
PROTOCOLS FOR LEUKOCYTE LOCOMOTION ASSAY AND VALUES

Test number	PMN cells + serum (upper compartment)	Chemotactic factor (lower compartment)	Locomotion index,[a] LI_{20} (μm) (mean ± SD)	To test or screen for
1	Normal + none	Medium	33.8 ± 5.5 (87)	Unstimulated migration
2	Patient + none	Medium		
3	Normal + none	ZymNS	44.5 ± 10.1 (87)	Chemotaxis, generation of complement-derived chemotactic factors, and CFI
4	Patient + none	ZymNS		
5	Normal + none	ZymPS		
6	Patient + none	ZymPS		
7	Normal + normal	None	57.1 ± 5.5 (87)	Chemokinesis, cell-directed inhibitor, and leukokinesis-enhancing factor
8	Patient + normal	None		
9	Normal + patient	None		
10	Patient + patient	None		

[a] The locomotion index (LI_{20}) is the average distance migrated by the PMNs when cells are counted at 10-μm intervals beginning at the 20-μm level. ZymNS, Zymosan-activated normal serum; ZymPS, zymosan-activated patients' serum. Numbers in parentheses represent number of subjects tested.

bath, the tubes are centrifuged and the PMNs resuspended in 0.7 ml Medium 199. The entire contents of the tube are transferred to the upper compartment of a chamber which has the lower compartment filled with 3% heat-stabilized zymosan-activated normal serum [activated serum from the C5a fragment preparation (above), which was heated but not chromatographed]. The chambers and filters are processed as previously described[28,29] (see also this volume [3]) and the %I calculated as described above.

The %I values are plotted against the concentration of sample on the x axis. The data follow the form of the Michaelis–Menten equation, and the value of K_i, the sample concentration giving 50% inhibition, is calculated by nonlinear regression. Other methods of solving the Michaelis–Menten equation may also be used. The sample concentration giving 50% of maximum inhibition is defined as one K_i unit of CDI activity. Because patients' sera, as well as sera from normal individuals, consist of a complex mixture of PMN inhibitors and stimulators, it is difficult or impossible to quantitatively estimate the CDI activity of patient sera directly. Therefore, 50 μl of normal serum is added to each assay chamber, which is a sufficient serum volume to produce near-maximal stimulation of PMN locomotion. Patient sera that contain CDI activity reverse the stimulation produced by the added normal serum, and the percent reversal is related to the concentration of the CDI activity.

To determine the K_i values for purified materials, such as immunoglobulins or immunoglobulin fragments (Table II), PMN are preincubated with various concentrations of the CDI preparations in the absence of serum, and the contents of the tubes, including the CDI, are transferred to the upper compartment of the chamber and the lower compartment filled with Medium 199. The control chambers have Medium 199 instead of CDI. The inhibitory effects reported in Table II are evaluated in two different manners,[28,29] the first based on the *average distance* migrated by the leukocytes and the second based on the *total number* of leukocytes that migrate into the filter. For both methods of assessment, cell counts are started at the upper surface of the filter and continued at each 10-μm level throughout the filter. The average distance migrated, *LI*, is calculated as follows:

$$LI = \Sigma\,(ND)/\Sigma\,N$$

where N is the number of cells counted at each 10-μm increment from the upper surface of the filter and D the distance in micrometers from the upper surface of the filter. At each concentration of inhibitor, the percent change in average distance migrated is calculated relative to a control containing Medium 199 in place of the inhibitor. In the second method, the percent change in the total number of migrating leukocytes is calculated relative to the number migrating into the control filter. K_i and maximum $\%I$ values are calculated from the Michaelis–Menten equation using nonlinear regression.

In addition to the isolated cases mentioned above,[24–26] various clinical conditions have been reported to be associated with increased serum levels of CDI or CDI-like activity. These include patients following thermal injury[30] or blunt trauma[31] and patients with paraproteinemia or cryoglobulinemia,[32] other neoplasms,[33] and liver cirrhosis.[34] In patients with liver cirrhosis, serum CDI activity is attributable to monomeric, dimeric, or aggregated IgA, since it can be removed with an anti-IgA immunoadsorbent column.[34]

[30] G. D. Warden, A. D. Mason, Jr., and B. A. Pruitt, Jr., *Ann. Surg.* **181,** 363 (1975).
[31] N. V. Christou and J. L. Meakins, *J. Surg. Res.* **26,** 355 (1979).
[32] D. E. Van Epps and R. C. Williams, Jr., *J. Exp. Med.* **144,** 1227 (1976).
[33] E. G. Maderazo, T. Anton, and P. A. Ward, *Clin. Immunol. Immunopathol.* **9,** 166 (1978).
[34] D. E. Van Epps, R. G. Strickland, and R. C. Williams, Jr., *Am. J. Med.* **59,** 200 (1975).

[22] Leukocyte Chemoattractant Receptors

By MARILYN C. PIKE and RALPH SNYDERMAN

Introduction

The accumulation of inflammatory cells is vital for many immunologically mediated functions as well as for wound healing. In the late nineteenth century, Metchnikoff postulated that chemical signals produced at sites of microbial invasion caused direct leukocyte migration.[1] It was not until 1962 that a quantitative *in vitro* assay for leukocyte chemotaxis was developed by Boyden, who demonstrated that leukocytes were in fact capable of directed migration toward serum treated with antigen–antibody complexes.[2] The first chemical identification of an endogenous chemoattractant substance was made by Snyderman *et al.*[3] and Shin *et al.*,[4] who described the chemotactic activity of C5a, the cleavage product of C5 which is generated by activation of both the alternative and classic complement pathways.

Although many substances were subsequently described to be chemotactic for leukocytes, it was not until 1975 that insight regarding the precise chemical structure of a chemoattractant substance was gained. Schiffman *et al.* found that N-formylated methionyl di- and tripeptides were potent chemotactic factors for human polymorphonuclear lymphocytes (PMNs)[4a] and monocytes.[5] This discovery allowed the synthesis of highly purified, highly radioactive chemotactic substances which could be used to identify cell surface receptors for formylated oligopeptide chemoattractants on leukocytes.[6,7]

Over the past several years, specific cell surface receptors for a variety of chemotactic agents have been described including the N-formylated oligo-

[1] E. Metchnikoff, *Br. Med. J.* **1,** 213 (1891).

[2] S. Boyden, *J. Exp. Med.* **115,** 453 (1962).

[3] R. Snyderman, H. Gewurz, and S. E. Mergenhagen, *J. Exp. Med.* **128,** 259 (1968).

[4] H. S. Shin, R. Snyderman, E. Friedman, A. Mellors, and M. M. Mayer, *Science* **162,** 361 (1968).

[4a] Abbreviations: PMNs, polymorphonuclear lymphocytes; LTB_4, leukotriene B_4; HBSS, Hanks' balanced salt solution; GBSS, Gey's balanced salt solution.

[5] E. Schiffmann, B. A. Corcoran, and S. M. Wahl, *Proc. Natl. Acad. Sci. U.S.A.* **72,** 1059 (1975).

[6] L. T. Williams, R. Snyderman, M. C. Pike, and R. J. Lefkowitz, *Proc. Natl. Acad. Sci. U.S.A.* **74,** 1204 (1977).

[7] S. Aswanikumar, B. Corcoran, E. Schiffmann, A. R. Day, R. J. Freer, H. J. Showell, and C. B. Pert, *Biochem. Biophys. Res. Commun.* **74,** 810 (1977).

TABLE I
CHARACTERISTICS OF CHEMOATTRACTANT RECEPTORS ON LEUKOCYTES[a]

Agonist	K_D (n*M*)	Receptors/cell ($\times 10^{-3}$)
N-Formylated oligopeptides		
fMet-Leu-Phe		
Human PMNs	10.0–22.3	55
Human monocytes	30.2	84
Guinea pig macrophages	11.0	10
fNle-Leu-Phe		
Rabbit PMNs	1.5	100
fNle-Leu-Phe-Nle-Tyr-Lys		
Human PMNs	1.0	120
Human monocytes	1.7–2.7	10–18
C5a		
Human PMNs	3–7	100–300
Leukotriene B_4		
Human PMNs	10.8–200	26–386

[a] Adapted from R. Snyderman and M. C. Pike, *Annu. Rev. Immunol.* **2,** 257 (1984).

peptides,[6,7] C5a,[8] and leukotriene B_4 (LTB_4).[9,10] These receptors have been found on various types of normal cells including human, rabbit, and equine PMNs, human monocytes, guinea pig macrophages,[6-11] as well as a variety of differentiated human leukemic cell lines such as HL60 and U937.[12,13] Binding to both intact cells and isolated membranes has been demonstrated for some of these cell types. This chapter provides methodology for measuring chemoattractant receptors on various types of cells and membranes. Table I summarizes the characteristics of chemoattractant receptors which have been described thus far.

Cell and Membrane Preparations

Materials

Heparinized venous blood (10 heparin U/ml)
Beef lung heparin (Upjohn)
3% (w/v) dextran (high MW, Pharmacia) in 0.9% NaCl

[8] D. E. Chenoweth and T. E. Hugli, *Proc. Natl. Acad. Sci. U.S.A.* **75,** 3943 (1978).
[9] R. A. Kreisle and C. W. Parker, *J. Exp. Med.* **157,** 628 (1983).
[10] D. W. Goldman and E. J. Goetzl, *J. Immunol.* **129,** 1600 (1982).
[11] R. Snyderman and M. C. Pike, *Science* **209,** 493 (1980).
[12] J. Niedel, I. Kahane, L. Lachman, and P. Cuatrecasas, *J. Biol. Chem.* **255,** 7063 (1980).
[13] M. C. Pike, D. Fischer, H. S. Koren, and R. Snyderman, *J. Exp. Med.* **152,** 31 (1980).

Lymphoprep (Bionetics)
1.6% (w/v) NaCl
0.2% (w/v) NaCl
Incubation buffer: 140 mM NaCl, 1.0 mM KH_2PO_4, 5 mM Na_2HPO_4, 0.5 mM $MgCl_2$, 0.15 mM $CaCl_2$, pH 7.4
Membrane buffer: 50 mM Tris–HCl with 10 mM $MgCl_2$, pH 7.7
0.5% shellfish glycogen (Sigma) in 0.9% NaCl
Hanks' balanced salt solution (HBSS), pH 7.4
Male Hartley guinea pigs (250–500 g)
Albino rabbits (> 1 kg)
50-ml plastic conical tubes
50-ml plastic syringes
14-gauge spinal needle
19-gauge butterfly needles
35-ml round-bottom centrifuge tubes

Procedures

Cells Derived from Venous Blood. Blood is collected in plastic syringes containing 10 U/ml of heparin, mixed 1 : 1 (v/v) with dextran in 50-ml syringes and allowed to sediment for 25 min at room temperature. The leukocyte-rich supernatant is decanted through butterfly needles into 50-ml conical centrifuge tubes, and 12–15 ml of Lymphoprep is gently underlayered using a spinal needle attached to a 50-ml syringe. The tubes are centrifuged at 20° for 40 min at 2000 *g* using a swinging-bucket rotor. Mononuclear leukocytes containing approximately 75% lymphocytes and 25% monocytes are harvested from the plasma–Lymphoprep interface. Following centrifugation, the excess Lymphoprep is removed and the remaining erythrocytes lysed with 10 ml of 0.2% NaCl which is added to the cell pellets and vortexed for 20 sec prior to addition of 10 ml of 1.6% NaCl. The cells are centrifuged for 10 min at 600 *g*, and the lysis procedure is repeated up to 2 more times if necessary. Cell pellets containing 96–99% PMNs are resuspended in incubation buffer for intact cell assays[6] or membrane buffer, pH 7.7, for preparation of membranes.[14]

Crude PMN membrane fractions are prepared by disruption of cells (5×10^7/ml) using two 12-sec bursts of a tissue homogenizer (Tekmar Co., Cincinnati, OH), followed by centrifugation at 200 *g* for 10 min.[14] The supernatant is decanted into round-bottom tubes, centrifuged at 41,000 *g* for 15 min, and the pellet is resuspended in incubation buffer. Further purification of plasma membranes can be accomplished by using methods

[14] C. Koo, R. J. Lefkowitz, and R. Snyderman, *J. Clin. Invest.* **72,** 748 (1983).

described by Borregaard *et al.*[15] Human monocytes can be purified by counterflow elutriation[16] (see also this series, Vol. 108 [20]) or by Percoll density centrifugation[17] (see also this series, Vol. 108 [9]).

Peritoneal Cells. Guinea pig macrophages and rabbit PMNs are harvested from the animals' peritoneal cavities approximately 72 and 16 hr, respectively, after an i.p. injection of a glycogen solution[18] (see also this series, Vol. 108 [25]). Cells are harvested in HBSS, centrifuged at 600 *g* for 10 min, washed in incubation buffer, and resuspended in incubation buffer. Guinea pig macrophage membranes are prepared using the procedure described above for neutrophil membrane preparation.[18]

Chemoattractant Ligands

Oligopeptide Chemoattractants

Radiolabeled N-formylated oligopeptide chemoattractants are available from New England Nuclear (Boston, MA). The purity of the compound can be assessed prior to use by high-pressure liquid chromatography using a C_{18} column and 0.2 N acetic acid : methanol (1 : 1, v/v).

Formylnorleucylleucylphenylalanylnorleucyltyrosyllysine (formyl hexapeptide) (Bachem Fine Chemicals, Torrance, CA) can be radiolabeled with ^{125}I or fluoresceinated.[19] Radioiodination of the peptide can be accomplished using chloramine-T or iodogen[19] followed by purification on a BioGel P-2 column equilibrated with 25 mM NaOH. The purified radioiodinated peptide prepared in this fashion has a specific activity ranging from 800 to 2000 Ci/mmol.[19] The peptide is stable for 3 months at $-20°$ when stored in a neutral solution containing 0.1% bovine serum albumin.

The formylated hexapeptide can be fluoresceinated using 20 mM hexapeptide with fluorescein isothiocyanate, 20 mM (Molecular Probes, Inc., Eugene, OR) in dry dimethyl sulfoxide (Pierce Chemical Co., Rockford, IL).[20] The product is purified by silicic acid thin-layer chromatography (Analtech plates) and eluted with chloroform/methanol/acetic acid, 3 : 1 : 0.1 at an R_f of ~0.5. The peptide is scraped from the plate and extracted with dimethyl sulfoxide. Its purity is verified by HPLC using a

[15] N. Borregaard, J. M. Heiple, E. R. Simons, and R. A. Clark, *J. Cell Biol.* **97,** 52 (1983).
[16] T. Yasaka, N. Mantich, L. Boxer, and R. L. Baehner, *J. Immunol.* **127,** 1515 (1981).
[17] F. Colotta, G. Peri, A. Villa, and A. Mantovani, *J. Immunol.* **132,** 936 (1984).
[18] R. Snyderman, M. C. Pike, S. Edge, and B. C. Lane, *J. Cell Biol.* **98,** 444 (1984).
[19] J. E. Niedel, I. Kahane, and P. Cuatrescasas, *J. Biol. Chem.* **254,** 10700 (1979).
[20] L. A. Sklar, D. A. Finney, Z. G. Oades, A. J. Jesaitis, R. G. Painter, and C. G. Cochrane, *J. Biol. Chem.* **259,** 5661 (1984).

C_{18} column as previously described and eluting with a gradient from 20 to 75% acetonitrile in 0.25 *M* triethylammonium phosphate, pH 3.[20]

C5a (see also this volume [46])

Human C5a is isolated from yeast cell-activated serum treated with 1.0 m*M* ε-aminocaproic acid.[21] The protein is labeled with ^{125}I by the method of Bolton and Hunter.[22] C5a (5–10 μg) is suspended in 0.1 *M* sodium borate, pH 8.5, and added to 200–400 μCi of Bolton–Hunter reagent (Amersham–Searle) and allowed to react for 30 min at 4°.[8,22] The reaction mixture is dialyzed against HBSS at 4° until no ^{125}I is detected in the dialysate. The specific activity of the ^{125}I-C5a prepared in this manner ranges from 8 to 12 μCi/mg. Purity of the preparations is assessed by SDS-polyacrylamide gel electrophoresis.[20]

[^3H]Leukotriene B_4

[^3H]LTB_4 is available commercially from New England Nuclear. The ^3H-labeled compound can also be purified from activated [^3H]arachidonic acid labeled PMNs as previously described.[9,10]

Chemoattractant Receptor-Binding Assays

The incubation conditions for quantification of the various chemoattractant receptors vary and will be dealt with individually. The measurement of bound and free ligand can be performed using several separation techniques. A method for quantifying oligopeptide chemoattractant receptors without physical separation of bound and free ligand is also discussed.

Oligopeptide Chemoattractant Receptors

Materials

Radiolabeled N-formylated peptide
Unlabeled N-formylated peptide, 60 μ*M*, in incubation buffer
Incubation buffer, pH 7.4 (see above)
PMN or macrophage cell preparations ($0.8-1.0 \times 10^7$/ml) or
PMN or macrophage membrane preparations (~0.4 mg/ml protein)
12 × 75 mm polypropylene tubes

Incubation Procedure. Twenty-five microliters of incubation buffer containing the desired concentration of labeled chemotactic peptide is

[21] H. N. Fernandez and T. E. Hugli, *J. Immunol.* **117,** 1688 (1976).

[22] A. E. Bolton and W. M. Hunter, *Biochem. J.* **133,** 529 (1973).

added to duplicate tubes containing 25 μl of buffer or approximately 1000-fold excess unlabeled peptide.[6] Reactions are started by carefully adding 100 μl (total incubation volume 0.15 ml) of cell or membrane material directly to the ligand solution. The tubes are then vortexed and incubated in a Dubhoff shaking water bath at the desired temperature. Care should be taken that shaking is effective yet not so vigorous as to cause dispersion of the suspension to the sides of the tubes. Reactions are terminated as indicated below. Nonspecific binding is defined as that amount of residual radioactivity bound to cells or membranes in the presence of the unlabeled N-formylated peptide.

C5a Chemotactic Receptors

Materials

^{125}I-C5a
Unlabeled C5a
Human PMNs (2.3×10^7 cells/ml)
Hanks' balanced salt solution (HBSS) with 0.5% gelatin
1.5-ml conical polypropylene centrifuge tubes (Beckman)

Incubation Procedure. Human PMNs, $0.9-1.5 \times 10^6$, contained in HBSS with 0.5% gelatin are added to duplicate control centrifuge tubes containing concentrations of ^{125}I-C5a ranging from 0 to 750 ng/ml alone or in the presence of 2 μg/ml of unlabeled C5a.[8] Following incubation in a shaking water bath, the tubes are centrifuged at 11,000 *g* for 30 sec in a Beckman microfuge (Beckman Instruments, Wakefield, MA). After centrifugation, one-half of the supernatant (S) is removed, transferred to a separate tube, and the radioactivity is determined along with the original tube containing both pellet and residual supernatant (P). The amount of cell-bound radioactivity is determined by the following formula[7]:

$$\% \text{ of total C5a bound} = \frac{[1 - 2 = \text{cpm(S)}]}{\text{cpm(S)} + \text{cpm(P)}} \times 100$$

Specific C5a bound is the percentage of total C5a bound minus the percent nonspecific binding defined as the residual amount of bound radioactivity in the presence of a 1000-fold excess of unlabeled C5a. Physical separation of bound versus free ^{125}I-C5a can also be accomplished using the separation techniques described below.

LTB_4 Chemoattractant Receptors (see also this volume [7])

Materials

[^{3}H]LTB_4
Unlabeled LTB_4

Gey's balanced salt solution (Gibco, Long Island City, NY) with 25 m*M* HEPES, pH 7.4 (GBSS)
Human PMNs (1×10^7 cells/ml)
1.5-ml conical polypropylene centrifuge tubes (Beckman)

Incubation Procedure. An aliquot of $[^3H]LTB_4$ in methanol is evaporated in a glass tube under nitrogen, vigorously resuspended in GBSS, and diluted serially. Cells are diluted in GBSS and incubated with the desired concentrations of $[^3H]LTB_4$ in the presence and absence of 10 μM unlabeled LTB_4 in a total volume of 130 μl.[10] Nonspecific binding can also be calculated by a method which utilizes $[^{14}C]LTB_4$.[9] Incubations are generally carried out at 4° for 45 min to attain steady-state conditions.[9,10]

Following incubation, two separation techniques can be employed. First, 100 μl of sample may be layered onto Ficoll–Hypaque or phthalate oil and processed as described below.[9,10] In the second method, a portion of the incubation mixture is placed into 2.5 ml of acetone which has been chilled in a dry ice–propanol bath, and the resulting suspension is filtered as described below except that filters are washed with acetone instead of incubation buffer.[10] It should be noted that high amounts of nonspecific $[^3H]LTB_4$ binding, at times exceeding 75% of total binding, are encountered in this assay. The acetone extraction/filtration method of separation of bound versus free ligand has been shown to minimize nonspecific binding as compared to the phthalate oil cushion method.[9]

Separation Procedures

The technique most widely used to rapidly separate bound from free ligand in hormone and chemoattractant receptor studies is filtration onto glass fiber filters. Various filtration devices are commercially available (i.e., Amicon Corp. or Hoeffer Scientific Instruments), and the one used in the authors' laboratory consists of a manifold containing 12 filtration ports which are in communication with a tank connected to a vacuum source (Duke University Physiology Instrument Shop, Durham, NC). At the end of the binding assay, approximately 5.0 ml of ice-cold incubation buffer is added to the tube, and the incubation mixtures are poured onto a Whatman GF/C filter contained on each port of the vacuum apparatus. The tube and then the filters are then rapidly washed with an additional 15 ml of ice-cold incubation buffer (or acetone for LTB_4 assays) to remove nonspecifically bound or trapped ligand.[6] A potential drawback of the filtration system is that small membrane fragments can pass through the filters. We therefore recommend using the smaller pore size Whatman GF/F filters when binding to purified leukocyte membrane fractions is being quantified. Following the wash procedure, the filters are placed in plastic scintillation vials, air dried, and following the addition of 10 ml of detergent-containing scintillation fluid (Leukofluor, Research Products, Inc., Mount Pleasant, IL) the radioactivity is determined.

Rapid centrifugation to separate bound from free ligand has also been used successfully to study chemoattractant receptor binding.[9,10,23] In this method 100 μl of incubation mixture is layered over 200–300 μl of silicone oil, 30% Ficoll–Hypaque, or a mixture of *n*-butyl phthalate and dionyl phthalate (7:2, v/v) in 400-μl polyethylene microtest tubes (Bio-Rad Laboratories, Richmond, CA). Cells are pelleted for 1–2 min at 8000 *g* in a Beckman Model 152 microfuge (Beckman Instruments, Fullerton, CA).[9,10] The supernatant and gradient are then removed by aspiration, and the bottoms of tubes containing cell or membrane pellets are cut off with a razor blade. The radioactivity of the contents of the tube can then be determined directly following resuspension in 10 ml of detergent-based scintillation fluid. This method results in less than 2% contamination by free ligand in the gradient immediately above the cell or membrane pellet and has been used predominantly to quantify direct binding of LTB_4 to cellular fractions.[9,10]

Analysis of Chemoattractant Receptor-Binding Data Obtained Using Direct Binding Techniques

Data can be analyzed using classic Scatchard techniques or computer programs which utilize least-squares curve-fitting analysis. Both methods assume that binding is in equilibrium and that there is no significant destruction of the ligand. These parameters must be examined in each experimental situation especially when using intact leukocytes and temperatures above 4° for binding studies since ligands can be rapidly internalized or degraded.

The two most widely used computer programs are SCTFIT[24] and LIGAND.[25] The power of these programs lies in their ability to give rigorous statistical estimates of multiple binding affinities which may be present in a given receptor system. When using these programs for analysis of multiple affinity states, the number of variables being analyzed dictates the number of data points which are necessary for meaningful analysis in a given experiment. For example, Fig. 1 shows the quantitation of two affinities for the oligopeptide chemoattractant receptor in guinea pig macrophage membrane preparations.[18] The K_D values of the two sites are 1.6 and 38.0 n*M* with 21% high-affinity sites. In order to effect such analysis,

[23] W. M. Mackin, H. Chi-Kuang, and E. L. Becker, *J. Immunol.* **129,** 1608 (1982).
[24] A. DeLean, A. A. Hancock, and R. J. Lefkowitz, *Mol. Pharmacol.* **21,** 5 (1982).
[25] P. J. Manson and D. Rodbard, *Anal. Biochem.* **107,** 200 (1980).

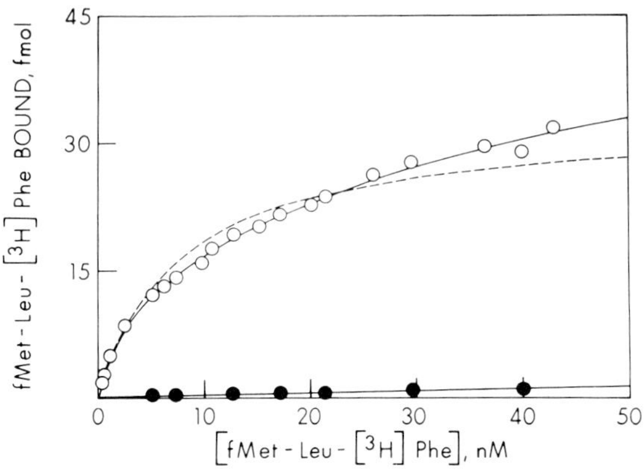

FIG. 1. Binding isotherm of fMet-Leu-[^{3}H]Phe to macrophage membrane preparations. Open circles represent actual data points for total binding, while filled circles represent nonspecific binding. The dashed line is a computer-generated isotherm assuming a homogeneous group of receptors with one affinity. The solid lines represent computer-generated isotherms based on a two-site model. The fit derived from a two-site model was significantly better ($P < 0.001$) than that derived from a one-site model. The two affinities (K_D values) derived from this experiment are 1.6 and 38.0 n*M* with 21% high-affinity sites. From Snyderman *et al.*[18]

approximately 20–25 data points are necessary for any given binding isotherm. Considerable skepticism should be exercised in evaluating the application of any modeling techniques to experiments where binding has not been performed under steady-state conditions or where insufficient amounts of data have been analyzed.

Analysis of Oligopeptide Receptor Binding Using Fluorimetric Techniques[19]

Materials

Fluoresceinated formylated hexapeptide
Antibody to fluorescein[26]
PMN suspension (10^7/ml)

[26] These antibodies are prepared by injecting rabbits with 50 μg of fluorescein isothiocyanate derivative of albumin or of keyhole limpet hemocyanin in Freund's complete adjuvant, followed by boosters (25 μg of protein) at 1 and 2 months and thereafter at two-month intervals. IgG is obtained by precipitation with 50% saturated ammonium sulfate, pH 7.0, and then dialysis versus phosphate-buffered saline or, when further purification is desired, versus 10 m*M* sodium phosphate buffer prior to chromatography on a DEAE-cellulose column.

Incubation buffer containing 0.1% bovine serum albumin, pH 7.4
Plastic cuvettes, 1-cm path length

Procedure. The cell suspensions (1.5 ml) contained in incubation buffer plus 0.1% bovine serum albumin are equilibrated in plastic cuvettes to the desired temperature by stirring in a thermostatted cuvette holder for equilibrium experiments or by placing them in the stirred and thermostatted sample turret of a spectrofluorometer (#4800, SLM Instruments, Inc., Urbana, IL). Cells are incubated with the fluoresceinated hexapeptide (0.5 – 10 n*M*) in the presence and absence of nonfluoresceinated hexapeptide for varying times.[20] Receptor-bound fluoresceinated peptide is quantified by measuring the residual fluorescence following the addition of 20 n*M* high-affinity antibody to fluorescein.[27] The antibody quenches the fluorescence of free hexapeptide, but receptor-bound ligand is inaccessible to quenching by the antibody until it is dissociated from the receptor.[20] The residual hexapeptide fluorescence in the presence of antibody is proportional to the amount of receptor-bound ligand since the fluorescence of the ligand is not affected by its association with the receptor. The spectroscopic conditions for the above assay are as follows: excitation at 490 nm through 8-nm slits and emission at 520 nm through 8-nm slits. Stray light is filtered with a three-cavity interference filter centered at 490 nm in the excitation path and a 3% glass filter in the emission path. External calibration of the degree of fluorescence is accomplished using Fluorotrol GF (Ortho Diagnostics, Westwood, MA) and BDRC Green Beads (Becton Dickinson FACS Systems, Sunnyvale, CA). The calibration of affinity contents and total numbers of receptor sites is described in Sklar *et al.*[20]

The above method of measuring oligopeptide receptors has allowed the real-time analyses of the oligopeptide chemoattractant receptor interaction and the relation of receptor occupancy to various evoked responses such as degranulation and superoxide production.[28]

[27] L. A. Sklar, Z. G. Oades, A. J. Jesaitis, R. G. Painter, and C. G. Cochrane, *Proc. Natl. Acad. Sci. U.S.A.* **78,** 7540 (1981).

[28] L. A. Sklar, P. A. Hyslop, Z. G. Oades, G. M. Omann, A. J. Jesaitis, R. G. Painter, and C. G. Cochrane, *J. Biol. Chem.* **260,** 11461 (1985).

[23] Contractile Proteins in Leukocytes

By MARK J. DINUBILE and FREDERICK S. SOUTHWICK

Introduction

Actin and the Cytoskeleton

The process of phagocytosis encompasses a broad spectrum of activities, including cellular locomotion, endocytosis, and secretion. How the phagocyte, e.g., a polymorphonuclear (PMN)[1] leukocyte or monocyte/macrophage, accomplishes these complex tasks is far from well defined. However, the key role of the actomyosin complex in skeletal muscle contraction and the discovery of filaments and subsequent identification of actin and myosin in a variety of motile cells have encouraged analogy between the contraction of muscle and the movement of phagocytes. The less orderly alignment of microfilaments in motile cells, the markedly reduced concentration of myosin relative to actin in phagocytes as opposed to muscle cells, and the differential cytoplasmic movement required for locomotion/endocytosis suggest that certain basic differences between the contractile systems of skeletal muscle and motile cells must exist.

The entangled appearance of filaments in the cytoplasm of motile cells provided a clue that actin polymers might be cross-linked. Such a network could amplify the contractile potential of the relatively limited supply of myosin. In addition, if the network structure could be independently regulated by changing actin filament length in localized regions of the cytoskeleton, differential movement of cytoplasm could be generated.

The Biochemistry of Actin

The cytoplasm of all motile cells contains large quantities of actin. This actin may exist as globular monomers (G-actin) or polymerize to form filaments (F-actin). The polymerization of actin has been likened to the condensation of a gas. When G-actin is added to a solution favorable for

[1] PMN, polymorphonuclear leukocyte; DFP, diisopropyl fluorophosphate; EGTA, ethylene glycol bis (β-aminoethyl ether)-N,N,N',N'-tetraacetic acid; S, supernatant; P, pellet; SDS, sodium dodecyl sulfate; PAGE, polyacrylamide gel electrophoresis; MES, 2-(*N*-morpholino)ethanesulfonic acid; EDTA, ethylenediaminetetraacetic acid; DTT, dithiothreitol; HPLC, high-pressure liquid chromatography; ABP, actin-binding protein (macrophage); PMSF, phenylmethylsulfonyl fluoride; NBD, 7-nitrobenz-2-oxa-1,3-diazole; FACS, fluorescence activated cell sorter.

polymerization, it remains in the monomeric state until a "critical monomer concentration" is exceeded; thereafter, any additional G-actin will be incorporated into filaments.

Kinetic studies indicate that there are two steps in actin filament assembly, nucleation and elongation. The rate of nucleation is slow and reflects the time required for three or four monomers to bind to each other. This is the rate-limiting step in filament formation. Once nuclei are formed, elongation begins.

The two ends of an actin filament are different. Myosin subfragments in the absence of ATP bind to actin filaments at 45° angles, defining barbed and pointed ends, which can be visualized by electron microscopy. Monomers assemble onto the barbed end (the preferred end for growth) more rapidly than onto the pointed end. In addition to different rate constants, each filament end may possess a different critical concentration in the presence of ATP and divalent cations. The apparent critical concentration for the entire filament (C_o) is intermediate between the lower critical concentration of the barbed end (C_o^B) and the higher critical concentration of the pointed end (C_o^P) and depends on the ionic conditions. At steady state when filament length is constant, there can be net growth of actin monomers onto the barbed end exactly matched by net disassembly from the pointed end. Under these circumstances, actin protomers can "treadmill" through a filament from the barbed to the pointed end.

Regulation of the Microfilament Network

Actin may be present in the cell as monomers, polymers, or cross-linked filaments. The state of actin is primarily determined by the intracellular ionic conditions and the concentration and availability of proteins which bind actin.

These regulatory proteins, often termed actin-binding or actin-modulating proteins, can profoundly influence the filament network in the cell cytoplasm. Based on their function they can be classified as (1) cross-linking proteins (e.g., α-actinin, macrophage actin-binding protein, filamin) which, depending on their size and flexibility, interact with polymerized actin to form bundles or gels; (2) end-blocking or capping proteins (e.g., gelsolin, 41,000-Da protein, acumentin) which selectively block filament ends and affect the polymer length and distribution (these proteins also "nucleate" actin assembly and some can sever preformed actin filaments); and (3) monomer-stabilizing proteins (e.g., profilin) which bind actin monomers and prevent their incorporation into filaments.

The effect of any actin-modulating protein on the cytoskeleton will be influenced by the presence of other actin-modulating proteins and may be different in different regions of the cytoplasm. The concentration of a crossing-linking protein necessary to gel actin is determined in part by the length distribution of actin filaments, which, in turn, is affected by capping (and severing) proteins that shorten actin filaments. Changes in ionic calcium concentration control the activity of some capping proteins, like gelsolin and 41,000-Da protein. Gelsolin and 41,000-Da protein bind only the barbed and acumentin the pointed ends of actin filaments, and these proteins therefore can change the critical monomer concentration necessary for actin polymerization. Through changes in critical concentration and other mechanisms, end-blocking proteins influence the amount of actin bound to profilin (profilactin). Obviously, complex interactions between actin and its regulatory proteins control the structure of the cytoskeleton in different cytoplasmic domains.

Detailed reviews of the contractile proteins of leukocytes are available.[1a,2] The following sections outline the methods required to produce cytoplasmic extracts from granulocytes and macrophages, to purify actin and actin-modulating proteins from these extracts, and to study the interactions of these various proteins. Finally we review some recently developed methods for examining the *in vivo* behavior of actin in phagocytes.

Preparation of Cytoplasmic Extracts from Leukocytes

Preparation of Human Granulocyte Extracts[3]

Human PMNs are obtained from discarded units of whole blood from a blood bank. Blood is processed immediately after withdrawal from the donor. Blood is dextran sedimented (3 parts blood to 1 part 2% T-500 dextran in 0.154 *M* NaCl solution), and the supernatant is diluted 1 : 1 with normal saline and centrifuged at 250 *g* for 30 min. This and all subsequent steps are performed at 4°. The resulting pellet is washed twice in 30 volumes of 0.154 *M* NaCl solution by centrifugation at 250 *g* for 10 min. This method yields white cell preparations containing 80–85% PMNs, the remainder of the cells being lymphocytes and monocytes. In some preparations 30-ml samples of blood are sedimented through a 13-ml layer of Ficoll–sodium diatrizoate solution according to the method of Bøyum[4] followed by dextran sedimentation. This method yields 95–98% PMNs (see also this series, Vol. 108 [9]).

[1a] F. S. Southwick and T. P. Stossel, *Semin. Hematol.* **20,** 305 (1983).
[2] T. P. Stossel, *J. Cell Biol.* **99,** 155 (1984).
[3] F. S. Southwick and T. P. Stossel, *J. Biol. Chem.* **256,** 3030 (1981).
[4] A. Bøyum, *Scand. J. Clin. Lab. Invest. Suppl.* **97,** 77 (1968).

Packed granulocytes are suspended in an equal volume of 0.154 *M* NaCl solution followed by the addition of 0.05 volumes of 0.1 *M* diisopropyl fluorophosphate (DFP, Sigma) (maintained as a stock solution in propylene glycol).[5] After 5 min, cells are washed with 10–15 volumes of deionized water mixed by inversion for 10 sec and immediately centrifuged in tightly sealed tubes. This step lyses most erythrocytes. The supernatant is discarded into a saturated NaOH solution in order to deactivate DFP. [Because DFP is a potent acetylcholinesterase inhibitor, this agent is carefully handled with double rubber gloves and in a fume hood. A 300-ml squeeze bottle with 5 *M* NaOH is always on hand to inactivate any spilled material rapidly. Atropine sulfate (1 mg/ml, Ivenex, Dexter Corp., Chagrin Falls, OH) should be available for immediate intramuscular injection in case of accidental exposure. The details are available in the product description.]

Packed granulocytes are mixed with 2–3 volumes of homogenization buffer containing 0.34 *M* sucrose, 5 m*M* EGTA, 20 m*M* imidazole–HCl, 1 mg% leupeptin, 7.8 mg% benzamidine, 4 mg% aprotinin, pH 7.5 (all available from Sigma Chemical Co.). Cells are broken in the cold by 60–120 strokes with a tight-fitting pestle in a glass Dounce homogenizer. This procedure ruptures greater than 90% of all cells as monitored by phase-contrast microscopy. Homogenates are then centrifuged at 12,000 *g* for 30 min yielding a clear supernatant (S) as the starting extract designated S_1. The final protein concentration varies from 5 to 8 mg/ml. To prepare actin-depleted extract (S_2), the solution is made 0.6 *M* with 3 *M* KCl and incubated for 2 hr at room temperature and then centrifuged at 100,000 *g* for 3.5 hr to sediment all actin filaments. The resulting supernatant contains approximately 40% of the actin contained in the S_1 extract. The pellet (P) is designated P_2.

Preparation of Rabbit Macrophage Extracts[6,7]

New Zealand white rabbits are injected in the marginal ear vein with 1–1.5 ml of Freund's complete adjuvant (Difco Laboratories, Detroit, MI) containing BCG (Bacillus Calmette–Guérin) vaccine (4% v/v) (University of Illinois Medical Center, Chicago, Illinois). The animals are sacrificed 2–3 weeks later, and the macrophages are isolated by tracheal lavage (see also this series, Vol. 108 [25]). Macrophages are then washed twice with ice-cold 0.15 *M* NaCl solution by centrifugation at 250 *g* at 4° for 10 min. They are then treated with DFP and homogenized as described for granu-

[5] P. C. Amrein and T. P. Stossel, *Blood* **56,** 442 (1980).
[6] J. H. Hartwig and T. P. Stossel, *J. Biol. Chem.* **250,** 5696 (1975).
[7] M. J. DiNubile and F. S. Southwick, *J. Biol. Chem.* **260,** 7402 (1985).

locytes. The homogenate is then centrifuged at 100,000 *g* for 1 hr. The resulting supernatant designated S_1 generally has a protein concentration of 6–10 mg/ml. To deplete this extract of actin, S_1 is made 0.1 *M* with 3 *M* KCl and incubated at 25° for 2 hr until the solution has gelled. The solution is then centrifuged at 12,000 *g* for 10 min to remove the cross-linked actin. The resulting supernatant S_2 contains approximately 50% of the actin found in S_1. The pellet is again designated P_2.

Comments. Actin is purified from S_1 extract while the purifications of profilin, gelsolin, and acumentin begin with S_2 extract. The preparation of actin-binding protein starts with the P_2 pellet. After collection and washing, macrophages are processed differently to purify α-actinin.

Purification of Actin from Leukocytes (see also this series, Vol. 85 [17] and [18])

Preparation of Macrophage Actin[8] *(using a modification of the method to purify actin from Acanthamoeba castellanii)*

Approximately 100 ml (wet volume) of DEAE-Sepharose CL-6B (Pharmacia) is suspended in 100 ml of buffer D (containing 10 m*M* imidazole–HCl, 0.1 m*M* $CaCl_2$, 0.75 m*M* 2-mercaptoethanol, 0.5 m*M* ATP, pH 7.5) made 0.1 *M* in KCl. ATP is added until its concentration (calculated from the absorbance at 260 nm) in the supernatant reaches 0.5 m*M* at equilibrium; the pH is readjusted to 7.5 with KOH before and after ATP is added. Generally, 0.5 g of ATP is sufficient to achieve saturation under these conditions, and equilibrium is complete within 15 min.

The ATP-saturated DEAE-Sepharose is packed into a column (2.5 × 15 cm) and equilibrated by washing with 300 ml of 0.1 *M* KCl in buffer D. Freshly prepared macrophage S_1 extract (within 2–4 hr of preparation) is applied to the column. Generally, 60–100 ml of a 6–10 mg/ml solution of S_1 extract containing 10% actin is used for a single purification. Fifty milliliters of buffer G (containing 3 m*M* imidazole base, 0.1 m*M* $CaCl_2$, 0.75 m*M* 2-mercaptoethanol, 0.5 m*M* ATP, pH 7.0) is applied to the DEAE-Sepharose column immediately *before* and *after* the sample to insure that the actin is not exposed to 0.1 *M* KCl. The column is next eluted with 200–300 ml of 0.1 *M* KCl in buffer D and 1 liter of a linear gradient from 0.1 to 0.5 *M* KCl in buffer D. Samples eluting between 0.19 and 0.24 *M* KCl are subjected to sodium dodecyl sulfate–polyacrylamide gel electrophoresis (SDS–PAGE) to confirm the presence of actin (molecular weight 42,000) as the major component of these fractions.

[8] D. J. Gordon, E. Eisenberg, and E. D. Korn, *J. Biol. Chem.* **251,** 1778 (1976).

The discontinuous 5–15% polyacrylamide gradient slab gel system of Laemmli[9] is used to determine the purity of the protein preparations. Protein samples are boiled for 1 min in 1% SDS, 10% sucrose, and 2% 2-merceptoethanol. The intensity of the Coomassie blue-stained protein bands is determined by scanning the gels with a densitometer (Zeineh soft laser densitometer, Fullerton, CA). Protein concentrations are measured by the method of Lowry *et al.*[10]

The fractions enriched with actin are pooled and dialyzed against buffer G containing 0.1 *M* KCl and 2 m*M* $MgCl_2$. The solution is then incubated for at least 30 min at 25° and centrifuged at 100,000 *g* for 3.5 hr at 20°. The clear gelatinous pellet is homogenized in 10–20 ml of buffer G containing 0.01% NaN_3 to retard the growth of microorganisms. The suspension is then dialyzed against buffer G with 0.01% NaN_3 for several days with frequent changes of buffer. Residual material that has not depolymerized should be removed by centrifugation at 100,000 *g* for 90 min. Actin is stored at 4° at a concentration of 5–10 mg/ml.

To remove actin aggregates and enhance the purity of G-actin, the depolymerized actin can be applied to a column of Sephadex G-150 and eluted with buffer G containing 0.01% NaN_3. Actin elutes as an asymmetric peak. The appropriate fractions may be pooled and concentrated to 5–8 mg/ml by ultrafiltration on an Amicon PM10 membrane. Gel-filtered actin should be used for kinetic analyses within 24 hr so that actin aggregates do not reform.

Preparation of Human Lymphocyte Actin[11]

Leukocytes can be harvested from normal human donors and patients with chronic lymphoid leukemia from heparinized blood or as a by-product of leukapheresis. Mononuclear cells are prepared by centrifugation on Ficoll–Hypaque gradients, and monocytes are removed by adherence to Falcon plastic culture dishes or by centrifugation through Percoll gradients (see also this series, Vol. 108 [9]). The degree of cellular purity can be assessed by Coulter Counter sizing. The preparations should contain less than 2% monocytes and 10% platelets. The lymphocytes are collected as pellets after centrifugation at 3,000 *g* for 10 min and stored at −70° until used.

[9] U. K. Laemmli, *Nature (London)* **227,** 680 (1970).

[10] O. H. Lowry, N. J. Rosebrough, A. L Farr, and R. J. Randall, *J. Biol. Chem.* **193,** 265 (1951).

[11] L. F. Liebes, R. Stank, D. Neurla, G. Grusky, D. Zucker-Franklin, and R. Silber, *Cancer Res.* **43,** 4966 (1983).

Actin can be purified from 5–7 g of packed lymphocytes by a method similar to that described above for macrophage actin.

Preparation of Guinea Pig Granulocyte Actin[12]

Adult male guinea pigs (500–800 g) are injected intraperitoneally with 20 ml of sterile 15% sodium caseinate in isotonic saline. Between 12 and 16 hr later, the guinea pigs are sacrificed by diethyl ether application, and the peritoneal exudate is removed by repeated lavage and aspiration with a solution consisting of 80% isotonic saline and 20% citrate–phosphate–dextrose (2.63% sodium citrate, 2.32% glucose, 0.327% citric acid, 0.25% sodium dihydrogen phosphate, adjusted to pH 7.4 with Tris).

The cell suspension is passed through two layers of muslin, and the cells are sedimented by centrifugation (100 *g*, 15 min) and washed 3 times with fresh aspirate solution and once with 0.34 *M* sucrose buffered to pH 7.4 with Tris–HCl. The cells are resuspended in 1.5 volumes of ice-cold 0.1 *M* 2-(*N*-morpholino)ethanesulfonic acid (MES) buffer, pH 6.8, containing 1 m*M* EGTA, 1 m*M* EDTA, and 0.3 *M* sucrose and are disrupted in a Potter–Elvehjem type homogenizer (side-to-wall clearance approximately 0.13 mm, 30 passes at 1400 rpm).

The homogenate is centrifuged at 1000 *g* for 4 min and the thick, creamy supernatant is carefully decanted. The loose pellet is resuspended in half the previous volume of sucrose-containing buffer and the homogenization repeated with 15 passes. After centrifugation, the supernatants are combined, yielding the nuclei-free cytoplasmic fraction. When observed under light microscopy there should be no detectable nuclear contamination in this fraction. The nuclei-free supernatant, after centrifugation at 100,000 *g* for 60 min at 4° yields a particulate phase (pellet) and a soluble phase (supernatant).

The supernatant is next diluted with an equal volume of 8 *M* glycerol, and the Mg^{2+} concentration is brought to 10 m*M* by addition from a 1 *M* stock $MgCl_2$ solution. This diluted soluble phase is then incubated at 37° for 20 min, and the flocculent precipitate which forms is collected by centrifugation at 100,000 *g* for 60 min at 20°. (Save the supernatant which still contains actin that can be precipitated by KCl; see next paragraph.) The pellet containing high concentrations of F-actin is resuspended by gentle hand homogenization in 5 ml of 10 m*M* Tris–HCl, 7.4, 2 m*M* ATP, 0.5 m*M* 2-mercaptoephanol, pH 7.4, depolymerized by dialysis against this solution overnight at 4°, and then clarified by centrifugation at 100,000 *g* for 30 min. This solution contains the now solubilized magnesium-precipitated actin.

[12] N. Crawford, H. Chahal, and P. Jackson, *Biochim. Biophys. Acta* **626,** 218 (1980).

The supernatant remaining after removal of the magnesium-precipitated actin is dialyzed to lower the glycerol content, then made 0.1 *M* by the addition of solid KCl and stirred gently at room temperature for 45 min. This solution is then made 0.6 *M* KCl by further addition of solid KCl and allowed to stand for 30–60 min, after which the solution is centrifuged at 100,000 *g* for 100 min at 20°. This pellet contains the potassium-precipitated actin, which can be solubilized in the same way as the magnesium-precipitated actin.

Comments. The actin found in leukocytes may not be homogeneous. The last method described above for actin purification yields only that actin which polymerizes in the granulocyte extract. Actin which forms filaments less readily is discarded. Similarly, actin purified by DEAE ion exchange and gel filtration may not be representative of the entire actin pool, since final yields represent only 3–5% of the actin found in the starting cytoplasmic extract.

ATP and a divalent cation are necessary to maintain G-actin in its native monomeric form. Since DEAE-Sepharose is an anion exchanger which binds ATP avidly, it is necessary to saturate the resin with ATP in order to obtain a detectable level of ATP in the eluate. Muscle G-actin is more than 50% denatured by conventional DEAE-cellulose chromatography. There is no denaturation of actin, however, in ATP-saturated DEAE-cellulose columns prepared freshly for each actin preparation.[8]

Methods for purifying actin from skeletal muscle are also well described.[13,14] Most investigators have used skeletal muscle actin to study the function of actin-modulating proteins purified from mammalian nonmuscle cells. Ideally, however, the function of these proteins should also be investigated using actin purified from the same nonmuscle cell. Some of the nonmuscle cell actin-binding proteins may have different affinities for nonmuscle actin as compared to skeletal muscle actin. For example, profilin purified from *Acanthamoeba castellanii* has a 10-fold higher affinity for *Acanthamoeba* actin than for skeletal muscle actin.[15]

Actin has been purified from many other nonmuscle sources.[16,17] The proteases of different cell types may necessitate certain modification, e.g., the avoidance of dithiothreitol (DTT) which apparently activates the proteases of *Entamoeba histolytica* despite the presence of inhibitors.[17]

In experiments assessing kinetic parameters, gel-filtered actin must be used (as described at the end of the section on purification of macrophage

[13] J. A. Spudich and S. Watt, *J. Biol. Chem.* **246,** 4866 (1971).

[14] J. D. Pardee and J. A. Spudich, this series, Vol. 85, [18].

[15] L. S. Tobachman and E. D. Korn, *J. Biol. Chem.* **257,** 4166 (1982).

[16] D. J. Gordon, J. L. Boyer, and E. D. Korn, *J. Biol. Chem.* **252,** 8300 (1977).

[17] I. Meza, M. Sabanero, F. Casares, and J. Bryan, *J. Biol. Chem.* **258,** 3936 (1983).

actin) to insure that actin aggregates do not act as nuclei for filament growth.

Purification of Macrophage Actin-Modulating Proteins

Preparation of Profilin[7]

The S_2 extract (300–600 mg total protein) is dialyzed against 30 volumes of a solution containing 1 m*M* EGTA, 1 m*M* DTT, 0.5 m*M* ATP, and 10 m*M* imidazole–HCl, pH 7.8 (buffer S). The dialyzed sample is applied to a column (2.5 × 30 cm) containing DEAE-Sepharose CL-6B anion-exchange resin (Pharmacia) equilibrated with 500 ml of buffer S. The column is then percolated with 200 ml of buffer S made 60 m*M* in KCl. Proteins which elute during this step are pooled and fractionally precipitated with ammonium sulfate between 1.2 and 2.4 *M*. The precipitate is resuspended in a small volume (<2 ml) of solution containing 10 m*M* imidazole, 0.01% NaN_3, pH 7.6 (buffer P), and dialyzed overnight in 400 volumes of this same buffer. The sample is then applied to a Sephadex G-75 (Pharmacia) gel filtration column (2 × 90 cm). Fractions are eluted with buffer P. Fractions which reduce the viscosity of F-actin as assessed by the falling ball assay (see below) and correspond to the enrichment on SDS–PAGE of a polypeptide with a molecular weight of 15,500 contain profilin.

An affinity chromatography method for purifying profilin has recently been described. This method is based on the observation that poly(L-proline) specifically binds profilin and profilin–actin complexes with high affinity.[18] Poly(L-proline) (M_r 1000–10,000, Sigma) (5 mg/ml of packed beads) is conjugated with cyanogen bromide-activated Sephadex beads (CNBr-activated Sepharose 4B, Pharmacia), using the protocol provided by the manufacturer. Macrophage S_1 extract, prepared as described above, is first made 0.1 *M* with 3 *M* KCl and 2 m*M* with 100 m*M* $MgCl_2$ and then incubated at 25° for 2 hr to polymerize actin in the extract. Incubation is followed by centrifugation at 100,000 *g* for 2 hr. We have found that this step removes high concentrations of macrophage actin. This and all subsequent steps are performed at 4°.

The resulting supernatant is next passed through a column (0.5 × 5 cm) containing the Sephadex beads conjugated with poly(L-proline). After being washed with 2 volumes of buffer containing 1 m*M* DTT, 0.02% NaN_3, and 10 m*M* imidizone–HCl, pH 7.4 (imidizole buffer), 1 column volume of 5 mg/ml poly(L-proline) suspended in imidizole buffer

[18] M. Tanaka and H. Shibata, *Eur. J. Biochem.* **151,** 291 (1985).

is percolated into the column and allowed to incubate for 90 min to dissociate profilin from the beads. The column is then eluted with 2 volumes of 0.1 *M* KCl in imidizole buffer. These 3 column volumes are saved and concentrated using a positive nitrogen pressure filtration with an Amicon PM10 membrane at 30 psi. The concentrate is then passed thru a Sephadex G-75 gel filtration column (1 × 90 cm) using imidizole buffer containing 0.1 *M* KCl. Elution fractions are monitored by absorbance at 280 nm. An early absorption peak corresponding to the void volume contains 42,000- and 15,500-Da polypeptides (profilin–actin complex). The second absorption peak consistently contains a single polypeptide of 15,500 Da (>98% purity as assessed by Coomassie blue-stained SDS–PAGE) which markedly inhibits actin nucleation. Profilin prepared by this method has a K_D value (2–3 μM) similar to profilin prepared from macrophages using ion-exchange chromatography and gel filtration.

Comments. The ion-exchange method purifies only profilin which is not bound to (or easily dissociable from) actin. A second fraction of profilin elutes from the ion-exchange column at a higher KCl concentration (0.19 *M*). This fraction of profilin is tightly bound to actin, generally requiring urea treatment for dissociation.[19] The harsh conditions required to derive free profilin from this fraction tend to denature the protein, resulting in some variation in its ability to modulate actin function. Comparisons of *Acanthamoeba* profilin derived from these two fractions, however, have revealed no functional or structural differences.[20] At present there is no definitive explanation for the very high affinity of profilin for actin under certain conditions. A putative cofactor which enhances the binding of actin by profilin has not been purified.

Affinity chromatography results in much higher yields (60%)[18] than the ion-exchange chromatography method (1–2%).[7] Purity as assessed by SDS–PAGE is also higher; however, profilin prepared by affinity chromatography is contaminated with poly(L-proline). We have found that the presence of this polyamino acid does not affect profilin's ability to inhibit actin nucleation or actin monomer addition to the pointed ends of actin filaments. To eliminate poly(L-proline) contamination, profilin can be dissociated from the affinity column by elution with 6 *M* urea and the urea subsequently removed by dialysis.[21]

[19] L. Carlsson, L. E. Nystrom, U. Lundberg, K. K. Kannan, H. Cid-Dresser, S. Longren, and H. Jormroll, *J. Mol. Biol.* **105,** 353 (1976).

[20] P. C.-H. Tseng, M. S. Runge, J. A. Cooper, R. C. William, Jr., and T. D. Pollard, *J. Cell. Biol.* **98,** 214 (1984).

[21] S. E. Lind, P. A. Jammey, C. Chaponnier, T.-J. Herbert, and T. P. Stossel, *J. Cell Biol.* **105,** 833 (1987).

Preparation of 41,000-Da Protein

The most recent actin-regulatory protein to be purified from macrophages is the 41,000-Da protein.[22] After loading the macrophage extract on the DEAE anion-exchange column as described above, the column is eluted with 1 liter of a linear 0–0.45 *M* KCl gradient in buffer S. Fractions are dialyzed in buffer S and then assayed for their ability to decrease the viscosity of polymerized actin in the presence of micromolar calcium (final $CaCl_2$ concentration 1 m*M* in buffer S) using the falling ball assay described below. There are two peaks of activity, the first eluting at a KCl concentration of 60 m*M* and corresponding to enrichment of a 41,000-Da polypeptide and the second eluting at a KCl concentration of 0.17 *M* corresponding to enrichment of 91,000-Da protein (gelsolin, see below). The samples representing the first activity peak are pooled and concentrated approximately 50-fold using a Collodion UH 100/10 vacuum dialysis membrane (Schleicher & Schuell, Keene, NH) in buffer S containing 0.1 *M* KCl. The sample is then applied to a series of two Protein Pak 125 (molecular mass range, native globular proteins, 2,000–80,000 Da) and two Protein Pak 300sw (molecular mass range, 10,000–400,000 Da) silica high-pressure gel filtration columns (Waters Associates, Milford, MA). Columns are eluted at 0.5 ml/min with buffer S containing 0.1 *M* KCl, pH 7.3. The fractions are analyzed for purity by SDS–PAGE and for activity in calcium-containing buffer by the falling ball assay as described below.

Preparation of Gelsolin[23,24]

Samples from the ion-exchange chromatography step corresponding to the second peak of calcium-sensitive activity described above are also pooled. Solid ammonium sulfate is then added to the solution to achieve a final concentration of 306 mg/ml. The precipitated proteins are collected by centrifugation at 15,000 *g* for 10 min at 4°, dissolved in 1 ml of a solution containing 0.8 *M* KCl, 1 m*M* EGTA, 5 m*M* DTT, and 10 m*M* imidazole–HCl, pH 7.5, and dialyzed against the same solution. This solution is also used for equilibration and elution of a Sephadex G-150 column (1.6 × 90 cm) to which the sample is applied. The eluted fractions containing gelsolin are assayed as described for the DEAE fractions.

Comments. All solutions used in purifying gelsolin contain very low concentrations of calcium (nanomolar range). When the ionized calcium

[22] F. S. Southwick and M. J. DiNubile, *J. Biol. Chem.* **261,** 14191 (1986).

[23] H. L. Yin and T. P. Stossel, *Nature (London)* **281,** 583 (1979).

[24] H. L. Yin and T. P. Stossel, *J. Biol. Chem.* **255,** 9490 (1980).

concentration is raised to the micromolar range this protein will irreversibly bind actin, making separation of gelsolin from actin difficult.

Preparation of Acumentin[25]

The fractions eluted from the DEAE anion-exchange column are also assayed for calcium-independent actin filament-shortening activity. After dialyzing elution fractions in buffer S, those samples containing a protein of molecular weight 65,000 on SDS–PAGE with activity which decreases the final viscosity of actin polymerized in 0.1 *M* KCl in buffer S (see below) are pooled and concentrated to a volume of 2–3 ml in a nitrogen pressure concentrator using an Amicon PM10 ultrafiltration membrane. The concentrated sample is applied on a Sephadex G-200 column (1.2 × 95 cm) equilibrated and eluted with 0.1 *M* KCl in buffer S. To enhance purity, this step can be followed by hydroxyapatite chromatography using a column (1 × 8 cm) eluted with 120 ml of a linear gradient of 0.05–0.4 *M* potassium phosphate buffer, pH 6.5. SDS-PAGE and falling ball viscometry are performed to determine the purity and activity of the eluted fractions.

The ability of acumentin to shorten actin filaments rapidly diminishes after the protein is purified to near homogeneity. Often the specific activity of this protein decreases during overnight gel filtration. More recently, we have employed gel filtration with two I-125 and two 300sw high-pressure liquid chromatography (HPLC) columns (Waters) in series. Purification by HPLC results in comparable purity and requires only 1 hr. Active samples are immediately diluted 1 : 1 with glycerol and stored at −20°. Under these storage conditions, acumentin remains active for several months.

Comments. Acumentin is a relatively labile protein which is very susceptible to proteolytic degradation. Use of fresh DFP during preparation of the starting extract is critical for preventing proteolytic cleavage to lower molecular weight by-products. This protein should be purified and stored in glycerol within 4–5 days of initial homogenization of the cells. Longer purification times result in rapid loss of acumentin activity.

Preparation of Actin-Binding Protein[5,26] *(see also this series, Vol. 85 [45])*

The P_2 pellet derived during preparation of the S_2 extract is resuspended in and then dialyzed against KI solution (1.2 *M* KI, 5 m*M* ATP, 5 m*M* DTT, 1 m*M* EGTA, 10 m*M* imidazole–HCl, pH 7.5). The dialyzate is clarified by centrifugation at 17,000 *g* for 10 min, and the superna-

[25] F. S. Southwick, N. Tatsumi, and T. P. Stossel, *Biochemistry* **22,** 6321 (1982).

[26] H. L. Corwin and J. H. Hartwig, *Dev. Biol.* **99,** 61 (1983).

tant is applied to a column (2.5 × 120 cm) of agarose (BioGel A-15, 200–400 mesh, Bio-Rad Laboratories). The column is equilibrated with KCl buffer (0.6 *M* KCl, 0.5 m*M* ATP, 0.5 m*M* DTT, 1 m*M* EGTA, 10 m*M* imidazole–HCl, pH 7.5). A volume of KI solution (described above) equal to approximately 10% of the volume of the column is applied immediately before the sample. Elution is with the 0.6 *M* KCl buffer. The eluted fractions are dialyzed into 0.1 *M* KCl, 1 m*M*, EGTA, 10 m*M* imidazole–HCl, pH 7.5.

The purity of the actin-binding protein (ABP) (two subunits, each with a molecular weight of 265,000) preparation can be determined by SDS–PAGE. The activity of this protein is measured by actin sedimentation activity or gel point determinations by falling ball microcapillary viscometry (see below).

Comments. ABP is also very susceptible to proteolysis. The importance of DFP treatment of cells and rapid purification followed by storage at −20° in 1 : 1 glycerol again needs to be emphasized.

Preparation of Macrophage α-Actinin[27] *(see also this series, Vol. 85 [29])*

Rabbit macrophages are obtained as described earlier in the preparation of macrophage extracts (see above). The cells are washed twice in 0.15 *M* NaCl and treated with 5 m*M* DFP. After 2 additional normal saline washes, 20 ml of packed cells are resuspended in 180 ml of ice-cold 0.1 *M* KCl, 10 m*M* EGTA, 2 m*M* $MgCl_2$, and 10 m*M* imidazole hydrochloride, pH 7.5, and 20 ml of 10% (w/v) Triton X-100 is added slowly while stirring; this buffer is designed to maintain the integrity of the cytoskeleton after solubilization. After 5 min at 0°, the solubilized cells are centrifuged for 20 min at 12,000 *g* at 4°, and the pellets containing the cytoskeletal proteins are resuspended in a mixture containing 2 m*M* EGTA, 1 *M* $MgCl_2$, and 10 m*M* imidazole–HCl, pH 7.5, and incubated for 30 min at 37° in a shaking water bath.

After centrifugation at 12,000 *g* at 4° for 20 min, the supernatant is applied directly to a 20 ml column of hydroxyapatite (BioGel HTP) equilibrated in the resuspension buffer. The column is washed with 100 ml of 25 m*M* potassium phosphate, pH 7.5, and then eluted with 100 ml of 90 m*M* pottassium phosphate, pH 7.5. Solid ammonium sulfate (30 g/100 ml) is added to the eluate, and the precipitated protein is collected by centrifugation at 12,000 *g* for 20 min at 4°. The protein is resuspended in 10 ml of buffer consisting of 0.1 *M* KCl, 0.1 m*M* EGTA, 2 m*M* NaN_3, and

[27] J. P. Bennett, K. S. Zaner, and T. P. Stossel, *Biochemistry* **23,** 5081 (1984).

10 mM imidazole–HCl, pH 7.5. After clarification by centrifugation at 12,000 g for 20 min at 4°, the sample is applied to a column (90 × 2.5 cm) of agarose (BioGel A-5m, 200–400 mesh) and chromatographed in the same buffer. α-Actinin elutes as the principal protein peak and generally is separated from other contaminants as assessed by Coomassie blue-stained SDS–PAGE. The "bundling" of actin filaments by this 103,000-Da protein is most simply measured by the falling ball assay (see below).

Preparation of Myosin[6] *(see also this series, Vol. 85 [7]–[12]).*

The macrophage S_2 extract is centrifuged at 80,000 g for 3 hr at 4°. The sediment is dissolved in ice-cold 0.6 M KCl solution with the aid of a glass tissue grinder and then applied to a column (2 × 80 cm) of Bio-Gel A-15m, 200–400 mesh (4% agarose), and eluted at 4° with 0.6 M KCl solution. Eluted fractions with potassium- and EDTA-activated ATPase activity are pooled, concentrated in a dialysis bag in dry Sephadex-G200, and dialyzed against a solution containing 0.6 M KI, 10 mM sodium thiosulfate, 5 mM ATP, 5 mM DTT, 20 mM Tris–maleate buffer, pH 7.0 (KI solution). The dialyzed protein solution is next applied to a column (2 × 80 cm) of Bio-Gel A-15m, 200–400 mesh, equilibrated with KI solution. Immediately before the addition of the sample, a volume of KCl solution equal to approximately 15% of the volume of the column is applied. Fractions are eluted with KCl solution at 4°, and those with potassium- and EDTA-activated ATPase activity, containing the purified myosin, are pooled and concentrated as described above and stored in 50% saturated ammonium sulfate solution at 4°.

Comments. Myosin undoubtedly plays an important role in cell motility. Myosin has been purified from leukocytes.[6,12,28] More recently two single-headed monomeric myosins (myosins IA and IB) have been purified from *Acanthamoeba castellanii*.[29,30]

Final Comments on Purification. With the exception of α-actinin, all of the above actin-modulating proteins can be purified from a single preparation of cells. Because of significant differences in their isoelectric points, each protein elutes from the DEAE ion–exchange column at a relatively discrete concentration of KCl. Acumentin and actin-binding protein should be purified first, followed by actin, profilin, 41,000-Da protein, and gelsolin. The structure and function of these latter four proteins remain relatively stable for at least 1 week when stored at 4°.

[28] L. A. Boxer and T. P. Stossel, *J. Clin. Invest.* **57,** 964 (1976).

[29] J. P. Albanesi, J. A. Hammer III, and E. D. Korn, *J. Biol. Chem.* **258,** 10176 (1983).

[30] H. Maruta, H. Gadasi, J. H. Collins, and E. D. Korn, *J. Biol. Chem.* **254,** 3624 (1979).

Study of the Properties of Purified Actin in the Presence of Purified Actin-Modulating Proteins

A comprehensive review of the commonly used methods for assessing actin filament formation is available in this series.[31] We discuss below only those methods which are most frequently used by investigators to study actin-modulating protein function.

Sedimentation Assay[3,32]

Samples are centrifuged at 160,000 *g* for 30 min in an air-driven centrifuge (Airfuge, Beckman Instruments, Inc., Palo Alto, CA). To determine the amount of actin which has sedimented, protein concentrations of the starting solution and of the supernatant fluid after centrifugation are measured, and the fraction of total protein representing actin is determined by quantitative densitometry of Coomassie blue-stained polyacrylamide gels after electrophoresis of the solutions in SDS.

Comments. The sedimentation assay has been used to study actin filament formation in cytoplasmic extracts from human granulocytes.[3] By this method we find that making the S_1 extract 0.6 *M* in KCl results in polymerization of approximately 60% of the total actin.

Under the conditions of this assay (160,000 *g* for 30 min), all molecules with a molecular weight greater than or equal to 600,000–700,000 would be expected to sediment. Therefore, actin filaments containing 14–16 monomers (approximate length 40–60 nm) would remain in the supernatant solution and could be misinterpreted as being in the monomeric state. Sedimentation is only a crude assay for distinguishing monomeric from filamentous actin.

Actin-sedimenting activity can also be used to monitor cross-linking of actin filaments.[32] The amount of actin remaining in the supernatant fluid after sedimentation can be compared in the presence and absence of a cross-linking protein.

Viscosity[3,33]

The viscosity of actin-containing solutions is measured in semimicroviscometers (Cannon–Manning, University Park, PA), extra low charge size 100 and 50, with buffer flow times of 48–50 and 95–100 sec, respectively. The flow time of buffer is compared to that of the actin solution.

[31] J. A. Cooper and T. D. Pollard, this series, Vol. 85, p. 182.

[32] E. A. Brotschi, J. H. Hartwig, and T. P. Stossel, *J. Biol. Chem.* **253,** 8988 (1978).

[33] H. L. Yin, K. S. Zaner, and T. P. Stossel, *J. Biol. Chem.* **255,** 9494 (1980).

Specific viscosity is calculated using the formula

$$\text{Specific viscosity} = \frac{\text{sample flow time (sec)}}{\text{buffer flow time (sec)}} - 1$$

Low shear viscosity of actin solutions can be measured by falling ball viscometry. One hundred-microliter capillary tubes (VWR, San Francisco, CA) are filled with the actin solution, one end sealed with tube sealer (Seal-ease, Clay Adams, Parsippany, NJ), and the tubes are set at a 45° angle. The time required for a 0.025-inch-diameter (grade 10) steel ball (Winsted Precision Ball Co., Winsted, CT) to roll the length of the tube is measured.

The usual concentration of actin used to measure viscosity by either method is 0.65 mg/ml. By standing the capillary tubes upright (90°) and using 1 mg/ml as the concentration of actin, the falling ball assay can be used to measure the gel point of F-actin (the concentration of cross-linking protein which causes the viscosity of actin-containing solution to increase abruptly) in the presence of cross-linking proteins and the solation of the actin gel by severing proteins.[33]

Comments. The viscosity of a solution containing actin will depend on many factors, including the critical actin monomer concentration, the total concentration of actin, the actin filament length distribution, and the extent and nature of cross-linking of actin filaments. Viscosity more closely reflects average filament length rather than absolute F-actin concentration.

Cannon–Manning viscometers subject actin filaments to high but constant shear which results in filament breakage. This technique is generally not as sensitive to small changes in F-actin concentration when compared to the falling ball capillary assay which can be performed at lower shear. However, shear in the latter assay is not constant but varies with the velocity of the falling ball. The falling ball assay tends to accentuate very small changes in filament length. Relative differences can be measured, but absolute viscosity values are difficult to accurately obtain by the falling ball method. The falling ball assay is useful for monitoring relative activity in column eluates while Cannon–Manning viscometers are most useful for quantitating the specific activities of actin filament-shortening proteins.

DNase I Inhibition[34]

Noncrystalline DNase I (type I, bovine pancreas, Sigma) is dissolved at a concentration of 0.1 mg/ml in 50 m*M* Tris–HCl, 0.25 m*M* $CaCl_2$, 0.01 m*M* phenylmethylsulfonyl fluoride (PMSF), 0.01% NaN_3, pH 7.5.

[34] I. Blikstad, F. Markey, L. Carlsson, T. Persson, and U. Lindberg, *Cell* **15,** 935 (1978).

Calf thymus DNA (final concentration 40 μg/ml) (type I, Sigma) is cut into fine pieces and dissolved in 0.1 mM Tris–HCl, 1.8 mM $CaCl_2$, 4 mM $MgSO_4$, 0.02% NaN_3, pH 7.5. The DNA is solubilized by stirring slowly at room temperature for 24–48 hr. The solution is then filtered through Whatman #1 filter paper. Absorbance at 260 nm should be 0.50–0.65. Ten microliters of DNase solution is added to 3 ml of DNA solution, and the slope of the increase in optical density at 260 nm is measured as the control. Ten microliters of DNAase solution is then mixed with 5–30 μl of sample, immediately (less than 5 sec) followed by the addition of 3 ml of DNA solution. The slope of the change in optical density at 260 nm is measured at the volume(s) of sample resulting in 25–75% reduction in slope compared to the control. If necessary, samples are diluted in 10 mM imidazole, 0.01% NaN_3, pH 7.6. A standard curve is derived by measuring the DNase I inhibition at known concentrations of G-actin. Using this curve, the apparent actin monomer concentrations of the samples can be determined.

Performance of the DNase I inhibition assay after the addition of guanidine–HCl, which depolymerizes actin, allows measurement of the total actin concentration in the sample. To measure the total actin, the sample is mixed with an equal volume of a mixture containing 1.5 M guanidine–HCl, 1 M sodium acetate, 1 mM $CaCl_2$, 1 mM ATP, and 20 mM Tris–HCl, pH 7.5. The sample is then incubated at 4° for 5–10 min to allow depolymerization of actin before adding DNase.

Comments. The purpose of the DNase assay is to determine the actin monomer and total actin concentration. The measurement of the critical actin monomer concentration under different ionic conditions can be easily performed.[7] Its advantage over the previously described assays is that DNase I inhibition is not greatly influenced by actin filament length. However, the number of actin filaments may affect the assay.[35] Actin-modulating proteins which bind to filaments, especially the barbed ends, or complex actin monomers may interfere with this assay. Guanidine may reduce the activity of DNase I.[36] In addition, the assay is difficult to perform reproducibly, and recorded values should reflect the mean of at least three measurements.

Light Scattering[15]

Samples are clarified by centrifugation for 10 min at 45,000 *g*; 400 μl of supernatant is placed into cuvettes in a fluorescence spectrophotometer with excitation and emission wavelengths of 400 nm. Bandpasses are 1.5

[35] J. C. Pinder and W. B Gratzer, *Biochemistry* **21,** 4886 (1982).

[36] K. M. K. Rao and J. Varami, *J. Immunol.* **129,** 1605 (1982).

and 2.0 nm for excitation and emission, respectively. Light scattering is then measured at an angle of 90° in arbitrary units.

Comments. Right-angle light scattering is based on the assumption that F-actin scatters the same amount of light regardless of filament length while G-actin scatters negligible light. This assumption is violated when the number of small filaments (approximately equal to or less than the wavelength of light used) is a significant fraction of the total filament population.[37] Thus, light scattering may overestimate the actin monomer concentration when the number of actin oligomers is relatively high.

Pyrenylactin Fluorescence[38-40]

Pyrenylactin is prepared by a modification of the procedure of Kouyama and Mikashi,[38] as described by Brenner and Korn.[39] Pyrene-conjugated G-actin increases its fluorescence by approximatley 20-fold when incorporated into filaments. This assay therefore can serve to distinguish monomeric from filamentous actin.

G-actin (1–2 mg/ml) is polymerized for 2 hr at room temperature in 5 m*M* Tris–HCl, 0.1 *M* KCl, 2 m*M* $MgCl_2$, 0.2 m*M* ATP, 0.1 m*M* $CaCl_2$, 0.01% NaN_3, pH 7.8. A 0.20-ml aliquot of freshly prepared *N*-pyrenyliodoacetamide (Molecular Probes, Junction City, OR) in dimethylformamide (14 mg/ml) is slowly added to 10 ml of the F-actin solution while rapidly vortexing. Actin which is exposed to high concentrations of dimethylformamide may be denatured. Alternatively, pyrenyliodoacetamide may be dissolved in dimethyl sulfoxide (DMSO) to minimize this problem. The mixture is then incubated overnight at room temperature in the dark. The reaction is quenched the next morning by the addition of 0.05 ml of 0.2 *M* DTT. Excess dye is removed by centrifugation at 2,500 *g* for 15 min. The labeled F-actin is pelleted at 100,000 *g* for 2.5 hr at 4°. The pellet is homogenized in 5 ml of buffer G (5 m*M* Tris–HCl, 0.2 m*M* DTT, 0.2 m*M* ATP, 0.1 m*M* $CaCl_2$, 0.01% NaN_3, pH 7.8) and dialyzed against this buffer for 72 hr at 4°, with frequent changes of buffer. The cycle of polymerization and depolymerization is then repeated. The labeled G-actin is clarified by centrifugation at 100,000 *g* for 2.5 hr at 4° and stored at this temperature in the dark.

The emission at 384 nm (and 407 nm) is measured in a fluorescence spectrophotometer after excitation at 365 nm. Bandpasses are 3 nm for emission and excitation.

[37] A. Wegner and J. Engel, *Biophys. Chem.* **3,** 215 (1975).

[38] T. Kouyama and K. Mihashi, *Eur. J. Biochem.* **114,** 33 (1981).

[39] S. L. Brenner and E. D. Korn, *J. Biol. Chem.* **258,** 5013 (1983).

[40] J. A. Cooper, S. B. Walker, and T. D. Pollard, *J. Muscle Res. Cell Motil.* **4,** 253 (1983).

Comments. The intensity of the fluorescence of F-actin may be reduced by the binding of some actin regulatory proteins (e.g., gelsolin, tropomyosin) and by an increase in the number of very short filaments.[41] Also, certian actin-modulating proteins may have a lower affinity for pyrene-conjugated actin, e.g., profilin isolated from *Acanthamoeba castellanii*[42] and human spleen.[43] In addition to measuring the rate and extent of actin polymerization, the fluorescence assay can also be used to measure the kinetics of F-actin depolymerization and the ability of actin-modulating proteins to sever performed actin filaments.[41]

Electron Microscopy (using acrosomal bundles from Limulus polyphemus sperm)

Actin filaments can be observed with the electron microscope, and the basic methodology has been reviewed elsewhere.[31] We describe below only the use of actin filament bundles from the acrosomal process of *Limulus* sperm as a "seed" for nucleated filament assembly.[44,45]

Limulus polyphemus can be obtained from the Marine Biological Laboratory, Woods Hole, MA. They are induced to discharge sperm by gentle rubbing near the gonadopores. Semen is immediately collected by a Pasteur pipet. Several crabs must be used to obtain sufficient sperm, which are washed in seawater and pelleted at 750 *g* for 5 min. The washing is again repeated and the pellet processed as detailed below.

The pellet is resuspended in greater than 3 volumes of 1% Triton X-100, 0.1 m*M* EGTA, 3 m*M* $MgCl_2$, and 30 m*M* Tris–HCl, pH 8.0. The sperm is then repeatedly pipetted to liberate the core filament bundles. The suspension is centrifuged at 750 *g* for 5 min. The supernatant is centrifuged again and resuspended. Then the bundles are pelleted from the supernatant by centrifugation at 12,000 *g* for 10 min, washed briefly in 10 m*M* Tris–HCl, pH 7.5, and pelleted again. The pellet is resuspended in 10 m*M* Tris–HCl, pH 7.5, containing 0.5% sodium laurylsarcosine (sarkosyl, Geigy Chemical Corp., Ardsley, NY). After 30 min, the suspension is centrifuged at 12,000 *g* for 10 min. The pellet is resuspended in 10 m*M* Tris–HCl, pH 7.5, with 0.5% sodium laurylsarcosine and pelleted again. The pellet is resuspended in appropriate ionic solution containing 0.5–6.0 *M* KCl and/or 0.5–2.0 m*M* $MgCl_2$ with actin and allowed to reach

[41] P. A. Jammey, C. Chaponnier, S. E. Lind, K. S. Zaner, T. P. Stossel, and H. L. Yin, *Biochemistry* **24,** 3714 (1985).

[42] A. A. Lal and E. D. Korn, *J. Biol. Chem.* **260,** 10132 (1985).

[43] B. Malm, *FEBS Lett.* **173,** 399 (1984).

[44] L. G. Tilney, *J. Cell Biol.* **64,** 289 (1975).

[45] E. M. Bonder and M. S. Mooseker, *J. Cell Biol.* **96,** 1097 (1983).

steady state. Electron microscopy can then be performed at various time intervals, counting the number of bundles with no growth, exclusive growth at the barbed or the pointed end, and bidirectional growth (Fig. 1). Such tabulation can be done in the presence and absence of millimolar divalent cations and ATP as well as after addition of profilin and/or actin-capping proteins (acumentin, gelsolin).

Comments. The previously described assays sample the entire population of actin filaments, while electron microscopy allows measurement of individual filaments and plotting of the percentage of filaments of various lengths. *Limulus* bundles allow observation of growth at each end of an actin filament, which can be easily identified as the barbed or pointed end because the bundles are tapered. The smaller or tapered end of the *Limulus* bundle corresponds to the barbed end of the actin filaments. The bundles can thereby serve the same function as actin filaments decorated with heavy meromysin subfragments. Since filaments within the bundle are packed in a paracrystalline array and are slightly thicker in diameter than pure actin filaments, it is claimed that new filaments as short as 10 monomers can be measured.

Study of the Behavior of Actin in Leukocytes

Triton-Insoluble Cytoskeleton-Associated Actin[46-48]

Human PMNs are purified by Ficoll–Hypaque or dextran sedimentation and then suspended in buffer containing 138 m*M* NaCl, 6 m*M* KCl, 0.64 m*M* Na_2HPO_4, 0.66 m*M* KH_2PO_4, 1.2 m*M* $MgSO_4$, 5.6 m*M* glucose, and 20 m*M* HEPES, pH 7.4. We find that maintaining the cells at room temperature in this low-calcium buffer minimizes cell activation. Just before the cells are warmed and exposed to the stimulus, a final concentration of 1.2 m*M* $CaCl_2$ is added to the solution. The final concentration of PMNs should be approximately 1.5×10^7 cells/ml. The cells are incubated at 37° for 10 min, then 500 μl of the PMN solution is added to a prewarmed Eppendorf tube containing fMet-Leu-Phe or other chemotactic stimulus.

The reaction is stopped by the addition of 55–65 μl (one-tenth of total volume) of Triton stop solution containing 10% Triton X-100 (Sigma), 7.5

[46] J. R. White, P. H. Naccache, and R. I. Sha'afi, *Biochem. Biophys. Res. Commun.* **108,** 1144 (1982).

[47] J. R. White, P. H. Naccache, and R. I. Sha'afi, *J. Biol. Chem.* **258,** 14041 (1983).

[48] R. Yassin, J. Shefcyk, J. R. White, W. Tao, M. Volpi, T. F. P. Molski, P. H. Naccache, M. B. Feinstein, and R. I. Sha'afi, *J. Cell Biol.* **101,** 182 (1985).

mg% benzamidine, 4 mg% apoprotinin, 1 mg% leupeptin, 1 m*M* PMFS, 54 m*M* PIPES, 22.5 m*M* HEPES, 9 m*M* EGTA, and 1.8 m*M* $MgCl_2$. Leupeptin and PMSF are particularly important for reducing proteolysis by lysosomal enzymes released during Triton treatment. PMNs may also be incubated with 5 m*M* DFP[5] prior to prewarming and exposure to the stimulus.

After mixing by a single gentle inversion, the solution is left for 2 min at 25°, and then centrifuged for 2 min at 12,000 *g* in an Eppendorf microfuge. We have found that longer centrifugation times (up to 8 min) do not significantly affect the concentration of actin found in the Triton-insoluble pellet. Centrifugation for 45 min at 150,000 *g* also yields similar results.[45]

The pellet (the Triton-insoluble fraction) is then solubilized by the addition of 100 μl of gel sample buffer containing 8 *M* urea, 1 g% SDS, 2 g% 2-mercaptoethanol, 10 g% sucrose, and 60 m*M* Tris–HCl, pH 6.8. The sample is next sonicated twice for 10 sec using a Branson sonifier, Model 200, with tapered microtip (Danbury, CT). The output should be set at 3. Immediately after sonication the solution is boiled for 5 min and electrophoresed into a 5–15% polyacrylamide slab gel.

Coomassie blue-stained gels are scanned using a Zeineh soft laser densitometer interfaced with an Apple II Plus computer. The area under the polypeptide band which comigrates with skeletal muscle actin (42,000 Da) and just below the ovalbumin standard (45,000 Da) is determined by integration for each Triton-insoluble extract. The effects of various stimuli on actin associated with the cytoskeleton is most simply expressed as a ratio of the actin concentration (expressed in integration units/100 μl) in stimulated cells divided by the actin concentration in unstimulated PMNs. Values between 1.5 and 2.0 are generally observed 10–40 sec after 5×10^{-8} *M* fMet-Leu-Phe stimulation at 37°. The percentage of the total cellular actin associated with the Triton-insoluble cytoskeleton can be determined by treating PMNs as described above except a stop solution containing no Triton is added to the cells after stimulation. The areas of the actin peaks in these whole cell preparations can be compared to Triton-treated pellets.

Comments. The Triton assay is a reasonably reproducible way of assessing *in vivo* actin filament assembly. The assay assumes that actin

---

FIG. 1. Gel-filtered G-actin (final concentration 0.5 μ*M*) was added to the *Limulus* bundles in 75 m*M* KCl, 3 m*M* $MgCl_2$, and 10 m*M* imidazole at pH 7.5, incubated for 6 min, and negatively stained. The figure is a composite micrograph of the two ends from a single *Limulus* bundle. The thicker end (bottom) corresponds to the pointed or low-affinity end of actin filaments, and the thinner end (top) corresponds to the barbed or high-affinity end. Note that actin filament assembly occurred off both ends of the bundle, with more growth off the thinner end. Courtesy of Dr. Lewis G. Tilney.

associated with the Triton-insoluble PMN fraction is filamentous actin which becomes trapped in or bound to the Triton-insoluble core. The assay probably cannot detect the presence of short actin filaments, since most short filaments, if not bound, should elute from the Triton-insoluble matrix. During the 4 min when the cells are being extracted with Triton and centrifuged it is also possible that some actin filaments depolymerize.

The pellet produced after Triton treatment and centrifugation must be completely dissolved in the gel sample buffer. We have found that extensive sonication is the best method of achieving this goal. Others have used incubation in a boiling water bath while vigorously vortexing for 10–15 min.[48] If the sample has not been completely solubilized, the density of the actin polypeptide band, after Coomassie blue staining of the polyacrylamide gel, may be considerably lower than expected. The other weakness of the assay is its dependence on consistent migration of the protein mixtures on SDS–PAGE. Diffusion of the proteins horizontally can also artificially lower the density of the actin peak.

NBD–Phallacidin Staining Assay[49–53]

NBD (7-nitrobenz-2-oxa-1,3-diazole)–phallacidin is a fluorescent analog of phallacidin, an acidic phallatoxin. By specifically binding to F-actin, this compound allows quantitation of the polymerized actin in the cell under different conditions.

The same modified Hanks'/HEPES buffer described for the preparation of PMNs for measurement of actin associated with the Triton-insoluble cytoskeleton may be used. Eight hundred fifty microliters of a PMN suspension containing 1.17×10^6 [for fluorescence activated cell sorter (FACS) analysis] or 2.35×10^6 cells/ml (for the extraction assay) is prewarmed for 10 min at 37° and then added to a 50 μl solution containing the stimulus. The reaction is then stopped, and the cells are fixed by the addition of 100 μl of 37% formalin (w/w) which is rapidly mixed by vortexing and allowed to incubate at 25° for 15 min.

Next, 50 μl of modified Hanks'/HEPES containing 3.46×10^{-6} *M* (for FACS analysis) or 6.93×10^{-6} *M* (for the extraction assay) NBD–

[49] T. H. Howard and W. H. Meyer, *J. Cell Biol.* **98,** 1265 (1984).

[50] T. H. Howard and C. O. Oresajo, *J. Cell Biol.* **101,** 1078 (1985).

[51] T. H. Howard and C. O. Oresajo, *Cell Motil.* **5,** 545 (1985).

[52] P. J. Wallace, R. P. Wersto, C. H. Packman, and A. M Lichtman, *J. Cell Biol.* **99,** 1060 (1984).

[53] T. P. Walsh, A. Weber, J. Higgins, E. M. Bonder, and M. S. Mooseker, *Biochemistry* **23,** 2613 (1984).

phallacidin and 2.1 mg/ml lysophosphatidylcholine (Sigma) is added to the fixed cell preparation and vortexed. NBD–phallacidin is shipped in absolute methanol at known concentrations. The desired volume of methanol containing NBD–phallicidin is air dried at 22° and then resuspended in modified Hanks'/HEPES containing the appropriate concentration of lysophosphatidylcholine. After addition of the stain solution, samples are incubated at 37° for 10 min. Cells are then centrifuged at 12,000 *g* for 1 min at room temperature in 1.5-ml Eppendorf tubes using an Eppendorf microfuge. Supernatant is removed, and the pellet is either resuspended in modified Hanks'/HEPES for analysis by FACS or resuspended in absolute methanol and macerated to perform the extraction assay.

For FACS analysis (see also this series, Vol. 108 [19]) resuspended cells are filtered through a 50-μm nylon mesh filter to remove large aggregates and then analyzed on a FACS (Becton Dickinson and Co., Oxnard, CA). Cells are excited with an argon laser at 488 nm and emission read at 522 nm with a long-pass filter. Results are stored and displayed as histograms of fluorescence [cell number versus fluorescence channel (0–250)]. Since the fluorescence histogram generally yields a normal distribution, the fluorescence can be recorded as the peak fluorescence channel number. Relative F-actin content is then expressed as the ratio of the stimulated cells' peak channel number to the unstimulated cells' peak channel number.[47] Alternatively, cell number can be plotted versus log fluorescence intensity. The mean and mode channel number, coefficient of variation, and standard deviation can be determined by computer analysis.[52]

For measurement of the relative fluorescence of extracted NBD–phallacidin, the macerated cell pellet is allowed to extract in methanol at room temperature in the dark for 1 hr. Cells are then again recentrifuged at 12,000 *g* for 1 min and the relative fluorescence intensity of the supernatant measured using a spectrofluorometer. Excitation and emission wavelengths are 465 and 535 nm, respectively. Both excitation and emission slit widths are set at 2 or 3 nm.[50]

Comments. Quantitation of the concentrations of F-actin in PMNs before and after fMet-Leu-Phe stimulation using this assay have yielded findings similar to those derived from the Triton extraction method. Results are readily reproducible, and this method lacks many of the pitfalls of the Triton-insoluble extract assay. The methanol extraction technique obviates the need for an expensive FACS.

Like the Triton-insoluble extract method, this assay quantitates the F-actin which remains adherent to lysophosphatidylcholine-insoluble cytoskeleton. The method assumes that formalin-treated actin binds NBD–phallicidin in the same way as native F-actin. Second, the assay assumes that NBD–phallicidin binding does not affect the state of actin. This phallotoxin in concentrations 100-fold higher than used in this assay have been shown to stimulate actin filament growth in nonmuscle cells.[54]

DNase I Inhibition Assay in Cell Lysates[36,55,56]

The DNase I inhibition assay (described earlier[34]) can be modified to quantitate the amount of G-actin and total actin in cell lysates. The protocol described here has been used by Fechheimer and Zigmond to study cytoskeletal changes in PMNs induced by chemotactic peptides.[55]

DNase I can be stored frozen at a concentration of 1 mg/ml in 50 mM Tris–HCl, 0.1 mM PMSF, and 1 mM $CaCl_2$, pH 7.5. Immediately before use it is diluted 1 : 100 with glass-distilled water. Calf thymus DNA is prepared as described previously[34] and diluted so that the optical density at 285 nm is approximately 1.7. The activity of the DNase is determined by mixing 70 μl of the DNase solution with an appropriate volume of Hanks' medium containing the Triton–lysis mixture (see below) and 2 ml of the DNA solution prewarmed to 30°. The change in optical density at 285 nm is recorded at 30° in a spectrophotometer.

PMNs (5×10^6/ml) are incubated for at least 7 min at 37° in Hanks' balanced salt solution without calcium, magnesium, or phenol red (Grand Island Biological Co.) to which EDTA (final concentration 10 mM) is added just before the start of the experiment. The inhibition of DNase activity of cell extracts prepared in the presence of EDTA and stored on ice is stable for at least 15 min. Cells are lysed by addition of 0.1 volume of Triton–lysis mixture which contains 5% Triton X-100, 6 mM ATP, 1 mM PMSF, and 2 mM DTT. The amount of the lysed cell extracts which gives approximately 50% inhibition of the DNase activity is determined.

To examine the effect of a chemotactic peptide (or other stimulus) on the state of actin, 0.1 volume of 10^{-6} M peptide or an equal volume of buffer is added to cells maintained at 37°. At various times following this addition, the Triton–lysis mixture is added. Control and stimulated cells are analyzed in parallel.

The cell lysates can be mixed with an equal volume of 1.5 M guanidine–HCl, 1 M sodium acetate, 1 mM $CaCl_2$, 20 mM Tris–HCl, and 1 mM ATP, pH 7.5, in order to assess the total amount of intracellular actin. The mixture is incubated on ice for 5–10 min to allow the F-actin to depolymerize. The DNase activity is measured in the presence of the guanidine, and the volume of the lysate required to give approximately 50% inhibition is determined. The percentage of the actin in the extract

[54] J. Wehland, M. Osborn, and K. Weber, *Proc. Natl. Acad. Sci. U.S.A.* **74,** 5613 (1977).

[55] M. Fechheimer and S. H. Zigmond, *Cell Motil.* **3,** 349 (1983).

[56] F. Lamb, M. Kaplan, and C. Gitler, *FEBS Lett.* **124,** 35 (1980).

which is polymerized is calculated from the inhibition of DNase activity by lysates prepared in the absence and presence of guanidine.

Acknowledgments

The authors thank Dr. Thomas Howard and Dr. John Hartwig for their help with the NBD–phallacidin assay and the preparative method for ABP, respectively; Dr. Lewis Tilney for the photomicrograph of a *Limulus* bundle; Dr. John Riefler for his careful review of the manuscript; and Ms. Patricia Cromartie for her invaluable secretarial assistance. Supported by National Institutes of Health Grant 3RO1 AI 23262 and a Multiple Dystrophy Basic Research Grant.

[24] Chemoattractant-Induced Membrane Phenomena of Phagocytes

By CHARLES D. SMITH and RALPH SNYDERMAN

Introduction

Cellular activation by exogenous stimuli requires that extracellular signals be translated into intracellular signals. Therefore, events occurring at the plasma membrane of the cell play important roles in stimulus–response coupling. Exposure of phagocytes to chemoattractants initiates a number of biological responses including directed migration, secretion of hydrolytic enzymes, and production of cytotoxic oxygen species.[1] These processes are activated by second messengers which are produced upon interaction of chemoattractants with their receptors.

This chapter is divided into three sections. The first describes methods for the isolation and subcellular fractionation of human polymorphonuclear leukocytes (PMNs).[1a] The second deals with methods for studying chemoattractant-induced changes in phospholipid metabolism. The third

[1] R. Snyderman and E. J. Goetzl, *Science* **213,** 830 (1981).

[1a] Abbreviations: PMNs, polymorphonuclear leukocytes; LSM, Lymphocyte Separation Medium; HEPES, *N*-2-hydroxyethylpiperazine-*N'*-2-ethanesulfonic acid; HHBSS, Hanks' balanced salt solution in HEPES; EGTA, ethylene glycol bis(β-aminoethyl ether)-*N,N,N',N'*-tetraacetic acid; IP_3, inositol 1,4,5-trisphosphate; PIP_2, phosphatidylinositol 4,5-bisphosphate; PIP, phosphatidylinositol 4-phosphate; IP, inositol phosphate; IP_2, inositol bisphosphate; PT, *Bordetella pertussis* toxin; SDS–PAGE, sodium dodecyl sulfate–polyacrylamide gel electrophoresis; EDTA, ethylenediaminetetraacetic acid; GTPγS, guanosine 5'-*O*-(3-thiotriphosphate).

deals with investigation of the role of a guanine nucleotide regulatory (N) protein in phagocyte activation by chemoattractants. Methods described herein have been used predominantly for studying responses of human PMNs to the oligopeptide chemoattractant fMet-Leu-Phe but are also applicable to other chemoattractants.

Preparation of Human PMNs and Their Subcellular Fractions

Preparation of Human PMNs (see also this series, Vol. 108 [9])

Heparinized blood (10 units of heparin/ml; Upjohn) from healthy donors is mixed with an equal volume of 3% dextran T500 (Pharmacia Fine Chemicals) in 0.9% NaCl. Erythrocytes are allowed to settle, and the leukocyte-rich plasma is transferred to a 50-ml conical centrifuge tube. This cell suspension is gently underlayed with a 15-ml cushion of Lymphocyte Separation Medium (LSM, Bionetics Laboratory Products) and centrifuged at 400 *g* for 40 min at room temperature. Leukocytes are resolved into two fractions: the mononuclear layer, which bands at the serum–LSM interface and which contains monocytes and lymphocytes, and the pellet, which contains PMNs and residual erythrocytes. For the isolation of PMNs, the serum and LSM layers are withdrawn by aspiration. Erythrocytes in the pellet are removed by 3 cycles of hypotonic lysis, each of which consists of resuspending the cells in 5 ml of ice-cold 0.2% NaCl for 15 sec followed by addition of 5 ml of cold 1.6% NaCl to restore isotonicity and centrifugation at 600 *g* for 10 min at 4° The isolation PMNs (approximately 10^9 cells/450 ml of blood which are >95% pure) are typically resuspended in Hanks' balanced salt solution (Gibco) containing 10 m*M* *N*-2-hydroxyethylpiperazine-*N*′-2-ethanesulfonic acid (HEPES) and 4.2 m*M* $NaHCO_3$, pH 7.4 (HHBSS).

Isolation of Plasma Membranes

Isolated PMNs are pelleted by centrifugation at 600 *g* for 10 min at 4° and resuspended to a concentration of 2×10^7 cells/ml of relaxation buffer (100 m*M* KCl, 3 m*M* NaCl, 3.5 m*M* $MgCl_2$, and 10 m*M* HEPES, pH 7.3). Aliquots of 15 ml are placed in a Cell Disruption Bomb (Parr Instrument Co.) and pressurized to 500 psi with nitrogen. After 20 min, the cell lysate is collected and combined with 300 μl of 100 m*M* neutralized ethylene glycol bis(β-aminoethyl ether)-*N,N,N′,N′*-tetraacetic acid (EGTA) to give a final concentration of 2 m*M* EGTA. The lysate is layered over a 15-ml cushion of relaxation buffer containing 41% sucrose in a polycarbonate ultracentrifuge tube (e.g., for a Beckman SW28 rotor) and centrifuged at

90,000 *g* for 60 min.[2] Plasma membranes band at the sucrose interface and are pelleted by centrifugation at 100,000 *g* for 45 min. The membranes are then resuspended in 10 m*M* HEPES, pH 7.0, containing 0.25 *M* sucrose (sucrose/HEPES) and used immediately or frozen at −80°.

An alternate procedure which allows simultaneous isolation of specific granules and azurophilic granules involves fractionation of the PMN lysate on self-forming gradients of Percoll[3] (Pharmacia). The cell cavitate is collected with EGTA as described above and centrifuged at 600 *g* for 10 min at 4° to pellet unbroken cells and nuclei. The resulting supernatant is placed in a 45-ml centrifuge tube (e.g., for a Sorvall SS34 rotor) over a 30-ml gradient of Percoll [15 ml of 54% Percoll in relaxation buffer (density 1.050 g/ml) overlayed on 15 ml of 88% Percoll in relaxation buffer (density 1.120 g/ml)]. Following centrifugation at 40,000 *g* for 25 min at 4°, the cell lysate is resolved into three prominent fractions. The least dense band contains plasma membranes (enriched 16-fold in alkaline phosphatase activity), the intermediate band contains specific granules (enriched 7-fold in vitamin B_{12}-binding protein), and the dense fraction contains azurophilic granules (enriched 5-fold in myeloperoxidase). These fractions are recovered and centrifuged at 100,000 *g* for 45 min at 4°. The organelles collect at the top of the Percoll pellet and can be recovered with a Pasteur pipet. The membranes are then diluted with sucrose/HEPES and centrifuged again at 100,000 *g* for 45 min at 4°. The membranes are recovered as above and are resuspended in sucrose/HEPES and frozen at −80°.

Measurement of Chemoattractant-Induced Phospholipid Metabolism

Modification of the phospholipid composition of plasma membranes is a rapid response to many hormonal stimuli, including chemoattractants. Alterations of phospholipid metabolism can have pronounced effects on the activities of membrane-associated enzymes or produce second messengers which act to regulate cell responses. Methods used in our laboratory for measuring chemoattractant-induced changes in phospholipid metabolism are described below.

Phospholipid Methylation

Hormone-induced changes in methylation of membrane phospholipids, e.g., conversion of phosphatidylethanolamine to phosphatidylcho-

[2] T. Maeda, K. Balakrishnan, and S. Q. Mehdi, *Biochim. Biophys. Acta* **731,** 115 (1983).

[3] N. Borregaard, J. M. Heiple, E. R. Simons, and R. A. Clark, *J. Cell Biol.* **97,** 52 (1983).

line, may alter the structure of the membrane sufficiently to provide a mechanism for regulating membrane function. Lipid methylation can be measured as the incorporation of ^3H from *S*-adenosyl-L-[*methyl*-^3H]methionine, which is synthesized by incubating cells with L-[*methyl*-^3H]methionine, into phospholipids.[4] PMNs (5×10^6 cells in 1 ml of HHBSS) are incubated at 37° for 15 min before the addition of 10 μCi of L-[*methyl*-^3H]methionine (New England Nuclear) and various stimuli, e.g., fMet-Leu-Phe. The cells are incubated up to an additional 60 min, pelleted, resuspended in 0.1 ml of HHBSS, and dissolved in 0.5 ml of chloroform : methanol (2 : 1, v/v). The tubes are then vortexed and centrifuged at 2000 *g* for 5 min. The aqueous upper phase is then removed by aspiration, and 0.1 ml of the organic phase is spotted onto a thin-layer cellulose plate containing a fluorescent indicator (Eastman chromogram), along with 10 μg of phosphatidylcholine. The plates are developed with chloroform : methanol : water (75 : 18 : 2, v/v/v) to separate the phospholipids from free L-[*methyl*-^3H]methionine. The phospholipids are visualized under ultraviolet light, scraped into scintillation vials, and mixed with 10 ml of Lefkofluor scintillation fluid (Research Products International Corp.) for determination of radioactivity. Incubation of guinea pig macrophages with chemotactic peptides results in a dose-dependent suppression of phosphatidylethanolamine methylation.[4]

Phosphoinositide Metabolism

Occupancy of many receptors results in the elevation of intracellular Ca^{2+} concentrations and activation of protein kinase C. These effects are mediated by inositol 1,4,5-trisphosphate (IP_3) and 1,2-diacylglycerol, respectively, which are produced by receptor-stimulated hydrolysis of phosphatidylinositol 4,5-bisphosphate (PIP_2) by phospholipase C.[5] Using radiolabeled PMNs (described below), chemoattractant-stimulated polyphosphoinositide hydrolysis can be measured as the decrease in radioactivity in phosphatidylinositol 4-phosphate (PIP) and PIP_2 or the increase in radioactivity in inositol phosphates. Radioactivity in PIP and PIP_2 is measured by liquid scintillation counting after isolation of these lipids by affinity chromatography using immobilized neomycin.[6] Inositol phosphates are analyzed by anion-exchange chromatography followed by liquid scintillation spectrometry.[7]

[4] M. C. Pike, N. M. Kredich, and R. Snyderman, *Proc. Natl. Acad. Sci. U.S.A.* **76,** 2922 (1979).

[5] M. J. Berridge and R. E. Irvine, *Nature (London)* **312,** 315 (1984).

[6] J. Schacht, this series, Vol. 72, p. 626.

[7] C. P. Downes and R. H. Michell, *Biochem. J.* **198,** 133 (1981).

PMN Labeling with $^{32}PO_4$. PMNs are isolated as described above and resuspended to 2×10^7 cells/ml of phosphate-free HHBSS (137 mM NaCl, 5.4 mM KCl, 1 mM $CaCl_2$, 4.3 mM $NaHCO_3$, 0.4 mM $MgCl_2$, 0.4 mM $MgSO_4$, 10 mM glucose, and 10 mM HEPES, pH 7.4). Neutralized ortho[^{32}P]phosphoric acid (New England Nuclear) is added to 40 μCi/ml, and the suspensions are incubated at 37° for 90–120 min. The cells are washed with HHBSS 3 times and then resuspended in HHBSS to 10^8 cells/ml, divided into 0.18 ml aliquots, and kept on ice. After 1–5 min of preincubation at 37°, the ^{32}P-labeled cells are stimulated by exposure to a chemoattractant for 15–300 sec. Reactions are terminated by the addition of 2 ml of chloroform:methanol (1:1, v/v), and [^{32}P]PIP and [^{32}P]PIP_2 levels are quantitated as described below.

PMN Labeling with myo-[2-^{3}H]Inositol. Isolated PMNs are resuspended to 2×10^7 cells/ml of Medium 199 (Hazelton Dutchland Corp.) containing 2% fetal calf serum. *myo*-[2-^{3}H]Inositol (New England Nuclear) is then added to 40 μCi/ml, and the cells are incubated for 18–24 hr at 37°. The cells are harvested, washed twice with HHBSS, resuspended to 10^8 cells/ml of HHBSS containing 2 mM LiCl, and divided into 0.18 ml aliquots on ice. Cells are stimulated as above, and the levels of [^{3}H]PIP, [^{3}H]PIP_2, and [^{3}H]inositol phosphates are determined as described below.

Analysis of Polyphosphoinositides. Reactions (volume $\leq$200 μl) are terminated by the addition of 2 ml of chloroform:methanol (1:1, v/v). Phospholipids are extracted by the addition of 0.5 ml of 2.4 N HCl, thorough mixing, and centrifugation at 1000 g for 10 min.[6] The organic lower phase is removed, and 1 ml of chloroform is added to the aqueous phase, which is mixed and centrifuged. The lower phase is combined with the previous extract and washed twice with 2 ml of methanol:1 N HCl (1:1, v/v) by mixing and centrifuging as above, each time removing the aqueous phase. The resulting organic phase is combined with 2 ml of 200 mM ammonium acetate in absolute methanol and subjected to affinity chromatography on 0.5 ml of neomycin covalently bound to glass beads as described by Schacht.[6] All the phospholipids except PIP and PIP_2 are washed from the column with 7 ml of chloroform:methanol:water (3:6:1, v/v/v) containing 150 mM ammonium acetate. PIP is eluted by washing with 7 ml of chloroform:methanol:water (3:6:1, v/v/v) containing 600 mM ammonium acetate. PIP_2 is eluted by washing with 7 ml of chloroform:methanol:14 N ammonium hydroxide (3:6:1, v/v/v). The amount of ^{32}P in the phospholipids can be directly analyzed as Cerenkov radiation in a liquid scintillation spectrometer (this typically produces approximately 40% of the cpm value determined in the presence of scintillation fluid). ^{3}H-Labeled phospholipids must be first dried (e.g., by heating in the scintillation vials overnight) before fluid is added for counting.

Analysis of Inositol Phosphates. Reactions are terminated, and the samples are acidified as described above. After centrifugation, the aqueous phase is removed and combined with 15 ml of 0.1 *N* NaOH to neutralize the extract and to reduce its ionic strength. The samples are then chromatographed on a 0.5-ml column of Dowex-1 (formate) equilibrated with 50 m*M* ammonium formate.[7] The column is washed with 10 ml of 50 m*M* ammonium formate to remove neutral compounds (e.g., inositol). Inositol phosphate (IP) is eluted into a scintillation vial with 10 ml of 0.2 *M* ammonium formate/0.1 *N* formic acid. Inositol bisphosphate (IP_2) is eluted with 10 ml of 0.4 *M* ammonium formate/0.1 *N* formic acid. IP_3 is eluted with 10 ml of 1.0 *M* ammonium formate/0.1 *N* formic acid. The eluants are then mixed with Lefkofluor scintillation fluid for determination of radioactivity. These chromatographic procedures do not resolve possible isomers of inositol phosphates (e.g., inositol 1,4,5-trisphosphate and inositol 1,3,4-trisphosphate); however, these may be separated using high-pressure liquid chromatography.[8]

Role of an N Protein in Receptor-Induced Polyphosphoinositide Breakdown

N proteins play pivotal roles in regulating the activity of adenylate cyclase and hence cAMP levels in cells. Accumulating evidence suggests that N proteins may play similar roles in coupling phospholipase C activation to receptors which stimulate polyphosphoinositide breakdown.[9] We have used two approaches to investigate N protein involvement in phospholipase activation by chemoattractants. The first involves use of *Bordetella pertussis* toxin (PT),[10] and the second involves direct activation of the phospholipase in isolated plasma membranes.[11,12]

PT as a Probe for N Protein Involvement

Bacterial toxins have provided powerful tools for analyzing the involvement of N proteins in regulating adenylate cyclase activity. For example, cholera toxin catalyzes the transfer of the ADP-ribose moiety of NAD to the α subunit of N_s. This modification results in the persistent activation of

[8] R. F. Irvine, E. E. Anggard, A. J. Letcher, and C. P. Downes, *Biochem. J.* **229,** 505 (1985).

[9] C. D. Smith, M. W. Verghese, and R. Snyderman, *in* "Sensing and Response in Microorganisms" (M. Eisenbach and M. Balaban, eds.), p. 215. Elsevier, New York, 1985.

[10] M. W. Verghese, C. D. Smith, and R. Snyderman, *Biochem. Biophys. Res. Commun.* **127,** 450 (1985).

[11] C. D. Smith, B. C. Lane, I. Kusaka, M. W. Verghese, and R. Snyderman, *J. Biol. Chem.* **260,** 5875 (1985).

[12] C. D. Smith, C. Cox, and R. Snyderman, *Science* **232,** 97 (1987).

the N protein and therefore of adenylate cyclase. In contrast, PT catalyzes the ADP-ribosylation and consequent inactivation of N_i and transducin.[13] Incubation of human PMNs with PT inhibits chemoattractant-induced cellular responses by inhibiting the activation of phospholipase C-mediated hydrolysis of PIP_2.[10,11]

Incubation Conditions. Pertussis toxin (List Biological Products Campbell, CA 95008) is incubated with unlabeled, $^{32}PO_4$-labeled, or [^{3}H]inositol-labeled PMNs (2×10^7 cells/ml of HHBSS) at concentrations of 0.1–1 μg/ml of HHBSS for up to 4 hr at 37°. The cells are pelleted, washed twice with HHBSS, and then assayed for chemoattractant-stimulated PIP_2 breakdown or production of inositol phosphates. By varying the dose of PT and the time of incubation, conditions can be defined which will cause inhibition of N protein function without inhibiting nonreceptor-mediated mechanisms.[10] For example, treating PMNs with 1 μg of PT/ml for 90 min at 37° produces an 84% inhibition of [^{3}H]IP_3 production induced by 10^{-6} *M* fMet-Leu-Phe, without inhibiting concanavalin A (100 μg/ml)-induced [^{3}H]IP_3 production, phorbol ester-induced superoxide production, or Ca^{2+} ionophore-stimulated secretion. Increasing the incubation time with PT to 4 hr produces even more complete inhibition of [^{3}H]IP_3 production induced by chemoattractants. For an accurate reflection of the effects of PT on receptor-mediated activation of phospholipase C, it is essential that the conditions used do not cause general cytotoxicity.

Identification of a PT-Sensitive N Protein. PT-sensitive N proteins in isolated membranes can be demonstrated by *in vitro* ribosylation using [*adenylate*-^{32}P]NAD and product analysis by sodium dodecyl sulfate–polyacrylamide gel electrophoresis (SDS–PAGE) and autoradiography. For studies with PMNs, plasma membranes (approximately 25 μg of protein) are incubated with 25 m*M* Tris–HCl, 4 m*M* $MgCl_2$, 1 m*M* EDTA, 1 m*M* dithiothreitol, 10 m*M* nicotinamide, 10 m*M* thymidine, 0.5 m*M* ADP, 10 μ*M* NAD with 1 μCi of [^{32}P]NAD (New England Nuclear), and 2 μg of PT in a volume of 50 μl for 30 min at 30°. The samples are then dissolved in an equal volume of electrophoresis sample buffer (125 m*M* Tris–HCl, 4% SDS, 20% glycerol, 10% 2-mercaptoethanol, and 0.002% bromphenol blue, pH 6.8) and analyzed by SDS–PAGE using a 10% polyacrylamide gel. The gels are dried and exposed to X-ray film for autoradiography. We have observed a PT-specific ADP-ribosylation product in PMN plasma membranes which migrates with an apparent molecular weight of 41,000. This protein therefore appears to be similar to the α subunit of N_i. Ribosylation of this protein by PT treatment of intact cells is indicated by a reduction in label incorporated into this protein during *in vitro* ribosylation of membranes from the PT-treated cells.

[13] A. G. Gilman, *Cell* **36,** 577 (1984).

N Protein-Mediated PIP_2 Breakdown in Isolated Plasma Membranes

Purified preparations of plasma membranes provide the most definitive system for analyzing N protein involvement in phospholipase C activation since possible regulatory compounds (e.g., Ca^{2+} and guanine nucleotides) can be directly added. Under physiological conditions, N proteins are activated through interaction with occupied receptors. This interaction promotes the dissociation of GDP bound to the N protein, thereby freeing it to bind GTP. Binding of GTP, or guanosine 5′-*O*-(3-thiotriphosphate) (GTPγS), activates the N protein until hydrolysis of the terminal phosphate of GTP regenerates the inactive GDP–N protein complex.[13] Therefore, activation of the N protein, and subsequently its effector enzyme, requires both hormone and GTP. Nonhydrolyzable analogs of GTP, e.g., GTPγS, directly bind to and activate N proteins. Therefore, receptor-mediated reactions which result in the activation of N proteins can often be mimicked by GTPγS.

An efficient method for producing radiolabeled PIP_2 in isolated membranes involves phosphorylation of the inositol phospholipids with [γ-^{32}P]ATP.[11] Isolated plasma membranes (10–25 μg of protein) in 25 μl of sucrose/HEPES (see above) are incubated for 60 sec at 37° with an equal volume of 100 m*M* HEPES, 10 m*M* $MgCl_2$, 1 m*M* spermine (Sigma), and 200 m*M* [γ-^{32}P]ATP (New England Nuclear) (800–1000 cpm/pmol), pH 7.0. With PMN plasma membranes, this results in the synthesis of phosphatidic acid, PIP, and PIP_2 which contain approximately 40, 50, and 30 pmol of ^{32}P/mg of protein, respectively.[11] These ^{32}P-labeled phospholipids can be conveniently analyzed by affinity chromatography using immobilized neomycin as described above. Appearance of ^{32}P-labeled PIP and PIP_2 demonstrates that the isolated plasma membranes possess functional phosphatidylinositol kinase and PIP kinase. The amount of ^{32}P-labeled PIP_2 reaches a maximum after 60 sec and remains stable for at least an additional 2 min. Therefore the change in [^{32}P]PIP_2 levels between 1 and 2 min is compared in samples incubated with $CaCl_2$ (0.1–1000 μM) and guanine nucleotides.

In the PMN plasma membrane system, PIP_2 hydrolysis can be activated by exposing the labeled membranes to $\geq 100\ \mu M$ $CaCl_2$, which directly activates phospholipase C. Addition of GTP causes a decrease in the Ca^{2+} requirement so that PIP_2 hydrolysis can be activated with an EC_{50} (concentration required for half-maximal PIP_2 breakdown) of approximately 4 μM Ca^{2+}. Most importantly, addition of a combination of fMet-Leu-Phe (10^{-7} *M*) and GTP (10 μM) further reduces the Ca^{2+} dependence so that PIP_2 hydrolysis occurs with as low as 0.1 μM Ca^{2+}.[12] In contrast to those from control cells, plasma membranes isolated from PT-treated

PMNs do not demonstrate stimulation of PIP_2 breakdown upon treatment with fMet-Leu-Phe plus GTP. GTPγS, either alone or in the presence of fMet-Leu-Phe, also activates the hydrolysis of PIP_2 at low concentrations of Ca^{2+}.[12] Therefore, the use of the ^{32}P-labeled plasma membrane system has provided strong support for the proposal that the fMet-Leu-Phe receptor is coupled to phospholipase C through an N protein. Similar systems should be generally applicable in studying the coupling of other receptors to PIP_2 breakdown.

[25] Biochemical Changes in Leukocytes in Response to Chemoattractants

By MARGRITH W. VERGHESE and RALPH SNYDERMAN

Introduction

Activation of phagocytic leukocytes by chemoattractants and most other inflammatory stimuli is associated with rapid increases in intracellular $[Ca^{2+}]$.[1,2] Optimal stimulation of the secretory and respiratory burst responses of these cells to chemoattractants requires Ca^{2+} in the extracellular medium.[3] Chemoattractants stimulate the release of the Ca^{2+} mobilizer inositol 1,4,5-trisphosphate (IP_3)[3a] and 1,2-diacylglycerol in leukocytes as a consequence of phospholipase C (PLC) activation and hydrolysis of phosphatidylinositol 4,5-bisphosphate (PIP_2).[4] The products resulting from PIP_2 hydrolysis stimulate Ca^{2+}/phospholipid- and/or calmodulin-dependent protein kinases which apparently leads to cellular activation.[5,6] Doses of chemoattractants which stimulate the respiratory burst and lysosomal

[1] T. Pozzan, D. P. Lew, C. B. Wollheim, and R. Y. Tsien, *Science* **221,** 1413 (1983).

[2] H. M. Korschak, K. Vienne, L. E. Rutherford, C. Wilkenfeld, M. C. Finkelstein, and G. Weissmann, *J. Biol. Chem.* **259,** 4076 (1984).

[3] J. E. Smolen and G. Weissmann, *Biochim. Biophys. Acta* **720,** 172 (1982).

[3a] Abbreviations: IP_3, inositol 1,4,5-trisphosphate; PLC, phospholipase C; PIP_2, phosphatidylinositol 4,5-bisphosphate; EGTA, (β-aminoethyl ether)-*N,N,N′,N′*-tetraacetic acid; DMSO, dimethyl sulfoxide; IBMX, isobutylmethylxanthine; 2-ME, 2-mercaptoethanol; PMSF, phenylmethylsulfonyl fluoride; HHBSS, Hanks' balanced salt solution with HEPES; TCA, trichloroacetic acid.

[4] F. Di Virgilio, L. M. Vicentini, S. Treves, G. Riz, and T. Pozzan, *Biochem. J.* **229,** 361 (1985).

[5] Y. Nishizuka, *Nature (London)* **308,** 693 (1984).

[6] M. J. Berridge and R. F. Irvine, *Nature (London)* **312,** 315 (1984).

enzyme secretion also result in redistribution of protein kinases in human polymorphonuclear leukocytes (PMNs).[7] Additionally, stimulation of PMNs with chemoattractants induces a rapid, but transient elevation in cAMP levels which may provide an inhibitory feedback mechanism.[8] Measurements of Ca^{2+} mobilization, cAMP levels, and protein kinase redistribution in PMNs have provided useful tools to study the response of these cells to mediators of inflammation. Measurement of phospholipid metabolism is covered elsewhere in this volume [24].

Methods

Methods for measuring Ca^{2+} redistribution commonly utilize the Ca^{2+}-sensitive fluoroprobes Quin 2-tetraacetomethoxy ester (Quin 2-AM)[1,2] and chlortetracycline.[3,9] The former is thought to monitor free cytosolic [Ca^{2+}], while the latter records membrane-associated [Ca^{2+}]. These methods are described elsewhere in this volume [26] and will not be duplicated here. Measurements of protein kinase activities are based on the incorporation of ^{32}P into histone from [γ-^{32}P]ATP[10] while cAMP elevations are quantified with commercially available radioimmunoassay kits.[8,11]

Materials

Hanks' balanced salt solution with Ca^{2+} (1.2 m*M*) and Mg^{2+} (0.8 m*M*)
Phosphate-buffered (10 m*M*), pH 7.4, isotonic saline without Ca^{2+}, Mg^{2+}
4-(2-Hydroxyethyl)-1-piperazineethanesulfonic acid (HEPES)
$NaHCO_3$
Triton X-100
Ethylene glycol bis(β-aminoethyl ether)-*N,N,N′,N′*-tetraacetic acid (EGTA)
Dimethyl sulfoxide (DMSO)
Isobutylmethylxanthine (IBMX)

[7] M. C. Pike, J. Jakoi, L. C. McPhail, and R. Snyderman, *Blood* **67,** 909 (1986).

[8] L. Simchowitz, L. C. Fischbein, I. Spilberg, and J. P. Atkinson, *J. Immunol.* **124,** 1482 (1980).

[9] A. S. Schneider, R. Herz, and M. Sonenberg, *Biochemistry* **22,** 1680 (1983).

[10] M. Wolfson, L. C. McPhail, V. N. Narsallah, and R. Snyderman, *J. Immunol.* **135,** 2057 (1985).

[11] M. W. Verghese, K. Fox, L. C. McPhail, and R. Snyderman, *J. Biol. Chem.* **260,** 6769 (1985).

2-Mercaptoethanol (2-ME)
Histone H1 (Sigma Type III-S)
ATP
Phosphatidylserine
1,2-Diolein
Phenylmethylsulfonyl fluoride (PMSF)
[γ-^{32}P]ATP
cAMP radioimmunoassay kit

The first two and the last two items are purchased from Gibco (Grand Island, NY) and New England Nuclear (Boston, MA), respectively. Sigma (St. Louis, MO) is the source of all other reagents escept for IBMX which is from Pharmacia (Piscataway, NJ).

Equipment

Microfuge (e.g., Fisher Centrifuge Model #59)
Liquid scintillation counter
Gamma counter, Vacuum filtration system

cAMP Measurement in Intact Cells

PMNs or mononuclear leukocytes (lymphocytes plus monocytes) are isolated from heparinized (10 U/ml) human blood by dextran sedimentation, followed by centrifugation over a lymphocyte-separating medium as described in this volume [24]. Monocytes can be purified from the mononuclear leukocyte fraction by counterflow elution (see this series, Vol. 108 [20]) or Percoll density centrifugation (see also this series, Vol. 108 [9]).

The cells are resuspended to 2×10^7/ml in Hanks' balanced salt solution with HEPES (HHBSS) and kept on ice. Immediately before assay, cells are prewarmed for 5 min at 37°. A stimulatory agent is then added, and the sample is incubated with shaking. At the appropriate time intervals, 200-μl aliquots of the cell suspension are withdrawn and immediately added to 1.0 ml of boiling phosphate-buffered saline containing 0.5 m*M* IBMX. Boiling is continued for exactly 5 min, and 5 ml of ice-cold 50 m*M* sodium acetate (pH 6.2) is added. Cells are sonicated for 10 sec and centrifuged at 3000 *g* for 15 min. The cAMP-containing supernatants (100 μl per sample) are transferred to glass tubes (12×75 mm) and are acetylated and immunoprecipitated exactly as described by the manufacturers of the cAMP radioimmunoassay kits. Extra care is needed to pellet the immunoprecipitates which should not be disturbed during the removal of the supernatants. The radioactivity is determined in a gamma counter, and cAMP contents can be estimated from the standard curve included in each assay.

Protein Kinase C Assay and Subcellular Redistribution

Cells are prepared as above but resuspended to 5×10^7/ml in HHBSS and prewarmed for 5–10 min at 37°. Stimuli are added at the desired concentrations for various times, the reactions are stopped by rapid addition of 5- to 10-fold excess ice-cold HBSS, followed by centrifugation. The pellet is resuspended to 5×10^7 cells/ml in extraction buffer consisting of 50 m*M* Tris–HCl (pH 7.5), 10 m*M* 2-ME, 1 m*M* PMSF, and 2 m*M* EGTA plus any desired concentration of Ca^{2+}.[12,13] Cells can be disrupted by sonication in an ice bath for 25 sec at 15–30 W with an Artek sonic dismembranator equipped with a microtip, or by nitrogen cavitation in a cell disruption bomb (Parr) for 20 min at 350 psi.[10] Nuclei and unbroken cells are removed by low-speed centrifugation (200 *g*, 10 min), and cytosolic and particulate fractions can be separated by ultracentrifugation (100,000 *g*, 60 min). The pellet is resuspended to the initial volume of extraction buffer, and, if necessary, the fractions can be stored at 0° for 2–3 days before assaying for protein kinase C. Freezing destroys protein kinase C activity!

Enzyme activity is assayed in a total volume of 0.25 ml containing 35 m*M* Tris–HCl (pH 7.5), 10 m*M* 2-ME, 0.2 m*M* PMSF, 0.01% Triton X-100, 0.4 m*M* EGTA, 10 m*M* $MgCl_2$, 0.6 m*M* $CaCl_2$, 20 μg/ml phosphatidylserine, 2 μg/ml 1,2-diolein, 5–30 μg protein from appropriate subcellular fractions, 50 μg Histone H1, and 50 μM ATP including approximately 2×10^6 cpm of [γ-^{32}P]ATP.[10] Phosphatidylserine and diolein are stored in chloroform (−20°), and appropriate aliquots are evaporated under nitrogen and resuspended in water by sonication. Protein kinase-mediated incorporation of ^{32}P into histone is allowed to proceed for 5–60 min at 30°, and reactions are stopped by the addition of 1 ml of ice-cold 25% trichloroacetic acid (TCA), followed by 0.5 mg bovine serum albumin as a carrier. Precipitates are collected on Whatman GF/F filters by vacuum filtration and are washed twice with 5 ml of TCA. The filters are allowed to dry and the radioactivity is determined in a liquid scintillation spectrometer. Net protein kinase C activity is expressed as picomoles ^{32}P incorporated per minute per 10^7 cell equivalents after subtracting activity observed in the absence of Ca^{2+} and lipids.

[12] D. M. Helfman, B. D. Appelbaum, W. R. Vogler, and J. F. Kuo, *Biochem. Biophys. Res. Commun.* **111,** 847 (1983).

[13] M. Wolf, P. Cuatrecasas, and N. Sahyoun, *J. Biol. Chem.* **260,** 15718 (1985).

[26] Ionic Events Relevant to Neutrophil Activation

By PAUL H. NACCACHE and RAMADAN I. SHA'AFI

Introduction

A variety of ionic events have been described in neutrophils in response to stimulation by physiological and nonphysiological agonists. Increased rates of transport of sodium, potassium, calcium, and hydrogen, together with transient changes in the intracellular levels of these ions, have been reported. In addition, stimulated changes in the membrane potential of neutrophils have also been described.[1-3] The advent of intracellularly trapped fluorochromes sensitive to concentrations of calcium and hydrogen has allowed the direct, real-time monitoring of their internal levels. These various ionic events are generally assumed to be relevant to the mechanism of activation of neutrophils by receptor-dependent soluble (and possibly also nonsoluble, i.e., phagocytic) stimuli.

Measurement of Ionic Transport

General Principles

The measurement of the rates of transport of various ions in neutrophils, as in other cells, relies on the general principles of compartmental analysis.[4,5] In view of the potential complexity of the behavior of multicompartment systems, the general outline of the latter will now be sketched. In practice, the following parameters are measured: (1) the initial rate of uptake or influx, (2) the initial rate of efflux, in which case, one monitors the exit from the cells of a labeled ion with which the cells had previously been loaded, and (3) steady-state measurements.

Measurements of influx and efflux rates are done primarily to derive information about the permeability of the cells to the desired ion. They are,

[1] R. I. Sha'afi and P. H. Naccache, *Adv. Inflam. Res.* **2,** 115 (1981).

[2] J. I. Gallin and B. E. Seligman, *in* "Regulation of Leukocyte Function" (R. Snyderman, ed.), p. 83. Plenum, New York, 1984.

[3] P. Wilkinson, "Chemotaxis and Inflammation." Churchill-Livingstone, Edinburgh, Scotland, 1982.

[4] A. K. Solomon, *in* "Mineral Metabolism" (C. L. Comar and F. Bronner, eds.), Vol. 1A, p. 119. Academic Press, New York, 1960.

[5] A. B. Borle, *Cell Calcium* **2,** 187 (1981).

however, determined by both the electrochemical gradient for that particular species and its permeability characteristics. Thus, caution needs to be exercised to insure that the measurements are rapid enough to be an accurate reflection of the initial rates, and that changes in concentrations, if any, are properly taken into account (see the section on interpretation of calcium flux data).

Analysis of the influx and efflux data is performed using standard compartmental analysis. The fundamental equation that governs the movement of a given substance in a closed, two-compartment system is

$$\ln(1 - S/S_{oo}) = -Kt \tag{1}$$

in which S and S_{oo} refer to the concentration of radioactivity (counts per unit volume) in the cells at any given time and at equilibrium, respectively. K is a first-order rate constant in reciprocal time and t is time. One should be aware that one of the basic assumptions in this analysis is that the concentrations of the ion under study are constant both in the cells as well as in the bathing media. The linearity between $(1 - S/S_{oo})$ and time is the basic characteristic of two-compartment systems. Fluxes are obtained by multiplying the rate constants by the intracellular concentrations, per liter of cell, of the respective cation.

Another important parameter is steady-state measurements. In steady-state experiments one allows the desired compound to equilibrate across the plasma membrane of the cells before examining the potential effect of the agonist under examination. Changes in the steady-state distribution thus reflect changes in the level (as opposed to the rate of transport or permeability characteristics) of the compound of interest.

The basic condition governing these experiments is that, at steady-state, the specific activities of the cation under study are the same in the cell and in the extracellular medium, i.e.,

$$[(\mathrm{dpm})_{i,oo}/(\mathrm{Ca})_{i,e}] = [(\mathrm{dpm})_{o,oo}/(\mathrm{Ca})_{o,t}] \tag{2}$$

where $(\mathrm{dpm})_{i,oo}$ and $(\mathrm{dpm})_{o,oo}$ are the steady-state disintegrations per minute in the cells (dpm/mg protein) and medium (dpm/ml), respectively, and $(\mathrm{Ca})_{i,e}$ and $(\mathrm{Ca})_{o,t}$ are the concentrations of cell-exchangeable (mol/mg protein) and total extracellular calcium (mol/ml), respectively. Thus, the size of the exchangeable internal compartment can be deduced from the changes in the cell-associated radioisotopic label assuming steady state has been reached. It should be emphasized that the above analysis does not imply anything by itself about the origin of any increase in the size of the internal exchangeable compartment.

Separation Method

Because of the rapidity of ionic movements across neutrophil membranes, separation of the cells from the bathing media is usually accomplished by centrifugation through water-immiscible media of densities intermediate between those of the cells and of the suspending buffers.[6] Several separation media can be used. The most common is the Versilube F50 silicone oil (General Electric Co., Waterford, NY) which has a density of 1.05 mg/ml.

Cell concentrations between 1 and 2×10^7 cells/ml are routinely used. The cells are separated from the suspending media by centrifuging a sample of cells (0.5–1.0 ml) through a layer of silicone oil (0.3–0.5 ml) for about 30 sec at 8,000–10,000 rpm in a microcentrifuge. The cells are separated from the suspending media within 5 sec. The cell pellet can be excised and, if necessary (i.e., for ^{45}Ca measurements), solubilized with 88% formic acid (0.2–0.5 ml/pellet, 30 min at room temperature). Counting efficiencies need to be monitored especially with tritiated isotopes. The extracellular water space can be estimated by measuring the amount of [^{14}C]inulin (New England Nuclear, Boston, MA) that is recovered in the cell pellets. Inulin, being impermeant, is an extracellular space marker. The amount of trapped extracellular space can thus be calculated from the ratio of the cell-associated inulin radioactivity (counts/cell pellet) to that of the suspension inulin radioactivity (counts/ml suspension). This ratio yields the volume of the trapped extracellular space (in ml). This value routinely averages about 10% of the total pellet water as measured with ^3H_2O,[7] a marker of the total volume of the cell pellet as it distributes equally in the intra- and extracellular spaces. The cell pellet volume can then be estimated, if desired, from the difference between the total cell pellet water volume and the extracellular space volume.

Special Applications

Sodium and Potassium. The measurement of the fluxes of sodium and potassium across the plasma membrane of neutrophils is facilitated by the fact that the cells' interior behave as a single compartment of homogeneous specific activity.[6,8] Thus, in the absence of changes in the intracellular concentration of these ions, direct conclusions can be derived from the determination of the initial rates of uptake and efflux and from changes in

[6] P. H. Naccache, H. J. Showell, E. L. Becker, and R. I. Sha'afi, *J. Cell Biol.* **73,** 428 (1977).
[7] L. Simchowitz and A. Roos, *J. Gen. Physiol.* **85,** 443 (1985).
[8] L. Simchowitz, I. Spilberg, and P. De Weer, *J. Gen. Physiol.* **79,** 453 (1982).

the steady-state values. The requirement for constant intracellular concentrations dictates that the cells need to be thermally equilibrated at the desired temperature (usually 37°, 10–15 min). In practice, 1–3 μCi of radioactivity (either ^{42}K or ^{22}Na) (New England Nuclear) are added per milliliter of neutrophil suspension at the desired cell density ($1-2 \times 10^7$ cells/ml). A zero-time datum is taken as quickly as possible. This sampling is repeated as many times as required for the particular experiment. The contribution of the extracellular trapped space can be estimated either from a parallel [^{14}C]inulin determination or by extrapolating the control uptake curve to the zero-time intercept.

Efflux experiments are performed by first loading the cells with the desired isotope until radioisotopic equilibration is reached (30–40 min). The cells are then washed free of extracellular isotope by 2–3 rounds of centrifugation in a microcentrifuge with 1.5 ml of the desired, thermally equilibrated, buffered solution each time. The cells are then resuspended at $1-2 \times 10^7$ cells/ml in isotope-free buffered solution, and the time course of the appearance of radioactivity in the extracellular medium, or, alternatively, the loss of label from the cells, is followed.

Calcium. Interpretation of the data. The measurement of calcium influx and efflux is performed essentially as just described for sodium and potassium. The interpretation of the data is complicated by the knowledge that the size of the exchangeable internal compartment does change during stimulation of the cells with physiological agonists. Thus increases in the apparent rates of influx and efflux may be accounted for by increases in permeability and/or intracellular levels of exchangeable calcium. In addition, one should remember that exchangeable and free calcium are not synonymous concepts; in fact, the size of the exchangeable calcium pool may exceed that of the free calcium by one or even two orders of magnitude. The clarification of the meaning of experimentally obtained increased rates of transport of calcium requires, at the very least, measurements under influx, efflux, and steady-state conditions (see above).[1,5,9]

There are several potential outcomes of steady-state experiments that depend, among other things, on whether the stimulation of the cells causes (1) a change in permeability, (2) a net uptake of calcium from the extracellular medium, or (3) a cytoplasmic release of calcium from previously unexchangeable stores with different specific activities than that of the prestimulation cytoplasmic exchangeable calcium. A few examples of physiologically relevant responses to stimulation of cell-associated counts are discussed in the following paragraphs.

[9] R. J. Petroski, P. H. Naccache, E. L. Becker, and R. I. Sha'afi, *Am. J. Physiol.* **237,** C43 (1979).

Cell stimulation may, in the simplest cases, be associated with selective increases in inward or outward (the calcium pump) permeability to calcium. These would result in increases and decreases in cell-associated counts, respectively. A change in bidirectional permeability, in and by itself and as long as it does not grossly short-circuit the ability of the cells to maintain their calcium levels within a physiologically relevant range, will not cause any change in the steady-state level of cell-associated ^{45}Ca as the specific activities of the internal and external compartments are by definition equal under these conditions. Thus, there will be no net gain or loss of radioactivity.

An internal release of calcium may induce a variety of changes in the steady-state level of ^{45}Ca depending on the relative efficiencies of the calcium-extruding mechanisms and of the permeability of the cells and on the specific activity of the pool of calcium that is released. Were, for some reason, the specific activities of the cytoplasmic calcium and of the mobilized pool to be the same (a condition that is physiologically unreasonable), then the release of the latter will result in an increase in the size of the internal exchangeable calcium, and thus will cause a net loss of ^{45}Ca as the cells restore their cytoplasmic free calcium concentration by activating the outwardly directed calcium pump.

In practical terms, one deals usually with a stimulus-sensitive pool of low specific activity compared to that of the cytoplasm, i.e., with a pool of calcium that was poorly exchangeable prior to stimulation. The release of the latter will thus increase the size of the pool of intracellular exchangeable calcium and concomitantly decrease the specific activity of the latter. The former event will result in the activation of the calcium-extruding mechanisms of the cells. The latter will stimulate the uptake of ^{45}Ca from the extracellular medium as the specific activities of the two pools of calcium will tend toward equilibrium again. Thus several potential outcomes on the amount of cell-associated counts are theoretically possible depending on the relative rates of extrusion versus uptake of calcium. When the latter predominates, then the net result of the internal release of calcium will be to cause a (transient) increase in cell-associated counts. Conversely, when the cell is capable of extruding calcium faster than it takes it up [e.g., in the presence of low ($<10\ \mu M$) concentrations of external calcium], then the release of the pool of internal calcium will result in a transient decrease in the cell-associated counts.

The possible outcomes discussed above of stimulated changes in membrane permeability and/or size of the internal exchangeable compartment on the rates of influx and efflux and on the steady-state distribution of a hypothetical cation are illustrated in Figs. 1 through 4. The cases selected in the preceding discussion should illustrate the pitfalls associated with a

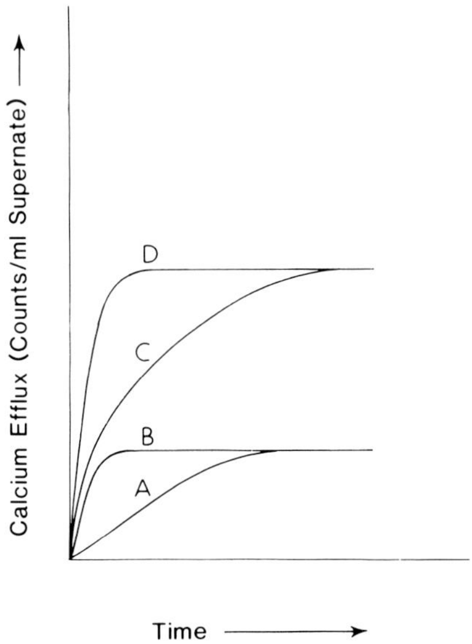

FIG. 1. Diagrammatic representation of the possible outcomes of an influx experiment. A, Control, unstimulated cells; B, increase in membrane permeability; C, increase in the size of the intracellular exchangeable compartment in the absence of a change in membrane permeability; D, increases in both the permeability of the membrane and the intracellular exchangeable level of the cation.

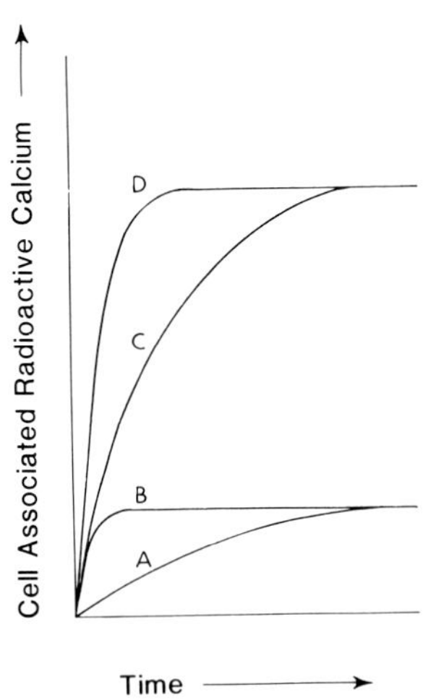

FIG. 2. Diagrammatic representation of the possible outcomes of an efflux experiment. The conditions are the same as those in Fig. 1.

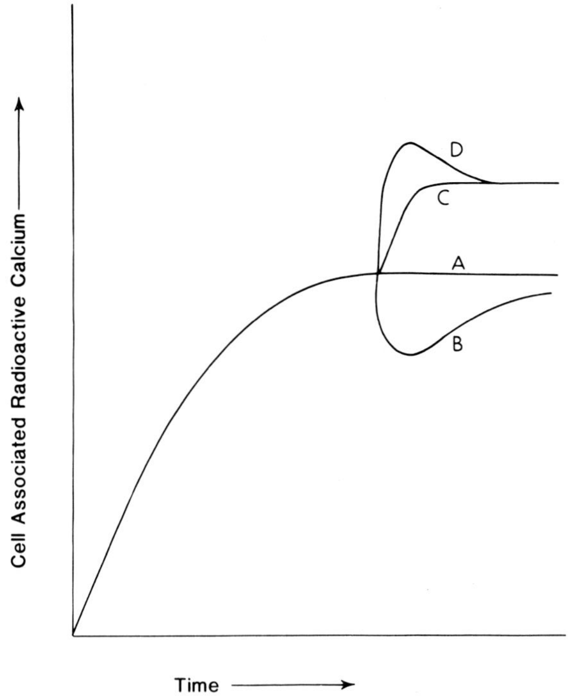

FIG. 3. Diagrammatic representation of some of the possible outcomes of a steady-state experiment in which an agonist is added in the presence of high extracellular concentrations of the cation. A, No change in the steady-state level; B, an increase in the level of intracellular exchangeable cation in the absence of any increase in membrane permeability; C, an increase in the level of intracellular cation as a result of net uptake from the extracellular medium without any increase in membrane permeability; D, increases in membrane permeability and in the level of intracellular exchangeable cation. Note that the increase in the level of exchangeable cation may come from the extracellular medium as well as from an internal source.

literal interpretation of the radioactivity data generated during calcium flux experiments. It is clear that changes in the internal levels of exchangeable calcium and of the cell-associated counts may be translated equally well into proportional, or inversely proportional, relationships.

Measurement of calcium fluxes. ^{45}Ca influx and efflux experiments are usually performed in the presence of 0.5 m*M* external calcium in order to decrease the amount of radioactivity required. For influx experiments, 0.5 μCi/ml of ^{45}Ca (New England Nuclear) is added to thermally equilibrated neutrophil suspensions (10^7 cells/ml). The accumulation of the label within the cells is then monitored under the desired experimental condi-

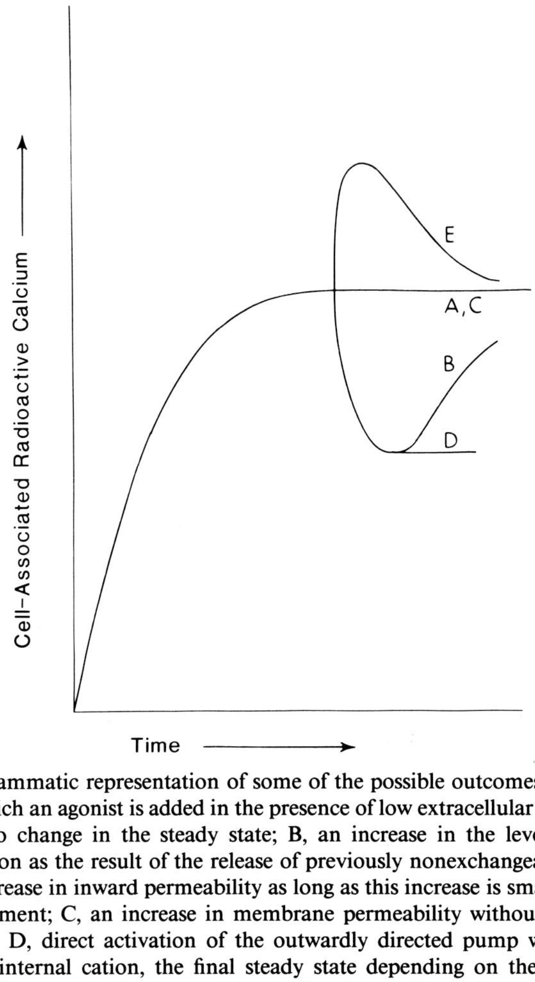

FIG. 4. Diagrammatic representation of some of the possible outcomes of a steady-state experiment in which an agonist is added in the presence of low extracellular concentrations of the cation. A, No change in the steady state; B, an increase in the level of intracellular exchangeable cation as the result of the release of previously nonexchangeable calcium with or without an increase in inward permeability as long as this increase is smaller than the rate of outward movement; C, an increase in membrane permeability without a change in the steady-state level; D, direct activation of the outwardly directed pump with or without a redistribution of internal cation, the final steady state depending on the degree of pump activation; E, an increase in the internal level of exchangeable cation as the result of the release of previously unexchangeable cation and an increase in the inward permeability as long as this increase in membrane permeability is much greater than the rate of outward movement of the cation.

tions. For efflux experiments, the cells (10^7 cells/ml) are loaded in the presence of ^{45}Ca (1 – 1.5 μCi/ml) for 50 – 60 min. The cells are then washed and handled as described above for sodium and potassium.

Steady-state experiments are performed with cells exposed to ^{45}Ca for 50 – 60 min, i.e., until equilibration of the label is achieved. The stimuli are then added to the cells in the presence of the extracellular label, and the changes, if any, in the cell-associated counts are monitored. Steady-state experiments can be performed under two experimental protocols: (1) in the presence of 0.5 m*M* Ca^{2+} or (2) in the absence of added calcium to the suspending buffer (i.e., in the presence of 1 – 10 μM calcium). The rationale for experiments performed in the absence of external calcium is that, under these conditions, the contribution of the external calcium to any detected changes will be minimized and, thus, that of the internal stores will be made more apparent (see Fig. 4 above).

A significant percentage of the ^{45}Ca counts that are found in the pellets of the cells represents calcium bound to the external surface of the cells. The contribution of the latter can be minimized by including a layer of EGTA (0.2 ml of a 2.2 *M* solution, pH 7.3) on top of the silicone oil (0.3 ml) before separating the cells from the unbound radioactivity. This procedure has been found to significantly decrease the background level of counts in the calcium flux experiments without modifying the qualitative nature of the effects of neutrophil agonists on calcium homeostasis.[10]

It should be emphasized that, in view of the presence of multiple unidentified internal compartments of undefined, and changing, specific activities, the interpretation of the ^{45}Ca data depends critically on the examination of as many parameters as possible. The resulting interpretation needs to satisfy, at the very least, the data obtained under the conditions outlined above, i.e., influx, efflux, and steady state (in the presence and absence of external calcium) measurements.

Hydrogen. Principle of the method. The internal pH (pH_i) of the neutrophils can be monitored by following the distribution of weak acids such as 5,5-dimethyloxazolidine-2,4-dione (DMO) (New England Nuclear) or weak bases such as trimethylamine (TMA). The method rests on the assumption that only the uncharged form of the indicator is permeant, and thus that their distribution is pH dependent.[11] Since the pK_a value of DMO (6.13 at 37°) is much closer to that of the biological buffers used than that of TMA (9.62 at 37°), the former distributes much more rapidly across the plasma membranes of neutrophils. In addition, weak acids tend to be excluded from the lysosomal compartments and to indicate more accurately the cytoplasmic pH. DMO has thus been the usual compound of choice.[12] The internal pH (pH_i) can be calculated as described in Roos and

[10] T. F. P. Molski, P. H. Naccache, and R. I. Sha'afi, *Cell Calcium* **4,** 57 (1983).

[11] A. Roos and W. F. Boron, *Physiol. Rev.* **61,** 296 (1981).

[12] R. I. Sha'afi, P. H. Naccache, T. F. P. Molski, and M. Volpi, *in* "Intracellular pH: Its Measurement, Regulation and Utilization in Cell Functions" (R. Nuccitelli and D. W. Deamer, eds.), p. 513. Liss, New York, 1982.

Boron[11] by using the following modified Henderson–Hasselbach equation[13]:

$$pH_i = pK + \log\{[(C_t/C_e)(1 + V_e/V_i) - V_e/V_i][1 + 10^{(pH_e - pK_a)}] - 1\} \quad (3)$$

where pH_e is the pH of the extracellular medium; pK_a, the dissociation constant of DMO (6.2); C_t, counts taken up by the cells in cpm/ml cell H_2O; C_e, counts in the extracellular medium in cpm/ml H_2O; V_e, the volume of trapped extracellular space; and V_i, the volume of cell water.

Application to neutrophils. Using the DMO method, the internal pH of neutrophils can be monitored as follows. [^{14}C]DMO is dried under nitrogen, resuspended in the desired buffered solution, and added (0.1 – 1 μCi/ml, 1 – 10 $\times 10^{-6}$ *M*) to thermally equilibrated cell suspensions (10^7 cells/ml). In view of the small magnitude of the changes in pH_i, great care has to be exercised to maintain the external pH constant to within less than 0.05 pH units during the course of the experiment. This can be accomplished by blowing carbon dioxide (5% CO_2, 95% air) over the samples or, alternatively, by removing the sodium bicarbonate from the suspending buffer and increasing, if necessary, the concentration of the organic buffers used such as Tris or HEPES. The various stimuli are added 1 min after DMO, and the radioactive contents of the cells monitored with the silicone rapid sampling method.

It should be remembered that the values taken for intracellular trapped space, fraction of responding cells,[14] and cell water all constitute potential sources of systematic errors in the determination of the absolute values of pH_i. Changes in pH_i, on the other hand, are to a great extent free from these constraints, are experimentally easier to define, and, more importantly, are more reliable.[12]

Fluorescence Assays

Intracellular Free Calcium

General Principles. The fluorescence assays depend on the development by R. Y. Tsien of fluorescent calcium-chelating indicators and of a method for loading them into intact cells.[15,16] The best known indicators are Quin 2 and the more recently introduced Fura 2.

The basic principle of the use of these compounds is as follows. They

[13] W. J. Waddell and T. C. Butler, *J. Clin. Invest.* **38,** 720 (1959).

[14] L. Simchowitz, *J. Biol. Chem.* **260,** 13248 (1985).

[15] R. Y. Tsien, T. Pozzan, and T. J. Rink, *Trends Biochem. Sci.* **9,** 263 (1984).

[16] T. J. Rink and T. Pozzan, *Cell Calcium* **6,** 133 (1985).

are presented to the cells as their respective tetraacetoxymethyl esters, which, because of the hydrophobicity thus confered, are freely permeable across cell membranes. Once in the cytoplasm, one of several unidentified esterases hydrolyzes the hydrophobic esters and liberates the free indicator. The latter is trapped inside the cells. It should be emphasized that it is only the hydrolyzed compounds (i.e., Quin 2 and/or Fura 2) which act as calcium indicators and not their esterified counterparts (i.e., Quin 2-AM and Fura 2-AM). This mechanism allows for significant accumulation of the indicator within the cell under examination. The exact magnitude of the concentration reached depends on the individual cell, and it is affected by, among others, the activity of the intracellular esterases, the amount of extracellular esterases, and the permeability of the membranes to the free acid form of the indicators. Conversely, the internal concentration of these compounds determines to a large extent whether they are acting as indicators or intracellular calcium buffers.

The chemical structures of Quin 2 and Fura 2 and their physical and fluorescence characteristics have been detailed by Tsien *et al.*[15,17] Grynkiewicz *et al.*,[18] and Rink and Pozzan.[16] Their most notable characteristic is that their fluorescence intensity increases upon binding calcium. Fura 2 possesses several advantages over the earlier probes such as Quin 2 including brighter fluorescence and excitation spectral shifts upon binding of calcium. This last property, the fluorescence excitation shift, enables one to determine directly the amount of intracellular calcium by measuring the ratio of the fluorescence of Fura 2 at two wavelengths (340 and 380 nm), thus bypassing the need to permeabilize the cells to determine the actual dye content of the cells as is required for Quin 2 (see below). For historical reasons, however, Quin 2 is the probe that has been the most widely used in neutrophil studies up to now.

Application to Neutrophils. Indicator loading is accomplished by incubating cell suspensions ($1-2 \times 10^7$ cells/ml) with $1-2 \times 10^{-5}$ *M* Quin 2-AM or $1-2 \times 10^{-6}$ *M* Fura 2-AM for 30 min at 37°. These probes are available from, among others, Molecular Probes (Junction City, OR) and Calbiochem (San Diego, CA). The cells are then washed free of the extracellular indicator with the desired buffered solution. Usually 2 washes are sufficient. An additional, 10-min incubation at 37° before the start of the actual measurements helps stabilize the baseline fluorescence in rabbit neutrophils possibly by allowing the hydrolysis of the internalized tetraacetoxymethyl esters to proceed to completion.

Fluorescence of Quin 2 and Fura 2 is monitored at the following wave-

[17] R. Y. Tsien, T. J. Rink, and M. Poenie, *Cell Calcium* **6,** 145 (1985).

[18] G. Grynkiewicz, M. Poenie, and R. Y. Tsien, *J. Biol. Chem.* **260,** 3440 (1985).

lengths: excitation 339 and 340 nm, emission 492 and 510 nm, respectively. As mentioned above, direct calibration of the Fura 2 signal can be accomplished by measuring the 340/380 ratio. Calibration of the Quin 2 signal, and, if so desired, also of the Fura 2 signal, is performed by measuring the signal generated by a lysed cell preparation (0.1% Triton X-100) in the presence of saturating (~1 mM) and extremely small (~10 nM) amounts of calcium. The latter can be achieved by adding excess amounts of calcium chelators and by increasing the pH of the medium to above 8. The free calcium can then be calculated as indicated by Tsien *et al.*,[19] using the following formula:

$$Ca^f = K_d[(F - F_{min})/(F_{max} - F)] \quad (4)$$

where Ca^f represents the concentration of free calcium (in nM); F, the fluorescence of the individual samples; F_{max}, the fluorescence of the released indicator in the presence of saturating amounts of calcium; F_{min}, the fluorescence of the released indicator in the presence of vanishingly small concentrations of calcium; and K_d, the affinity constant of the probe for calcium (115 nM for Quin 2, 224 nM for Fura 2).

Chlortetracycline Fluorescence

General Principle. Chlortetracycline is a hydrophobic fluorescent antibiotic.[20] Its spectral characteristics (quantum yield and spectral maxima) depend on the hydrophobicity of the environment and on the level of calcium and magnesium in its immediate surroundings. The fluorescence of chlortetracycline increases upon partitioning into the lipid bilayer of biological membranes. As the divalent cation chelate has a net positive charge, its partitioning is itself divalent cation dependent, as more probe will accumulate in calcium-containing hydrophobic environments than in calcium-depleted membranes. Conversely, the release of calcium from membranes leads to the subsequent loss of chlortetracycline. The fluorescence of chlortetracycline increases upon binding calcium and/or magnesium. In aqueous solution chlortetracycline binds to magnesium and to calcium equally well, while the order of affinity constants favors calcium over magnesium in hydrophobic environments (K_d 9 μM for Ca^{2+} and 25 μM for Mg^{2+} in 70% methanol[20]). The spectral maxima of the calcium and the magnesium chelates of chlortetracycline differ enough to allow one to distinguish between the two forms of the probe. Thus chlortetracycline has been used in several cell types to monitor the state of "membrane-as-

[19] R. Y. Tsien, T. Pozzan, and T. J. Rink, *J. Cell Biol.* **94,** 325 (1982).
[20] A. H. Caswell, *Int. Rev. Cytol.* **56,** 145 (1979).

sociated" calcium. The quotation marks are meant to indicate that serious questions remain about the exact nature of the cellular events which underlie the commonly observed changes in chlortetracycline that follow stimulation by a variety of agonists. The calibration of the chlortetracycline signal in whole cells is a similarly complicated problem that has not been systematically studied to date.[21]

Application to Neutrophils. The effects of the desired stimulus on the fluorescence of neutrophil-associated chlortetracycline can be monitored using one of two experimental protocols. The two methods differ only in that in the first the measurements are made in the presence of external chlortetracycline while in the other the cells are washed free of extracellular probe before monitoring the fluorescence of the cells.

In the first protocol,[22] neutrophil suspensions (5×10^6 cells/ml) are incubated with 2×10^{-5} *M* chlortetracycline (Sigma Chemical Co., St. Louis, MO) in a buffered salt solution that contains neither magnesium nor proteins and whose calcium concentration is adjusted to 0.5 m*M*. The lack of proteins and magnesium is necessary to minimize the contribution of extracellular chlortetracycline to the overall signal. The uptake of the probe into the cells can be monitored by observing the increase in fluorescence of the cells (excitation wavelength 390 nm, emission wavelength 530 nm). The cells are incubated with the probe until the rate of increase of the fluorescence diminishes considerably, i.e., about 30 min. The cells can then be transferred serially to the spectrofluorophotometer and their fluorescence and the effect of the desired stimuli monitored.

Alternatively,[23,24] the cells can be washed free of the extracellular chlortetracycline immediately prior to transfer to the spectrofluorophotometer. The composition of the media during the loading period in this procedure does not require the absence of magnesium and proteins. It is only following the washing step that these precautions need to be observed. The advantage of this protocol is that it minimizes the contribution of the extracellular probe and allows for easier measurements of the spectral characteristics (excitation and emission maxima) of the probe before and after stimulation. Its disadvantage is that it involves an additional manipulation of the cells and thus introduces mechanical and thermal disturbances, the contributions of which need to be assessed.

The major difference between these two protocols is thus the presence

[21] N. Gains, *Eur. J. Biochem.* **111,** 199 (1980).

[22] P. H. Naccache, H. J. Showell, E. L. Becker, and R. I. Sha'afi, *J. Cell Biol.* **83,** 179 (1979).

[23] P. H. Naccache, M. Volpi, H. J. Showell, E. L. Becker, and R. I. Sha'afi, *Science* **203,** 461 (1979).

[24] J. E. Smolen and G. Weissmann, *Biochim. Biophys. Acta* **720,** 172 (1982).

or absence of chlortetracycline during the measurement phase. The release of calcium, as indicated by the decrease in fluorescence of the cell-associated probe, can thus be monitored equally well using either procedure. However, as mentioned in the preceding section, calcium release is accompanied by the displacement of chlortetracycline from the membranes. In the absence of external probe, the released chlortetracycline is thus diluted in the suspending medium, and the fluorescence of the cells never recovers to its prestimulation level. In the presence of chlortetracycline in the buffer, the decrease in fluorescence is transient. The mechanism underlying the return of the fluorescence to its basal level remains to be rigorously studied. It is likely to involve both the reuptake of calcium into the pool that is displaced by stimulation of the cells and the repartitioning of the probe within the same environment. It is thus conceivable that this latter protocol may provide a means of monitoring, and understanding, the mechanisms responsible for the turning off of the calcium signal.

Finally, one should remain aware of the potential inhibitory effects of chlortetracycline on neutrophil functions. The latter will be more evident at relatively high concentrations of the probe ($>5 \times 10^{-5}\ M$) and low extracellular calcium.[25]

Intracellular pH

General Principles. Fluorescent indicators with the properties of the calcium indicators described above have also recently become available. They, as the calcium indicators, are presented to the cells as their tetraacetoxymethyl esters which are freely permeable to the cells and get trapped following the hydrolysis of the ester bonds. The most common fluorescein derivative used for the measurement of intracellular pH is 2,7-bis(carboxyethyl)-5(6)-carboxylfluorescein (BCECF).[26]

Application to Neutrophils. Loading of the tetraacetoxymethyl ester of BCECF (BCECF-AM) (HSC Research Development Corp., Toronto, ON, or Molecular Probes, Junction City, OR) is accomplished as follows.[27] Neutrophil suspensions (2×10^7 cells/ml) are incubated with 3 μg/ml BCECF-AM for 30 min at 37°. They are then washed twice in the desired buffered solution and resuspended at the same cell density.

The excitation wavelength is 500 nm and the emission wavelength 530 nm. If required, a high-pass filter with a cutoff of 515 may be used on the emission side of the cell compartment. The calibration of the signal can be performed in the following ways. The cells are lysed with Triton X-100,

[25] J. G. R. Elferink, *Biochem. Pharmacol.* **33,** 3667 (1984).

[26] T. J. Rink, R. Y. Tsien, and T. Pozzan, *J. Cell Biol.* **95,** 189 (1982).

[27] S. Grinstein and W. Furuya, *Biochem. Biophys. Res. Commun.* **122,** 755 (1984).

and the fluorescence of the released probe can then be measured at various pH values. A correction factor needs to be applied to the value so determined because the excitation peak of the intracellular dye is red-shifted about 5 nm, thus slightly reducing the signal for any given pH.[26] Alternatively, the internal hydrogen ion concentration can be equalized with that of the outside medium by using the H^+/K^+-antiporter nigericin[26] (Calbiochem), using the procedure of Thomas *et al.*[28]

The fluorescence of the BCECF-loaded cells is first measured in the buffered solution that is being used. The cells are then resuspended in the presence of buffered solutions containing 130 m*M* potassium (i.e., concentrations approximating the internal level of this cation). Nigericin (10 μg/ml) is added, and the fluorescence is measured at various values of the external pH. Under these conditions, the internal and external pH are equalized, and a calibration curve relating the fluorescence of the cells loaded with BCECF to various values of the external pH can be constructed. The latter allows the value of the original internal pH to be deduced. It should be remembered that BCECF is not totally impermeant and that a certain amount will diffuse out of the cells with time. Thus one needs to ascertain the magnitude of the contribution of the external probe. This can be done simply by measuring the amount of fluorescence of the cell supernates as a function of time.

Membrane Potential

General Considerations. Measurements of the membrane potential of neutrophils do not directly involve the examination of ionic events. However, as the latter are determined by both the level and the permeability of the plasma membrane to the different permeant ions, changes of the membrane potential do indeed indirectly result from altered ionic homeostasis.

The membrane potential of cells in suspensions can be estimated using one of a series of cyanine dyes[29] the most commonly used being diS-C_3(5). The fluorescence response of these dyes results from the potential-dependent partition of the dye between the cells and the external medium. The uptake of the dye increases as the cell hyperpolarizes and conversely decreases during cell depolarization. The emission from the cell-associated dye decreases during a hyperpolarization response as the dye aggregates and becomes less fluorescent. A depolarization is associated with increases

[28] J. A. Thomas, R. N. Buchsbaum, A. Zimniak, and E. Racker, *Biochemistry* **18,** 2210 (1979).

[29] A. S. Waggoner, *Annu. Rev. Biophys. Bioeng.* **8,** 47 (1979).

in fluorescence while a hyperpolarization is indicated by decreased levels of fluorescence. The ratio of dye to membrane that produces the optimal fluorescence changes must therefore be experimentally determined.

Application to Neutrophils. Neutrophils ($1-2 \times 10^6$ cells/ml) are incubated with $1-2 \times 10^{-6}$ *M* diS-C_3-5 (Molecular Probes) until a stable baseline fluorescence is established (3–6 min). Caution should be exercised to maintain the stock solution of the probe in the dark. The excitation and emission wavelengths are 620 and 670 nm, respectively. Adsorption of the dye to glass can be minimized by the addition of bovine serum albumin (1 mg/ml).[8] The presence of catalase (550 IU/ml) and/or superoxide dismutase (25 μg/ml) to avoid quenching of the dye by reactive oxygen metabolites is necessary in order to accurately determine the shape of the membrane potential changes at times longer than 1–2 min after the addition of neutrophil agonists.[30] The resting membrane potential can be estimated by the null point method in the presence of the potassium-specific ionophore valinomycin, i.e., by determining the concentration of external potassium at which the addition of valinomycin (Sigma) (1×10^{-5} *M*) produces no changes in fluorescence.

Acknowledgments

This chapter was supported in part by grants from the National Institutes of Health (AM 31000 and AI 13734), the Medical Research Council of Canada (DG-337), and the Fonds de la Recherche en Sante du Quebec (850032).

[30] J. C. Whitin, R. A. Clark, E. R. Simons, and H. J. Cohen, *J. Biol. Chem.* **256,** 8904 (1981).

Section II

Inflammation

A. Methods for the Study of the Cellular Phenomena of Inflammation and Experimental Models of Inflammation
Articles 27 through 44

B. Methods for the Study of Complement in Chemotaxis and Inflammation
Articles 45 through 49

[27] Cellular and Vascular Phenomena in Inflammation

By ANDREW C. ISSEKUTZ and THOMAS B. ISSEKUTZ

Introduction

Inflammation is the response of body tissues to noxious or injurious stimuli. It is a complex series of events resulting in cellular and vascular alterations. The response is usually quite similar whether the stimuli are chemical, physical, microbial, or immunological in nature. Acute inflammation is characterized by swelling and redness or erythema as a result of plasma and protein leakage from blood vessels into the interstitial space, usually accompanied by an increase in the local blood flow (hyperemia). These are the earliest visible responses, which may be followed by platelet deposition and polymorphonuclear leukocyte (PMNL)[1] margination in the microvasculature and subsequent PMNL infiltration into the tissues. If the reaction proceeds to tissue and vascular injury, fibrin and additional platelet deposition may develop (thrombosis) and red cell extravasation (hemorrhage) may occur. In the later stages of acute inflammation, and certainly during delayed hypersensitivity of T lymphocyte-mediated inflammatory reactions, infiltration of the involved tissue with blood monocytes and lymphocytes also occurs. The aim of this chapter is to outline in detail methods which are currently in use to measure these various aspects of the inflammatory reaction. Obviously, space does not allow a comprehensive review of all methodologies available, and, therefore, only brief reference will be made to alternative techniques which are also applicable.

Methods for the Study of Inflammation

General Considerations

The skin is the most readily accessible and convenient tissue in which to study the inflammatory reaction. Inflammatory stimuli may be administered with a minimum of trauma and physical artifact by intradermal injection of the test material through a very thin needle, e.g., 30-gauge.

[1] Abbreviations: PMNL, polymorphonuclear leukocyte; fNLP, *N*-formylnorleucylphenylalanine; ZAP, zymosan-activated plasma; PBS, phosphate-buffered saline; ACD; acid–citrate–dextrose; PRP: platelet-rich plasma; PPP; platelet-poor plasma; RBC; red blood cell; LRP; leukocyte-rich plasma; DTH; delayed-type hypersensitivity; KLH; keyhole limpet hemocyanin; IUDR; 5-iodo-2-deoxyuridine; FCS; fetal calf serum.

Multiple skin test sites may be studied, usually on the back of the same animal, as long as proper randomization of sites is used to eliminate regional effects and proper controls are included to rule out the effects of lesion numbers or timing on the course of individual reactions. In some species such as the rabbit, the intradermal injection of up to 0.2 ml of control substances such as pyrogen-free saline induces virtually no inflammatory response by itself. In other species such as the rat, however, the intradermal injection of as little as 0.05 ml induces a transient increase in vascular permeability. Furthermore, rabbits are very placid and usually require no restraint during intradermal injections. In contrast, rats and guinea pigs are usually much more combative, and rats may even need to be anesthetized.

A wide variety of agents can be used to induce inflammation in the skin using intradermal injection. Specifically, saline-washed, live or formalin (0.5%)-killed bacteria such as *Escherichia coli* [10^5–10^8 colony-forming units (CFU) skin site], *E. coli* endotoxins (1–100 ng/site) (Difco Laboratories, Detroit, MI), or boiled, washed zymosan A (0.2–2 mg/site) (Sigma Chemical Co., St. Louis, MO) induce inflammation in the skin of rabbits which is maximal within 2 hr of injection. An immune complex vasculitis can be induced with the dermal Arthus reaction. For this, 5–10 mg of an antigen such as human IgG (Sandoglobulin, Sandoz Pharmaceuticals) is injected intravenously and, 60 min later, hyperimmune antibody containing serum to this antigen is injected intradermally. The reaction will evolve over 4–6 hr into a hemorrhagic vasculitis if the antibody titer in the serum is adequate. Injection of PMNL chemotactic factors such as *N*-formylnorleucylphenylalanine (fNLP, 10^{-8}–10^{-5} M) (Sigma) or zymosan-activated plasma (ZAP) as a source of $C5a_{desArg}$ are potent acute inflammatory stimuli. The fNLP is dissolved as a stock solution (10^{-2} *M*) in dimethyl sulfoxide and then diluted in saline. ZAP is made by incubating 9 parts plasma (containing 5 U heparin/ml) with 1 part boiled (30 min), washed zymosan A (50 mg/ml) for 60 min at 37° with mixing. The zymosan particles are removed by centrifugation (2000 *g* for 15 min) prior to injection of ZAP. Finally, in addition to intradermal injection of inflammatory agents, topical application of stimuli capable of penetrating the stratum corneum and epidermis, such as arachidonic acid (0.1–10%) in acetone,[1a] or contact sensitizing agents (e.g., picryl chloride)[2] (see below—lymphoblast migration), especially when applied to the ears of mice, may be used to elicit acute or delayed inflammatory reactions, respectively.

[1a] J. M. Young, D. A. Spires, C. J. Bedford, B. Wagner, S. J. Ballaron, and L. M. DeYoung, *J. Invest. Dermatol.* **82,** 367 (1984).

[2] M. L. Rose, D. M. V. Parrott, and R. G. Bruce, *Cell. Immunol.* **27,** 36 (1976).

Although the skin is a convenient tissue in which to elicit inflammation, it does not allow for the recovery of cells and of *in vivo* generated mediators. If the study requires the latter, it is most advisable to induce inflammation in a serous cavity such as the peritoneal or the pleural space. Our preference is the pleural cavity of rabbits because it allows sequential sampling in the conscious animal without interference by bowel or ommentum as is often the case in the peritoneum.

Under local or general anesthesia, with aseptic measures, a small-bore needle, e.g., 20-gauge, is carefully advanced at an oblique angle through an intercostal space of the right hemithorax until the parietal pleura is traversed. At this point, the lung on that side will collapse. Then a large-bore, e.g., 12- or 13-gauge, needle is inserted through the same space, and a thin catheter made of Tygon or a 5 French pediatric feeding tube, bearing side holes as well as an end hole, is advanced through the needle for about 10–15 cm. The needle may then be withdrawn and the intrapleural position of the catheter verified by instillation and ready withdrawal of pyrogen-free saline. Any residual fluid and air is aspirated through the catheter, it is then filled with 50 units of heparin/ml in saline and sealed with a clamp (such as is used by blood banks to seal blood collection bags). The catheter is secured into position with sutures at the skin and protected by wrapping with an elastic bandage around the thorax. This procedure is very well tolerated as long as the lung is not lacerated. These pleural catheters can usually be used for at least 5–7 days following placement. Prior to testing an inflammatory stimulus either directly after insertion or, preferably, 24–48 hr later, the pleural space is washed with 2–3 exchanges of 10 ml each of a pyrogen-free buffered physiological medium such as Tyrode's solution (see below). The inflammatory stimulus may be introduced in sufficient volume, e.g., 5–10 ml, so that later either the entire exudate or sequential timed samples of 1–3 ml may be collected for analysis of cells and cell-free supernatant. Collection of the samples in anticoagulant, e.g., 2–4 units heparin/ml of exudate or 0.38% (final) sodium citrate, is recommended.

It should be stressed that no matter which route is used to inject inflammatory stimuli, all test materials should be carefully prepared to prevent gram-negative endotoxin (pyrogen) contamination. Endotoxin in a variety of species including the rat, rabbit, sheep, and human is an extremely potent inflammatory stimulus inducing both acute and delayed inflammatory reactions at doses as low as 0.1 ng. Therefore, it is suggested that media and buffered salt solutions for *in vivo* use be made with pyrogen-free water or saline which has been approved for human use. This is available from a number of pharmaceutical companies or from hospital suppliers.

In addition to the methodology described, which requires no special equipment or expertise, two specialized techniques which allow for direct visualization of the microvasculature and blood cells *in vivo* should be mentioned. These are the hamster cheek pouch and rabbit ear chamber preparations. These are described in detail elsewhere.[3,4]

Measurement of Plasma Exudation

Plasma exudation is the result of enhanced leakage of plasma water and macromolecules (proteins) primarily from postcapillary venules due to the widening of endothelial cell gap junctions. This response is triggered by a variety of vasoactive inflammatory mediators. The method of Udaka *et al.*[5] is perhaps the simplest to employ when plasma exudation in a solid tissue is to be measured. This technique requires ^{125}I-labeled human serum albumin or autologous serum albumin which may be radiolabeled by any of a number of techniques including chloramine-T or insolublized glucose oxidase–lactoperoxidase (Enzymobeads, Cat. No. 170-6001, Bio-Rad Laboratories, Richmond, CA). Alternatively, ^{125}I-labeled human serum albumin may be purchased from Amersham (Oakville, ON) or Merck Frosst, although this tends to be much less cost efficient than preparing the material. There are two important considerations with regard to the preparation. First, the iodinated albumin must be virtually free of unbound ^{125}I. This can be assessed by repeated dialysis of the preparation to ensure that less than 2% of the total radioactivity is in the dialyzate. Second, in our experience in a rabbit model of dermal inflammation, iodinated human, bovine, or rabbit albumin yield similar results. However, if one is studying Arthus reactions induced by the *in situ* formation of bovine albumin–anti-bovine albumin immune complexes, ^{125}I-labeled *autologous* albumin must be used to prevent cross-reactions between the anti-bovine albumin antibodies and human as well as bovine albumin.

In order to quantitate plasma protein exudation at a given point in time in solid tissues such as the skin or lungs, the procedure is to inject intravenously a dose of 10–20 μCi/kg of the ^{125}I-labeled albumin 20–30 min prior to sacrificing the animal. In the case of dermal inflammatory reactions in the rabbit, rat, or guinea pig, sets of duplicate to quadruplicate skin sites may be injected with the same inflammatory stimulus at various times prior to the injection of ^{125}I-labeled albumin, e.g., 0.5, 1, 2 hr, etc., so that the kinetics of plasma protein exudation during the inflammatory reaction

[3] N. Leaf and H. A. Zarem, *Microvasc. Res.* **2,** 77 (1970).

[4] E. Svensjo, K.-E. Arfors, G. Arturson, and G. Rutili, *Upsala J. Med. Sci.* **83,** 71 (1978).

[5] K. Udaka, Y. Takeuchi, and H. Z. Movat, *Proc. Soc. Exp. Biol. Med.* **133,** 1384 (1970).

may be assessed. Immediately following sacrifice, a 0.5- to 2-ml sample of blood is collected from the heart into anticoagulant. This sample is then centrifuged (1000 *g* for 15 min) to harvest the plasma. In the case of dermal inflammatory reactions, the skin is removed and excess blood in the vessels is manually expressed toward the periphery of the pelt. It is desirable to freeze the skin at -20 to $-70°$ and then to punch out a fixed area of the skin test site using a leather punch (e.g., diameter 15–19 mm in the rabbit) available from most leather craft suppliers. The skin lesions and a fixed volume of plasma, e.g., 100 μl are then analyzed in a gamma counter for ^{125}I content.

In the event that plasma protein exudation is measured in a vascular organ such as the lungs, the residual blood containing a large amount of ^{125}I-labeled albumin in the plasma fraction cannot be merely expressed but must be removed by perfusion of the organ. If this is to be performed, it is best to anticoagulate the animal with heparin (e.g., 500 U/kg i.v.) immediately before sacrifice. Using the lung as an example, during intense pulmonary inflammation the albumin may leak into the interstitial compartment, the alveoli, and the bronchial tree. Measurement of extravasated albumin in this organ may be performed by using a two-stage method in which the lung is first lavaged *in situ,* at least twice, by the instillation of phosphate-buffered saline (PBS) via a polyethylene tracheal cannula (e.g., PE 260, Becton Dickinson Co., Parsippany, NJ), taking care not to rupture the parenchyma. This lavage fluid is then saved for ^{125}I determination. The lung is then perfused *in situ* by securing a cannula (e.g., PE 200, Becton Dickinson, or a 5 French feeding tube, Argyle Division of Sherwood Medical, St. Louis, MO) into the main pulmonary artery and into the ascending aorta. The lung is maintained in moderate inflation by injection of air via the tracheal cannula and perfused via the pulmonary artery at physiological pressures (e.g., 20 cm water) with a balanced salt solution such as Tyrode's containing either 0.5–1% albumin or 20% autologous heparinized plasma.

Calculation of the degree of plasma protein exudation into tissues such as the skin or the lung is then made by determining the radioactivity in the tissue and relating this to the level of radioactive albumin in the plasma at the time of sacrifice. The following equation shows the calculation of the amount of plasma albumin equivalents (ul) that have extravasated into the tissues:

$$\mu\text{l plasma extravasated} = \frac{\text{cpm in tissue}}{\text{cpm}/\mu\text{l plasma}}$$

If desired, one can standardize the values for the wet or dry weight of the tissues.

Prior to the use of ^{125}I-labeled albumin, investigators used Evan's blue or a similar dye administered intravenously to quantitate albumin extravasation.[5] This dye binds with high affinity to albumin and its accumulation in inflamed tissue or exudate fluid was used as a measure of albumin extravasation. This technique requires extraction of the dye from the tissue to allow spectrophotometric quantitation of the Evan's blue–albumin complex; it is therefore more tedious.

Quantitation of Hyperemia

Inflammation is often accompanied by vasodilatation and increased blood flow in the inflamed tissue, especially during the acute phase. This response can be measured in one of several ways. Hay *et al.*[6] applied the radiolabeled microsphere technique to the measurement of inflammatory hyperemia. This technique requires that a catheter be implanted into the aortic arch or the left ventricle of the experimental animal. In the author's opinion, it is preferable to perform this 1 or more days prior to induction of inflammation so that the animal need not be anesthetized at the time that the blood flow measurements are made. If such is the case, the catheters should be filled with a solution of 3.8% sodium citrate containing 100 units of heparin/ml and sealed. Alternately, the aortic catheter, usually inserted retrograde via the right carotid artery, may be secured in position while the inflammatory reaction is underway. Blood flow in this model is measured using radiolabeled microspheres 15 μm in diameter, which impact in the terminal arterioles and do not recirculate. These are available from New England Nuclear bearing a variety of isotopes. At the time of measurement, 5 μCi/kg of the microsphere suspension is injected into the aortic catheter. If serial measurements of blood flow are required, microspheres carrying different gamma isotopes may be injected at different times as long as the gamma emission spectra of the various labels can be separated by the gamma spectrometer in use, e.g., ^{57}Co, ^{51}Cr, ^{85}Sr. The tissue must be excised and radioactive content determined in a gamma spectrometer. The results of these experiments may be expressed as a percent increase in the radioactivity of the tissue, and thus in the blood flow to the inflamed tissue relative to control (noninflamed) tissue, or the results may be calculated as a percentage of the total cardiac output being delivered to the tissue, determined from the formula:

$$\text{\% of cardiac output} = \frac{\text{cpm in tissue}}{\text{cpm injected}} \times 100$$

[6] J. B. Hay, M. G. Jonhston, B. B. Hobbs, and H. Z. Movat, *Proc. Soc. Exp. Biol. Med.* **150,** 641 (1975).

An alternative and simpler method of measuring inflammatory hyperemia, one that the authors employ regularly, is basically a modification of that verified by Mendell and Hollenberg.[7] This technique employs rubidium-86 (^{86}RbCl) as an isotope and relies on the rapid transport of ^{86}Rb into tissue cells by their sodium–potassium exchange pump. This isotope is injected intravenously as a bolus of 15 μCi/kg. If cardiac arrest is achieved within 60 sec of this bolus injection, the rate of ^{86}Rb accumulation in organs and tissues is directly proportional to the rate of blood flow. We routinely sacrifice the animal 45 sec after ^{86}RbCl injection using an overdose of pentobarbitol followed immediately by saturated KCl solution. The inflamed tissue is excised, and the radioactivity is determined in a gamma spectrometer. As with the microsphere technique or ^{125}I-labeled albumin permeability measurements, in dermal inflammatory reactions, the kinetics of the blood flow response in lesions of different ages (i.e., in lesions initiated at different times prior to sacrifice) can be determined. The clear advantage of the ^{86}Rb technique is that one need not anesthetize the animal or have ready access to the systemic arterial circulation. However, unlike the microsphere technique, blood flow to the inflammatory tissue can be measured only once, and the results can only be expressed in a comparative fashion such as a percent increase or fold increase of blood flow relative to control noninflamed sites, based on the relative ^{86}Rb content.

A third technique for measuring local blood flow was introduced by Williams[8] and employs xenon-133. However, it is beyond the scope of this chapter to discuss this method in detail, and the reader is referred to the original publication.

Measurement of Polymorphonuclear Leukocyte Accumulation

Characteristic of acute inflammatory reactions is the striking PMNL infiltration of the involved tissues within a few hours of the initiation of a reaction. This infiltration, when induced in a body cavity, can be quantitated by doing serial leukocyte counts on the exudate fluid. In a solid tissue, serial thin histological sections may be made and stained for microscopic quantitation of PMNLs. This procedure tends to be extremely tedious and requires counting of many microscopic fields to determine the average number of leukocytes in a fixed area of tissue (e.g., PMNL/mm^2). Two alternative approaches are (1) to homogenize the tissue and assay the supernatant for the presence of PMNL-associated proteins such as myelo-

[7] P. L. Mendell and N. K. Hollenberg, *Am. J. Physiol.* **221,** 1617 (1971).

[8] T. J. Williams, *J. Physiol. (London)* **254,** 4 (1976).

peroxidase[9] or (2) to measure the accumulation in the tissue of radiolabeled PMNLs. We have used the latter technique employing ^{51}Cr to label blood PMNLs or a mixed blood leukocyte suspension. When analyzing a large number of samples (e.g., 35–45 lesions) in a solid tissue such as the skin of rabbits, we have found this method not only to be rapid and convenient but to also correlate well with histological measurements.[10] The described technique is slightly modified from Ref. 10. It is usually used in NZW rabbits weighing 3–4 kg. However, it can be performed with smaller, younger animals of 800–1000 g and potentially also in newborn rabbits. The blood sample sizes are given for the 3- to 4-kg animals and would need to be adjusted proportionately when smaller rabbits and lesser amounts of blood are used.

Materials

1. Tyrode's solution, Ca^{2+}, Mg^{2+}-free, made up as a 10 times concentrated (10×) stock solution in sterile 50-ml polypropylene centrifuge tubes containing the following: 4 g NaCl, 0.1 g KCl, 0.025 g $NaH_2PO_4 \cdot H_2O$, 0.5 g $NaHCO_3$, and 0.5 g glucose made up to 50 ml with pyrogen-free water (Baxter Travenol, Malton, ON). This solution is stored frozen and can be thawed repeatedly prior to use. Isotonic 1× Tyrode's solution is made by diluting 5 ml of 10× to 50 ml with pyrogen-free water in a 50-ml polypropylene tube. One 5-ml aliquot is removed, and the amount of 0.3 *N* HCl required to adjust the pH to 7.2 is determined. This aliquot is discarded. Then 9 times this amount of 0.3 *N* HCl is also added to the remaining 45 ml.

2. Hydroxyethylcellulose (Polysciences Inc., Cat. No. 5570, Warington, PA). This is a 1% solution made in a 50-ml polypropylene centrifuge tube as follows. One-half gram of hydroxyethylcellulose is washed twice with *cold,* sterile, pyrogen-free saline by mixing and centrifuging the poorly soluble hydroxyethylcellulose at 1,000 *g* for 15 min at 4°. The washed hydroxyethylcellulose is then resuspended to 50 ml with sterile, pyrogen-free *saline* and mixed at 37° for 4–5 hr until dissolved. This solution is stored frozen and can be thawed repeatedly at 37° prior to use; however, thorough mixing after thawing is recommended prior to use. A *cautionary note* is that the supplier of hydroxyethylcellulose (Polysciences Inc.) appears to have changed the formulation of this compound in the past few years, although its viscosity and catalogue number have remained the same. The material on the market at present does not differentially sedi-

[9] P. P. Bradley, D. A. Priebat, R. D. Christensen, and G. Rothstein, *J. Invest. Dermatol.* **78,** 206 (1982).

[10] A. C. Issekutz and H. Z. Movat, *Lab. Invest.* **42,** 310 (1980).

ment rabbit red blood cells as well as the one marketed 4–5 years ago. Preliminary results indicate that hydroxyethylcellulose (Cat. No. 542900) manufactured by Fluka Chemical Corp. (Hauppauge, NY) is a suitable alternative; however, this material is quite soluble and therefore the washing step, alluded to above, must be omitted. [Anyone interested in this procedure is advised to contact the author (A.I.) for information.]

3. Sodium [^{51}Cr]chromate (New England Nuclear, Cat. No. NEZ030S, supplied as 1 mCi/ml in sterile saline).

4. Acid–citrate–dextrose (ACD formula A, Travenol Labs., Division of Baxter Travenol)

5. Percoll (Pharmacia Fine Chemicals, Uppsala, Sweden) is made up as an isotonic stock solution as follows: to 9 ml of Percoll, 1 ml of 10× stock Ca^{2+},Mg^{2+}-free Tyrode's solution and 5–10 μl of phenol red (1 mg/ml) are added. The pH of this solution is then adjusted by gradually adding a few microliters of 1–3 *N* HCl *while* vortex mixing, until the phenol red indicator turns orange (pH 7–7.2). This solution is referred to as 100% Percoll. It should be made up and stored under sterile conditions.

6. Ammonium chloride, 0.84% in water.

7. Sterile plasticware (tubes, pipets, tips, etc.) is used throughout, and contact between glass or other potentially pyrogenic material and leukocytes or solutions should be avoided.

Procedure. Thirty milliliters of blood is collected from the central ear artery of an awake 3- to 4-kg rabbit into a 50-ml sterile polypropylene tube containing 5 ml of ACD anticoagulant. Eight milliliters of the collected blood is centrifuged (300 *g* for 10 min at room temperature) to obtain platelet-rich plasma (PRP). The plasma is replaced by 1× Tyrode's solution, Ca^{2+},Mg^{2+}-free, and the resuspended cells are combined with the rest of the blood. PRP is further centrifuged (800 *g* for 20 min, 4°) to obtain platelet-poor plasma (PPP) for later use. One volume of 1% hydroxyethylcellulose is then mixed with 4 volumes of ACD–blood at 37° in order to sediment the red blood cells (RBCs). Blood adhering to the cap of the tube should be removed, and any bubbles or foam forming on the surface should be carefully aspirated to enhance separation of red cells from the leukocyte-rich plasma (LRP). The tube is incubated in a 37° bath at a 60° angle. While the red cells are sedimenting, the LRP is harvested with a sterile plastic disposable pipet at 10- to 15-min intervals. The process is usually complete (i.e., approximately half of the total volume is recovered as the LRP phase) in 45–60 min.

Once the LRP is obtained, one of two procedures may be adopted depending on whether a mixed leukocyte suspension or purified PMNLs are to be labeled. In order to achieve the latter, the entire amount of LRP is carefully layered onto a 10-ml cushion of 66% plasma Percoll (6.6 ml 100% Percoll, 1 ml PPP, and 2.4 ml isotonic 1× Ca^{2+},Mg^{2+}-free Tyrode's solution) in a 50-ml polypropylene centrifuge tube. The tube is then centrifuged (400 *g* for 30 min, room temperature), and the plasma, the plasma–Percoll interface (consisting of platelets and mononuclear cells), and as much of the Percoll layer as possible are discarded. The pellet contains PMNLs and some red cells, which cannot be readily removed without compromising the *in vivo* function of the PMNLs. This pellet is gently resuspended in 0.3 ml PPP (saved from the initial 8 ml of blood) and 1.7 ml of isotonic 1× Ca^{2+},Mg^{2+}-free Tyrode's solution. Then 100 μCi of $Na_2{}^{51}CrO_4$ are added and the resuspended cells incubated for 30 min at 37°. Resuspension and dispersion of cells must *always* be done with *gentle mixing,* or at most by repeated aspiration and drainage through a large-bore plastic pipet. Vortex mixing will damage the cells for *in vivo* purposes.

We have observed that in acute inflammatory reactions, in which the leukocytic infiltrate is composed of 90–95% PMNLs, one obtains identical results by using the shorter procedure of omitting the Percoll gradient centrifugation and directly centrifuging the LRP (150 *g* for 10 min, room temperature) to obtain a mixed leukocyte–red cell pellet. This pellet usually contains 35–40% PMNLs and 60–65% mononuclear cells. These mixed cells are resuspended in PPP, 1× Ca^{2+},Mg^{2+}-free Tyrode's solution, 100 μCi $Na_2{}^{51}CrO_4$ and labeled in an identical fashion to the cell pellet from the Percoll gradient above.

In either case, after 30 min of incubation with ^{51}Cr, 2.5–3 ml of PPP is added and the total volume made up to 25–30 ml with 1× Ca^{2+},Mg^{2+}-free Tyrode's solution. The cells are then centrifuged (150 *g* for 10 min, room temperature), and the button is resuspended in approximately 0.5 ml PPP diluted to 2.5–3 ml with 1× Ca^{2+},Mg^{2+}-free Tyrode's. Two 30-μl aliquots of the labeled cell suspension are removed and added to 4 ml of 0.84% NH_4Cl prewarmed to 37°. After 3 min, when the red cells have lysed, the tubes are centrifuged (200 *g* for 10 min, 4°). The supernatant is then discarded, and the cell pellet, which contains the labeled leukocytes, can be used to assess the degree of labeling.

The remaining radiolabeled cell suspension (i.e., 2.5–3 ml), usually containing $1-2 \times 10^6$ cpm and $50-100 \times 10^6$ PMNLs, is injected intravenously into the test rabbit. The labeled cells may be injected at the same time that the skin sites are injected with the inflammatory stimulus. The accumulation of ^{51}Cr-labeled PMNLs in the lesions would then be measured in a cumulatative fashion usually over 2–4 hr. Alternatively, the labeled cells may be injected into the rabbit usually 60 min prior to sacrifice at a time when there are already lesions of different ages (e.g., 1, 2, 4, 8 hr, etc.). In this way, the ^{51}Cr content of the skin lesions is a measure of the *rate* of leukocyte accumulation per 60 min. The latter method is preferred if lesions of 4 hr or more are studied because, with prolonged reactions, the half-life of rabbit PMNLs (3–3.5 hr), becomes a significant factor influencing ^{51}Cr accumulation. With a brief, 60-min pulse labeling of the circulating blood PMNLs and thus of the lesions with ^{51}Cr PMNLs, this limitation is overcome. Using this labeling protocol >95% of the ^{51}Cr remains cell associated, 35–50% of the ^{51}Cr PMNLs are in the circulating pool, and the blood PMNLs have a normal half-life of 3–3.5 hr in rabbits. The ^{51}Cr associated with circulating red cells due to the combined labeling of PMNLs and RBCs is insufficient to influence the results in the skin even when moderately hemorrhagic reactions are studied.[10,11]

The actual number of PMNLs accumulating in skin sites or other solid tissue (e.g., the lung), may be determined by collecting a 5-ml blood sample into 0.2% (final) EDTA when half of the total time of ^{51}Cr–leukocyte circulation in the blood has elapsed, e.g., 30 min after ^{51}Cr–leukocyte injection when these cells are left to circulate for 60 min. This sample is then combined with 1 ml of 1% hydroxyethylcellulose in a 14-ml polystyrene tube to sediment the red cells, and the LRP is centrifuged through 3 ml of 68% Percoll (400 *g* for 30 min at 4°). The plasma, mononuclear cell layer, and Percoll are removed, and the cell pellet is resuspended in 12 ml of 0.84% NH_4Cl *prewarmed* to 37°. After 3 min of mixing, the tube is centrifuged (150 *g* for 10 min at 4°). The cell pellet, which consists of 95% PMNLs, is resuspended in 2 ml of 1 × Tyrode's solution. The *total* cell number in 2 ml is determined, and the ^{51}Cr content of the cells is measured in a gamma counter. From these values, one can calculate the number of PMNLs per cpm in the blood. Using this value, the ^{51}Cr in the lesions or tissue may be converted to numbers of PMNLs using the following formula: Number of PMNLs/tissue sample = cpm/tissue sample × PMNLs/cpm in blood. If one is quantitating the PMNL accumulation in a vascular organ, such as the lung, one needs to remove excess blood by perfusing the organ in the same manner as described for the removal of ^{125}I-labeled albumin containing plasma during the quantitation of protein exudation (see above).

Quantitation of Monocyte Infiltration (see also this series, Vol. 108 [9])

The infiltration of solid tissue during inflammatory reactions by blood monocytes can be quantitated histologically or with radioisotope methods. Since compared to PMNLs, monocytes are a minor component of the cellular infiltrate during the acute phase of inflammation, the leukocyte

[11] T. B. Issekutz, A. C. Issekutz, and H. Z. Movat, *Am. J. Pathol.* **103,** 47 (1981).

labeling technique described above requires certain modifications. First of all, a larger volume of blood, e.g., 45–50 ml, is collected from one rabbit into 0.2% EDTA or ACD (7.5–8 ml) anticoagulant. The mononuclear cells are separated by layering 20–25 ml of blood onto each of two 15-ml cushions of 64% Percoll. After centrifugation (400 *g* for 40 min at room temperature) the plasma (PPP) is removed and saved and the mononuclear cells at the plasma–Percoll interface are harvested in a minimum of volume. This cell suspension is then *diluted at least 5-fold* with 1× Ca^{2+},Mg^{2+}-free Tyrode's solution. The cells are sedimented at 200 *g* for 12 min at room temperature and incubated in 0.2 ml of PPP, 1.8 ml of 1× Ca^{2+},Mg^{2+}-free Tyrode's solution and 100 μCi of $Na_2{}^{51}CrO_4$ for 30 min at 37°. To the labeled cells, 3–4 ml of plasma and 30–40 ml of Ca^{2+},Mg^{2+}-free Tyrode's solution is added, and the cells are sedimented (150 *g* for 10 min, room temperature). Following resuspension in 20% PPP–Ca^{2+},Mg^{2+}-free Tyrode's solution, the labeled cells are injected into relatively small rabbits (1–1.5 kg) in order to maximize the circulating blood leukocyte radioactivity. The blood half-life of the labeled monocytes is 39 hr.[11] As with the PMNLs, the accumulation of ^{51}Cr-labeled mononuclear cells in inflamed tissue can be measured in a cumulatative fashion, in which the lesions are induced at the time the labeled cells are administered intravenously, or in a kinetic fashion in which the labeled cells are injected 1 hr prior to sacrifice at a time when lesions in various stages of inflammation are present.

Quantitation of Platelet Deposition and Thrombosis in the Microvasculature

Platelets accumulate in inflammatory reactions both at an early stage prior to any evidence of vessel disruption or thrombosis and during the later phases when vessel disruption is evident. For the measurement of platelet deposition, ^{51}Cr- or ^{111}In-labeled platelets have been employed. We have preferred to use ^{111}In because this isotope gives a much greater degree of cell labeling than ^{51}Cr and has excellent gamma photon emission properties for gamma counting or external imaging. This greater degree of labeling is preferable in platelet work because, unlike PMNLs, which accumulate in massive numbers in the tissues and can easily be detected with ^{51}Cr-labeled PMNLs, platelets usually accumulate in the microvasculature in relatively small quantities.[12]

Using the rabbit as a model, we suggest the collection of 20–30 ml of blood into a 50-ml polypropylene tube containing 4–5 ml of ACD anticoagulant. The platelets should then be separated by differential centrifuga-

[12] A. C. Issekutz, M. Ripley, and J. R. Jackson, *Lab. Invest.* **49,** 716 (1983).

tion, first at 300 *g* for 10 min at room temperature to obtain PRP, and then the PRP is centrifuged at 800 *g* for 20 min at room temperature to sediment the platelets and obtain PPP. The *walls* of the polypropylene tube and the *surface* of the platelet pellet are washed *very gently* with 2–3 ml of Ca^{2+},Mg^{2+}-free Tyrode's solution adjusted to *pH 6.5*. This procedure is repeated 3 times in order to minimize the presence of plasma proteins, which inhibit the labeling. The platelets are then resuspended in 2 ml of Ca^{2+},Mg^{2+}-free Tyrode's solution (pH 6.5), warmed for 2 min to 37°, and then 10–15 μCi of aqueous ^{111}In-oxine solution (Amersham) is added. The platelets are incubated with gentle swirling for 75–90 sec at 37° after which time labeling is terminated by the addition of 8–10 ml of PPP. The total volume is made up to 50 ml with Ca^{2+},Mg^{2+}-free Tyrode's solution (pH 6.5), and the labeled platelets are sedimented (800 *g* for 20 min, room temperature). The supernatant is then discarded, and the labeled platelets are resuspended in 1 ml of PPP and 2 ml of Ca^{2+},Mg^{2+}-free Tyrode's solution. Small aliquots may be taken to assess labeling efficiency, and the remaining suspension is injected intravenously into the rabbit bearing inflammatory reactions.

The deposition of platelets in inflammatory tissues can be quantitated in the same manner as PMNL accumulation, i.e., in a cumulative fashion (by injecting the platelets at the same time that inflammation is initiated) or in a kinetic fashion (by administering the platelets at a time when inflammatory reactions are already in progress, e.g., 1 hr prior to sacrificing of the animal). If one wishes to relate the ^{111}In content of the inflammatory lesions to the actual number of platelets deposited, a blood sample (e.g., 1 ml) is collected in 0.2% EDTA (final concentration) anticoagulant 30–60 min following platelet administration. On this sample, a platelet count is performed and the ^{111}In content of a measured volume of blood (e.g., 1 ml) is determined. From these values one can calculate the platelet specific activity in the blood at the time platelet deposition is occurring, i.e., the number of platelets/cpm in the blood. By multiplying the radioactive content of the lesions by this platelet specific activity factor, the number of platelets present in the lesions can be calculated.

Platelets must be handled gently at all times; vigorous shaking or vortexing must be avoided. Resuspension may be facilitated by gentle aspiration through a large-bore plastic pipet, but this must be done gently in order to avoid shearing forces. Foaming of the suspension must also be minimized, and all centrifugation must be at room temperature or 37° If these precautions are followed, greater than 95% of the radiolabel remains platelet associated *in vivo*, and 85–90% of the platelets are in the circulation within 30 min of administration. The circulating half-life of rabbit platelets is 48 hr.[12]

In addition to platelet deposition during inflammation, fibrin thrombi can also form. Labeled platelets can be used to measure this thrombus formation. An alternative, simpler technique utilizes iodinated fibrinogen. Unlike platelet deposition, however, fibrin deposition during clotting can occur not only intravascularly by also in extravascular sites. Two conditions in the use of fibrinogen as an indicator of thrombosis have to be met. First, the fibrinogen radioactivity should be at least 90% clottable upon treatment with thrombin or upon addition to clotting whole blood. In our experience, radioiodination with the insolubilized glucose oxidase–lactoperoxidase kit (Enzymobead Reagent) available from Bio-Rad Laboratories gives satisfactory results. Generally, lyophilized human or bovine fibrinogen (Sigma) is used at a concentration of 0.5 mg/ml and iodinated according to the manufacturer's suggested protocol. A second precaution is that ^{125}I-labeled fibrinogen may accumulate not only in lesions with fibrin deposition but also, like albumin, during increased vascular permeability. For this reason, it is advisable to measure simultaneously protein exudation and fibrin deposition using ^{131}I-albumin and ^{125}I-fibrinogen. These two proteins should be administered intravenously at the same time. The ratio of ^{125}I to ^{131}I in the plasma and in the inflammed tissue should then be determined to verify that accumulation of ^{125}I-fibrinogen is not merely a reflection of protein exudation, but is in fact due to the deposition of fibrin clot.

Quantitation of Hemorrhage

Many inflammatory reactions are characterized by vascular disruption and red cell extravasation or hemorrhage. There have been few techniques to measure this parameter, and most have been fairly tedious, requiring homogenization of the inflammatory lesions. Several years ago, we developed a technique employing heavily labeled rabbit RBCs in order to quantitate hemorrhage in dermal inflammatory reactions.[13] The RBCs are labeled *in vivo* by i.v. injection of 300 μCi of [^{59}Fe]ferrous citrate (New England Nuclear) in 2–3 ml of rabbit serum. By 48–72 hr up to 9×10^5 cpm is present per ml of blood and greater than 99% of the radioactivity is bound to RBCs. This rabbit can then be used as a "donor" of labeled RBCs for the purpose of labeling the circulating RBC pool of another rabbit in which inflammatory reactions are induced (e.g., in the skin). In a typical experiment, $10-15 \times 10^6$ cpm on RBCs (about 12–15 ml of heparinized blood) are transferred into the test animal, usually 16 hr prior to initiating the inflammatory reactions. No major blood group incompatibilities seem

[13] M. M. Kopaniak, A. C. Issekutz, C. E. Burrowes, and H. Z. Movat, *Proc. Soc. Exp. Biol. Med.* **163**, 126 (1980).

to exist between NZW rabbits. The inflammatory reactions are then initiated and allowed to develop while the labeled RBCs are in the circulation. A blood sample is obtained to quantitate the amount of ^{59}Fe per ml of blood, and the hematocrit is determined so that standardization of the RBC fraction, for example at 40% hematocrit, may be performed. The amount (μl) of blood equivalents in the lesions can be determined by dividing the cpm ^{59}Fe in the lesions by the cpm of $^{59}Fe/\mu l$ of blood, adjusted to a 40% hematocrit.

The ^{59}Fe-labeling method has also been used by administering smaller doses of [^{59}Fe]ferrous citrate such as 50 μCi to label the RBCs of the *test* rabbit and thereby avoid the need for a transfusion. This dose of ^{59}Fe will result in $1.5-2 \times 10^5$ cpm/ml of blood on RBCs. We have found, however, using the skin, that a small amount of ^{59}Fe is incorporated into this tissue, resulting in a relatively high radioactive content in control, nonhemorrhagic skin sites. This is a significant problem when mild to moderate reactions are studied, and therefore the transfusion method is preferable for these. With the transfusion method, the hemorrhage can also be studied in a kinetic fashion by administering the labeled RBC, for example, 1-2 hr prior to sacrifice. As with protein exudation, PMNL, monocyte, platelet, or fibrinogen accumulation, RBC extravasation and hemorrhage may be measured in a highly vascular organ as long as the intravascular, ^{59}Fe-bearing RBCs are removed by thorough perfusion prior to radioactive analysis of the organ.

In many types of inflammatory reactions, it may be desirable to quantitate several parameters of inflammation simultaneously. This can be done using the above described radioisotope techniques as long as the gamma spectrometer in use can distinguish the emission spectra of the various isotopes and make the necessary spillover corrections. Figure 1 illustrates the results of simultaneous measurements of plasma exudation, blood flow, leukocyte accumulation, and hemorrhage in acute dermal inflammatory reactions in the rabbit induced by killed *E. coli* and the reversed Arthus reaction.

Measurement of Lymphocyte Accumulation

One of the major histological features of chronically inflamed tissues is the accumulation of large numbers of lymphocytes and monocyte-derived macrophages in the inflammatory site. Both of these types of cells reach the inflamed tissue by migrating out of the blood through postcapillary venules. Most of the work examining the migration of lymphocytes into inflammatory sites has used cutaneous contact sensitivity or delayed-type hypersensitivity (DTH) reactions. Oxazolone and picryl chloride have

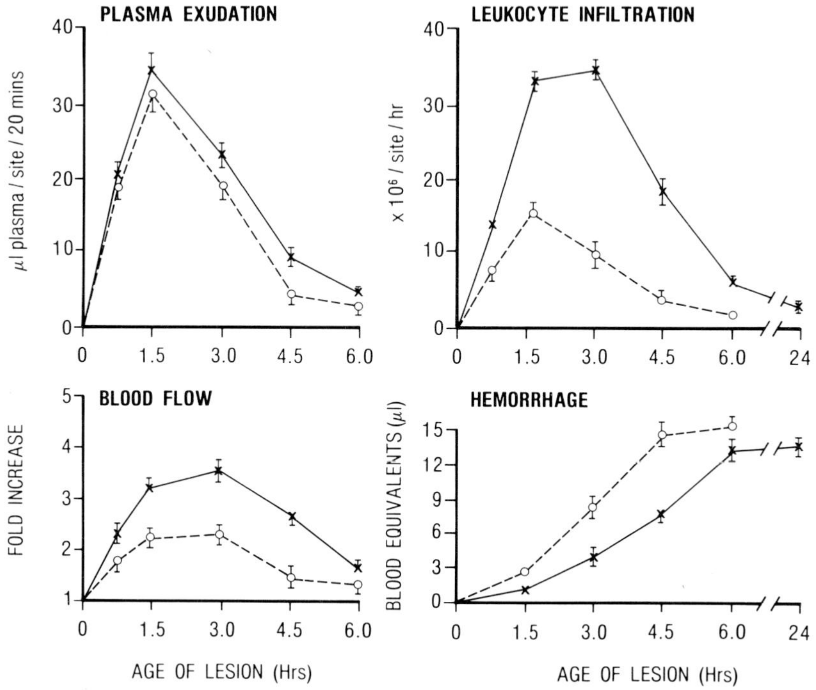

FIG. 1. Simultaneous quantitation of the dermal inflammatory response induced by *E. coli* and the reversed Arthus reaction. Rabbits were injected intradermally with either 1×10^8 formalin-killed *E. coli* (X) or rabbit anti-human IgG serum (reversed Arthus reaction) (O) in 0.2 ml at various times prior to sacrifice. For the reversed Arthus reaction, the animals were given intravenously 10 mg of purified human IgG (Sandoglobulin, Sandoz Pharmaceuticals) at least 1 hr prior to injection of skin sites with antisera. For measurement of hemorrhage, the rabbits were transfused 24 hr prior to sacrifice with 12×10^6 cpm of ^{59}Fe incorporated into RBCs (see text). Subsequently, ^{51}Cr-labeled blood leukocytes were injected intravenously 1 hr prior to sacrifice, ^{125}I-labeled rabbit albumin was injected 20 min prior to sacrifice, and ^{86}RbCl was injected 45 sec prior to sacrifice. The absolute values in *control* (saline-injected) skin averaged 5 μl plasma/site/20 min (=250 cpm ^{125}I) for plasma exudation, 200 cpm of ^{86}Rb/site for blood flow, 0.3×10^6 leukocytes/site/hr (=75 cpm ^{51}Cr) for leukocyte infiltration, and 1.1 μl blood/site (=120 cpm ^{59}Fe) for hemorrhage. These control values were subtracted from those for the inflammatory lesions, but the reader may calculate the raw cpm for each reaction at each time point from the given control values. Values are means ± SEM of triplicate lesions in at least six animals per stimulus.

been widely used to induce contact sensitivity, and a variety of antigens including tubercle protein, keyhole limpet hemocyanin (KLH), and sheep erythrocytes have been used to induce DTH reactions. The most widely used models of lymphocyte migration have used dividing lymphoblasts obtained from antigen-stimulated subcutaneous lymph nodes.[2,14]

[14] C. A. Ottaway and D. M. V. Parrott, *J. Exp. Med.* **150,** 218 (1979).

These cells are radiolabeled *in vitro* with 5-[^{125}I]iodo-2-deoxyuridine ([^{125}I]IUDR). In our laboratory, we have also investigated the migration of small nondividing T lymphocytes into cutaneous DTH reactions.[15] Our studies have made it possible to examine the migration of these small lymphocytes, which form the majority of the lymphoid cells in chronic inflammatory tissues. Both of these procedures are described.

Materials

1. Animals. Many inbred strains of mice can be used. Our preference is 6- to 12-week-old BALB/c mice. Inbred rats are also useful for these experiments, because of the larger amount of lymphoid tissue available and the large number (16–20) of skin sites for injection on each rat. Our preference is 200-g animals of the AO or Fischer strain.

2. Chemicals. The contact sensitizing agents 2-ethoxy-4-methyl-oxazolone and picryl chloride are available from British Drug House (Toronto, ON). The radiochemicals 5-[^{125}I]iodo-2-deoxyuridine and ^{111}In-oxine are available from Amersham (Arlington Heights, IL). RPMI 1640 medium and fetal calf serum (FCS) are available from most suppliers of tissue culture materials. KLH can be purchased from Calbiochem (San Diego, CA).

Procedures. Lymphoblast migration. The usual procedure for obtaining lymphoblasts is to stimulate subcutaneous lymph nodes with a contact sensitizing antigen. The skin on the flanks of mice is shaved and painted with 10 mg of 2-ethoxy-4-methyloxazolone dissolved in ethanol (10%, w/v). Three days later, the animals are sacrificed by cervical dislocation, and the axillary and brachial lymph nodes are removed. A cell suspension is prepared from these lymph nodes by gently teasing the nodes apart with forceps and a scalpel in RPMI 1640 medium containing 5% FCS. Cells are diluted to 10 ml in a 15-ml polystyrene tube, and debris is removed by sedimentation at 1 *g* for 5 min. The supernatant is collected, centrifuged at 300 *g* for 10 min, and the cells are washed twice more in RPMI 1640 medium. After the number of viable cells in the suspension is determined by Trypan blue dye exclusion, the cells are resuspended in RPMI 1640 with 5% FCS at a concentration of 10^7 cells/ml, 0.5 μCi/ml[^{125}I]IUDR is added, and the cells are incubated for 60 min at 37° in a shaking water bath. After incubation, the cells are washed 3 times in medium and suspended at a concentration of 2.5×10^7 cells/ml for injection.

A sample of the labeled cells is retained for gamma counting of the injected dose of radioactivity, and 0.4 ml of the labeled cell suspension is injected into the tail vein of each recipient mouse. If lymphoblast migration into cutaneous contact sensitivity sites is studied, one or both ears of

[15] T. B. Issekutz, D. M. Webster, and J. M. Stoltz, *Immunology* **58,** 87 (1986).

the recipient mice are painted on both sides with 0.05 ml 10% 2-ethoxy-4-methyloxazolone or 10% picryl chloride in olive oil either at the time of the injection of labeled cells or, preferably, 1 day before the injection of labeled cells since this results in greater accumulation of labeled lymphoblasts. Alternatively, animals can be presensitized to the agent by painting the shaved abdomens with 10 mg of 2-ethoxy-4-methyloxazolone dissolved in ethanol (10%, w/v) 4–14 days earlier. The ears are then painted with the same sensitizing agent, at the same dose as stated above, in order to elicit a secondary response at the time of injection of labeled cells. The animals are usually sacrificed 24 hr after the intravenous injection and the accumulation of labeled lymphocytes in the inflamed tissue determined by measuring the radioactivity of the ears.

There are several possible modifications of the procedure described above. Any subcutaneous lymph node stimulated with an antigen that induces large numbers of lymphoblasts after 3–5 days can be used to produce cells that will take up [^{125}I]IUDR and be appropriate for assessing migration into cutaneous inflammatory sites. However, intestinal lymph nodes are not suitable as a source of lymphoblasts for studies of cutaneous inflammatory sites. Lymphoblast migration into cutaneous DTH sites can also be studied using cells labeled with [^{125}I]IUDR. In the rat, the footpads can be injected with an antigen, such as KLH, and the lymphoblasts in the axillary and popliteal lymph nodes labeled 4 days later.

Measurement of small lymphocyte accumulation. Most of the small lymphocytes obtained from subcutaneous lymph nodes, both in mice and rats, do not migrate very well from the blood into cutaneous inflammatory sites.[15,16] Therefore, these cells are not useful in quantitating the migration of small lymphocytes into inflammatory reactions in the skin. However, our work[15] and that of others[16] has demonstrated that small lymphocytes, obtained from an inflammatory site, can migrate rapidly out of the blood into a cutaneous inflammatory reaction when injected intravenously. In our experience, the rat has been an especially useful model in this regard, and the details of the procedure will be described for this animal model. The principles employed in this model can also be used for studies in mice.

Inbred rats are injected intraperitoneally with any of a variety of agents to induce a peritoneal exudate rich in lymphocytes. In our experience, 5×10^7 plaque-forming units of live vaccinia virus (ATCC, Rockville, MD) appears to be superior to most other agents for inducing suitable exudates,[15] although 10^7 killed *Listeria monocytogenes* (ATCC),[17] 10 ml

[16] G. L. Asherson and G. G. Allwood, *Immunology* **22,** 493 (1972).

[17] M. J. Lefford, *Immunology* **41,** 643 (1980).

5% casein (Sigma),[18] or even bacterial culture[17] broths, such as 3 ml Bacto-peptone (British Drug House), can be used. Exudates induced by the latter agents have greater numbers of macrophages and fewer lymphocytes than those induced with virus.

Five days after the intraperitoneal injection of virus, or 3–4 days after injection of any of the other agents, the animals are sacrificed by an overdose of ether, and the skin on the abdomen is disinfected with alcohol and peeled away from the underlying musculature. Thirty milliliters of PBS containing 5 units of heparin/ml is injected carefully to avoid damage to the bowel. The abdomen is opened by an incision along the linea alba, and the peritoneal fluid containing the exudate cells is collected with a pipet. Thirty milliliters more of heparinized PBS is injected, and as much as possible is again recovered from the peritoneal cavity. Approximately 50 ml of fluid is usually obtained. The cells present in the peritoneal fluid are isolated by centrifugation, washed twice in culture medium, and counted by Trypan blue exclusion. In virus-induced exudates, these cells consist of approximately 40–50% macrophages, 40–45% small to medium lymphocytes, and about 5% lymphoblasts.

For most studies of lymphocyte migration into inflammatory tissues, depletion of macrophages is sufficient to provide a population of lymphocytes for labeling and migration studies.[15] To this end, the cells are suspended in RPMI 1640 and 10% FCS at a concentration of 10^7 cells/ml, and 8 ml of the cell suspension is added to each 100-mm-diameter petri dish. After incubation at 37° in humidified 5% CO_2 for 2 hr, the nonadherent cells are recovered from these petri dishes, counted, and washed twice in RPMI 1640 medium before being resuspended to a concentration of 10^8 cells/ml (see also this series, Vol. 108 [27]). These cells can then be radiolabeled with $Na_2{}^{51}CrO_4$ at a dose of 100 μCi/ml in RPMI 1640 plus 10% FCS for 45 min at 37° or, as we prefer, with ^{111}In-oxine. If the latter isotope is used, 7 μCi (1.6×10^7 cpm) is added per 10^8 cells in RPMI 1640 without any serum, and the cells are incubated for 10 min at room temperature. The cells are then washed twice in RPMI 1640 plus 10% FCS and resuspended at a concentration of $1-2 \times 10^7$ cells/ml in RPMI 1640 and 10% rat serum. Each rat is injected with 0.8 ml of labeled cells carrying $0.5-1.0 \times 10^6$ cpm.

For the study of cutaneous DTH reactions, we have found KLH and ovalbumin particularly satisfactory in the rat. Two weeks prior to use, animals are sensitized by the injection of 0.1 ml of antigen (50 μg), emulsified in Freund's complete adjuvant (Difco), in each of their four footpads.

[18] S. H. E. Kaufmann, R. Schauer, and H. Hahn, *Immunobiology* **160**, 184 (1981).

On the day of the experiment, the animals are anesthetized with ether, and the labeled cells are injected via the tail vein. The back of each animal is shaved, and an appropriate dose of the antigen and control medium is injected in three or four replicate sites in a volume of 0.1 ml/site. After 24 hr, the animals are sacrificed, the skin on the back removed, and the excess blood in the superficial veins squeezed out. The skin is frozen, and the injected areas punched out with a 12-mm-diameter leather punch. The radioactivity in the tissues is determined in a gamma counter. In a typical experiment in which 10^6 cpm on 10^7 peritoneal lymphocytes were injected intravenously, DTH sites induced by the injection of 50 μg KLH had 3,240 $\pm$ 142 (SEM) cpm, control injected sites had 152 $\pm$ 20 cpm, and uninjected skin had 77 $\pm$ 10 cpm.

The procedure described above will label the lymphoblasts as well as the small lymphocytes in the peritoneal exudate population. The lymphoblasts, however, form only a minority of the labeled cells, and our previous studies have shown that most of the labeled cell migration into the DTH lesions is attributable to the small lymphocytes,[15] so most of the radioactivity in the lesions is due to the accumulation of these small cells. Finally, it is also possible to examine the kinetics of the migration of lymphocytes into the inflammatory sites by injecting animals intradermally at various times (e.g., from 0 to 96 hr) and giving labeled cells intravenously 2 hr before sacrifice.[19] This allows the comparison of the rate of lymphocyte migration at any given time with measurements of the other inflammatory parameters described above, such as blood flow and vascular permeability.

[19] T. B. Issekutz, D. M. Webster, and J. M. Stoltz, *Cell. Immunol.* **99,** 322 (1986).

[28] Carrageenan Foot Edema Test

By Ivan G. Otterness and Peter F. Moore

Introduction

Paw swelling is a convenient *in vivo* method for assessing inflammation in response to a variety of irritants, phlogistics, and immune challenges. The use of carrageenan as an irritant was introduced empirically as a method of detecting clinically useful antiinflammatory drugs[1] and has

[1] C. A. Winter, E. A. Risley, and G. W. Nuss, *Proc. Soc. Exp. Biol. Med.* **111,** 544 (1962).

proven most useful to date for studying inflammation involving mediation by prostaglandins.[2,3] As other mediators are proven to be of clinical relevance, other phlogistics may be utilized in the foot edema test as a rapid assay of antiinflammatory properties.

Method

Preparation of Carrageenan

The amount of swelling depends on the carrageenan preparation, particularly the amount of λ-carrageenan in the preparation,[4] and the amount of carrageenan injected subplantarly. We are currently using a carrageenan preparation (Viscarin, Marine Colloids, Springfield, NJ) which results in approximately 1 ml of swelling in Sprague–Dawley-derived rats (CD rats, Charles River) after an injection of 50 μl of a 1% solution in water. A uniform solution of carrageenan is obtained when the carrageenan is prepared 2 weeks in advance of its use and allowed to slowly solubilize. This appears to improve the consistency of edema formation. Nonetheless, a useful preparation can be made by grinding or homogenizing the carrageenan shortly before use.

Animals

Good health and husbandry of animals is important since infection alters the inflammatory response. Although rat strain, sex, and age have been declared insignificant by some authors,[5,6] we have noted strain differences in the response to carrageenan. This is of even more concern in mice than in rats because of the availability of widely variant inbred strains. We have consistently used 160- to 200-g male Sprague–Dawley outbred rats with good results. We have used inbred 25- to 30-g (12-week-old) C3H/HeN or C3H/HeJ mice. Outbred Swiss–Webster-derived mice also have given good results.

[2] J. R. Vane, *Nature (New Biol.)* **231,** 232 (1971).

[3] I. G. Otterness and M. L. Bliven, *in* "Nonsteroidal Antiinflammatory Drugs" (J. G. Lombardino, ed.), p. 111. Wiley, New York, 1985.

[4] E. W. Moore and R. W. Trottier, Jr., *Res. Commun. Chem. Pathol. Pharmacol.* **7,** 625 (1974).

[5] E. Arrigoni-Martelli and I. Conti, *Farmaco, Ed. Prat.* **19,** 134 (1964).

[6] S. Garattini, A. Jori, D. Bernardi, C. Carrara, S. Paglialunga, and D. Segre, *in* "Nonsteroidal Antiinflammatory Drugs" (S. Garrattini and M. N. G. Dukes, eds.), p. 151. Excerpta Medica, Amsterdam, 1965.

Injection of Carrageenan

In the rat, 50 μl of a 1% carrageenan solution is injected subplantarly using a $\frac{5}{8}$-inch 25-gauge needle on a 100-μl glass syringe. The needle is inserted bevel down through the callus (see Fig. 1) at an angle of approximately 5–20° for approximately 6 mm in order to deliver the carrageenan uniformly into the same subplantar site of each animal. Generally this step is best carried out by two people, one person to hold the rat and extend the paw, the other to inject the carrageenan.

In the mouse, a 25-μl volume of a 1% solution of carrageenan is injected into the subplantar surface of the mouse paw. A 100-μl gas-tight syringe (Hamilton Co., Reno, NV) with a 25-gauge needle is used to deliver the carrageenan. The injection procedure is similar to the rat except it can be done easily by one person. The mouse is placed under a small, weighted mortar or breaker and the paw extended out through the small triangular opening. The injection is made in the center of the paw similarly to the rat, although there is no callus to make the site exactly reproducible.

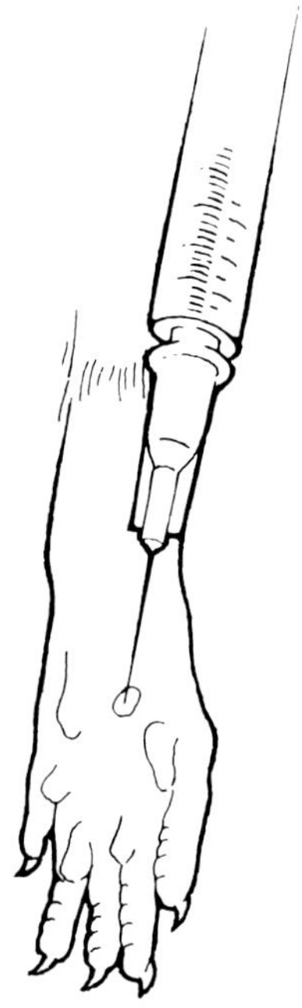

FIG. 1. The rat paw in perspective showing the syringe needle as it is inserted through the callus on the plantar surface in order to define a reproducible injection site. The syringe needle is inserted bevel down at an angle of 5–10° with the surface and to a depth of around 6 mm.

Measurement in the Rat

The volume of the rat paw can be determined by the technique of fluid displacement plethysmography, i.e., measuring the weight or the pressure of the displaced fluid. Most investigators use mercury as the fluid to increase the sensitivity of the method. A line is first drawn at the level of the lateral maleolus prior to injection of the carrageenan solution. This delineates the paw volume to be included in the measurement. Typically, the volume of the paw is measured by dipping it in fluid up to the level of the drawn line. It is imperative that there be a foot pedal or other device to trigger the recording of the volume of the paw. If the operator has to look away from the rat to read the measuring device, there is frequently an

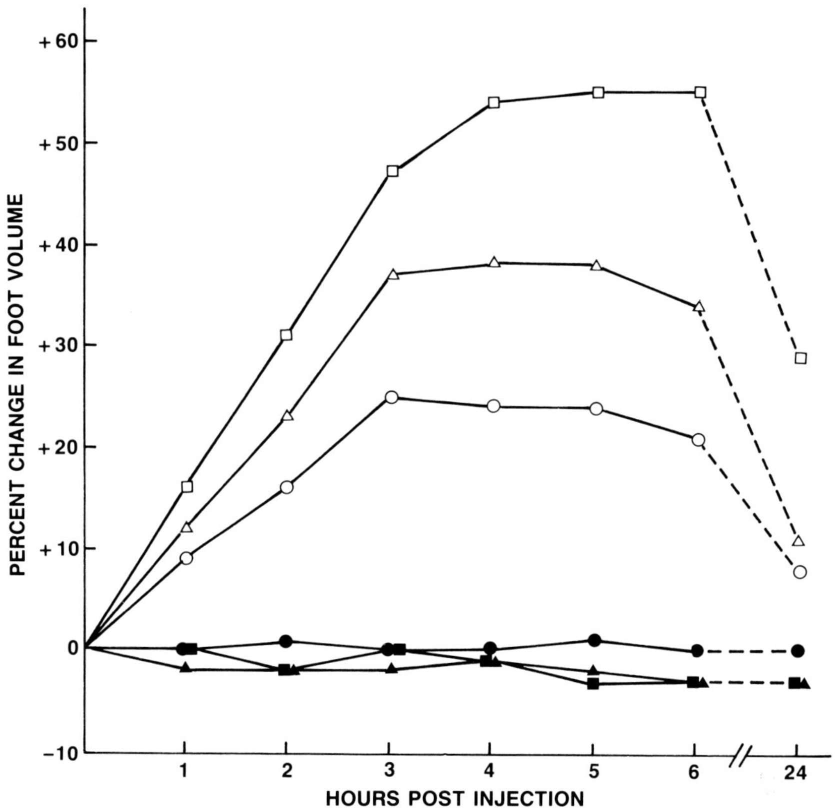

FIG. 2. The effect of the amount of carrageenan injected into the rat paw on the amount of swelling developing with time. The swelling is given as the mean percent increase in foot volume. The mean control paw volume was 1.77 ml. The paw was uninjected (●) or injected with 50 μl of distilled water (▲), saline (■), or carrageenan at 0.1% (○), 0.3% (△), or 1% (□).

increase in variability of the measurements. In our standard assay, the volume of the injected paw is read immediately after (V_0) and 3 hr after (V_3) carrageenan injection. The change in volume of the paw is then $V_3 - V_0$. Other investigators have used the volume of the contralateral, noninjected paw instead of V_0 as the control volume to determine the degree of swelling. If that is done, care must be taken to insure that there is not a bias in volume measurement between readings taken on right and left paws.

Pressure Measurement. This is the commonly used method of determining the paw volume because it is based on the original publication by Winter *et al.*[1] Volume is measured by displacement of mercury when the paw is dipped to a fixed depth, here the level of the line drawn at the lateral maleolus. A pressure transducer coupled to a digital output is used to read the pressure upon activation of a foot pedal. A commercial apparatus is availablc (Buxco, Sharon, CT). Since the actual measurement is the increase in pressure caused by the upward displacement of mercury, the relationship of the pressure to the displaced volume must be calibrated for the apparatus. We use ball bearings of known volume for the calibration. The ball bearings can be easily submerged and removed from the mercury using a magnet. Figure 2 shows a representative set of curves indicating the dependency of swelling on carrageenan concentration and time for the CD rat.

Weight Measurement. A beaker containing water is placed on an automatic balance. As before the paw is dipped into the fluid to the marked level. In this case the volume is determined as change in weight due to the displacement of fluid by the paw. Otherwise the procedure is unchanged including the use of a foot-activated pedal for initiating automatic recording of the weight.

Measurement in the Mouse

Although measurement of the mouse paw by volume displacement or weight can be done as in the rat, it is far more convenient to simply measure paw thickness using a caliper. A Schnelltaster caliper (Schlesinger's, Brooklyn, NY) which is spring loaded and measures from 0 to 10 mm works very well. Paw thickness is measured just prior to injection of carrageenan (T_0) and 3 hr later (T_3). Alternatively the thickness of the uninjected paw can be measured at 3 hr and used as a control. The change in paw thickness is recorded as $\Delta T = T_3 - T_0$. The injection volume causes an immediate increase in paw thickness of about 0.8 mm which largely disappears within the first hour. However, it means that, unlike the

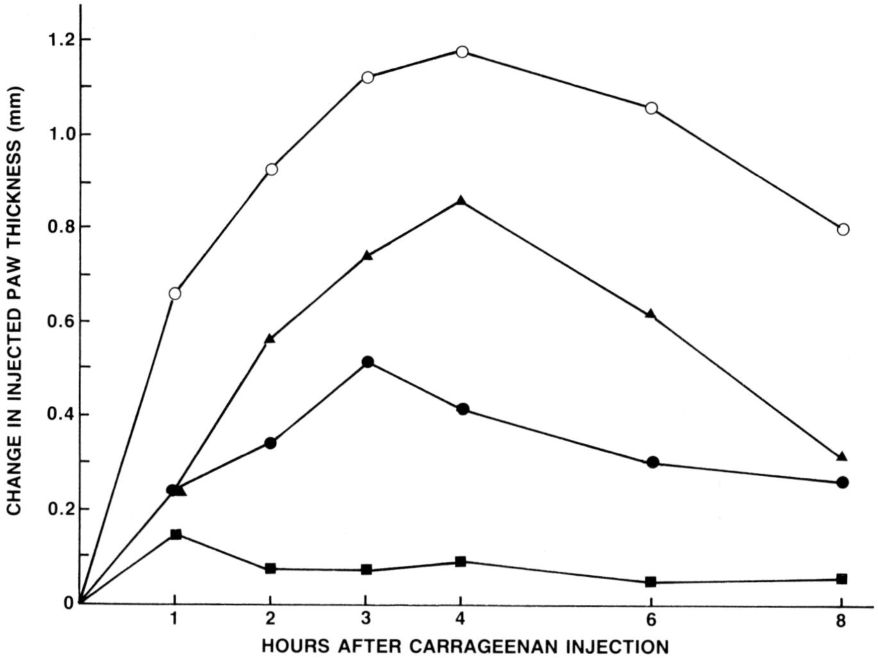

FIG. 3. The effect of the amount of carrageenan injected into the mouse paw on the amount of swelling developing with time. Swelling is given as the mean increase in paw thickness. The mean preinjection paw thickness was 2.04 mm. The paw was injected with 25 μl of saline (■) or carrageenan at 0.1% (●), 0.3% (▲), or 1% (○).

rat, a paw thickness measured immediately after injection is not a suitable control measurement. A typical response curve for varying doses of carrageenan which shows the development of swelling with time in C3H/HeN mice is presented in Fig. 3.

Computations

The change in volume due to the injection is then $\Delta V = V_3 - V_0$. The percent inhibition of swelling due to a treatment is then simply 1 minus the swelling of the treated group divided by the swelling of an untreated group times 100 to convert to percentage.

$$\text{Percent inhibition} = (1 - \Delta V_{\text{treatment}}/\Delta V_{\text{control}}) \times 100$$

In the case of measurements made in mice, the change in thickness, ΔT, is used instead of ΔV.

Other Experimental Considerations

Paw swelling is sensitive to temperature, and thus it is important that the laboratory temperature be kept constant.[7] Low temperatures in particular inhibit the responsiveness of the assay to nonsteroidal antiinflammatory drugs. Stress should be minimized as adrenal stimulation inhibits paw swelling.[8,9] The time of day for the assay should be standardized as there is a circadian rhythm in adrenal output that leads to small but significant differences in swelling.[10,11] The use of anesthetics to facilitate the handling of animals has also been shown to inhibit the induction of paw swelling.[12]

Applications

The carrageenan edema assay has been useful for the study of the mediators found in the developing edema associated with inflammation. There has been considerable study of the roles of histamine, serotonin, kinins, complement factors, prostaglandins, and leukocytes in various stages of the developing reactions.[3] A recent report indicates that platelet-activating factor (PAF) or a PAF-like material is released and contributes to the early edematous response induced by carrageenan.[13] Volume measurements in conjunction with specific antagonists, histological evaluation, and biochemical or biological assay of mediators will continue to be fruitful avenues of investigation. Other edema-inducing substances such as kaolin, nystatin, and yeast have been substituted for carrageenan.[14]

The most common use of this assay is for the evaluation of drugs which may be effective in the treatment of human arthritis. Most of the nonsteroidal antiinflammatory classes of drugs have been discovered by this methodology.[15] Typically, drugs are administered 1 hr prior to subplantar injection of carrageenan. The administration of drugs, computations of effective drug doses, and comparative potency determinations have been discussed in detail elsewhere.[3,14] The doses of nonsteroidal antiinflammatory drugs required to inhibit carrageenan-induced rat foot edema have

[7] A. Y. Green, D. Green, P. A. Murray, and A. B. Wilson, *Br. J. Pharmacol.* **41,** 132 (1971).

[8] C. A. Winter, E. A. Risley, and R. H. Silber, *J. Pharmacol. Exp. Ther.* **162,** 196 (1968).

[9] G. Labrecque, F. Doré, and P.-M. Bélanger, *Life Sci.* **28,** 1337 (1981).

[10] N. Loubbaris, G. Cros, J. J. Serrano, and M. Boucard, *Life Sci.* **32,** 1349 (1983).

[11] G. Labrecque, P.-M. Bélanger, and F. Doré, *Pharmacology* **24,** 169 (1982).

[12] D. E. Griswold, S. Alessi, E. F. Webb, and D. T. Walz, *J. Pharmacol. Methods* **8,** 161 (1982).

[13] S.-B. Hwang, M.-H. Lam, C.-L.Li, and T. Y. Shen, *Eur. J. Pharmacol.* **120,** 33 (1986).

[14] C. G. van Arman, *Handb. Exp. Pharmacol.* **50,** 75 (1979).

[15] K. F. Swingle, *in* "Antiinflammatory Agents, Chemistry and Pharmacology" (R. A. Scherrer and M. W. Whitehouse, eds.), Vol. 13, p. 33. Academic Press, New York, 1974.

been shown to correlate well with clinically useful doses in man.[16,17] The assay in the mouse has not been well characterized. Two published papers[18,19] and some limited drug studies of the authors suggest that the response to antiinflammatory drugs in the murine carrageenan assay is not as closely related to human clinical effects as the carrageenan assay in the rat.

[16] J. G. Lombardino, I. G. Otterness, and E. H. Wiseman, *Arzneim.-Forsch.* **25**, 1629 (1975).
[17] I. G. Otterness, E. H. Wiseman, and D. Gans, *Agents Actions* **9**, 177 (1979).
[18] J. Weis, *Med. Exp.* **8**, 1 (1963).
[19] L. Levy, *Life Sci.* **8**, 601 (1969).

[29] Sponge Implants as Models

By PHILIP J. BAILEY

Introduction

The inflammatory process is a response to tissue injury involving the infiltration of cells, the accumulation of fluids, the production of mediators, and the release of hydrolytic enzymes, all of which take place in a sequential and well-ordered series of events. To study this phenomenon in tissues of the whole animal is a very desirable but extremely complicated procedure. Tissue fluids are not always accessible for biochemical examination, and the cellular infiltration is most commonly assessed by time-consuming histologic methods. Accordingly, methods have been developed using sponge matrices *in vivo,* impregnated with a wide variety of inflammatory stimuli. The use of such sponge implants allows them to be removed at various times after implantation where the cellular infiltration can be assessed and fluids collected for biochemical analysis. Using the sponge implantation technique, quantitative data on the mediators present, the number and types of cells accumulating, and the effect of drugs on these processes can be readily obtained.

Methods

Preparation and Implantation of Sponges

A number of different sponge matrices have been used in this model, and very often the material from which the sponge is made influences the nature and intensity of the inflammatory response. Ford-Hutchinson *et al.*[1] described the use of sponges cut from polyvinyl foam sheets, extracted with 70% (v/v) ethanol. Higgs *et al.*[2] used polyester sponges impregnated with carrageenan, and Parnham *et al.*[3] have described the use of carrageenan-impregnated polyester sponges fitted with an indwelling catheter. A very useful material is Cellpress (Kongsfoss, Fabrikker A/S, Bygdoalle 5, Oslo 2, Norway) which is a spongy matrix made from specially treated cellulose intended for use as a matrix anchor for allografts. This material is nontoxic and induces only a modest cellular infiltration unless impregnated with inflammatory materials such as carrageenan or zymosan. Thus the inflammatory response observed using this type of sponge matrix results largely from the particular type of inflammatory stimulus which is included within the sponge and is not simply due to an effect of the material of the sponge itself.

The sponge implant is cut from blocks of the Cellpress material into cubes 1.0 cm square and approximately 5 mm thick (weight approximately 30 mg). No pretreatment of this material is required prior to implantation except that the sponges are sterilized by autoclaving. Antibiotics may be added (100 units penicillin and 100 μg streptomycin), but this is not usually necessary. Just prior to implantation, the inflammatory material is added to the sponge at room temperature as a solution or suspension in 0.9% NaCl, each sponge receiving 0.2 ml. Good stimulation of cell infiltration has been obtained with 3.0 mg of zymosan and 2.0 mg of carrageenan per sponge. It has been found that it is better to add a measured amount of the inflammogen rather than to soak the sponges in a solution of the irritant, a fixed volume of irritant/sponge resulting in a more consistent cellular infiltration.

Sponges are implanted subcutaneously in male Sprague–Dawley rats (140–165 g). The abdomen is washed thoroughly with 70% (v/v) ethanol in water. Rats are lightly anesthetized using Metophane inhalation anesthesia. A small incision is made in the midline of the ventral surface, and a pocket is created by inserting a blunt-ended pair of scissors into the incision and gently prying the skin loose from the underlying body wall, taking care that no bleeding occurs. The sponge implant is then placed into the pocket taking care that it is located at least 2 cm from the initial incision. Using rats of approximately 150 g body weight, up to four sponges can be implanted into each animal, two each side of the midline incision. The incision is then closed with a single stainless steel clip, and the animals are

[1] A. W. Ford-Hutchinson, J. R. Walker, and M. J. H. Smith, *J. Pharmacol Methods* **1,** 3 (1978).

[2] G. A. Higgs, R. J. Flower, and J. R. Vane, *Biochem. Pharmacol.* **28,** 1959 (1979).

[3] M. J. Parnham, I. L. Bonta, M. J. P. Aldofs, and P. Bragt, *Agents Actions* **7,** 539 (1977).

returned to cages where they are maintained at constant temperature and allowed access to food and water.

Sponge implants can be left *in situ* for periods ranging from several hours to days depending on the type of experiment. In experiments designed to study, for example, the early infiltration of polymorphonuclear leukocytes (PMNs)[3a] and the production of arachidonic acid oxygenation products, the sponges are left in the animal for 6 hr. If chronic granuloma formation is studied the implants may be left in the animal for several days.[4] After the desired period of time the animals are killed by asphyxiation in carbon dioxide, the abdominal skin reflected (care being taken to avoid bleeding), and the sponge removed with forceps taking care to avoid excessive squeezing which could result in loss of accumulated fluid.

Estimation of Leukocyte Accumulation

Sponge implants treated with various stimuli become rapidly infiltrated with cells. The early cellular infiltrate consists of PMNs, whereas later infiltrates (over a period of several days) consist of mainly mononuclear cells and fibroblasts. Numerous methods have been described for measuring the accumulation of cells within the sponge. The most obvious method is to fix the sponge implant in formol-buffered saline and perform standard wax-embedded sections as described by Boyle and Mangan.[5] This technique is tedious and time consuming, however, and demands accurate serial sectioning. Treatment of the sponge with 0.5 ml of 0.5% w/v trypsin (at pH 7.4 for 20 min at 37°) has also been described.[1] The enzyme treatment loosens the cells from the sponge matrix. The cells are then squeezed out into balanced salt solution and differential cell counts performed on the exudate. The enzyme treatment, however, often fails to completely free the cells from the sponge matrix so that the number of cells obtained is not an accurate measure of the total cells infiltrating the sponge. A more accurate method for measuring the early infiltration of PMNs is to measure the levels of myeloperoxidase (MPO) extractable from the sponge. This enzyme is present at a high specific activity within PMNs. On the other hand, the later infiltration of mononuclear cells can be quantitated by measuring the levels of the lysosomal enzyme *N*-acetylglucosaminidase (NAG), present in high levels in activated macrophages.

[3a] Abbreviations: PMNs, polymorphonuclear lymphocytes; MPO, myeloperoxidase; NAG, *N*-acetylglucosaminidase; HTAB, hexadecyltimethylammonium bromide; LTB_4, leukotriene B_4.

[4] G. A. Higgs, K. G. Mugridge, S. Moncada, and J. R. Vane, *Proc. Natl. Acad. Sci. U.S.A.* **81,** 2890 (1984).

[5] E. Boyle and F. R. Mangan, *Br. J. Exp. Pathol.* **61,** 351 (1980).

Although MPO and NAG measurements provide reliable quantitative data, their activity can vary on the degree of activation of the cell type.

Procedure for Measuring Leukocyte Accumulation Using MPO Levels of Sponge Implants

Reagents

0.5% (w/v in water) hexadecyltrimethylammonium bromide (HTAB) (Fisher Scientific Co., Fair Lawn, NJ)

30% hydrogen peroxide (Fisher) diluted 1 : 100 (v/v) with water; dilute a further 1 : 100 with 10 m*M* KH_2PO_4 (pH 6.0) just before use

o-Dianisidine dihydrochloride (3,3′-dimethoxybenzidine dihydrochloride, Eastman Kodak Co., Rochester, NY), 12.9 mg dissolved in 1.0 ml water

Procedure. All procedures are performed at room temperature. Sponges are removed from the animal and placed in 10 ml 0.5% HTAB. The whole sponge implant is then ground in a Polytron homogenizer (Kinematica GMBH, Luzeon, Switzerland) at the highest setting for 10 sec. Debris is removed by centrifugation at 800 *g* for 10 min. The supernatant is used for MPO determination, and activity is stable for several days at room temperature.

The enzyme assay incubation mixture consists of 25 μl dianisidine solution, 3.0 ml H_2O_2 solution, and 0.1 ml sponge homogenate supernatant. Incubation is for 5 min at room temperature, and the absorption is read at 460 nm. The amount of dianisidine oxidized is calculated using an extinction coefficient of 1.13×10^4.

The type of cells infiltrating the sponge change with the age of the implant. The first cells to infiltrate the sponge are the PMNs, which accumulate in the sponge in the first 1–12 hr. If the implant is left *in situ* for 24 hr or more the cell infiltration changes to predominantly macrophages and fibroblasts. An estimate of macrophage infiltration is given by the level of NAG which is present in high specific activity in activated macrophages.

Procedure for Measuring NAG Levels of Sponge Implants

Reagents

0.9% saline containing 0.1% (v/v) Triton X-100 (New England Nuclear, Boston, MA)

Citrate/phosphate buffer: 100 ml 0.1 *M* citric acid, 155 ml 0.1 *M* Na_2HPO_4, adjusted to pH 4.5

0.2 *M* glycine buffer: mix equal volumes of 0.8 *M* glycine (Baker Chemical Co., Phillipsburg, NJ), 0.8 *M* NaCl, and 0.8 *M* NaOH adjusted to pH 10.6

2.24 m*M* p-nitrophenyl-*N*-acetyl-β-D-glucosaminide (Sigma Chemical Co., St. Louis, MO) prepared in citrate/phosphate buffer

Procedure. The sponge implant is removed from the rat and placed in 10.0 ml ice-cold saline/Triton solution and ground in a Polytron disintegrator (Kinematica) for 10 sec at maximum speed. The debris is removed by centrifugation at 1500 *g* for 10 min. If the NAG assay is not carried out immediately, the supernatant fractions can be kept frozen until used.

NAG activity is based on the hydrolysis of *p*-nitrophenyl-*N*-acetyl-β-D-glucosaminide by a modification of the procedure of Woollen *et al.*[6] To 0.1 ml sponge extract is added 1.5 ml citrate/phosphate buffer. The reaction is started by the addition of 1.5 ml 2.24 m*M* *p*-nitrophenyl-*N*-acetyl-β-D-glucosaminide. The mixture is incubated at 37° for 10 min and the reaction terminated by the addition of 1.6 ml 0.2 *M* glycine buffer. Hydrolysis of the substrate is determined by measuring the color absorption at a wavelength of 400 nm, assuming an extinction coefficient for *p*-nitrophenol of 9.95×10^3.

The method can also be adapted for the Technicon Autoanalyzer as follows. The sample aliquot (6.4–100 μl) is picked up by the probe and mixed in a 3-way cactus with a stream of substrate (2.24 m*M* *p*-nitrophenyl-*N*-acetyl-β-D-glucosaminide), and simultaneously percolated with a stream of air. The mixture of substrate, sample, and air is thoroughly mixed by passing it along a coil of 14 turns. The mixture enters the heating bath at 37° where it is incubated for 15 min. At the exit from the heating bath the stream is joined by a stream of 0.2 *M* glycine buffer (pH 10.4) which stops the reaction. The mixture proceeds to the flow cell of the spectrophotometer where the amount of *p*-nitrophenol liberated by the action of the enzyme on the substrate is measured at 400 nm. A tracing proportional to the amount of *p*-nitrophenol released is described on the recorder. From the height of the tracing and extinction coefficient (9.95×10^3) of *p*-nitrophenol, enzyme activity is calculated and expressed as micromoles *p*-nitrophenol liberated per hour.

Using the marker enzyme technique of MPO for the PMN infiltration and NAG for the macrophage infiltration, it has been shown that in a carrageenan-impregnated sponge the maximum infiltration of PMNs takes place at approximately 12 hr whereas the mononuclear cell infiltration undergoes a slower increase over 5–14 days, reflecting the development of

[6] J. W. Woollen, R. Heyworth, and P. G. Walker *Biochem. J.* **78,** 111 (1961).

TABLE I
ACCUMULATION OF CELLS IN CARRAGEENAN-IMPREGNATED SPONGES AS REFLECTED BY ENZYME MARKER ACTIVITY

Myeloperoxidase levels (MPO)		N-Acetylglucosaminidase levels (NAG)	
Age of implant (hr)	MPO activity ± SEM (μmol oxidized/hr/sponge)	Age of implant (days)	NAG activity ± SEM (μmol/hr/sponge)
1	1.3 ± 0.3	1	8.0 ± 0.2
2	1.8 ± 0.2	5	12.2 ± 1.9
4	20.1 ± 2.8	8	22.6 ± 2.5
6	61.7 ± 8.9	12	27.9 ± 0.9
8	85.9 ± 6.7	14	29.6 ± 2.1
12	164.3 ± 8.3		
24	146.0 ± 16.2		

an encapsulated granuloma. Typical data relating to a carrageenan-impregnated sponge are shown in Table I. Note the time scale difference of the MPO assay (which is measured in hours), illustrating the very rapid infiltration of PMNs compared with the NAG determinations (which are measured in days) reflecting mononuclear cell infiltration. The maximum rate of NAG accumulation in the sponge is seen between 5 and 8 days.

Determination of Mediators in the Fluid of Sponge Implants

The accessibility of the fluid phases of sponge implants makes this model very useful for the determination of mediators generated *in situ* in response to various stimuli. These mediators include plasma proteins, lysosomal enzymes, complement-derived products, vasoactive amines, and products of arachidonic acid oxidation. Drug levels within the sponge can also be estimated. The recently discovered leukotriene pathway of arachidonic acid activation has prompted speculation as to the role of these mediators in the inflammatory response. The sponge technique has been used to measure the levels of these products, particularly leukotriene B_4 (LTB_4) (a potent chemotactic agent for leukocytes). The following procedure describes a method for determining the levels of LTB_4 in zymosan-impregnated sponges.

Reagents

Freshly distilled methanol to which is added an equal volume of 0.1 *M* sodium acetate (pH 4.4)

Freshly distilled diethyl ether High-performance liquid chromatography (HPLC) solvent system consisting of methanol/water/acetic acid, 63:37:0.08 (v/v/v), pH 5.7

Procedure. The Cellpress sponges (see above) are treated at room temperature with 0.2 ml of a solution of 1.5% w/v zymosan in 0.9% saline (3 mg zymosan/sponge) and implanted into rats as described above. Six hours later the rats are killed by asphyxiation in carbon dioxide, and the sponges are removed and immediately placed in 2 ml of cold methanol. The sponges are squeezed with forceps while immersed in the cold methanol, the sponges are then removed, excess methanol is squeezed out, and the sponges are discarded. The methanol extract is then centrifuged at 10,000 *g* at 4° for 10 min. LTB_4 is extracted by shaking the methanol supernatant with 6 ml diethyl ether, which is evaporated to 100 μl under reduced pressure. Four hundred microliters of methanol is then added to the ether concentrate, and the mixture is concentrated to a volume of 70 μl under vacuum. Thirty-five microliters of this extract is then added to an HPLC reversed phase column (250 × 2 mm) packed with Nucleasil (C_{18} Bondapak). The solvent system consists of methanol/water/acetic acid (63:37:0.08, v/v/v, pH 5.7) and the flow rate maintained at 1 ml/min. The absorbance of the column effluent is monitored at 270 nm using a spectroflow absorbance detector. The amount of LTB_4 is determined from the area of the peak as calculated by the integrator and compared with authentic LTB_4 standards. Alternatively, a radioimmunoassay can be performed on the fractions or directly on the sponge exudates as described by Ford-Hutchinson *et al.*[7] It is interesting that, although a large amount of LTB_4 can be recovered from sponges impregnated with zymosan, very little is recovered from sponges treated with carrageenan. This seems to be due to the fact that the PMNs infiltrating a carrageenan-treated sponge are killed on contact with the carrageenan, whereas cells in the zymosan sponge remain viable.

Comments

The techniques of sponge implantation described in this chapter represent a convenient system for providing ready access to the cells and fluids of an ongoing inflammatory response. Using these techniques it has been shown[8] that indomethacin reduces leukocyte migration into subcutaneously implanted sponges but that this effect was distinct from the effect of indomethacin in reducing the content of prostaglandin in the exudate. It was found[9] that the dual 5-lipoxygenase/cyclooxygenase inhibitor BW 755C reduced the level of LTB_4, and prostaglandin E_2 as well as leukocyte

[7] A. W. Ford-Hutchinson, G. Brunet, P. Savarel, and S. Charleson, *Prostaglandins* **28,** 13 (1984).

[8] J. R. Walker, M. J. H. Smith, and A. W. Ford-Hutchinson, *Agents Actions* **6,** 602 (1976).

[9] J. A. Salmon, P. M. Simmon, and S. Moncada, *J. Pharm. Pharmacol.* **35,** 808 (1983).

migration in carrageenan-impregnated sponges, although cell influx still occurred at levels of BW 755C which completely inhibited LTB_4 synthesis. Ford-Hutchinson *et al.*[7] concluded that LTB_4 can induce leukocyte migration into sponge exudate but that measurements of LTB_4 in such exudates could not be correlated with the degree of leukocyte infiltration. These examples serve to illustrate the usefulness of the sponge model in dissecting the events in the inflammatory response. One major disadvantage of this model is the absence of a vasculature in the early phase of the model. The extravasation of cells from the blood vessels into the tissues is a major initiator of the inflammatory response, and this process obviously cannot be studied in the sponge model. Nevertheless the ready accessibility of the cells and fluids in the sponge system makes this a useful tool in the study of the biochemical events in the inflammatory process.

[30] Carrageenan Pleurisy

By MARCIA L. BLIVEN and IVAN G. OTTERNESS

Introduction

Inflammation in most tissues can be measured by gross changes such as edema, erythema, or irritancy (pain). However, these gross measures give no indication of the role infiltrating leukocytes play in the inflammation. Traditionally, histologic sectioning has been the principal means of measuring infiltration of inflammatory cells. Histologic procedures are time consuming and, because of sampling difficulties, often best suited to qualitative rather than quantitative evaluation. Furthermore, functional studies cannot be done with the inflammatory cells. And, if potential inflammatory mediators are to be studied, tedious extraction procedures are required for their recovery from the solid tissue. Several models are available, however, in which it is possible to collect cells and fluid from the inflammatory exudate. Cells can be recovered for determination of their numbers and type and for measurement of their functional capacity. Similarly, the exudate fluid can be quantitated and can be used for the identification and measurement of inflammatory mediators. The most widely accepted model which fits these criteria is the carrageenan pleurisy model. It has been extensively studied[1] and has the advantage of long use and standardization of conditions.

[1] R. Vinegar, J. F. Truax, J. L. Selph, and F. A. Voelker, *Fed. Proc., Fed. Am. Soc. Exp. Biol.* **41,** 2588 (1982).

Methods

Preparation of Carrageenan

Carrageenan (Viscarin, Marine Colloids, Springfield, NJ) containing high amounts of λ-carrageenan is prepared the day before use by grinding or homogenizing with normal saline (0.9 gm NaCl/100 ml in distilled water). Carrageenan is generally suspended at a concentration of 2.5 mg/ml for standard assay conditions.

Animals

Outbred male Sprague–Dawley rats of good health weighing approximately 200 $\pm$ 20 g have been routinely used by most investigators. The rats are normally fed *ad libitum.* For reproducibility between experiments, room temperature, type of handling, caging, and time of initiating of the study should be held constant.

Injection of Carrageenan

Rats are lightly anesthetized with ether and placed on their left sides. An injection of 0.2 ml of normal saline containing carrageenan is made into the pleural cavity on the right side of the mediastinum between the third and fifth ribs using a 26-gauge $\frac{1}{2}$-inch needle. A blunt-tipped needle is used to minimize the possibility of puncturing the lung. A highly reproducible cell infiltration and fluid accumulation is obtained when 500 μg of carrageenan is used to induce pleurisy, but a dose–response to carrageenan is observed over the range of 100–2000 μg.

Measurements

Four hours after carrageenan injection, the rats are sacrificed with ether and the peritoneum opened to expose the diaphragm. A small incision is made near the sternum, and 1 ml of heparinized (10 mg%) phosphate-buffered saline is used to rinse out the pleural cavity through the opening. The exudate and cells are aspirated through the blunt tip of a 1-ml plastic syringe. The volume of exudate is measured and recorded. The total leukocyte number is determined manually using a hemacytometer; this procedure may be performed automatically with a particle counter.

Cell smears are made to obtain differential counts of the types of the infiltrating cells. Ideally, cell smears should be prepared in a cytocentrifuge.[2] Smears prepared in this manner retain their cellular morphology

[2] M. Martimazzi, *Am. J. Clin. Pathol.* **56,** 719 (1971).

better than standard air-dried smears. Although this is not critical for cells from blood, cells from inflammatory fluids are more fragile and easily lose their characteristic morphology upon air drying. Wright stain is suitable for routine differentiation of infiltrating cells. Cells separated from the fluid samples are also suitable for functional studies, i.e., chemotaxis, phagocytosis, and mediator release.

The cell-free fluid may be used for measurements of known mediators such as histamine, serotonin, prostaglandins, leukotrienes, thromboxanes, complement split products, and kinins. It is also a source of fluid for examination of less well-defined or novel mediators.

Discussion

Pleural exudate formation is biphasic with the first phase developing between 1 and 3 hr and the second phase between 5 and 16 hr. Typically, 1×10^8 cells in approximately 0.75 ml of fluid can be expected to be recovered 4 hr after an injection of 500 μg of carrageenan (Fig. 1). The amount of fluid accumulation and the number of infiltrating cells are directly related to the amount of carrageenan injected[3,4] and the time of collection.[4,5] A typical dose–response curve 4 hr after carrageenan injection is shown in Fig. 1. With 500 μg of carrageenan, the fluid volume peaks at 16 hr and total cell accumulation peaks at 4 hr. Leukocyte populations measured at 4 hr after carrageenan injection routinely consist of 90–95% neutrophils.[6] However, if the inflammation is allowed to proceed for another 20 hr the predominant cell population is the monocyte.[1]

Both histamine and serotonin have been identified in pleural exudate fluids with levels peaking at 1 hr after carrageenan injection[7]; however, it is unlikely that histamine[5,8] or serotonin[5] are significant mediators of the reaction. Levels of the arachidonic acid metabolites PGE_2,[7] TXB_2,[9] 6-keto-$PGF_{1\alpha}$ [9] and $PGF_{2\alpha}$[7] can be measured in exudate fluids. PGE_2 peaks at around 3 hr after carrageenan injection, while TXB_2 rises to a plateau at

[3] H. Gilchrest and A. S. Watnick, *Agents Actions* **7,** 529 (1977).

[4] A. P. Almeida, B. M. Bayer, Z. Horakova, and M. A. Beaven, *J. Pharmacol. Exp. Ther.* **214,** 74 (1980).

[5] R. Vinegar, J. F. Truax, and J. L. Selph, *Fed. Proc., Fed. Am. Soc. Exp. Biol.* **35,** 2447 (1976).

[6] I. G. Otterness, D. L. Larson, and J. G. Lombardino, *Agents Actions* **12,** 308 (1982).

[7] F. Capasso, C. J. Dunn, S. Yamamoto, D. A. Willoughby, and J. P. Giroud, *J. Pathol.* **116,** 117 (1975).

[8] Z. Horakova, B. M. Bayer, A. P. Almeida, and M. A. Beaven, *Eur. J. Pharmacol.* **62,** 17 (1980).

[9] M. Tissot, C. Bonne, B. Martin, M. Solier, and J. P. Giroud, *Agents Actions* **14,** 76 (1984).

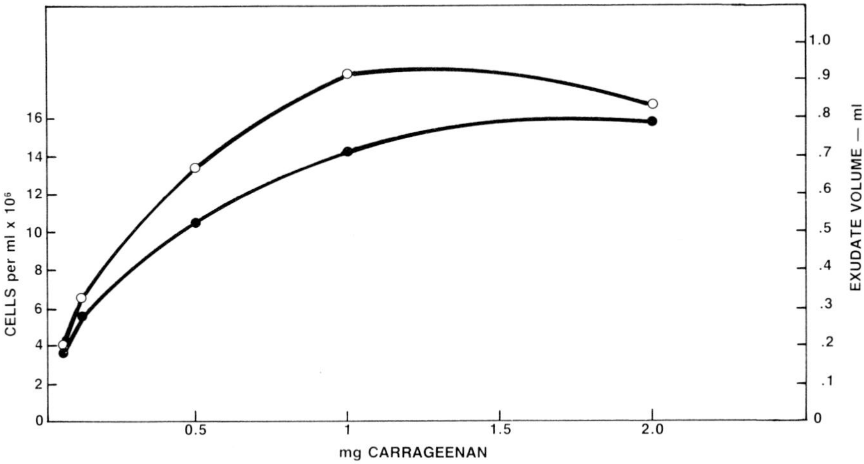

FIG. 1. Effect of the amount of carrageenan on the numbers of infiltrating cells and the amount of fluid measured 4 hr after intrapleural injection of carrageenan. Recovered cell numbers (●) are expressed as cells per ml, and collected exudate volume (○) is measured in ml.

3 hr and begins to decrease at 7 hr.[10] Peak intrapleural PGE_2 coincides with the maximum number of cells in the exudate[11] while peak $PGF_{2\alpha}$[11] appears to coincide with monocyte migration (24 hr). LTB_4 has been identified in exudate fluid after 5 hr but only in small quantities, while LTC_4 and LTD_4 have been identified in more significant amounts.[12]

Cells isolated from pleural exudates have been shown to release several mediators of inflammation[9,13,14] *in vitro* under varying stimulatory conditions. Their ability to release other mediators to phagocytosis[15] and chemotaxis[16] has not been examined closely.

A primary use for the carrageenan pleurisy model is the study of antiinflammatory drugs. Nonsteroidal, as well as steroidal, antiinflamma-

[10] M. Katori, Y. Harada, K. Tanaka, H. Miyazaki, M. Ishibashi, and Y. Yamashita, *Adv. Prostaglandin Thromboxane Res.* **8,** 173 (1980).

[11] F. Capasso, C. J. Dunn, S. Yamamoto, D. A. Deporter, J. P. Giroud, and D. A. Willoughby, *Agents Actions* **5,** 528 (1975).

[12] A. Ueno, K. Tanaka, and M. Katori, *Prostaglandins* **26,** 493 (1983).

[13] H. L. White, L. D. Faison, J. F. Truax, J. L. Selph, and R. Vinegar, *Prostaglandin Leukotriene Med.* **20,** 1 (1985).

[14] M. I. Siegel, R. T. McConnell, R. W. Bonser, and P. Cuatrecasas, *Prostaglandins* **21,** 123 (1981).

[15] J. Bird and J. P. Giroud, *J. Pharmacol. Methods* **14,** 305 (1985).

[16] M. Roch-Arveiller, G. Roblin, M. Allain, and J.-P. Giroud, *J. Pharmacol. Methods* **14,** 313 (1985).

tory drugs have been reported to inhibit the development of the pleurisy.[1,17,18] Therefore, nonsteroidal antiinflammatory drugs can be studied for their effects on fluid volumes and cell numbers, and for their influence on the presence/release of arachidonate metabolites.

However, the inflammatory reaction elicited by doses of carrageenan greater than 500 μg is difficult to inhibit with pharmacologically relevant drugs.[17] Often nonspecific doses of nonsteroidal antiinflammatory drugs are required to effect significant reductions in cell infiltration and edema formation under these conditions. Results with pharmacologic agents are dependent on the time of measurement. Both the appearance and disappearance of cell populations and mediators follow a distinct time course which alters sensitivity to pharmacologic manipulation.

Comparison to Other Exudative Models (see also this volume [10] and [29])

Two other models fulfill the criteria for ease of sampling of cells and fluid from the site of inflammation: the air bleb model and the sponge implantation model. As with the pleurisy model, the favored irritant has been carrageenan, although other agents can be used to induce cell infiltration and edema. The pleurisy model is generally favored because it requires no manipulation to construct the collection site. In the sponge model, a small carrageenan-impregnated sponge must be surgically implanted beneath the skin of rats. It is used primarily for measuring migration of cells into the sponge cavities as the exudate volume is more difficult to quantitate. Surgical stress is also unavoidable at the time of implantation. The sponge may act as a barrier to the entrance of cells into the sponge. Histologically, we generally observe a much greater density of cells in the immediate surrounding tissue than in the sponge itself. Moreover, the type of sponge is very crucial. Some sponges affect the morphology of the infiltrating cells. Some sponges lack good intercommunication between the pores of the sponge, thereby limiting cell infiltration largely to the outer edges of the sponge. Other sponges seem to induce fibrin deposition or strong cell adherence, making the cells difficult to recover for counting and functional studies. Thus, we consider it imperative to do good histologic studies on control and implanted sponges so that the sponge's role in the reaction can be well understood.

[17] R. Vinegar, J. F. Truax, J. L. Selph, A. Lea, and P. R. Johnston, *Eur. J. Rheum. Inflammation* **1,** 204 (1978).

[18] I. G. Otterness and M. L. Bliven, *in* "Nonsteroidal Antiinflammatory Drugs" (J. G. Lombardino, ed.), p. 166. Wiley (Interscience), New York, 1985.

In the air bleb model, a subcutaneous pouch is formed by injection of a large (usually 50-ml) bolus of air. Subsequently, an inflammation is induced by injection of carrageenan into the air bleb. Carrageenan causes the accumulation of fluid and formation of a granuloma over a 3- to 6-day period allowing the quantitative measurement of the exudative reaction and leukocyte immigration. One hundred times more carrageenan must be used to induce significant accumulation of cells and fluids in early, less than 3-day, air pouches.[19] This appears to be related to the low vascularity and low cellularity of the newly formed pouch lining.

Because of its reproducibility and ease of standardization, the carrageenan pleurisy model remains the most widely used exudative model for evaluation of drug effects. However, it must be kept in mind that drug effects in the carrageenan pleurisy,[18] sponge,[20] and air pouch[21,22] models often differ. This is not surprising since a comparison of the time course of the inflammatory events show that cell numbers and fluid accumulation peak at different times in each model. The mechanisms and timing of inflammatory events of each model must be carefully examined in order to directly compare the effects of drugs in one model with those in another.

[19] A. D. Sedgwick, Y. M. Sin, J. C. W. Edwards, and D. A. Willoughby, *J. Pathol.* **141,** 483 (1983).

[20] I. G. Otterness and M. L. Bliven, *in* "Nonsteroidal Antiinflammatory Drugs" (J. G. Lombardino, ed.), p. 159. Wiley (Interscience), New York, 1985.

[21] S. Tsurufuji, H. Sato, K. R. Min, and K. Ohuchi, *J. Pharm. Dyn.* **1,** 8 (1978).

[22] H. Sato, M. Hashimoto, K. Sugio, K. Ohuchi, and S. Tsurufuji, *J. Pharm. Dyn.* **3,** 345 (1980).

[31] Adjuvant Arthritis

By JOEL D. TAUROG, DENNIS C. ARGENTIERI, and RICHARD A. MCREYNOLDS

Introduction

Adjuvant arthritis (AA, also referred to as adjuvant polyarthritis, adjuvant-induced arthritis, or adjuvant disease) is an experimental disorder unique to rats that was first described in detail by Pearson in 1956.[1,2] Two

[1] C. M. Pearson, *Proc. Soc. Exp. Biol. Med.* **91,** 95 (1956).

[2] C. M. Pearson, *Ann. Rheum. Dis.* **15,** 379 (1956).

years earlier, Storek *et al.*,[3] had reported the phenomenon in an abstract. The fundamental process requires the inoculation of an oily emulsion or suspension of material possessing Freund's type adjuvant activity into an area of potent lymphatic drainage in a susceptible rat. Characteristically, 10–14 days after the injection of adjuvant, swelling, warmth, erythema, and tenderness appear in distal joints and tendons of affected animals. Maximum severity usually occurs between postadjuvant days 16 and 21, and the acute inflammation then gradually subsides over the ensuing weeks. In some cases, the arthritis can follow an undulating course for many weeks.[4,5] AA is widely used for studying the physiology, biochemistry, and pharmacology of inflammation and as a model of cell-mediated autoimmune disease, human arthritis, and chronic pain.[6–10]

Methods

Animals

Depending on experimental requirements, either outbred or inbred strains of rats can be used. Wistar, Sprague–Dawley, Long–Evans, Holtzman, and Lobund are among the many susceptible outbred strains that have been used. Inbred strains are now more commonly used than outbred strains. LEW, DA, PVG, WF, BN, and WKA are a few of the highly susceptible inbred strains; F344 rats have variable susceptibility, and the BUF strain is generally resistant.[11–13] For many types of experiments, LEW rats offer the convenience of being 100% susceptible to moderate to severe disease, as well as being docile and widely available.

Rats younger than 5 weeks are relatively resistant to AA.[14] Six- to 10-week-old rats, 180–250 g, are optimal. In commonly used rat strains,

[3] H. C. Stoerk, T. C. Bielinski, and T. Budzilovich, *Am. J. Pathol.* **30,** 616 (1954).

[4] C. M. Pearson, *in* "Mechanisms of Hypersensitivity" (J. H. Shaffer, G. A. LoGrippo, and M. W. Chase, eds.), p. 647. Little–Brown, Boston, 1959.

[5] C. M. Pearson, *J. Chronic Dis.* **16,** 863 (1963).

[6] M. E. J. Billingham and G. E. Davies, *Handb. Exp. Pharmacol.* **50,** 108 (1979).

[7] C. M. Pearson and Y.-H. Chang, *Ann. Pheum. Dis. Suppl.* **38,** 102 (1979).

[8] J. T. Rosenbaum, *Ann. Intern. Med.* **94,** 261 (1981).

[9] M. E. Rosenthale and R. J. Capetola, *Fed. Proc.* **41,** 2577 (1982).

[10] B. H. Newbould, *Br. J. Pharmacol.* **21,** 127 (1963).

[11] Z. Zidek and F. Perlik, *J. Pharm. Pharmacol.* **23,** 389 (1971).

[12] O. Kahashi, C. M. Pearson, F. J. W. Beck, and M. Alexander, *Infect. Immun.* **17,** 244 (1977).

[13] J. R. Battisto, R. N. Smith, K. Beckman, M. Sternlicht, and W. L. Welles, *Arthritis Rheum.* **25,** 1194 (1982).

[14] C. M. Pearson and F. D. Wood, *Arthritis Rheum.* **2,** 440 (1959).

there is no major sex difference in regard to either incidence or severity of the disease. It is advisable to use only specific pathogen-free rats maintained under barrier conditions; for example, our data have suggested that subclinical pulmonary infection with *Mycoplasma pulmonis,* which is quite common in nonbarrier-housed rats, suppresses the severity of AA, at least in LEW rats.[15]

Genetics

A clear picture of the genetic control of susceptibility and resistance to AA in rats has yet to emerge. This is unlike the situation in, for example, experimental autoimmune encephalomyelitis or collagen-induced arthritis, susceptibility to both of which in rats is largely controlled by genes of the major histocompatibility complex.[16-18] Although such linkage for AA has also been claimed,[13] the available data suggest at least a very significant influence of non-MHC genes.[19,20]

Adjuvants

The term Freund's (type) adjuvant generally refers to those adjuvants composed of either dried bacteria or bacterial cell wall constituents in an oily vehicle, with or without an emulsifier. Such adjuvants enhance both humoral and cell-mediated responses to a variety of antigens. It is adjuvants of the Freund's type that induce chronic polyarthritis in rats; other types of immunological adjuvants such as mineral earths, endotoxins, polynucleotides, methylated albumin, and calcium alginate are not by themselves arthritogenic.[21]

Although mycobacteria are by far the most commonly used organisms in preparations of Freund's adjuvants,[22] other types of gram-positive bacteria, or preparations of their cell wall peptidoglycan, have been shown to be arthritogenic when injected into rats in an oily vehicle. Such bacteria include *Corynebacteria rubrum, Nocardia asteroides, Lactobacillus plan-*

[15] J. D. Taurog, S. L. Leary, M. A. Cremer, M. L. Mahowald, G. P. Sandberg, and P. J. Manning, *Arthritis Rheum.* **27,** 943 (1984).
[16] R. M. Williams and M. L. J. Moore, *J. Exp. Med.* **138,** 775 (1973).
[17] M. M. Griffiths, E. J. Eichwald, J. H. Martin, C. B. Smith, and C. W. DeWitt, *Arthritis Rheum.* **24,** 781 (1981).
[18] M. M. Griffiths and C. W. DeWitt, *J. Immunol.* **133,** 3043 (1984).
[19] J. D. Taurog and S. L. Leary, *Arthritis Rheum.* **26,** 815 (1983).
[20] A. R. MacKenzie, P. R. Sibley, and B. P. White, *Br. J. Exp. Pathol.* **60,** 507 (1979).
[21] M. W. Whitehouse, K. J. Orr, F. W. J. Beck, and C. M. Pearson, *Immunology* **27,** 311 (1974).
[22] J. Freund, *Adv. Tuberc. Res.* **7,** 130 (1956).

tarum, Bacillus megaterium, and several *Streptococcus* and *Staphylococcus* species.[21,23,24] It appears that in all of these cases, arthritogenicity and adjuvanticity reside in the same cell wall components; the synthetic compound *N*-acetylmuramyl-L-alanyl-D-isoglutamine (MDP), which is generally considered the minimal structure required for adjuvant activity, is also arthritogenic in rats.[25-27] In addition, compounds which have no strong structural resemblance to bacterial cell wall peptidoglycan can have Freund's type adjuvant activity and can induce arthritis in rats.[28]

For most purposes, intradermal injection of heat-killed mycobacteria suspended in oil remains the most convenient and reproducible method for inducing adjuvant arthritis. *Mycobacterium butyricum* and *M. tuberculosis* strain H37Ra, commercially available from Difco Laboratories (Detroit, MI), are widely used; in addition, a mixture of three strains of *M. tuberculosis,* C, DT, and PN, is available from the Central Veterinary Laboratory (Weybridge, England). Bacillus Calmette–Guérin, available from a variety of sources (e.g., Pasteur Institute, Paris, France), can also be used.

Up to several hundred milligrams of the heat-killed organisms are ground to a fine, pale powder with a mortar and pestle; the powder is then weighed and suspended at 3–10 mg/ml in light mineral oil, pristane, or squalane (all available from Sigma Chemical Co.)[21] Such a suspension can be stored indefinitely at 4° without loss of arthritogenicity. Just before each group of animals is to be injected, the suspension is dispersed evenly either manually or by sonication, and the total amount needed is aliquoted with a plastic syringe into sterile 5-ml tubes. For injection, a 1-ml plastic tuberculin syringe is inserted into the bottom of the tube, and the suspension is repeatedly withdrawn and expressed to ensure even dispersion. The syringe is then filled and fitted with a 23- or 25-gauge needle. Ten rats are then quickly given intradermal injections of 0.1 ml of the suspension, and the process is repeated for the next group of rats.

Some drug-screening protocols call for comparison between the acute inflammation at the injection site and the delayed arthritis at distant sites.[6] In such cases, the intradermal injection is given into one hind footpad, with the animal either restrained or anesthetized. In most other cases, the injection is best given at the base of the tail; this leaves both hind paws free

[23] O. Kahashi, C. M. Pearson, Y. Watanabe, and S. Kotani, *Infect. Immun.* **16,** 861 (1977).
[24] T. Koga, K. Maeda, K. Onoue, K. Kato, and S. Kotani, *Mol. Immunol.* **16,** 153 (1979).
[25] S. Nagao and A. Tanaka, *Infect. Immun.* **28,** 624 (1980).
[26] Y.-H. Chang, C. M. Pearson, and L. Chedid, *J. Exp. Med.* **153,** 1021 (1981).
[27] O. Kohashi, K. Aihara, A. Ozawa, S. Kotani, and I. Azuma, *Lab. Invest.* **47,** 27 (1982).
[28] Y.-H. Chang, C. M. Pearson, and C. Abe, *Arthritis Rheum.* **23,** 62 (1980).

for evaluation of the delayed arthritis. No special apparatus is required; the tail can simply be pulled firmly out of the cage, leaving the rat braced between the wall and the top of the cage for the few seconds required to inject the adjuvant.

In all intradermal injections, it is best for the needle to be almost parallel to the skin surface as it enters the skin, and then to travel within the dermis for about 10–15 mm before the material is injected. The tissue turgor of the dermis serves as a valve to prevent material from leaking out of the needle tract. The injected material should significantly distend the skin to form a compact bleb; if the injection is too superficial, or the needle tract is too short, the material is likely to leak through the skin. If the injection is too deep, i.e., subcutaneous instead of intradermal, there will not be an adequate bleb. In either case, the arthritogenic potency of the adjuvant is reduced. Necrosis and ulceration of the skin at the injection site are common but rarely cause significant morbidity.

When water-soluble materials such as MDP or larger peptidoglycans are to be used, a water-in-oil emulsion must be made. The water-soluble material is dissolved in a volume of phosphate-buffered saline (PBS) such that 0.05–0.1 ml of a 1 : 1 emulsion of PBS and oil will deliver the desired amount of adjuvant-active material. For most purposes, the most convenient and effective oil is Freund's incomplete adjuvant (Difco).

If the emulsion is prepared manually, the aqueous and oily components are separately drawn up into glass, Luer-lock syringes. The loaded syringes are attached to either a three-way stopcock or to a double-headed small-bore needle, care being taken to eliminate all bubbles. The aqueous component is forcefully injected into the oily component, and the mixture is then rapidly and vigorously passed back and forth between the two syringes. The emulsion is ready for intradermal injection when a drop of it floats intact when dropped onto a beaker of water. Water-in-oil emulsions can also be prepared with an electric blender (e.g., Virtis). A thorough discussion of the preparation of water-in-oil emulsions of Freund's complete adjuvant (CFA) has been published elsewhere.[29]

In cases in which the arthritogenic material is precious, severely limited in quantity, or of doubtful potency, maximum effect can be obtained by injecting the emulsion directly into the rats' inguinal lymph nodes.[23,24,30] This method is described in detail in the original report of Newbould.[31]

[29] B. B. Mishell and S. M. Shiigi (eds.), "Selected Methods in Cellular Immunology," p. 487. Freeman, San Francisco, California, 1980.

[30] B. H. Newbould, *Ann. Rheum. Dis.* **23,** 392 (1964).

[31] B. H. Newbould, *Innumology* **9,** 613 (1956).

Clinical Evaluation of Adjuvant Arthritis

Methods for the serial, *in vivo* evaluation of arthritis in rodents can be broadly grouped into two classes, one based on visible inspection and the other based on measuring devices. Of the first class, the scoring system of Wood *et al.*,[32] or variants of it, is the most commonly used. In this system, each paw, and sometimes also the tail, is given a score between 0 and 4, based on the severity and extent of the erythema and edema of the joints and periarticular tissues; usually the injection site is not counted, and thus the maximum score is usually 16. We have recently described a greatly expanded scoring scale, from 0 to 156, that allows separate evaluation of swelling and erythema in different regions of each paw.[33] It is not yet clear if this system will prove to have broad applicability.

Two common types of measuring devices are used for the "objective" assessment of joint swelling. Micrometer calipers can be used to measure either the bimaleolar diameter of the hind paw or the dorsal–ventral thickness of either hind or front paws. An estimate of actual paw volume can be made by displacement of a column of mercury.[34] A potential disadvantage of the reliance on size measurements is that one may be left with no good assessment of the activity of the inflammatory process, since permanent swelling due to bony enlargement is a common sequela of adjuvant arthritis. However, paw volume measurement has proved to be the most sensitive and reproducible method for quantifying both the inflammatory response and the antiinflammatory activity of drugs in adjuvant arthritic rats, and is described in detail below.

Whatever method is used for assessing the acute arthritis, the peak severity generally occurs between days 16 and 21 following the injection of adjuvant. Whether or not one reports a time course or peak severity, it is customary to report a mean score for each treatment group of rats recorded during this period.

Paw Volume Measurement as a Method of Quantifying the Inflammatory Response of Adjuvant Arthritic Rats

The method used today does not deviate appreciably from that first described by Winter *et al.*[34] in 1962, when the carrageenan hind paw edema assay was introduced. The apparatus used consists of a mercury-filled plethysmograph connected to a pressure tranducer by water-filled

[32] F. D. Wood, C. M. Pearson, and A. Tanaka, *Int. Arch. Allergy* **35,** 456 (1969).

[33] J. D. Taurog, S. S. Kerwar, R. A. McReynolds, G. P. Sandberg, S. L. Leary, and M. L. Mahowald, *J. Exp. Med.* **162,** 962 (1985).

[34] C. A. Winter, G. A. Risley, and G. W. Nuss, *Proc. Soc. Exp. Biol. Med.* **111,** 544 (1962).

tubing. Displacement of the mercury causes an increase in pressure to register in the tranducer. The signal is then amplified electronically and displayed on a meter or printer.

The present technology is microprocessor based; a good example is the Edema Computer system made by Buxco Electronics (Sharon, CT) (Fig. 1). This system consists of a self-contained plethysmograph table/platform, a Statham PM 131TC differential tranducer, an electronic edema computer, and a printer. Measurements are actuated by a foot switch. The latest model has the ability to store right and left paw volume data from different days and calculate the changes in paw volume, as well as group mean changes and standard errors. This system allows the investigator to process a large number of animals without the burdensome task of performing mathematical computations for each sampling. The system can be interfaced with a personal or main-frame computer for additional data handling and statistical evaluation.

Once the system has been calibrated with objects of known volume and exhibits good linearity, paw volumes can be measured in a fast and reproducible manner. The hind paw is immersed in the mercury to a predetermined anatomical site, usually the hairline on the dorsal surface. The paw volume is recorded by depressing the foot switch; care must be taken to hold the rat steady during measurement. A typical time course following injection of adjuvant into one hind footpad is shown in Fig. 2.

Radiographic Evaluation of Adjuvant Arthritis

Roentgenography provides a permanent and objective record of the pathologic changes of experimental arthritis that offers the advantage of serial sampling from an individual animal. We have found it quite satisfactory to use a Siemens portable dental X-ray unit, together with Kodak high-speed occlusal film. Rats are anesthetized with ether or chloral hydrate and positioned so that both hindlimbs are X-rayed together, followed by both forelimbs. We have recently described a scoring system,[33] which in practice is best applied blindly. Separate scores of 0 to 3 are assigned to each extremity for periosteal reaction, bony erosions, joint space narrowing, and joint space destruction. It is also possible to take radiographs postmortem, either of intact animals or of extremities fixed in formalin; however, in our experience, the resoluation in radiographs of fixed tissues is often suboptimal.

Magnification radiography has recently been applied to collagen-induced arthritis,[35] but is equally applicable to AA.

[35] T. W. Jamieson, A. A. DeSmet, M. A, Cremer, K. L. Kage, and H. B. Lindsley, *Invest. Radiol.* **20,** 324 (1985).

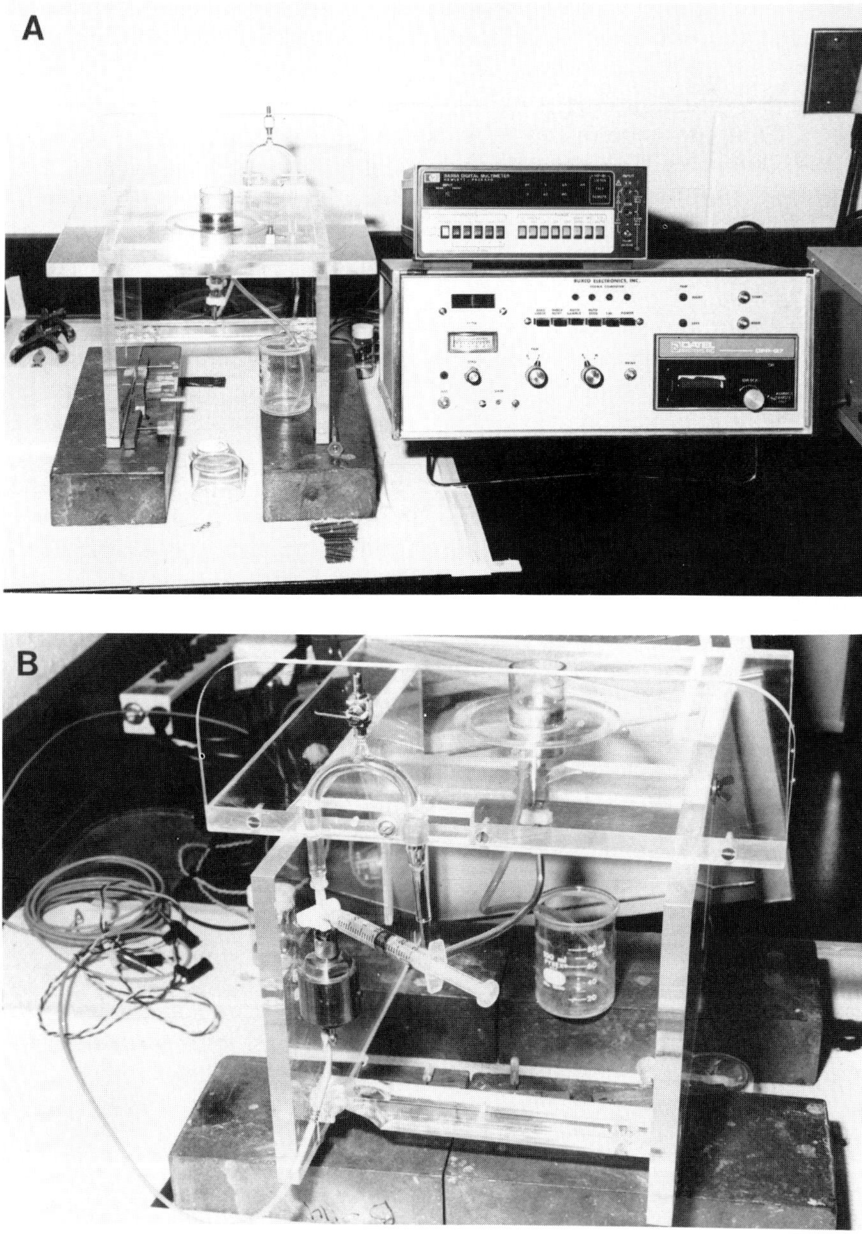

FIG. 1. Buxco edema computer system. (A) Front view showing plethysmograph table, mercury reservoir, a 50-ml glass syringe cut down to 35 ml, and electronic edema computer. (B) Rear view of the table showing connections of tubing to transducer. Note that the glass U tube and transducer connection are equipped with a stopcock to vent air from the system.

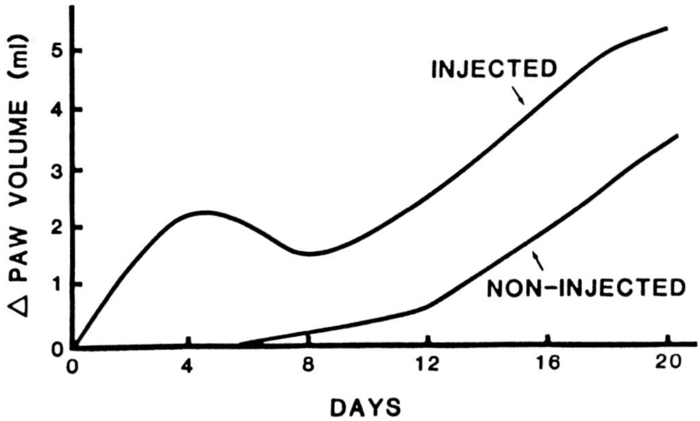

FIG. 2. Time course of paw volume changes of both the injected and noninjected paws in adjuvant arthritic rats. Δpaw volume is the change in paw volume from day 0, the day of adjuvant injection into one hind paw.

Histologic Evaluation of Adjuvant Arthritis

The articular and extraarticular histologic changes in AA have been well characterized.[4,5,14,36–39] The most convenient site for light microscopic examination is the tarsal region, although with care any joint can be processed satisfactorily. Animals are sacrificed, and the extremities to be studied are quickly disarticulated and stripped of skin and muscle. Bone cutters are used to cut as close to the relevant joint as possible; for example, if the tarsal joints are to be examined, cuts are made through the metatarsals and through the distal tibia. The specimen is placed in a porous cloth bag with a pullstring, which is then pulled tight, and the bag is labeled with a Sharpie marker. (Point bags with a double drawstring, measuring 6.5 × 10 cm, are obtained from Chase Bag Co., P.O. Box 1677, Reidsville, NC 27320.) A large number of bags can be put into a single large reservoir of neutral-buffered formalin (pH 7.0, with 0.2 *M* sodium phosphate buffer); we have found it convenient to use plastic 1- or 5-gallon containers obtained from the hospital food service.

After a few days of fixation, the tissues are ready for decalcification. Although this can be accomplished in a few days in formic acid, better resolution is usually obtained from tissues decalcified for 4–8 weeks in EDTA. A large reservoir of neutral-buffered formalin is prepared to contain 10% disodium EDTA, and the cloth bags containing the fixed speci-

[36] C. M. Pearson, B. H. Waksman, J. T. Sharp, *J. Exp. Med.* **113,** 485 (1961).
[37] C. M. Pearson and F. D. Wood, *Am. J. Pathol.* **42,** 73 (1963).
[38] N. A. Burstein and B. H. Waksman, *Yale J. Biol. Med.* **37,** 177 (1964).
[39] B. H. Waksman and S. J. Bullington, *Arch Ophthalmol.* **64,** 751 (1960).

mens are transferred into it. Decalcification then proceeds for 4–8 weeks; it is best if the EDTA solution is subjected to gentle agitation throughout this period, e.g., on a platform rocker.

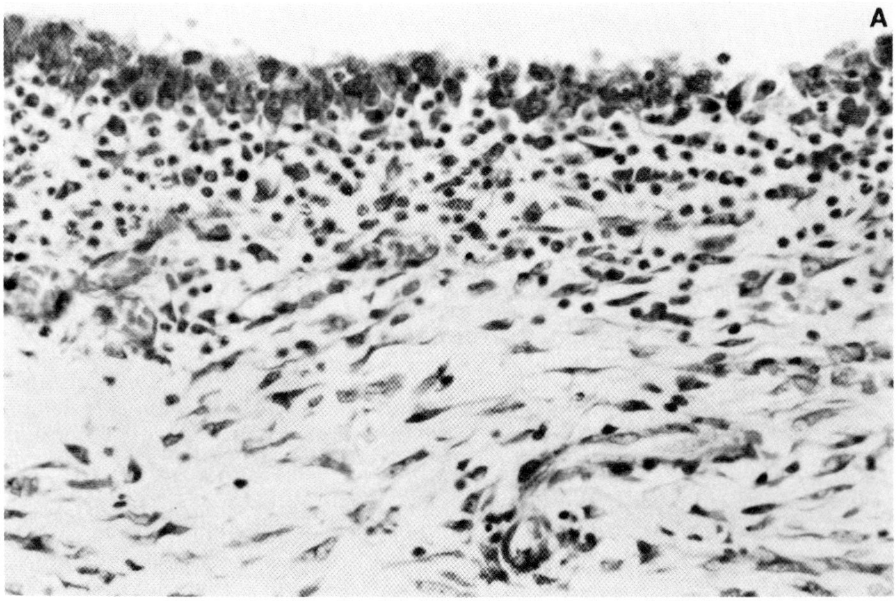

FIG. 3. Histology of adjuvant arthritis. (A) Acute synovitis. Acute inflammatory changes are present, including edema and the infiltration of polymorphonuclear neutrophils. Neutrophils accumulate in the synovium nearest the joint space, as shown here, and penetrate the discontinuous synoviocyte lining to enter the joint space. Mild synoviocytic hyperplasia is present. Giemsa stain. ×100. (B) Acute synovitis and intraarticular exudate. Acute synovitis leads to the accumulation within the joint space of polymorphs, cell debris, and macrophages enmeshed in a fibrin web. E, Exudate within joint space. Giemsa stain. ×100. (C) Periarticular granuloma. Unique to adjuvant arthritis is the formation of small granulomas in periarticular soft tissues. These granulomas contain macrophages and occasional giant cells. Arrow indicates a macrophage in the granuloma. Giemsa stain. ×100. (D) Pannus. Ingrowth of granulation tissue from the synovium leads to the formation of pannus, which covers and adheres to the articular surfaces. C, Cartilage; P, pannus. Note the numerous small blood vessels in the pannus. At this stage (day 21 after adjuvant injection), few inflammatory cells remain in the pannus. Giemsa stain. ×100. (E) Pannus with invasion of cartilage. Adherent pannus may exhibit vascular buds penetrating into cartilage, with destruction of the cartilage. Arrows mark penetration of pannus into cartilage. C, Cartilage; P, pannus. Giemsa stain. ×100. (F) Destruction of bone. Marrow elements are replaced by granulation tissue; stem cells are lost. Remodeling of intraarticular bone is a prominent feature of AA; shown here is resorption of the subchondral bony plate by osteoclasts (arrow). C, Cartilage; P, pannus. Giemsa stain. ×100.

When the specimens are sufficiently decalcified to feel rubbery and pliable, they are ready for embedding and sectioning. Although most histology laboratories routinely embed specimens in paraffin, better sections can be obtained if the specimens are embedded in methacrylate. In practice, the investigator is usually limited by the capability of the services available. Sections of 3–4 μm are stained with Giemsa or hematoxylin–eosin.

Joint specimens are evaluated for the extent of inflammatory changes in the synovium and joint space, for pannus formation and erosion of articular cartilage and subchondral bone, and in late specimens for fibrous or bony ankylosis and osteoneogenesis. The progression of histologic changes typical of AA is shown in Fig. 3. Although most of the changes are also common in other experimental arthritides, granuloma formation in periarticular soft tissue is unique to AA (Fig. 3C). The macrophages in these granulomas have been shown by immunofluorescence to express class II major histocompatibility antigens (R. A. McReynolds, unpublished observations).

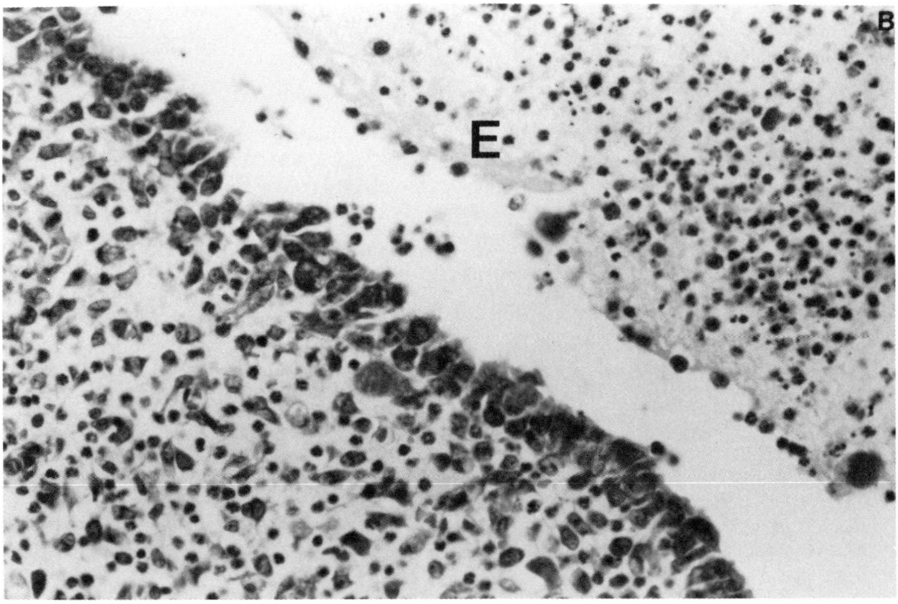

FIG. 3. *(Continued)*

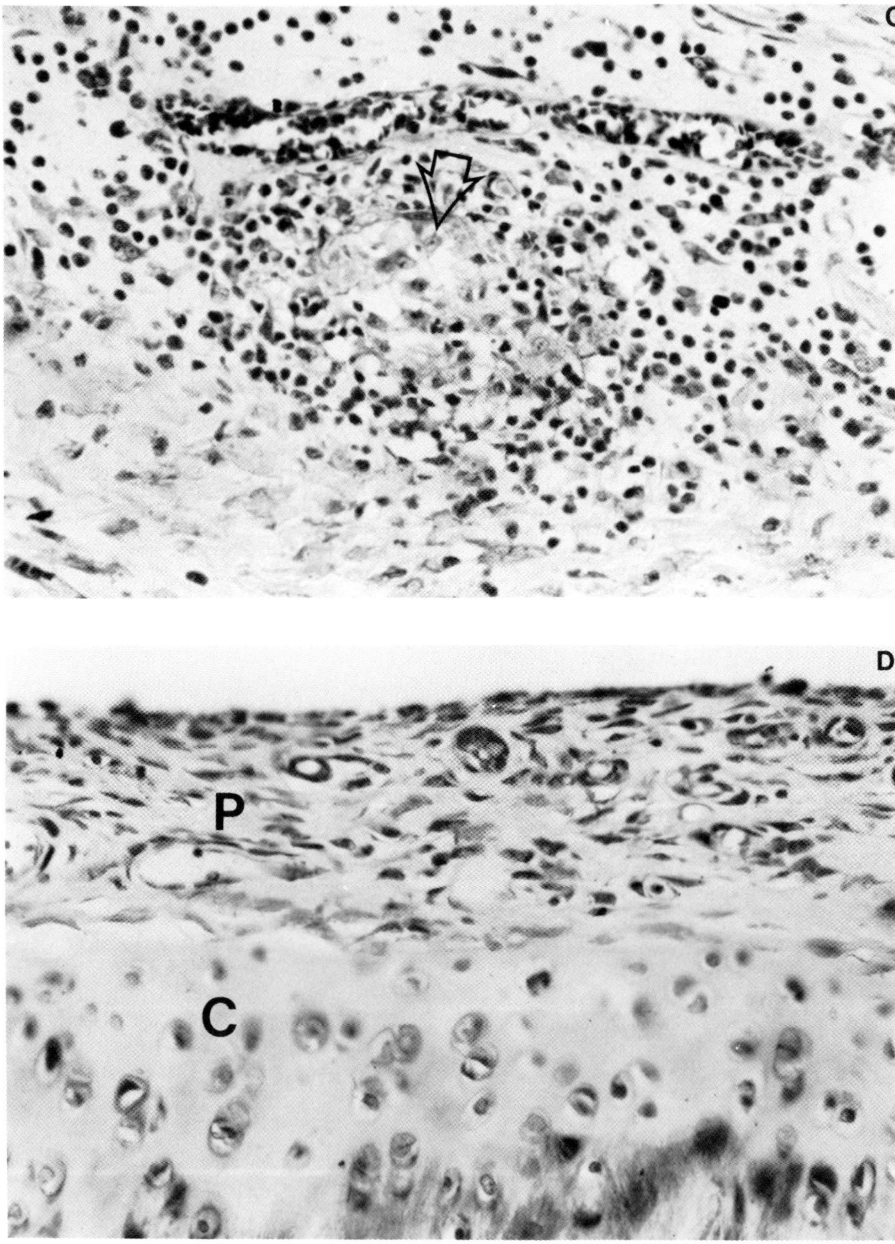

FIG. 3. *(Continued)*

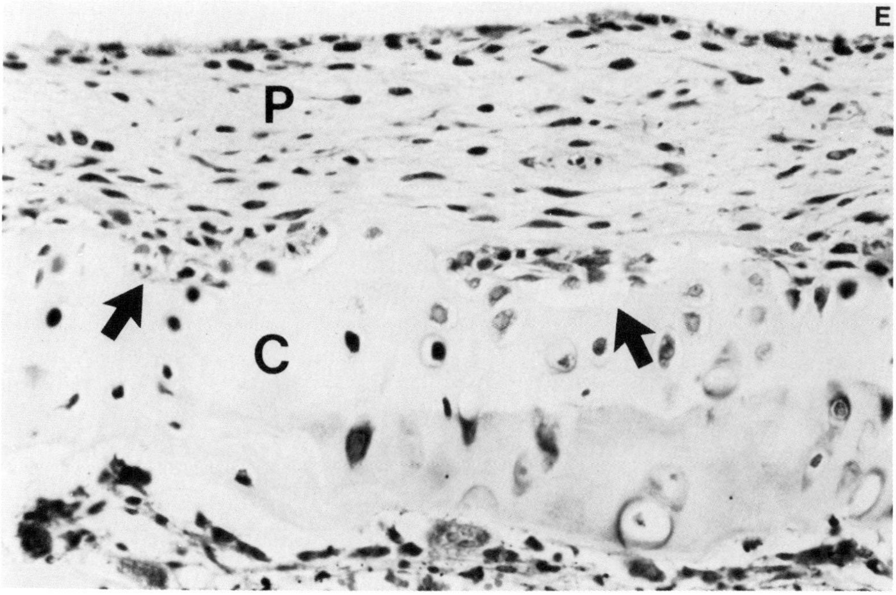

FIG. 3. *(Continued)*

Applications

Effect of Pharmacologic Agents on Adjuvant Arthritis

The adjuvant arthritic rat has long been used as an assay for detecting the antiinflammatory activity of drugs.[6] Figure 2 illustrates the change in volumes of both the injected and uninjected paws of the adjuvant arthritic rat with respect to time. One can use this time course to differentiate between the type of antiinflammatory activity a drug exhibits. Daily oral drug treatment can begin at various stages of the inflammatory response. Prophylactic treatment usually begins upon injection of the adjuvant (day 0) and continues until the chronic inflammatory response is established (day 16). Nonsteroidal antiinflammatory drugs (NSAIDs) and the corticosteroids inhibit the acute inflammation of the injected paw, which peaks around days 4–5, as well as the chronic inflammation of both the injected and uninjected paws at day 16. The NSAIDs are particularly good inhibitors of the synthesis of proinflammatory prostaglandins and are effective in reducing the edematous response initiated by prostaglandins.

Immunosuppressive agents are ineffective against the early inflammation but are quite effective in reducing the chronic inflammation of the

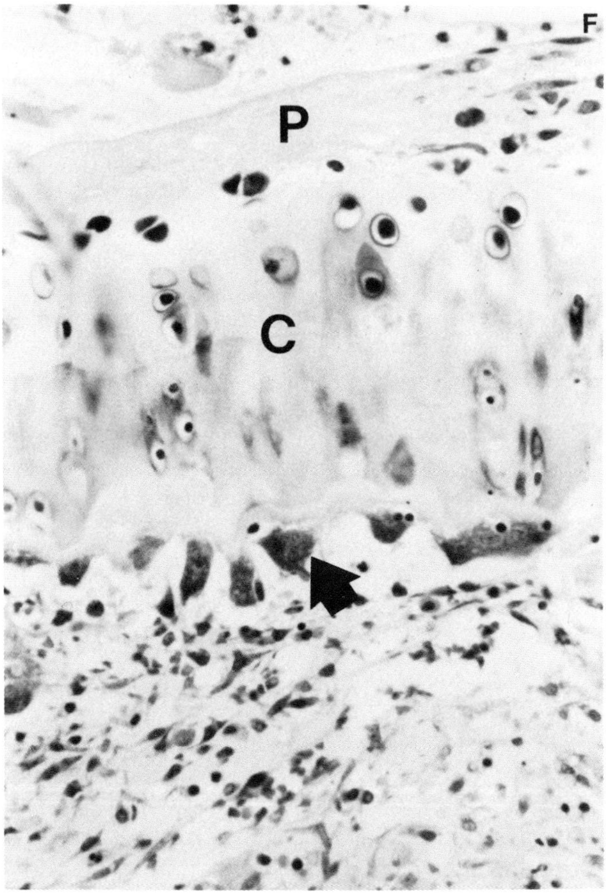

FIG. 3. *(Continued)*

disease. If paw volumes are measured after drug therapy has been discontinued, the differences in the mechanisms of action of these three drug classes can be easily demonstrated. Once therapy with NSAIDs is discontinued, prostaglandin formation will proceed, the inflammatory response rebounds, and paw volumes increase steadily. Thus NSAIDs provide symptomatic relief by reducing the edema, without altering the course of the disease. The corticosteroids and immunosuppressive agents, on the other hand, are disease modifying, causing a permanent attenuation of the disease as indicated by the lack of significant change in paw volumes once therapy is withdrawn. These points are summarized in Table I.

TABLE I
EFFECT OF PHARMACOLOGIC AGENTS ON AA

Drug class[a]	Day 4, injected paw	Day 16: Injected	Day 16: Uninjected	Day 28: Injected	Day 28: Uninjected
	Activity[b]				
NSAIDs (e.g., indomethacin)	+	+	+	−	−
Corticosteroids	+	+	+	+	+
Immunosuppressives (e.g., cyclophosphamide)	−	+	+	+	+

[a] Compounds administered daily from day 0 (day of adjuvant injection) through day 16.
[b] Inhibition of paw volume increases.

Cell-Mediated Passive Transfer of AA

The clinical, histologic, and radiographic features of AA can be transferred to naive syngeneic rats with lymphoid cells harvested from donors given CFA. This observation suggests strongly that AA is primarily a cell-mediated autoimmune disease. Two passive transfer protocols have recently been described. Holoshitz *et al.* have produced a cloned T-cell line reactive with *M. tuberculosis* that induces arthritis in lethally irradiated, but not normal, syngeneic recipients.[40–42] Taurog *et al.* have shown that mitogen-stimulated lymph node, spleen, or peripheral blood mononuclear cells from donors with AA can transfer arthritis to normal syngeneic recipients.[33,43,44] LEW rats were used in all of these experiments.

The following is a protocol for the passive transfer of AA with mitogen-stimulated lymphoid cells. Six- to 10-week-old LEW rats are given a single injection into the base of the tail of 0.6 mg of *M. butyricum* or *M. tuberculosis* in oil, as described above. Nine to 12 days later, the donors are sacrificed, and sterile single cell suspensions are prepared from the pooled inguinal and lumbar lymph nodes and from the spleens. Erythrocytes are removed by either Ficoll–metrizamide centrifugation (specific gravity

[40] J. Holoshitz, A. Matitiau, and I. R. Cohen, *J. Clin. Invest.* **73,** 211 (1984).
[41] J. Holoshitz, A. Matitiau, and I. R. Cohen, *J. Clin. Invest.* **75,** 472 (1985).
[42] W. van Eden, J. Holoshitz, Z. Nevo, A. Frenkel, A. Klajman, and I. R. Cohen, *Proc. Natl. Acad. Sci U.S.A.* **82,** 5117 (1985).
[43] J. D. Taurog, G. P. Sandberg, and M. L. Mahowald, *Cell. Immunol.* **75,** 271 (1983).
[44] J. D. Taurog, G. P. Sandberg, and M. L. Mahowald, *Cell. Immunol.* **80,** 198 (1983).

> 1.09) or by ammonium chloride lysis (see also this series, Vol. 108 [6]). The cells are then cultured for 2 days in a humidified 5% CO_2 atmosphere at 2×10^8 cells in 100 ml of medium per recumbent 150-cm^2 flask. Medium consists of Dulbecco's minimal essential medium containing 4.5 g/liter glucose, 10 mM HEPES, 5×10^{-5} M 2-mercaptoethanol, antibiotics, 2 μg/ml concanavalin A (Con A), and a pretested serum source. We have successfully used 5% horse serum, 5% fetal bovine serum, and 2.5% fresh normal rat serum, horse serum being the most convenient and economical.

Following 2 days of culture, the cells are counted, centrifuged, and resuspended at $2-5 \times 10^7$ viable cells/ml of serum-free medium; the entire procedure is conducted at room temperature. The cells are taken up into a plastic syringe and passed once through a 25- or 27-gauge needle into a fresh tube; they are then taken up into a fresh syringe and injected through the same caliber needle intravenously into syngeneic recipient rats. Up to 5 ml can be safely injected into a 150- to 200-g rat if cell clumps have been removed by prepassage through a small needle, if the cell concentration is 5×10^7/ml or less, and if the injection is given over several minutes.

(Con A)-stimulated AA lymph node cell cultures usually show a high percentage of blast cells and good viability; in contrast, the spleen cell cultures usually show poor viability (<50%) with few apparent blasts. Nonetheless, both transfer arthritis effectively; our standard dose per recipient is $4-5 \times 10^7$ viable spleen cells or 1×10^8 viable lymph node cells. Arthritis or tendinitis first appears in 4–8 days, usually in the hind paws. The clinical manifestations of the passive arthritis are usually milder than those induced directly by intradermally administered CFA.

Induction of Tolerance to AA

Although the evidence is still largely circumstantial, AA behaves like an antigen-specific, T-cell-mediated autoimmune response.[42] As would be expected for such a response, AA is susceptible to tolerance phenomena. Administration of 10–100 μg of dead mycobacteria in an aqueous suspension renders a rat relatively resistant to the arthritogenic effect of subsequent CFA given intradermally.[45,46]

Recently, Kaibara *et al.* described a protocol in which rats were given the immunosuppressive agent cyclosporin A in conjunction with CFA immunization; these rats failed to develop AA, and following cessation of the cyclosporin A treatment, they remained resistant to the effects of

[45] B. H. Waksman, C. M. Pearson, and J. T. Sharp, *J. Exp. Med.* **85,** 403 (1960).

[46] E. M. Eugui and R. H. Houssay, *Immunology* **28,** 703 (1975).

subsequent CFA.[47] This tolerization was antigen-specific, since there was no cross-tolerance between the arthritogenic effects of CFA and Type II collagen.

Holoshitz *et al.* have shown that an antimycobacterial cloned T-cell line, which is arthritogenic when given to irradiated rats, is tolerogenic when given to normal rats, such that the normal recipients are resistant to subsequent AA induction by CFA.[41] Complete protocols for these tolerance systems are contained in the references cited.

[47] N. Kaibara, T. Hotokebuchi, K. Takagishi, I. Katsuki, M. Morinaga, C. Arita, and S. Jinguishi, *J. Exp. Med.* **159,** 1388 (1984).

[32] Type II Collagen-Induced Arthritis in Rats

By SUSAN C. RIDGE, ARNOLD L. ORONSKY, and S. S. KERWAR

Introduction

Intradermal administration of native type II collagen emulsified in Freund's incomplete adjuvant induces a hind paw arthritis in approximately 40% of the immunized rats.[1] This model of arthritis, which can also be induced in mice,[2] shares serological, histological, and radiological features with human rheumatoid arthritis.[3] Sera of some patients with rheumatoid arthritis contain large amounts of anticollagen IgG.[4] Histological studies have shown that, similar to the human lesion, joint of rats or mice with collagen arthritis have extensive synovitis with a proliferative pannus overlaying cartilage, destruction of cartilage and bone, and a massive infiltrate of mononuclear cells in the synovium.[5] In addition, a large number of neutrophils are present in the joint space. Radiologic studies in rats with collagen arthritis indicate extensive destruction of cartilage and bone. Although this model is not an exact counterpart of the human disease,[3] the above similarities suggest that the model is relevant to the study of the human lesion.

[1] D. E. Trentham, A. S. Townes, and A. H. Kang, *J. Exp. Med.* **146,** 857 (1972).

[2] J. S. Courtenay, M. J. Dallman, A. D. Dayan, A. Martin, and B. Mosedale, *Nature (London)* **283,** 666 (1980).

[3] D. E. Trentham, *Arthritis Rheum.* **25,** 911 (1982).

[4] D. Collier, S. S. Kerwar, M. Garavoy, K. Fye, and J. D. Stobo, *Arthritis Rheum.* **27,** 1201 (1984).

[5] J. P. Caulfield, A. Hein, R. Dynesius-Trentham, and D. E. Trentham, *Lab Invest.* **46,** 321 (1982).

TABLE I
TYPE II COLLAGEN ARTHRITIS: SUSCEPTIBLE AND RESISTANT STRAINS

Strain	Genetic linkage	Susceptible	Resistant
Mice			
DBA1/J;B10G;B10Q	$H2^q$	+	
DBA/Z;B10D2	$H2^d$		+
B10K;CBA/N;B10BR	$H2^k$		+
B10	$H2^b$		+
Rats			
Outbred Royalhart		+	
Wistar Furth	$RT1^u$	+	
DA; Lewis Inbred	$RT1^a$	+	
Buffalo	$RT1^b$		+
AUG	$RT1^c$		+
MAXX:LEW.B3	$RT1^n$		+

Like the human lesion, this model responds clinically to several agents used in the treatment of rheumatoid arthritis.[6] These include nonsteroidal antiinflammatory agents, steroids, immunosuppressive drugs and D-penicillamine. However, unlike human disease, this model does not respond to gold salts.[7,8] Both human rheumatoid arthritis and Type II collagen arthritis are under genetic control. Studies[9,10] have identified strains of rats and mice that are susceptible and resistant to the development of the lesions (Table I).

Mechanistic studies related to the induction of arthritis have been conducted. These studies have shown that the induction of arthritis is a complement-dependent lesion.[11,12] Thus, if rats are decomplemented with cobra venom factor, the onset of the disease is delayed.

Cell-mediated immunity to Type II collagen has been implicated in the pathogenesis of this disease.[13,14] Transfer of arthritis was achieved in recipients by the intravenous administration of sensitized T cells.[14] Evidence

[6] S. S. Kerwar, A. E. Sloboda, J. E. Birnbaum, and A. L. Oronsky, *J. Rheum.* **8,** 84 (1981).

[7] A. E. Sloboda, J. E. Birnbaum, A. L. Oronshy, and S. S. Kerwar, *Arthritis Rheum.* **24,** 616 (1981).

[8] W. J. McCune, D. E. Trentham, and J. R. David, *Arthritis Rheum.* **23,** 932 (1983).

[9] M. M. Griffiths, E. J. Eichwald, D. B. Martin, and C. W. DeWitt, *Arthritis Rheum.* **24,** 781 (1981).

[10] P. H. Wooley, H. S. Luthra, J. M. Stuart, and C. S. David, *J. Exp. Med.* **154,** 688 (1981).

[11] K. Morgan, R. B. Clague, M. J. Shaw, S. A. Firth, and T. M. Towse, *Arthritis Rheum.* **24,** 1356 (1981).

[12] S. S. Kerwar, N. Bauman, A. L. Oronsky, and A. E. Sloboda, *J. Immunopharmacol.* **3,** 323 (1982).

[13] L. R. Klareskog, R. Holmdahl, E. Larsson, and H. Wigzell, *Clin. Exp. Immunol.* **51,** 117 (1983).

[14] D. E. Trentham, R. A. Dynesius, and J. R. David, *J. Clin. Invest.* **62,** 359 (1978).

supporting the role of humoral immunity in the pathogenesis of collagen arthritis has also been provided. It has been shown[15] that rats which develop arthritis have large amounts of anticollagen IgG in their sera whereas immunized rats which do not develop the disease have minimal amounts of antibody. The arthritis can be transferred to recipients by the intravenous administration of immune sera[16] or affinity-purified anticollagen IgG.[17] Passive arthritis induced by anticollagen IgG is also a complement-dependent lesion.[17] Similar studies indicating the critical role of antibody in the pathogenesis of collagen arthritis in mice have been published.[18]

Major differences between active arthritis (antigen induced) and passive arthritis (antibody induced) have been demonstrated.[19] Unlike the active disease, which is a chronic lesion, passive arthritis is transient and lacks a proliferative pannus in the joint space.[19] It has therefore been suggested that both cell-mediated and humoral immunity are involved in the pathogenesis of collagen arthritis. In this mechanism,[19] responder rats develop a robust humoral response to Type II collagen (anticollagen IgG_{2a}).

Some of these serum antibodies bind to the Type II collagen present in the articular cartilage of the hind paws[20] and activate the complement system. This activation leads to the generation of complement components, C3a and C5a. C3a can increase vascular permeability and allow for trafficking of sensitized lymphocytes into the joint space. These sensitized lymphocytes can proliferate in the joint space and secrete lymphokines[21] that may induce proliferation of synovial tissue. C5a can induce migration of inflammatory cells into the joint space. These inflammatory cells contain connective tissue-degrading enzymes (collagenase, elastase, neutral and acid cathepsins) that can induce joint destruction. In this mechanism, anticollagen IgG plays a critical role in initiating the disease and cell-mediated immunity is required for sustaining the lesion. Evidence supporting this hypothesis has been presented.[22]

[15] J. M. Stuart, M. A. Cremer, A. H. Kang, and A. S. Townes, *Arthritis Rheum.* **22,** 1344 (1979).

[16] J. M. Stuart, M. A. Cremer, A. S. Townes, and A. H. Kang, *J. Exp. Med.* **155,** 1 (1982).

[17] S. S. Kerwar, M. E. Englert, R. A. McReynolds, M. J. Landes, J. M. Lloyd, A. L. Oronsky, and F. J. Wilson, *Arthritis Rheum.* **26,** 1120 (1983).

[18] J. M. Stuart and F. J. Dixon, *J. Exp. Med.* **158,** 378 (1983).

[19] M. E. Englert, R. A. McReynolds, M. J. Landes, A. L. Oronsky, and S. S. Kerwar, *Cell. Immunol.* **90,** 258 (1985).

[20] S. S. Kerwar, S. Gordon, R. A. McReynolds, and A. L. Oronsky, *Clin. Immunol. Immunopathol,* **29,** 318 (1983).

[21] S. M. Helfgott, D. Dynesius-Trentham, E. Brahn, and D. E. Trentham, *J. Exp. Med.* **162,** 1531 (1985).

[22] J. D. Taurog, S. S. Kerwar, R. A. McReynolds, G. P. Sandberg, S. L. Leary, and M. L. Mahowald, *J. Exp. Med.* **162,** 962 (1985).

Methods

Preparation of Bovine Native Type II Collagen

Three hundred grams of bovine nasal septum is cut into small pieces ($4\ mm^3$) and washed in 2 liters of 50 m*M* Tris–HCl, pH 7.4, containing 2 *M* $MgCl_2$ for 48 hr at 4°. The cartilage pieces are then rinsed several times with cold water (2 liters rinse). The cartilage pieces are suspended in 600 ml of 0.5 *M* acetic acid, and the pH of the suspension is adjusted to 2.5 with formic acid. Crystalline pepsin (Boehringer Mannhein) is added (1/50th wet weight—6 g pepsin), and the digestion is allowed to proceed for 72 hr at 4° with stirring. The undigested residue is separated by centrifugation (20,000 *g*, 20 min), and the viscous supernatant is dialyzed for 48 hr in the cold against 50 m*M* Tris/0.2 *M* NaCl, pH 7.6 (10 liters with 3 buffer changes). This viscous material is passed through a DEAE-cellulose column (8 × 20 cm) and eluted with the same buffer. (Type II collagen does not bind to DEAE-cellulose under these conditions, and this column chromatography is used to remove any contaminating proteoglycans present in the preparations.) Collagen is precipitated from the DEAE eluate by the addition of NaCl to 3 *M*. The precipitate is dissolved in 1 liter of 50 m*M* Tris/0.2 *M* NaCl, pH 7.6, and passed through a second DEAE-cellulose column (8 × 20 cm) under the conditions detailed above. After precipitation with NaCl (3 *M*) the collagen is dissolved in 0.5 *M* acetic acid (1 liter) and dialyzed against 10 m*M* Na_2HPO_4 for 48 hr (10 liters with 3 changes). The collagen precipitate is harvested by centrifugation and dissolved in 0.5 *M* acetic acid (1 liter). Collagen is reprecipitated by the addition of NaCl to 1 *M*, and the precipitate is redissolved in 0.5 *M* acetic acid (1 liter). This material is dialyzed for 48 hr against 0.1 *M* acetic acid (10 liters with 3 changes) and lyophilized.

Immunization and Clinical Assessment of Arthritis

Type II collagen is solubilized in 1 m*M* acetic acid at a concentration of 1.5 mg/ml and then emulsified with an equal volume of Freund's incomplete adjuvant (Difco Labs., Midland, MI). Outbred male Royalhart Wistar rats (160–180 g; Royal Hart Lab Animals, New Hampton, NY) are immunized with native Type II collagen. Each rat is immunized with 0.1 ml of the emulsion (75 μg collagen), which is administered intradermally at the base of the tail. Immunized animals are boosted on day 7 with 75 μg of collagen as specified above. Hind paw diameters are measured around the ankle joint with a vernier caliper. Rats with paw diameters of 8 mm or greater are considered arthritic (uninjected rats have a paw diameter of about 7 mm).

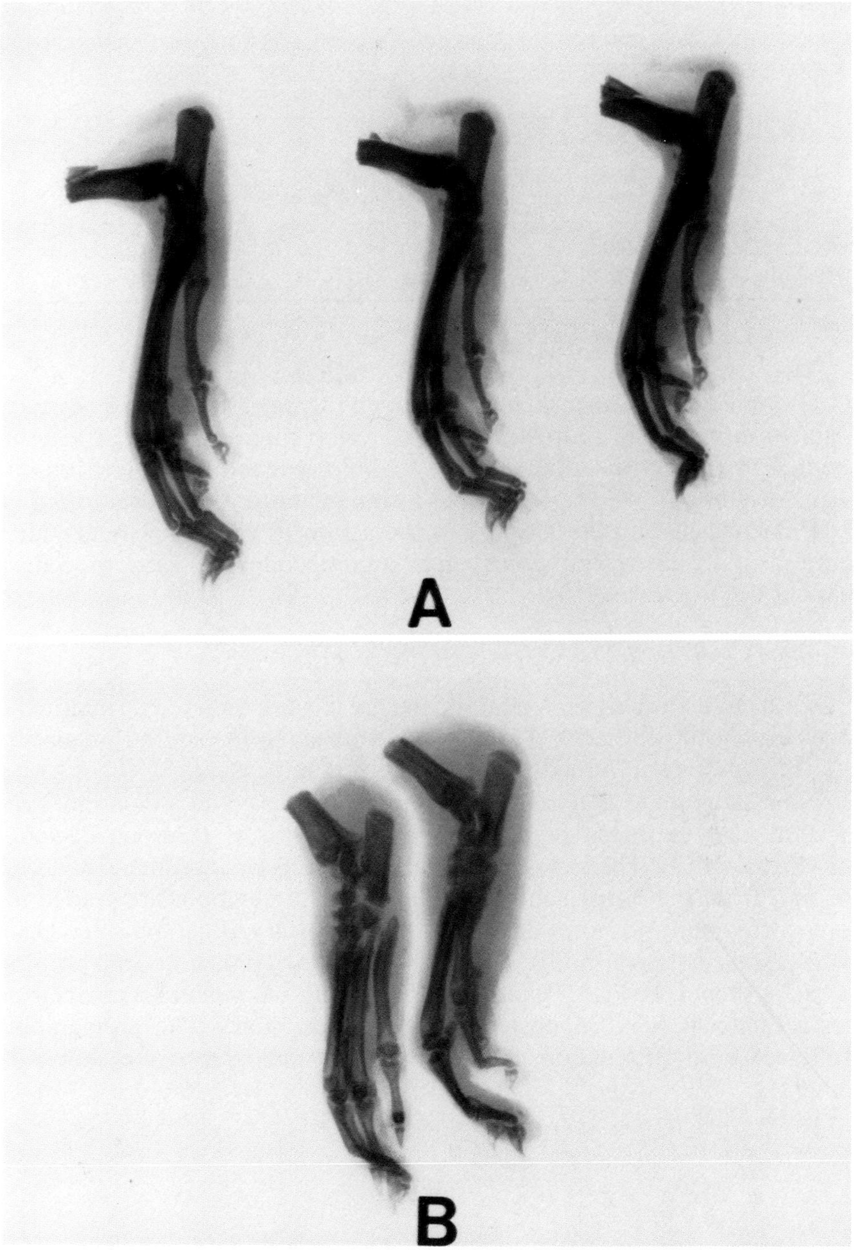

FIG. 1. Radiographs of hind paws of rats with collagen arthritis. (A) Normal, (B) arthritic.

TABLE II
RADIOGRAPHIC EVALUATION OF THE JOINTS OF HIND PAWS OF COLLAGEN ARTHRITIC RATS

	Radiographic grade[a]				
Joint	Bone demineralization	Erosions	Periostitis	Cartilage space narrowing	Distal phalanges
Normal	0	0	0	0	0
Collagen arthritis	2.0	3.0	2.0	3.0	2.0

[a] Grading: 0 = normal, 4 = maximum destruction.

The incidence of arthritis is 40–60% and peaks about 3 weeks after initial immunization. Animals begin to develop clinical signs of arthritis in about 2 weeks, but maximal damage (radiological and histological assessment) occurs at 5 weeks after initial immunization. Antibodies to native Type II collagen can be detected in the serum as early as 2 weeks after immunization. These antibodies can be quantified by an ELISA procedure that has been described.[17]

Histologic and Radiologic Assessment of Arthritis

For histologic analysis, the hind paws are fixed in buffered formalin and decalcified in formic acid. The tissue is embedded in paraffin or plastic, and the sections are stained with hematoxylin and eosin.[17]

For radiographic analysis, the hind paws are placed on a Polaroid Type 55 film and exposed to X-rays using a Faxitron (Hewlett-Packard, McMinville, OR). The exposure is for 5 min at 60 kV, and the distance of the film from the X-ray source is 45 cm. The radiographs are graded for several parameters of joint destruction.[23] A typical radiograph of collagen arthritic rats is shown in Fig. 1. In Fig. 1A, radiographs of normal rat hind paws are shown. In Fig. 1B, radiographs of hind paws of collagen arthritic rats are shown (day 35 postcollagen immunization). The radiographic grades of joint destruction assigned to these radiographs are shown in Table II.

[23] R. L. Clark, J. T. Cutrino, S. K. Aderle, W. J. Cromartie, and J. H. Schwab, *Arthritis Rheum.* **22,** 25 (1979).

[33] Collagen-Induced Arthritis in the Mouse

By PAUL H. WOOLEY

Introduction

Type II collagen-induced arthritis (CIA)[1] was first described in the mouse by Courtenay *et al.*,[1a] following the discovery of the experimental disease model in the rat. There are many clinical and immunological similarities between the models; however, there are also a number of important differences. The most apparent difference is the time course of the two diseases: rat CIA has a relatively short duration (3 weeks) between immunization and the appearance of arthritis, whereas mouse CIA shows a mean onset date around 6 weeks postimmunization. The progression of the arthritis in mice is also slower: the affected joints in rats are frequently committed within 1 week after onset, while new joints may become involved in mouse CIA after 9 weeks of disease duration. Genetic susceptibility has been demonstrated in both the rat[2] and the mouse,[3] but the more defined genetic systems of the mouse have permitted a more detailed analysis. It is apparent that the disease is more restricted in mice, since most inbred rat strains are CIA susceptible, while most inbred mouse strains are CIA resistant. There are important immunological variations between the species: adoptive transfer has been reported in rat CIA, whereas this has proved unsuccessful in the murine model. Finally, mice are not susceptible to adjuvant-induced arthritis, making complete adjuvant preferable for the induction of disease in this species. The majority of published studies of pharmacological intervention in CIA have been conducted in the rat; however, some recent data suggest that the rat and murine models may respond in a similar manner to both nonsteroidal antiinflammatory drugs and most disease modifying antirheumatic drugs.[4]

[1] Abbreviations: CIA, collagen-induced arthritis; PMNs, polymorphonuclear leukocytes; MPO, myeloperoxidase; NAG, *N*-acetyl-β-D-glucosaminidase; SAP, serum amyloid P; RIA, radioimmunoassay; ELISA, enzyme-linked immunoabsorption assay; PBS, phosphate-buffered saline; BSA, bovine serum albumin; DTH, delayed-type hypersensitivity; StA, *Staphylococcus* protein A.

[1a] J. S. Courtenay, M. J. Dallman, A. D. Dayan, A. Martin, and B. Mosedale, *Nature (London)* **283,** 666 (1980).

[2] M. M. Griffiths, E. J. Eichwald, J. H. Martin, C. B. Smith, and C. W. DeWitt, *Arthritis Rheum.* **24, 781** (1981).

[3] P. H. Wooley, H. S. Luthra, J. M. Stuart, and C. S. David, *J. Exp. Med.* **154,** 688 (1981).

[4] K. Phadke, R. L. Fouts, J. E. Parrish, and L. D. Butler, *Immunopharmacology* **10,** 51 (1985).

TABLE I
INCIDENCE OF ARTHRITIS (%) IN $H\text{-}2^q$ AND $H\text{-}2^r$ INBRED MOUSE STRAINS USING VARIOUS SPECIES OF TYPE II COLLAGEN

		Type II Collagen			
Strain	*H-2*	Chick	Bovine	Porcine	Deer
DBA/1 J	*q*	71	62	70	ND[a]
DBA/1 Lac	*q*	89	65	90	ND
B10.Q	*q*	84	53	11	60
B10.G	*q*	41	15	17	18
BUB/J	*q*	77	ND	60	ND
BZH	*q*	0	52	43	ND
SWR/J	*q*	0	0	0	0
B10.RIII	*r*	0	73	90	75
RIII/J	*r*	0	0	0	0

[a] ND, Not done.

Strain Selection

The selection of mouse strains for the investigation of CIA is restricted mostly by the genetic regulation of disease susceptibility. Arthritis has been induced only in mice bearing one of two independent haplotypes, $H\text{-}2^q$ and $H\text{-}2^r$. The regulation is effected through Class II major histocompatibility complex genes, demonstrated through recombinant haplotype analysis.[3] Recently, the susceptibility has been specifically associated with the *I-A* subregion, but this susceptibility is also influenced by background genes. The species source of Type II collagen also influences strain susceptibility, and it is postulated that this may be due to several immunogenic epitopes present on the Type II collagen molecule,[5] possibly on the cyanogen bromide peptide fragment 11.[6] Table I shows the $H\text{-}2^q$ and $H\text{-}2^r$ inbred mouse strains studied to date, using several species of Type II collagen. The most universally susceptible strain appears to be the DBA/1 Lac mouse, currently available from Jackson Laboratories (Bar Harbor, ME) as the substrain DBA/1 Lac J. This strain, although originally bred from DBA/1 J, is consistently more susceptible.[1] DBA/1 is the source of the $H\text{-}2^q$ haplotype of B10.Q mice, and this strain is more susceptible than B10.G mice, where the $H\text{-}2^q$ haplotype was translocated from the GL mouse.

[5] P. H. Wooley, H. S. Luthra, M. M. Griffiths, J. M. Stuart, A. Huse, and C. S. David, *J. Immunol.* **135,** 2443 (1985).

[6] K. Terato, K. A. Hasty, M. A. Cremer, J. M. Stuart, A. S. Townes, and A. S. Kang, *J. Exp. Med.* **162,** 637 (1985).

B10 congenic $H\text{-}2^q$ mice show some regulation by the species source of collagen, being susceptible to most Type II collagen species but relatively resistant to porcine collagen. SWR/J mice are consistently resistant to the induction of CIA using any species of Type II collagen, despite strong similarities in the immune response when compared to DBA/1 Lac J mice. BUB mice demonstrate a high disease incidence to both chick and porcine collagen, while BZH mice are susceptible only to bovine collagen and develop an atypical joint disease and frequent ear lesions. It is possible that the disease seen in BZH mice represents a model of relapsing polychondritis. The $H\text{-}2^r$ strain B10.RIII develops a high incidence of CIA in response to bovine, porcine, and deer Type II collagen but is resistant when injected with chick, rat, or human collagen. Background gene influence is evident since the source of the $H\text{-}2^r$ MHC, the RIII/J inbred strain, is resistant to CIA. Insufficient recombinant haplotype mice are available to permit the complete analysis of the MHC control of disease susceptibility in the $H\text{-}2^r$ haplotype.

Induction and Assessment of Collagen Arthritis in Mice

Selection and Maintenance of Mice

Several general considerations, in addition to strain selection, are required to achieve maximal disease incidence in the collagen arthritis model. Mice should be a minimum of 7 weeks of age, although no upper age restriction is detected.[7] The animal must be in good health, since viral diseases in particular cause a marked decrease in disease incidence. Male mice develop CIA earlier after immunization than female mice, although the overall final disease incidence between sexes is not statistically different. This may be due to a hormonal influence, since CIA is affected by pregnancy.[8] Diet is also a consideration, and the disease incidence varies in mice fed several commercially available rodent chows (Table II). The highest disease incidence is observed in mice fed a diet designed for breeders (Purina Mouse Chow 5015), and studies with defined diets have indicated that the major dietary influence may be the fat component.[9]

[7] P. H. Wooley, A. M. Dillon, H. S. Luthra, J. M. Stuart, and C. S. David, *Transplant. Proc.* **15,** 180 (1983).

[8] F. Hirahara, P. H. Wooley, H. S. Luthra, C. B. Coulam, M. M. Griffiths, and C. S. David, *Am. J. Reprod. Immunol. Micro.* **11,** 44 (1986).

[9] K. Leslie, W. A. Gonnerman, M. D. Ullman, K. C. Hayes, C. Franzblau, and E. S. Cathcart, *J. Exp. Med.* **162,** 1336 (1985).

TABLE II
INFLUENCE OF COMMERCIAL RODENT DIETS ON TYPE II COLLAGEN-INDUCED ARTHRITIS IN B10.Q MICE[a]

Purina diet	CIA incidence (%)	CIA onset (day)	CIA severity (clinical score)
5001	65	48.8	6.1
5008	47	35.5	5.8
5012	47	41.3	5.5
5015	71	33.7	6.6
5025	28	64.0	3.8

[a] Mice immunized with chick collagen.

Preparation of Type II Collagen

Type II collagen is most frequently prepared from chick sternae (of birds made lathyritic by the addition of 0.3% β-aminopropionitrile to the drinking water for 2 weeks)[10] or bovine articular cartilage.[2] Tissue is divided into small pieces and extracted overnight with 50 m*M* Tris–HCl buffer (pH 7.5) containing 4 *M* guanidine–HCl. The residue is washed in water, and collagen is extracted by the addition of 0.5 *M* acetic acid containing pepsin at a 1 : 50 (pepsin : cartilage) weight ratio at 4° for 48 hr. Solid NaCl is added to a final concentration of 0.9 *M*, and after overnight precipitation the resulting collagen is collected by centrifugation (20,000 *g* for 1 hr) and dissolved in acetic acid. The solution is dialyzed against 50 m*M* Tris–HCl buffer containing 0.2 *M* NaCl (Tris–NaCl) and passed through a DEAE column equilibrated with the same buffer. The effluent is collected and dialyzed against 0.2 *M* Na_2HPO_4, and the precipitated collagen is collected by centrifugation and dissolved in Tris–NaCl. An equal volume of 4.4 *M* NaCl is added to the mixture which is left at room temperature for 48 hr. The precipitated Type I collagen is removed by centrifugation. Type II collagen is precipitated by dialysis versus 0.1 *M* Na_2HPO_4, dissolved in 0.1 *M* acetic acid, lyophilized, and stored desiccated at −20°.

Induction of Disease

Lyophilized Type II collagen is cut into small pieces and dissolved at 2 mg/ml in cold 0.1 *M* acetic acid by rolling on an end-over-end rotator for 24 hr. The solution should appear clear and viscous prior to preparation of

[10] M. A. Cremer, J. M. Stuart, A. S. Townes, and A. H. Kang, *J. Immunol.* **124,** 2912 (1980).

the emulsion. Equal volumes of cold collagen solution and Freund's complete adjuvant (Ra37, Difco) are mixed vigorously and maintained at 4°. Mice are injected with 100 μl of freshly prepared cold emulsion intradermally at a single site at the base of the tail. Glass syringes and 26-gauge $\frac{1}{2}$-inch hypodermic needles are preferable for immunization, to ensure the accuracy of delivery. Particular attention should be paid to achieving an intradermal injection. The dose of Type II collagen is dependent on the mouse strain and the species source of collagen.[5] A boost injection is not necessary for CIA to develop; however, it may increase the disease incidence in some cases (e.g., DBA/1 Lac J mice immunized with bovine Type II collagen). Boost injections consist of Type II collagen emulsified in Freund's incomplete adjuvant and are administered intradermally at the base of the tail 7 days after the primary injection. Boost injections given intraperitoneally after 21 days delay the onset of disease in some mice.[7]

Assessment of Collagen-Induced Arthritis

Clinical Assessment. CIA appears in mice approximately 6 weeks after the initial immunization, depending on the strain of mice and the species of collagen. Early onset is observed in DBA/1 Lac J mice immunized with chick collagen (4–5 weeks), B10.RIII mice immunized with porcine collagen (4–5 weeks), or either strain immunized with deer collagen (3–4 weeks). Onset is most frequently characterized by erythema and edema in one limb. Simultaneous onset in two or more limbs occurs in about 10% of mice. Front and hind limbs are affected with equal frequency. Measurements with constant tension calipers show increases from the normal paw width of 2.1 mm to a range from 3.0 to 4.5 mm. Maximal joint swelling is most often observed in the affected metatarsal/metacarpal region. Individual limbs achieve peak swelling a mean of 6 days after onset. In the majority of mice, the disease progresses to involve two or more limbs.

After the inflammatory stage of the arthritis, a minority (5%) of the limbs show a spontaneous remission, resulting in a return to apparently normal paws. The majority of affected limbs progress to joint distortion, due to ulnar deviation and subluxation of the metacarpophalangeal joints or the metatarsalphalangeal joints. Many paws remain at this clinical stage; however, 20–35% progress to joint ankylosis, which is detected by manipulation of the limb. Ankylosis is easy to detect in the affected ankle, and with practice it may be felt in the wrists. "Sausage digits" are also seen in a number of paws. Knees are involved in CIA, but this is an infrequent occurrence and difficult to detect due to fur coverage and the minor detectable changes in knee width using caliper measurements.

Radiological and Histological Assessment. CIA in the mouse is difficult to assess by radiographic techniques, due to the size of the joints. Radiographs are obtained using dental X-ray cassettes, and portable X-ray generators are convenient for this purpose. However, only gross changes are seen on the films, unless a magnifying system is employed.

The severity of the arthritis is scored by histological assessment. The peripheral limbs are removed at the hair line, and the skin removed to the digits. After fixation in 10% formalin solution for a minimum of 24 hr, the joints are decalcified in Surgipath decal for 18 days. Limbs are processed through alcohol and xylene and embedded in paraffin with an orientation to allow a lateral section from the ankle or wrist to the digits. Serial sections of 4–8 μm are cut to provide histology from the tarsal and metatarsal joints of the rear paw and the carpal and metacarpal joints of the front paw. Assessment of arthritis is most accurate on approximately medial sections.

Histological assessment is also carried out on the knee joints. After the joint region is dissected free of skin, the knee is excised by cutting through the femur and tibia/fibula approximately $\frac{1}{2}$ inch from the knee. The bulk of the large attached muscles (superficial gluteus, vastus lateralis, gastrocnemius, tibialis) is removed close to the knee joint to aid fixation and decalcification. The joint is orientated in the embedding medium to cut lateral sections, including the patella,

Sections cut in the described manner are stained with hematoxylin and eosin. The following scoring system is used to assess the arthritis[3]:

0 Normal synovial membrane (2–3 cells thick), smooth articulating cartilage surfaces
1 Synovial membrane hypertrophy, cellular infiltrate into the synovial fluid
2 Grade 1 plus pannus formation, with superficial cartilage erosions
3 Grade 2 plus major erosion of the cartilage and subchondral bone
4 Loss of joint integrity through erosion, massive cellular infiltrate, presence of ankylosis

An example of mouse CIA histology is shown in Fig. 1.

Biochemical Assessment (see also this volume [40] and [44]). We have recently shown that the inflammatory response in CIA may be assessed by the use of enzyme markers for polymorphonuclear leukocytes (PMNs) and macrophages. The paws are dispersed into 5 ml of cold sodium phosphate buffer (50 m*M*, pH 6.0) using a Polytron homogenizer for 30 sec. One-half milliliter of 5% hexadecyltrimethylammonium bromide (Sigma) solution is added to the sample, and the preparation is vortexed. The supernatant is removed after centrifugation, and the resulting solution is assayed for

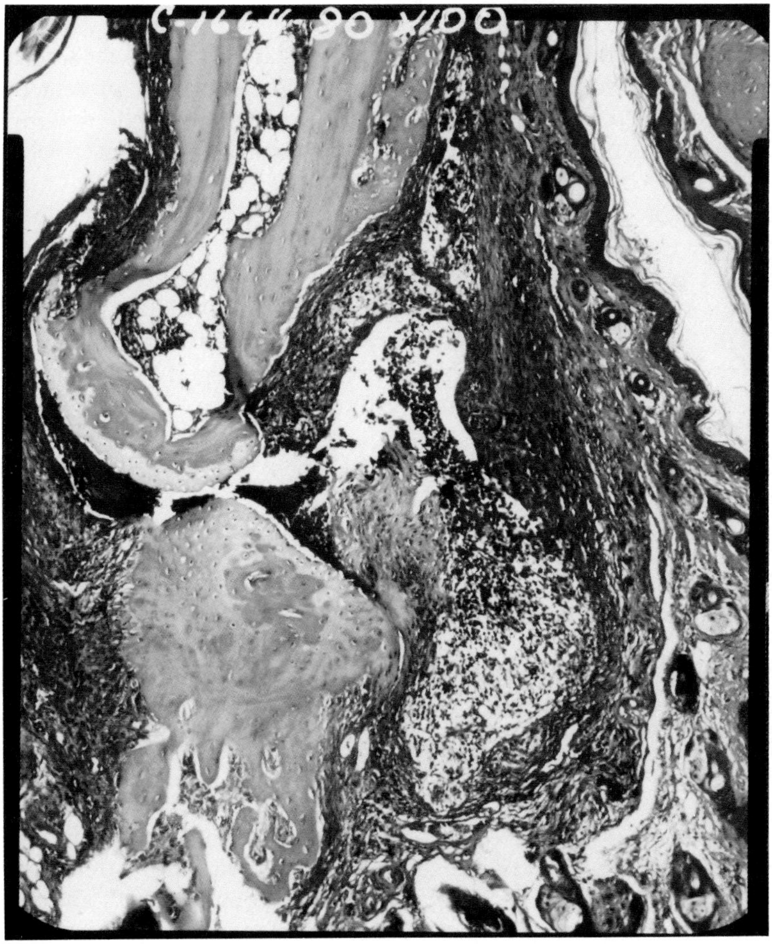

FIG. 1. Histological appearance of Type II collagen-induced arthritis. Synovial hypertrophy and pannus eroding both cartilage and subchondral bone are seen. Massive proliferation of mononuclear cells is observed within the synovial cavity. The appearance of the section indicates that joint subluxation has occurred, resulting in deformation of the articulating surfaces.

protein content[11] and the presence of myeloperoxidase (MPO)[12] and *N*-acetyl-β-D-glucosaminidase (NAG)[13] activity. Significant increases in the

[11] J. Pierce and C. H. Suelter, *Anal. Biochem.* **82,** 478 (1977).

[12] K. Suzuki, H. Ota, S. Sasagawa, T. Sakatani, and T. Fujikura, *Anal. Biochem.* **132,** 345 (1983).

[13] H. U. Schorlemmer, P. Davies, W. Hylton, M. Gugig, and A. C. Allison, *Br. J. Esp. Pathol.* **58,** 315 (1977).

levels of MPO, the enzyme marker for PMNs, are seen in affected paws during the inflammatory phase of the disease, with levels of MPO raised approximately 5-fold over levels in the normal paw. This technique has proved sensitive enough to detect the mild inflammation present during the passive transfer of disease. Less dramatic changes in the macrophage enzyme marker (NAG) are seen in the arthritic paw extract, in which NAG activity rises to double the levels seen in the normal paw.

The murine acute phase reactant, serum amyloid P (SAP), has also been studied in CIA. The SAP level is measured by electroimmunoassay[14] in sera taken from B10.RIII and B10.Q mice at weekly intervals during the induction of arthritis. A transient increase (up to 10-fold normal) in SAP level is detected prior to the appearance of clinically detectable disease, and this change in SAP may have either prediction properties or pathological significance for the development of arthritis.[15]

Immune Response to Type II Collagen

Antibody Response to Type II Collagen

High levels of circulating antibody to Type II collagen are found to be an invariable phenomenon in CIA. Several assays have been used to assess the anticollagen titer, the most frequently reported being the solid-phase radioimmunoassay (RIA) and the enzyme-linked immunoabsorption assay (ELISA). In both tests, Type II collagen is nonspecifically adsorbed on the surface of polystyrene wells or tubes using a 10–30 μg/ml solution in phosphate-buffered saline (PBS). The remaining protein binding sites on the plastic are blocked by the addition of 0.5% bovine serum albumin (BSA) solution in PBS. Test sera are added at appropriate dilutions in BSA/PBS and incubated overnight at room temperature. After washing 6 times with PBS, the bound immunoglobulin is detected with either radio-labeled anti-mouse immunoglobulin or *Staphylococcus* protein A in the RIA or similar reagents coupled to an enzyme marker in the ELISA.

Anti-Type II collagen antibody is proportional to the amount of radioactivity bound after washing (RIA) or the amount of color developed in the enzymatic reaction (ELISA). Actual antibody levels may be derived by regression analysis from a control curve of a known quantity of antibody in both the RIA and ELISA. The RIA may be modified to determine the isotype of the antibody response by using labeled isotype-specific reagents

[14] M. B. Pepys, M. Baltz, K. Gomer, A. J. S. Davies, and M. Doenhoff, *Nature (London)* **278,** 259 (1979).

[15] M. L. Bliven, P. H. Wooley, M. B. Pepys, and I. G. Otterness, *Arthritis Rheum.* **29,** 1131 (1986).

as the probe. Whereas a high antibody titer to Type II collagen antibody is a consistent finding in mice with CIA, high antibody titers are also detected in susceptible mice that failed to develop CIA and in CIA-resistant strains of mice.[1] There are reports of both IgG_1 and IgG_{2a} as the major isotype in the immune response.[16,17] It may be significant that both of these isotypes have the ability to fix complement, since the current view implies that the primary lesion in collagen-induced arthritis arises due to antibody-mediated damage to cartilage. However, the specificity of the antibody response is probably the most critical determinant in the development of arthritis.[5,6]

Cellular Response to Type II Collagen

The T-cell proliferative response to Type II collagen is less marked than most cell-mediated immune responses in mice. Responses may be detected *in vitro* in both spleen and lymph node cells cultured in the presence of collagen by [^{3}H]thymidine incorporation using standard cell culture techniques[18,19] (see also this series, Vol. 163 [54]). The peak dose–response to Type II collagen is approximately 50 μg/ml, and *in vitro* denatured collagen stimulates a higher response than the native molecule. No differences are seen between collagen-immunized mice with and without arthritis in either the strength or the specificity of the proliferative response.[18] The peak T-cell response occurs 2–3 weeks after immunization,[18] although responses are detected in arthritic mice over 7 weeks from the primary immunization, by the administration of booster immunizations with Type II collagen.[19]

The DTH response to collagen is detected by a radiometric ear assay.[20] Type II collagen at 1 mg/ml in 0.1 *M* acetic acid is dialyzed against PBS at 4° with two buffer changes. A collagen-immunized mouse is anesthetized, and 10 μl of the collagen solution is injected intradermally in the pinna of one ear using a 50-μl Hamilton syringe and a 30-gauge needle. A control injection of PBS alone is given in the opposite ear. To block the synthesis of native deoxyuridine, 200 μl of $10^{-7}M$ fluorodeoxyuridine (Sigma) is injected intraperitoneally, and after a 20-min interval 2 μCi of ^{125}I-labeled deoxyuridine (New England Nuclear) is injected intraperitoneally in

[16] P. H. Wooley, H. S. Luthra, A. Huse, and C. S. David, *Arthritis Rheum. Suppl.* **28,** s74 (1985).

[17] W. C. Watson and A. S. Townes, *J. Exp. Med.* **162,** 1878 (1985).

[18] J. M. Stuart, A. S. Townes, and A. H. Kang, *J. Clin. Invest.* **69,** 673 (1982).

[19] M. Dallman and C. G. Fathman, *J. Immunol.* **135,** 1113 (1985).

[20] P. H. Wooley, H. S. Luthra, W. P. Lafuse, A. Huse, J. M. Stuart, and C. S. David, *J. Immunol.* **134,** 2366 (1985).

200 μl of 0.8% NaCl. The mouse is sacrificed after 24–48 hr, and the ears are removed at the hair line. The time depends on the mouse strain and severity of the response. The radioactivity in the ear is measured by a gamma counter. The delayed-type hypersensitivity (DTH) reaction is reported as the ratio of cpm in the collagen-injected ear/cpm in the control ear. A typical time course of the immune response and appearance of arthritis is shown in Fig. 2. It is apparent that the peak DTH response (and the reported peak T-cell proliferative response) occurs substantially earlier than the arthritis, at around 10 days postimmunization, while the antibody level is maximal at the onset of disease.

Passive Transfer of Collagen-Induced Arthritis

The passive transfer of collagen arthritis in mice has been performed using polyclonal anti-type II collagen immunoglobulin[21,22] or monoclonal anti-type II collagen antibody.[23] The specificity of the immunoglobulin is important in the passive transfer, and polyclonal antisera are more successful in this technique than most monoclonal antibodies to Type II collagen. The transfer is accomplished by intravenous or intraarticular injection of the immunoglobulin. Serum pools are obtained from mice with recent onset arthritis, and a crude immunoglobulin fraction is prepared by precipitation with an equal volume of saturated ammonium sulfate solution. After resuspension and dialysis against PBS, this solution is used for the passive transfer or for the preparation of more highly purified immunoglobulin. Immunosorption of the immunoglobulin on Sepharose 4B bound to *Staphylococcus* protein A (StA) (Pharmacia) is utilized to prepare highly purified IgG. Small (5–10 ml) StA columns may be prepared by swelling the gel in PBS and pouring the desired volume into a plastic syringe. The immunoglobulin solution is passed over this column (10 ml/hr) at room temperature. The column is then washed with excess PBS until the eluate is protein free. The bound IgG is then eluted from the StA column by the passage of 0.6% acetic acid, which is immediately neutralized by the addition of 2 *M* Tris and dialyzed against PBS.

Antibody specific for Type II collagen may be obtained if Type II collagen is bound to cyanogen bromide-activated Sepharose 4B (CNBr) (Pharmacia). One gram of CNBr-Sepharose is swollen and washed in 200 ml of 1 m*M* HCl, and the acid is removed by filtration on a sintered

[21] J. M. Stuart and F. J. Dixon, *J. Exp. Med.* **158,** 378 (1983).

[22] P. H. Wooley, H. S. Luthra, C. J. Krco, J. M. Stuart, and C. S. David, *Arthritis Rheum.* **27,** 1010 (1984).

[23] J. M. Stuart and F. J. Dixon, *Arthritis Rheum. Suppl.* **26,** s14 (1983).

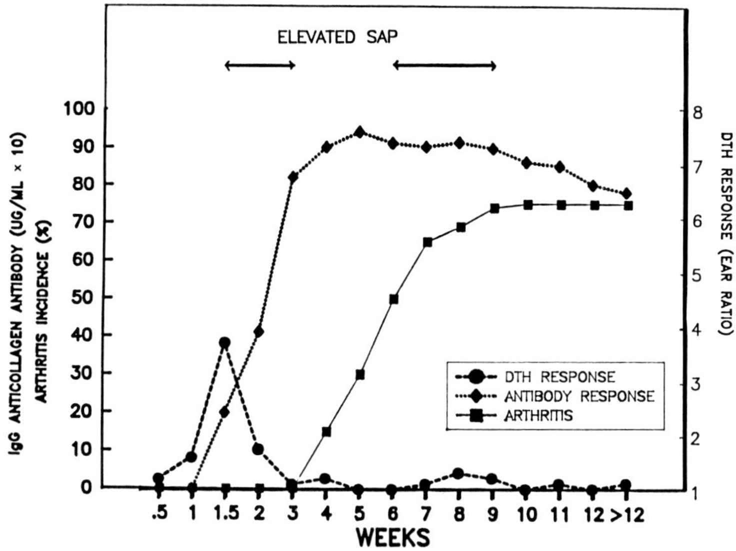

FIG. 2. The time course of the immunological features of Type II collagen-induced arthritis. An early, short-lived DTH reaction to collagen is observed, which is essentially absent by the time of disease onset. IgG anticollagen antibodies are present by day 10 postimmunization and continue to increase to a peak at 5 weeks, after which a slowly declining plateau phase is observed. Arthritis appeared around the period of peak anticollagen antibody levels and rose to maximum incidence after 8 weeks. Two periods of elevated serum SAP levels are observed: one shortly after immunization and the other corresponding to the development of clinical signs.

glass. The beads are left to react with 10 ml of Type II collagen at 1 mg/ml in 0.4 phosphate buffer (pH 7.6) at 4° overnight. Remaining reaction sites are blocked by the addition of 0.2 *M* glycine, and the beads are washed alternatively with 0.1 *M* acetate buffer (pH 4) and 0.1 *M* $NaHCO_3$ buffer (pH 8). Columns may be prepared as described above, and immunoglobulin of all isotypes specific for Type II collagen may be eluted from this immunosorbant as described above.

Mice are injected intravenously with immunoglobulin solutions in volumes of up to 1 ml via the tail vein. High concentrations of anticollagen antibody are required to transfer disease. The minimum required dose is 2 mg,[22] and the highest incidence is reported for mice given two injections on consecutive days.[21] If mice are injected intraarticularly, using a 50-μl Hamilton syringe and 30-gauge needle, 20 μl of a 600 μg/ml anticollagen antibody solution is sufficient to induce arthritis.[24] The clinical appearance

[24] P. H. Wooley, H. S. Luthra, S. Singh, A. Huse, J. M. Stuart, and C. S. David, *Mayo Clin. Proc.* **59,** 737 (1984).

of the transferred arthritis resembles a mild inflammatory stage of CIA. This usually appears 48 – 72 hr after immunoglobulin injection, never progresses to a chronic arthritis, and resolved within 48 hr. On histological examination, synovial hypertrophy is the most frequently observed change, frequently accompanied by early pannus formation. The deposition of IgG and C3 on the articular surface of the recipient mouse joints may be demonstrated by immunofluorescence,[21] and the intravenous injection of radiolabeled anticollagen antibodies results in high uptake in the joints of transferred mice.[22,24]

CIA as a Model for Experimental Immunotherapy and Pharmaceutical Research

Murine CIA has a number of immunological and pathological features in common with rheumatoid arthritis and, therefore, has been used for both experimental immunotherapy and pharmacological treatment of autoimmune disease. If preventative protocols are employed, significant decreases in the arthritis incidence and/or significant delays in the onset of disease are the most pertinent parameters of study. In curative protocols, mice are usually divided into treatment or control groups at disease onset. Then changes in the progression of the disease (joint count and clinical score) and histological index are monitored. If immunomodulatory therapies are investigated, the humoral and cell-mediated responses to Type II collagen are also followed.

Immunotherapy

Antibodies to Ia antigens (available from the Transplantation Immunology Section, NIH, Bethesda, MD) are administered to B10.Q and B10.RIII mice prior to the injection of Type II collagen. Milligram quantities of immunoglobulin per mouse are used for this experimental intervention, and the procedure is similar to that described for the suppression of experimental autoimmune encephalomyelitis,[25] namely intraperitoneal injection of anti-Ia antibody on 3 consecutive days, before, at, and after collagen immunization. Significant decreases in the resulting incidence of CIA are seen if polyclonal anti-Ia reagents are used. However, if monoclonal anti-Ia antibodies are utilized, protection depends on the fine specificity of the antibody. Only one in four monoclonal antibodies which reacted with the *I-A* subregion of *H-2^q^* is capable of suppressing CIA in B10.Q

[25] L. Steinman, J. T. Rosenbaum, S. Sriram, and H. O. McDevitt, *Proc. Natl. Acad. Sci. U.S.A.* **78,** 7111 (1981).

mice, while only monoclonals reacting with epitope I of the Ia.7 molecule are effective in suppressing arthritis in B10.RIII mice. The decrease in disease incidence is also accompanied by a fall in the level of early IgG antibody to Type II collagen, but the fall did not affect the final maximum titer. The peak DTH response to Type II collagen is also unaffected by the anti-Ia treatment.[20]

Monoclonal antibody to L3T4 (available from Becton Dickinson), the phenotypic marker for murine T helper cells, is also effective in preventing the onset of CIA. Injection of 100 μg of antibody the day prior to and the day after Type II collagen immunization, followed by similar doses at 2-week intervals, results in a significant decrease in the incidence of CIA. This protocol is also shown to deplete L3T4$^+$ T cells from the spleen and lymph nodes and to cause a significant decrease in the antibody production against Type II collagen.[26]

Acknowledgments

The author wishes to thank Dr. Joan Chapdelaine, Janey Whalen, and Marcia Bliven for assistance in the preparation of the manuscript.

[26] G. E. Ranges, S. Sriram, and S. M. Cooper, *J. Exp. Med.* **162,** 1105 (1985).

[34] Streptococcal Cell Wall-Induced Arthritis in Rats

By SUSAN C. RIDGE, JOHN B. ZABRISKIE, ARNOLD L. ORONSKY, and S. S. KERWAR

Introduction

Intraperitoneal administration of sonicated cell walls prepared from group A streptococci induces an acute inflammation in the hind paws of inbred female Lewis rats.[1] This acute inflammation, which occurs within 72 hr, resolves in a few days and is followed by a chronic erosive arthritis that persists for several weeks. Histologic and radiologic analysis of the hind paws has shown that the lesion is, in many respects, similar to human rheumatoid arthritis.[1,2] Other features of this model that are similar to

[1] W. J. Cromartie, J. G. Craddock, J. H. Schwab, S. K. Anderle, and V. H. Yang, *J. Exp. Med.* **146,** 1585 (1977).

[2] R. L. Clark, J. T. Cuttino, S. K. Anderle, W. J. Cromartie, and J. H. Schwab, *Arthritis Rheum.* **22,** 25 (1979).

rheumatoid arthritis are complement dependence,[3] endocrine and sex factors,[4,5] genetic linkage,[6] interleukin-2 deficiency,[7] and the remittive and relapsing nature of the lesion.[1]

Mechanistic studies with this cell wall model have shown that the chronic arthritis is a T-cell-dependent lesion.[8,9] Thus athymic inbred Lewis rats *(rnu/rnu)* develop the acute inflammatory lesion when they are treated with cell walls. However these athymic rats do not develop the chronic lesion unless they are treated with spleen cells derived from euthymic (*rnu*/+) littermates. These observations indicate that the acute arthritis induced by cell walls is a T-cell-independent lesion and the chronic disease is T-cell dependent. Since the acute lesion is complement dependent,[3] it has been proposed[8] that cell walls activate the alternate pathway of the complement system,[10] *in situ,* allowing for the generation of complement peptides, C3a and C5a. These complement components can increase vascular permeability (C3a) and induce chemotaxis (C5a) of inflammatory cells to sites of complement activation. Enzymes released by these inflammatory cells (collagenase, elastase, neutral proteases) can degrade the connective tissue matrix of the cartilage.[11]

Cell walls prepared from group A, B, and C streptococci can induce arthritis in rats.[1] A property shared by these cell walls is the resistance of the cell wall peptidoglycan to digestion by lysozyme.[12] Resistance to lysozyme degradation allows for the cell walls to persist in tissues for extended periods of time.[1,6,13] Molecular weight analysis of the peptidoglycan has shown that an average molecular weight of 5×10^6 or greater is required for disease induction. Peptidoglycans of molecular weight of 5×10^6 or lower fail to elicit the disease. Recently, it has been shown that if rats

[3] J. H. Schwab, J. B. Allen, S. K. Anderle, F. Dalldorf, R. Eisenberg, and W. J. Cromartie, *Immunology* **46,** 83 (1982).

[4] J. B. Allen, D. Blatter, G. B. Calandra, and R. L. Wilder, *Arthritis Rheum.* **26,** 560 (1983).

[5] R. L. Wilder, G. B. Calandra, A. J. Garvin, K. D. Wright, and C. T. Hansen, *Arthritis Rheum.* **25,** 1064 (1982).

[6] R. L. Wilder, J. B. Allen, L. M. Wahl, G. B. Calandra, and S. M. Wahl, *Arthritis Rheum.* **26,** 1442 (1983).

[7] S. C. Ridge, J. Zabriskie, T. Diamantstein, A. L. Oronsky, and S. S. Kerwar, unpublished (1986).

[8] S. C. Ridge, J. Zabriskie, A. L. Oronsky, and S. S. Kerwar, *Cell. Immunol.* **96,** 231 (1985).

[9] J. B. Allen, D. G. Malone, S. M. Wahl, G. B. Calandra, and R. L. Wilder, *J. Clin. Invest.* **26,** 1042 (1985).

[10] J. Greenblatt, R. J. Boackle, and J. H. Schwab, *Infect. Immun.* **19,** 296 (1978).

[11] E. D. Harris, Jr., *in* "Joints and Synovial Fluid" (L. Sokoloff, ed.), Vol. 1, p. 243. Academic Press, New York, 1978.

[12] R. L. Wilder, *Ann. Intern. Med.* **101,** 820 (1984).

[13] R. Eisenberg, A. Fox, J. J. Greenblatt, S. K. Anderle, W. J. Cromartie, and J. H. Schwab, *Infect. Immun.* **38,** 127 (1982).

treated with cell walls are administered mutanolysin (a murimidase preparation isolated from *Streptomyces globisporus*), they fail to develop the chronic lesion.[14] It has been proposed that mutanolysin cleaves the cell wall peptidoglycan to fragments of a molecular weight less than 5×10^6.

Methods

Group A streptococci, type 3, strain D58, or type 12 (American Type Culture Collection, Rockville, MD), are grown in Todd Hewitt broth (BBL Microbiology Systems, Cockeysville, MD) and harvested in late log phase. The cells are washed in phosphate-buffered saline, and a 10% cell suspension (wet weight/volume) is prepared in phosphate-buffered saline. To the cell suspension a few drops of 1-octanol are added. The cell suspension is mixed with an equal amount (volume) of dry glass beads (0.1 mm diameter, B. Braun, San Mateo, CA), and 100 ml of this cell–glass bead suspension is homogenized in a Bead Beater (Biospecific Products, P.O. Box 772, Bartlesville, OK). The container with the cell suspension is surrounded by ice. The cell suspension is homogenized for a total period of 45 min. Since the cell suspension warms up, homogenization is conducted in 5-min intervals with a 10-min interval of cooling. Cell disruption is monitored by using gram staining[15] of the suspension. If cocci are present, homogenization is continued until all of the cocci are disrupted.

When cell disruption is complete, the homogenate is filtered through a glass filter and the glass beads are washed on the filter with cold phosphate-buffered saline. The filtrate and the wash are mixed, and the material is centrifuged at 10,000 *g* for 20 min. The pelleted cell walls are washed once with 3 volumes of cold phosphate-buffered saline, centrifuged at 10,000 *g* for 20 min, and the pellet lyophilized.

The lyophilized cell walls are suspended in cold phosphate-buffered saline at a concentration of 5 mg/ml. Ribonuclease (Sigma Chemical Co., St. Louis, MO) is added to a concentration of 0.25 mg/ml, and the mixture is incubated at 37° for 4 hr. The cell walls are pelleted by centrifugation at 10,000 *g* for 20 min. The pellet is washed once with cold phosphate-buffered saline. After centrifugation, the cell walls are suspended to their original volume (see above) in phosphate-buffered saline. Trypsin is added to a final concentration of 0.25 mg/ml. The cell wall suspension is incubated at 37° for 4 hr. The cell walls are pelleted by centrifugation at

[14] M. J. Janusz, C. Chetty, R. A. Eisenberg, W. J. Cromartie, and J. H. Schwab, *J. Exp. Med.* **160,** 1360 (1984).

[15] P. Gerhard, *in* "Manual of Methods for General Bacteriology," p. 26. Am. Soc. Microbiol., Washington, D.C., 1982.

10,000 *g* for 20 min, washed in cold phosphate-buffered saline, and the pellet is lyophilized again.

To prepare sonicated cell walls, the lyophilized material is suspended in cold phosphate-buffered saline at a concentration of 2% (weight/volume). The suspension is sonicated intermittently using a Braun Sonifier (Braun Sonic 2000) for a total time of 60 min. The beaker containing the cell wall suspension is immersed in a dry ice–acetone bath. Sonication is stopped when the temperature of the suspension rises to about 15°.

After sonication, the material is centrifuged at 10,000 *g* for 10 min. The supernatant is filtered through a Nalgene filter (0.45 μm), and the sonicate (filtrate) is assayed for its rhamnose content.[16] The pellet is resuspended in phosphate-buffered saline and is sonicated again for 60 min, as detailed above. After centrifugation at 10,000 *g* for 10 min, it is filtered as detailed above and assayed for rhamnose.

Induction and Assessment of Arthritis

Inbred female Lewis rats (100–120 g, Charles River Breeding Labs., Wilmington, MA) are used. Rats are injected intraperitoneally with sonicated cell walls (Nalgene filtrate) at a rhamnose equivalent of 6 mg/100 g rat. Hind paw diameters (both paws) are measured around the ankle joint with a vernier caliper.

Hind paw inflammation (acute lesion) appears within 72 hr. Paw diameters increase from approximately 6.5 mm to about 8–9 mm. This acute inflammation resolves within 7 days, and chronic inflammation appears between days 10 and 15. This chronic inflammation (hind paw diameters 8–9 mm) is remittive and relapsing and persists for at least 120 days.

Histologic and Radiologic Analysis

Rats are sacrificed at various periods of time, and their hind paws are subjected to histological and radiological analysis. For histological analysis, the paws are fixed in buffered formalin and decalcified in formic acid. After embedding in paraffin, sections are prepared and stained with hematoxylin and eosin or Geimsa. For radiological analysis, the hind paws are placed on a Polaroid Type 55 film and subjected to X-ray exposure in a Hewlett-Packard radiographic apparatus (Faxitron, Hewlett-Packard, McMinville, OR). The focus-to-film distance is 45 cm, and the exposure is for 5 min at 60 kV. Each negative radiograph is graded for several parameters of joint destruction.[2]

Representative radiographs of the hind paws are shown in Fig. 1. Figure

[16] A. Dische and L. B. Shettles, *J. Biol. Chem.* **175,** 595 (1948).

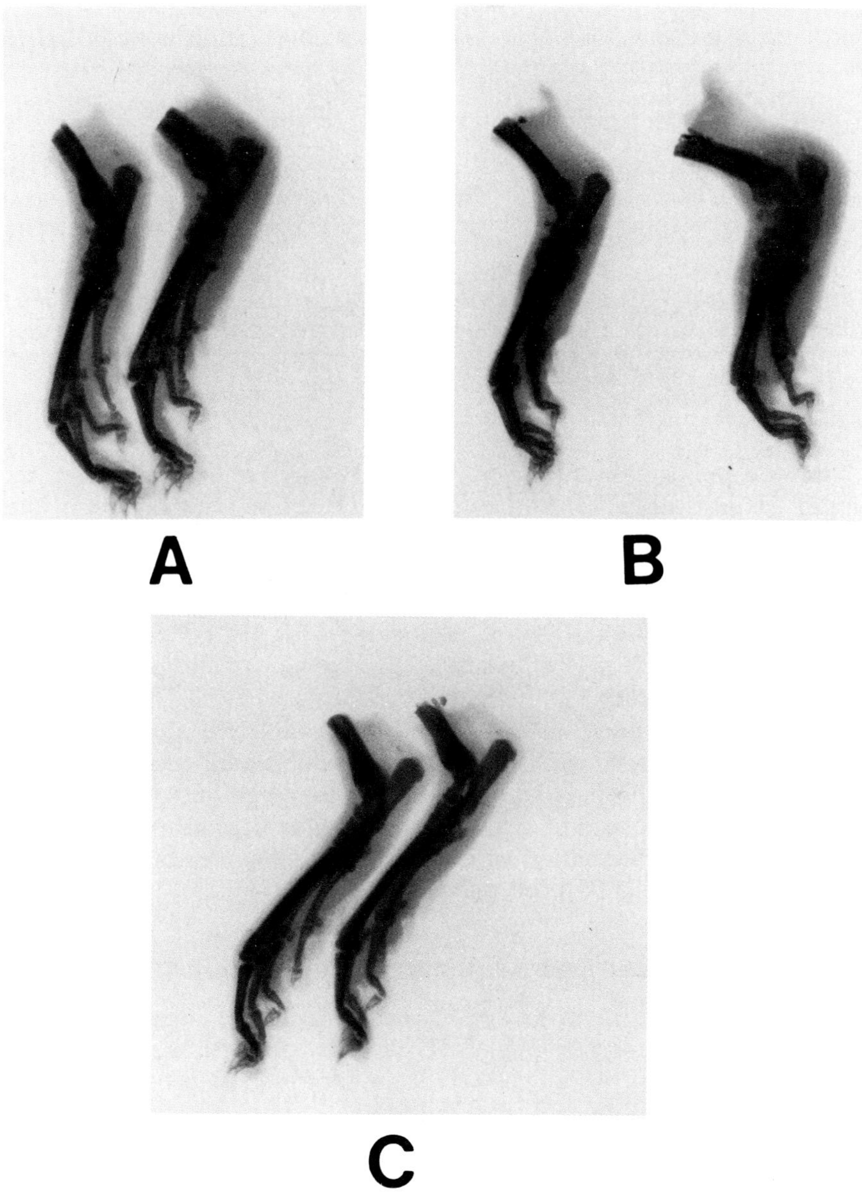

FIG. 1. Radiographs of hind paws of rats with cell wall arthritis. (A) Radiograph prepared on day 21 post cell wall treatment, (B) radiograph prepared on day 35 post cell wall treatment, and (C) radiograph of normal rat.

TABLE I
EVALUATION OF RADIOGRAPHS OF THE HIND PAWS OF RATS WITH CELL WALL ARTHRITIS[a]

Condition	Radiographic grade				
	Bone demineralization	Erosions	Periostitis	Cartilage space narrowing	Distal phalanges
Cell wall arthritis					
Day 22 (Fig. 1A)	2.0	2.0	0	0.5	2.0
Day 35 (Fig. 1B)	3.0	4.0	2.0	3.5	2.0
Normal (Fig. 1C)	0	0	0	0	0

[a] See Fig. 1. Grading: 0 = normal, 4 = maximum destruction.

1A shows a radiograph of the hind paws of cell wall arthritic rats X-rayed on day 21 post cell wall treatment. Figure 1B shows a radiograph of the hind paws of cell wall arthritic rats X-rayed on day 35 post cell wall treatment. In comparing these radiographs, it is apparent that joint destruction is greater on day 35. For comparative purposes, a radiograph of the hind paws of normal rats is also shown (Fig. 1C). Radiographs of arthritic rats also show the extent of soft tissue inflammation that is observed in this disease.

Table I shows an evaluation of these radiographs. In our experience, rats at 35 days post cell wall administration exhibit joint destruction that can be assessed by histological and radiographic techniques. Thus, if one wishes to assess the effects of clinically used nonsteroidal antiinflammatory agents or antirheumatic drugs with this model, it is preferable to conduct assessments on day 35 post cell wall treatment.[17]

Relevance of the Cell Wall Model to the Human Lesion

Rheumatoid arthritis is a complicated immunologic disease,[18] and the lesion induced in rats by streptococcal cell walls is, at best, merely a model of the human lesion. Although several similarities exist between the cell wall model and the human disease, additional studies are required before

[17] S. C. Ridge, J. J. Gibbons, Jr., J. Lucas, J. Zabriskie, A. L. Oronsky, and S. S. Kerwar, *in* "Biologically Based Immunomodulators in the Therapy of Rheumatic Diseases" (S. Pincus, D. S. Pisetsky, and L. J. Rosenwasser, eds.), p. 97. Elsevier, New York, 1987.

[18] N. J. Zvaifler, *in* "Rheumatoid Arthritis, Etiology, Diagnosis, Management" (P. Utsinger, N. J. Zvaifler, and G. E. Ehrlich, eds.), p. 151. Lippincott, Philadelphia, Pennsylvania, 1985.

one can conclude that this model is an exact counterpart of rheumatoid arthritis.

Caution

Streptococci can induce infections of the throat, and it is possible that the organisms can induce other lesions. Therefore, care should be taken when growing and processing streptococci cells and cell walls.

[35] Chronic Arthritis in Rabbits

By HUGO E. JASIN

Introduction

Advances in our knowledge of the pathogenic mechanisms operative in rheumatoid arthritis have been hampered by the unavailability of spontaneous or induced animal models of arthritis that closely resemble this chronic inflammatory disease. Acquisition of experimental data pertinent to the human disease is made more difficult by the rather sketchy and incomplete knowledge available of the pathophysiology of this disease. Moreover, the pathologic anatomy of connective tissues affected with chronic inflammatory reactions tends to show a similar appearance regardless of the mechanisms underlying the model under study. These considerations suggest that the transfer of knowledge from the animal to the human counterpart should be made with full awareness of the limitations inherent to the experimental conditions associated with the collection of data. Moreover, generalizations derived from work on experimental models can be meaningful only if they are actually or potentially ascertainable from studies of the human disease.

The first attempts to produce an experimental arthritis by means of intraarticular injection of antigen date from over 70 years ago,[1] but this model attracted the attention of investigators only after Dumonde and Glynn[2] described a chronic inflammatory arthritis in rabbits sensitized to autologous or heterologous fibrin and injected intraarticularly with the insoluble antigen. The features that made this model particularly suitable for the study of pathogenic mechanisms in rheumatoid arthritis were its

[1] D. L. Gardner, *Ann. Rheum. Dis.* **19,** 297 (1960).
[2] D. C. Dumonde and L. E. Glynn, *Br. J. Exp. Pathol.* **43,** 373 (1962).

chronicity, the striking histologic similarity with its human counterpart, and the fact that it was immunologically induced by well-defined antigens. Although the original model was produced by intraarticular injection of fibrin, an insoluble antigen, soluble antigens such as ovalbumin,[3,4] bovine serum albumin,[4] ferritin,[5,6] and horseradish peroxidase[7-9] injected in rabbits previously immunized to the same antigens in Freund's complete adjuvant also induced a chronic arthritis lasting in some cases over 6 months. A rabbit model closely resembling antigen-induced arthritis produced by injection of viable *Mycoplasma arthritidis* or *Mycoplasma pulmonis* may involve immune mechanisms similar to those operative in the model produced by protein antigens.[10]

Methods

Rabbits of either sex weighing more than 2 kg are suitable for the induction of arthritis. The immunization procedure is designed to generate a maximal or near maximal immune response to the antigen used. Antigens are usually dissolved in 0.15 M NaCl solution in concentrations ranging from 5 to 20 mg/ml. After micropore sterilization, the solutions are emulsified with equal volumes of Freund's complete adjuvant containing 2 mg/ml heat-killed *Mycobacterium tuberculosis* (Difco Laboratories, Detroit, MI). The animals are injected with 2-ml volumes of the emulsion as follows: 0.05 ml is injected in each footpad, 0.5 ml in each gluteus muscle, and the remaining 0.8 ml given intradermally into 4–6 sites in the nape of the neck. Although controlled studies are not available, it is our impression that the yield and severity of chronic arthritis are increased when the animals are reinjected in the same fashion 2–3 weeks following the primary immunization. Alternatively, some investigators[2,11] select the rabbits with hyperreactivity to the immunizing antigen by assessment of the inflammatory reaction developing 4, 24, and 48 hr after intradermal injections of 10–100 μg of antigen. We have found a good correlation

[3] R. Consden, A. Doble, L. E. Glynn, and A. P. Nind, *Ann. Rheum. Dis.* **30,** 307 (1971).

[4] T. D. Cooke and H. E. Jasin, *Arthritis Rheum.* **15,** 327 (1972).

[5] H. Tateishi, H. E. Jasin, and M. Ziff, *Arthritis Rheum.* **16,** 133 (1973).

[6] O. Ohno, H. Tateishi, and T. D. Cooke, *Arthritis Rheum.* **21,** 81 (1978).

[7] A. Doble, J. Dorling, L. E. Glynn, J. A. Webb, and J. H. Wilcox, *Br. J. Exp. Pathol.* **54,** 646 (1973).

[8] R. C. Graham and S. L. Shannon, *Am. J. Pathol.* **67,** 69 (1972).

[9] P. Hasselbacher, J. L. Nacht, D. A. Labosky, and M. E. Steinberg, *J. Rheumatol.* **7,** 596 (1980).

[10] L. R. Washburn, B. C. Cole, and J. R. Ward, *Arthritis Rheum.* **23,** 837 (1980).

[11] M. E. Steinberg, C. R. McRae, L. D. Cohen, and H. R. Schumacher, *Clin. Orthop. Relat. Res.* **97,** 248 (1973).

between the serum antibody levels of each individual animal and the degree of severity of the chronic synovitis.[4]

Arthritis is induced by intraarticular injections of antigen in sterile isotonic saline solution given 2–3 weeks after the immunization procedure. It is difficult to decide, on the basis of the available literature, on the optimal doses of antigen capable of inducing maximal inflammatory responses. However, most investigators have found that antigen amounts ranging from 1 to 20 mg/joint appear to yield significant degrees of chronic synovitis. The choice of antigen depends on the purpose of the study to be carried out. Horseradish peroxidase[7–9] or similar enzyme antigens, for instance, offer a convenient and sensitive method for the detection of antigen and specific antibodies. Ferritin[5,6] has been used for studies involving localization of this iron-laden molecule by electron microscopy techniques.

In order to minimize trauma to the injected joints it is desirable to anesthetize the animals with small doses of intravenous sodium pentobarbital or equivalent anesthetic agents. The knees are carefully shaved and swabbed with iodine solution or Merthiolate. The knee is flexed to 120° preferably by an assistant who simultaneously exerts moderate pressure on the caudal segment of the limb in an attempt to separate the distal ends of femur and tibia. A $\frac{1}{2}$-inch 26-gauge needle is inserted at a point equidistant between the inferior border of the patella and the edge of the tibial plateau, parallel to the sagital plane of the limb and bisecting the angle formed by femur and tibia. The joint cavity is usually reached by inserting the needle 5–10 mm from the skin surface depending on the size of the animal to be injected. Volumes ranging from 0.5 to 2.0 ml can be injected without difficulty. In most cases, proper needle placement can be ascertained by palpation of the suprapatellar pouch engorgement as the antigen solution fills the joint cavity.

Pathology[2,11,12]

Local injection of large amounts of antigen in animals with high titers of precipitating antibodies give rise to acute Arthus reactions, and the joint cavity is no exception.[13] From 1 to 4 hr following intraarticular challenge, acute inflammation associated with thick, purulent exudate, severe edema, and hemorrhagic synovitis develops. Microscopic examination shows diffuse polymorphonuclear heterophile exudate, capillary congestion and thrombosis, and severe edema. This acute reaction reaches its peak be-

[12] E. P. Gall and E. A. Gall, *J. Rheumatol.* **7,** 13 (1980).

[13] C. V. DeShazo, P. Henson, and C. G. Cochrane, *J. Clin Invest.* **51,** 50 (1972).

tween 24 and 48 hr after antigen injection. The acute inflammatory response subsides slowly over a period of 7–10 days, being replaced by chronic inflammatory changes. At the end of the first week, a visible joint exudate is present only in a small proportion of animals. At this time, microscopic examination of the synovium shows vasodilation interspersed with organizing thrombii in medium and small size vessels. There is abundant fibrinous exudate on the synovial surface which shows moderate cell infiltration and early evidence of collagen deposition. The synovial lining cell layer shows hyperplastic and hypertrophic changes reaching a thickness of 4 to 8 cells. In the subsynovial tissues, the heterophile infiltration is largely replaced by mononuclear cells, predominantly macrophages.

By 2–4 weeks, the synovium appears thickened and slightly congested. On gross inspection, a small proportion of injected animals may show pannus tissue invading articular cartilage which may be seen more frequently in animals injected with insoluble antigens (e.g., fibrin).[2] At this time, histologic examination shows a typical picture of chronic inflammation. The synovial lining cell layer is uniformly hypertrophic. The subsynovium shows areas of diffuse lymphocytic and plasma cell infiltration with a tendency to localize around small blood vessels. In scattered areas, the round cell infiltrates are reminiscent of the primary lymphoid follicles found in lymph nodes. The articular cartilage also shows evidence of progressive damage. Early in the development of the chronic arthritis, ultrastructural studies show evidence of chondrocyte death in the superficial layers. There is progressive increase in fibrillation and degenerative changes of the collagen fibers. Histochemical stains reveal profound depletion of acidic proteoglycan with consequent decrease in cartilage thickness.

The gross and histologic picture described above may vary considerably from animal to animal. In general, the severity of the acute Arthus reaction tends to be reproducible. However, the degree of chronic progressive synovitis may range from complete recovery, 6–8 weeks after intraarticular challenge with antigen, to severe irreversible damage with complete resorption of articular cartilage and development of secondary osteoarthritis. It is for this reason that this experimental model has been used only infrequently in pharmacologic studies involving antiinflammatory or immunosuppressive agents.

Immunopathology

Antigen-induced arthritis is a particularly attractive model for the study of chronic inflammation induced by immunologic mechanisms. One of the hallmarks of the inflammatory cellular infiltrate in rheumatoid synovitis is the presence of large numbers of plasma cells that synthesize immunoglobulins[14,15] in amounts comparable to those measured in spleen or lymph nodes. Thus, it has been suggested that the inflamed synovial tissue behaves as an autonomous lymphoid organ. Similarly, in antigen-induced arthritis, the inflamed synovial tissues secrete immunoglobulins to the same extent as the central lymphoid organs of the same animals.[4] Moreover, 30–40% of these immunoglobulins represent specific antibody to the antigen injected up to 8 weeks previously, indicating that the sustained local antibody response may be driven by antigen persisting locally. Indeed, when the fate of radioactive antigen injected intraarticularly is followed using an external probe, appreciable amounts are retained locally only in animals previously immunized to the same antigen. In addition, retained antigen is released slowly with a half-life of 20 days or more.[4,16] More accurate estimation of the retained antigen can be obtained by quantitation of the radioactivity present in tissues dissected from arthritic and control joints. In these type of experiments, from 3 to 6 times as much antigen is retained in arthritic joints as compared to uninflammed control joints 6 weeks after intraarticular injection.[4]

An important feature of this model is the localization of the retained antigen mainly in relatively avascular intraarticular collagenous tissues such as articular cartilage, menisci, and ligaments. Identification of the retained antigen by immunofluorescence and autoradiography indicates that a major portion is retained in the form of complement-fixing immune complexes in the superficial areas of the intraarticular tissues, close to the joint cavity.[17] It is likely that these immune complexes contain antibody in excess since it can be shown that tissues obtained from arthritic joints specifically bind additional antigen when incubated *in vitro* with solutions containing the homologous antigen. Thus, this experimental model is particularly suitable for studies on mechanisms of trapping of immune complexes in connective tissue,[16,18] their possible pathogenic role in chronic inflammation,[19] and the processes that influence their retention and release from these tissues.

Pathogenesis

The immune mechanisms involved in the development of a protracted chronic inflammatory synovitis in this model are not clearly understood. It

[14] J. D. Smiley, C. Sachs, and M. Ziff, *J. Clin Invest.* **47,** 624 (1968).

[15] A. S. Sliwinsky and N. J. Zvaifler, *J. Lab. Clin. Invest.* **76,** 304 (1970).

[16] J. R. Hollister and M. Mannik, *Clin. Exp. Immunol.* **16,** 615 (1974).

[17] T. D. Cooke, E. R. Hurd, M. Ziff, and H. E. Jasin, *J. Exp. Med.* **135,** 323 (1972).

[18] H. E. Jasin, *Clin. Exp. Immunol.* **22,** 473 (1975).

[19] H. E. Jasin and T. D. Cook, *Clin. Exp. Immunol.* **33,** 416 (1978).

is likely that the magnitude of the immune response following immunization with Freund's complete adjuvant may determine the severity and chronicity of the arthritis. However, long-term maintenance of the inflammatory reaction may depend on immunologic mechanisms other than the immune response to the original antigen because the synovitis is occasionally observed long after the disappearance of detectable antigen within the injected joint. Therefore, studies addressing the role of the initial antigen and the local immune responses should be conducted within 6–8 weeks after intraarticular injection, the interval in which the specificity of the local antibody response to the injected antigen has been clearly documented.[4]

The findings discussed in the previous section indicate that long-term retention of antigen in avascular collagenous structures depends on its trapping in the form of insoluble immune complexes maintained in antibody excess by the high antibody concentration found in the joint cavity. It is likely that such an environment is provided in part by the active local synthesis of antibody. This feature may be of paramount importance for the retention of antigen because both antigen and antibody are free to diffuse out of the tissue when strong chaotropic solutions or antigen in excess are used to solubilize the sequestered immune complexes,[17] suggesting that they are retained within the collagen fiber meshwork mainly on the basis of their size and poor solubility. It should be emphasized that the above discussion on sequestered antigens pertains only to such molecules that are irreversibly trapped within the tissues. Depending on the experimental design used, a variable proportion of the radioactive antigen present in connective tissue is free to diffuse out. Therefore, studies dealing with trapped antigens in tissues should be conducted only after exhaustive washing for several days with large volumes of solutions containing a mixture of tissue protease inhibitors to minimize proteolytic breakdown of the sequestered immune complexes.

Evaluation[4,12]

The wide range in the severity of the synovitis observed between individual animals suggests that any experimental design necessitating quantitation of the inflammatory reaction should include at least four rabbits per group and blind reading of histology sections by at least two observers.

Quantitation of the cellular exudate within the joint cavity provides a simple and objective method of assessing the magnitude of the inflammatory response. As mentioned above, soon after the initial acute Arthus reaction abates, the synovial cavity may contain very small amounts of synovial fluid, precluding direct cell quantitation. Accurate measurements can be achieved, however, by washing the suprapatellar pouches with 1 ml of normal saline solution containing 1 unit/ml sodium heparin. The joint should be dissected by sectioning the infrapatellar fat pad and releasing the capsule from either side of the patella, taking care not to pierce the synovial membrane which provides the lateral "walls" of the pouch. In this manner it is possible to flush the intact pouch several times and still recover most of the fluid injected. Cell counts can be made without further dilution using a hemocytometer. Typical cell counts 1–2 weeks after antigenic challenge may range from 2 to 10×10^4 cells per joint.

Synovium specimens for histology studies can be obtained from the lateral aspects of the suprapatellar pouch, the infrapatellar fat pad, and the posterior compartment. The latter is easily reached after sectioning the cruciate ligaments. It is important to obtain histology sections from several areas of each joint because the degree of synovitis varies not only between animals but also between different areas of the same joint. Quantitative studies should include assessment of (1) lining cell hyperplasia, (2) mononuclear cell infiltration of the subsynovium, (3) degree of polymorphonuclear heterophile infiltration, and (4) presence of invasive pannus and erosive changes of articular cartilage. The latter should combine naked-eye examination of the affected joints with histology studies of articular cartilage specimens selected from areas showing erosive changes and/or pannus invasion. If the studies deal with healing and repair mechanisms, it may be desirable to include scores for fibroblast infiltration and collagen deposition.

[36] Murine Lupus

By SYAMAL K. DATTA

Introduction

Systemic lupus erythematosus (SLE)[1] is a complex syndrome of autoimmune disorders. Genetic, immunologic, virologic, and environmental factors have all been implicated in SLE, but there is no understanding of its

[1] Abbreviations: SLE, systemic lupus erythematosus; NZB, New Zealand Black; NZW, New Zealand White; PBS, phosphate-buffered saline; FCS, fetal calf serum; goat anti-Rauscher gp70, goat antiserum to the envelope glycoprotein antigen of the Rauscher murine leukemia virus; dsDNA, double-stranded DNA; ssDNA, single-stranded DNA; zDNA, DNA in the left-handed helical configuration; poly(I), polyinosinic acid; poly(dT), polydeoxythymidine; BUN, blood urea nitrogen; PAS, periodic acid–Schiff; Ig, immunoglobulin;

fundamental mechanism.[1a,2] It is not clear if some of the factors implicated in SLE are the cause or consequence of the disease or are merely associated with it. Because of this complexity, genetically uniform inbred mouse strains that spontaneously develop a disease similar to human SLE are essential tools to define the etiology and pathogenesis of the disease.[3,4]

Mouse Strains Used in Lupus Research

The New Zealand strains of mice, notably New Zealand Black (NZB) mice and the (NZB × NZW)F_1 hybrids that are produced by mating NZB with New Zealand white (NZW) mice, were the first described models of spontaneous SLE.[5,6] When NZB were crossed with NZW mice, the hemolytic disease pattern characteristic of NZB mice changed to one of severe glomerulonephritis in the F_1 hybrids. A similar shift in disease pattern occurs when the NZB mice are crossed with a normal mouse strain called SWR.[7,8] Recently the MRL and BXSB strains have been added as models of SLE.[3,9] Two congenic MRL strains were developed: one strain of mice with massive lymphadenopathy and severe lupus is called MRL-*lpr/lpr* because the lymph node enlargement is determined by an autosomal recessive gene termed lymphoproliferation *(lpr),* and the other strain without lymphadenopathy is called MRL-+/+ whose genome is identical (<0.1% difference) to the MRL-*lpr/lpr* strain except for lacking the *lpr* gene. The MRL-+/+ mice develop a milder form of lupus late in life. The lupus-prone BXSB is a recombinant inbred line derived from two progenitor strains, a C57BL/6 female and a SB/Le male.[3,9] Genes linked to the Y chromosome of the BXSB strain that were inherited from the SB/Le

GαMIg, goat antibody to mouse immunoglobulin; RαMIg, rabbit antibody to mouse immunoglobulin; Ab, antibody; MAb, monoclonal antibody; RαId, rabbit antiidiotypic antibody; MEM, Eagle's minimal essential medium; HAT, hypoxanthine–aminopterin–thymidine; RIA, radioimmunoassay; ELISA, enzyme-linked immunosorbent assay; FITC, fluorescein isothiocyanate conjugate.

[1a] R. S. Schwartz, *Prog. Allergy* **35,** 1 (1984).

[2] G. C. Tsokos and J. E. Balow, *Prog. Allergy* **35,** 93 (1984).

[3] A. N. Theofilopoulos and F. J. Dixon, *Adv. Immunol.* **37,** 269 (1985).

[4] H. R. Smith and A. D. Steinberg, *Annu Rev. Immunol.* **1,** 175 (1983).

[5] M. Bielschowsky, B. J. Helyer, and J. B. Howie, *Proc. Univ. Otago Med. Sch.* **37,** 9 (1959).

[6] J. B. Howie and B. J. Helyer, *Adv. Immunol.* **9,** 215 (1968).

[7] S. K. Datta, N. Manny, C. Andrzejewski, J. Andre-Schwartz, and R. S. Schwartz, *J. Exp. Med.* **147,** 854 (1978).

[8] J. W. Eastcott, R. S. Schwartz, and S. K. Datta, *J. Immunol.* **131,** 2232 (1983).

[9] E. D. Murphy and J. B. Roths, *in* "Genetic Control of Autoimmune disease" (N. E. Rose, P. E. Bigazzi, and N. L. Warner, eds.), p. 207. Elsevier/North-Holland, Amsterdam, 1978.

background were found to markedly accelerate autoimmune disease in these mice.

The lupus strains of mice can be obtained from the Jackson Laboratories (Bar Harbor, ME). Among these strains, the MRL-*lpr/lpr* mice, both males and females, develop severe disease at an earlier age. Autoantibodies are detectable by 1–2 months, and death from lupus nephritis occurs between 6 and 12 months of age.[3,9] In contrast, the MRL-+/+ strain develops autoantibodies and mild lupus in the second year of life. Lupus is also accelerated in BXSB male mice due to Y-chromosome linked genes, whereas female BXSB develop a milder form of disease in the second year of life.[3] The (NZB × NZW)F_1 and (NZB × SWR)F_1 hybrids resemble human SLE; early and severe lupus nephritis occur in females whereas males have a milder disease late in life.[6,8] This difference in the onset and severity of lupus is due to sex hormones, in contrast to the genetically determined differences in the BXSB strain where lupus is accelerated in male mice.

The consistent pathologic lesion in lupus mice is immune complex-mediated glomerulonephritis, although its severity and incidence varies.[3-9] Lupus mice spontaneously produce autoantibodies to nucleic acids, nuclear proteins, and histones. Among these, autoantibodies to DNA play a major role in lesions of SLE.[3,10] IgG antibodies to DNA are deposited in glomeruli, complexed with DNA. Anti-DNA autoantibodies in the renal lesions are concentrated 30 to 1000 times more than in serum, in some of these animals. In addition, antibody to gp70 immune-complexed with gp70, a retroviral envelope glycoprotein, is found in renal lesions of the majority of lupus strains.[3] The immune complex deposits in the kidney activate the complement system resulting in glomerulonephritis and proteinuria.

Unlike the disease in MRL and BXSB strains and the crosses of NZB with NZW or SWR mice, in the case of the NZB strain itself, the major manifestation of autoimmune disease is hemolytic anemia.[5] The incidence of lupus nephritis is very low in the NZB mice although they do produce anti-DNA antibodies.[6-8]

Among the lupus mouse strains described above, the MRL-*lpr/lpr* mice possess a recessive gene, *lpr,* that accelerates the development of SLE by causing proliferation and accumulation of a subset of T lymphocytes.[9] These T cells produce certain factors that induce polyclonal activation of B lymphocytes including those capable of secreting autoantibodies.[11] No

[10] D. Koffler, R. Carr, V. Agnello, R. Thoburn, and H. G. Kunkel, *J. Exp. Med.* **134,** 294 (1971).

[11] G. J. Prud'homme, T. M. Fieser, F. J. Dixon, and A. N. Theofilopoulos, *Immunol. Rev.* **78,** 160 (1984).

human counterpart of such a model exists. Moreover, the *lpr* gene produces full-blown SLE only in the genetic background of its congenic partner, the MRL-+/+ mice, and not in the background of other normal nonautoimmune mouse strains.[12] However, the congenic MRL-+/+ mice themselves develop nephritis and produce high levels of autoantibodies later in life.[13,14] Thus, the MRL-*lpr/lpr* strain is a model for the study of secondary factors that can accelerate and increase the severity of SLE. The primary mechanism of autoimmunity in the MRL model lies in the congenic MRL-+/+ background, and this strain has not been systematically studied. Similarly, in the BXSB model, Y-linked genes *accelerate* the development of SLE in mice with a lupus-prone genetic background. Many studies on the secondary effects of the Y-chromosome accelerating genes have been published, but nothing is known about the primary underlying defects in the background genome of this model.[15]

The NZB × SWR cross, on the other hand, has unique features that enable a genetic dissection of various factors which may be primarily involved in autoimmune disease. The F_1 hybrids of NZB and the normal SWR mouse strain uniformly develop lethal glomerulonephritis even without the superimposition of any accelerating genes like *lpr.* As in human SLE, the incidence of lupus nephritis is higher in the female (NZB × SWR)F_1 hybrids.[7,8] The other cross of NZB that develops a high incidence of lupus nephritis is the (NZB × NZW)F_1 hybrid. The NZW parents of this cross, however, are not normal; they produce anti-DNA antibodies and develop nephritis later in life.[16] They also share background genes with the NZB strain, and both the NZB and NZW strains are similar in virologic phenotype, i.e., they produce exceptionally high levels of retroviruses and retroviral antigen (gp70) throughout life.[17] In contrast, the SWR strain of mice do not produce any infectious retroviruses and have extremely low serum levels of gp70.[7,18] Moreover, the SWR mice are free of autoimmune disease, and they do not produce any autoantibodies spontaneously.[7,8] Therefore, because of their unique virologic and immunologic phenotypes, differing from both the NZB and NZW strains, the SWR mice are valuable genetic tools that can be used to define the role of various

[12] V. E. Kelley and J. B. Roths, *Clin. Immunol. Immunopathol.* **37,** 220 (1985).

[13] D. S. Pisetsky, S. A. Caster, J. B. Roths, and E. D. Murphy, *J. Immunol.* **128,** 2322 (1982).

[14] B. P. Giroir, E. C. Raps, R. M. Lewis, and Y. Borel, *Cell. Immunol.* **75,** 337 (1983).

[15] R. T. Steinberg, M. L. Miller, and A. D. Steinberg, *Clin. Immunol. Immunopathol.* **35,** 67 (1985).

[16] V. E. Kelley and A. Winkelstein, *Clin. Immunol. Immunopathol.* **16,** 142 (1980).

[17] T. Yoshiki, R. C. Mellors, M. Strand, and J. T. August, *J. Exp. Med.* **140,** 1011 (1974).

[18] S. K. Datta, P. J. McConahey, N. Manny, A. N. Theofilopoulos, F. J. Dixon, and R. S. Schwartz, *J. Exp. Med.* **147,** 872 (1978).

NZB abnormalities in the etiology of autoimmune disease. Furthermore, the NZB × SWR crosses are also suitable for identifying the genetic factors contributed by a completely normal strain in the development of lupus nephritis.

NZB and NZB × NZW mice have two features: uniform development of autoimmune disease and production of xenotropic retroviruses in high levels throughout life. Viral antigens (the viral envelope glycoprotein, gp70) and antiviral antibodies together with DNA complexed with anti-DNA antibodies are deposited in renal lesions of NZB × NZW mice.[17] These observations suggested a cause-and-effect relationship between the virus and the autoimmune disease.[19–24] The virus was thought both to trigger the disease and to provide antigens that generated nephritogenic immune complexes. The xenotropic variety of retroviruses expressed by NZB mice cannot infect mouse cells; it can infect only cells of heterologous species.[19] However, with the NZB × SWR cross, the relevance of the virus to autoimmunity could be ascertained by genetic techniques.

The genetic analysis became feasible by the demonstration that the expression of xenotropic viruses in NZB mice is a genetically determined trait.[25,26] A highly sensitive and quantitative assay was developed to detect xenotropic viruses. The assay reflected the *in vivo* virologic status of the animals. The segregation ratios of the virologic phenotype of the progeny of NZB × SWR crosses led to the identification of two autosomal dominant genes that governed the expression of xenotropic virus in NZB mice.[25,26] One of these genes, *Nzv-1,* specified high-grade expression of the virus, and the other, *Nzv-2,* specified low-grade expression of the virus. Homozygosity for recessive alleles at both loci resulted in a virus-negative mouse strain, such as SWR.[25,26] Matings between NZB and virus-negative SWR mice resulted in three kinds of progeny in F_2 and (F_1 × SWR) backcross generations: high-virus, low-virus, and virus-negative phenotypes. The virologic phenotypes of the progeny mice were stable upon testing by splenic biopsy and thus helped to establish that development of autoimmunity did not require the expression of xenotropic virus. Moreover, the other NZB traits, namely, high levels of circulating gp70 antigen and gp70 immune complexes, could also be dissociated from the develop-

[19] J. A. Levy, *Biomedicine* **24,** 84 (1976).
[20] N. Talal, *Arthritis Rheum.* **13,** 887 (1970).
[21] P. H. Lambert and F. J. Dixon, *Clin. Exp. Immunol.* **6,** 829 (1970).
[22] R. S. Schwartz, *N. Engl. J. Med.* **293,** 132 (1975).
[23] R. C. Mellors and J. W. Mellors, *Proc. Natl. Acad. Sci. U.S.A.* **73,** 233 (1976).
[24] M. Strand and J. T. August, *J. Virol.* **14,** 1584 (1974).
[25] S. K. Datta and R. S. Schwartz, *Nature (London)* **263,** 412 (1976).
[26] S. K. Datta and R. S. Schwartz, *Virology* **83,** 449 (1977).

ment of glomerulonephritis in the NZB × SWR crosses.[18] The dissociation between virus expression and development of autoimmune disease was not a peculiar feature of NZB × SWR crosses, similar results were found in crosses between NZB and other nonautoimmune strains, such as B10.A, C57B1/6, and AKR mice, and in recombinant inbred lines derived from NZB and C58 pro-genitor strains.[7,27]

Experimental Procedures for the Study of Retroviruses and Autoimmunity in Murine Lupus

A list of commercially available reagents for these studies is given in Table I.

Mice

Genetically uniform inbred strains of mice, such as the NZB and the SWR parental strains, may be purchased from the Jackson Laboratories. Reciprocal crosses are made to produce the F_1 and backcross progeny mice since the maternal direction of the matings does not affect results. Thus the F_1 progeny mice consist of (NZB × SWR)F_1 and (SWR × NZB)F_1; first backcross to SWR are (F_1 × SWR) and (SWR × F_1); first backcross to NZB are (F_1 × NZB) and (NZB × F_1); and the F_2 progeny are derived by mating F_1 mice.

Splenic Biopsy

Mice are anesthetized by intraperitoneal injection with 0.1 ml of Nembutal (1 : 10 dilution in normal saline) per 10 g of body weight. The mouse to be operated on is then ear-tagged and laid down left side up. The hair is shaved off with a hair clipper, and the skin is wiped with Betadine followed by 70% ethanol. The mouse is then placed on a dissecting board inside a bench-top hood. The mouse's tail and hind limbs are taped down on the board, and its head and chest are also loosely taped down. A 1-inch-long incision is made in the prepared skin area (over the spleen). The skin flaps are retracted and pinned down to the board with needles bent to an L shape. A small incision is then made in the abdominal wall without damaging any internal organs. The tip of the spleen with adjoining mesenteric tissue is gently pulled out through the opening. A noose of 4-O silk surgical suture is tied around the spleen tip. The portion of the exposed spleen tip beyond the suture is cut and placed in a sterile tissue culture dish. The remainder of the spleen with the tied end is placed back in the

[27] S. K. Datta, F. L. Owens, J. E. Womack, and R. J. Riblet, *J. Immunol.* **129,** 1539 (1982).

TABLE I
COMMERCIAL ANTIBODIES AND CELL LINES USED IN THE STUDY OF MURINE LUPUS

Reagent	Source[a]
Polyclonal antibodies	
Rabbit anti-mouse Ig	Cappel Laboratories, Cooper Biomedical, Malvern, PA
Goat anti-mouse Ig	Cappel Laboratories
Rabbit anti-chicken egg albumin	Cappel Laboratories
FITC–rabbit anti-goat IgG	Cappel Laboratories
FITC–goat anti-mouse IgM	Meloy Laboratories Inc., Springfield, VA
Goat anti-mouse IgM, IgG_{2a}, IgG_{2b}, IgG_1, IgA	Meloy Laboratories
Rabbit anti-mouse IgG_3, κ, λ	Litton Bionetics, Kensington, MD
Rabbit anti-mouse IgM	Litton Bionetics
Alkaline phosphatase conjugated goat anti-mouse Ig, goat anti-rabbit Ig	Boehringer Mannheim, Indianapolis, IN
FITC–goat anti-Moloney MuLV, FITC–goat anti-Rauscher gp70	Office of Program Resources, Virus and Antisera Distribution Center, Biological Carcinogenesis Branch, Division of Cancer Etiology, National Cancer Institute, NIH, Bethesda, MD; Officer in charge: Dr. John S. Cole
Monoclonal antibodies	
MAR 18.5 (mouse anti-rat κ)	Becton Dickinson, Mountain View, CA
Bet 1 (rat anti-mouse IgM)	American Type Culture Collection (ATCC), Rockville, MD
Mouse myeloma proteins	
MOPC 104E, MOPC 195, UPC 10, TEPC 183, J 606	Litton Bionetics
Cell lines	
Mink cell line (CCL64) for xenotropic virus assay	ATCC

[a] Other commercial sources for these reagents are also available.

peritoneal cavity, and the abdominal wall is sutured. The skin flaps are then apposed with surgical suture clips. The mouse is then injected subcutaneously with 0.5 ml of normal saline to compensate for the loss of body fluids during the operation. Mice are placed in a clean cage near a warm lamp for an hour before returning to the animal housing area. One month after biopsy the skin clips are removed.

Preparation of Spleen Cell Suspensions (see also this series, Vol. 67, p. 108)

Preparation of spleen cell suspensions is carried out under a laminar flow hood with sterile equipment. The aseptically removed splenic biopsy specimen is placed in a tissue culture dish (60 × 15 mm, Corning, NY) containing 4 ml of McCoy's 5A medium (Gibco, Grand Island Biological Co., NY) supplemented with 10% fetal calf serum and antibiotics (120 U/ml penicillin and 120 μg/ml streptomycin). The spleen fragment is squashed between the frosted ends of a pair of glass slides. The cells so obtained are harvested and centrifuged at 250 *g* for 10 min at 4°, washed once, and resuspended in the same medium. Clumps are removed by filtration through a small plug of nylon wool (Leukopak, Fenwal Co., Morton Grove, IL) packed loosely in a Pasteur pipet. Cell viability is determined by Trypan blue (0.1%) dye exclusion and concentration of cells adjusted as required.

Xenotropic Retrovirus Assays

Serially diluted spleen cell suspensions are tested for virus production by an infectious center assay.[7,25,26] In this method, the spleen cells are cultured with mink lung indicator cells (ATCC, CCL64) as described by Hartley and Rowe.[28] The day before the experiment, the stock mink cell line is trypsinized (0.25% trypsin–EDTA, Gibco), and 2×10^5 cells in 4 ml of supplemented McCoy's 5A medium are seeded in 60-mm tissue culture dishes containing sterile glass coverslips. The dishes are placed in an incubator at 37°, 6% CO_2 humidified atmosphere. The next day the cells are treated for 1 hr with 25 μg/ml of diethylaminoethyl–dextran (Sigma Chemical Co., St. Louis, MO), in 2 ml of serum-free McCoy's medium. The cultures are then rinsed, and the subconfluent monolayer of mink cells is inoculated with serial dilutions of spleen cell suspensions (10^7, 10^6, and 10^5 cells) in 4 ml supplemented medium. After incubation for 2 days the medium is removed and the cultures are fed with 4 ml of fresh medium. Two days later the cultures are examined by an inverted phase microscope to determine confluency. The coverslips are removed from the dishes, fixed in cold acetone, air dried, and stored at −20°. The mink cells attached to the coverslips are later stained with a fluorescein isothiocyanate-conjugated goat antiserum prepared against Tween–ether disrupted Moloney leukemia virus (Office of Program Resources, Virus and Antisera Distribution Center, Officer in charge, Dr. John S. Cole, National Cancer Institute, Bethesda, MD), which is specific for retroviral antigens. The antiserum is diluted 1 : 40 in a rhodamine contrast solution (1 : 50 in phosphate-buffered saline) and used to stain the coverslips for 1 hr at 37°. The coverslips are

[28] J. W. Hartley and W. P. Rowe, *J. Virol.* **19,** 19 (1976).

then washed and mounted on slides. Foci of cells showing cytoplasmic fluorescence are counted under a fluorescence microscope.

Virus titer is expressed as $\log_{10}$ infectious centers per 10^7 spleen cells. Progeny of the NZB × SWR crosses have three virologic phenotypes: high virus, low virus, and virus negative. The mean log titer of virus per 10^7 spleen cells in all high-virus animals is 3.49 ± 0.43, and for low-virus mice 0.76 ± 0.44. These values segregate into two distinct populations that do not overlap by ± 3SD. This justifies the selection of 2.0 log titer (100 infectious centers/10^7 cells) as the upper and lower limits defining the two groups. All NZB mice tested between 1 month and 2.5 years of age have high titers of virus whereas all age-matched SWR mice are virus negative. All (NZB × SWR)F_1 and (SWR × NZB)F_1 hybrids and (F_1 × NZB) backcross progeny express high titers of virus like the NZB parents. In a test of 72 first backcross to SWR (F_1 × SWR) progeny the ratio of virus-positive to virus-negative mice was 3 : 1, and among 80 (SWR × NZB)F_2 mice it was 15 : 1. Moreover, among the (F_1 × SWR) progeny 37 had high virus titers, 17 had low titers, and 18 were virus negative, yielding a segregation ratio for high to low to negative virus phenotypes of 2 : 1 : 1. Similarly, among the F_2 mice 61 had high titers, 14 had low titers, and 5 were virus negative, giving a segregation ratio of 12 : 3 : 1. Maternal direction of the cross and sex of the animals did not affect the results.

The segregation ratios are compatible with two autosomal dominant loci controlling the expression of xenotropic virus. One of these loci (termed *Nzv-1*) is responsible for high-grade expression of xenotropic virus, and the other *(Nzv-2)* specifies low titers of xenotropic virus expression. The high-virus phenotype is either homozygous or heterozygous for *Nzv-1;* the low-virus phenotype lacks *Nzv-1* and is either homozygous or heterozygous for *Nzv-2;* and the virus-negative mice are homozygous for the recessive alleles at both *Nzv-1* and *Nzv-2.* The second backcross of the low- or high-virus (F_1 × SWR) progeny with the virus-negative SWR parent or intercrosses with each other confirm this finding.[26]

Tests for Autoantibodies

Antierythrocyte Autoantibodies. A microtiter direct Coombs test is carried out as follows. Blood is collected by tail bleeding into heparinized hematocrit tubes. After centrifugation in a Damon/IEC hematocrit centrifuge the hematocrit is determined, the tube is cut at the plasma/red cell junction, and the packed red cells are ejected with a squeeze bulb into test tubes containing warm (37°) phosphate-buffered saline (PBS, 10 m*M*, pH 7.2). Erythrocytes are washed 3 times with warm PBS, and 1 μl of the red cell suspension is mixed with 20 μl of 1:5, 1:125, and 1:150 dilutions of rabbit anti-mouse immunoglobulin (Cappel Laboratories) in a U-well microtiter plate (Dynatech Laboratories Inc., Alexandria, VA). The rabbit anti-mouse Ig had been preabsorbed with normal mouse red cells. After 1 hr at room temperature the sedimentation pattern is read and graded from 1+ to 5+.[29] All mice should be tested at least twice during the study.

Anti-DNA Autoantibodies. Serum samples are obtained from the test animals by allowing blood to clot at room temperature and separating by centrifugation. Serum antibodies to double-stranded (dsDNA) and single-stranded DNA (ssDNA) are measured by a radioimmunoprecipitation assay as described.[30]

For the dsDNA assay, [^{3}H]- or [^{14}C]dsDNA (New England Nuclear, Boston, MA) is treated with S_1 nuclease (Worthington, Cooper Biomedical, Malvern, PA) specific for ssDNA (see below, Nucleic Acid Preparations). ssDNA is prepared by boiling the radiolabeled dsDNA for 15 min and then chilling immediately in ice. Heat-inactivated test sera are incubated with the radiolabeled DNA and then precipitated with affinity-purified rabbit anti-mouse immunoglobulin (RαMIg) antibody (Cappel Laboratories). Ten microliters of serum sample is added to 90 μl of phosphate–EDTA buffer (60 mM phosphate buffer, 30 mM sodium EDTA, pH 8.2) in Eppendorf microfuge tubes (Brinkman Instruments, Sybron, NY), mixed, and heat inactivated at 56° for 20 min in a water bath. Four such tubes are prepared for each test serum. To each of duplicate tubes containing the heat-inactivated sera, either 50 μl radiolabeled dsDNA (0.5–1 μg, ~5,000 cpm) or 50 μl of labeled ssDNA is added, and the tubes are mixed and incubated for 1 hr at 37°. One hundred microliters of the RαMIg antibody (1:20 dilution) is added to each tube; tubes are mixed and then incubated again at 37°. The tubes are centrifuged at 12,000 g for 2 min in a microfuge (Beckman Instruments, Fullerton, CA), the supernatant is aspirated off, and the pellet in each tube is washed with 1 ml of buffer. The tubes are centrifuged and washed once more. Radioactivity is measured in a scintillation counter after dissolving the precipitates in 100 μl of 0.1 N NaOH. Results are expressed as percentage of the total amount of radioactivity present in the precipitates. The mice are tested twice during the study and are considered positive if at least once they show levels of DNA-binding antibody greater than 17.5% for dsDNA and 14% for ssDNA. These values are 2SD higher than the mean levels in control mice. Sera from age-matched SWR, BALB/c, and AKR mice are used as controls.

[29] E. Gold, *Vox. Sang.* **15,** 222 (1968).

[30] M. Papalian, E. Lafer, R. Wong, and B. D. Stollar, *J. Clin Invest.* **65,** 469 (1980).

Tests for Lupus Nephritis

Proteinuria. Urines are tested repeatedly (weekly to biweekly) for proteinuria by Albustix (Miles Laboratories, Elkhart, IN). A 2+ reading means 100 mg protein/dl, and 3 to 4+ is equivalent to 350–2000 mg protein/dl. Up to 2+ may be found in normal male mice.

Blood Urea Nitrogen. The blood urea nitrogen (BUN) test is carried out with Azostix (Miles). Normal mice (SWR) have levels of 10–20 mg/dl. Values of 40 mg/dl or above are abnormal.

Immunofluorescence of Kidney Sections. Immunoglobulin Deposits. Cryostat sections (4 μm) are made on snap-frozen kidney tissue. Sections are air dried, washed in PBS, and fixed in ethanol: ether (1 : 1) for 10 min and then in 95% ethanol for 20 min. The sections are then stained with fluorescein-conjugated anti-mouse IgG (Meloy Laboratories Inc., Springfield, VA) and examined under a fluorescence microscope. The sections are graded on a 0 to 4+ scale by estimating both the brightness and the extent of fluorescent staining.[7,8] In addition to an overall score for the entire glomerulus, separate scores should be assigned to the mesangium and capillary loops. All slides should be read on coded specimens by two observers. Kidneys from 9-month-old (NZB × NZW)F_1 mice may be used as positive controls and those from 3- and 12-month-old SWR or BALB/c mice as negative controls.

Retroviral Antigens. Direct Fluorescence: Sections of the kidneys are stained with a 1 : 40 dilution of fluorescein isothiocyanate-conjugated goat antiserum prepared against Tween–ether disrupted Moloney leukemia virus (see above, Xenotropic Retrovirus Assays). A 1 : 50 dilution of rhodamine solution is used as a counterstain.

Indirect Immunofluorescence: Kidney sections are overlaid with an anti-Rauscher gp70 serum made in goats (Office of Program Resources, National Cancer Institute). After incubation and washing, the slides are stained with a fluorescein-conjugated rabbit anti-goat IgG (preabsorbed with normal mouse spleen cells and serum). A 1 : 40 dilution of the anti-Moloney leukemia virus serum usually stains with 3+–4+ intensity mink (ATCC, CCL 64), rabbit (SIRC), human (RD), and rat (NRK) cell lines infected with xenotropic viruses derived from NZB, NIH Swiss (AT 124), and BALB/c (BALB:virus-2) mice. It also stains with equal intensity at 1 : 40 dilution NIH 3T3 cells infected with ecotropic (N-tropic) AKR Gross virus and BALB 3T3 cells infected with ecotropic (B-tropic) virus from B10.A mice. A 1 : 20 dilution of this serum does not stain uninfected NIH and BALB 3T3 cells or uninfected rat embryo, mink, and rabbit cells, as well as cell lines derived from spleens and embryos of SWR mice. The goat anti-Rauscher gp70 serum has similar properties. These two antisera consistently stain glomeruli in kidney sections obtained from 9-month-old (NZB × NZW)F_1 mice but do not stain glomeruli in kidney sections from 3-month-old SWR or BALB/c and 12-month-old SWR mice.

Histopathology. Glomerular lesions are graded as 2+, 3+, or 4+.[31] A 2+ lesion corresponds to focal or focal and diffuse thickening of the capillaries of 30–60% of the glomeruli; in a 3+ lesion the capillaries of all the glomeruli are diffusely and focally thickened with hypercellularity of glomeruli; 4+ lesions are characterized by the preceding together with sclerosis of glomeruli, massive proliferation of epithelial cells, and numerous tubular casts (Fig. 1). A grade of 0 is given to kidneys without glomerular lesions. All specimens should be coded before examination.

Summary

Genetic tests with the NZB × SWR cross indicate that the xenotropic virus and gp70 antigen are *endogenous* and thus potential autoantigens in NZB mice. Like other autoantigens, such as erythrocytes and DNA, they may play secondary roles, but they are not required primarily for autoimmunization. Whether an autoimmune reaction occurs against xenotropic viral antigens or other autoantigens depends on genes ("autoimmunity genes") that are distinct from viral genes.[7,18,25,26]

Methods for Studying Nonspecific Immunologic Defects in Murine Lupus

Polyclonal hyperactivity of B lymphocytes with increased immunoglobulin production has been observed in all the murine models as well as in human SLE.[2,3,27,32] This trait could be secondary to a decreased suppressor T-cell or increased helper T-cell function, or it could be due to a primary B-cell defect. In NZB mice these generalized B-cell abnormalities are complex and are influenced by unlinked sets of genes.[27] One of these B-cell hyperactivity traits is manifested by spontaneous hypersecretion of IgM.[33] This abnormality is expressed in NZB mice starting in fetal life independent of any T-cell influence.[32] Since the SWR strain is normal for this trait, by using the NZB × SWR crosses (F_1, backcross, and F_2 progeny) it was found that two sets of autosomal genes cause hyperactivity of a subset of B cells in NZB mice.[27,32] It was postulated that the intrinsic hyperactivity of this subpopulation of B cells in NZB mice also results in

[31] C. L. Manaligod, C. L. Pirani, F. Miysasta, and V. E. Pollak, *Nephron* **4,** 215 (1967).

[32] N. Manny, S. K. Datta, and R. S. Schwartz, *J. Immunol.* **122,** 1220 (1979).

[33] H. M. Moutsopoulos, M. Boehm-Truitt, S. S. Kassan, and T. M. Chused, *J. Immunol.* **119,** 1639 (1977).

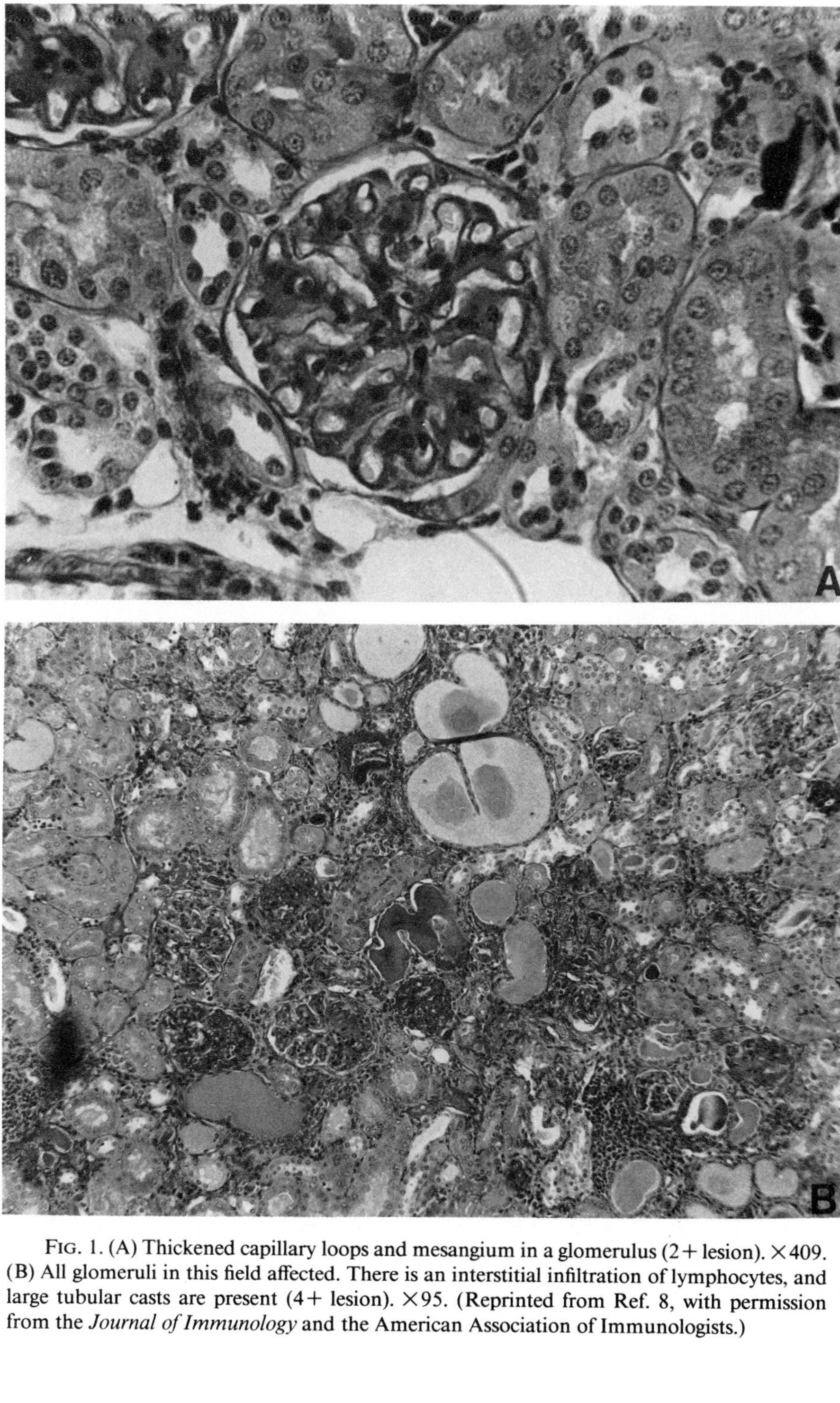

FIG. 1. (A) Thickened capillary loops and mesangium in a glomerulus (2+ lesion). ×409. (B) All glomeruli in this field affected. There is an interstitial infiltration of lymphocytes, and large tubular casts are present (4+ lesion). ×95. (Reprinted from Ref. 8, with permission from the *Journal of Immunology* and the American Association of Immunologists.)

the activation of autoreactive B cells, causing autoimmune disease.[4,11,32,33] However, genetic segregation analysis in the NZB × SWR crosses and in NX8 (NZB × C58) recombinant inbred lines showed that inheritance of the spontaneous B-cell hyperactivity trait could influence the severity of SLE but that the trait was not necessary for the development of autoantibodies and autoimmune disease.[8,27]

Tests for Polyclonal B-Cell Hyperactivity

Spontaneous IgM Secretion in 4-Hr Cultures. Single cell suspensions are obtained from splenic biopsy fragments in leucine-free Dulbecco's modified Eagle's minimal essential medium (MEM) (Gibco) supplemented with 2 m*M* glutamine, 10 m*M* HEPES buffer, pH 7.2, penicillin, streptomycin, and 5% dialyzed fetal calf serum (FCS) (Gibco). The cells are washed twice in ice-cold medium, and viable cells are counted by Trypan blue dye exclusion. Usually 2×10^6 cells are cultured in 1 ml of leucine-free medium in Falcon tissue culture tubes (16 × 125 mm, No. 3033, Falcon Plastics, Oxnard, CA) with 25 μCi (0.1 ml) of [4,5-^{3}H(*N*)]leucine (New England Nuclear) for 4 hr at 37° in a 6% CO_2 atmosphere. The cells are centrifuged (850 *g* for 15 min), and the supernatants are harvested to assay for IgM production as described below.[8,27,32]

A high-titer monospecific antiserum to IgM is prepared by repeated injection of rabbits with 100 μg of mouse IgM (light chain class λ_1, from MOPC 104E mouse plasmacytoma, Litton Bionetics) emulsified with Freund's complete adjuvant. The antiserum should be monospecific by Ouchterlony gel analysis and by the demonstration that it reacts only with mouse IgM in immunoelectrophoresis. The cell-free culture supernatants are centrifuged at 50,000 × *g* for 30 min and then divided into equal aliquots. Two hundred microliters of the supernatant is incubated with 0.2 ml of the rabbit anti-mouse IgM and 25 μg of mouse myeloma IgM at 37° for 1 hr and overnight at 4°. The titer of rabbit anti-mouse IgM used in the coprecipitation radioimmunoassay should be slightly in excess of that which at equivalence precipitated 25 μg of IgM. A nonspecific serologic blank to control for any radiolabeled material that might be trapped in the rabbit antiserum–antigen matrix must be set up simultaneously. This control consists of 0.2 ml of the supernatant, 12.5 μg of chicken egg albumin, and 0.2 ml of rabbit anti-chicken egg albumin serum (Cappel Laboratories) at equivalence. This control yields a precipitate similar in size and protein content to that of the specific serologic reaction. After 3 washings with PBS at 4°, the precipitates are dissolved twice in 0.1 *N* acetic acid and reprecipitated each time with equal volumes of freshly made 0.1 *N* NaOH. This procedure minimizes the amount of radiolabeled material trapped nonspecifically in the immune precipitate. The precipitates are then dissolved in 0.1 *N* NaOH, and their radioactivity is determined in 10 ml of scintillation fluid (HP Ready-Solv, Beckman). Net serospecific IgM is calculated by subtracting total nonspecific counts per minute (cpm) from the specific counts. The results are expressed as IgM cpm secreted by the number of cultured cells. The radiolabeled material in the specific immune precipitates in this coprecipitation assay is shown to be IgM by sodium dodecyl sulfate–polyacrylamide gel electrophoresis analysis.

Staining for Cytoplasmic Immunoglobulin M. To enumerate cytoplasmic IgM-containing cells aliquots of spleen cell suspension obtained by biopsy (see above) are collected by centrifugation, resuspended in 8 ml of Dulbecco's MEM, and layered over 3 ml of FCS in a 15 ml tube. The cells are then centrifuged through the FCS gradient (400 *g* for 10 min), followed by a wash in PBS with 1% FCS. The cells are suspended at a density of 4×10^5 cells/ml in 1% FCS in PBS, and 0.3 ml of the suspension is deposited on slides by centrifugation in a cytocentrifuge (Shandon Scientific Co., London, England). Smears are briefly air dried and fixed for 12 min in 95% cold methanol. The slides are then washed in PBS (pH 7.4), and the smears are stained with class-specific fluorescein isothiocyanate-conjugated goat anti-mouse IgM (1 : 10 dilution, specific for mouse IgM μ chain, Meloy Laboratories) for 45 min at 37° in a moist chamber. The slides are then washed 3 times in PBS, mounted, and examined under a fluorescence microscope with selective filters and a Ploem epiillumination system. Thymocytes and fibroblast smears serve as controls. At least 1000 cells with cytoplasmic IgM are counted. Those smears that do not show any IgM-containing cells are scanned in entirety ($\sim 1.2 \times 10^5$ cells per smear).

Summary

As stated above, the autoimmune NZB mice have a primary B-cell abnormality manifested by spontaneous hypersecretion of IgM in short-term cultures. Compared to the normal SWR strain the NZB spleen cells secrete 15 to 22 times more IgM spontaneously.[8,32] The purpose of the procedures described above is to characterize the genes specifying the intrinsic B-cell defect and to determine their relationship to "autoimmunity genes." The increase in IgM secretion results from two independently segregating genetic traits: (1) an increased number of IgM-containing cells and (2) an increased rate of secretion of IgM per cell.

Methods for Studying Anti-DNA Autoantibodies in Murine Lupus

Autoantibodies to DNA play a major role in the pathogenesis of lupus nephritis.[10] Normal mice can produce such antibodies, however, and lupus anti-DNA antibodies share idiotypes with antibodies that are widely conserved in normal mice.[34] Furthermore, both NZB and (NZB × SWR)F_1 mice produce anti-DNA antibodies, and yet only F_1 hybrids develop a high incidence of severe lupus nephritis.[7,8] Therefore, qualitative differences among anti-DNA antibodies are important in determining their pathogenicity, and genes contributed by the normal SWR parent probably determine the expression of such pathogenic antibodies in the F_1 hybrids.[35,36] Indeed, in contrast to their autoimmune NZB parents, the (NZB × SWR)F_1 hybrids produce highly cationic IgG anti-DNA antibodies that have distinct antigenic specificity patterns and the allotype of the normal SWR parent.[35,36] IgG anti-DNA antibodies with positive charge are especially pathogenic in SLE.[37] Moreover, since DNA, like the glomerulus, is negatively charged, the probability of its deposition and persistence in glomeruli will increase markedly if it is complexed with such cationic anti-DNA antibodies produced by the F_1 mice.[38,39]

Nucleic Acid Preparations

Native double-stranded DNA is prepared by dissolving calf thymus DNA (Worthington) in SSC buffer (150 m*M* sodium chloride, 20 m*M* sodium citrate, pH 7.5), extracting with chloroform : isoamyl alcohol (24 : 1), and then ethanol precipitating and twirling on a glass rod.[30] The ethanol precipitate is redissolved in sodium citrate buffer, dialyzed into 60 m*M* NaCl, 10 m*M* $ZnCl_2$, 0.1 *M* sodium acetate, pH 5, and digested with *Aspergillus* S_1 nuclease (Worthington, 100 U/mg DNA) for 45 min at 37°. The reaction is stopped by adding EDTA (10 m*M*). The DNA is then ethanol precipitated, redissolved in SSC buffer, and chromatographed over a Sephacryl S-200 column (Pharmacia, Piscataway, NJ). The final A_{260} to A_{280} ratio of the preparation is greater than 1.85. Denatured single-stranded DNA is prepared from the dsDNA by boiling for 15 min and then immediately cooling on ice. Polyinosinic acid [poly(I)] is obtained from P.L. Biochemicals (Milwaukee, WI), polydeoxythymidine [poly(dT)] from Collaborative Research, Inc. (Cambridge, MA), and cardiolipin from Supelco, Inc. (Bellefonte, PA) or Sigma; brominated poly(dG–dC)·poly(dG–dC)

[34] S. K. Datta, B. D. Stollar, and R. S. Schwartz, *Proc. Natl. Acad. Sci. U.S.A.* **80,** 2723 (1983).

[35] S. K. Datta, *UCLA Symp. Mol. Cell. Biol.* **18,** 877 (1984).

[36] J. Gavalchin, J. A. Nicklas, J. W. Eastcott, M. P. Madaio, B. D. Stoller, R. S. Schwartz, and S. K. Datta, *J. Immunol.* **134,** 885 (1985).

[37] F. Ebling and B. H. Hahn, *Arthritis Rheum.* **23,** 392 (1980).

[38] Y. Lawrence, C. Agodoa, V. J. Gauthier, and M. Mannik, *J. Immunol.* **134,** 880 (1985).

[39] J. L. Barnes and M. A. Venkatachalam, *J. Exp. Med.* **160,** 286 (1984).

(zDNA) is prepared by the method described by Lafer *et al.*[40] Briefly, poly(dG–dC) (P.L. Biochemicals) in SSC buffer is treated with bromine water until the $OD_{260/295}$ value is 0.35 indicating that the DNA is in a left-handed configuration. Tritiated *Escherichia coli* DNA is obtained from New England Nuclear.

Derivation of Anti-DNA Antibody-Producing Hybridomas

Hybridomas producing anti-DNA antibodies are obtained from NZB and (NZB × SWR)F_1 mice by following standard procedures described in detail elsewhere (see this series, Vol. 73 [1]). Female NZB (7- to 13-month-old) and F_1 (4- to 8-month old) mice are used. Spleen cells from unimmunized mice are fused with SP2/0-Ag-14 using PEG (polyethylene glycol 1000, Baker Chemicals, Phillipsburg, NJ) at a concentration of 44.4%. After a 2-min treatment, a 3-min centrifugation (500 *g*), and a wash, the cells are taken up in Dulbecco's minimal essential medium (DMEM, 4.3 g glucose/liter, Gibco) supplemented with 15% FCS, 2 m*M* L-glutamine, 1 m*M* sodium pyruvate, and antibiotics and incubated overnight at 37° in a 6% CO_2 humidified atmosphere. The cells are harvested, placed in selective HAT media, and then aliquoted into 96-well flat-bottom microtiter plates (Costar, Cambridge, MA) together with a feeder layer of 5×10^3 mouse thymocytes. The cultures are incubated at 37° in a 6% CO_2 humidified atmosphere and fed weekly. Wells with growth are tested for anti-DNA antibody production. Those producing anti-DNA Ab are subcloned and then expanded in flasks using DMEM with 15% FCS, 2 m*M* L-glutamine, 125 μg/ml Pen–Strep, 10 m*M* HEPES, and 0.5 ml nonessential amino acids/100 ml.

Selection of Anti-DNA Antibody-Producing Hybridomas. The screening of the hybridomas is carried out by a solid-phase radioimmunoassay (RIA). ssDNA (2.5 μg) in 100 μl of sodium citrate buffer is placed in polystyrene tubes (12 × 75 mm) for a minimum of 16 hr. The tubes are rinsed twice with washing buffer (0.1 *M* K_2HPO_4, pH 7, 0.025% heat-inactivated FCS, 0.02% sodium azide), and then 0.1 ml of hybridoma supernatant is added and the mixture incubated for 1 hr at room temperature. The tubes are washed 3 times with washing buffer, and 100 μl of ^{125}I-labeled goat anti-mouse IgM and IgG (~50,000 cpm, Cappel Laboratories) is added and incubated for 1 hr at room temperature. The tubes are rinsed 3 times, and the radioactivity is determined. Hybridomas whose supernatants yield counts of at least twice the negative control (supernatant from the SP2/0 parent) are then subcloned through two cycles of limiting dilu-

[40] E. M. Lafer, R. P. C. Valle, A. Moller, A. Nordheim, P. S. Schur, A. Rich, and B. D. Stollar, *J. Clin. Invest.* **71**, 314 (1983).

tion. The cells are then seeded into flasks for antibody production. Hybridomas are also selected for binding to dsDNA-coated microwell plates by an enzyme-linked immunosorbent assay (ELISA) (see below, Analysis of Specificity of Hybridoma Antibodies).

Purification of Monoclonal Antibodies. The antibodies are affinity purified on goat anti-mouse Ig–Sepharose columns (Pharmacia; prepared according to the manufacturer's suggestions). The antibodies are eluted with 0.5 *M* acetic acid, dialyzed versus PBS (10 m*M* sodium phosphate buffer, pH 7.2, 150 m*M* NaCl), and stored at 4° with 0.02% azide and 2 m*M* phenylmethylsulfonyl fluoride (PMSF).

Determination of Isotype of the Monoclonal Anti-DNA Antibodies

Ouchterlony double diffusion in agar is used for isotype analysis (see this series, Vol. 73 [1]). Goat anti-mouse IgM, IgG_{2a}, IgG_{2b}, IgG_1, and IgA are obtainable from Meloy Laboratories or Litton Bionetics; rabbit anti-mouse IgG_3 and rabbit anti-mouse κ and λ light chains from Litton Bionetics.

Determination of Allotype.

The Ig heavy chain allotypes of (NZB × SWR)F_1 hybridoma-derived anti-DNA antibodies may be determined by one of the three assays. Antiallotype sera are prepared by reciprocal immunization of SWR and A/J mice (A/J have the same allotypic determinants as NZB).[41] In this protocol, allotype donor mice are hyperimmunized with repeated injections of heat-killed *Bordetella pertussis* and then bled. The pooled antipertussis hyperimmune sera are complexed with the bacteria, and the complexes are used to immunize the recipient strain for production of anti-allotype serum.

The immunization protocol is as follows: The allotype donor mice are first immunized with heat-killed *B. pertussis* (Massachusetts Department of Public Health, Jamaica Plain, MA). On day 1, 2×10^9 bacteria in 0.2 ml saline are injected intraperitoneally and in four places subcutaneously. On days 3 and 5 the same amount of bacteria is injected S.C. in four places. On days 21 and 37 bacteria are injected as in day 1. The immunized mice are bled on days 25, 32, and 45 and the sera are pooled. The recipient mice are then immunized for the production of antiallotype serum as follows: On day 1, 10^9 bacteria plus 20 μl of antiserum from the immunized donor mice are mixed in 0.2 ml saline and incubated at room temperature for 1 hr to allow complex formation. The mixture is then injected i.p. into the

[41] L. A. Herzenberg and L. A. Herzenberg, *Handb. Exp. Immunol.* **1,** 12.1 (1978).

recipient strain of mice. These mice are boosted with the immune complexes, as in day 1, on days 3, 5, 22, 26, 50, and 57. On day 64 and then weekly, each recipient is bled and tested for antiallotype titer. The high-titered sera are pooled. The recipient mice can be boosted if titer goes down, and 10 days after boost they may be bled and tested.

The antiallotype serum is then absorbed to isolate specific antibody by using Sepharose-conjugated immunoabsorbents. Cyanogen bromide-activated Sepharose beads (Pharmacia) are conjugated to immunoglobulins of SWR and A/J mice. The SWR and A/J mice are first hyperimmunized with heat-killed *B. pertussis* (as above), and the Ig in the hyperimmune sera are precipitated by 50% saturated ammonium sulfate, dialyzed into PBS (10 m*M* sodium phosphate, pH 7, 150 m*M* NaCl), and then conjugated to CNBr-Sepharose according to the manufacturer's instructions. The Sepharose–Ig columns (5 ml column bed with 10 mg of bound Ig) are washed with 0.1 *M* acetic acid to remove any extraneous protein and then with 50 m*M* Tris (pH 7) to neutrality. To block nonspecific absorption 2 ml of normal serum from A/J and SWR mice are first run through the SWR Ig–Sepharose and A/J Ig–Sepharose columns, respectively, followed by 50 m*M* Tris buffer washing until no more 280-nm-absorbing material appears in the effluent. The anti-SWR allotype serum made in A/J mice is loaded on to the SWR Ig–Sepharose column, and the anti-A/J allotype serum made in the SWR strain is absorbed on the A/J Ig column. The columns are then washed with Tris buffer until no more 280-nm-absorbing material comes off. To recover the bound antiallotype antibodies, the columns are eluted with 0.1 *M* acetic acid. Fractions (2.5 ml) are collected into tubes containing 2.5 ml of 2× Tris buffer for neutralization. The protein peaks are pooled and dialyzed into PBS. The purified anti-allotype antibodies should be absorbed with normal sheep red blood cells.

This protocol does not produce antibodies to IgG_{2b} (*Igh-3* locus) or IgM (*Igh-6* locus) allotypic determinants in good titers. BALB/c mice, in contrast to the NZB strain, share many IgG_{2b} allotypic determinants with the SWR strain.[41,42] Therefore, an antiallotypic serum reactive with the IgG_{2b} allotype of SWR mice is prepared by immunizing young (2- to 3-month-old) NZB mice with a BALB/c IgG_{2b} myeloma protein, MOPC 195, according to the method of Mage *et al.*[43] The immunization is accomplished by a total of eight s.c. injections, each with 75 μg of MOPC 195 (Litton Bionetics) in 0.05 ml saline, into four sites including two footpads, axillary,

[42] C. M. Huang, M. Parsons, V. T. Oi, H. S. Huang, and L. A. Herzenberg, *Immunogenetics* **18,** 311 (1983).

[43] R. Mage, R. Lieberman, M. Potter, and W. D. Terry, *in* "The Antigens" (M. Sela, ed.), Vol. 1, p. 299. Academic Press, New York, 1973.

and inguinal regions. The first injection is given with the myeloma protein mixed with 0.05 ml Freund's complete adjuvant; the second (4 days later) is mixed with 0.05 ml incomplete Freund's adjuvant; and the third to eighth injections are given with the myeloma Ig only in saline at weekly intervals. Mice are bled weekly for 4 weeks starting 7 days after the last injection, and 1.5–2 ml of antiserum is collected from each mouse.

The indirect hemagglutination assay described by Mage *et al.*[43] may be used to allotype the monoclonal IgG anti-DNA antibodies derived from (NZB × SWR)F_1 mice. In this assay, sheep red blood cells, (SRBC) coupled with mouse serum Ig or myeloma protein are agglutinated by antiallotype sera. This agglutination can be inhibited by addition of Ig of the corresponding allotype. Thus, if the particular allotype is lacking in the added sample the agglutination test will be positive. SWR, A/J, and NZB mice are hyperimmunized with heat-killed *B. pertussis* (as above).[41] Immunoglobulins in the hyperimmune sera are precipitated with 50% saturated ammonium sulfate, dialyzed versus PBS, and subsequently coupled covalently to SRBC by glutaraldehyde.[44] One-half milliliter of packed SRBC is washed 3 times in normal saline and suspended to 8 ml in saline. Mouse Ig or myeloma protein (3 mg) is added to the cells in 1 ml saline, and then the cells are continually stirred while 3 ml of saline containing 25% glutaraldehyde is added over a 15-min period. The cells are mixed for another 45 min, then washed 3 times in saline, and resuspended in 100 ml of PBS–BSAE buffer (10 m*M* phosphate buffer, pH 7.2, 150 m*M* NaCl, 0.1% bovine serum albumin, 0.02% sodium azide, 10 m*M* sodium EDTA). The SRBC conjugate is stable at 4° for more than 1 year. If the cells do not settle in a clean button without added antibody, the BSA concentration in the buffer may be raised to 1% to give clear buttons.

The hemagglutination assay for allotyping is done in polystyrene microtiter V bottom plates (Dynatech) using PBS–BSAE buffer as the antiserum diluent. First the antiallotype serum is titrated alone to determine its end point. A dilution of antiallotype corresponding to 1–2 hemagglutinating units (i.e., the end point or twice that concentration) is chosen for the inhibition assay; this is often around a 1 : 1000 dilution of the antiallotype serum. Samples (1 μl = 1 μg) of each of the anti-DNA antibodies to be allotyped are added to 50 μl of the appropriate antiallotype serum dilution in the microtiter wells. Next 25 μl of the 0.5% v/v indicator SRBC conjugate is added, and the well contents are mixed by agitating the plate. The plates are covered, incubated at room temperature for a period ranging from 3 hr to overnight, and then scored for agglutination.[29] Positive agglu-

[44] J. Evens, M. Steel, and E. Arthur, *Cell* **3**, 153 (1974).

tination means that the test anti-DNA antibody does not bear the allotype, and inhibition of agglutination indicates the presence of the allotype. Anti-DNA antibodies inhibiting the agglutination of SWR Ig-coupled SRBC by the A/J anti-SWR allotype serum are designated to bear the SWR allotype; anti-DNA antibodies with the NZB allotype inhibit the agglutination of either A/J or NZB Ig-coupled SRBC by the SWR anti-A/J allotype serum. It should be established in preliminary tests that the SWR anti-A/J allotype serum does not agglutinate SWR Ig-coated SRBC, and conversely, the A/J anti-SWR allotype serum does not react with A/J or NZB Ig-coupled SRBC. Hyperimmune serum Ig from SWR, A/J, or NZB mice may be also used as competitive inhibitors in the allotype assays as positive and negative controls. Purified hybridoma anti-DNA antibodies derived from NZB mice may be also used as competitors in the allotyping assays as an additional control.

The (NZB $\times$ SWR)F_1-derived monoclonal antibodies of IgG_{2b} class have the SWR allotype if they inhibit the agglutination of either SWR Ig-coupled or MOPC 195-coupled SRBC by the NZB anti-MOPC 195 serum. They have the NZB allotype if they do not inhibit the agglutination. The NZB anti-MOPC 195 serum does not agglutinate SRBC coupled with hyperimmune serum Ig from NZB mice or a monoclonal IgG anti-DNA antibody derived from NZB mice. This excludes the presence of any rheumatoid factor antibody in the NZB anti-MOPC 195 serum.

Anti-DNA antibodies of the IgM class are allotyped by an enzyme-linked immunosorbent assay (ELISA), using a monoclonal rat anti-mouse IgM antibody, Bet 1, which defines an allotypic determinant on IgM called Igh 6.5.[45] SWR serum, like BALB/c, inhibits the binding of Bet 1 to MOPC 104E myeloma protein very strongly, whereas NZB serum or monoclonal IgM anti-DNA antibodies derived from NZB mice inhibit very weakly. By using a range of dilutions of the competitor sera, significant differences in the inhibition curves of SWR and NZB sera (or NZB monoclonal IgM antibodies) can be obtained. The Bet 1 assay is carried out as described, with minor modifications.[45] Briefly, 75 μl of serial dilutions of monoclonal anti-DNA antibody are incubated with 75 μl of 1:20 dilutions of Bet 1 for 3 hr and then plated into wells coated with 0.5 μg of MOPC 104E. After a 3-hr incubation, the wells are washed 4 times with PBS, 1% BSA. Then 75 μl of a 1:10 dilution of MAR 18.5, a mouse anti-rat κ antibody of IgG_{2a} class,[46] is added to each well, and the wells are incubated for 1 hr. The wells

[45] J. T. Kung, S. O. Sharrow, D. G. Sieckmann, R. Liebermann, and W. E. Paul, *J. Immunol.* **127,** 873 (1981).

[46] L. L. Lanier, G. A. Gutman, D. E. Lewis, S. T. Griswold, and N. L. Warner, *Hybridoma* **1,** 125 (1982).

are then washed 4 times with PBS, 1% BSA. The amount of bound MAR 18.5 is detected by addition of 50 μl of a 1:600 dilution of protein A conjugated with β-galactosidase (Zymed Laboratories, San Francisco, CA). After an overnight incubation at 4°, the wells are washed 4 times, 50 μg of *p*-nitrophenyl-β-galactopyranoside substrate (Zymed Laboratories) is added to each well, and the color development is read at 405 nm on a Dynatech reader. The Bet 1 antibody, MOPC 104E, and the competing monoclonal IgM anti-DNA antibodies do not bind to protein A. The Bet 1 hybridoma cells and monoclonal antibody may be obtained from the American Type Culture Collection, and the MAR 18.5 antibody from Becton Dickinson (see Table I).

Isoelectric Focusing of Anti-DNA Antibodies

Ampholine PAG plates (5% acrylamide, 2.4% ampholines, pH 3.5–9.5) from LKB Inc. (Rockville, MD), may be used to determine the p*I* of anti-DNA antibodies with IgG isotypes. Focusing is carried out on a flat-bed LKB apparatus as described.[47] The anodic buffer is 1 *M* H_3PO_4, and the cathodic buffer is 1 *M* NaOH. The pH gradient is measured immediately after focusing.[47] The focused antibodies are fixed and stained with silver nitrate according to the procedure of Morrissey.[48]

The p*I* of anti-DNA antibodies of IgM class are determined by using a composite agarose–acrylamide gel system as described by Jackson *et al.*[49] The pH gradient in the gel is determined as for the isoelectric focusing of IgG antibodies. The gel is stained in 0.25% Coomassie blue for 30 min and is destained in methanol:acetic acid:distilled H_2O (5:1:5) overnight.

Analysis of Specificity of Hybridoma Antibodies for Various Nucleic Acids and Phospholipids

Competitors [ssDNA, dsDNA, zDNA, poly(I), poly(dT), and cardiolipin] are used to inhibit the binding of the monoclonal antibodies to ssDNA. The analysis may be performed by using a solution RIA and/or a solid-phase ELISA, as described.[30,36,50,51] In the RIA the monoclonal antibody is first incubated with inhibitor (competing antigen) and then with

[47] V. J. Gauthier, M. Mannik, and G. E. Striker, *J. Exp. Med.* **156,** 766 (1982).

[48] J. H. Morrissey, *Anal. Biochem.* **117,** 307 (1981).

[49] D. E. Jackson, C. A. Skandera, J. Owens, E. T. Lalley, and P. C. Montgomery, *J. Immunol. Methods* **36,** 315 (1980).

[50] Y. Shoenfeld, J. Rauch, H. Massicote, S. K. Datta, J. Andre-Schwartz, B. D. Stollar, and R. S. Schwartz, *N. Engl. J. Med.* **308,** 414 (1983).

[51] Y. Shoenfeld, D. A. Isenberg, J. Rauch, M. P. Madaio, B. D. Stollar, and R. S. Schwartz, *J. Exp. Med.* **158,** 718 (1983).

[^{3}H]ssDNA, followed by rabbit anti-mouse Ig (see the section on tests for assaying serum anti-DNA antibodies, above). The precipitate is centrifuged, washed, and resuspended in 0.1 *N* NaOH, and the radioactivity is determined. The percent inhibition is calculated as [(counts with no inhibitor) − (counts with inhibitor)]/(counts with no inhibitor) × 100 and plotted against the amount of PO_4 (nmol) in the inhibitor solution on a log scale.

In the ELISA assay, 96-well Dynatech Immulon plates are coated with 150 μl/per well of a poly(L-lysine) solution [50 μg/ml poly(L-lysine) hydrobromide, MW 150,000, Sigma] for 30 min. The plates are washed 3 times with TBS (30 m*M* Tris, 140 m*M* NaCl, pH 7.4) using a Dynatech washer, and then 150 μl of ssDNA solution (3 μg/ml in TBS) is added and incubated for 1 hr. The plates are washed 2 times with TBS and then 150 μl/well of a poly(L-glutamate) solution [50 μg/ml poly(L-glutamine) sodium, MW 50,000, Sigma] are added. After 2 hr the plates are washed 2 times with TBS and left with 150 μl of TBS/well at 4° until use. One hour before use the TBS is aspirated off, and 150 μl of phosphate–EDTA blocking buffer (60 m*M* K_2HPO_4, pH 8, 30 m*M* EDTA, 3% BSA) is added following which the plates are washed just before the assay. The assay is performed by preincubating the inhibitor and monoclonal antibody for 1 hr in the phosphate–EDTA buffer. One hundred fifty microliters of the inhibitor–antibody solution is placed in each well (duplicates are run for each concentration of inhibitor used) and incubated for 1 hr. The plates are washed 2 times with PBS plus 1% Tween (Sigma), and then 150 μl of a 1:2000 dilution of goat anti-mouse Ig (IgG and IgM) conjugated to alkaline phosphatase (Boehringer Mannheim, Indianapolis, IN) in TBS is added to each well; the plates are incubated overnight. The plates are washed once with PBS–1% Tween and then twice with TBS before addition of 150 μl of *p*-nitrophenyl phosphate substrate (Sigma, 1 tablet in 5 ml of 25 m*M* $NaCO_3$ buffer, pH 9.5, with 2 m*M* MgCl) per well. The color developed in the plates is read on a Dynatech microELISA reader, using wells with no added hybridoma antibody as the blank. Results are plotted as the percent inhibition calculated as [(OD with no inhibitor) − (OD with inhibitor)]/(OD without inhibitor) × 100 versus the amount of PO_4 (nmol) in the inhibitor solution.

In these assays, each purified monoclonal antibody is used at a dilution giving 50% maximum binding of the respective antibody to ssDNA. Therefore, inhibition by the competing antigens depends on the relative avidity of each antibody for the respective antigen. Serial 5-fold dilutions of each competitor are used in the assays. A plot of percent inhibition versus the quantity of competing antigen (expressed as nanomoles of phosphate in the competing antigen) used is then made.[30,36,50,51] The amount of PO_4 of the competitor needed to cause 50% inhibition of binding of the monoclonal antibody to ssDNA is used to compare relative affinities for the antigens. There is no competition by yeast RNA (Sigma) in either type of assay, and there is no significant binding of the monoclonal anti-DNA antibodies to plates coated with yeast RNA in the solid-phase ELISA.

The competition assays test the relative avidity of the monoclonal antibodies for the nucleic acids and synthetic polynucleotides as well as for cardiolipin. The two assays (RIA and ELISA) give similar results for a single monoclonal antibody. All of the monoclonal anti-DNA antibodies are polyspecific; they bind to three or more of the competing antigens with considerable variation in avidity (Fig. 2). They are classified into three categories based on their antigenic specificity patterns.[36] Category I antibodies have relatively strong affinity for poly(I) and ssDNA and variable affinity for zDNA. These antibodies do not bind dsDNA, poly(dT), or cardiolipin. Category II antibodies bind strongly to both dsDNA and ssDNA (in contrast to Categories I and III) and have varying affinities for poly(I), poly(dT), and zDNA. These antibodies also do not react with cardiolipin. Category III antibodies bind strongly to poly(dT) and ssDNA (unlike Category I and II) and have varying affinities for poly(I) and zDNA. They do not bind dsDNA (unlike Category II) or cardiolipin. Category I antibodies are exclusively derived from the (NZB $\times$ SWR)F_1 hybrids wheras both NZB- and F_1-derived antibodies are found in the binding Categories II and III.[36]

Idiotypic Analysis of Anti-DNA Antibodies

Production and Purification of Antiidiotypic Antibodies. Antiidiotypic sera are produced by immunization of female NZW rabbits (2.5–2.7 kg in weight) with purified monoclonal anti-DNA antibodies.[51,52] A *Staphylococcus aureus*-monoclonal antibody sandwich is prepared for immunization. For monoclonal IgG anti-DNA antibodies, 250 μl of a 10% Staph A suspension (Ig Sorb, Biopure Enzyme Center, Boston, MA) is washed 3 times with PBS (10 mM NaH_2PO_4, 150 mM NaCl, pH 7.2), and the pellet is resuspended and incubated with 500 μg of the monoclonal antibody (MAb) in 1 ml PBS on a rocker at room temperature for 1 hr. The Staph A conjugate is centrifuged, washed thrice with PBS, resuspended in 1 ml PBS, and emulsified with Freund's complete adjuvant (Difco Laboratories, Detroit, MI). In the case of monoclonal IgM anti-DNA antibodies, first 150 μg of rabbit anti-mouse IgM (RαMIgM, Litton Bionetics) is incubated

[52] C. Andrejewski, J. Rauch, E. Lafer, B. D. Stollar, and R. S. Schwartz, *J. Immunol.* **125,** 226 (1981).

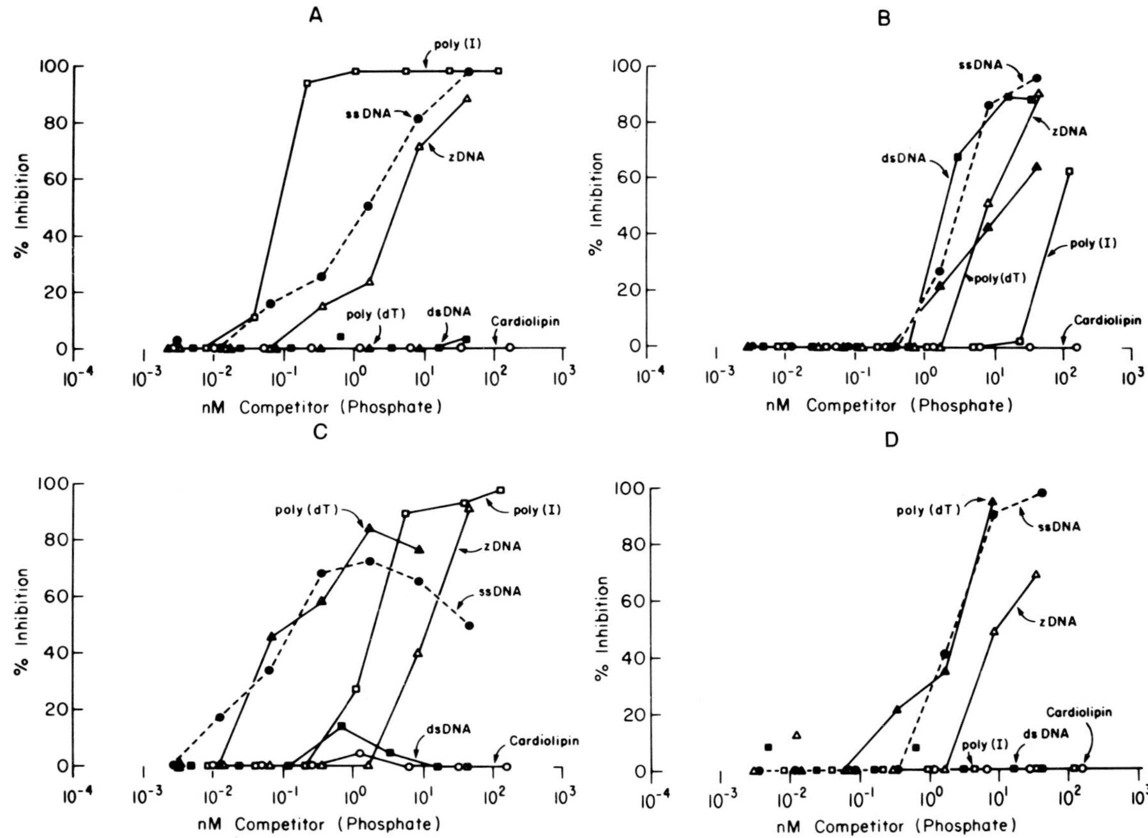

FIG. 2. Competitive immunoassays showing antigenic specificity of monoclonal anti-DNA antibodies. Competition by six different antigens to inhibit binding of antibodies to ssDNA is shown. (A) Antibody of the antigen binding Category I; (B) monoclonal antibody with antigenic specificity pattern of Category II; (C and D) antibodies with binding pattern of Category III. (Reprinted from Ref. 36, with permission from the *Journal of Immunology* and the American Association of Immunologists.)

with washed Staph A in 1 ml PBS at room temperature for 1 hr. The Staph A–RαMIgM conjugate is washed 3 times with PBS and then incubated with 500 μg of the IgM MAb in 1 ml PBS overnight at 4°. The next day the conjugate is washed 3 times with PBS, resuspended in 1 ml of PBS, and emulsified with 1 ml of Freund's complete adjuvant in a pair of Luer-locked syringes. A rabbit is immunized subcutaneously at approximately 10 sites on the back and hind-quarters. The rabbit is boosted monthly with 150–250 μg of MAb conjugated to Staph A in Freund's incomplete adjuvant. Seven days after the second boost the rabbit is bled, and the serum is tested against mouse Ig in an Ouchterlony plate. If the anti-mouse Ig has a titer of at least 1:16, the rabbit is bled, otherwise the rabbit is boosted again. The rabbit is then boosted monthly and bled 7–10 days later once the 1:16 titer of anti-mouse Ig reactivity has been attained.

Purification of Rabbit Antiidiotypic Antibody (RαId). The immune rabbit serum is applied to a protein A–Sepharose column (Pharmacia), and the bound antibodies are eluted with 3 *M* potassium thiocyanate. The eluted fractions are neutralized immediately with Tris to pH 7.0 and dialyzed extensively against PBS. The rabbit serum is run over the protein A–Sepharose column repeatedly until no more protein is eluted. The dialyzed eluate is then absorbed on a mouse myeloma Ig–Sepharose column [MOPC 195 (IgG_{2b}, κ), UPC 10 (IgG_{2a}, κ), TEPC 183 (IgM, κ), J 606 (IgG_3, κ), and MOPC 104E (IgM, λ), Litton Bionetics] exhaustively until no protein is eluted by 0.5 *M* acetic acid. [The myeloma immunoglobulins are conjugated to CNBr-activated Sepharose (Pharmacia) according to the manufacturer's instructions.] The flow-through fractions are pooled and concentrated using a pro-Dicon concentrator-dialyzer (Biomolecular Dynamics, Beaverton, OR). A pool of preimmune normal rabbit serum may be used as a control.

Determination of Specificity of RαId. Immulon I microtiter well plates (Dynatech) are coated with the MAb (1 μg/ml in 50 m*M* borate buffer, pH 8.6, overnight at 4°) used to immunize the rabbit, as well as control myeloma antibodies (MOPC 195, MOPC 104E, TEPC 183, J 606, UPC 10) in separate wells. The plates are washed twice with PBS–1% Tween, then twice with PBS and blocked with 60 m*M* phosphate buffer, pH 8.2, containing 30 m*M* EDTA and 0.1% BSA for 1 hr at room temperature. Dilutions of RαId or absorbed normal rabbit serum in 0.1% Tween–PBS are added, and the plates are incubated at room temperature for 2 hr. The plates are then washed as indicated above, and 150 μl of goat anti-rabbit Ig–alkaline phosphatase conjugate (Boehringer Mannheim) diluted in 0.1% Tween–PBS at 1:2000 is added to the wells. The plates are incubated overnight at 4° and the next day are washed as indicated above. After the plates are washed, bound alkaline phosphatase conjugate is detected by the addition of 150 μl *p*-nitrophenyl phosphate. The reaction is stopped at 1 hr with 5 *N* NaOH, and optical density is read at 405 nm in a Dynatech Model MR 580 microELISA reader. If there is no nonspecific binding of the antiidiotypic antibody to the myelomas, one can proceed to the competition assay. If there is nonspecific binding, the RαId should be absorbed again on the myeloma columns.

We deliberately do not absorb the antiidiotypic reagents with serum Ig from normal SWR or young preautoimmune NZB and (NZB $\times$ SWR)F_1 mice. This strategy is necessary because in contrast to antibody idiotypes that are induced by deliberate immunization with antigens or haptens, certain spontaneously produced lupus anti-DNA autoantibody idiotypes are shared by naturally occurring antibodies that are widely conserved in normal mice.[34,35] The majority of these "natural" antibodies are not directed against DNA, although they share the idiotypic markers of lupus anti-DNA autoantibodies. Similar to other conserved germ-line idiotypes that are expressed by preimmune normal animals, these natural antibodies sharing lupus idiotypes may be directed against common environmental pathogens, or they may have an important physiologic role in the early development of the normal immune system.[34,35] Therefore absorption with serum Ig from normal or young preautoimmune animals might remove important idiotypic reactivities from the rabbit antiidiotypic antibodies. The absorption steps with the myeloma Ig are adequate to render the rabbit antiidiotypic antibodies truly idiotype specific.[53,54] The myeloma proteins used for absorption of the rabbit antiidiotypic sera share isotypic and allotypic determinants of the monoclonal anti-DNA antibodies that are used to immunize the rabbits, and therefore they can be used to remove any anti-constant region reactive antibody from the rabbit antiidiotypic sera.[53,54] The idiotypic specificity of the purified rabbit antiidiotypic antibodies is established when they fail to react with many F_1- and NZB-derived anti-DNA MAb that are matched in isotype, allotype, and antigenic specificity patterns to those autoantibodies that cross-react with the respective rabbit antiidiotype antibodies.[53,54]

Competitive Inhibition Assays for Detection of Cross-Reactive Idiotypes. The affinity-purified monoclonal anti-DNA antibody is diluted in 50 m*M* borate buffer, pH 8.6. Concentrations of monoclonal antibody giving 50% of maximum binding to the homologous antiidiotype antibody are incubated in wells of polystyrene plates overnight at 4°. The idiotype-coated plates are washed 3 times with TPBS (phosphate-buffered saline containing 1% Tween 20) and then with PBS. The dilution of the RαId

[53] J. Gavalchin, R. Seder, and S. K. Datta, *J. Immunol.* **138,** 128 (1987).

[54] J. Gavalchin and S. K. Datta, *J. Immunol.* **138,** 138 (1987).

that gives 50% maximal binding to the immunizing MAb in the solid phase is determined by the direct binding ELISA described above. To achieve proper dilution in the reaction mixture, twice this concentration of RαId in 200 μl vol is used to incubate with an equal volume of the competing anti-DNA MAb or control myeloma Ig. The competing antibodies are diluted 10-fold serially from 10 μg to 0.01 ng in PBS, 1% BSA, 0.05% Tween, and the RαId is also diluted in the same buffer. After a 2-hr incubation at room temperature, 150 μl of the mixture is added in duplicate to the idiotype-coated plates for 2 hr. After 3 washes with TPBS and PBS, 150 μl of goat anti-rabbit immunoglobulin conjugated to alkaline phosphatase is added, and the plates are incubated overnight at 4°. The final steps of the assay are identical to those described above for determination of specificity of RαId.

The antiidiotypic antibodies can thus be used to detect cross-reactive idiotype families among the monoclonal anti-DNA autoantibodies in relation to their cationic or anionic charge, antigenic specificity patterns, and animal of origin.[53,54] In addition, the respective RαId can be used to study the production and regulation of autoantibodies bearing the cross-reactive idiotypes, either in lymphocyte cultures *in vitro* or in serum *in vivo*.[34,35,53,54] The pathogenic role of autoantibodies sharing the cross-reactive idiotypes can also be established by analyzing Ig deposited in the renal lesions of lupus mice with nephritis, using the respective rabbit antiidiotypic antibodies.[54]

Summary

In mice with lupus nephritis qualitative changes in anti-DNA antibodies occur, such as IgG switch and increased cationic charge, to render these antibodies pathogenic. Pathogenic anti-DNA idiotypes can be encoded by genes of a normal mouse strain such as SWR, where they remain dormant. When the normal mice are crossed with an autoimmune strain like NZB, the F_1 hybrids express these idiotypes owing to defects in immunoregulation, resulting in the development of lethal glomerulonephritis.

Acknowledgments

I thank Drs. Robert Schwartz and David Stollar for stimulating collaboration in different portions of the work summarized here, and Ms. Barbara Lewis for processing the manuscript. The studies were supported by National Institutes of Health Grant RO1 CA 31789.

[37] Experimental Allergic Encephalomyelitis

By ROBERT H. SWANBORG

Introduction

Experimental allergic encephalomyelitis (EAE)[1] is a T-cell-mediated autoimmune central nervous system (CNS) disease, characterized clinically by ascending paralysis and histologically by perivascular mononuclear cell inflammation of the brain and spinal cord.[1a] The disease can be readily induced in susceptible strains of laboratory animals by a single injection of CNS extract emulsified in Freund's complete adjuvant (CFA).[1a] The major EAE-inducing antigen in CNS tissue is myelin basis protein (BP), which represents about 30% of the total myelin protein and is an integral structural constituent of the myelin sheath.[2] EAE can be adoptively transferred from immunized donors to syngeneic recipients with viable lymphocytes, and transfer is enhanced dramatically by culturing the donor cells *in vitro* in the presence of the T-cell mitogen concanavalin A[3] or BP.[4-7]

Methods

Preparation of Myelin Basic Protein

The most economical source of BP is bovine brain or spinal cord (which can usually be obtained from a local slaughterhouse). Although bovine BP is highly encephalitogenic for guinea pigs, it is relatively inactive in rats.[8] In contrast, BP prepared from guinea pig CNS tissue is encephali-

[1] Abbreviations: EAE, experimental allergic encephalomyelitis; CNS, central nervous system; CFA, Freund's complete adjuvant; BP, myelin basic protein; C/M, chloroform/methanol; HBSS, Hanks' balanced salt solution.

[1a] P. Y. Paterson and R. H. Swanborg, *in* "Immunological Diseases" (M. Samter, ed.), 4th Ed., Chap. 76, p. 1877. Little, Brown, Boston, 1988.

[2] M. W. Kies, *Ann. N.Y. Acad. Sci.* **122,** 161 (1965).

[3] H. S. Panitch and D. E. McFarlin, *J. Immunol.* **119,** 1134 (1977).

[4] J. R. Richert, B. F. Driscoll, M. W. Kies, and E. C. Alvord, Jr., *J. Immunol.* **122,** 494 (1979).

[5] J. H. Holda, A. M. Welch, and R. H. Swanborg, *Eur. J. Immunol.* **10,** 657 (1980).

[6] D. J. Hinrichs, C. M. Roberts, and F. J. Waxman, *J. Immunol.* **126,** 1857 (1981).

[7] H. Ovadia and P. Y. Paterson, *Cell. Immunol.* **65,** 66 (1981).

[8] R. E. Martenson, G. E. Deibler, M. W. Kies, S. Levine, and E. C. Alvord, Jr., *J. Immunol.* **109,** 262 (1972).

togenic for rats and guinea pigs of genetically susceptible strains at the 10–25 μg dosage level[8] and is routinely utilized. Although either brain or spinal cord can be employed, we find that better yields with fewer protein contaminants are obtained when BP is extracted from spinal cord tissue. Our method is adapted directly from the procedure originally described.[2]

Frozen guinea pig spinal cords (150–250 g, which represents approximately 150–250 spinal cords) are homogenized in a Waring Blendor with cold chloroform/methanol (2/1; C/M) then extracted overnight in 3 liters of C/M in the cold room with constant stirring using a propeller mixer. The residue is vacuum-filtered through Whatman grade 90 filter paper or equivalent (but is not taken to complete dryness) then is reextracted with 3 liters of cold C/M for 1 hr at 4°. The residue is refiltered, then extracted with 900 ml cold distilled water for 1 hr at 4° on a magnetic stirrer, refiltered, and suspended in 300 ml 0.01 *N* HCl.

The pH is maintained between 3.0 and 3.2 for 1 hr, by dropwise addition of 1 *N* HCl (as needed) at room temperature with constant stirring. Then the acid extract is centrifuged (10 min at 48,000 *g*, 4°), and the supernatant is filtered through Whatman No. 4 filter paper. At 4°, an equal volume of saturated ammonium sulfate is added dropwise to the supernatant, with constant stirring, and the precipitate is allowed to settle overnight. The supernatant is discarded, and the precipitate is dissolved in 50 ml 0.01 *N* HCl and dialyzed (in preboiled dialysis tubing) against 6 liters of distilled water at 4° for 24 hr, with one change of the dialysis bath after about 6 hr.

The contents of the dialysis bag are centrifuged (10 min at 48,000 *g*, 4°) to remove the trace precipitate, and the slightly acidic supernatant is adjusted to pH 7.0 with 0.25 *N* NaOH, with constant stirring. A precipitate forms. The precipitate is removed by centrifugation (10 min at 48,000 *g*, 4°), and the supernatant is lyophilized and stored at 4° as crude BP. The yield is usually 600–800 mg. This material is enriched in BP (Fig. 1) and can be used to elicit EAE, or it can be further purified by gel filtration chromatography.

Purification of Myelin Basic Protein

The BP is purified by gel filtration at 4° through a column (2.6 × 95 cm) of Sephadex G-50 Superfine (Pharmacia), equilibrated with 0.01 *N* HCl.[9] Approximately 150 mg BP dissolved in 1.5 ml of 0.01 *N* HCl is applied to the column, and 5-ml fractions are collected. The eluate is

[9] R. H. Swanborg, J. E. Swierkosz, and R. G. Saieg, *J. Immunol.* **112,** 594 (1974).

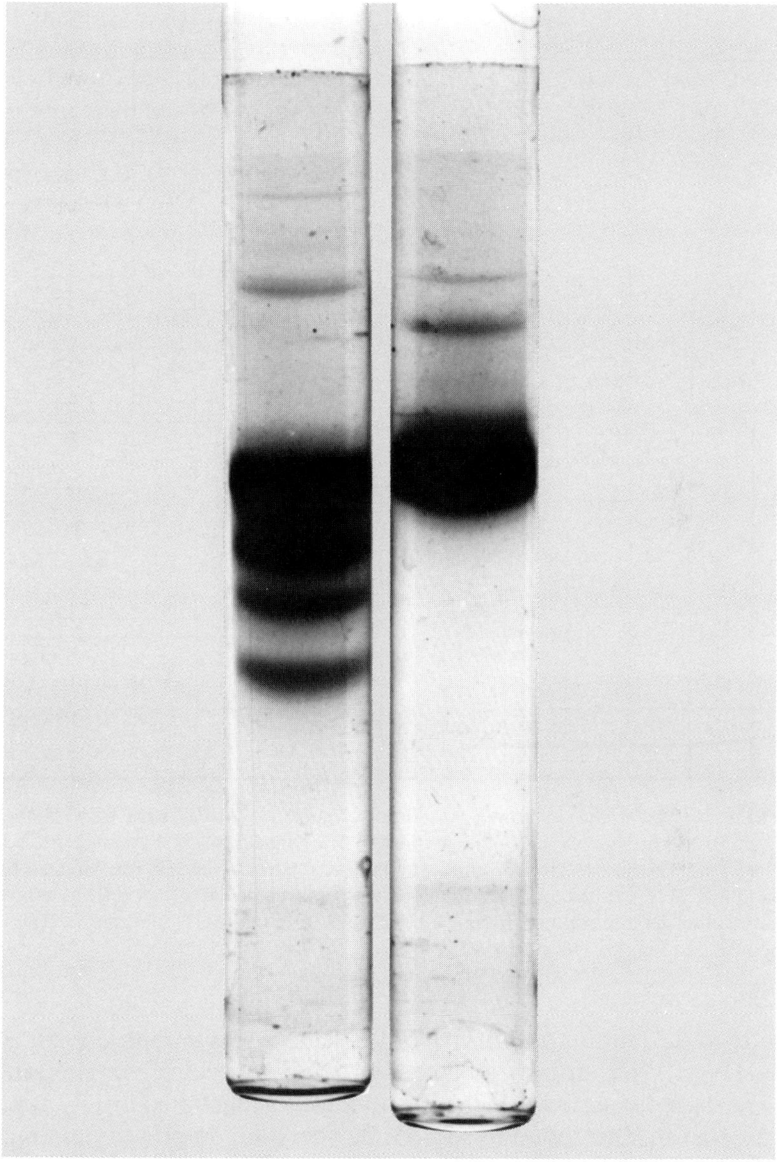

FIG. 1. Polyacrylamide gel electrophoresis of crude guinea pig myelin basic protein (left) and BP purified by gel filtration (right). The protein band above the major BP component in the right-hand gel represents aggregated BP.[10] Migration is toward the cathode (bottom).

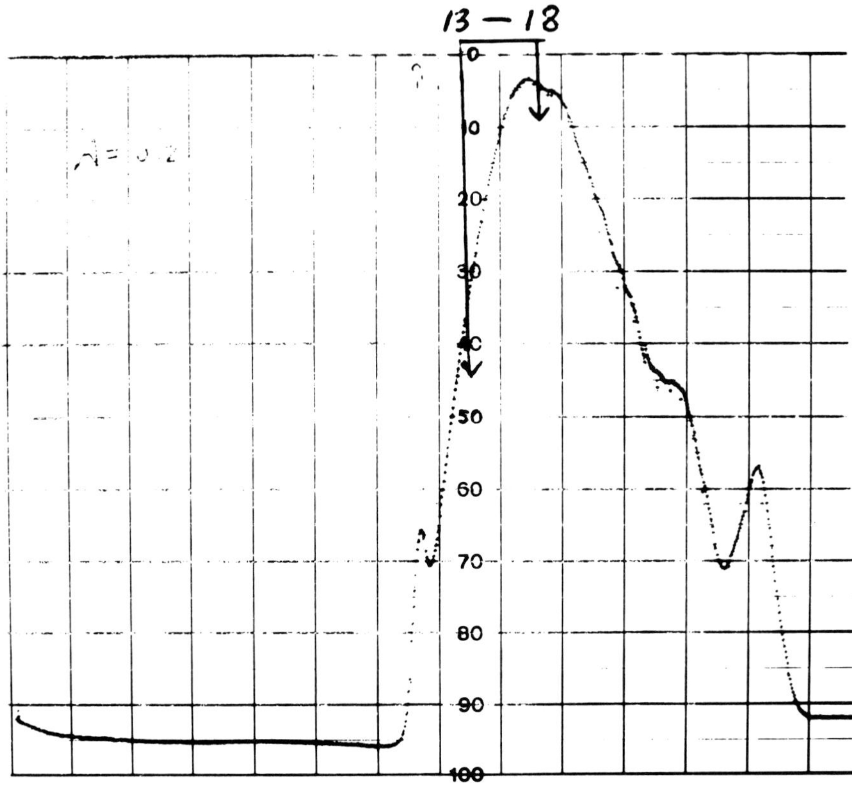

FIG. 2. A typical elution profile at 280 nm of guinea pig BP. Crude BP (150 mg) was dissolved in 1.5 ml 0.01 *N* HCl and applied to a Sephadex G-50 Superfine gel column (2.6×95 cm). The flow rate was 25 ml/hr, and the void volume was 168 ml. After a 5-hr wait period (LKB 2111 MultiRac), 5-ml fractions were collected. Each fraction was assayed by polyacrylamide gel electrophoresis, and tubes 13–18 were pooled, lyophilized, and stored as purified BP (see Fig. 1). The yield was 35 mg.

monitored at 280 nm (LKB Uvicord S), and an asymmetric peak is obtained (Fig. 2). The upper leading edge of the peak, which contains purified BP (Fig. 1), is lyophilized (yield 20–40 mg). As illustrated in Fig. 1, purity is evaluated by electrophoresis in 5% polyacrylamide gels containing 8 *M* urea in 1 *M* acetic acid, with migration toward the cathode (1 hr, 2.5 mA/gel), according to the method described by Martenson *et al.*[10]

[10] R. E. Martenson, G. E. Deibler, and M. W. Kies, *Biochim. Biophys. Acta* **200,** 353 (1970).

Active Induction of EAE

The purified guinea pig BP induces EAE in Lewis rats and Hartley guinea pigs at 10 μg. However, we usually use 25 μg as a routine challenge dose (25 μg BP plus 100 μg *Mycobacterium butyricum*).

Approximately 1 mg guinea pig BP is dissolved in 1 ml distilled water. To confirm weighing accuracy, the concentration of the BP solution is checked spectrophotometrically at 280 nm. The absorbance should be approximately 0.6. Freund's complete adjuvant is prepared by adding 4 mg/ml *M. butyricum* to Freund's incomplete adjuvant. Both components may be purchased from Difco (Detroit, MI).

Disperse the mycobacteria throughout the adjuvant using a vortex mixer and pipet 0.5 ml CFA into a 12 × 75 mm test tube. Add 0.3 ml BP solution and mix repeatedly, drawing and expelling the emulsion in a 1 ml Luer-Lok tuberculin syringe (Becton Dickinson, No. 2027) fitted with an 18-gauge needle. When thickened, add 0.2 ml BP solution and continue emulsifying until a drop of the emulsion is stable when dropped on an aqueous surface. This is a 1/1 (v/v) BP–CFA emulsion. Inject 0.05 ml of the emulsion into one hind footpad (using the tuberculin syringe fitted with a 26-gauge needle) of a susceptible Lewis rat or Hartley guinea pig. This dosage contains 25 μg BP and 100 μg *M. butyricum,* which is sufficient to induce severe EAE in the immunized animal.

EAE develops in 9–12 days. In rats, the severity of clinical disease is graded as follows: 1, loss of tail tonicity; 2, definite hindquarter weakness; and 3, hind leg paralysis. Rats with scores of 2–3 are often incontinent (Fig. 3). Most Lewis rats recover from EAE by day 18. In contrast, most guinea pigs with EAE die by day 14–18.

Rats are sacrificed, and the entire spinal column is placed in 10% formalin–saline (100 ml formaldehyde plus 900 ml 0.15 *M* NaCl). After 5 days, the spinal cord has been fixed and can be easily removed from the spinal column. This can be further facilitated by soaking the spinal column overnight in a decalcifying solution (e.g., J. T. Baker, No. G161-2). Longitudinal sections are prepared (10 μm thick), embedded in paraffin, and stained with hematoxylin–eosin.

In rats, inflammatory lesions are common in the spinal cord (Fig. 4) but are only rarely found in brain sections. In contrast, lesions are found in both the brain and spinal cord of guinea pigs with EAE. Sections are examined microscopically under low power (10× objective) without knowledge of source, and graded as described by Levine and Wenk[11]: 0, no

[11] S. Levine and E. J. Wenk, *Proc. Soc. Exp. Biol. Med.* **114,** 220 (1963).

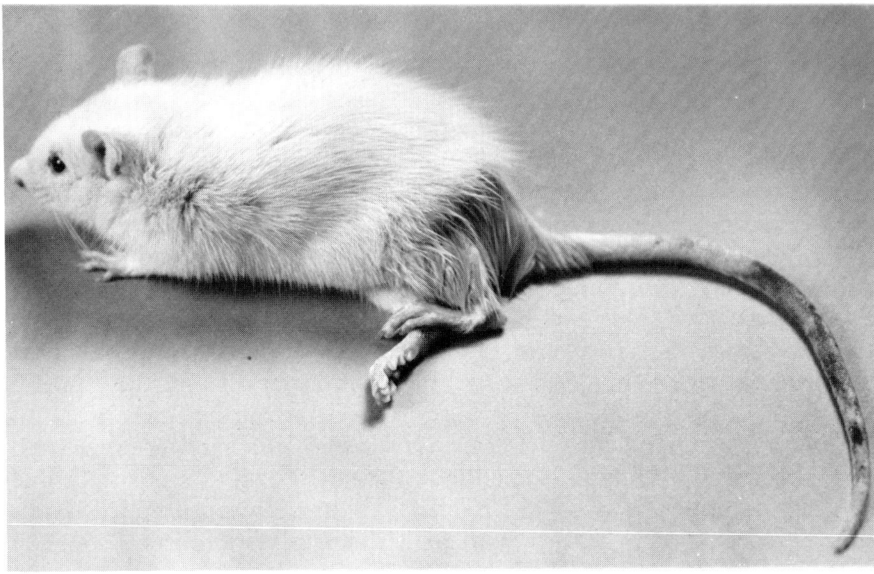

FIG. 3. A Lewis rat with grade 3 EAE, exhibiting hind limb paralysis and incontinence.

perivascular infiltration; 1, approximately five lesions in an entire section; 2, a few lesions in each of a few fields; 3, lesions in many different fields; 4, extensive lesions in almost all fields.

Adoptive Transfer of EAE

The adoptive transfer of EAE experiments are conducted in susceptible isohistogeneic Lewis rats, which are commercially available (e.g., Harlan Sprague Dawley). *Direct* transfer of EAE with lymph node cells from immunized donors requires large numbers of cells ($2-5 \times 10^8$/recipient).[12] Recently, however, it has become possible to transfer severe EAE with sensitized donor cells by the simple expedient of culturing donor spleen cells with concanavalin A[3] or BP[4] for 24–72 hr prior to transfer. This procedure activates the EAE effector cells and results in successful adoptive transfer of disease with $1-5 \times 10^7$ donor cells.

Donor Lewis rats (we routinely use 8-week-old females) are immunized with BP–CFA as described in the previous section. Ten to 12 days later the donor(s) is sacrificed, and the spleen is aseptically removed and placed into a small petri dish containing Hanks' balanced salt solution (HBSS, Gibco, 310-4020). Using sterile instruments and aseptic technique, the spleen is

[12] P. Y. Paterson, *J. Exp. Med.* **111,** 119 (1960).

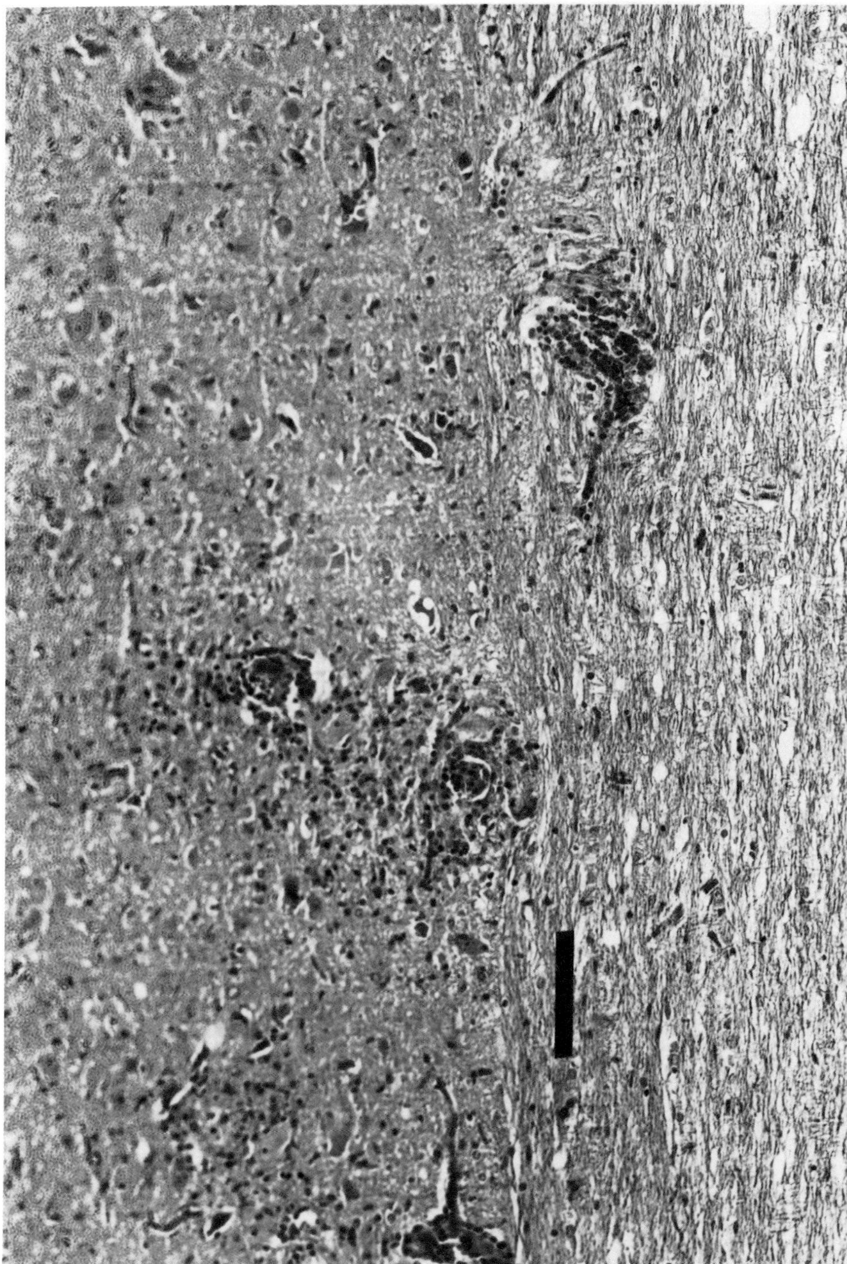

FIG. 4. A hematoxylin–eosin-stained section of spinal cord from a Lewis rat with EAE, showing perivascular and parenchymal mononuclear cell infiltration. Original magnification $\times 100$; bar, 100 μm.

cut into small pieces which are placed into a sterile plastic tube (17 × 100 mm) (Falcon 2001) containing HBSS. The spleen fragments are then homogenized in the plastic tube using the pestle from a Teflon Potter–Elvehjem tissue grinder (American Scientific Products, 08-414-16C). Subsequently, the spleen cell suspension is filtered through cotton in a Pasteur pipet into a 15-ml conical tube and centrifuged for 15 min at 470 *g*. The cells are resuspended in 8 ml Tris–NH_4Cl [9 parts 0.83% NH_4Cl plus 1 part 0.01 *M* Trizma Base (Sigma), pH 7.2] and refrigerated for 10 min. Then, HBSS is added to fill the tube, and the cells are centrifuged for 15 min at 470 *g*, resuspended in 6–10 ml HBSS, and counted in a hemacytometer. Each spleen yields approximately 2×10^8 nucleated cells; viability should exceed 95%.

The cells are cultured with BP in tissue culture medium RPMI 1640, prepared as follows: To 100 ml RPMI 1640 (containing 25 m*M* HEPES buffer; M. A. Bioproducts, 12-115-B) add 1 ml L-glutamine (200 m*M*) (Gibco, 320-5030), 1 ml penicillin–streptomycin solution (Gibco, 600-5140), 5 ml fetal bovine serum (Gibco, 200-6140) [this is heat inactivated at 56° for 1 hr; individual lots must be checked to ensure that they do not inhibit effector cell activation], 0.1 ml 2-mercaptoethanol (Eastman 4196) [1/1000 dilution in phosphate-buffered saline, pH 7.2, which is sterilized by passage through a Millex-GS 0.20-μm filtered unit (Millipore 10432)], and 0.2 mg BP dissolved in 1 ml HBSS. The final concentration of the supplements are as follows: 2 m*M* L-glutamine, 100 units/ml penicillin, 100 μg/ml streptomycin, 5% fetal bovine serum, 2×10^{-5} *M* 2-mercaptoethanol, 2 μg/ml BP.

The spleen cells are suspended in the supplemented RPMI 1640 medium at a concentration of 2×10^6/ml and cultured for 72 hr at 37°, 5% CO_2, in sterile tissue culture flasks. The cells, harvested by centrifugation for 15 min at 470 *g*, are washed twice in HBSS and counted in a hemacytometer. The concentration is adjusted to 5×10^7 viable cells/ml. Typically, the viability of BP-cultured spleen cells is 30–40%.

Viable cells ($1-5 \times 10^7$, 0.2–1 ml of the cell suspension) are injected intraperitoneally into each female Lewis rat recipient. Clinical signs of EAE develop 4–6 days later. The clinical and histologic manifestations of the disease are identical to those seen in actively immunized rats (Figs. 3 and 4).

Discussion

The methods described in this chapter have provided us with a highly reproducible model to investigate the cellular mechanism of autoimmune encephalomyelitis.[5,13–17] In particular, the discovery that EAE can be transferred with relatively small numbers of activated donor lymphocytes[3–7] will permit evaluation of the cellular interactions involved in the induction of this autoimmune disease. It has already been demonstrated that antigen-presenting cells,[13] helper T cells,[14,15] and interleukin 2[17] play a role in the induction of EAE. Clearly, future investigations will provide further insight into the pathogenesis of this autoimmune disease which has served as an experimental model of human demyelinating diseases.[1]

Acknowledgments

Work carried out in the author's laboratory has been supported by National Institutes of Health Grant NS-06985-19 and National Multiple Sclerosis Society Grant RG1073-C-6. I thank Mrs. Chuty Erves for preparing the manuscript.

[13] J. A. Killen and R. H. Swanborg, *J. Immunol.* **129,** 759 (1982).
[14] R. H. Swanborg, *J. Immunol.* **130,** 1503 (1983).
[15] C. B. Pettinell and D. E. McFarlin, *J. Immunol.* **127,** 1420 (1981).
[16] J. H. Holda, D. Silberg, and R. H. Swanborg, *J. Immunol.* **130,** 732 (1983).
[17] L. Ortiz-Ortiz and W. O. Weigle, *J. Immunol.* **128,** 1545 (1982).

[38] Experimental Glomerulonephritis

By DAVID J. SALANT and ANDREY V. CYBULSKY

Introduction

There now exist several animal models that closely resemble various forms of human glomerulonephritis. The resemblance includes, in most instances, morphological, immunohistological, and pathophysiological similarities. While classic studies of acute serum sickness and antiglomerular basement membrane (anti-GBM)[1] antibody disease (nephrotoxic

[1] Abbreviations: GBM, glomerular basement membrane; IF, immunofluorescent microscopy; C, complement; MN, membranous nephropathy; EM, electron microscopy; PHN, passive Heymann nephritis; HN, Heymann nephritis; BN, Brown Norway; SD, Sprague–Dawley; PBS, phosphate-buffered saline; DDW, distilled deionized water; CFA, Freund's complete adjuvant; IFA, Freund's incomplete adjuvant; FITC, fluorescein isothiocyanate; DOC, sodium deoxycholate; MVBS, modified isotonic veronal-buffered saline; ANA, antinuclear antibody; TBM, tubular basement membrane; cBSA, cationized bovine serum albumin; CVF, cobra venom factor; BPB, *p*-bromophenacyl bromide; PVP, polyvinylpyrrolidone.

serum nephritis) led to the concept that the immunopathogenesis of glomerular injury was separable into one of two categories distinguishable by immunofluorescent microscopy (IF),[1a] namely, immune complex glomerulonephritis characterized by granular deposits of immunoglobin and anti-GBM nephritis represented by linear staining of the GBM, recent years have witnessed a change in thinking.[2] Although circulating immune complexes may be trapped in glomeruli, especially in a mesangial or subendothelial location, such deposition generally does not cause morphological or functional evidence of glomerular injury. On the other hand, it is now apparent from animal studies that immune complexes often form *in situ* from the binding of circulating antibodies to fixed or planted glomerular antigens and that such locally formed complexes regularly produce glomerular damage. Fixed antigens may include intrinsic components of glomerular epithelial, endothelial, or mesangial cells or matrix constituents of the GBM or mesangium. Planted antigens may localize in glomeruli by the interaction of circulating endogenous or exogenous substances with cell surface or matrix components. The IF staining pattern (linear versus granular) of such locally formed immune deposits and the likelihood of their being visible as electron-dense aggregates thus depend more on the distribution, density, and mobility of the antigens than on the mechanism of deposit formation.

Mediators of the glomerular injury induced by such immune deposits vary widely in different models. They include the complement (C) system acting either through its terminal, membranolytic pathway or via the inflammatory action of leukocytes; antibody-directed influx of leukocytes; and antibody alone.[3] It has been shown that host species, the nature and dose of antibody, and the distribution of antigen influence which mediators will operate in the various models studied to date. In this chapter, we describe methods for producing and studying three models with which we have experience and which cover the spectrum of fixed and planted antigens and various mediator systems.

Active and Passive Heymann Nephritis

Heymann nephritis[4] (HN) is induced in Lewis and other susceptible[5] strains of rats by immunization with a proximal renal tubular epithelial cell

[1a] C. B. Wilson and F. J. Dixon, *in* "The Kidney" (B. M. Brenner and F. C. Rector, eds.), 3rd Ed., p. 800, Saunders, Philadelphia, Pennsylvania, 1986.

[2] W. G. Couser and D. J. Salant, *Kidney Int.* **17**, 1 (1980).

[3] W. G. Couser and D. J. Salant, *in* "Nephrotic Syndrome: Contemporary Issues in Nephrology" (B. M. Brenner and J. H. Stein, eds.), Vol. 9, p. 47. Churchill-Livingstone, New York, 1982.

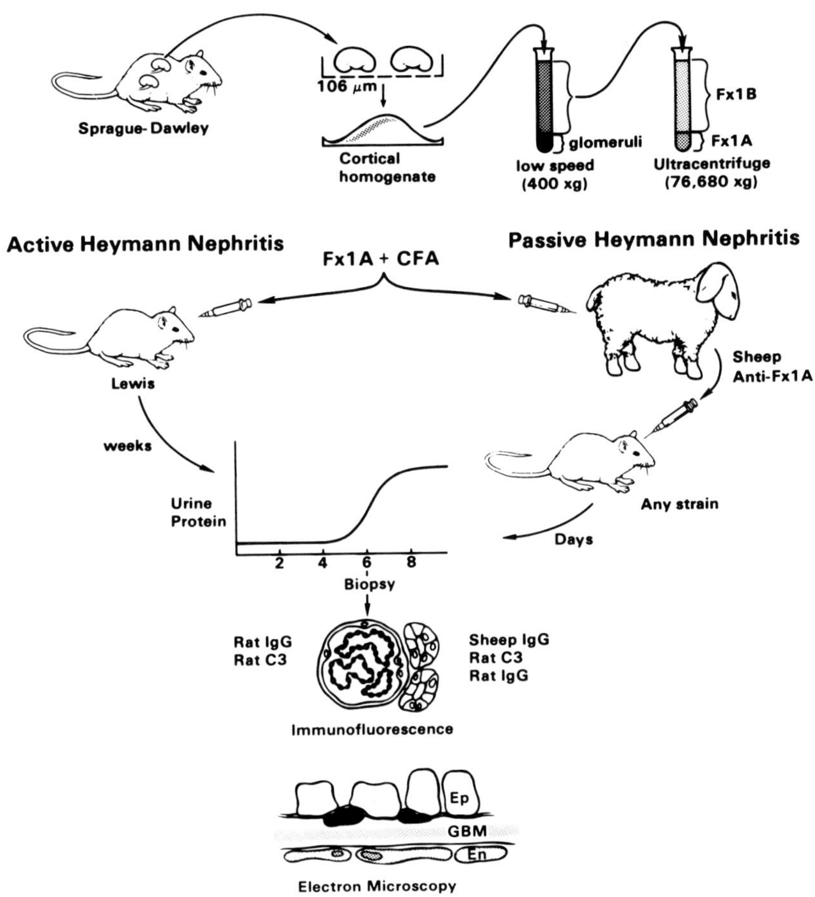

FIG. 1. Protocol for producing active and passive Heymann nephritis. See text for details. CFA, Freund's complete adjuvant; Ep, glomerular visceral epithelial cell; GBM, glomerular basement membrane; En, endothelial cell.

brush border fraction (Fx1A)[6] as illustrated in Fig. 1. This results, within 4–8 weeks, in an immune complex glomerulonephritis that closely resembles human membranous nephropathy (MN). On IF, all animals have staining for rat IgG in a granular pattern decorating the glomerular capillary wall. On electron microscopy (EM), this staining is seen to be due to

[4] W. Heymann, D. B. Hackel, S. Harwood, S. G. F. Wilson, and J. L. P. Hunter, *Proc. Soc. Exp. Biol. Med.* **100,** 660 (1959).

[5] B. Stenglein, G. H. Thoenes, and E. Gunther, *Clin. Exp. Immunol.* **33,** 88 (1978).

[6] T. S. Edgington, R. J. Glassock, J. I. Watson, and F. J. Dixon, *J. Immunol.* **99,** 1199 (1967).

electron-dense deposits in an exclusively subepithelial location on the outer aspect of the GBM. Abnormal proteinuria occurs within 8 weeks in about 60% of immunized rats and is associated with glomerular deposits of C in a very similar distribution to IgG. Reimmunization of persistently nonproteinuric rats will induce C deposition and proteinuria in about 60% of those reimmunized.[7] Proteinuria persists for the life of the rats despite cessation of new immune deposit formation. Lifespan is not substantially decreased compared to unimmunized, caged rats.

Passive Heymann nephritis (PHN) is induced with heterologous anti-Fx1A[8,9] as depicted in Fig. 1. Within minutes of injection of anti-Fx1A, deposits of the heterologous IgG are detectable in glomeruli by IF, and, after 3–5 days, one can visualize subepithelial electron-dense deposits that are very similar to those seen at 4–6 weeks in HN and in very early human MN. Moreover, with a suitable C-fixing antiserum, glomerular capillary wall deposits of C are found after about 3 days followed by heterologous-phase proteinuria on the fourth or fifth day postinjection.[9] Glomerular deposits of rat IgG, signaling the onset of an autologous (anti-IgG) reaction, are usually visible about 3 days after injection. In the absence of heterologous-phase proteinuria, the autologous phase does not manifest with overt proteinuria until about 2 weeks have elapsed (see below). When superimposed on heterologous-phase injury, the autologous phase causes an increase in proteinuria after about 7–10 days. Proteinuria in both the heterologous and autologous phases is C mediated but leukocyte independent.[10,11] Rats with PHN that become severely proteinuric remain so for life. As in HN, new immune deposit formation ceases within a few weeks of the onset of proteinuria and lifespan is not appreciably shortened.

Originally, both HN[6] and PHN[8] were believed to be the result of nonspecific trapping of circulating immune complexes of antibody and small quantities of soluble tubular brush border antigen. We now recognize that the glomerular deposits actually form *in situ* as the result of antibody binding to fixed glomerular antigens.[12,13] These antigens, common to

[7] B. Noble, J. B. Van Liew, G. Andres, and J. R. Brentjens, *Clin. Immunol. Immunopathol.* **30,** 241 (1984).

[8] A. Z. Barabas and R. Lannigan, *Br. J. Exp. Pathol.* **55,** 47 (1974).

[9] D. J. Salant, S. Belok, M. M. Stilmant, C. Darby, and W. G. Couser, *Lab. Invest.* **41,** 89 (1979).

[10] D. J. Salant, S. Belok, M. P. Madaio, and W. G. Couser, *J. Clin. Invest.* **66,** 1339 (1980).

[11] S. Adler, D. J. Salant, J. E. Dittmer, H. G. Rennke, M. P. Madaio, and W. G. Couser, *Kidney Int.* **23,** 807 (1983).

[12] B. J. C. van Damme, G. J. Fleuren, W. W. Bakker, R. L. Vernier, and P. J. Hoedemaker, *Lab. Invest.* **38,** 502 (1978).

[13] W. G. Couser, D. R. Steinmuller, M. M. Stilmant, D. J. Salant, and L. M. Lowenstein, *J. Clin. Invest.* **62,** 1275 (1978).

proximal tubular epithelial cells, are also produced, transported, and expressed on the surface of glomerular visceral epithelial cells.[14] It is currently believed that redistribution of such glomerular epithelial cell antigens follows antibody binding; subsequent shedding of the resulting complexes is thought to account for the observed subepithelial electron densities.[15]

Selection of Species

Not all rat strains are equally susceptible to HN. Susceptibility appears to be multifactorial and includes genetic and environmental factors.[5] Strains carrying the Lewis, Fischer, AS, BDV, and AS2 genetic backgrounds are most susceptible, whereas Brown Norway (BN), AVN, and DA are resistant. Susceptibility or resistance to HN is also conferred, at least in part, by major histocompatibility *(RT-1)* genes. Thus, strains with the Lewis background also carry a *RT-1*-linked susceptibility gene *($RT\text{-}1^l$)*, whereas strains with the BN genetic background carry $RT\text{-}1^n$, an HN resistance gene. Close but incomplete linkage between HN susceptibility genes and RT-1 is suggested by RT-1 segregation studies involving Lewis × BN and Lewis × L.BN hybrids.[5] The means by which genetic factors confer susceptibility or resistance to HN is unknown.

PHN, in contrast to HN, can be induced in all rat strains with heterologous anti-Fx1A. Using sheep antiserum to Sprague–Dawley (SD) Fx1A (described below) we have observed identical IF and ultrastructural glomerular immune deposits and similar quantities of heterologous-phase proteinuria in SD, Lewis, BN, and Munich–Wistar rats.[16]

Species other than rats are resistant to HN and PHN. Although considerable cross-species reactivity exists with regard to proximal tubular brush border antigens,[16] only in rats are certain of these antigens expressed on glomerular epithelial cells. Thus glomerular IF deposits typical of PHN can be produced in rats with various heterologous antisera to rat, human, and guinea pig Fx1A but cannot be induced in the species (other than rat) from which the Fx1A is derived.[2,16,17] However, when normal mice are injected with a heterologous antiserum against a Pronase digest of mouse Fx1A,

[14] D. Kerjaschki and M. G. Farquhar, *J. Exp. Med.* **157,** 667 (1983).

[15] G. Camussi, J. R. Brentjens, B. Noble, D. Kerjaschki, F. Malavasi, O. A. Roholt, M. G. Farquhar, and G. Andres, *J. Immunol.* **135,** 2409 (1985).

[16] W. G. Couser and D. J. Salant, unpublished observations (1978).

[17] K. J. M. Assman, R. Ronco, M. M. Tangelder, W. P. H. Lange, P. Verroust, and R. A. P. Koene, *Am. J. Pathol.* **121,** 112 (1985).

granular glomerular capillary wall immune deposits are seen on IF.[18] These IF deposits are more homogeneous than those seen in rat PHN, and their ultrastructural appearance is different.[17] Furthermore, some antigens detected on rat glomerular epithelial cells by immunohistochemistry are not detectable on mouse glomeruli.[17]

Rabbit and sheep anti-Fx1A are most often used for inducing PHN. We prefer sheep antiserum because (1) it produces heterologous-phase proteinuria more consistently than rabbit anti-Fx1A; (2) large volumes of antiserum can be obtained from one adequately immunized animal; (3) it may be advantageous to be able to separate the immune IgG into C-fixing and non-C-fixing subclasses.

Induction of HN

Preparation of the Immunogen (Fx1A) (Fig. 1). Fx1A is isolated from Sprague–Dawley (SD) rat kidneys by sieving and ultracentrifugation as described by Edgington *et al.*[6]

Materials

SD rats (250–300 g) or frozen SD rat kidneys. Twenty kidneys will usually yield enough antigen to immunize at least 20 rats.

Stainless steel ASTME sieve #140 (106-μm mesh opening, 8-inch diameter, W. S. Taylor, Inc., Mentor, OH) and collecting pan; single-edged razor blade; small curved scissors; fine forceps; petri dish; 25-ml Erlenmeyer flask; 250-ml plastic squirt bottle.

Phosphate-buffered saline (PBS) (20 m*M* sodium phosphate, 150 m*M* NaCl, pH 7.2), 6 liters at 4°; ice-filled bucket.

Ether or chloral hydrate (3.6%, 1.2 ml/100 g intraperitoneally) for anesthesia.

Refrigerated centrifuge; ultracentrifuge; lyophilizer.

Procedure

1. Under anesthesia, SD rat kidneys are excised and placed in cold PBS. If the Fx1A is to be used to prepare heterologous antiserum, the kidneys should first be perfused with about 50 ml of cold, heparinized PBS (1 U heparin/ml PBS) via an 18-gauge needle placed in the aorta to free them of blood. Renal perfusion is facilitated by placing a ligature around the aorta above the renal arteries and exsanguinating the rat via a 16-gauge needle inserted into the heart.
2. The kidneys are decapsulated, adherent tissue is dissected free, and

[18] K. J. M. Assman, M. M. Tangelder, W. P. J. Lange, T. M. Tadema, and R. A. P. Koene, *Kidney Int.* **24**, 303 (1983).

they are sliced sagitally into 3–4 pieces with a sharp blade. The cortex is trimmed free from each piece with curved dissecting scissors, and the pooled cortices are washed with PBS, placed in a petri dish on ice, and minced with a razor blade to a fine paste.

3. The cortical paste is gently pressed through a moistened 140-mesh sieve with the undersurface of a 50-ml Erlenmeyer flask using a circular motion. The whole surface of the sieve should be used; passage through the sieve is aided by generous spraying with cold PBS from a plastic squirt bottle. Material passing through the sieve, containing glomeruli and tubular fragments, is collected in a stainless steel collecting pan fitted to the underside of the sieve and placed on ice. No more than 200–300 ml of PBS (squirted and poured) is required for this step. Sieving is continued until only white fibrous material remains on the surface.
4. The sieved material is centrifuged at 400 *g* for 10 min at 4° to sediment glomeruli. The supernatant is recentrifuged once or twice until no sediment remains. The supernatant (fraction 1, Fx1) is rich in tubular epithelium.
5. Fx1 is ultracentrifuged at 78,680 *g* for 45 min at 4°. The sediment is washed with distilled and deionized water (DDW) 3 times, recentrifuging at 78,680 *g* for 45 min after each wash. The final sediment (Fx1A), enriched for proximal tubular brush border membranes, is frozen, lyophilized, and stored at −70° or at 4° in a desiccator. The supernatant (Fx1B) is discarded. A typical yield is 300 mg Fx1A from 20 kidneys.

Immunization. Sufficient Fx1A to provide 10 mg/rat is suspended in PBS to a concentration of 100 mg/ml. This is emulsified with an equal volume of Freund's complete adjuvant (CFA) comprising 4 mg of pulverized *Mycobacterium tuberculosis* H37Ra (Difco Laboratories, Detroit, MI) per milliliter of incomplete Freund's adjuvant (IFA) (Difco Laboratories). (Note that CFA as supplied by Difco Laboratories contains *M. butyricum,* 0.5 mg/ml; it can be used if supplemented with additional *M. butyricum* or *M. tuberculosis* to 4 mg/ml.) Anesthetized Lewis (or other susceptible) rats weighing 60–120 g are injected with 0.1 ml of this emulsion into each of the hind footpads.

Using this single-dose protocol, proteinuria exceeding 50 mg/24 hr (normal, <5 mg/24 hr) usually develops within 8–10 weeks in about 60% of rats; all demonstrate typical glomerular capillary wall deposits of rat IgG after 4 weeks.[19] Some authors favor a booster injection of identical compo-

[19] W. G. Couser, M. M. Stilmant, and C. Darby, *Lab. Invest.* **34,** 23 (1976).

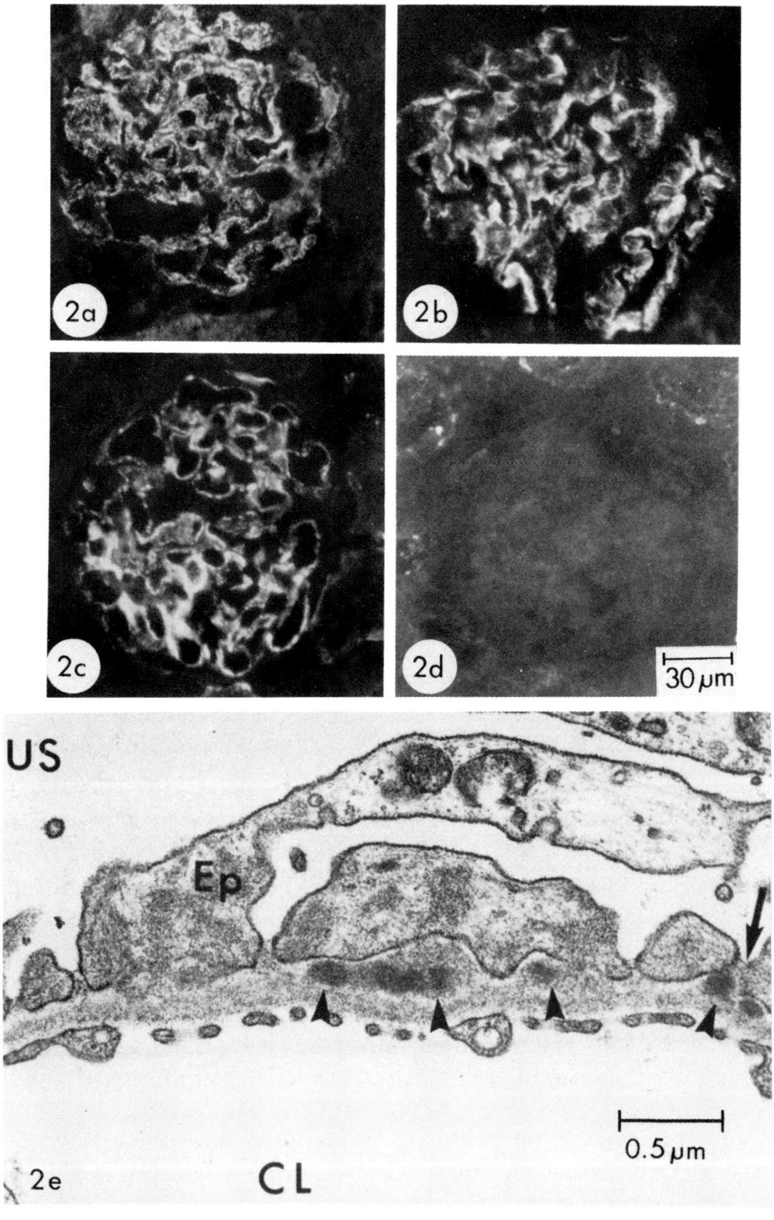

FIG. 2. Immunofluorescent and electron micrographs of representative glomeruli in renal biopsies of rats with passive (PHN) and active Heymann nephritis (HN). (a) PHN rat glomerulus (5 days) stained for sheep IgG. Note the finely granular staining of glomerular capillary walls. (b) HN rat glomerulus (8 weeks) stained for rat IgG. (c) Glomerulus from a

sition in the front footpads[20] or a smaller dose in IFA given subcutaneously[21] (s.c.) after 4 weeks. *Bordetella pertussis* vaccine (Eli Lilly and Co., Indianapolis, IN), 0.05 ml s.c. of a suspension containing 2.0×10^{10} bacteria/ml, at the time of initial immunization has also been used by some investigators.[5,21] It should be emphasized that there is wide variability in the frequency with which different batches of Lewis rats obtained from the same supplier and immunized according to the same protocol develop proteinuria.[7] It is therefore uncertain if these additions to the immunization protocol increase the frequency of proteinuria.

Characteristics of HN

Serum titers of anti-brush border antibody, as measured by indirect IF on normal rat kidneys, have usually peaked (1:32 to 1:64) by 6–8 weeks and then decline.[20] Urine is collected for 18–24 hr from rats by housing them in individual metabolic cages (Fenco Cage Co., Boston, MA) with free access to water but not food. Urine protein excretion, as measured turbidometrically by a salicylsulfonic acid method[22] using Labtrol protein standards (Dade Diagnostics, Inc., Aguada, Puerto Rico), begins to increase after about 4 weeks. From a normal value of less than 5 mg/24 hr, it usually reaches a maximum of 200–300 mg/24 hr after 10 weeks and then plateaus. Histological and immunohistological features vary according to the time of renal biopsy but are most characteristic at about 6 weeks postimmunization. At this time, glomeruli appear normal by light microscopy but show bright granular capillary wall staining for rat IgG (Fig. 2) and less intense staining for rat C3 on IF. Small electron-dense deposits in a subepithelial location are first seen by EM after 4 weeks. By 6 weeks,

[20] D. L. Mendrick, B. Noble, J. R. Brentjens, and G. A. Andres, *Kidney Int.* **18,** 328 (1980).

[21] A. K. Bhan, E. E. Schneeberger, L. G. Baird, A. B. Collins, K. Kamata, D. Bradford, M. E. Erikson, and R. T. McCluskey, *Lab. Invest.* **53,** 421 (1985).

[22] J. Davidson and J. B. Henry, *in* "Clinical Diagnosis by Laboratory Methods," 14th Ed., p. 48. Saunders, Philadelphia, Pennsylvania, 1969.

5-day PHN rat injected with γ_1 sheep anti-Fx1A (C-fixing) and stained for rat C3. The staining pattern is similar to that of sheep IgG but of lesser intensity. (d) Glomerulus from a 5-day PHN rat injected with γ_2 sheep anti-Fx1A (non-C-fixing) and stained for rat C3. The glomerulus is negative for C3 but was positive for sheep IgG (as in Fig. 2a). Similar findings are obtained in rats depleted of complement and given γ_1 anti-Fx1A. (e) Electron micrograph showing a section of glomerular capillary wall from a rat with PHN (10 days). Note the immune deposits (arrowheads) in the lamina rara externa of the glomerular basement membrane with effacement of adjacent epithelial cell (Ep) foot processes and displacement of filtration slit diaphragms (arrow). CL, Capillary lumen; US, urinary space. Composed, in part, from Salant *et al.*[36]

these deposits are larger and are associated with changes in epithelial cell morphology, including podocyte effacement and displacement of slit diaphragms. Proximal tubular cell injury, due to filtration and binding of circulating anti-brush border antibodies, is seen only if serum antibody titers are boosted just prior to the onset of proteinuria.[20]

In severely proteinuric rats, hypoalbuminemia occurs but edema and ascites are unusual. Glomerular filtration rate and renal plasma flow are similarly reduced by about 25–30%, and some authors,[23,24] though not all,[25] have noted considerable heterogeneity of function among nephrons.

Induction of PHN

Preparation and Characterization of Heterologous Anti-Fx1A (Fig. 1). This is a protocol for preparing sheep anti-Fx1A, but rabbit antiserum is produced in the same way except that the dose of Fx1A is reduced to 20 mg/rabbit.

Procedure. Rat Fx1A is prepared as described above. Seventy-five to 100 mg of Fx1A is suspended in 20 ml of PBS and emulsified in an equal volume of CFA as supplied by Difco Laboratories (contains *M. butyricum,* 0.5 mg/ml). The immunogen is injected into multiple subcutaneous sites over the back, thighs, and shoulders of an adult sheep. Four to six boosters of 75 mg Fx1A in IFA are given at monthly intervals, and test bleeds are obtained 1 week after the fourth through sixth boosters. At least 4 months of immunization are necessary to obtain a nephritogenic C-fixing antiserum that causes heterologous-phase proteinuria.

The antiserum is heat inactivated (56°, 30 min) and tested as follows:

1. Indirect IF on normal rat kidney.[26] Serial dilutions of antiserum are applied to ether–ethanol-fixed cryostat sections of normal rat kidney. After washing, bound sheep IgG is detected with an appropriately diluted fluorescein isothiocyanate (FITC)-conjugated anti-sheep IgG such as rabbit anti-sheep IgG (~20 mg/ml, diluted 1:8; Cappel Laboratories, Cochranville, PA). Typically, staining of proximal tubular brush border is seen at anti-Fx1A dilutions of 1:10,000 to 1:64,000 glomeruli remain unstained.
2. Micro-Ouchterlony (see also this series, Vol. 70 [9]). Anti-Fx1A usually forms three precipitin lines at titers up to 1:16 using insolu-

[23] M. E. M. Allison, C. B. Wilson, and C. W. Gottschalk, *J. Clin. Invest.* **53,** 1402 (1974).
[24] I. Ichikawa, J. R. Hoyer, M. W. Seiler, and B. M. Brenner, *J. Clin. Invest.* **69,** 185 (1982).
[25] M. J. Zamlauski-Tucker, J. B. Van Liew, and B. Noble, *Kidney Int.* **28,** 504 (1985).
[26] W. G. Couser and E. J. Lewis, *Am. J. Clin. Pathol.* **61,** 873 (1974).

ble Fx1A at 10 mg/ml as antigen. If Fx1A is solubilized with 1% sodium deoxycholate (DOC), five lines at titers up to 1 : 16 are seen.

3. *In vivo* nephritogenicity. Incremental doses of 0.5 – 1.5 ml are given by slow intravenous (i.v.) injection to 140 to 180-g anesthetized rats. Urine protein excretion is measured (as described above), and renal cortical tissue is obtained by biopsy or at sacrifice on the fifth day and snap-frozen in dry ice – isopentane for IF microscopy. A good antiserum will produce proteinuria (20 – 100 mg/24 hr) after 5 days and show moderately intense granular staining of the glomerular capillary walls by the heterologous IgG and faint to moderate staining for rat C3[9] (Fig. 2) (for details of reagents see below). Faint staining for rat IgG (using a noncross-reactive fluoresceinated antiserum such as goat anti-rat IgG, ~24 mg/ml, diluted 1 : 16, from Cappel Laboratories) is generally visible in a similar distribution. Early adverse effects of the antiserum include some respiratory difficulty and fluffing of the hair on the back. These effects usually resolve within 1 or 2 hr. Single intravenous or intraperitoneal (i.p.) doses of antiserum greater than 1.8 ml often kill rats of this size. Excessive mortality at lower doses may be due to antibodies to rat blood cells or serum proteins. The former can be removed by twice absorbing the antiserum with washed rat blood cells at a ratio of 1 : 1 (v/v); the latter can be detected by Ouchterlony immunodiffusion against rat serum and removed by absorption with CNBr-activated Sepharose-linked rat serum proteins (100 mg serum : 20 ml antiserum). The immunoabsorbent is prepared as described by the manufacturer (Pharmacia) and can be reused after regeneration with 0.1 *M* glycine – HCl, pH 2.5.

Once the antiserum meets these criteria, the sheep is bled via the external jugular vein. One liter of blood can be removed safely at each bleeding from an adult sheep and collected into 0.15 *M* EDTA (60 ml EDTA/liter of blood). The plasma is separated and coagulated with calcium chloride – magnesium chloride (0.3 *M* Ca^{2+}, 0.1 *M* Mg^{2+}), 1 ml/15 ml plasma, for 1 hr at 37° followed by dialysis against PBS. After removing the fibrin clot that forms during dialysis, the defibrinated antiserum is heat inactivated (56°, 30 min), absorbed as necessary with rat cells and serum, filtered (0.45 μm), and lyophilized or stored frozen at −70°.

Characteristics of PHN

After a single nephritogenic i.v. dose of sheep anti-Fx1A, rats have faint IF deposits of sheep IgG in glomeruli within minutes; these increase in intensity over 3 – 5 days, becoming more granular with time (Fig. 2). When a C-fixing antiserum is used, rat C appears in glomeruli after 3 days (Fig. 2) and C-mediated, leukocyte-independent proteinuria (20–100 mg/24 hr) occurs after 5 days[10] (see below for methods of leukocyte and C depletion). Terminal components of complement, including the C5b-9 membrane attack complex, can be demonstrated in glomeruli in the same distribution as sheep IgG and rat C3.[27] Although rat IgG is visible in glomeruli after about 3 days, autologous-phase injury becomes apparent only after about 7–10 days, when proteinuria begins to increase further, reaching about 150–250 mg/24 hr after 14 days. The severity of both heterologous- and autologous-phase injury is dose dependent. Early autologous-phase proteinuria is also C mediated and leukocyte independent.[11] Glomerular histology on light microscopy during the first few weeks after immunization is normal and shows no inflammatory cell infiltration. In addition to subepithelial electron densities, EM shows visceral epithelial cell changes that resemble those in early HN described above (Fig. 2). Depending on the dose and nephritogenicity of the antiserum, glomerular filtration rate and renal plasma flow are reduced by about 20% during the period of maximal proteinuria.[24] In contrast to HN, these effects are more homogeneous from one glomerulus to another.[24]

Rats followed for extended periods show persistent but less severe proteinuria.[28] Heterologous and host IgG are present on IF for several months but do not represent new immune deposits. After about 3 months, GBM thickening and prominence of the mesangial matrix become apparent on light microscopy, and focal sclerosing lesions of glomeruli may appear.[28] EM reveals progressive GBM thickening around immune deposits with ultimate dissolution of the latter leaving lucent zones within the thickened GBM.[28]

Isolation of Sheep Anti-Fx1A IgG Subclasses

Both subclasses of IgG obtained from sheep hyperimmunized with rat Fx1A contain antibodies that form glomerular deposits. However, the ability to cause heterologous-phase proteinuria resides with the C-fixing γ_1 subclass.[10] Thus, qualitatively and quantitatively identical subepithelial immune deposits can be produced with γ_1 and γ_2 anti-Fx1A, but only the former activate C *in vivo* and cause proteinuria. It is therefore often useful to isolate anti-Fx1A IgG subclasses, a procedure that is facilitated by

[27] D. T. Perkinson, P. J. Baker, W. G. Couser, R. J. Johnson, and S. Adler, *Am. J. Pathol.* **120,** 121 (1985).

[28] B. M. Iverson, R. Matre, and J. Ofstad, *Acta. Pathol. Microbiol. Immunol. Scand.* **90,** 241 (1982).

differences in electrical charge of γ_1 and γ_2 and is readily accomplished by ion-exchange chromatography on DEAE-Sephacel (Pharmacia).

Materials

Saturated and 50% ammonium sulfate.
Sodium phosphate, 0.5 *M*, pH 7.4, 2 liters.
Sodium phosphate, 17.5 m*M*, pH 7.4 (starting buffer), 12 liters.
Sodium chloride, 150 m*M* in starting buffer (elution buffer), 2 liters.
DEAE-Sephacel (Pharmacia), 215 ml.
Glass chromatography column (2.6 × 40 cm).
High-speed refrigerated centrifuge; UV monitor (optional); fraction collector; gradient mixer.

Procedure. The procedure is described for 100 ml of starting material and is carried out at 4°.

1. Saturated ammonium sulfate (100 ml) is added dropwise over several hours to the absorbed and heat-inactivated antiserum (100 ml) under continuous stirring to produce a 50% ammonium sulfate precipitate.[29] After washing with 50% ammonium sulfate, the precipitate is dissolved in 30–40 ml PBS and then dialyzed against a 2000:1 excess of PBS followed by dialysis against starting buffer. Ample space must be allowed in the dialysis tubing to allow for sample expansion on transfer from PBS to the hyposmolar starting buffer. A euglobulin precipitate that forms during dialysis should be removed by centrifugation and discarded. An aliquot of the dialyzed sample is tested with a drop of 10% barium chloride to ensure that all sulfate has been removed.
2. DEAE-Sephacel (210 ml) is equilibrated with 0.5 *M* sodium phosphate (four 500-ml changes over 24–48 hr) at room temperature followed by starting buffer (four 500-ml changes over 12–24 hr) at 4°. The final pH should be 7.4. The gel is poured and packed into a glass column (2.6 × 40 cm) with about 500–800 ml of starting buffer and the flow rate adjusted to about 0.5–1.0 ml/min.
3. After applying the sample, about 300 ml of starting buffer is run through to elute the γ_2 subclass. Fractions of 5–7.5 ml are collected and OD_{280} values monitored (continuously with an in-line UV monitor or spectrophotometrically on collected fractions). γ_2 IgG elutes as a broad single peak in the void volume; thereafter OD_{280} returns to baseline (Fig. 3).
4. The γ_1 subclass is then eluted by applying a linear sodium chloride

[29] P. D. Gorevic, F. C. Prelli, and B. Frangione, this series, Vol. 116, p. 6.

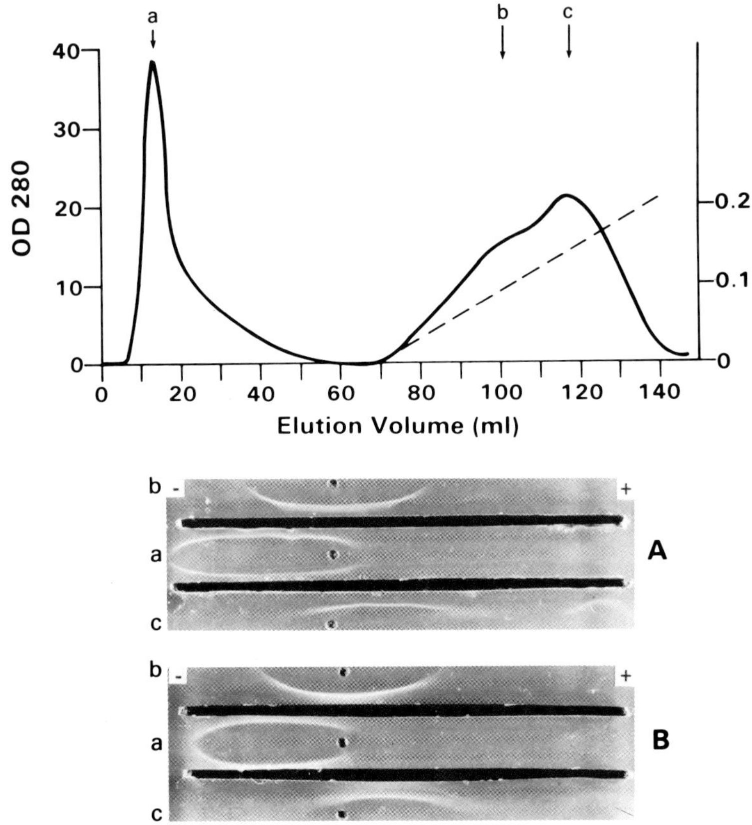

FIG. 3. Purification of sheep anti-Fx1A subclasses by ion-exchange chromatography (DEAE-Sephacel) and their characterization by immunoelectrophoresis (IEP). The first peak (containing γ_2) was obtained with 17.5 mM phosphate starting buffer; γ_1 was eluted with a linear gradient (dashed line) of NaCl in the starting buffer. Samples obtained at arrows a, b, and c were examined by IEP against (A) rabbit anti-sheep serum proteins (Miles Laboratories, Inc., Elkhart, IN) and (B) rabbit anti-sheep IgG (Miles Laboratories). Fraction a was pure IgG that migrated cathodally; fraction b contained mostly IgG that straddled the well and a very small amount of β-globulin; fraction c contained IgG that migrated anodally and was contaminated with other serum proteins. Electrophoresis was performed in 1% agarose for 1 hr at 50 V.

gradient using equal volumes (250 ml) of starting and 150 mM elution buffer. Most of the γ_1 IgG bound to the ion exchanger is eluted as a broad "peak" (Fig. 3) that includes a small amount of contaminating non-IgG globulins. Pure γ_1 is found in the early part of the gradient. Monitoring successive fractions by immunoelectrophoresis with an antiserum to sheep serum proteins (see Fig. 3 for details) allows one to assess the fractions for purity of IgG. Alternatively, a higher yield of impure γ_1 may be obtained by stepwise elution with 0.3 *M* sodium chloride in starting buffer. The material can be further purified by gel filtration on Sephacryl S-200 (Pharmacia) to obtain the 150-kDa protein peak.

5. Fractions containing the isolated subclasses are pooled, concentrated, dialyzed against PBS, filtered (0.45 μm), and stored at $-70°$. Typically, about 500 mg of γ_2 and 2000 mg of γ_1 are isolated from 100 ml of hyperimmune sheep serum.

Characterization of Sheep Anti-Fx1A IgG Subclasses

Immunoelectrophoresis (Fig. 3). On immunoelectrophoresis in 1% agarose at pH 8.2 against anti-sheep IgG, γ_2 makes a single arc that extends cathodally from the well, while γ_1 makes a single arc that straddles the well but extends anodally.

Isoelectric Focusing. Electrofocussing has shown the following isoelectric (p*I*) ranges[10]: γ_1 5.3–7.8; γ_2 8.1–9.6.

In Vitro Fixation of C.[10,30] Normal mouse (or rat) kidney is incubated with each IgG subclass of anti-Fx1A to produce proximal tubular brush border deposits of IgG in the presence of serum as a source of C. Positive C fixation is then detected by straining for C3. Mouse kidney is the preferred substrate as it produces less nonspecific background staining. *In vitro* C fixation should be tested in the presence of fresh and heat-inactivated serum from the species in which studies are to be done (e.g., rat in the case of PHN). However, we have found that fresh human serum is a convenient and satisfactory source to screen for C fixation.

Materials

Acid-cleaned glass slides (1 × 3 inch).
Cryotome and fluorescent microscope.
Modified isotonic veronal-buffered saline (MVBS): barbital (3 m*M*)–barbital sodium (1.8 m*M*)–NaCl (140 m*M*), with Ca^{2+} (0.15 m*M*) and Mg^{2+} (0.5 m*M*), pH 7.4 (freshly prepared).[31]
PBS.
Fresh or fresh-frozen human serum (about 5 ml). Half of the serum is heat inactivated at 56° for 30 min.
γ_1 and γ_2 anti-Fx1A (from heat-inactivated antiserum as above).
Nonimmune sheep IgG.

[30] P. M. Burkholder, *J. Exp. Med.* **114,** 606 (1961).

[31] M. M. Mayer, *in* "Experimental Immunochemistry" (E. A. Kabat and M. M. Mayer, eds.), 2nd Ed., p. 149. Thomas, Springfield, Illinois, 1961.

High-titer human antinuclear antibody (ANA) serum (positive control) from a patient with systemic lupus erythematosis.

Monospecific goat anti-human C3–FITC, ~30 mg/ml (Cappel Laboratories), absorbed twice with mouse liver powder (Cappel Laboratories).

Monospecific rabbit anti-sheep IgG-FITC (H and L chain-specific, Cappel Laboratories) and rabbit anti-human IgG–FITC, ~17 mg/ml (Cappel Laboratories).

Procedure

1. Cryostatic sections of normal mouse kidney on eleven slides (Table I) are fixed (10 min) in ether–95% ethanol (1 : 1), and washed in PBS.
2. Sections are then incubated at room temperature for 30 min with γ_1 and γ_2 anti-Fx1A, each diluted to 0.5 mg/ml in fresh MVBS containing fresh or heat-inactivated human serum (1 : 20) (Table I).
3. After washing with PBS (5 min, 3 times), sections are stained for 30 min at room temperature with goat anti-human C3–FITC (1 : 8) or with rabbit anti-sheep IgG–FITC (1 : 8).
4. Sections incubated with nonimmune sheep IgG and stained exactly as above serve as negative controls.

TABLE I

In Vitro AND *In Vivo* COMPLEMENT FIXATION OF SHEEP ANTI-Fx1A IgG SUBCLASSES ON NORMAL MOUSE OR RAT KIDNEY

Incubation components[a]	Stain[b]	Result[c]
γ_1 anti-Fx1A + HIS	Anti-sheep IgG	PCT
γ_1 anti-Fx1A + HIS	Anti-human C3	—
γ_1 anti-Fx1A + FHS	Anti-human C3	PCT
γ_2 anti-Fx1A + HIS	Anti-sheep IgG	PCT
γ_2 anti-Fx1A + HIS	Anti-human C3	—
γ_2 anti-Fx1A + FHS	Anti-human C3	—
Nonimmune sheep IgG + FHS	Anti-sheep IgG	—
Nonimmune sheep IgG + FHS	Anti-human C3	—
ANA + HIS	Anti-human IgG	Nuclear
ANA + HIS	Anti-human C3	—
ANA + FHS	Anti-human C3	Nuclear

[a] Cryostat sections of normal kidney are incubated with γ_1 or γ_2 anti-Fx1A, nonimmune sheep IgG, or antinuclear antiserum (ANA) in the presence of fresh (FHS) or heat-inactivated (HIS) human serum.

[b] Sections are stained with fluoresceinated anti-sheep IgG and anti-human IgG and C3.

[c] Staining of proximal tubular brush borders (PCT) or cell nuclei.

5. Additional sections, incubated with ANA diluted 1:20 (or to another predetermined dilution that gives positive nuclear staining for human IgG with minimal background staining) in MVBS containing fresh or inactivated human serum (1:20) and stained for human C3 and IgG as above, are included as positive controls.

Table I indicates the expected results with the various test and control reagents. Positive staining for IgG on proximal tubular brush borders is observed with γ_1 and γ_2, whereas only γ_1 anti-Fx1A produces positive brush border staining for C3.

In Vivo Fixation of Complement. Renal tissue is obtained 4 or 5 days after a 15-mg i.v. injection of γ_1 or γ_2 anti-Fx1A and is stained with rabbit anti-sheep IgG–FITC and goat anti-rat C3–FITC (Cappel Laboratories). Both antisera produce granular glomerular capillary staining for sheep IgG, but only kidneys from rats injected with γ_1 anti-Fx1A have positive glomerular staining for rat C3. (Fig. 2). Furthermore, when sections of kidneys from rats injected with γ_1 or γ_2 anti-Fx1A are tested for *in vitro* C fixation in the presence of fresh human serum (as above), only those from γ_1-injected rats have positive glomerular staining for human C3.

Autologous Phase of PHN

In PHN rats with heterologous-phase proteinuria, autologous-phase injury is marked by worsening proteinuria, but its onset is obscured by the former. By using non-C-fixing γ_2 anti-Fx1A, one can create pure autologous-phase injury, the onset of which is usually observed after 15–16 days as severe proteinuria.

Accelerated Autologous Phase. To accelerate the onset of autologous phase, rats are preimmunized with sheep IgG (1 mg in CFA s.c.) followed 5 days later by 8–10 mg i.v. of γ_2 sheep anti-Fx1A. This procedure results in no heterologous injury but advances the onset of autologous-phase proteinuria to 7–8 days post anti-Fx1A.[11] Whereas heterologous-phase proteinuria is unaffected by immunosuppression, autologous injury is prevented by sublethal total body irradiation (658 rads at 129 rads/min) or L-asparaginase (2000 U/kg/day starting on the day before anti-Fx1A). The disadvantages of this model is that preimmunization produces circulating anti-sheep IgG antibodies which partly complex with and clear the subsequently administered sheep anti-Fx1A IgG. This results in fewer glomerular immune deposits and less severe injury (mean proteinuria 28 mg/day).

Transplant Model of Autologous-Phase Injury. The transplant model is used to hasten the onset of severe autologous-phase injury while avoiding the disadvantages mentioned above in native kidneys. Pure autologous-phase proteinuria that is C mediated and leukocyte independent[11] can be induced by transplanting rat kidneys containing subepithelial deposits of sheep IgG into recipients actively or passively immunized against sheep IgG. Sheep IgG is immunologically planted in donor glomeruli by injecting anti-Fx1A. To avoid heterologous injury the γ_2 subclass of anti-Fx1A is injected, and kidneys are transplanted 3 days later. γ_2 sheep IgG thus severe as a nonnephritogenic antigen for subsequent *in situ* immune complex formation.[11,32] Three days postinjection is selected for transplantation because kinetic studies show approximately 70% of maximum glomerular binding of antibody at this time[33] but no detectable autologous disease.

Active model. Inbred (Lewis) rats (250–300 g) are used as donor and recipient to avoid the problem of transplant rejection as glomerular injury is only fully developed 5 days after transplantation.[11]

Procedure

1. Recipient rats are immunized with 1 mg of sheep IgG in CFA s.c. 6 days before transplantation.
2. Donors are given 6–8 mg γ_2 sheep anti-Fx1A i.v. 3 days prior to transplantation. (This dose may require adjustment according to potency of the antiserum to achieve the desired outcome.)
3. Urine protein excretion is measured 24 hr before transplantation in both donor and recipient to confirm the absence of proteinuria and hence the lack of nephritogenicity of γ_2 anti-Fx1A or preimmunization.
4. The left kidneys of donor rats are orthotopically transplanted into recipients using a modification of the technique of Lee.[34] Donor kidneys are gently perfused with 3 ml of chilled, heparinized RPMI medium (Gibco, Grand Island, New York) prior to removal. A short segment of donor aorta and a vena cava cuff are left attached to the renal artery and vein, respectively. End-to-side anastomoses are then performed to the recipient aorta and inferior vena cava. Ischemia time is about 30–40 min. After removing the vascular clamp, most kidneys reperfuse immediately. The donor and recipient ureters are anastomosed end-to-end over a short piece of 0.025-inch o.d. silastic tubing (Dow Corning Corp., Midland, MI) as suggested by Daniller *et al.*[35] Most kidneys produce urine soon thereafter.

   Depending on the purpose of the study, the recipient is unilaterally or bilaterally nephrectomized. Leaving a single host kidney provides a useful control for comparative purposes but precludes

[32] L. A. van Es, A. P. R. Blok, L. Schoenfeld, and R. J. Glassock, *Kidney Int.* **11,** 106 (1977).
[33] D. J. Salant, C. Darby, and W. G. Couser, *J. Clin. Invest.* **66,** 71 (1980).
[34] S. Lee, *Surgery* **61,** 771 (1967).
[35] A. Daniller, R. Buchholz, and R. A. Chase, *Surgery* **63,** 956 (1968).

using whole animal clearance techniques or serum creatinine (Worthington Diagnostics, Freehold, NJ) for assessing function of the renal isograft. Other controls should include normal kidneys transplanted into preimmunized recipients and antigen-containing kidneys transplanted into nonimmunized recipients.

Passive Model. Inbred or outbred (e.g., Sprague–Dawley) rats can be used for acute studies in this model as severe injury (manifesting as proteinuria) is apparent within 48 hr of transplantation.

Procedure

1. Rat anti-sheep IgG is prepared by repeated immunizations (3–4 times at monthly intervals) of 20 SD (150-g) rats with pure sheep IgG (2 mg) injected at multiple s.c. and intramuscular (i.m.) sites. The initial immunization is given in CFA and subsequent boosters in IFA. Randomly selected rats are test bled 1 week after each booster, and, when most demonstrate a titer of 1:4 or greater on micro-Ouchterlony against sheep IgG (1 mg/ml), serum is collected by exsanguination, pooled, heat inactivated, and stored at $-70°$.
2. Donor rats are given γ_2 anti-Fx1A (6–8 mg) i.v. 3 days before transplantation as above.
3. The left donor kidney is orthotopically transplanted (as above) into unilaterally or bilaterally nephrectomized naive recipients. Immediately thereafter, 1.5–2.0 ml of rat anti-sheep IgG is injected i.v. (This dose may need adjustment if the desired outcome is proteinuria.)
4. The same controls should be included as in the active model.

With suitable antisera, 6 mg γ_2 anti-Fx1A, trace labeled with ^{125}I, usually produces 9 μg of glomerular-bound sheep IgG per kidney, and 15 mg of the IgG fraction of rat anti-sheep IgG, trace labeled with ^{131}I, produced 31 μg of glomerular-bound rat antibody per kidney 48 hr after transplantation.[36] This combination induced moderate proteinuria after 48 hr. More severe proteinuria is produced with 8 mg γ_2 anti-Fx1A and the same dose of rat anti-sheep IgG.

Characteristics of the PHN Autologous-Phase Models

The active and passive transplant models are similar to each other as well as to the accelerated autologous-phase model in native kidneys with respect to the presence of electron-dense immune deposits located exclu-

[36] D. J. Salant, S. Adler, C. Darby, N. J. Capparell, G. C. Groggel, I. D. Feintzeig, H. G. Rennke, and J. E. Dittmer, *Kidney Int.* **27**, 938 (1985).

sively in the subepithelial space, noninflammatory glomerular histology, granular glomerular capillary wall deposits of sheep and rat IgG and rat C3 on IF, and C-mediated leukocyte-independent proteinuria.[11,36] The main difference lies in the rate of onset and magnitude of proteinuria. In the active transplant model, moderate proteinuria (mean 33 mg/day, normal <4 mg/day) occurs within 5 days. In the passive model, heavy proteinuria (mean 102 mg/day) develops within 48 hr, which greatly facilitates manipulations designed to elucidate mediation mechanisms, such as depletion of C and leukocytes, and pharmacological maneuvers.

Antiglomerular Basement Membrane (Anti-GBM) Nephritis

Anti-GBM antibody is broadly defined as any antibody that reacts with the GBM in a linear, capillary wall, pattern on IF (Fig. 5). Active anti-GBM nephritis (Steblay's, autoimmune nephritis) is produced by active immunization of the host species with a glomerular preparation and has been best described in sheep[37] and guinea pigs[38]; it will not be discussed further here. Rather, we describe methods for producing the more widely employed passive model, induced with a single i.v. injection of heterologous anti-GBM antiserum. This antiserum is classically raised against crude preparations of glomeruli or GBM and is probably reactive with multiple GBM, endothelial, and possibly epithelial cell wall constituents.[39] The actual antigens recognized have not yet been characterized. Antisera have also been raised to purified components of the capillary wall, including Type IV collagen,[40] proteoglycan,[41] and laminin.[42] While injection of these antisera results in linear GBM staining on IF, in general, their capacity to induce disease appears to be limited.

In classic anti-GBM nephritis, heterologous-phase proteinuria develops within the first 24 hr and resolves after 48–72 hr. Reappearance of proteinuria after 8–10 days represents autologous-phase injury. The evolution of histopathological and pathophysiological changes, as well as the variety of mechanisms that mediate injury during the heterologous and autologous phases, were recently reviewed in depth.[1,3] Suffice it to say that heterologous-phase injury is produced by humoral mediators, including antibody

[37] R. W. Steblay, *J. Exp. Med.* **116,** 253 (1962).

[38] W. G. Couser, M. M. Stilmant, and E. J. Lewis, *Lab. Invest.* **29,** 235 (1973).

[39] D. L. Mendrick, H. G. Rennke, R. S. Cotran, T. A. Springer, and A. K. Abbas, *Lab. Invest.* **49,** 107 (1983).

[40] M. Yaar, J. M. Foidart, K. S. Brown, S. I. Rennard, G. R. Martin, and L. Liotta, *Am. J. Pathol.* **107,** 79 (1982).

[41] H. Makino, J. T. Gibbons, M. K. Reddy, and Y. S. Kanwar, *J. Clin. Invest.* **77,** 142 (1986).

[42] D. R. Abrahamson and J. P. Caulfield, *J. Exp. Med.* **156,** 128 (1982).

alone,[43] antibody and C,[44] and antibody, C, and leukocytes.[45] The autologous phase is mediated by macrophages[46,47] or possibly neutrophils[48] independently of C. Depending on the host species, origin of the antiserum, and its C-fixing properties, glomerular histology may appear normal or demonstrate sequential infiltration of lymphocytes, neutrophils, and macrophages during the heterologous phase and even crescentic proliferation during the autologous phase. As in PHN, the active autologous phase can be accelerated by preimmunization of the host animal with heterologous IgG[46] (see below). Passive autologous models of anit-GBM nephritis have also been utilized on occasion[36,47]; these involve initial, passive immunization with anti-GBM antibody that serves as a planted antigenic target for subsequently administered isologous anti-IgG.

Selection of Species

Anti-GBM nephritis has been studied in a variety of animals, including rats, rabbits, mice, and guinea pigs. The selection of species depends primarily on the purpose of the study, and, in many instances, the choice may be solely dependent on the preference of the investigator. Certain unique features of the various species are worth pointing out. The quantity of kidney-bound antibody required to induce proteinuria in rats (75 μg/g kidney weight) is relatively high as compared to rabbits or guinea pigs (15 μg/g kidney weight).[1,3] Techniques for measuring renal physiological processes (e.g., micropuncture and clearance studies) have been best developed for the rat, while studies in the complement field are facilitated by strains of guinea pigs deficient in C4,[43] mice deficient in C5,[49] or rabbits deficient in C6.[44] In contrast to rats, rabbits manifest crescentic glomerulonephritis during the autologous phase, which may implicate different inflammatory mechanisms. Although mice are less suitable than rats for physiological, surgical, and quantitative studies, the availability of highly inbred, genetically defined strains, including athymic, nude mice,[50] as well as many mouse-specific immunological reagents, makes them a highly suitable species in which to study cellular as well as humoral immunity in

[43] W. G. Couser, M. M. Stilmant, and N. B. Jermanovich, *Kidney Int.* **11,** 170 (1977).

[44] G. C. Groggel, D. J. Salant, C. Darby, H. G. Rennke, and W. G. Couser, *Kidney Int.* **27,** 643 (1985).

[45] C. G. Cochrane, E. R. Unanue, and F. J. Dixon, *J. Exp. Med.* **122,** 99 (1965).

[46] G. F. Schreiner, R. S. Cotran, V. Pardo, and E. R. Unanue, *J. Exp. Med.* **147,** 369 (1978).

[47] S. R. Holdsworth, T. J. Neale, and C. B. Wilson, *J. Clin Invest.* **68,** 686 (1981).

[48] P. F. Naish, N. W. Thomson, I. J. Simpson, and D. K. Peters, *Clin. Exp. Immunol.* **22,** 102 (1975).

[49] E. R. Unanue, M. R. Mardiney, Jr., and F. J. Dixon, *J. Immunol.* **98,** 609 (1967).

[50] Y. Kusuyama, T. T. Nishihara, and K. Saito, *Clin. Exp. Immunol.* **46,** 20 (1981).

renal injury. A recently described method for inducing heterologous- and autologous-phase anti-GBM nephritis in the mouse[51] is described below.

Immunogens used to produce nephrotoxic anti-GBM antisera include whole kidney homogenate,[52] isolated glomeruli,[43] and isolated GBM.[51,53] Most appear to generate antisera that are predominantly reactive with glomerular capillary wall antigens *in vivo,* with an insignificant degree of cross-reactivity to other renal structures such as tubular basement membrane (TBM) and proximal tubular brush border. The widest experience has been gained using isolated whole glomeruli for immunization. While substantial species cross-reactivity is often present, the glomeruli are generally prepared from the species in which the anti-GBM nephritis will eventually be studied. A protocol for the isolation of rat glomeruli and GBM is described, but rabbit, mouse, guinea pig, or human glomeruli can be isolated in a similar manner.

Preparation of the Immunogen

Isolation of Glomeruli

Material

SD rats (250–300 g) or frozen rat kidneys. Thirty kidneys produce a final yield of about 50–75 mg of lyophilized glomeruli.

Stainless steel ASTME sieves #80 (180 μm), #140 (106 μm), and #200 (75 μm) and collecting pan (all 8-inch diameter, W. S. Tyler).

Other materials are the same as those required for Fx1A isolation (see above) except for the ultracentrifuge.

Procedure

1. Using the same procedure described in steps 1–3 for the preparation of Fx1A (see above), renal cortices are gently pressed through the #140 sieve into the collecting pan.
2. The #80 sieve is placed on top of the #200 sieve, and the material in the collecting pan is poured over the sieves. Large contaminants are caught in the #80 sieve, glomeruli on the #200 sieve, and smaller tubular fragments are washed through. Both sieves are rinsed with PBS, and the glomeruli are collected at one edge of the #200 sieve by rinsing it with a sharp stream of PBS from a wash bottle. Finally the glomeruli are washed from the sieve into a centrifuge tube.

[51] K. J. M. Assman, M. M. Tangelder, W. P. J. Lange, G. Schrijver, and R. A. P. Koene, *Virchows Arch.* **406,** 285 (1985).

[52] D. K. Hammer and F. J. Dixon, *J. Exp. Med.* **117,** 1019 (1963).

[53] S. R. Holdsworth, A. M. Thomson, E. F. Glasgow, J. P. Dowling, and R. C. Atkins, *J. Exp. Med.* **147,** 98 (1978).

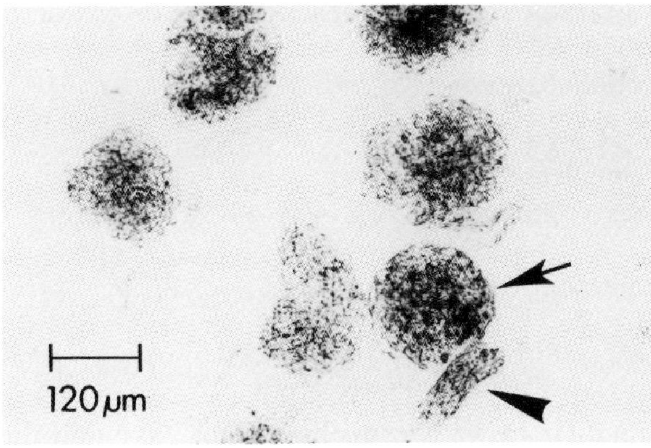

FIG. 4. Phase-contrast micrograph of a wet preparation of rat glomeruli isolated by sieving (see text for details). One encapsulated glomerulus (arrow), one tubular fragment (arrowhead), and several decapsulated glomeruli are visible.

3. The glomeruli are washed several times with PBS (centrifuged at 60–100 *g* for 5 min at 4°) until soluble protein is not detected in the supernatant (measured at OD_{280}). At this point the glomeruli are suspended and washed with 2–3 changes of DDW (1:100 v/v, 4°) to remove PBS and lyse cells. The final pellet is then lyophilized and stored.

The yield and purity of the glomerular preparation can be ascertained by microscopy after the PBS washing steps (Fig. 4). The best yield (about 50%, assuming 38,000 glomeruli per kidney) and highest purity (fewer than 10% tubular fragments) are obtained by using large volumes of PBS to wash through the #140 sieve (about 800 ml) and #200 sieve (1–2 liters). To isolate mouse glomeruli, the #200 sieve is replaced with a #325 ASTME (45 μm, W. S. Tyler) sieve.[51] Tubular contamination may be as high as 35%.

Preparation of GBM

There are two methods to free GBM of cell membranes: sonic disruption[54] or detergent solubilization.[55] The two methods have occasionally been combined.[51]

Sonic Disruption. Glomeruli are isolated as described above and, after osmotic lysis in DDW, are sonically disrupted for 1 min in two 30-sec

[54] R. G. Price and R. G. Spiro, *J. Biol. Chem.* **252,** 8597 (1977).

[55] E. Meezan, J. T. Hjelle, and K. Brendel, *Life Sci.* **17,** 1721 (1975).

bursts from a Branson sonifier (Branson Sonic Power Co., Danbury, CT) at an output of 30 W.

Detergent Solubilization

Materials

Deoxyribonuclease I (Sigma).
Sodium deoxycholate (DOC), 4%, 40 ml containing 0.1% sodium azide.
Sodium chloride, 1 *M*, 40 ml.

Procedure

1. After osmotic lysis with DDW (100 : 1 v/v with 0.1% sodium azide), the glomerular pellet is centrifuged at 8000 *g* and the supernatant discarded.
2. DNA released from lysed cells tends to gel on exposure to DOC, which precludes adequate washing. Therefore, the glomerular pellet is suspended in 40 ml of 1 *M* sodium chloride containing 2000 Kunitz units of DNase for 1–2 hr at 27° with constant stirring followed by centrifugation at 8000 *g*.
3. The pellet is then suspended in 40 ml of 4% DOC and stirred for 2–4 hr at 27° followed by centrifugation at 8000 *g*.
4. The DOC-treated pellet is washed by extensive suspension and centrifugation in DDW followed by lyophilization.

The detergent method produces intact acellular ghosts of glomeruli composed of GBM and mesangial matrix. Sonication results in acellular fragments of GBM. Subtle alterations in the biochemical composition of the GBM may be incurred by the different methods (see Kefalides *et al.*[56] for an extensive review of GBM composition) but do not preclude production of a nephritogenic antiserum.[51]

Preparation of Anti-GBM Antibody

Rabbits and sheep have been most commonly used for immunization, and both species are capable of producing nephritogenic C-fixing anti-GBM antibody after 8 weeks.[44,52] However, sheep appear to be more susceptible than rabbits to the development of autoimmune nephritis, which may lead to death early after onset of immunization, thereby limiting the period available to produce C-fixing antibody. Also to be taken into consideration, as discussed above, are differences in immunoglobulin chemistry between species.

[56] N. A. Kefalides, R. Alper, and C. C. Clark, *Int. Rev. Cytol.* **61,** 167 (1979).

Procedure. A variety of immunization protocols have been used to generate anti-GBM antibody. Thus, no uniformity exists in doses of immunogen or immunization schedule. For example, reported doses of antigen have ranged from 1 to 300 mg protein per injection for rabbits[52] and from 25 to 75 mg for sheep.[44,57] Similarly, antibody has been observed as early as 1 week after the initiation of immunization,[52] but C-fixing antibody more commonly appears from 8 weeks to over 6 months. We have routinely used the following protocol for producing sheep anti-rat and anti-rabbit GBM antisera with whole glomeruli.

Immunization with glomeruli. Sheep are initially immunized with 40–75 mg of whole glomeruli suspended in 8 ml of PBS and emulsified with CFA exactly as described above for production of sheep anti-Fx1A. Booster injections of 15–50 mg of glomeruli in IFA are given every 4 weeks. Blood is obtained 1 week after each booster and tested, as described below. With this protocol, some sheep generate nephritogenic, C-fixing anti-GBM antibody by 10 weeks.[44] Others make nephritogenic anti-GBM antibodies that cause proteinuria largely independently of C fixation.[57] Also, the immunization protocol may lead to the early development of autoimmune nephritis and terminal renal failure, which may preclude the generation of C-fixing antibody.

Immunization with GBM. The procedure has been successfully used to produce nephritogenic goat and rabbit anti-mouse GBM.[51] Adult goats or rabbits are immunized with 1 mg of GBM in CFA at multiple s.c. sites followed after 4 and 6 weeks by boosters of 0.2 and 1 mg of GBM s.c., respectively. The animals are bled 10 days after the second booster injection. A similar approach has been used to produce anti-rabbit GBM, using 25 mg of particulate rabbit GBM to immunize sheep at biweekly intervals.[53]

Characteristics of Anti-GBM Antiserum

Indirect IF on Normal Kidney. The main purpose of indirect IF on normal kidney is to ensure that the antiserum reacts only with GBM. Faint, linear, TBM staining may also be seen with some antisera and is acceptable for most purposes. Tubular brush border staining implies that there was substantial tubular contamination of the immunogen. Therefore, if the antiserum is to be used in rats for studies lasting more than 24–48 hr, it should be absorbed repeatedly with Fx1A (prepared as described above) until such staining disappears, to avoid the occurrence of PHN.

[57] W. G. Couser, C. Darby, D. J. Salant, S. Adler, M. M. Stilmant, and L. M. Lowenstein, *Am. J. Physiol.* **249,** F241 (1985).

Heterologous Phase. Immunized animals are bled 1 week after every boost and the antiserum is tested *in vivo.* Adult test animals are injected i.v. with incremental amounts of antiserum (0.05–0.5 ml of sheep antiserum or 0.5–1.0 ml of rabbit antiserum per rat; 0.5–2.0 ml of sheep antiserum per rabbit; 0.03–0.5 ml of sheep antiserum per guinea pig), and urine is collected for determination of protein excretion during the first 24 hr (usually overnight for 18 hr). Kidney biopsies are performed on the next day to verify the presence of antibody by IF (as described above). A good antiserum will produce proteinuria of over 100 mg/24 hr in rats and over 1000 mg/24 hr in rabbits during the first day. Excessively high doses, if not lethal, will often result in acute failure accompanied by anuria during the first 24 hr, with recovery of urine output at 24–48 hr. IF should demonstrate intense, linear, glomerular capillary wall staining for heterologous IgG (Fig. 5). Fainter staining may also be present on the TBM. If the antibody is C fixing, host C3 should also be seen in glomeruli in a similar pattern, but of lesser intensity (Fig. 5). Proteinuria and glomerular deposits of IgG unaccompanied by C3 staining suggests the likelihood of C-independent, antibody-induced injury. The presence of C3 staining does not imply that proteinuria is wholly C mediated, however. This can be confirmed by depletion of C as described below. Histological sections of glomeruli may also be examined for the presence of infiltrating leukocytes (Fig. 6), but their pathophysiological role should be assessed by depletion studies, as outlined below. Once the antiserum satisfies the necessary

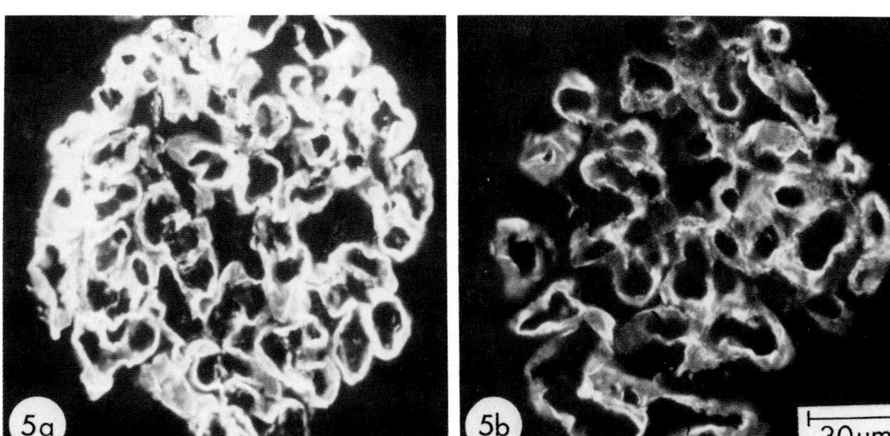

FIG. 5. Immunofluorescent micrographs of representative rat glomeruli in a renal biopsy obtained 24 hr after injecting sheep anti-rat GBM antiserum: (a) stained for sheep IgG; (b) stained for rat C3.

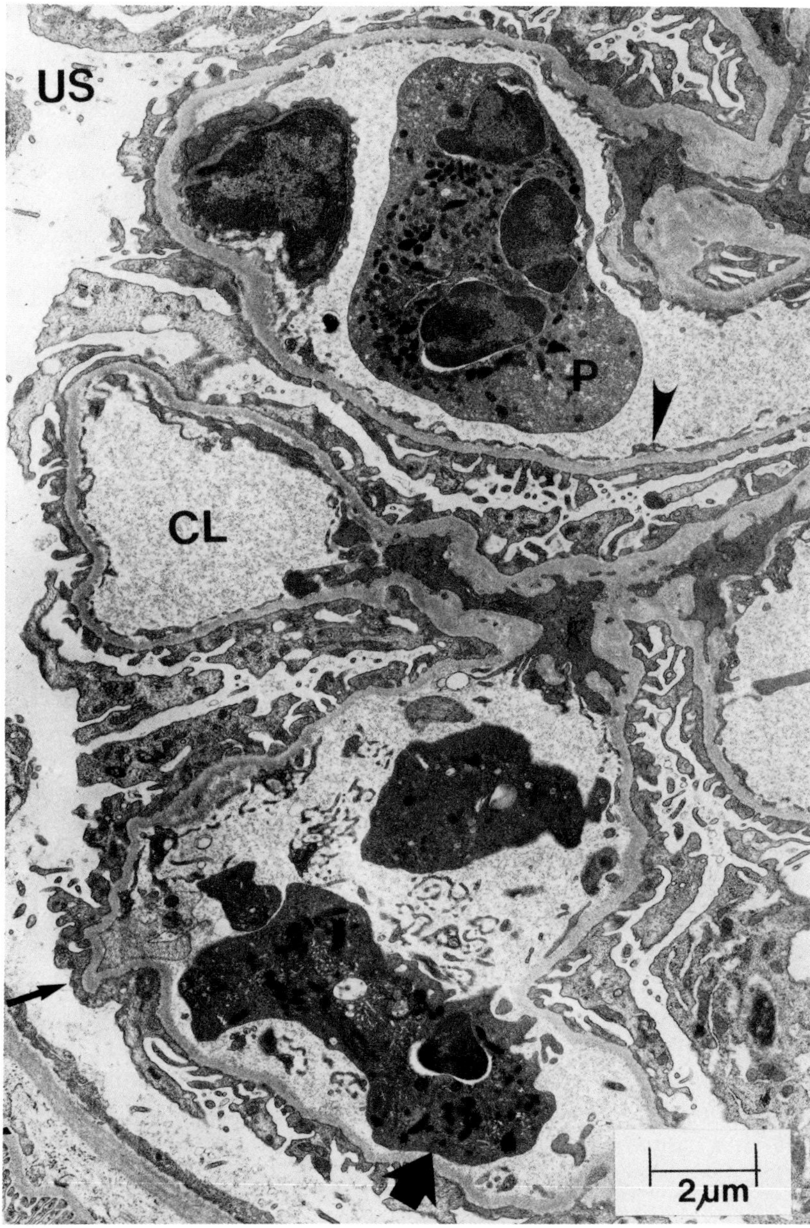

FIG. 6. Low-power electron micrograph of a representative rat glomerulus in a renal biopsy obtained 24 hr after injecting sheep anti-rat GBM antiserum (0.05 ml). Note the infiltration of polymorphonuclear leukocytes (P), one of which is adjacent to an area of basement membrane denuded of endothelium (broad arrow). In some areas, the endothelium is normally fenestrated and intact (arrowhead), but in other areas it is detached. Focal regions of epithelial cell foot process effacement (thin arrow) are also present. CL, Capillary lumen; US, urinary space. Micrograph kindly prepared by Dr. Dale Abrahamson, University of Alabama at Birmingham.

requirements, animals are exsanguinated. Further absorption of the antiserum or isolation of IgG is carried out if needed (see above). The following are typical nephritogenic doses of anti-GBM IgG: sheep anti-rat, 5–10 mg/100 g[57]; γ_1 sheep anti-rabbit, 12 mg/kg[44]; goat anti-mouse, 4 mg/mouse[51]; sheep anti-guinea pig, 0.5 mg/100 g.[43]

In addition to proteinuria, profound dose-dependent renal pathophysiological changes are observed in the heterologous phase. These include reductions in renal plasma flow, glomerular filtration rate, and ultrafiltration coefficient, and have been ascribed to stimulated sympathomimetic activity[58] and altered glomerular eicosanoid production.[59] In those models in which leukocytes are implicated, proteinuria appears to be due, in part, to the release of reactive oxygen species,[60] while leukocyte-independent, C-mediated proteinuria depends on an intact terminal C pathway.[44]

Autologous Phase

As indicated above, most animals that have developed heterologous anti-GBM nephritis, including rats, mice, and rabbits, will progress to an autologous phase of injury occurring at 8–10 days.[1a,3] The amount of heterologous antibody that fixes in the kidney to produce glomerular injury is also sufficient to elicit a secondary antibody response. Guinea pigs, while producing autologous antibody, do not develop proteinuria during the autologous phase and thus would appear to be an exception.[43]

Accelerated Autologous Phase. Preimmunization of susceptible species will accelerate the onset of autologous-phase injury. Furthermore, in this situation, the heterologous phase of injury can be abolished if only a small, "subnephritogenic" dose of anti-GBM antibody is administered. For example, in rats, administration of a 1 mg of rabbit IgG in CFA i.p. 5 days prior to administration of a subnephritogenic dose of rabbit anti-rat GBM antibody (i.e., one that does not normally cause proteinuria) results in the early development of proteinuria that peaks after 2–4 days.[46] A similar protocol can be adapted for use with different primary antibodies or in other species.

Passive Autologous Phase. In the following protocol, described by

[58] R. C. Blantz, T. H. Hostetter, and B. M. Brenner, *in* "Contemporary Issues in Nephrology" (C. B. Wilson, B. M. Brenner, and J. H. Stein, eds.), Vol. 3, p. 122. Churchill-Livingstone, New York, 1979.

[59] E. A. Lianos, *Nephron* **37,** 73 (1984).

[60] A. Rehan, K. J. Johnson, R. C. Wiggins, R. G. Kunkel, and P. A. Ward, *Lab. Invest.* **51,** 396 (1984).

Holdsworth *et al.*,[47] rabbits are given an i.v. subnephritogenic dose of either horse[47] or sheep[61] anti-GBM antibody, which serves as "planted" GBM antigen. Fifteen hours later, a nephritogenic dose of rabbit anti-horse or anti-sheep IgG is administered i.v. Within the next 12 hr rabbits develop heavy proteinuria in association with histological findings of exudative glomerulonephritis. Both anti-GBM and anti-IgG antisera may be prepared by methods similar to those described above for sheep anti-GBM and rat anti-sheep IgG, respectively.

Autologous-phase injury, as defined by the accelerated and passive models described above, is C independent and is mediated by macrophages, possibly via Fc receptor adhesion.[62]

Glomerulonephritis Induced with Planted Antigens

Several investigators recently have taken advantage of the anionic nature of heparan sulfate proteoglycans in the laminae rara of the GBM and sialoglycoproteins in the glycocalyxes of glomerular epithelial and endothelial cells to "plant" a variety of cationic proteins in the glomerular capillary wall.[63-66] There they serve as planted antigens for subsequent *in situ* immune complex formation with circulating antibodies induced by active[65] or passive[64] immunization. The lectin-binding properties of cell surface glycoproteins have also been similarly used to fix concanavalin A on the glomerular endothelial cell surface.[67] In most instances, these models have required that the antigens be perfused *ex vivo* through a kidney temporarily isolated from the circulation to avoid diffuse binding to circulating and cell-surface anionic residues or glycoproteins and to allow sufficient glomerular binding before exposure to antibody. Several cationized proteins have been used for this purpose including ferritin, IgG, ovalbumin, and both human and bovine serum albumin (BSA)[63-66] The degree of glomerular injury depends on the doses of antigen and antibody, the extent of antigen cationization, and the time between antigen deposition and exposure to antibody.

[61] S. R. Holdsworth and R. Bellomo, *Kidney Int.* **26,** 162 (1984).

[62] S. R. Holdsworth, *J. Immunol.* **130,** 735 (1983).

[63] A. Vogt, R. Rohrbach, F. Shimizu, H. Takamiya, and S. Batsford, *Kidney Int.* **22,** 27 (1982).

[64] T. Oite, S. R. Batsford, M. J. Mihatsch, H. Takamiya, and A. Vogt, *J. Exp. Med.* **155,** 460 (1982).

[65] T. Oite, F. Shimizu, I. Kihara, S. R. Batsford, and A. Vogt, *Am. J. Pathol.* **112,** 185 (1983).

[66] W. A. Border, H. J. Ward, E. S. Kamil, and A. H. Cohen, *J. Clin. Invest.* **69,** 451 (1982).

[67] S. M. Golbus and C. B. Wilson, *Kidney Int.* **16,** 148 (1979).

Glomerulonephritis Induced with Cationized BSA (cBSA) in Rabbits (see also this volume [41])

This model is a variant of chronic serum sickness[68] and was first described and characterized by Border *et al.*[66] It does not require renal perfusion of the cationized antigen but relies on recurrent cycles of antigen and antibody excess induced by single daily injections of antigen to preimmunized animals. Subsequent studies confirmed that the glomerular immune deposits produced with cBSA form *in situ.*[69] The main advantage of this model over conventional chronic serum sickness in rabbits is that a fairly uniform glomerular lesion, characterized by predominantly subepithelial immune deposits, is produced with a fixed dose of immunogen in all immunized animals.

Preparation of the Immunogen

BSA is cationized by activating carboxyl groups with carbodiimide in the presence of the nucleophile, ethylenediamine, according to a modification of the method of Hoare and Koshland.[70] By substituting amino groups for carboxyl groups, the net molecular charge is changed from negative to positive; the degree of change is proportional to the degree of substitution. Cross-linking of intra- and interchain amino and carboxyl groups with subsequent polymerization is largely prevented by using a dilute solution of BSA.

Materials

Crystalline fraction V bovine serum albumin (Miles Laboratories), 5 g.
Anhydrous ethylenediamine (Sigma), 67 ml.
1-Ethyl-3-(3-dimethylaminopropyl)carbodiimide hydrochloride (Sigma), 1.8 g.
Hydrochloric acid, 6 *N*, 400 ml.
Sodium acetate–acetic acid buffer, 4 *M*, pH 4.75, 50 ml.

Procedure. To 500 ml of DDW in a 1-liter flask, 67 ml of ethylenediamine is added. After cooling the solution to 25° and adjusting the pH to 4.75 with about 350 ml of 6 *N* HCl, 25 ml of a 20% solution of BSA (w/v) in DDW is added. While maintaining constant pH and temperature, 1.8 g of carbodiimide is added with constant stirring. The reaction is continued

[68] B. Noble and J. R. Brentjens, this volume [41].

[69] S. G. Adler, H. Wang, H. J. Ward, A. H. Cohen, and W. A. Border, *J. Clin. Invest.* **71,** 487 (1983).

[70] D. G. Hoare and D. E. Koshland, Jr., *J. Biol. Chem.* **242,** 2447 (1967).

for 2 hr and then is quenched by adding 30 ml of 4 *M* acetate buffer, pH 4.75. The reaction mixture is dialyzed extensively against DDW, lyophilized, and stored at −70° until used.

Horizontal isoelectric focusing of the cationized product in 5% thin-layer polyacrylamide gel using ampholytes with a pH range of 3.5–9.5 (Bio-Rad Laboratories, Richmond, CA), reveals a p*I* of 8.8–9.3 as compared to native BSA with a p*I* of 4.5–5.3.[71] Sodium dodecyl sulfate polyacrylamide gel electrophoresis (SDS–PAGE) using a 5% vertical gel or gel filtration chromatography on Sepharose 6B (Pharmacia) should demonstrate negligible polymerization, except for a small population of dimers normally found in BSA preparations.[66]

Induction of Glomerulonephritis with cBSA[66]

Male New Zealand white rabbits (1.5–2.0 kg) are immunized with 1 mg of cBSA and 1 μg endotoxin (Difco Laboratories) in 1 ml of PBS, i.v. Daily i.v. injections of 25 mg of cBSA are begun 1 week later and continued until completion of the study. Before injection, cBSA is dissolved in PBS (12.5–25 mg/ml) and centrifuged for 30 min at 27,000 *g*.

Characteristics of the Model

After 1 week of daily cBSA, 70% of rabbits have proteinuria that exceeds the upper 99% confidence limit of normal (18 mg/24 hr).[71] In about 30% of animals, values in excess of 1 g/24 hr are found. On IF microscopy, linear granular deposits of rabbit IgG, C3, and BSA are seen to decorate all glomerular capillary walls (Fig. 7). Light microscopy reveals mild hypercellularity with occasional neutrophils occupying capillary lumena. On EM, the subepithelial region of the GBM contains numerous, small electron densities that are often located proximal to the filtration slit diaphragm (Fig. 7). Occasional subendothelial and mesangial deposits may also be noted. At this stage, during which glomeruli are relatively free of infiltrating leukocytes, the lesion resembles early human membranous nephropathy, and proteinuria is dependent on an intact terminal C pathway.[71] Thus, C6-deficient rabbits have normal urine protein excretion despite qualitatively and quantitatively similar glomerular immune deposits as C-replete animals.[71]

After 2 weeks of cBSA, glomerular deposits of rabbit IgG and C3 assume a more granular appearance and appear to partly mask the detection of BSA on IF. Light microscopy and EM show greater hypercellularity

[71] G. C. Groggel, S. Adler, H. G. Rennke, W. G. Couser, and D. J. Salant, *J. Clin. Invest.* **72,** 1948 (1983).

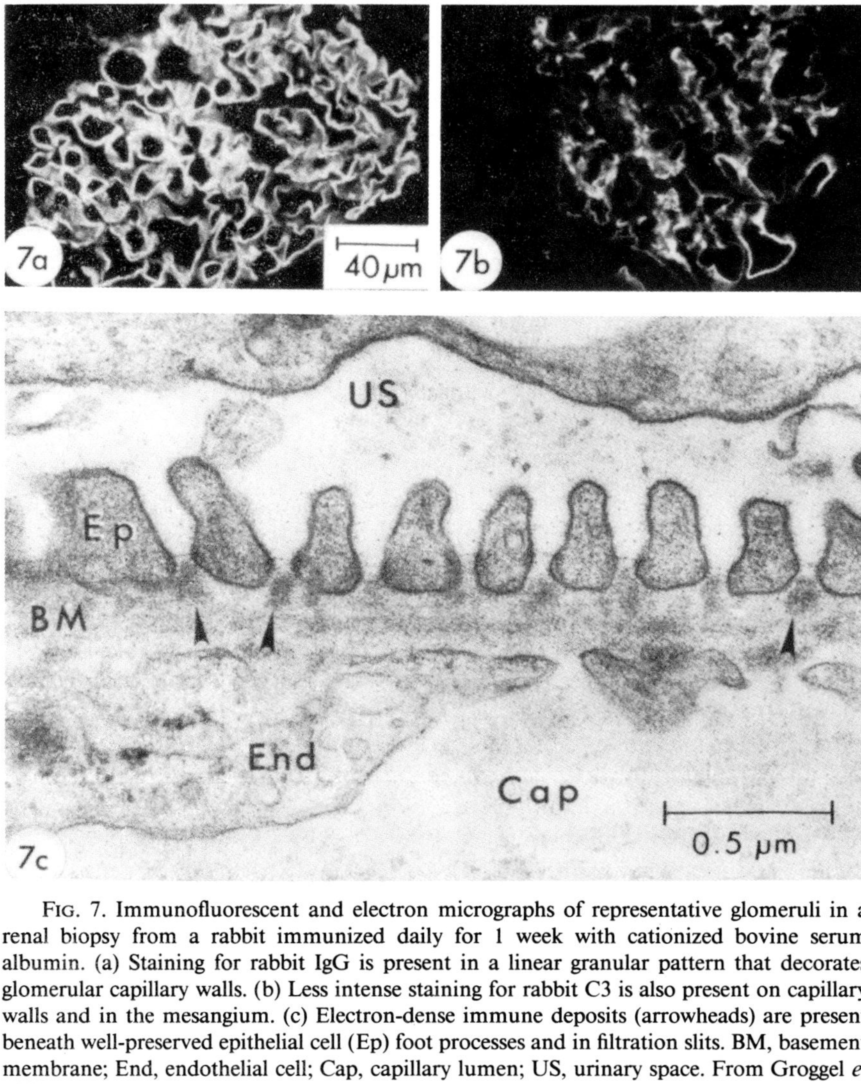

FIG. 7. Immunofluorescent and electron micrographs of representative glomeruli in a renal biopsy from a rabbit immunized daily for 1 week with cationized bovine serum albumin. (a) Staining for rabbit IgG is present in a linear granular pattern that decorates glomerular capillary walls. (b) Less intense staining for rabbit C3 is also present on capillary walls and in the mesangium. (c) Electron-dense immune deposits (arrowheads) are present beneath well-preserved epithelial cell (Ep) foot processes and in filtration slits. BM, basement membrane; End, endothelial cell; Cap, capillary lumen; US, urinary space. From Groggel *et al.*[71]

of glomeruli arising from infiltration of mononuclear and polymorphonuclear leukocytes. Electron-dense deposits remain in a predominantly subepithelial location but appear larger and more flocculent. Coincident with the influx of inflammatory cells, proteinuria is found to develop with equal severity in C6-deficient and C-replete rabbits[71]; however, the mediation of injury in this phase has not yet been established.

In Vivo Complement and Leukocyte Depletion

The definition of mediator systems, including complement, coagulation factors, leukocytes, and platelets, in experimental glomerulonephritis primarily relies on selective mediator depletion *in vivo.* Outlined below are methods to deplete animals of C and leukocytes.

Complement Depletion

Pretreatment with cobra venom factor (CVF) on a schedule of 300 units/kg, divided into three equal i.p. doses on the day prior to antibody administration, has been found to adequately deplete animals of C3 for at least 24–48 hr in our own experience and in that of others.[72] Should it be necessary to maintain depletion of C beyond 48 hr, a maintenance dose of 100 units/kg, i.p., must be given every day after the loading dose.[10] It is rarely possible to sustain adequate depletion for more than 6–8 days due to immunological neutralization of the CVF. Purified CVF from *Naja naja kaouthia* or *Naja haje* is available from Cordis Laboratories, Inc. (Miami, FL) and is claimed to be free of phospholipase A_2 activity. However, if the presence of small amounts of phospholipase activity could interfere with the interpretation of results, phospholipase A_2 should be inactivated. This is carried out by incubating the CVF with 0.1 m*M* *p*-bromophenacyl bromide (BPB) (Aldrich Chemical Co., Inc., Milwaukee, WI) at 37° and pH 8.0 for 30 min.[73] BPB is removed on a Sephadex G-25 minicolumn (PD 10, Pharmacia). CVF can also be purified directly from *Naja naja* venom (Sigma) by sequential ion-exchange and gel filtration chromatography.[74,75] Adequacy of C depletion is determined by measuring the total hemolytic activity of serum after CVF treatment and comparing it to pretreatment values.[31] Alternatively, the C3 concentration in serum may be assessed by radial immunodiffusion.[10] Hemolytic activity or C3 levels below 15% of normal are indicative of adequate C depletion.

Leukocyte Depletion

Neutrophil depletion is accomplished in rats with antineutrophil serum that may be prepared as follows.[10,45] Rats are injected with 10 ml of 0.1% oyster glycogen (Sigma), i.p., on 2 consecutive days. The cellular exudate (consisting of 90–100% neutrophils) is then collected, washed, and emulsi-

[72] C. G. Cochrane, H. J. Müller-Eberhard, and B. S. Aikin, *J. Immunol.* **105,** 55 (1970).

[73] J. O. Shaw, M. F. Roberts, R. J. Ulevitch, P. Henson, and E. A. Dennis, *Am. J. Pathol.* **91,** 517 (1978).

[74] M. Ballow and C. G. Cochrane, *J. Immunol.* **103,** 944 (1969).

[75] C.-W. Vogel and H. J. Müller-Eberhard, *J. Immunol. Methods* **73,** 203 (1984).

fied in CFA. Rabbits are immunized once with 2×10^8 cells in 5 ml CFA in multiple s.c. and i.m. sites, and antineutrophil serum is collected after 3 weeks. Early response serum is more effective at depleting rat neutrophils than that collected after longer immunization periods.[45] Antiserum is absorbed with rat plasma, leukocyte-free erythrocytes, and lymph node cells (using the mesenteric and popliteal lymph node cells of one 300-g rat for 20 ml antiserum). A globulin fraction is prepared by precipitation with 50% ammonium sulfate[29] and is reconstituted with PBS in half of the original volume. A good antiserum will cause rats to become neutropenic (<250 neutrophils/mm^3) within hours of an intravenous or intraperitoneal dose of <1.5 ml of globulin and will maintain depletion for 12–24 hr. Twice daily doses may therefore be necessary for continuous depletion. Efficacy is generally lost after 4–5 days.

Rats are rendered pancytopenic, especially neutropenic and thrombocytopenic, by a single i.p. dose of cyclophosphamide (Cytoxan, Mead Johnson and Co., Evansville, IN), 150 mg/kg, or by lethal whole-body irradiation with 860 rads at a dose rate of 140 rads/min. Such rats will usually survive for at least 5–6 days under normal laboratory conditions.

Rabbits can be depleted of neutrophils by i.p. injection of nitrogen mustard (Mustargen hydrochloride, Merck, Sharpe and Dohme, West Point, PA), 1.75 mg/kg. By the third day postinjection neutrophil counts have been reported[45] to fall below 200/mm^3. Nitrogen mustard is less specific than antineutrophil serum and causes some degree of pancytopenia.

Quantitation of Glomerular Deposits of Exogenous Antibody

Experimental conditions frequently demand an accurate measurement of the amount of antibody bound in glomeruli. For example, before concluding that blockade of a given mediator system *in vivo* is responsible for abrogating glomerular injury, it is necessary to ensure that the manipulation has not affected the amount of antibody deposited.[10,44] Another important use of quantitative techniques is to assess the effects of various factors on the glomerular binding of passively administered antibodies or immune complexes.[10,76-78] Such factors include properties of the immune reactants (charge, size, avidity, etc.), hemodynamic determinants, systemic clearing mechanisms (e.g., complement receptors, mononuclear phagocyte

[76] M. P. Madaio, D. J. Salant, S. Adler, C. Darby, and W. G. Couser, *Kidney Int.* **26,** 397 (1984).

[77] I. D. Feintzeig, J. E. Dittmer, A. V. Cybulsky, and D. J. Salant, *Kidney Int.* **29,** 649 (1986).

[78] L. Raij, R. K. Sibley, and W. F. Keane, *J. Lab. Clin. Med.* **98,** 558 (1981).

system), and others.[79] While IF microscopy can detect gross changes in the amount of glomerular immune deposits, it is insensitive to small differences in binding that could nevertheless account for variations between groups of differently treated animals. Quantitative techniques have been best developed to study the deposition of exogenous immune reactants,[33,80] and these will be described here. The reader is referred elsewhere for a description of the indirect methods necessary to quantitate endogenous components of glomerular immune deposits.[71] A full discussion of the uses, advantages, and shortcomings of quantitative techniques is provided in a recent review.[79]

Principle

To quantitate glomerular binding of radiolabeled exogenous antibody, it is necessary to isolate glomeruli from the rest of the renal cortex because a variable proportion of kidney-bound antibody is usually present in extraglomerular locations, a fact not readily appreciated by IF microscopy.[33,79] It is also important to include a means of correcting for, or eliminating nonspecific glomerular contamination with unbound circulating IgG. This is best achieved by means of the dual-isotope technique in which a plasma marker (e.g., host albumin or IgG), labeled with a second isotope, is injected prior to glomerular isolation.[44,79,80] Alternatively, nonspecific contamination of isolated glomeruli can be virtually eliminated[33] by perfusing the kidneys with heparinized saline (as described earlier under the procedure for isolating Fx1A). Last, one should distinguish between tracer studies—injection of a small (nanogram to microgram) dose of labeled antibody to determine the effect of a given manipulation on antibody binding in the absence of a nephritogenic response—and studies in which a nephritogenic dose of antibody is injected after mixing it with a small quantity of labeled antibody. What follows is a description of the technique used to quantitate the glomerular binding of a nephritogenic dose of heterologous anti-rat kidney IgG antibody (e.g., sheep anti-GBM or anti-Fx1A), using the dual-isotope method to correct for nonspecific binding.

Radiodination of IgG

Several methods have been described to safely iodinated IgG without destroying antibody activity,[79] including chloramine-T,[81] lactoperoxi-

[79] D. J. Salant, A. V. Cybulsky, and I. D. Feintzeig, *Kidney Int.* **30,** 255 (1986).
[80] C. B. Wilson and F. J. Dixon, *J. Immunol.* **105,** 279 (1970).
[81] P. J. McConahey and F. J. Dixon, this series, Vol. 70, p. 210.

dase,[82] and iodine monochloride.[83] We have found that the following solid-phase chloramine-T method provides safe, simple and efficient iodination of IgG.

Materials

High specific activity ^{125}I and ^{131}I as sodium iodide in 0.1 *N* NaOH, carrier-free (New England Nuclear).
Iodobeads (Pierce Chemical Co., Rockford, IL).
PBS.
IgG fraction of antibody preparation, purified as described above (see Isolation of Sheep Anti-Fx1A IgG Subclasses).
BSA (or nonimmune IgG), also to be iodinated and serve as a plasma marker.
AG1-X8 anion-exchange resin (Bio-Rad Laboratories), or Sephadex G-25 (Pharmacia).
10-ml disposable syringe or 10-ml glass serological pipet with glass wool to plug the outlet.

Procedures

1. The desired amount of protein (usually 0.1–5.0 mg IgG at about 10 mg/ml in PBS) is placed in a glass test tube (12×75 mm) together with 4–7 Iodobeads.
2. The ^{125}I or ^{131}I in the carrier vial is diluted with 600 μl of PBS, and 0.5–1.0 mCi per milligram of protein to be labeled is transferred to the test tube.
3. Additional buffer is added if necessary to just cover the Iodobeads.
4. The vial is stoppered and the reaction allowed to proceed at room temperature for 15 min. The reaction is terminated simply by removing the sample from the Iodobeads.
5. A carrier protein is added at this point to protect the sample from radiation damage and to facilitate recovery during subsequent steps to remove free iodine. For tracer studies, we usually add an excess of "cold" nonimmune IgG or BSA. For studies in which the labeled IgG is to serve as a marker for a nephritogenic dose, we add unlabeled antibody IgG.
6. Two "desalting" steps are almost always necessary to ensure that 98–99% of the isotope is protein bound (i.e., is precipitable with 10% trichloroacetic acid). We have used the following combinations of chromatography and dialysis with equal success.

[82] M. Morrison, this series, Vol. 70, p. 214.
[83] M. A. Contreras, W. F. Bale, and I. L. Spar, this series, Vol. 92, p. 277.

6a. The sample is applied to an AG1-X8 anion-exchange column packed in a 10-ml syringe. It is eluted with PBS, and about 20 0.75 to 1.0-ml samples are collected. An aliquot of 1 μl from each sample is counted in a gamma scintillation counter for 0.1 min, and all fractions exceeding 1×10^4 cpm are pooled. The pool is then concentrated by vacuum dialysis against PBS in a collodion bag (UH 100/25, Schleicher and Schuell, Inc., Keene, NH) concentrating apparatus (UH 100/1, Schleicher and Schuell, Inc.).

6b. A small amount of Blue Dextran solution (Pharmacia) is added to the sample to turn it faintly blue. The sample is applied to a Sephadex G-25 (Pharmacia) column packed in a 10-ml serological pipet and eluted with PBS. All the effluent that emerges in the void volume with Blue Dextran is collected, pooled, and dialyzed against 200 times the sample volume of PBS for 48 hr.

With this procedure we have obtained 24–47% utilization of isotope and specific activities of 50–800 μCi/mg. Trichloroacetic acid (10%) precipitability with either "desalting" procedure is 95–99%. At these levels of radioactivity, we have not found evidence of damage to antibody activity; however, this can be ascertained for individual preparations as described elsewhere.[79] Most unbound isotope is readily disposed of as dry waste after allowing the Iodobeads and columns to dry in a fume hood.

Quantitation of in Vivo Glomerular Binding of Antibody[33]

Materials

Materials for isolating glomeruli as listed above (see Isolation of Glomeruli).
Disposable, plastic 1 and 3-ml syringes and 30-gauge needles.
Conical polystyrene disposable 15-ml centrifuge tubes.
Narrow-stem 65-mm polypropylene funnels.
Glass test tubes (10 × 75 mm) (lead-free).
Aqueous polyvinyl pyrrolidone (PVP), 30%. Make up 30 g (Sigma, PVP-40T, MW 40,000) in 100 ml DDW with 0.1% sodium azide and filter (0.45 μm) into a sterile bottle.
Gamma counter.
Fuchs–Rosenthal hemocytometer (0.2-mm deep, Hausser Scientific, Blue Bell, PA).

Procedure

1. Rats (or other experimental animals) are given 0.45 g/100 ml NaCl and 5 mg/100 ml KI in their drinking water for 24–48 hr before

injecting antibody. This facilitates urinary excretion and blocks thyroid uptake of free iodine.

2. Sufficient unlabeled and labeled antibody IgG are mixed prior to injection to provide each rat with a nephritogenic dose of antibody and about $1-2 \times 10^6$ cpm of protein-bound isotope. The final specific activity is determined on the day of injection from the protein-bound counts and protein concentration of the mixture. The volume injected intravenously should not be more than 1.5 ml in a 200-g rat. To ensure accuracy when tracer doses are used, the labeled antibody should be diluted, if necessary, with PBS to inject a volume of 1 ml. The exact dose delivered may not be that shown on the syringe barrel and can be predetermined by injecting the contents of the syringe and 30-gauge needle into a counting tube and determining the radioactivity. The average of three replicates is taken to represent the administered dose. The contents of the tubes are recovered and returned to the sample.
3. The ^{125}I-labeled antibody is injected intravenously, and the rats are maintained on saline–KI drinking water, as above, until sacrifice. Blood samples (100 μl) can be drawn at intervals from the retroorbital plexus or tail vein if it is necessary to follow the plasma disappearance kinetics.[10,76–79] Note that heterologous IgG is rapidly catabolized over the course of several days and that free iodine is excreted in the urine. Therefore precautions should be taken handling the cage bedding and fur of the rats as both become contaminated during this time.
4. About an hour before sacrifice, 1×10^6 cpm of the ^{131}I-labeled plasma marker (e.g., BSA) is injected intravenously. Immediately prior to sacrifice a 100-μl sample of blood is obtained, and the radioactivity is determined in a dual-channel, well-type gamma counter with window settings for ^{125}I (15–80 keV) and ^{131}I (260–470 keV). The percent "spillover" of ^{131}I counts into the ^{125}I channel should be predetermined in the range of activity anticipated in the samples by counting some diluted ^{131}I-labeled material in both channels. Spillover counts in the ^{125}I channel are typically about 20% of counts in the ^{131}I channel.
5. The kidneys are removed, decapsulated, trimmed free of extrarenal tissue, and rinsed with PBS.
6. Glomeruli are isolated from the combined kidneys of individual rats by sieving on #140, #80, and #200 ASTME sieves, as described above for the production of anti-GBM antibody. Glomeruli on the #200 sieve are washed into a 15-ml conical centrifuge tube using a

funnel. After 2 washes with PBS, the glomeruli are sedimented by centrifugation at 60–100 *g* for 5 min, the supernatant is aspirated off, and the glomerular pellet is transferred to a lead-free glass test tube (12 × 75 mm) using 1 ml of 30% PVP. (PVP helps suspend glomeruli for visual counting; see below.) Glomeruli tend to stick to test tubes and pipets. Those adherent to the sides of the conical centrifuge tube should be ignored, but one should take care to transfer those suspended in PVP to the bottom of the glass test tube. The tubes are placed in a gamma counter, and ^{125}I and ^{131}I are determined.

7. Glomeruli are suspended by brisk agitation on a Vortex Genie mixer (Scientific Products, Evanston, IL). Aliquots of 12.5 μl are applied into each chamber of a Fuchs–Rosenthal hemocytometer (glomeruli are too large for conventional clinical hemocytometers). Five replicate aliquots are counted, the median value determined, and the number of glomeruli in the 1-ml sample calculated by dividing the median by the combined volume of both counting chambers (6.4 μl). When the yield of glomeruli is high it may be necessary to increase the volume of PVP to 2 ml to avoid excessive clumping. The median value is then multiplied by 2.
8. Radioactive counts in all blood and glomerular samples are corrected for background activity spillover and isotope decay. Specific glomerular binding of ^{125}I-labeled antibody (Ab) is then calculated from Eq. (1).

$$\text{Binding of } {}^{125}\text{I-Ab} = \text{glomerular } {}^{125}\text{I} - \frac{\text{glomerular } {}^{131}\text{I} \times \text{blood } {}^{125}\text{I}}{\text{blood } {}^{131}\text{I}} \text{ (cpm)} \tag{1}$$

From the specific activity (SA) of the ^{125}I-labeled antibody, the number of glomeruli in the sample (N), and the average number of glomeruli in a rat kidney (38,000),[84,85] one can calculate the amount of IgG antibody bound in the glomeruli of one kidney as follows:

$$\text{IgG(Ab)} = \text{Binding of } {}^{125}\text{I-Ab [Eq. (1), (cpm)]} \times \frac{38{,}000}{N} \times \frac{1}{\text{SA (cpm/mg)}} \quad \text{(mg/kidney)}$$

[84] J. Strassberg, J. Paule, H. C. Gonick, M. H. Maxwell, and C. R. Kleeman, *Nephron* **4**, 384 (1967).

[85] L. J. Barratt, J. D. Wallin, F. C. Rector, Jr., and D. W. Seldin, *Am. J. Physiol.* **224** (Suppl. 3), 643 (1973).

Application of Quantitative Techniques

Two examples of the use of quantitative techniques in experimental glomerulonephritis are presented in Tables II and III. Several other examples are described or cited in Ref. 78.

Table II shows the results of an experiment in which two groups of rats are injected with anti-Fx1A IgG to determine if there is any difference in the nephritogenicity of the γ_1 (C-fixing) and γ_2 (non-C-fixing) subclasses.[10] The IgG subclasses are isolated (as described above) and labeled with ^{125}I. Preliminary studies are carried out to ensure that the doses injected produce similar quantities of glomerular-bound antibody after 5 days, the time at which urine is collected to measure protein excretion. In this study, the dual-isotope method is not used to correct for nonspecific binding but kidneys are perfused *in vivo* prior to removal.[33] The results demonstrated considerable overlap between the groups with regard to glomerular binding of antibody, and yet only rats given the C-fixing γ_1 anti-Fx1A are proteinuric. IF staining for sheep IgG is the same in both groups, but only the group given γ_1 anti-Fx1A had glomerular staining for rat C3 (Fig. 2). Quantitative techniques allow the conclusion that a difference in protein excretion occurs despite equivalent glomerular deposits of the two antibody subclasses.

Table III presents an experiment in which normal and C6-deficient rabbits are injected with an identical nephritogenic dose of ^{125}I-labeled sheep anti-GBM IgG to assess the role of terminal C components in the

TABLE II

GLOMERULAR ANTIBODY BINDING (GAB) AND URINE PROTEIN EXCRETION AFTER 5 DAYS' ADMINISTRATION OF γ_1 AND γ_2 ANTI-FX1A TO RATS[a]

Antibody injected[b]	GAB[c] (μg/2 kidneys)	Urine protein excretion (mg/24 hr)
γ_1 anti-Fx1A (6)[d]	91 ± 31.1[e]	41 ± 22.0[e]
γ_2 anti-Fx1A (6)	66 ± 12.8	5 ± 1.6

[a] From Salant *et al.*[10]

[b] Rats are intravenously injected with either γ_1 or γ_2 sheep anti-rat Fx1A labeled with ^{125}I. In preliminary studies, the dose of γ_1 anti-Fx1A (15 mg) regularly produced proteinuria; the dose of γ_2 anti-Fx1A (18–22 mg) was adjusted to produce GAB levels within the range induced by γ_1 anti-Fx1A.

[c] GAB is measured in glomeruli isolated from the saline-perfused kidneys of individual rats.

[d] Number of rats studied is given in parentheses.

[e] Values are means ± SD.

TABLE III
GLOMERULAR ANTIBODY BINDING (GAB) AND URINE PROTEIN EXCRETION 24 HR AFTER ADMINISTRATION OF SHEEP ANTI-RABBIT GBM TO NORMAL AND C6-DEFICIENT (C6D) RABBITS[a]

Group[b]	GAB[c] (μg IgG/10^4 glomeruli)	Urine protein excretion (mg/g creatinine)
Normal (5)[d]	15.8 ± 2.71[e]	42.0 ± 26.3[e]
C6D (6)	18.7 ± 1.99	5.1 ± 5.5

[a] From Groggel *et al.*[44]
[b] Rabbits are intravenously injected with sheep anti-rabbit GBM IgG (12 mg/kg) labeled with ^{125}I.
[c] Since kidneys are not perfused with saline, GAB is corrected for nonspecific binding by the double-isotope technique using nonimmune sheep ^{131}I-IgG as a plasma marker.
[d] Number of rabbits studied is given in parentheses.
[e] Values are means ± SD.

development of heterologous-phase proteinuria.[44] Urine protein excretion is measured for 24 hr after antibody injection, and glomerular binding is assessed by the dual-isotope method at the end of the urine collection period. The results show a lower rate of urine protein excretion in the C6-deficient group, which cannot be attributed to their having less glomerular antibody.

Acknowledgments

This work was supported by U.S. Public Health Service Grant RO1 DK30932 from the National Institute of Diabetes, Digestive and Kidney Diseases, by an American Heart Association Established Investigator Award (D.J.S.), and by the Kidney Foundation of Canada and Medical Research Council of Canada (A.V.C.).

[39] Autoimmune Orchitis and Thyroiditis

By PIERLUIGI E. BIGAZZI

Introduction

Studies of organ-specific autoimmune diseases occurring in laboratory animals have provided useful and often illuminating information regarding the etiologic and pathogenetic mechanisms of human autoimmune disease. Autoimmune disease can be induced in animals by the administration of drugs and other chemicals[1] or X-irradiation and by surgical manipulations (e.g., thymectomy). More commonly, such experimental models of autoimmunity are obtained by the immunization of laboratory animals with homologous or heterologous organ-specific antigens, usually emulsified in Freund's complete adjuvant or in association with other adjuvants (e.g., *Bordetella pertussis*).[2,3]

However, as Burnet once noted, "An elderly woman with Hashimoto's disease of the thyroid has not been injected with rabbit thyroid in Freund's complete adjuvant."[4] In other words, there is a need for animal models of autoimmune disease that are not dependent on the artificial administration of antigens and adjuvants and that resemble more closely the situation occurring in humans. For this reason, we have suggested that careful investigation of spontaneous organ-specific autoimmune disease of animals may provide additional insights into human illness.[5] In this chapter, we review both an experimentally induced and a spontaneous model of organ-specific autoimmune disease.

Postvasectomy Autoimmune Orchitis

Autoimmune orchitis is induced experimentally in a variety of animal species by immunization with spermatozoa or testicular antigens, emulsified with Freund's complete adjuvant or associated with *B. pertussis*.[6] However, there is one model of autoimmune orchitis occurring in rabbits, initiated by occlusive surgery of the vas deferens (vasectomy) and *not* dependent on the administration of exogenous adjuvants.[7-10] In the following section we describe in detail the procedures employed in our labora-

[1] P. E. Bigazzi, *in* "Immunotoxicology and Immunopharmacology" (J. Dean, M. I. Luster, A. E. Munson, and H. Amos, eds.), p. 277. Raven, New York, 1985.

[2] P. E. Bigazzi and N. R. Rose, *in* "The Autoimmune Diseases" (N. R. Rose and I. R. Mackay, eds.), p. 161. Academic Press, Orlando, Florida, 1985.

[3] P. E. Bigazzi, *in* "Autoimmunity and Endocrine Disease" (R. Volpe, ed.), p 345. Dekker, New York, 1985.

[4] F. M. Burnet, "Auto-immunity and Auto-immune Disease," p. 101. Davis, Philadelphia, Pennsylvania, 1972.

[5] P. E. Bigazzi and N. R. Rose, *Prog. Allergy* **19,** 245 (1975).

[6] K. S. K. Tung and A. C. Menge, *in* "The Autoimmune Diseases" (N. R. Rose and I. R. Mackay, eds.), p 537. Academic Press, Orlando, Florida, 1985.

[7] P. E. Bigazzi, L. L. Kosuda, L. H. Harnick, R. C. Brown, and N. R. Rose, *Clin. Immunol. Immunopathol.* **5,** 182 (1976).

[8] P. E. Bigazzi, L. L. Kosuda, K. C. Hsu, and G. A. Andres, *J. Exp. Med.* **143,** 382 (1976).

[9] P. E. Bigazzi, *in* "Vasectomy" (I. H. Lepow and R. Crozier, eds.), p. 339. Academic Press, New York, 1979.

[10] P. E. Bigazzi, *in* "Reproductive Immunology" (N. Gleicher, ed.), p. 461. Liss, New York, 1981.

tory to obtain and evaluate this animal model of organ-specific autoimmune disease.

Vasectomy

Materials

Rabbits, New Zealand White, male, 3.1–3.6 kg (Beckens Research Animal Farms, Sanborn, NY, or Jenks Rabbitry, Feeding Hills, MA)
Acepromazine maleate (Aveco, Fort Dodge, IA)
Ketamine hydrochloride (Ketaset, Bristol Laboratories, Syracuse, NY, or Ketalar, Parke-Davis, Morris Plains, NJ)
Surgical instruments (scissors, knives, retractors, etc.), sterile
Sutures (4-O silk, chromic) (Ethicon Inc., Somerville, NJ)
Furazolidone powder (Topazone, Norden, Smith Kline, Philadelphia, PA)
Syringes, needles (22-gauge), centrifuge tubes, etc.

Evaluation of Autoimmune Responses to Spermatozoa

Materials

Cytocentrifuge Cytospin, SCA-0030 (Shandon Instruments, Sewickley, PA)
Absolute methanol
Phosphate-buffered saline (PBS),[10a] pH 7.0–7.5
Vasectomized and control rabbit sera, stored at $-70°$
Fluorescein isothiocyanate (FITC)-conjugated goat antibodies to rabbit immunoglobulins (available from commercial sources, e.g., Sigma Chemical Co., St. Louis, MO, or Cappel, Cooperbiomedical Inc., Malvern, PA)
Fluorescence microscope (Leitz Dialux, equipped for transmitted light microscopy with an HBO 200-W mercury lamp, exciter filters BG12 and BG38, absorption filter K510)
Tissue culture medium RPMI 1640 (Associated Biomedic Systems, Buffalo, NY)
Buffered glycerol (1 volume PBS and 9 volumes glycerol)

[10a] Abbreviations: PBS, phosphate-buffered saline; FITC, fluorescein isothiocyanate; CFA Freund's complete adjuvant; BUF, Buffalo; RT, rat thyroglobulin; BSA, bovine serum albumin; ART, antibodies to rat thyroglobulin; IIF, indirect immunofluorescence; ELISA, enzyme-linked immunosorbent assay; SAT, spontaneous autoimmune thyroiditis; EAT, experimentally induced thyroiditis.

Immunohistopathology and Histopathology

Materials

Testes and kidneys from vasectomized and control rabbits (portions of specimens kept frozen at $-70°$, fixed in 10% buffered formalin, or fixed in a mixture of paraformaldehyde and glutaraldehyde)
Cryostat (Wide Range Microtome-Cryostat WRC, Harris LoTemp Products, North Billerica, MA)
Microtome
Ultramicrotome (Reichert OMU2)
FITC-conjugated goat antibodies to rabbit immunoglobulins (Sigma)
FITC-conjugated goat antibodies to rabbit C3 (Cappel)
Fluorescence microscope (Leitz Dialux with epiillumination)
Light microscope (Leitz Dialux)
Electron microscope (Philips EM 300)
Buffered glycerol (1 volume PBS and 9 volumes glycerol)

Methods

Vasectomy

New Zealand White rabbits are bilaterally vasectomized by means of the following procedure. Food, but not water, is withheld for 12 hr prior to the administration of anesthesia. Rabbits weighing 3.1 – 3.6 kg are tranquilized with a mixture of ketamine and acepromazine (1 ml acepromazine/ 10 mg ketamine) for a final dosage of 0.45 ml anesthetic per kg body weight. The anesthesized rabbit is secured to a board in a dorsally recumbent position, and the hair of the abdominal surface, inside of the hind legs, and around the penis and testes is carefully shaved. The area is prepared for surgery and draped. A vertical midline incision, about 2.5-cm long, is made in the skin of the suprapubic area (Fig. 1). With careful blunt dissection through this single mid-ventral incision, both spermatic cords are located and identified. The components of each cord are reached through a small cut in the cord wall, the vas deferens is carefully isolated, avoiding bleeding and trauma to the blood vessels, and ligated in two sites with silk sutures (Fig. 2). The part between the sutures, approximately 1 cm, is cut out (Fig. 3) and fixed in formalin for histological examination. The other components of the cord (arteries, veins, etc.), are gently pushed back into the cord, the small cut is sutured with silk, and the incision in the skin also sutured with silk.

After surgery the rabbits receive furazolidone powder topically and a combination of penicillin and streptomycin (0.5 ml per rabbit) intramus-

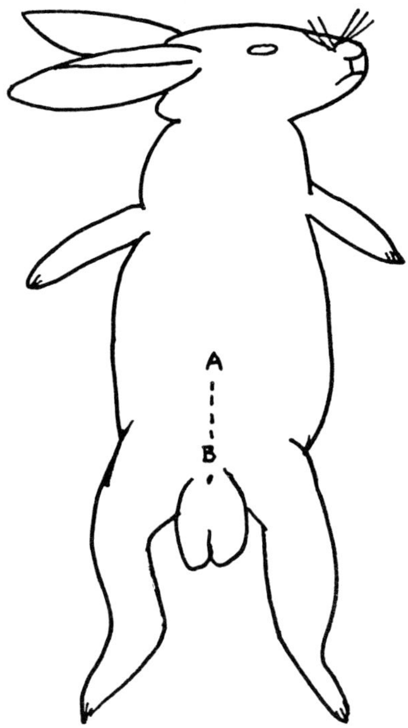

FIG. 1. Schematic outline of surgical incision for vasectomy in rabbits.

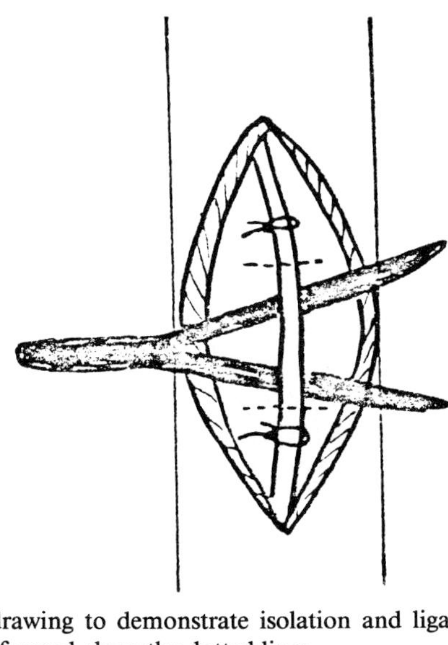

FIG. 2. Schematic drawing to demonstrate isolation and ligation of the vas deferens. Excision of the vas is performed along the dotted lines.

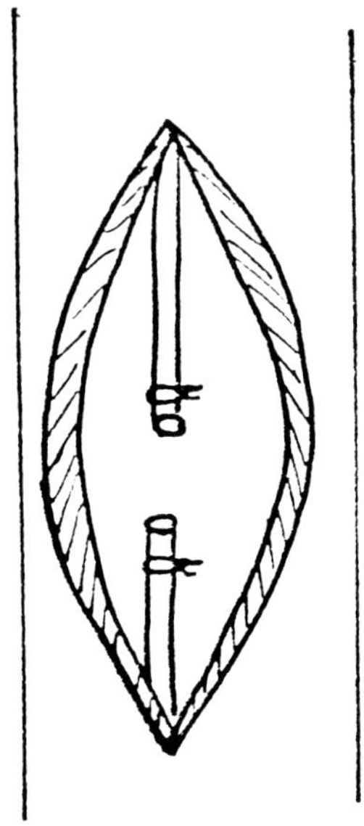

FIG. 3. Schematic drawing of the vas deferens after ligation and excision.

cularly for 5 days. Bleedings are taken from the central artery of the ear approximately every 15–30 days, and sera are stored at 70° until tested. Sham vasectomies are performed on rabbits of the same weight, anesthesized and treated following the procedure described above, with the exception that the vas deferens once isolated, is not ligated and cut out, but pushed back into the spermatic cord with the other cord components. Postsurgical treatment and trial bleedings are performed exactly as in the vasectomized group.

Detection of Antibodies to Spermatozoa by Indirect Immunofluorescence on Sperm Smears

Spermatozoa are obtained from the epididymis of normal rabbits after unilateral or bilateral orchiectomy. After washing the spermatozoa in medium RPMI 1640, smears are prepared on clean microscope slides by using a cytocentrifuge at 1500 rpm for 10 min. Sperm smears are dried, fixed in absolute methanol for 30 min at room temperature, washed in PBS, and incubated for 30 min with 2-fold serial dilutions of rabbit serum starting with a dilution of 1 : 10. They are then washed in PBS for 30 min, incubated for 30 min with the FITC-conjugated immunoglobulin fraction of goat antiserum to rabbit immunoglobulins, and washed for 30 min. Sera obtained at different intervals from vasectomized rabbits (including a prevasectomy bleeding), sham vasectomized rabbits, and normal animals are tested in each experiment. As an additional control, the goat anti-rabbit immunoglobulin conjugate is also used on sperm smears without any previous incubation with serum. The slides are mounted in buffered glycerol, examined under a fluorescence microscope, and reactions are read as positive (subjectively rated as 1+, 2+, or 3+ according to the brightness), weakly positive (±) and negative.

Immunohistopathology

Small fragments of testes and kidneys obtained from sham vasectomized and control rabbits are snap-frozen in liquid nitrogen and stored at −70°. Four-micrometer-thick sections are cut in the cryostat and air dried. The sections are stained using FITC-conjugated antibodies to rabbit IgG, C3, albumin, and fibrinogen. The specificity of the conjugates is tested by immunodiffusion in agar and immunoelectrophoresis. Sections are incubated for 30 min with the appropriate dilution of the conjugate, washed for 30 min in PBS, and then mounted in buffered glycerol. The sections are examined under a fluorescence microscope.

Evaluation of Animal Model

It can be expected that approximately 50% of vasectomized rabbits will develop circulating antibodies against spermatozoa. In our laboratory, none of the prevasectomy sera and none of the sera from sham vasectomized and normal untreated rabbits have been found to give positive reactions. The kinetics of the antibody response to spermatozoal antigens should show that the percentage of positive animals increases with time. For example, we have found that 2% of sera from vasectomized rabbits are positive 2 and 4 months following vasectomy, 8.7% are positive at 6 months, and 27% at 8 months. In the following months the percentage of positive animals varies between 25 and 37%. Positive sera stain the acrosomal region of rabbit spermatozoa (as shown in Fig. 4), with titers of acrosomal antibodies ranging from 10 to 320.

Rabbits with high and persistent levels of circulating antibodies to spermatozoa show thickening of basement membranes of seminiferous tubules and rare focal infiltrates of mononuclear cells in the interstitium. Occasionally, multinucleated spermatids, vacuolization of the basal portion of the tubules, as well as partial and total disappearance of Sertoli and spermatogonial cells may be observed. In thin sections, the main abnormalities of the basal lamina are small areas of increased contrasts, irregularities directed toward the epithelium, and accumulation of mononuclear

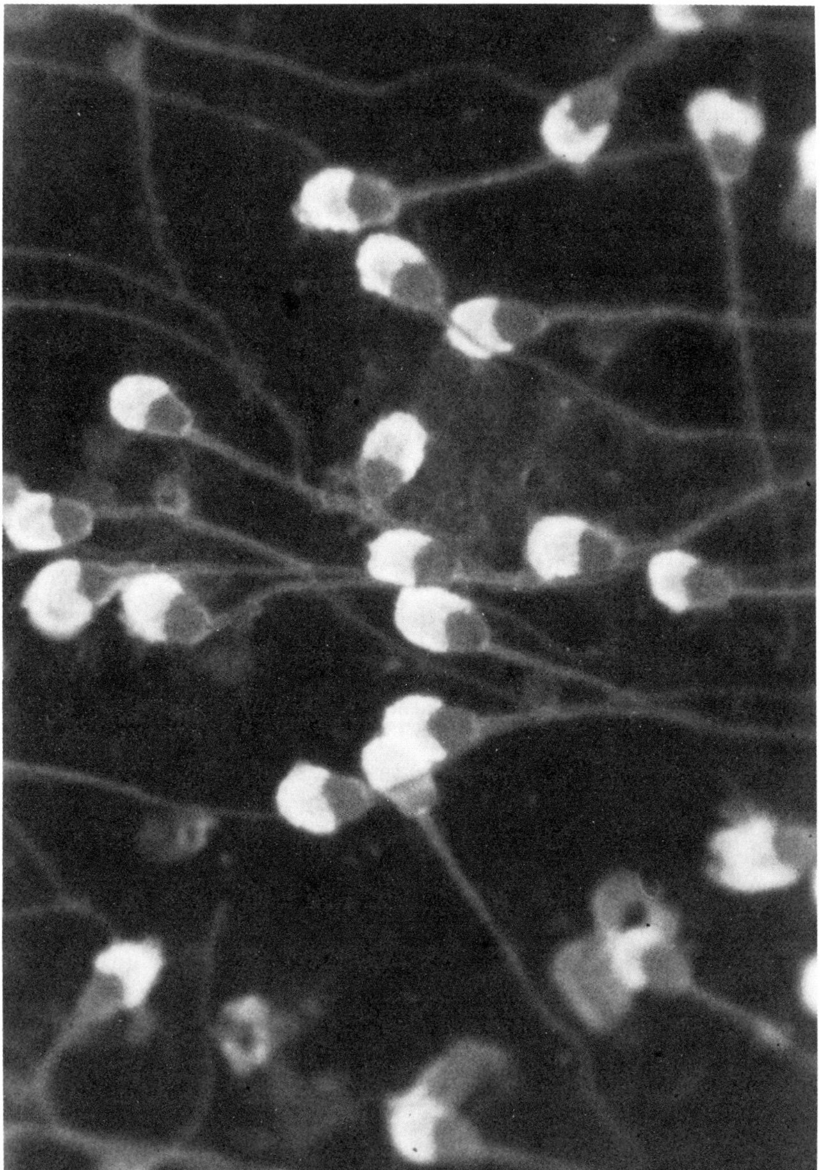

FIG. 4. Immunofluorescent staining of the acrosomal region of rabbit spermatozoa. Original magnification × 352.

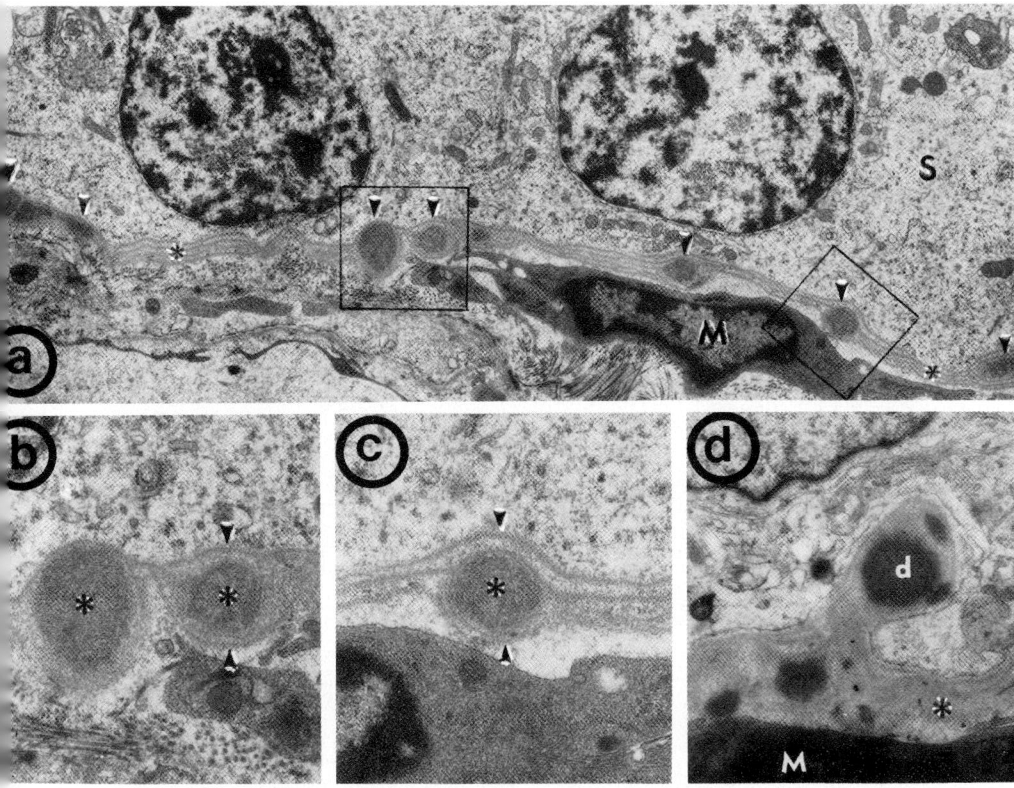

FIG. 5. (a) Electron-opaque deposits (arrowheads) within the basement membrane (asterisks) of a seminiferous tubule from a vasectomized rabbit. S, Spermatogonium; M, macrophage. The boxed areas are seen at higher magnification in Fig. 5b, c. Original magnification ×7,000. (b, c) Three electron-opaque deposits (asterisks) between the split layers (arrowheads) of the lamina densa. Original magnification ×14,000. (d) Deposits of marked electron opacity in the basement membrane (asterisk) and within a spike (d). M indicates part of a macrophage. Original magnification ×14,000. Reproduced from Bigazzi *et al.*[8]

cells and rarely, polymorphonuclear leukocytes. By electron microscopy, the most characteristic finding is the deposition of electron-opaque material in the basement membranes of seminiferous tubules (Fig. 5). In rabbits with less severe or initial lesions, the deposits are found within the basement membrane proper (lamina densa). In rabbits with more pronounced changes there is formation of new layers of epithelial basement membrane and development of "spikes" directed toward the epithelium. The electron-opaque deposits, frequently surrounded by newly formed layers of epithelial basement membrane, are localized within the "spikes," between

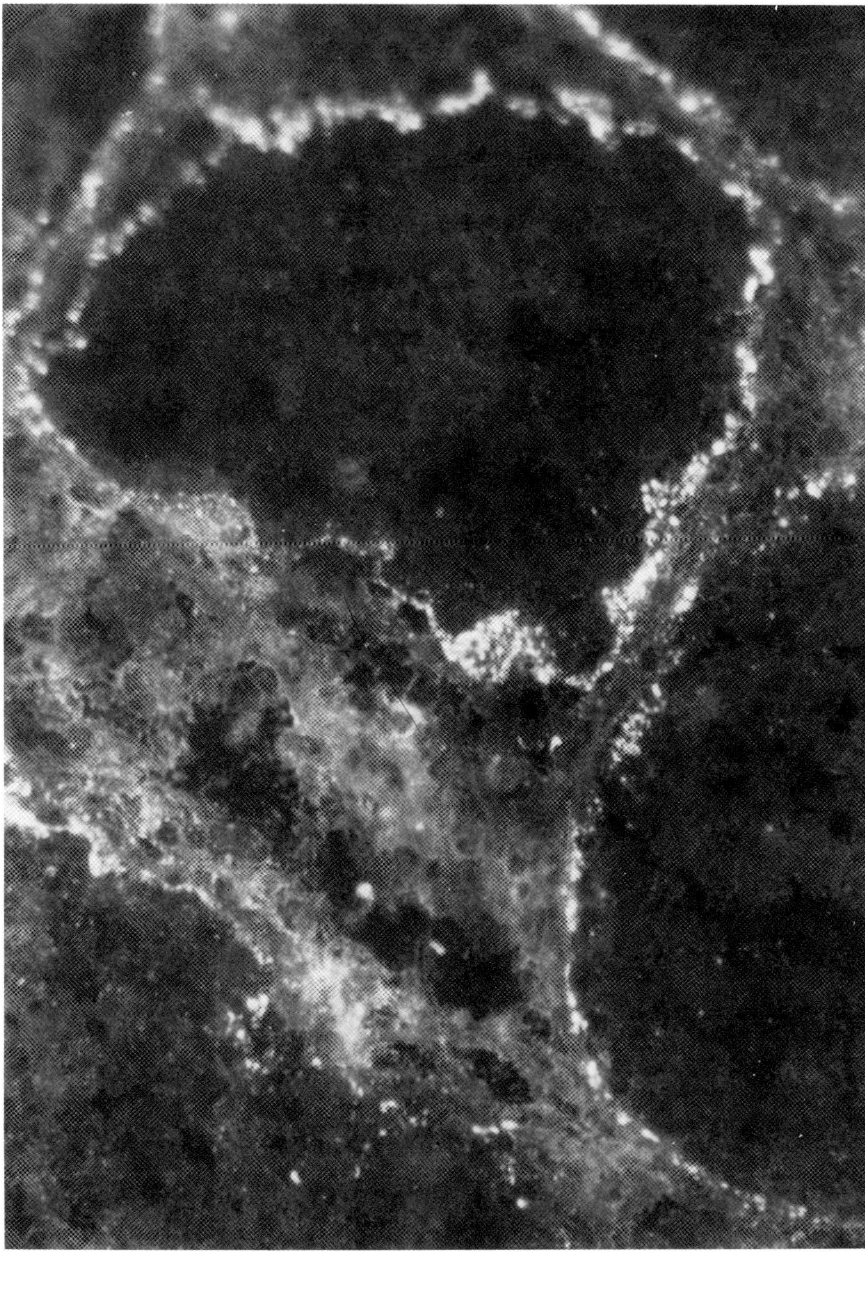

FIG. 6. Granular deposits of rabbit immunoglobulins along the basement membrane area of the seminiferous tubules of a vasectomized rabbit. Original magnification ×157.

the collagenous layers of the basal lamina, or in the basement membranes of interstitial vessels in minimal amount. Granular deposits of rabbit IgG and C3 can be demonstrated along the seminiferous tubules by immunofluorescence (Fig. 6). In rabbits with mild or initial changes, the deposits are seen along the lamina densa that appears in proximity to the collagenous basal lamina. In contrast, in severely damaged tubules the immune deposits are separated from the collagenous basal lamina by a space containing macrophages.

Spontaneous Autoimmune Thyroiditis

As mentioned previously, there are several experimental models of autoimmune thyroiditis. The majority of these models are obtained by the injection of thyroid extract or purified thyroglobulin emulsified in Freund's complete adjuvant (CFA).[2] Other models are induced by the combination of thymectomy and irradiation of experimental animals or by the administration of chemicals such as methylcholanthrene.[1] In pure-bred obese strain (OS) chickens[11] and in rats of the Buffalo (BUF) inbred strain, however, autoimmune thyroiditis develops spontaneously, i.e., without any apparent cause or manipulation.

Materials

Buffalo (BUF) rats (Simonsen Laboratories, Gilroy, CA), retired female breeders, i.e., females older than 1 year that have produced a number of litters

Rat thyroids, obtained at sacrifice of normal BUF rats or purchased from commercial sources (Pel-freeze Biologicals, Rogers, AK), kept frozen at $-70°$

^{125}I (Amersham-Searle, Arlington Heights, IL)

Lactoperoxidase (Calbiochem, La Jolla, CA)

Sephadex G-75 (Pharmacia Fine Chemicals, Uppsala, Sweden)

Gamma counter (Nuclear Chicago, Des Plaines, IL)

FITC-conjugated rabbit antibodies to rat immunoglobulins (available from commercial sources, e.g., Cappel)

Rabbit antibodies to rat immunoglobulins (Cappel)

Fluorescence microscope (Leitz Dialux equipped with an HBO 200-W mercury lamp and set for transmitted light microscopy, exciter filters BG12 and BG38, barrier filter K510)

Buffered glycerol (1 volume PBS and 9 volumes glycerol)

[11] G. Wick, J. Most, K. Schauenstein, G. Komer, H. Dietrich, A. Ziemiecki, R. Fassler, S. Schwarz, N. Neu, and K. Hala, *Immunol. Today* **6,** 359 (1985).

Light microscope (Leitz Dialux)
Electron microscope (Philips EM 300)

Methods

Purification and Radiolabeling of Rat Thyroglobulin

Rat thyroids are obtained from normal animals or purchased from commercial sources (Pel-freeze Biologicals). Thyroids are thawed, dissected from the trachea, cleaned from fat and connective tissue, and placed in a volume of PBS equal in milliliters to twice the weight of the tissue in grams. Tissues are ground in a Virtis homogenizer for 5 min. Centrifuge tubes containing the homogenate are placed overnight on wrist-action shaker. In the morning the thyroid homogenate is centrifuged for 30 min at 30,000 rpm in the No. 40 rotor of a Model L preparative ultracentrifuge (Beckman Instruments, Wakefield, MA). The pinkish supernatant, free of the dense pellet below and the lipid pellet above, is collected, the protein concentration determined by the biuret reaction, and the solution adjusted to 30 mg/ml. This material, which is designated "crude rat thyroid extract," is stored at 20°. Rat thyroglobulin (RT) is prepared from "crude rat thyroid extract" by passing 10 ml of the thyroid extract through a Sephadex G-200 (Pharmacia) column. The column is washed with PBS, and fractions are collected automatically at 10-min intervals and monitored at 280 nm with a spectrophotometer. RT is usually recovered in the first protein peak eluting from the column, and it sediments as a single peak in the analytical ultracentrifuge with a sedimentation coefficient of approximately 19 S. Column-purified RT with a concentration of 1 mg/ml is stored at $-20°$.

RT is labeled with ^{125}I using lactoperoxidase, as previously described.[12] Briefly, 1 mCi (37 MBq) of ^{125}I, lactoperoxidase (1% of the total protein), and 2 μl of H_2O_2 are added to 100 μg of RT in 10 μl of phosphate buffer, pH 7.3. The mixture is incubated for 15 min at room temperature, and the reaction stopped by addition of 2-mercaptoethanol. Next, 10 μl of 50 m*M* KI is added, and the mixture incubated for 15 min at room temperature. After the addition of 50 μl of bovine serum albumin (BSA) (40 mg/ml), the mixture is filtered through a Sephadex G-75 column. The efficiency of the radiolabeling procedure is controlled by precipitation with 20% trichloroacetic acid and determination of the radioactivity in the precipitate (usually, $>90\%$ of ^{125}I is recovered in the pellet). Radiolabeled RT (^{125}I-RT) is aliquoted and stored at $-70°$ until use.

[12] M. Zanetti and P. E. Bigazzi, *Eur. J. Immunol.* **11,** 187 (1981).

Detection of Circulating Antibodies to Rat Thyroglobulin

Indirect Immunofluorescence. As previously described,[13] sera are tested on rat thyroid cryostat sections, fixed in methanol at 56° for 3 min. Sera to be used are serially diluted in PBS, and dilutions are incubated on the sections for 30 min at room temperature in a humid chamber. Slides are rinsed, washed in PBS for 30 min, and then incubated with FITC-conjugated rabbit antisera to rat immunoglobulins for 30 min in a humid chamber. After an additional 30-min wash in PBS, sections are mounted in buffered glycerol and then examined under a fluorescence microscope. Sera containing antibodies to rat thyroglobulin (ART) give a floccular fluorescence of the thyroid colloid (Fig. 7). Reactions are scored on an arbitrary scale as 3+, 2+, 1+, and −.

Radioimmunoassay[12]. In the first step of the radioimmunoassay, 25 μl of ^{125}I-RT (10,000 cpm) is added to 25 μl of rat serum to be tested for ART, diluted 1 : 10 in 0.1% BSA in PBS. After an incubation of 30 min at 37°, the total cpm in each tube is determined using a gamma counter. Following the incubation, rat IgG is precipitated by the addition of 400 μl of rabbit antiserum to rat IgG at the appropriate dilution (the concentration of antiserum necessary to precipitate the rat Ig in the mixture must be predetermined). After an additional incubation at 37° for 30 min, the tubes are centrifuged at 2,000 *g* for 220 min, washed with cold PBS, centrifuged again, and the cpm in the pellet counted. Tests are performed in duplicate, and the percentage of ^{125}I-RT binding by the test serum is calculated from the following formula:

$$^{125}\text{I-RT Binding (\%)} = \frac{\text{cpm in the pellet}}{\text{total cpm}} \times 100$$

Histopathology

Thyroids, stained with routine hematoxylin and eosin stains, are evaluated according to the percentage of infiltration of the organ. The condition of the whole thyroid is inferred from observation of six to eight serial sections taken from at least four separate cutting levels of the fixed tissue. In most cases both lobes are examined. The final evaluation is usually based on observations of both lobes. These determinations are necessarily quite subjective and qualitative; therefore, only three separate categories of thyroid pathology are made. Thyroids exhibiting definite thyroiditis, but with less than 30% of the parenchyma replaced by mononuclear cells, are said to have "low infiltration." Thyroids exhibiting 30–70% loss of normal

[13] B. Noble, T. Yoshida, N. R. Rose, and P. E. Bigazzi, *J. Immunol.* **117,** 1447 (1976).

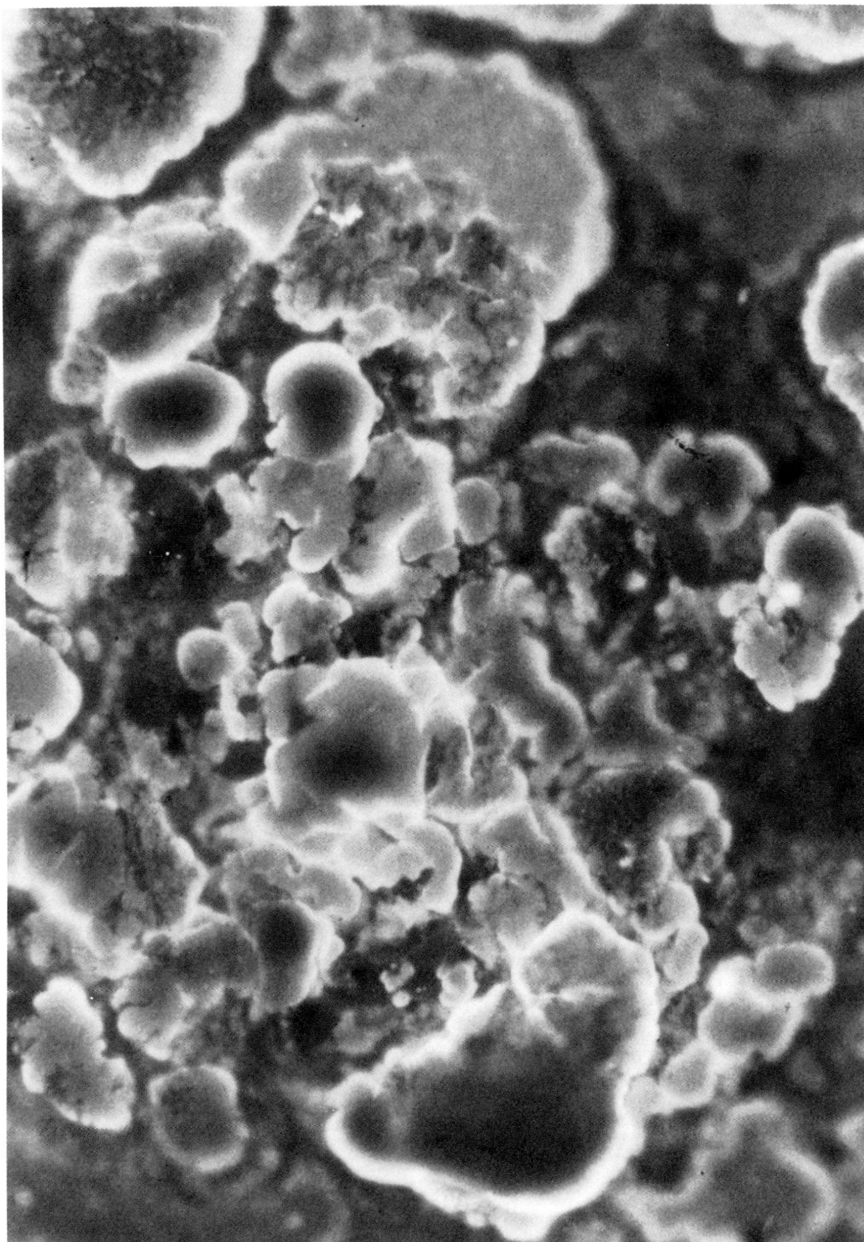

FIG. 7. Floccular pattern of immunofluorescence staining of rat thyroglobulin obtained with BUF rat serum containing autoantibodies to rat thyroglobulin. Original magnification ×100.

structure are labeled "intermediate." When more than 70% of the thyroid is replaced by the infiltrating cells, the condition is described as "high infiltration."

Evaluation of Animal Model

On the basis of our experience, it can be expected that approximately 24% of female retired breeders (older than 1 year) will have circulating ART, detectable by indirect immunofluorescence (IIF).[13] The presence of circulating ART, detected by IIF, correlates well with the incidence of inflammatory infiltration of the thyroid, present in 26% of the retired female breeders. Animals with low titers of serum ART detected by IIF have less infiltration of their thyroids, whereas rats with high antibody titers have intermediate levels of infiltration. On the other hand, rats with a high degree of infiltration have relatively low titers of ART by IIF.

More recently, we have found that a similar correlation exists between circulating ART detected by radioimmunoassay and thyroiditis (Bigazzi *et al.,* manuscript in preparation). In a group of 25 BUF rats, found positive for serum ART by radioimmunoassay, 96% had thyroiditis. On the other hand, 98% of another group of 52 BUF rats, negative for ART, had perfectly normal thyroids. Interestingly, similar studies, performed using an enzyme-linked immunosorbent assay (ELISA) to detect serum ART,[14] have failed to confirm this correlation between thyroiditis and circulating antibodies to thyroglobulin (Bigazzi *et al.,* manuscript in preparation).

The extent of thyroiditis may vary between affected animals from moderate (<30%) to intermediate (30–70%) and high (>70%) infiltration[13] (Fig. 8). The thyroid infiltrates observed in BUF rats are composed of lymphocytes, plasma cells, and macrophages, often with distinct germinal centers. No Hurthle cells are present.

The incidence of spontaneous autoimmune thyroiditis (SAT) in untreated female BUF rats is a function of age and parity.[1] No disease is found in female BUF rats less than 10 weeks of age, while 13% of rats aged 11–18 weeks show thyroid lesions. As previously mentioned, older female rats that have been bred repeatedly have a significantly higher incidence of thyroiditis than age-matched virgins. There is no difference between retired male breeders and nonbred males. In recent years, we have noticed a decrease in the incidence of thyroiditis in BUF female breeders (Bigazzi *et al.,* manuscript in preparation). This may be due to variations in environmental factors (especially viral and bacterial infections) or changes in the breeding practices of commercial suppliers.

[14] J. P. McCoy, Jr., J. H. Michaelson, and P. E. Bigazzi, *Life Sci.* **32,** 109 (1983).

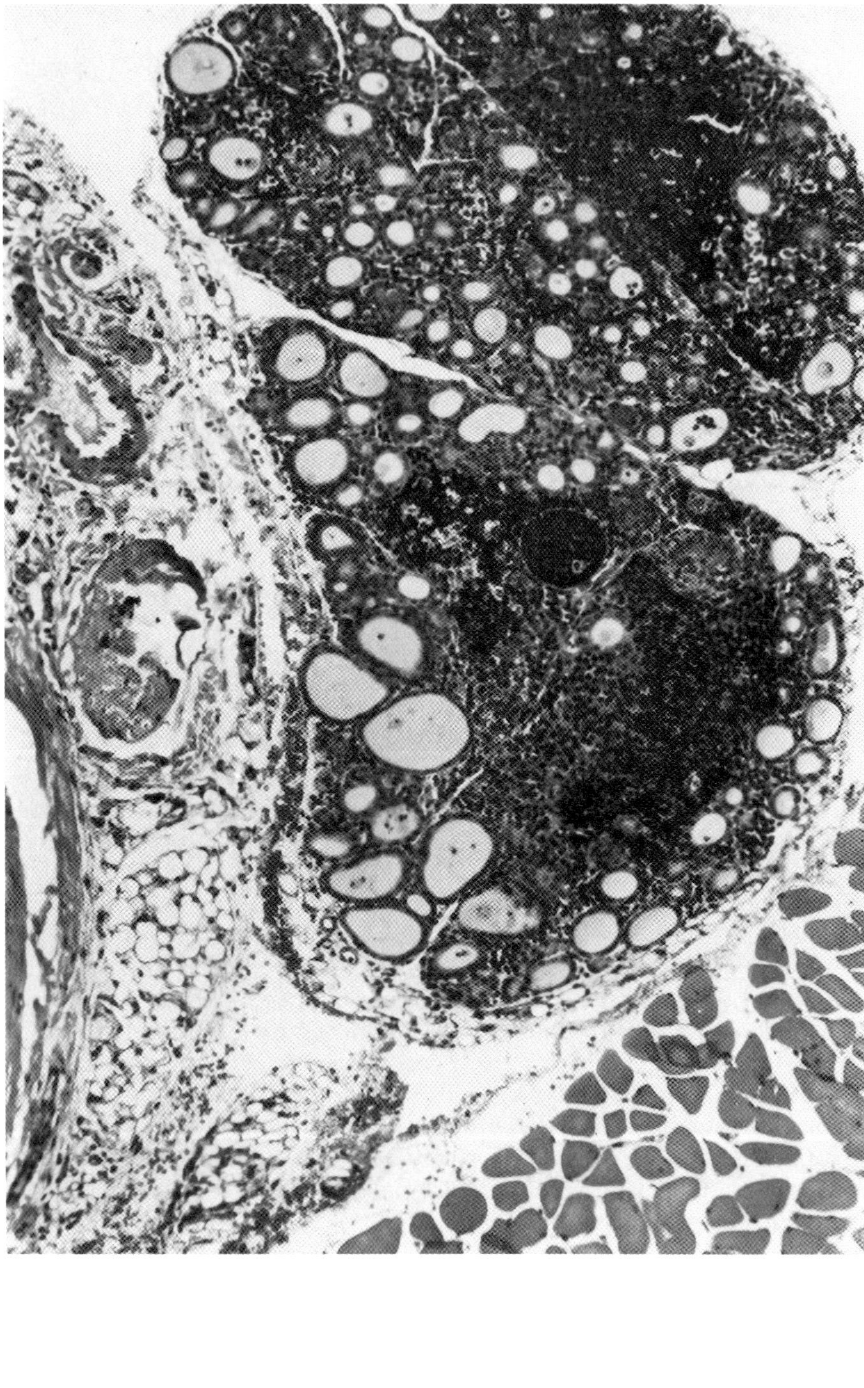

FIG. 8. Extensive mononuclear cell infiltration of a BUF rat thyroid. Original magnification ×40.

Neonatal thymectomy increases the incidence of thyroiditis in BUF rats.[1] Thyroid infiltration is found in over 80% of these animals. The percentage of rats with thyroiditis seems to be correlated with the time of thymectomy, since the highest values are observed when thymectomy is performed within 24 hr of birth. When surgery is performed at 48 hr, the percentage of animals with thyroiditis decreases to 26%, while thymectomies performed on 21-day-old BUF rats do not change the incidence of thyroiditis versus controls.

Conclusions

In this chapter we have reviewed two models of organ-specific autoimmune responses to spermatozoal antigens and a variety of immunohistotaneously. Both models can be very useful for investigations of the etiology and pathogenesis of organ-specific autoimmune disease. As an example, we have recently begun studies of the role of the idiotype–antiidiotype network in autoimmunity, utilizing BUF rats with SAT. [12,14–16]

The first model of organ-specific autoimmune disease has the advantage of being induced by a single surgical operation, vasectomy, without the need of any adjuvant. Interestingly, the findings in vasectomized rabbits resemble those of the experimental autoimmune orchitis obtained in rabbits by immunization with testicular antigens in CFA.[6] The autoimmune orchitis induced in rabbits by vasectomy is characterized by autoimmune responses to spermatozoal antigens and a variety of immunohistopathological and histopathological changes, with a picture very suggestive of *in situ* immune complex formation, a mechanism of organ-specific tissue damage previously observed at the level of the kidney or the thyroid in other experimental models of autoimmune disease.[8]

The second model described is the spontaneous autoimmune thyroiditis of BUF rats. It should be noted first that one of the major advantages of this model is its occurrence in animals of an inbred strain. There are very few models of organ-specific autoimmune disease that arise spontaneously (i.e., without experimental manipulations) in inbred animals. Second, BUF rat thyroiditis is similar to the condition observed in humans because of its predilection for older females and the presence of circulating autoantibodies to thyroglobulin. The pathogenesis of SAT in BUF rats is still unclear. We have noted an absence of delayed-type hypersensitivity reactions in animals with SAT and a correlation between circulating autoantibodies to thyroglobulin and thyroid infiltration with mononuclear cells.

[15] M. Zanetti, R. W. Barton, and P. E. Bigazzi, *Cell. Immunol.* **75**, 292 (1983).
[16] P. E. Bigazzi, *Ann. N.Y. Acad. Sci.* **475**, 66(1986).

These findings suggest that SAT of BUF rats may be autoantibody mediated in contrast to experimentally induced thyroiditis (EAT) that seems to be mediated by T cells. Neonatal thymectomy accelerates the onset and increases the incidence of SAT in BUF rats, while it abrogates EAT of rabbits, guinea pigs, rats, and mice, again suggesting that the experimentally induced disease has a different cellular basis from SAT.

Acknowledgments

These studies were supported by Grants HD17546 from NICHHD and ESO3684 from NIEHS, National Institutes of Health. We acknowledge the expert secretarial assistance of Ms. Ruth Conrod in the preparation of the manuscript.

[40] Arthus Phenomenon

By PHILIP J. BAILEY and DANIEL S. FLETCHER

Introduction

The Arthus reaction is an inflammatory response resulting from the deposition of immune complexes within tissues. Tissue damage occurs due to the activation of components of the complement system, the infiltration of polymorphonuclear leukocytes (PMNs),[1] the release of lysosomal enzymes[1a] together with an increase in vascular permeability generated by the local release of mediators, and, in extreme cases, hemorrhage resulting from damage to the vessel wall. A study of the inflammatory response and the mechanism of tissue damage occurring in the Arthus phenomenon would therefore seem to have some relevance to the events occurring in a variety of pathological situations such as rheumatoid arthritis, glomerular nephritis, and systemic lupus erythematosus where immune complexes are deposited.

The Arthus reaction can be elicited in a variety of animal species, rat, rabbit, and guinea pig being perhaps the most common, as well as in a number of different tissues including skin, the pleural space, synovial joints, and the paw. It can be induced basically in two ways: (1) as the

[1] Abbreviations: PMNs, Polymorphonuclear leukocytes; RPA, reverse passive Arthus reaction; BSA, bovine serum albumin; R-ABSA, rabbit anti-BSA; PBS, phosphate-buffered saline; MPO, myeloperoxidase; HTAB, hexadecyltrimethylammonium bromide; HSA, human serum albumin.

[1a] C. G. Cochrane and B. S. Aiken, *J. Exp. Med.* **124,** 733 (1966).

direct reaction, in which the animal is presensitized to a suitable antigen reaction elicited by injection of the antigen at a selected site, or (2) in a passive manner, in which the performed antibody is injected intravenously into a nonsensitized animal followed by injection of the antigen at a selected site.

To elicit the direct Arthus reaction the animal is first immunized with a soluble antigen (e.g., bovine serum albumin, egg albumin, hemocyanin) using standard immunization techniques (usually with Freund's complete adjuvant). Once precipitating antibody has appeared in the circulation the Arthus reaction can be elicited by injecting the specific antigen at the selected site. It is essential that the antibody be of the precipitating type because nonprecipitating antibody does not cause an Arthus response. For the same reason antigen excess should be avoided. This makes it virtually necessary to test a variety of antigen levels when eliciting the response in an actively immunized animal.

More commonly, the Arthus reaction is elicited as a reverse passive phenomenon in which the antibody is injected at the reaction site and the antigen is introduced intravenously. The passive reaction is probably the most commonly used method of eliciting the Arthus phenomenon because it avoids the often long sensitization procedure necessary to obtain sufficient antibody titers and is, in general, more reproducible because it avoids the fluctuation in antibody titer obtained in actively sensitized animals and thus allows greater reproducibility within groups of animals. In this chapter methods are described for elicitation of the reverse passive Arthus phenomenon in rat skin, but the methodology is readily applicable to other species and a wide variety of different anatomical sites.

Methods

Procedure Used to Elicit the Reverse Passive Arthus Reaction (RPA) in Rat Skin

The procedure which has been used successfully to elicit the Arthus phenomenon in rat skin is the bovine serum albumin (BSA)–antibovine serum albumin system,[2] although many differing antigen/antiserum combinations can be used. A system, described below, in which the primary use of the Arthus reaction is to study the mechanism of hemorrhagic lesion formation utilizes the ovalbumin–rabbit antiovalbumin combination. In many instances, however, where the Arthus reaction is being used to study the effect of drugs on plasma leakage and leukocyte accumulation, it is

[2] P. J. Bailey and A. Sturm, *Biochem. Pharmacol.* **32,** 475 (1983).

desirable to adjust conditions to result in a mild, essentially reversible Arthus response.

Reagents

BSA, crystallized (Miles Laboratories Inc., Elkhart, IN)
Rabbit antibovine serum albumin (R-ABSA) (lyophilized, Cappel Laboratories, Cochranville, PA)

Procedure. When developing the BSA–anti-BSA system to produce a mild reversible response to study the effect of pharmacologic agents, it was found that the injection of the undiluted R-ABSA obtained from Cappel Laboratories into the skin causes a nonspecific inflammatory response in the absence of antigen. The R-ABSA was therefore diluted 1:4 with phosphate-buffered saline (PBS) which in the skin was found to produce a minimal inflammatory response compared to saline in the absence of intravenously injected antigen but which resulted in a good edematous and cellular response when antigen was injected into the tail vein.

Male Sprague–Dawley rats weighing 130–160 g are separated into groups of six animals per group. Animals are lightly anesthetized (with Metofane inhalation anesthetic), and the hair is shaved from the middorsal region with electric clippers. Immediately after shaving, each animal is injected with 40 μl PBS at a site on the left side of the middorsal line. Forty microliters of R-ABSA (5.0 mg/ml antibody protein, diluted 1:4 with PBS) is injected intradermally into a skin site on the right side of the middorsal line, each skin site receiving 0.05 mg antibody protein. Immediately following the intradermal challenge, each rat receives 0.5 ml PBS containing 1.0 mg BSA injected directly into the tail vein.

In the skin the maximum edematous and cellular response occurs 4–6 hr postchallenge and thereafter shows a slow decline. At various times after intradermal challenge each group of animals is sacrificed by asphyxiation in carbon dioxide. The full thickness of skin is removed from the back of each animal and a disk 8 mm in diameter punched out with a metal punch (available at common hardware stores) at each intradermal injection site.

Measurement of Leukocyte Infiltration Using Myeloperoxidase (MPO) Determination

The main infiltrating cell in the RPA is the PMN, which in contrast to the macrophage and other local connective tissue cells contains myeloperoxidase at high specific activity. The amount of MPO in a skin sample is therefore a reflection of the number of infiltrating PMNs since noninflamed skin contains very little extractable MPO.

Reagents

Hexadecyltrimethylammonium bromide (HTAB), 0.5% (w/v) (Fisher Scientific Co., Fair Lawn, NJ), must be warmed slightly if precipitation occurs

Hydrogen peroxide, 30% H_2O_2 (Fisher), diluted 1:100 (v/v) with water: just before use it should be diluted a further 1:100 (v/v) with 10 m*M* KH_2PO_4 (pH 6.0)

o-Dianisidine dihydrochloride (3,3′-dimethoxybenzidine dihydrochloride, Eastman Kodak Co., Rochester, NY), 12.9 mg dissolved in water just before use

Procedure. MPO is extracted by placing each skin portion in 10 ml 0.5% (w/v) HTAB at room temperature. The skin is then ground with a Polytron homogenizer (Brinkman Instruments, Westbury, NY) at the highest setting for 15 sec. Tissue debris is removed by centrifugation at 800 *g* for 10 min at room temperature, and the supernatant fraction is used for MPO activity determination. Enzyme activity is stable for several days at room temperature.

The MPO assay is an adaptation of the method described by Bradley *et al.*[3] Three milliliters of 0.003% hydrogen peroxide is mixed with 0.2 ml skin extract, and the reaction is started by the addition of 25 μl of *o*-dianisidine. Incubation is performed at room temperature for 5 min, and then the absorption is measured at 460 nm. If desired, the reaction may be stopped by the addition of 0.2 ml 0.2% sodium azide. The spectrophotometric reading of the MPO assay can be used directly as an expression of the amount of enzyme per skin site, or it can be expressed as micromoles of hydrogen peroxide reduced using an extinction coefficient of 1.13×10^4 for oxidized dianisidine.

Measurement of Plasma Leakage

A cardinal feature of the Arthus phenomenon is an increase in vascular permeability leading to the accumulation of fluid and plasma proteins at the site of the reaction.[4] The simplest measure of edema formation is the increase in wet weight of the site injected with antibody as compared with the site injected with PBS.[2] After sacrificing the animal, an 8-mm disk of full skin is punched out with a metal punch and placed on filter paper moistened with PBS. The edema caused by the RPA reaction is then measured as the difference (expressed in milligrams) between the wet

[3] P. P. Bradley, D. A. Priebat, R. D. Christensen, and G. Rothstein, *J. Invest. Dermatol.* **78,** 206 (1982).

[4] Y.-H. Chang and I. G. Otterness, *Eur. J. Pharmacol.* **69,** 155 (1981).

weights of the antibody-injected site and the PBS-injected site. This method of measuring the extravasation of plasma proteins and fluids is rapid and convenient and has the added advantage that the weighed skin sites can then be used for measurement of leukocyte infiltration by determining MPO. A more quantitative method of determining plasma protein leakage from blood vessels is to inject intravenously an iodinated protein such as ^{125}I-labeled human serum albumin (HSA).[5] One microCurie is injected into the tail vein in 0.5 ml 0.9% (w/v) NaCl at any time during the development of the Arthus reaction. At specific times, 1.0 ml blood is removed by cardiac puncture and the animals sacrificed by asphyxiation in CO_2. Each reaction site is punched out from the skin and the isotope content determined in a gamma counter. The amount of accumulated plasma protein, expressed as microliter equivalents of plasma, is calculated by dividing the ^{125}I-HSA cpm of the skin site by the ^{125}I-HSA cpm of 1 μl of whole blood.

That simple wet weight measurements are indeed a true measure of extravasation of fluid from the blood vessels into the tissue has been confirmed using the ^{125}I-HSA technique described above. Leakage of the iodinated protein into the skin parallels the increase in wet weight. Also, the inhibition of ^{125}I-HSA leakage and wet weight determination obtained with a drug such as dexamethasone are of the same order, suggesting that the two measurements are a true reflection of edema formation. The disadvantage of wet weight measurements is that it does not allow a determination of the rate of protein leakage at any one time but is simply a measure of the total accumulation of fluid at the site.

Measurement of Hemorrhage in the RPA Reaction

When a severe RPA reaction occurs, damage to the blood vessels in the affected tissue is great enough to allow for the extravasation of erythrocytes as well as plasma fluid and proteins. This hemorrhage may be measured directly in the RPA skin sites if the hemoglobin of the rat's erythrocytes is labeled with ^{59}Fe prior to using the animals for experimentation.[6]

Reagents

Ovalbumin (5 × crystalline, Calbiochem, La Jolla, CA) Rabbit anti-ovalbumin serum (lyophilized, Cappel Laboratories)

[^{59}Fe]Ferrous citrate (1 μCi/μl, New England Nuclear, Boston, MA)

[5] M. M. Kapaniak, A. C. Issekutz, and H. Z. Movat, *Am. J. Pathol.* **98,** 485 (1980).

[6] M. M. Kapaniak, A. C. Issekutz, C. E. Burrowes, and H. Z. Movat, *Proc. Soc. Exp. Biol. Med.* **163,** 126 (1980).

Procedure. Male Fisher-344 rats (190–200 g) are anesthetized with Metofane, and 2.5 ml blood is withdrawn to stimulate erythropoiesis. Immediately afterward the rats are injected intravenously with 0.5 ml of saline containing 10 μCi [^{59}Fe]ferrous citrate. Rats are then used for the RPA experiments 3–4 days later. Analysis of blood samples shows that all (>99%) of the radioisotope is bound within the erythrocytes.

Induction of Reverse Passive Arthus Reaction

Rats (200–210 g), whose erythrocytes were previously labeled with ^{59}Fe, are used. Each is injected intradermally at four sites along the back with 50 μl of rabbit antiovalbumin serum (200 μg antibody protein/50 μl antiserum), followed immediately by an intravenous injection of 1 ml of saline containing 2 mg of ovalbumin. Control animals receive intravenous saline only. Hemorrhage is usually visible in the RPA sites by the fourth hour, and further erythrocyte accumulation ceases about 8 hr after initiation of the RPA. At the appropriate times, a 1-ml blood sample is taken by cardiac puncture, the animals are sacrificed by CO_2 asphyxiation, and the skin sites punched out with a metal punch as previously described. The radioactivity of the blood and skin samples is determined in a gamma counter. The accumulation of erythrocytes in the RPA skin sites is expressed in terms of microliter equivalents of blood. This value is obtained for each animal by dividing the average ^{59}Fe cpm of the skin sites by the average ^{59}Fe cpm of 1 μl of that animal's blood.

Comments

The ease with which the Arthus phenomenon can be induced in experimental animals makes this an attractive model to study the inflammatory process *in vivo.* The fact that the model involves deposition of immune complexes, the activation of complement, and is dependent on circulating neutrophils makes this phenomenon very relevant to several disease processes. It has been shown[2] that antiinflammatory steroids are very effective in reducing the edema and cellular infiltration in the Arthus reaction. On the other hand, nonsteroidal antiinflammatory drugs, while significantly reducing the levels of infiltrating cells, do not do this sufficiently to affect the edema. It would appear, therefore, in direct contrast to the inflammation induced by conventional agents (e.g., carrageenan or zymosan), that prostaglandins do not play a major role in the causation of the Arthus phenomenon. The use of this model of inflammation should lead to new insights into the pharmacologic modulation of the inflammatory response.

[41] Experimental Serum Sickness

By BERNICE NOBLE and JAN R. BRENTJENS

Introduction

Experimental serum sickness, resulting from an immune response against foreign serum proteins, is the classic example of tissue injury induced by immune complexes and is the model from which most has been learned about the potential of immune complexes to damage the kidney and other organs. Although the antigens used to elicit experimental serum sickness in laboratory animals are heterologous proteins, our present appreciation of the contribution of immune complexes to the pathogenesis of certain autoimmune diseases, particularly systemic lupus erythematosus, depends on insights gained from the study of serum sickness.

In all forms of experimental serum sickness, the clinical course and the severity of disease may be affected by the amount of antibody that is produced and by the quality of that antibody. Research into the pathogenesis of immune complex-mediated tissue injury has been aimed at identifying the qualitative and quantitative aspects of the immune response that promote inflammation in the kidney and in other organs. Other aspects of immune complex pathology that can be studied with one or another of the available animal models of experimental serum sickness include the role of soluble or cellular mediators of inflammation, the value of different approaches to therapy, and the relationship of functional impairment to immune complex formation.

Acute Serum Sickness Produced by Immunization of Rabbits with Native BSA

The majority of rabbits will react to an intravenous injection of foreign proteins with a specific antibody response. This feature has made the rabbit the most widely used animal in the study of acute as well as chronic experimental serum sickness. Although in the earliest investigations serum sickness was induced by the intravenous injection of whole serum, at the end of the 1940s Hawn and Janeway showed that purified serum proteins could be equally effective in this respect.[1] That observation was important because it provided the opportunity to analyze the pathogenesis of serum

[1] C. V. Z. Hawn and C. A. Janeway, *J. Exp. Med.* **85,** 571 (1947).

sickness by quantitative immunological techniques. Pioneering studies of this type were performed in the 1950s by Germuth[2] and by Dixon and collaborators.[3] Over the years, native bovine serum albumin (BSA)[3a] has become the protein traditionally used for the production of serum sickness.

Immunization Scheme

Crystallized BSA is dissolved in sterile 0.8% NaCl to form a 10% solution. New Zealand White rabbits, weighing 2–2.5 kg, receive 250 mg of BSA/kg body weight via a lateral ear vein. Nine to 15 days later, about 60% of the animals treated in this manner will develop acute serum sickness.

To increase the incidence and uniformity of the disease, the following procedures are employed. First, 3 days before the intravenous injection of BSA, each rabbit receives a total of 1 ml of an emulsion prepared by mixing equal parts of a BSA solution (6 mg BSA/ml isotonic saline) with Freund's complete adjuvant (Difco Laboratories, Detroit, MI); the 1 ml of emulsion is divided over multilple subcutaneous sites on the back of the animal.[4] Second, for each rabbit, 10 μg of *Escherichia coli* endotoxin (Sigma Chemical Co., St. Louis, MO) is added to the BSA solution that is administered intravenously.[5] When used in combination, these two procedures increase the incidence of acute serum sickness to over 90% of the animals.

Monitoring the Development of the Disease (see also this volume [36])

The most prominent manifestation of acute serum sickness is a reversible, acute proliferative glomerulonephritis. By immunofluorescence microscopy, scarce, finely granular deposits composed of BSA and rabbit Ig and complement are found along the glomerular capillary wall. The clinical signs of the renal disease are proteinuria and, in the most severely affected animals, a decrease in the glomerular filtration rate. The joints and the cardiovascular system are also damaged in acute serum sickness. The peak incidence of the lesions in the joints and vessels precedes those in the kidneys by a few days.[5] Full development of the glomerulonephritis of

[2] F. G. Germuth, Jr., J. Exp. Med. **97,** 257 (1953).

[3] F. J. Dixon, J. J. Vazquez, W. O. Weigle, and C. G. Cochrane, *Arch. Pathol.* **65,** 18 (1958).

[3a] Abbreviations: BSA, bovine serum albumin; PBS, phosphate-buffered saline; BW, body weight, HSA, human serum albumin; BGG, bovine γ-globulin.

[4] G. Camussi, C. Tetta, M. C. Deregibus, F. Bussolino, G. Segoloni, and A. Vercellone, *J. Immunol.* **128,** 86 (1982).

[5] W. T. Kniker and C. G. Cochrane, *J. Exp. Med.* **122,** 83 (1965).

acute serum sickness coincides with the end of the so-called "immune phase" of BSA elimination. (see below).

From this brief phenomenological description of acute serum sickness, it may be clear that monitoring the development of the disease may be achieved by measuring the urinary protein excretion, by examining kidney biopsies, or by following the disappearance of BSA from the circulation. The measurement of proteinuria does not appear to be an adequate parameter as increased glomerular permeability is usually present only for 1–2 days and may be easily missed. Furthermore, in an appreciable number of animals, renal immunopathology is not reflected in increased urinary protein excretion.[6] Seeking immunopathologic evidence of glomerulonephritis in kidney biopsies is not very practical as acute serum sickness may develop in individual rabbits at any time from 9 to 15 days after the intravenous injection of BSA.

The occurrence of acute serum sickness nephritis can most reliably and accurately be predicted by following the disappearance of BSA from the circulation.[6] A practical way to do so is to label the BSA that will be injected with radioactive iodine. A simple iodination procedure of proven value makes use of chloramine-T as an oxidizing agent.[7] Daily measurement of the levels of radiolabeled BSA in plasma reveals that the antigen is eliminated in three distinct phases.[2,3,8] In the first 2 days, 60–75% of the original BSA dose leaves the circulation. This is a reflection the equilibration of the protein between the intravascular and extravascular space. The second phase is characterized by a logarithmic rate of disappearance of BSA from the circulation and represents normal catabolism of BSA. Within 4–5 days, antibodies against BSA are produced, an event that marks the beginning of the third, or immune, phase of antigen elimination. After this phase of accelerated disappearance of BSA, free anti-BSA antibodies become detectable in the circulation. All animals in which more than 99% of the injected BSA dose disappears from the circulation between days 9 and 15 will have immunopathologic evidence of acute serum sickness glomerulonephritis. Animals that have eliminated BSA 15 or more days after the injection will have only very mild disease or, having not produced any specific antibodies, will not develop disease at all.[6]

Determination of Urinary Protein Excretion (see also this volume [36])

To collect urine, rabbits are placed for about 15 hr in a metabolism cage (Lab Products, Inc., Maywood, NJ) with free access to water but not

[6] C. B. Wilson and F. J. Dixon, *J. Immunol.* **105,** 279 (1970).

[7] P. J. McConahey and F. J. Dixon, *Int. Arch. Allergy* **29,** 1985 (1966).

[8] D. W. Talmage, F. J. Dixon, S. C. Bukantz, and G. J. Dammin, *J. Immunol.* **67,** 243 (1951).

food. Of several methods for protein determination in the urine,[4,6] we routinely use the technique described by Shevky and Stafford.[9] An aliquot of urine is centrifuged at about 600 *g* for 5 min. Four milliliters of the supernatant is pipetted into a McNaught and Mackay–Shevky–Stafford sedimentation tube (Catalog No. 46800, Kimble Products, Toledo, OH) and layered with 2.5 ml of Tsuchiya's reagent (prepared by adding 7.5 g phosphotungstic acid and 25 ml concentrated HCl to 500 ml 85% ethanol). After mixing, the sedimentation tube is centrifuged 4° for 30 min at 600 *g*. One-tenth milliliter of sediment equals 72 mg of protein per 100 ml of urine. Normal rabbits excrete less than 50 mg protein/24 hr. In acute serum sickness, daily protein excretion may vary widely but can exceed 1 g.

Preparation of Kidney Tissue for the Evaluation of Immunopathology

Samples of kidney tissue may be obtained by surgical biopsy or at necropsy. For immunofluorescence microscopy, the tissues are embedded in Optimal Cutting Temperature compound (OCT, VWR Scientific, Rochester, NY), snap-frozen in liquid nitrogen, and stored at −20° until sections are cut with a cryostat. The frozen tissue sections, approximately 4 μm thick, are air dried, rinsed several times in phosphate-buffered saline (PBS), pH 7.2, and incubated for 30 min with fluorescein isothiocyanate-conjugated antiserum to immunoglobulins, complement, or BSA (Cooper Biomedical Inc., Malvern, PA). After extensive rinsing with PBS, the tissue can be examined with a fluorescence microscope to ascertain the pattern and extent of distribution of immune reactants within the kidney.

For conventional light microscopy the kidney tissue may be fixed by immersion in 10% Millonig's modified phosphate-buffered formalin (Surgipath, Grayslake, IL) and embedded in paraffin. Staining with hematoxylin–eosin and periodic acid–Schiff reagent will suffice for most evaluations, but other commonly used histochemical stains may also be valuable, including the methenamine silver stain to detect "spike formation" in the basement membrane and the Masson trichrome stain for the evaluation of fibrous tissue formation (for details of these histological techniques, see Ref. 10). For routine electron microscopy minced kidney tissue can be fixed with 2% glutaraldehyde in 0.1 *M* phosphate buffer, pH 7.2, and postfixed in osmium tetroxide.[10] Study by light microscopy of tissue embedded in plastic, cut to 1 μm thickness, and stained with toluidine blue may also prove to be instructive.

[9] M. C. Shevky and D. D. Stafford, *Arch. Intern. Med.* **32,** 222 (1923).

[10] D. C. Sheehan and B. B. Hrapchak, "Theory and Practice of Histotechnology." Mosby, St. Louis, Missouri, 1980.

Chronic Serum Sickness Produced by Immunization of Rabbits with Native BSA

Immunization Scheme

To produce chronic serum sickness, rabbits are treated daily with one or more intravenous injections of BSA. Two approaches may be used, one in which the daily dose of BSA is kept constant,[11] another in which the dose is adjusted in each individual animal to the strength of its own antibody response.[12,13] In both protocols, a 250 mg% solution of crystallized BSA in sterile 0.8% NaCl is used. The appropriate dose is injected into a lateral ear vein of New Zealand White rabbits weighing 2–2.5 kg.

Upon daily challenge with BSA, a certain number of rabbits will never produce antibodies and, therefore, will never develop chronic serum sickness. In animals that do form antibodies, several distinct phases may be recognized during the course of daily immunization with BSA (Fig. 1).[11-13] After 2–3 weeks the immunization regimen produces an acute, reversible glomerulonephritis that may be manifested clinically by transient proteinuria. This phase is followed by one in which free anti-BSA antibodies may become detectable in the circulation and in which, despite the daily injection of BSA, only minimal, mainly mesangial, immune deposit formation can be observed in the kidney. The last phase, which coincides with a decay in the antibody response to BSA,[11,14] is characterized by the extensive accumulation of immune deposits in glomeruli and by clinical and histological evidence of a glomerulonephritis of variable severity. In the majority of animals, this kidney disease, called chronic serum sickness glomerulonephritis, is irreversible, progressive, and may lead to death from uremia.

If rabbits are given constant daily doses of relatively low amounts of BSA (from 5 to 25 mg), it may, especially in animals with a vigorous antibody response, take many months to induce chronic serum sickness. In contrast, if the initial dose of 12.5 mg/day is increased after 3 weeks to match the antibody response of so-called high responders, chronic serum sickness may be produced in 6–10 weeks (Fig. 1).[12,13] The adjustment of the daily dose of BSA is made every 7 or 10 days and is based on an

[11] F. G. Germuth, Jr., L. B. Senterfit, and A. D. Pollack, *Johns Hopkins Med. J.* **120,** 225 (1967).

[12] F. J. Dixon, J. D. Feldman, and J. J. Vazquez, *J. Exp. Med.* **113,** 899 (1961).

[13] J. R. Brentjens, D. W. O'Connell, I. B. Pawlowski, K. C. Hsu, and G. A. Andres, *J. Exp. Med.* **140,** 105 (1974).

[14] B. Albini, J. R. Brentjens, R. Olson, E. Ossi, and G. A. Andres, *Proc. Conv. Immunol. 6th,* p. 207 (1979).

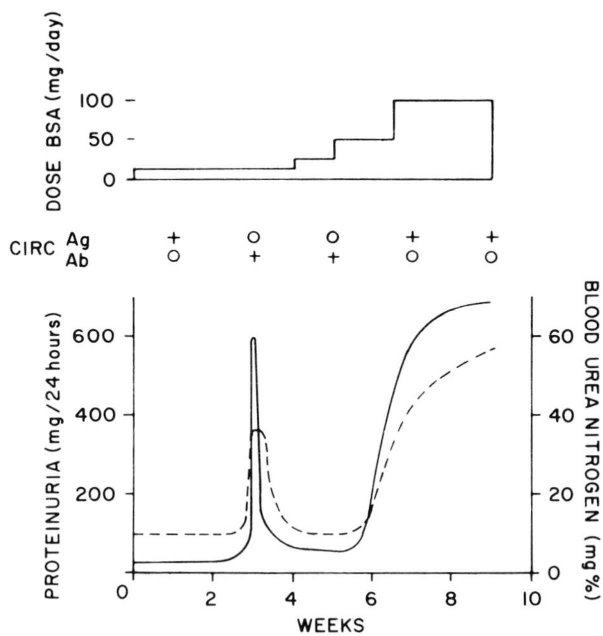

FIG. 1. Schematic representation of the induction of chronic serum sickness in rabbits immunized with native BSA. At the top of the diagram, the daily dose of BSA is indicated. The presence of free circulating (CIRC) antigen (Ag) or antibody (Ab) is recorded as +, the absence of either as 0. Protein excretion is indicated by a solid line, blood urea nitrogen by a dashed line.

estimation of the strength of the specific immune response (see below). Serological testing of sera of rabbits given daily intravenous injections of low doses (12.5 mg) of BSA for 3 weeks will give the following results: 10–15% of the rabbits will not have free antibody but will have free BSA in their circulation. These animals have failed to produce antibodies against BSA and will endure no ill effects of the chronic immunization. Another small group (10–15%) of the rabbits will have a relatively low antibody response. Some of these rabbits will show neither free antibody nor free BSA in their circulation. Continued injection of BSA at 12.5 mg/day may induce chronic serum sickness in these individuals. Other animals in the same group will have low titers of precipitating anti-BSA antibodies, and one may consider raising their daily BSA dose to 25 mg. The remaining rabbits (60–70%) will have high titers of precipitating anti-BSA antibodies in the circulation. In these rabbits the daily BSA dose can be increased from 12.5, to 25, to 50, to 75, to 100, and eventually to 150 mg until a decay in titer is observed; they can be maintained on the BSA dose at which this phenomenon occurs. Amounts of BSA greater than 50 mg must be injected very slowly, in two or more equal portions at intervals of 4–8 hr, to reduce the possibility of anaphylactic shock. However, despite all precautions, about 10% of the rabbits with a high antibody response will die from anaphylaxis.

Monitoring the Development of the Disease

Direct immunofluorescence tests on kidney tissue from rabbits with chronic serum sickness reveal the presence of mesangial and subepithelial immune deposits, composed of BSA, rabbit Ig, and complement. In addition, in the most severely affected animals, subendothelial immune deposits may also be found. The histological lesions are variable, ranging from mesangial to membranous to exudative endo- and extracapillary proliferative glomerulonephritis.[11,12,15] The incidence of membranous glomerulonephritis is especially high among rabbits with a low antibody response and among those treated with daily constant, low doses of BSA.[15,16] In contrast, practically all animals with a vigorous response given appropriately adjusted doses of BSA develop severe proliferative glomerulonephritis.[12,15] In high responder animals given daily high doses of BSA, immune deposits are also detected at extraglomerular sites in the kidney and, with variable frequency, in many organs beside the kidney (Table I).[17,18] Therefore, their disease may be characterized as a systemic immune complex disease.[18] The kidney lesions of chronic serum sickness are accompanied by heavy proteinuria and, frequently, by a decrease in glomerular filtration rate. Useful parameters for monitoring the development of chronic serum sickness include circulating anti-BSA antibody titer, urinary protein excretion, serum creatinine or blood urea nitrogen concentration, and kidney immunopathology (Fig. 1). To determine the antibody titer (see below) rabbits are bled weekly, 12–24 hr after the last BSA injection, from the central ear artery. This blood sample may also be used for the assessment of kidney function. Urinary protein excretion is also measured on a weekly basis (see above, Acute Serum Sickness). A biopsy is indicated when precise information of the lesions in the kidney is required (see above, Acute Serum Sickness).

[15] D. W. O'Connell, J. R. Brentjens, and G. A. Andres, *Nephron* **20,** 316 (1978).

[16] F. G. Germuth, Jr., J. J. Taylor, S. Y. Siddiqui, and E. Rodriguez, *Lab. Invest.* **37,** 162 (1977).

[17] J. R. Brentjens, D. W. O'Connell, B. Albini, and G. A. Andres, *Ann. N.Y. Acad. Sci.* **254,** 603 (1975).

[18] J. R. Brentjens and G. A. Andres, *Arthritis Rheum.* **25,** 880 (1982).

TABLE I
TISSUE LOCALIZATION OF IMMUNE COMPLEXES IN HIGH RESPONDER RABBITS WITH CHRONIC SERUM SICKNESS[a]

Organ/structure	Frequency of positivity (%)	Severity of inflammation[b]
Glomerulus	100	+++
Spleen	80–90	0
Ovary	80–90	+
Eye (ciliary processes, Bruch's membrane)	80–90	+ to ++
Lung	70–80	+ to +++
Heart	70–80	0 to +
Kidney, extraglomerular	60–70	+ to +++
Serous membranes	60–70	++
Salivary/lacrimal glands	60–70	+
Choroid plexus	50–60	0
Intestine	40–50	+ to ++
Endocrine glands	40–50	0
Spleen, sinusoidal walls	40–50	+ to ++
Synovia	40–50	+
Liver, sinusoidal walls	5–10	0
Dermal–epidermal junction	0	0

[a] When the daily dose of BSA is increased to match antibody production in rabbits with a vigorous immune response to BSA, immune complex deposits may be found in many organs beside the kidney.

[b] 0, Absent; +, slight; ++, moderate; +++, marked.

Detection of Antibodies to BSA

Any reliable technique for the detection of antibodies may be used. In our laboratory the strength of the precipitating anti-BSA antibody response in the rabbit is routinely screened by a semiquantitative immune precipitin test. After preparing serial dilutions (1/100 to 1/6400) of a 1% solution of BSA in borate-buffered saline, pH 8.4, prepared by mixing 95 parts of 0.85% NaCl with 5 parts of a borate buffer made by adding 6.2 g of boric acid, 9.5 g of borax (sodium tetraborate, $Na_2B_4O_7 \cdot 10H_2O$), and 4.4 g NaCl to 1 liter of distilled water, 2 drops of each dilution are mixed in small tubes with 2 drops of undiluted rabbit serum. The test is read after 24 hr and scored visually on a semiquantitative scale of negative to 3+. Free circulating BSA is detected in a similar manner by incubating a high titered anti-BSA serum with an equal volume of the rabbit serum being tested. Anti-BSA serum of high titer can be obtained from rabbits immunized for chronic serum sickness.

Chronic Serum Sickness Produced by Immunization of Rats with Native BSA

Unlike rabbits, rats do not respond to intravenous injections of native foreign serum proteins with a vigorous antibody response. As a result, acute serum sickness, resulting from a single high dose of antigen, is unknown in rats. Chronic serum sickness can be produced in rats, but only after antibody synthesis has been first stimulated by the subcutaneous administration of the antigen in an emulsion with Freund's adjuvant. Unfortunately, intravenous injections of antigen into animals with significant antibody titers may cause fatal anaphylactic reactions. The immunization protocol outlined below has been devised to minimize that difficulty. For the induction of chronic serum sickness in rats, BSA is the most widely used antigen,[19,20] although immunizations with human serum albumin and human γ-globulin have also been described.[21,22] Rat strains used successfully include Lewis (LEW), Fisher (F344), Wistar, SHR, Holtzmann, and Sprague–Dawley.[21–23] Brown Norway (BN) and Wistar King–Aptekman (WKA) rats may be less suitable.[21,23]

Immunization Scheme

The following scheme has been devised to produce chronic serum sickness with great reproducibility in young female LEW rats. Modifications may be required for animals of different age and sex. Strain differences in susceptibility will also alter the induction of disease. Female LEW rats, weighing approximately 175 g, receive injections of an emulsion prepared by mixing equal volumes of a BSA solution (30 mg BSA/ml isotonic saline) with Freund's incomplete adjuvant (Fig. 2). Each rat is given 0.2 ml of emulsion, containing 3.0 mg BSA. The injection is administered in the rear foot pads, 0.1 ml to each foot. The immunization is repeated 2 weeks later using the front feet. Four weeks after the first injection, blood is collected from the tail vein into heparinized hematocrit capillaries, and the plasma is tested by Ouchterlony immunodiffusion tests (see also this series, Vol. 70 [9]) for precipitating antibodies to BSA. If antibody titers are too low (less than 1:4) a third identical BSA injection is administered in several intradermal sites on the back near the neck.

Daily injections of BSA in a soluble form are initiated when 4-fold

[19] L. Arisz, B. Noble, M. Milgrom, J. R. Brentjens, and G. Andres, *Int. Arch. Allergy Appl. Immunol.* **60,** 80 (1979).

[20] W. K. Bolton and B. C. Sturgill, *Br. J. Exp. Pathol.* **59,** 167 (1978).

[21] K. Yamamoto, T. Oite, I. Kihara, and F. Shimizu, *Clin. Exp. Immunol.* 57, 575 (1984).

[22] R. H. Fennell and V. M. Pardo, *Lab. Invest.* **17,** 481 (1967).

[23] B. Noble, M. Milgrom, J. B. Van Liew, and J. R. Brentjens, *Clin. Exp. Immunol.* **46,** 499 (1981).

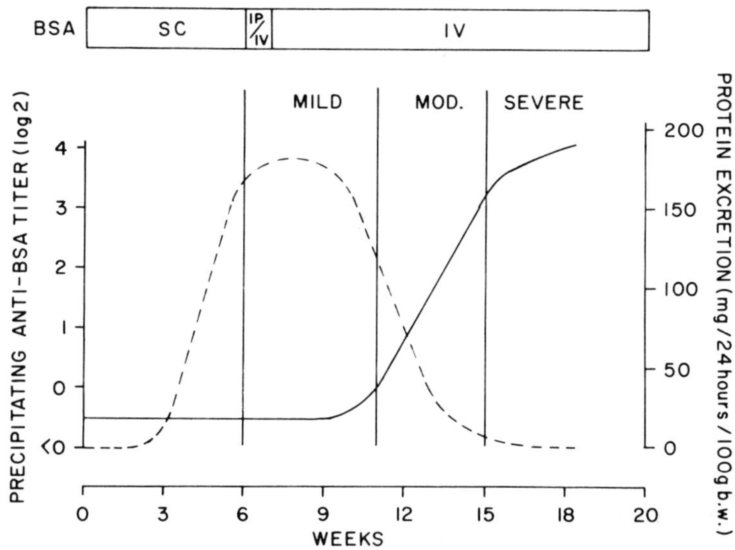

FIG. 2. Schematic representation of the induction of chronic serum sickness in rats immunized with native BSA. The mode of introduction of BSA is indicated at the top of the diagram (SC, subcutaneous; IP, intraperitoneal; IV, intravenous). Protein excretion is indicated by a solid line, anti-BSA antibody titer by a dashed line. Vertical lines show the division of the natural history of the disease into three discrete stages, mild, moderate, and severe.

dilutions of plasma form visible lines in Ouchterlony immunodiffusion plates (see below). On the first day, BSA is administered intraperitoneally. Early in the day, 0.5 mg BSA/0.5 ml saline is given. At hourly intervals the dose is increased in 0.5-mg increments until a total of 6.0 mg has been given. On the following day, 0.5 mg BSA/0.5 ml is administered by direct injection into a tail vein. On the third day 2.0 mg BSA/0.5 ml is given intraperitoneally in a single dose, on the fourth 1.0 mg BSA intravenously, on the fifth 2.0 mg BSA intraperitoneally, and on the sixth 2.0 mg BSA intravenously. From that time 2.0 mg BSA can be safely administered intravenously on a daily basis, provided that no more than 1 day is occasionally missed.

With this regimen most animals will become proteinuric within 4–6 weeks. The onset of proteinuria may be accelerated by increasing the daily BSA dose.[23] Giving daily tail vein injections over a period of many months can be technically demanding. It is recommended to warm the tail first in a 50° water bath and to use 25- or 26-gauge needles. Care should be taken not to introduce any antigen into extravascular sites in the tail. Local Arthus reactions may result in tissue necrosis and/or edema that will hinder subsequent intravenous injections.

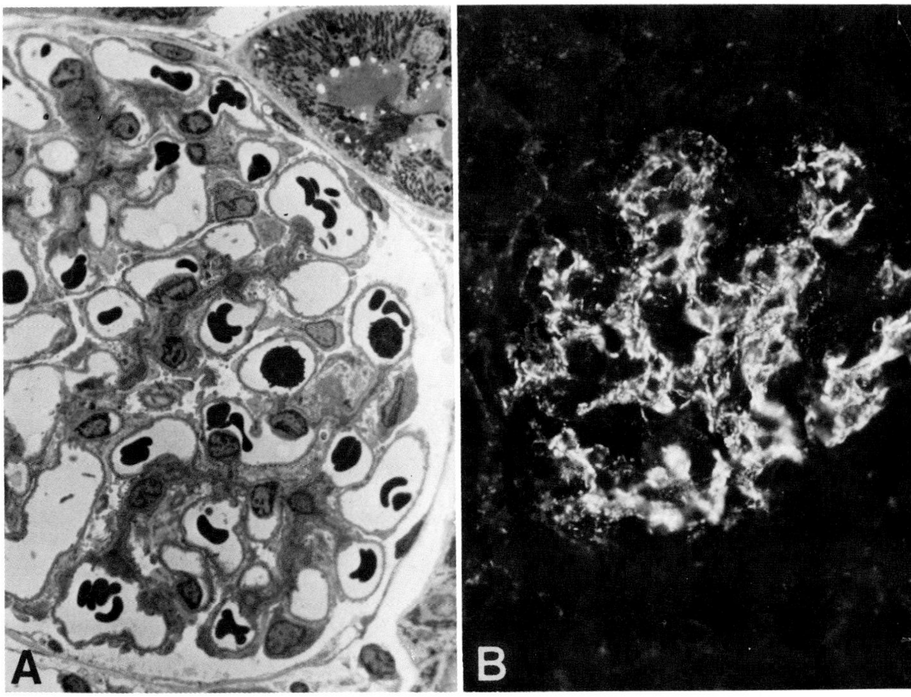

FIG. 3. (A) Light micrograph of part of a glomerulus of a rat in the mild phase of chronic serum sickness. There is a moderate expansion of the mesangial area. Toluidine blue. ×1,200. (B) A glomerulus of a rat in the mild phase of chronic serum sickness, stained for BSA by the direct immunofluorescence technique. BSA appears to be present predominantly in the mesangial area. ×400.

Monitoring the Progression of Disease (see also this volume [36])

The early development of chronic serum sickness can be monitored very precisely by regular determinations of anti-BSA titer and urinary protein excretion.[24] The onset of proteinuria is very closely correlated with the disappearance of precipitating antibodies from the circulation (Fig. 2). In the preproteinuric stage, designated the mild stage of chronic serum sickness, histopathology is minimal or absent (Fig. 3a). Direct immunofluorescence tests on kidney biopsies reveal only a mesangial accumulation of immune reactants (Fig. 3b). The second stage of glomerulonephritis, which is called moderate chronic serum sickness, has a rather abrupt onset, protein excretion in most animals increasing from normal [<1.0 mg/100 g

[24] J. B. Van Liew, J. R. Brentjens, and B. Noble, *Kidney Int.* **24,** 160 (1983).

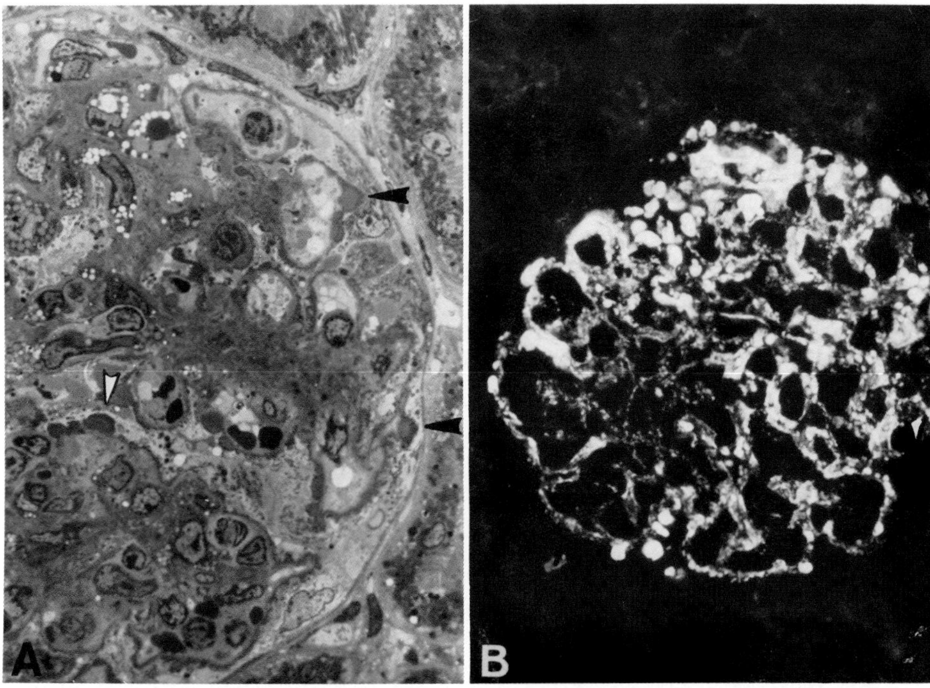

FIG. 4. (A) Light micrograph of part of a glomerulus of a rat with severe chronic serum sickness. There is excessive proliferation of endocapillary cells resulting in partial or complete occlusion of glomerular capillary lumina. Subepithelial deposits are clearly visible (arrowheads). Toluidine blue. ×1,200. (B) Glomerulus of a rat with severe chronic serum sickness, stained for BSA by the direct immunofluorescence technique. Granular deposits of BSA of varying size are seen along the glomerular capillary walls and in some mesangial areas. ×400.

body weight (BW) per day] to above 100 mg/100 g BW per day within 48 hr. Biopsies taken after proteinuria is established show proliferative, inflammatory lesions in most glomeruli (Fig. 4a), associated with peripheral capillary wall deposits of BSA, rat Ig, and complement (Fig. 4b). Animals in the third and final, or severe, stage of chronic serum sickness glomerulonephritis can be recognized by the formation of ascites fluid that occurs only in that stage.[25] The precise onset of severe chronic serum sickness can be detected by careful monitoring of sodium excretion and/or creatinine clearance, which remain normal throughout the mild and moderate stages of disease and drop sharply when the severe stage begins.

Creatinine clearance may be determined by analyzing matched plasma

[25] J. B. Van Liew, B. Noble, and J. R. Brentjens, *Nephron* **40,** 476 (1985).

and urine samples with the method of Hare.[26] Twenty-five microliters of plasma is added to 100 μl 10% trichloroacetic acid in a 400-μl mirocentrifuge tube. The supernatant obtained after centrifugation for 2 min at 7500 *g* in a Beckman Microfuge B (Beckman Instruments, Spinco Division, Palo Alto, CA) is removed. One milligram of Lloyd's reagent (hydrated aluminum silicate, Eastman Kodak Co., Rochester, NY) and 10 μl saturated oxalic acid are added to another 400-μl microcentrifuge tube. Appropriately diluted urine, the plasma supernatant described above, and standard solutions of creatinine are added in 100-μl aliquots to the Lloyd's reagent, mixed several times over 10 min, and centrifuged for 1 min, after which the supernatant is discarded. A freshly prepared solution of alkaline picrate (5 ml picric acid, 1 ml 10% NaOH, 12 ml H_2O) is added in 100-μl aliquots to each tube, mixed several times over 10 min, and incubated for 30 min in a water bath at 37°. After centrifugation for 2 min, the optical density of the supernatant is measured at 500 nm. Creatinine clearance (ml/min × 100 g BW) can be determined from the following relationship:

$$\text{Creatinine clearance} = \frac{\text{creatinine concentration of urine (mg/dl)}}{\text{creatinine concentration of plasma (mg/dl)}} \times \text{urine flow rate (ml/min} \times \text{100 g BW)}$$

In severe chronic serum sickness of rats, diffuse and exudative, proliferative glomerulonephritis is accompanied by segmental necrosis in many glomeruli, the formation of inspissated tubule casts, and some tubule atrophy. Large dense deposits, representing immune complex aggregates, are discernible in subepithelial sites along the glomerular basement membrane (Fig. 5a). Abnormalities of the glomerular capillary wall include foot process fusion, invasion of blood-borne cells, and disruption of the integrity of the basement membrane (Fig. 5b). To document these changes, biopsies may be taken and tissue prepared for study as mentioned above (Acute Serum Sickness).

The development of severe glomerulonephritis in rats immunized chronically with BSA is also associated with systemic immune complex deposition. Affected organs include lungs, heart, liver, spleen, choroid plexus, ovary, testes, peritoneum, and intestine.[19] Under certain conditions, intraperitoneal administration of BSA may produce extensive systemic immune complex deposition in the absence of severe kidney disease.[23]

Detection of Antibodies to BSA

Anti-BSA antibodies are readily measured in this model of serum sickness, although once daily intravenous antigen injections have been

[26] R. S. Hare, *Proc. Soc. Exp. Biol. Med.* **74,** 148 (1950).

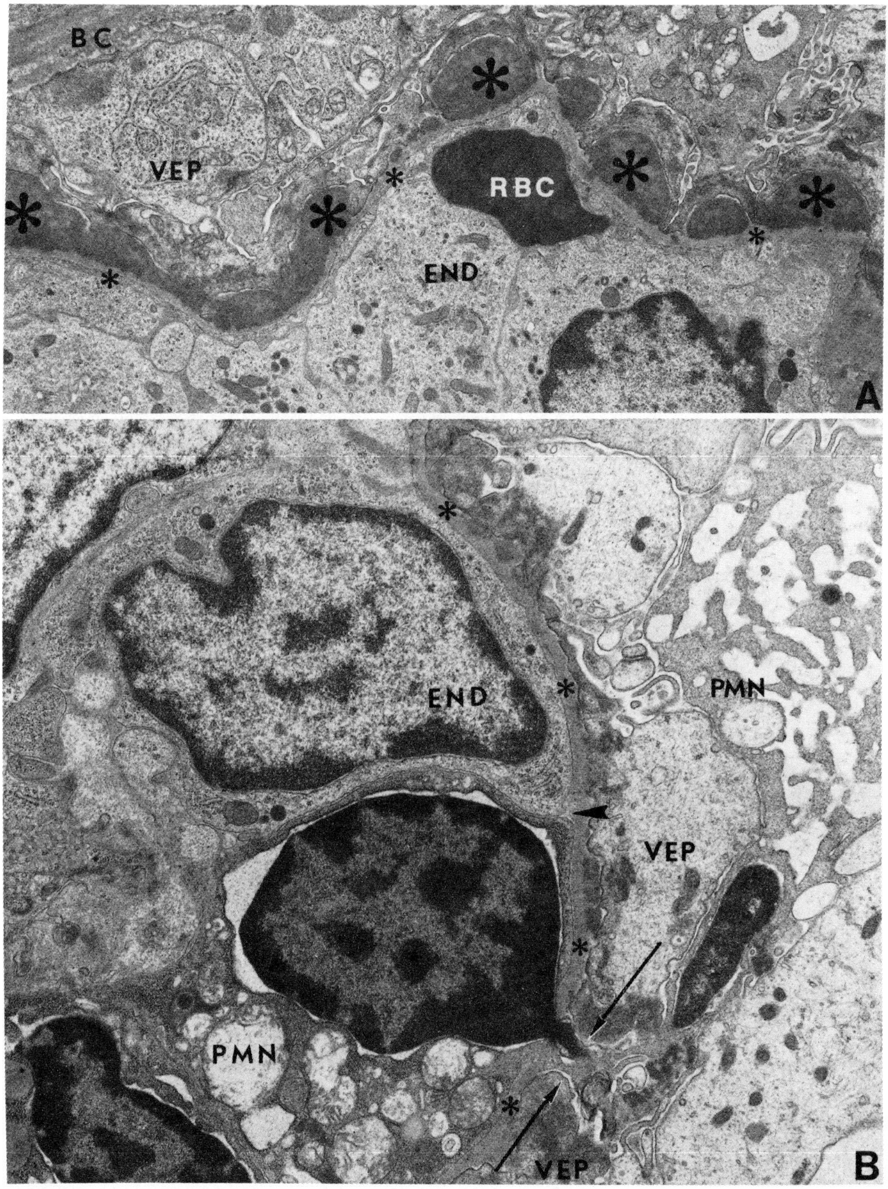

FIG. 5. (A) Electron micrograph of the glomerular capillary wall of a rat with severe chronic serum sickness, showing subepithelial electron-dense deposits (large asterisks). BC, Bowman's capsule; VEP, visceral epithelial cell; END, endothelium; RBC, part of a red blood cell. The small asterisks mark the glomerular basement membrane. ×6,000. (B) A polymorphonuclear leukocyte (PMN) in direct contact with the glomerular basement membrane (beginning of direct contact is indicated by the arrowhead) appears to be in the process of crossing the glomerular capillary wall. The glomerular basement membrane (marked by asterisks) and the continuity of the visceral epithelial cell lining are disrupted (arrows). VEP, Visceral epithelial cell; END, endothelial cell. ×6,000.

initiated care must be taken to collect blood at a fixed interval after the antigen injection. For routine screening the best time is 24 hr after the previous BSA dose, as the measurable titer is significantly reduced just after the injection of antigen and the return to maximum, preinjection values may take several hours. Antibodies to BSA may be measured by almost any kind of antibody test, including radioimmunoassay, enzyme-linked immunoassay, tanned cell hemagglutination, tube precipitation, and complement fixation (see also this series, Vol. 70 [9] and [10]). For samples to be taken frequently from the same rat, a method must be chosen that requires only a small amount of blood. For this reason the tube precipitation test, which proves so convenient for measuring anti-BSA antibodies in rabbits with serum sickness, cannot be used to monitor the disease in rats. Since the onset of proteinuria in this model is so well correlated with the disappearance of the precipitating antibodies that can be detected by means of Ouchterlony gel immunodiffusion, this test, which is quite simple to perform, is particularly valuable for monitoring disease progression in rats. Three milliliters of 1% agarose is poured into a small petri dish (9 × 50 mm). Wells of 3 mm diameter, separated by a distance of 3 mm, are cut into the gel. Each well is filled either with 4 μl of appropriately diluted plasma or with 4 μl of BSA, 0.2 mg/ml, using a 10-μl Hamilton syringe for precise delivery. The plates are incubated for 24 hr at 4° and examined with a bright light to detect lines of precipitation. The end point of the titration is the highest dilution giving an unequivocal positive reaction, i.e., formation of a precipitation line in the gel.

Determination of Urinary Protein Excretion (see also this volume [36])

Total daily protein excretion in the urine can be estimated by measuring the protein concentration in urine samples collected overnight for 15–18 hr. To prevent contamination of the urine sample, it may be necessary to deprive the animals of food during the collection period. To collect urine on consecutive days, a complete liquid diet should be provided, as diminished protein intake will also reduce protein excretion. Any convenient, rapid quantitative test for protein will serve to measure the urinary protein excretion of rats. The biuret test or the Bio-Rad test of Bradford[27] is adequate. In our opinion, both the Shevky–Stafford test (recommended for analysis of rabbit urine) and the paper "dipsticks" used in clinical laboratories lack sufficient precision, sensitivity, and reproducibility to be useful for the evaluation of the protein concentration of rat urine.

[27] M. M. Bradford, *Anal. Biochem.* **72**, 248 (1976).

Chronic Serum Sickness in Mice Produced by Immunization with Native Foreign Serum Proteins

Severe chronic serum sickness has been elicited in several mouse strains using immunization protocols similar to the one used for rats in that they depend on stimulating antibody production by preimmunization with antigen and adjuvant. Subcutaneous administration of adjuvant-emulsified BSA (0.5 mg, 3 times at 2-week intervals) to male and female C3H/NB mice, followed by daily intraperitoneal injections of soluble BSA (0.3–0.5 mg), leads to immune complex glomerulonephritis with proteinuria, the accumulation of electron-dense deposits in the peripheral capillary wall, and diffuse proliferative lesions in glomeruli.[28] The time of onset of proteinuria is closely linked to the disappearance of precipitating anti-BSA antibodies from circulation and is a function of the daily BSA dose. Kidney disease is accompanied by the accumulation of immune complex deposits in many other organs. A similar procedure, using human serum albumin as the antigen, has been described to produce chronic immune complex glomerulonephritis in female C57BL/6J mice.[29]

Several immunization schemes have been described to produce chronic immune complex glomerulonephritis in mice that do not depend on presensitization with antigen in adjuvant. Serum sickness develops in Swiss albino,[30] BALB/c, and C3H[31] mice given regular injections of ferritin or apoferritin for 1 or 2 months. In mice selectively bred to produce antibodies of low affinity, Steward[32] has shown that prolonged (up to 4 months) daily intraperitoneal injections of human serum albumin (1 mg HSA/100 g) cause a high incidence of severe glomerulonephritis with immune complex deposits in the peripheral capillary wall. When TO mice bred for high-affinity antibodies are exposed to a similar immunization regimen, milder kidney immunopathology develops, with immune complexes deposited predominantly in mesangial sites.

Serum Sickness Produced by Immunization with Cationic Proteins (see also this volume [38])

In an attempt to elucidate the role of electrical charge in nephritogenesis, the effects of immunization with cationic and anionic species of foreign proteins have been studied in rabbits, rats, and mice. Severe

[28] B. Noble, K. Olson, M. Milgrom, and B. Albini, *Clin. Exp. Immunol.* **42,** 255 (1980).

[29] A. O. Haakenstaad, G. Striker, and M. Mannik, *Lab. Invest.* **48,** 323 (1983).

[30] M. M. Stilmant, W. G. Couser, and R. S. Cotran, *Lab. Invest.* **32,** 746 (1975).

[31] S. S. Iskandar and J. C. Jennette, *Lab. Invest.* **48,** 726 (1983).

[32] M. W. Steward, *Clin. Exp. Immunol.* **38,** 414 (1979).

glomerulonephritis, with heavy proteinuria, elevated blood urea nitrogen, and the accumulation of immune deposits in subepithelial sites, develops in New Zealand White rabbits given a primary intravenous dose of highly cationic BSA (p*I* > 9.5) with 1 μg of endotoxin as adjuvant followed 1 week later by a 6-week regimen of daily intravenous injections of 25 mg cationic BSA.[33] Rabbits exposed to an identical immunization schedule with a highly anionic derivative of BSA (p*I* 3.5–4.5) or with native BSA (p*I* 4.5–5.0) will sustain a form of glomerular injury in which immune complexes are limited to mesangial areas, with mild mesangial hypercellularity and slight proteinuria.

A similar spectrum of glomerular immunopathology can be observed in mice immunized with cationic, anionic, or native bovine γ-globulin (BGG).[34] Daily intravenous injections to BALB/cAn NC/BR mice of cationic BGG (p*I* > 9.5, 0.2 mg/day) for 3 weeks leads to the formation of subepithelial immune deposits, membranous rather than proliferative glomerular pathology, and marked proteinuria. Slightly less cationic antigen (p*I* 8–10) produces milder disease with immune deposits located primarily in the mesangium and only focally in subepithelial sites. In contrast, when highly anionic BGG (p*I* 2–5) is administered in an identical fashion, no abnormalities of kidney structure or function are produced.

Cationic antigens have been used to produce a form of acute serum sickness in rats. Wistar rats are preimmunized with 500 μg native human IgG in Freund's complete adjuvant by introducing the emulsion in several subcutaneous and intramuscular sites.[35] Three weeks later 100 μg of cationic human IgG (p*I* > 9.5) is perfused directly into the left renal artery. This method leads to severe proliferative glomerulonephritis, accompanied by the extensive subepithelial accumulation of immune complex deposits. In this model immunopathology is limited primarily to the perfused left kidney.

Cationic derivatives of serum proteins may be prepared according to a method described originally by Danon and co-workers.[36] Five milliliters of H_2O is added to 250 μl *N,N*-dimethyl-1,3-propanediamine (Eastman Kodak), and the pH of the solution is adjusted to 6.5 with 1 *N* HCl. To this solution 200 mg of any serum protein is added. After gentle stirring, 375 mg 1-ethyl-3-(3-dimethylaminopropyl)carbodiimide (Sigma) is added.

[33] W. A. Border, H. J. Ward, E. S. Kamil, and A. H. Cohen, *J. Clin. Invest.* **69,** 451 (1982).

[34] G. R. Gallo, T. Caulin-Glaser, S. N. Emancipator, and M. E. Lamm, *Lab. Invest.* **48,** 353 (1983).

[35] T. Oite, F. Shimizu, I. Kihara, S. R. Batsford, and A. Vogt, *Am. J. Pathol.* **112,** 185 (1983).

[36] D. Danon, L. Goldstein, Y. Marikovsky, and E. Skutelsky, *J. Ultrastruct. Res.* **38,** 500 (1972).

The reaction mixture is maintained at pH 6.5 by the addition of 0.2 *N* HCl, as needed, during the following 6 hr. The reaction is allowed to continue at room temperature overnight, in a covered vessel. The product is dialyzed against PBS and filtered to sterilize.

To prepare anionic derivatives, serum proteins are succinylated.[37] One gram of protein is added to distilled water which is adjusted to pH 7. Solid succinic anhydride (Sigma) is introduced in small increments over a period of 15 min to 1 hr. The amount of succinic anhydride to be added depends on the desired extent of succinylation. The reaction mixture is maintained at pH 7 by the addition of NaOH or Na_2CO_3 as needed.

Passive Serum Sickness

The passive administration of immune complexes, formed *in vitro* under well-defined conditions, has also been used to identify determinants of immune complex pathogenicity. Although such preformed complexes usually accumulate in the mesangial areas of the kidney and produce little or no immunopathology, some factors have been recognized to influence the deposition of immune complexes in the peripheral capillary wall. Precipitating antigen–antibody systems,[38] cationic antigens,[39] and low-affinity antibodies[40,41] have been implicated in subepithelial immune deposit formation in passive serum sickness. Although, in some studies, mild histopathology has been described to be associated with the accumulation of preformed immune complexes in glomeruli, evidence of functional impairment, such as increased glomerular permeability, has not been reported.

[37] I. M. Klotz, this series, Vol. 11, p. 576.

[38] L. Agodoa, V. J. Gauthier, and M. Mannik, *J. Exp. Med.* **158,** 1259 (1984).

[39] G. R. Gallo, T. Caulin-Glaser, and M. Lamm, *J. Clin. Invest.* **67,** 1305 (1981).

[40] F. Germuth, F. Rodriguez, C. Lorelli, E. Trump, L. Milano, and O. Wise, *Lab. Invest,* **41,** 366 (1979).

[41] A. M. Lew, N. A. Staines, and M. W. Steward, *Clin. Exp. Immunol.* **57,** 413 (1984).

[42] Mast Cells and Mast Cell Products

By FRANK M. GRAZIANO

Introduction

In the century since the mast cell was first described by Paul Ehrlich, a great deal has been learned about this cell and how it can profoundly affect

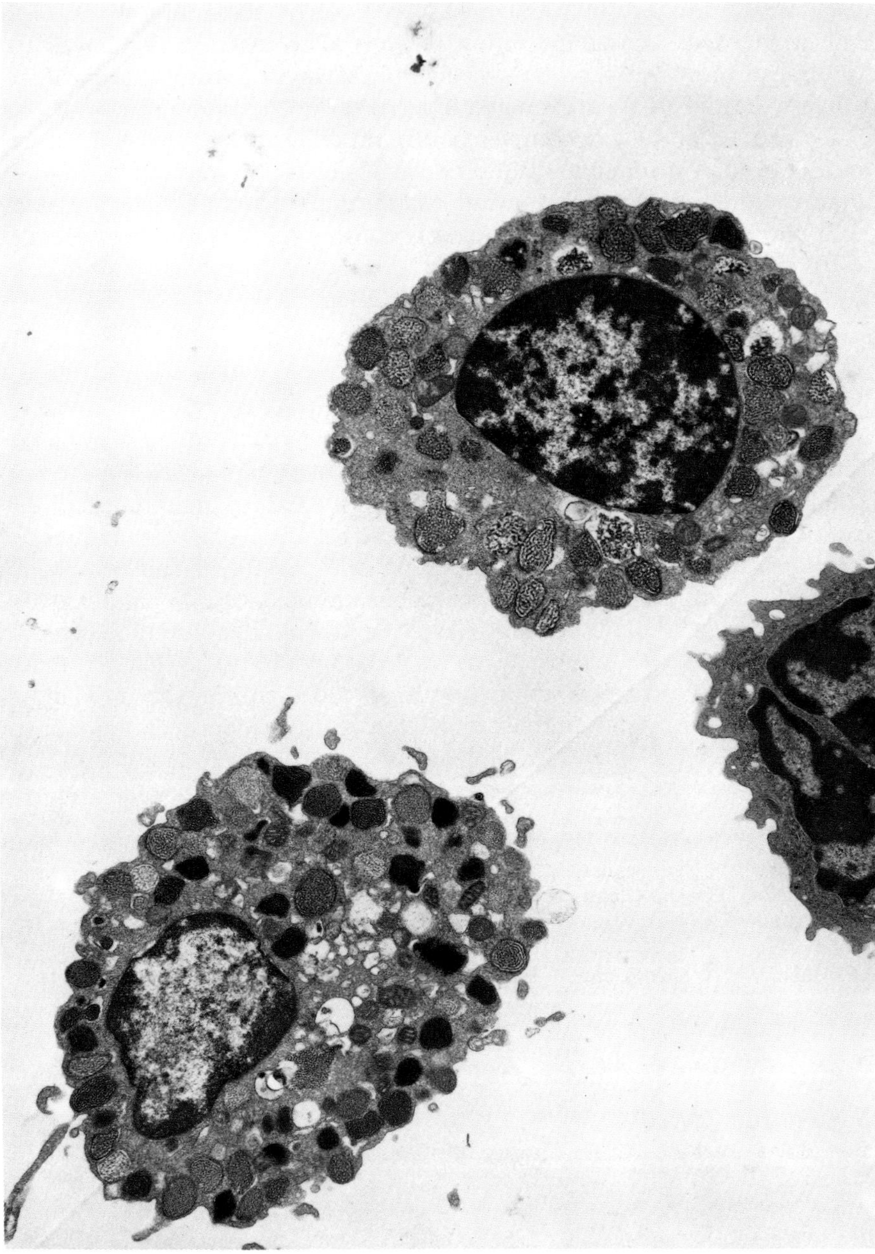

FIG. 1. Guinea pig pulmonary mast cells purified with a combination of enzyme digestion of lung tissue, centrifugal elutriation, and Percoll density gradient centrifugation. Granules have typical dense, diffuse, or whorl patterns and are surrounded by a perigranular membrane. ×4568.

the microenvironment in which it exists. In mammalian tissue, mast cells are generally found as ovoid or irregularly elongated cells. When isolated in suspension they measure 10–15 μm in diameter and contain the usual subcellular elements. The characteristic feature of the mast cell, however, is the presence of many dense membrane-bound secretory granules which measure 0.2–0.5 μm in diameter. In rodents, these granules are amorphous. In humans and guinea pigs they possess a definite subgranular architecture dominated by repeating subunits which give the granules a scroll or whorllike appearance (Fig. 1). Mast cells are widely distributed in the loose connective tissue surrounding blood vessels, nerves, and glandular ducts, and they appear under epithelial, serosal, and synovial membranes.

While it is not totally clear from which precursor the mast cell is derived, it is clear that this cell participates in a variety of pathologic processes by the release of preformed and newly generated chemical products (mediators) from the granules contained in its cytoplasm. Much of the data concerning the role of the mast cell in disease has in the past been based on study of this cell in chopped tissues. Since other cell types are present in these preparations, information concerning the mast cell has largely been indirect and circumstantial. With the advent of techniques to study tissue mast cells in monodispersed and purified preparations, our understanding of their role in health and disease should become more clear. Currently, mast cells can be obtained in monodispersed suspension from a number of tissues from several animal species. It is the major purpose of this chapter to review the techniques utilized to isolate and purify mast cells from tissues and animals in which the inflammatory process is critically being studied. Methods used to measure some of the chemical mediators contained in mast cell granules and shown to have importance in inflammation are also described.

Isolation of Rat Peritoneal Mast Cells

Prior to the development of techniques for isolation of mast cells from various tissues, the rat peritoneal mast cell (RPMC),[1] because it can be

[1] Abbreviations: RPMC, rat peritoneal mast cell; BSA, bovine serum albumin; MCM, mast cell medium; EDTA, ethylenediaminetetraacetic acid; MMC, mucosal mast cell; CTMC, connective tissue mast cell; HBSS, Hanks' balanced salt solution; HEPES, *N*-2-hydroxyethylpiperazine-*N*′-2-ethanesulfonic acid; PBS, phosphate-buffered saline; CMF-HBSS, calcium- and magnesium-free Hanks' balanced salt solution; T, Tyrode's buffer; CMF-T, calcium- and magnesium-free Tyrode's buffer; TGMD, Tyrode's buffer plus magnesium, gelatin, and DNase; TG, Tyrode's buffer plus gelatin; OPT, *o*-phthalaldehyde; HNMT, histamine *N*-methyltransferase; SAMe, *S*-adenosylmethionine; TLC, thin-layer chromatog-

isolated in suspension by simply irrigating the peritoneal cavity, served as a primary source of information concerning mast cell function. It has become clear, however, that distinct differences in morphology and function exist between RPMCs and human mast cells, and findings with RPMC cannot always be directly related to human (or other species) mast cell function. Despite this, study of the RPMC continues to add a great deal to our knowledge of mast cells, and isolation procedures are described below.

Procedure

Male or female rats (both Wistar and Sprague–Dawley rats have been used) weighing 200–400 g may be used in the procedure. Mast cells are harvested and processed in a balanced salt medium containing 150 m*M* NaCl, 3.7 m*M* KCl, 3.0 m*M* Na_2HPO_4, 3.5 m*M* KH_2PO_4, 0.9 m*M* $CaCl_2$, 5.6 m*M* dextrose, 0.1% (w/v) bovine serum albumin (BSA), 0.1% (w/v) gelatin, and heparin 10 units/ml, pH 6.8, adjusted with 5 *N* NaOH (mast cell medium, MCM). Animals are sacrificed by ether asphyxiation, put on dry ice, a 2 inch square of abdominal skin removed, and the peritoneum washed with water to remove remaining debris. A small incision is made in the peritoneum, and 20 ml of MCM is instilled into the peritoneal cavity via plastic tubing fitted to a syringe. The abdomen is massaged for 90 sec, then opened, and the lavage fluid drained (using a funnel) into a 50-ml plastic tube. The abdominal cavity is washed once with 10 ml of MCM, and this is added to the original collection (two-thirds of the total volume used for lavage is a usual return). Lavage fluid is filtered through gauze to remove any large debris and centrifuged at 150 *g* for 15 min at room temperature. Pelleted cells from each animal are pooled and washed 3 times (150 *g* for 5 min at room temperature) with heparin-free MCM. Approximately 20×10^6 cells with $0.5-2 \times 10^6$ of these being mast cells (average purity 5%, range 2–10%) can be obtained per animal using this procedure. It has been noted that a greater number of mast cells can be recovered from the lavage fluid of older (350–400 g) animals. Also, isolation of RPMCs is carried out at room temperature and the cells used immediately. Cold temperatures have been noted to cause histamine desensitization in these cells.

Substantial further purification of RPMCs has been obtained using

raphy; HETEs, monohydroxyeicosatetranoic acids; LT, leukotriene; SRS-A, slow reacting substance of anaphylaxis; RIA, radioimmunoassay; ECF-A, eosinophil chemotactic factor of anaphylaxis.

various materials including BSA,[1a,2,3] Ficoll,[4,5] Percoll,[6] and metrizamide.[7] Other material such as sucrose–gelatin has been used, but cells lose more than 70% of their total histamine content and their ability to respond to antigen and secretagogues.[8] Thirty eight percent (w/v) BSA, or 35% (w/v) Ficoll solutions are frequently used to purify mast cells from peritoneal lavage fluid. In the general procedure, BSA or Ficoll solutions are prepared in MCM without heparin, and 4 ml is carefully layered into 10 ml plastic tubes. Pooled peritoneal lavage cells are resuspended in MCM at 20–40 $\times 10^6$ cells per ml, and 2 ml of this suspension (per tube) is layered carefully onto the BSA or Ficoll solution. The tubes are centrifuged at 300 *g* for 30 min at room temperature. After centrifugation a distinct cell band is observed at the medium–gradient interface and a pellet is noted at the bottom of the tube. Mast cells at highest purity are found in the pellet. Medium, cell interface layer, and gradient material can be discarded to within 1 ml of the pellet. These cells are washed twice with 9 ml of MCM (250 *g* for 10 min, room temperature). If several tubes are run, the pellets are pooled and further washed (2 times) with MCM. Mast cell purities of 90–98% and yields of 50–75% of the starting mast cell concentration in peritoneal lavage fluid have been described with use of both BSA[1a,2,3] and Ficoll.[4,5] Similar purities and yields have been described for Percoll.[6]

Mast cell counts are performed with the specific mast cell dyes Toluidine blue[9] or Alcian blue.[10] Briefly, Toluidine blue staining of cell preparations is performed by adding 0.08 ml of solution I [0.1% ethylenediaminetetracetic acid (EDTA) in saline] to a plastic tube. This is followed sequentially by 0.02 ml of whole normal serum, 0.02 ml of sample cell suspension, and 0.1 ml of solution II [0.5% cetylpyridinium chloride (Sigma Chemical Co., St. Louis, MO), 25 ml; distilled water, 25 ml; 0.8% Toluidine blue in 5% aluminum sulfate, 20 ml]. Cells are mixed and allowed to stand for 2–3 min. Only mast cells will stain a purple color with this dye. In the Alcian blue method, a small sample of cells to be stained is centrifuged and the supernatant fluid removed. To the cells are added

[1a] S. Coutis, R. Nehring, Jr., and N. U. Jariwala, *J. Immunol.* **124,** 2309 (1980).
[2] T. Sullivan, K. Parker, W. Stenson, and C. W. Parker, *J. Immunol.* **114,** 1473 (1975).
[3] R. Lemanske, K. Joiner, and M. Kaliner, *J. Immunol.* **130,** 1881 (1981).
[4] D. Conrad, H. Bazin, A. Sehon, and A. Froese, *J. Immunol.* **114,** 1688 (1975).
[5] P. Cooper and D. Stanworth, *Prep. Biochem.* **4,** 105 (1974).
[6] L. Enerback and I. Svensson, *J. Immunol. Methods* **39,** 135 (1980).
[7] R. Yurt, R. Leid, J. Spragg, and K. F. Austen, *J. Immunol.* **118,** 1201 (1977).
[8] A. Johnson and N. Moran, *Proc. Soc. Exp. Biol. Med.* **123,** 886 (1966).
[9] R. Bray and P. Van Arsdel, *Proc. Soc. Exp. Biol. Med.* **106,** 255 (1961).
[10] J. Gilbert and L. Ornstein, *Blood* **46,** 279 (1975).

0.4 ml of solution I (0.1% EDTA in saline) and 0.45 ml solution III (0.076% cetylpyridinium chloride, 0.7% $LaCl_3 \cdot 6H_2O$, 0.9% NaCl, 0.21% Tween 20, 0.143% Alcian blue—filter through 0.45-μm filter before each use). The mixture is stirred gently for 1 min, then 0.05 ml of 1 *N* HCl is added and the solution stirred. Mast cells stain blue with this dye.

Purification of RPMCs through any density media may result in the loss of 50–80% of the IgE receptor binding activity.[1a] In one report, 50% of the IgE antibody receptors on RPMCs were observed to be lost with BSA gradient purification and 76% lost with Ficoll.[1a] Active receptors remaining on the cells, however, have the same affinity constant for IgE antibody as do those on cells not purified (mast cells in peritoneal lavage fluid) with density gradients. Mouse peritoneal mast cells have also been obtained using a combination of peritoneal lavage and Percoll density gradient centrifugation.[11]

Isolation of Tissue Mast Cells

Mast cells have been isolated from a variety of tissues (skin, lung, heart, stomach, intestine, mesentary) and from a number of species, including rats,[12–14] guinea pigs,[15–18] dogs,[19] sheep,[20] and humans.[21–25] Unlike the

[11] H. Dullens, C. Van Basten, R. DeWeger, and W. Den Otter, *J. Immunol. Methods* **40,** 367 (1981).

[12] A. D. Befus, F. Pearce, J. Gauldie, P. Horsewood, and J. Bienenstock, *J. Immunol.* **128,** 2475 (1982).

[13] T. Lee, F. Shanahan, H. Miller, J. Bienenstock, and A. D. Befus, *Immunology* **55,** 721 (1985).

[14] F. Shanahan, J. Denburg, J. Fox, J. Bienenstock, and A. D. Befus, *J. Immunol.* **135,** 1331 (1985).

[15] B. Undem, F. Green, T. Warner, C. Buckner, and F. M. Graziano, *J. Immunol. Methods* **81,** 187 (1985).

[16] F. Pearce, H. Behrendt, U. Blum, G. Poblete-Freundt, P. Pult, C. Stang-Voss, and W. Schmutzler, *Agents Actions* **2,** 45 (1977).

[17] H. Halim, B. Fischer, and W. Schmutzler, *Agents Actions* **12,** 189 (1982).

[18] K. Barrett, M. Ennis, and F. Pearce, *Agents Actions* **13,** 3122 (1983).

[19] A. Soll, K. Lewin, and M. Beaven, *Gastroenterology* **77,** 1283 (1979).

[20] J. Huntley, G. Wallace, and H. Miller, *Res. Vet. Sci.* **33,** 58 (1982).

[21] C. Fox, A. Dvorak, S. Peters, A. Kagey-Sobotka, and L. Lichtenstein, *J. Immunol.* **135,** 483 (1985).

[22] E. Schulman, D. MacGlashan, S. Peters, R. Schleimer, H. Newball, and L. Lichtenstein, *J. Immunol.* **129,** 2662 (1982).

[23] T. Ishizaka, D. Conrad, E. Schulman, A. Sterk, and K. Ishizaka, *J. Immunol.* **130,** 2357 (1983).

[24] D. MacGlashan, R. Schleimer, S. Peters, E. Schulman, G. K. Adams III, H. Newball, and L. Lichtenstein, *J. Clin. Invest.* **70,** 747 (1982).

[25] S. Peters, E. Schulman, R. Schleimer, D. MacGlashan, H. Newball, and L. Lichtenstein, *Am. Rev. Respir. Dis.* **126,** 1034 (1982).

rodent peritoneal mast cell, which is readily accessible in suspension by lavage of the peritoneal cavity, the tissue mast cell occurs in low number and is much less accessible. One exception is the dog mastocytoma where the tumor mass is rich in mast cells. Such cells have been isolated and studied,[26-28] but one must be cautious in extrapolating data from this tumor cell to normal cells. The cornerstone in isolating mast cells from normal tissue is digestion of the tissue with proteolytic enzymes. Once in a monodispersed suspension cells may be enriched in purity by more standard techniques. There is no one general enzyme digestion procedure that applies to the isolation of all tissue mast cells. Each tissue appears to require specific enzymes or combination of enzymes for digestion. It is not practical, therefore, to describe in detail each procedure. In the following discussion, techniques for mast cell isolation will be specifically described for rats, guinea pigs, and humans. Inflammation and the involvement of the mast cell in this process is being studied particularly in these species.

Intestinal Mast Cells

Gut mast cells have been isolated from a number of species including rats, guinea pigs, dogs, sheep, and humans. In the rat much interest has been generated by studies suggesting a heterogeneity of mast cell populations in the intestine. In this regard, a mucosal mast cell (MMC) isolated from the lamina propria of the intestine has been observed to be functionally and morphologically distinct from connective tissue mast cells (CTMCs). Procedures for the isolation of the MMC from both rat and human intestine are described below.

Rat Mucosal Mast Cell Isolation Procedure. To facilitate MMC isolation, outbred rats (no differences among strains have been noted for MMC isolation) are infected with 3000 third-stage larvae of the nematode *Nippostrongylus brasiliensis*[12] (Carolina Biological Supply Co., Burlington, NC). On or about day 35 after infection, the small intestine is removed from the animal, flushed of fecal material (saline, 200–400 ml), and trimmed of mesentery, adherent connective tissue, fat, and Peyer's patches (normally 20 present). The intestine is opened longitudinally, cut into 3- to 5-cm pieces, and washed well (200 ml) in Hanks' balanced salt solution (HBSS) containing 25 m*M* HEPES (*N*-2-hydroxyethylpiperazine-*N*′-2-ethanesulfonic acid), pH 7.4. The tissue then is cut further into 1- to 2-mm pieces and stirred in a beaker at 37° in 3 (10 min) changes of calcium- and magnesium-free HBSS containing 1.3×10^{-4} *M* EDTA to remove epithe-

[26] M. Phillips, W. Gold, and E. J. Goetzl, *J. Immunol.* **131,** 906 (1983).
[27] E. Goetzl, M. Phillips, and W. Gold, *J. Exp. Med.* **158,** 731 (1983).
[28] M. Phillips, P. Barnes, and W. Gold, *Am. Rev. Respir. Dis.* **132,** 1019 (1985).

lial cells. Between each 10-min incubation, HBSS washes (100 ml) of tissue are quickly performed to remove mucus and debris.

Enzyme digestion of the tissue follows and consists of incubation of the tissue in 100 ml of HBSS (plus 20% heat-inactivated fetal bovine serum) containing 25 U/ml of collagenase (EC 3.4.24.3, Grand Island Biological Co.). After a 60-min incubation at 37°, the supernatant is decanted, and cells are obtained by filtering the fluid through gauze and pelleting by centrifugation at 200 *g* for 8 min at room temperature. A second enzyme digestion (37° for 60 min, collagenase 25 U/ml) is performed on the remaining tissue and the freed cells collected as described above. The tissue remaining after this enzyme digestion is mechanically dispersed by repeated syringing (10-ml syringe, approximately 20 times) using 20 ml of RPMI 1640 containing 5% fetal bovine serum and 2 m*M* L-glutamine.

Cells from the second enzyme digestion and the mechanical disruption step are pooled (cells from the first digestion are discarded because of a large number of contaminating eosinophils), washed in RPMI, cleared of coarse debris by rapid passage (10–25 ml/min) through a nylon wool column (300 mg washed nylon wool in a 10 ml syringe) (see also this series, Vol. 108 [28]), and centrifuged at 100 *g* for 10 min at room temperature. Pooled cells have been shown to contain 30 ± 3% mast cells in a total of approximately 6×10^7 cells.[12,13] Major contaminating cells have been shown to be lymphocytes (30%) eosinophils (12%), and epithelial cells (7%). Further enrichment of MMCs obtained from enzyme digestion has been achieved with the sequential use of Percoll density gradient centrifugation and unit gravity velocity sedimentation in a continuous BSA gradient[13] (see also this series, Vol. 108 [8]).

MMCs are first enriched from enzyme digests of intestinal tissue using a discontinuous Percoll gradient. Stock Percoll solution is made by adding 9 parts Percoll (Pharmacia) to 1 part 10× RPMI (adjusted to pH 7.4 and 300 mos*M*/kg water). Percoll solutions of 30 and 80% are prepared by adding the appropriate amount of RPMI to the stock solution. Gradients are formed by carefully layering 15 ml of 80% Percoll in the bottom of a 50-ml clear polystyrene tube followed by layering of 20 ml of the 30% Percoll onto the 80% solution. A cell suspension of MMCs (up to 1×10^8 per gradient) in 10 ml of RPMI is layered onto this discontinuous gradient, and the tube is centrifuged at 600 *g* for 20 min (10°). Critical in this step for proper separation of cells is a slow acceleration of the centrifuge. MMCs of 65 ± 5.2% (20% recovery of MMCs loaded) have been shown to be present in the pellet at the bottom of the gradient tube.[13] The major contaminating cell is the small lymphocyte.

Pelleted cells are washed and resuspended in 10 ml of phosphate-buffered saline (PBS) containing 0.2% BSA and bottom loaded into a unit gravity velocity sedimentation apparatus (Sta-Put; Johns Scientific Co., Toronto, ON). A continuous BSA gradient is built from below using two connecting flasks containing 1% and 2% BSA. Gradient formation and sedimentation (sedimentation at unit gravity for 3 hr) is performed at 4°. Sixteen 25-ml fractions are collected from below and centrifuged at 200 *g* for 5 min (4°). The cells resuspended, and the fractions are pooled four at a time (giving four final fractions) in a balanced salt buffer. Mast cell counting is facilitated by Toluidine blue staining, and cell viability is assessed with Trypan blue dye exclusion (see also this series, Vol. 108 [6]). Fraction 1 obtained from this procedure contains greater than 95% MMCs (the small lymphocyte is the major contaminating cell) with 84% of the MMCs loaded recovered. These cells represent approximately 17% of the total mast cells obtained from enzyme digestion.

MMCs collected with this procedure have been demonstrated to be functionally intact in experiments examining mediator release,[12,13] and the enzyme digestion procedure has not been shown to affect the morphology or function of these cells.[14] Use of centrifugal elutriation (used successfully for mast cell isolation in other tissue) has not been successful with MMCs, since mast cells appear to be morphologically damaged (granule extrusion) using this procedure.[13]

Human Intestinal Mucosal Mast Cell Isolation Procedure. Most work with human gastrointestinal mast cells has involved *in situ* histochemical studies describing the distribution and characteristics of the cells in normal and diseased tissue. Recently, single-cell suspensions of mast cells from human intestinal mucosa have been obtained.[21] This procedure differs from that used for rat MMC isolation.

Human colonic tissue is obtained from grossly normal surgical specimens and transported in RPMI medium buffered with 25 m*M* HEPES containing 100 μg/ml penicillin and 100 μg/ml gentamycin. The mucosa is dissected from the underlying muscular layers, cut into pieces (1 × 1 cm) and washed for 5 min in calcium- and magnesium-free (CMF-HBSS) containing 25 m*M* HEPES and 1 m*M* dithiothreitol to remove mucus. A second wash in CMF-HBSS/HEPES without dithiothreitol follows, and the tissue is then incubated with stirring at 37° for 60 min in CMF-HBSS/HEPES containing 0.75 m*M* EDTA to separate epithelium from the rest of the mucosa. Epithelial cells, now in the supernatant fluid, are removed by filtration through Nytex mesh (150-μm pore size, available from Tetco, Inc., Elmsford, NY) and discarded. The remaining tissue is cut into 2 × 2 mm pieces, enzymatically digested and dispersed in six steps.

(1) Tissue is stirred for 60 min at 37° in 100 ml of HBSS/HEPES containing collagenase (30 U/ml) and 20% newborn calf serum. Freed cells are collected by filtration through Nytex mesh (Fraction 1). (2) A second

identical digestion procedure is performed, and cells are collected (Fraction 2). (3) Tissue remaining after the second enzyme digestion is disrupted by forceful syringing through a 30-ml syringe and freed cells collected (Fraction 3). (4) The remaining tissue is digested further (40 min at room temperature) with a combination of Pronase (1.5 mg/g wet weight tissue) and chymopapain (0.5 mg/g wet weight tissue), and freed cells are collected (Fraction 4). (5) A second, similar enzyme digestion is performed on remaining tissue and the cells again collected (Fraction 5). (6) Finally, the tissue remaining is disrupted by syringing as described above (Fraction 6). The greatest number of total cells and mast cells have been shown to be recovered from fractions 4, 5, and 6 with $3.1 \pm 2.1\%$, $2.1 \pm 1.6\%$, and $1.6 \pm 0.5\%$ mast cell purity obtained, respectively.[21] Using this procedure 4.5×10^5 mast cells per gram of starting tissue have been obtained.[21] These intestinal mast cells are indistinguishable from isolated human pulmonary mast cells in mediator release characteristics and morphologic appearance when viewed in the electron microscope.[21]

Human Pulmonary Mast Cells

Recently, purification of human lung mast cells to near homogeneity has been achieved.[22] This method is based on previous techniques in which human lung tissue was dispersed into a single cell suspension by the use of proeolytic enzymes. The lung mast cell has now become the most comprehensively studied human mast cell, and the procedure for isolation and purification involves three steps: enzyme digestion of tissue, centrifugal elutriation, and cell affinity column chromatography or Percoll density gradient centrifugation.

Procedure. Macroscopically normal human lung tissue has been obtained at thoracotomy for carcinoma of the bronchus or other neoplasms where normal appearing tissue could be dissected free of tumor, pleura, bronchi, and visible blood vessels.[22] Tyrode's buffer (T) in this work contains (in g/liter) 8 g NaCl, 0.2 g KCl, 0.05 g NaH_2PO_4, 0.26 g $CaCl_2 \cdot 2H_2O$, 0.25 g $MgCl_2 \cdot 6H_2O$, and 1 g glucose. CMF-T is Tyrode's buffer without calcium and magnesium, and TGMD is Tyrode's buffer without calcium but with magnesium, gelatin (1 g/liter) and DNase (Type 1, 15 mg/liter).

Surgical tissue is kept in CMF-T, cut into 5- to 20-mg fragments, washed extensively, and incubated overnight in the same buffer. The following day the tissue is again washed extensively in CMF-T, and enzymatic dispersion of cells is accomplished in four digestion steps. (1) One gram of lung per milliliter of CMF-T is incubated in a mixture of Pronase (2 mg/ml) and chymopapain (0.5 mg/ml) for 30 min at room temperature.

Freed cells are separated from residual lung fragments by filtration through Nytex mesh of 100-μm pore size (Fraction 1). (2) The Pronase–chymopapain digestion is repeated with the residual lung fragments from step 1 (Fraction 2). (3) Residual lung fragments from step 2 are now subjected to a 30-min incubation at room temperature with collagenase (Type 1, 1 mg/ml) and elastase (Type 2, 10 units/ml) in TGMD and freed cells harvested (Fraction 3). (4) The collagenase–elastase digestion is repeated with residual lung fragments from step 3 (Fraction 4). All harvested cells are washed (two times) in TGMD at 200 *g* for 10 min, and mast cell counts are performed with Alcian blue staining. Enzymatic digestion of lung tissue using this method yields $0.4-1.4 \times 10^6$ mast cells per gram of wet tissue, and Fractions 2–4 yield 28–33% of the total recovered mast cells at an average purity of 5%.[22]

Monodispersed lung cell Fractions 1–4 ($0.5-200 \times 10^9$ cells) are combined in 50–200 ml of TGMD and elutriated using a JE-6 elutriator rotor (Beckman Instruments) (see also this series, Vol. 108 [20]). Rotor speed is controlled with a 10-turn potentiometer which allows for accurate rpm selection (within 10 rpm). Before each run, the tubing and chamber are cleaned by sequential washes with 70% ethanol (200 ml), distilled water (500 ml), and TGMD (200 ml). Lung cells are loaded into the elutriation chamber (standard chamber) via a 50-ml syringe fitted with a 3-way valve.[29] Cells are loaded at a rotor speed of 1825 rpm and a flow rate of 11–12 ml/min. The flow of buffer is driven by a peristaltic pump and monitored with a flow meter. Rotor speed is maintained at 1825 rpm, and the cells are fractionated by incrementally increasing the flow rate to 14, 18, 20, 26, and 30 ml/min. With each change in flow rate, three fractions are collected (each 50 ml) and numbered consecutively. Fractions 13 to 15 generally have the greatest yield and purity of mast cells. With higher flow rates greater purity can be obtained, but with low cell yield. Pooled fractions containing the highest yield and purity of mast cells demonstrate $10.9 \pm 1.3 \times 10^6$ mast cells with an average purity of $29 \pm 3.5\%$.[22]

Further purification of elutriated mast cells has been obtained using cell affinity column chromatography (using benzylpenicilloyl-specific IgE antibody)[22] or Percoll density gradients.[23] The more convenient method of the two procedures is Percoll density gradient centrifugation. In this procedure, cells obtained from the sequential enzyme digestion of tissue and elutriation (fractions with greater than 20% purity from elutriation) are washed with Hanks' balanced salt solution and fractionated with a discontinuous Percoll gradient. Cells applied to the gradient ($1-4 \times 10^7$) are

[29] T. Contreras, J. Jemionek, H. Stevenson, V. Hartwig, and A. S. Fauci, *Immunology* **54,** 215 (1980).

suspended in 0.8 ml of Percoll stock solution (9 parts Percoll plus 1 ml of 10× HBBS) and placed at the bottom of a plastic tube (12 × 75 mm). On top of this cell suspension is carefully layered 0.8-ml aliquots of each of the following Percoll solutions: 80, 70, 60, 50, and 40% (prepared by adding the appropriate amount of HBSS to stock Percoll). The tubes are centrifuged at 100 *g* for 12 min, and the cells at each interface are collected and washed with CMF-T to remove Percoll. Mast cells of 70–95% purity have been obtained at the 60/70% and 50/60% interfaces.[23] No significant difference has been noted in mediator release characteristics between purified lung mast cells and chopped lung preparations.[25] Further, when the mediator release characteristics of purified lung mast cells obtained by enzyme digestion of pulmonary tissue or monodispersed lung cells containing mast cells obtained by mechanical disruption of pulmonary tissue (alone) were compared, again, no significant differences were noted.[30]

Guinea Pig Pulmonary Mast Cells

The guinea pig is commonly used as an animal model in the study of immediate hypersensitivity reactions in the lung, and guinea pig pulmonary mast cells have been isolated in monodispersed cell suspension.[15–18] We recently have been able to obtain guinea pig lung mast cells in high purity,[15] and these methods are described below.

Procedure. Outbred female Hartley guinea pigs weighing 250–350 g are used for the isolation procedure. Tyrode's buffer plus gelatin (TG), used throughout the isolation procedure (carried out at room temperature), 137 m*M* NaCl, 2.6 m*M* KCl, 0.35 m*M* NaH_2PO_4, 5.5 m*M* glucose, 11.9 m*M* $NaHCO_3$, and 1 g/liter gelatin with the pH adjusted to 7.4 with HCl. Percoll stock solution is prepared as described above (osmolality, 285 mos*M*/kg water). Desired densities of Percoll are prepared by adding HEPES-buffered saline (containing 10 m*M* HEPES, 137 m*M* NaCl, 5 m*M* KCl, 0.5 mg/ml gelatin, and 1 mg/ml dextrose, pH adjusted to 7.4 with NaOH) to the stock solution.

Animals are sacrificed by a sharp blow to the head and exsanguinated. The lungs are removed and perfused via the pulmonary artery and trachea with approximately 60 ml of TG. The lungs (approximately 3 g of tissue per animal) are trimmed of large airways and blood vessels, minced finely with scissors, chopped to 1-mm^3 fragment size with a McIlwain tissue chopper (Brinkman), washed by incubating in 50 ml of TG with vigorous shaking for 5 min, and filtered through a Nytex mesh of 100-μm pore size. Washed and filtered tissue is digested with enzymes in a five-step proce-

[30] M. Church, G. J.-K. Pao, and S. Holgate, *J. Immunol.* **129,** 2116 (1982).

dure. (1) Lung fragments are incubated with shaking for 30 min in TG (1 ml/g tissue) containing 125 U of collagenase (Type 1) and 10 U of elastase (Type 1) per gram of tissue. After this incubation period, 50 ml of TG is added to the preparation, and freed cells are separated from remaining tissue fragments by once again filtering through Nytex mesh. Steps 2–4 consist of a similar enzyme treatment of the tissue fragments remaining after each digestion. (5) After the four sequential enzyme treatments, loosely bound cells are jarred free by briskly shaking the remaining fragments in 50 ml of TG. All harvested cells are centrifuged at 300 *g* for 10 min and washed with TG.

Monodispersed lung cells containing the highest percentage of mast cells (usually from Steps 3, 4, and 5) are pooled and fractionated with a JE-6 elutriator rotor driven by a J-21B Centrifuge using system preparation and centrifuge controls as described above for human cells. Guinea pig lung cell suspensions ($20-200 \times 10^6$ nucleated cells), filtered twice through Nytex mesh to remove large aggregates, are loaded into the elutriation chamber (Sanderson chamber) via a 20-ml syringe fitted with a 3-way valve.[29] Cells are loaded at a flow rate of 28 ml/min and a rotor speed of 3950 rpm. The flow rate is then increased to and maintained at 30 ml/min, and the cells are fractionated by incrementally decreasing the rotor speed to 3800, 3650, 3380, 2900, and 0 rpm. The flow of buffer is driven by a peristaltic pump and monitored with a flow meter. After each adjustment in rotor speed, two to four 45-ml fractions are collected. An additional 90-ml collection is obtained at the end of each run with the centrifuge turned off. All fractions are centrifuged (300 *g* for 10 min) and analyzed for mast cell content using Alcian blue.

Continuous Percoll density gradients are prepared by centrifuging 10 ml of Percoll (1.080 g/ml) at 20,000 *g* for 15 min. Monodispersed lung cell fractions obtained at rotor speeds of 3380 and 2900 rpm in the elutriation process usually contain the highest percentage of mast cells. These fractions are pooled, washed, layered as a 1-ml suspension ($10-35 \times 10^6$ cells) onto the continuous Percoll gradient, and centrifuged at 400 *g* (22°) for 20 min. A second continuous Percoll gradient formed simultaneously is layered with density marker beads (Pharmacia) and centrifuged similarly. After centrifugation, cells are carefully removed from Percoll in three fractions according to the densities formed by the density marker beads. Fraction 1 consists of densities 1.049 g/ml to, but not including, 1.076 g/ml. This fraction may contain significant numbers of mast cells but is low in purity. Fraction 2 (1.076–1.088 g/ml) contains the enriched purity mast cells as does Fraction 3 (1.088–1.120 g/ml), but the latter fraction contains few cells and is largely contaminated by erythrocytes. Fraction 2 has been demonstrated to contain a mean of 65 ± 5%

purity mast cells (range 50–90%) with a 40 ± 7% yield of the mast cells present in the enzyme dispersed cell suspension.[15]

Comment

Success in using any procedure to isolate lung mast cells (or any other tissue mast cells) is predicated on the number of mast cells in the initial monodispersed cell suspension. While collagenase alone has been used by others to obtain guinea pig lung mast cells,[16–18] we have found that a combination of collagenase and elastase is synergistic and most successful in routinely obtaining lung cell suspensions containing 3–6% mast cells at $4-8 \times 10^6$ mast cells per 6 g of tissue.[15] We estimate this procedure releases 75–90% of the mast cells in the lung tissue (based on histamine content per cell and total lung tissue histamine). Combining elutriation and Percoll density gradient centrifugation of monodispersed lung cells proved to be more effective than either technique alone in the procedure. The elutriation step usually leads to a 5- to 7-fold purification of mast cells and Percoll density gradient centrifugation a 3- to 4-fold increase in purity of the elutriated cells. No difference in mast cell mediator release characteristics was demonstrated when mechanically dispersed lung cells (no enzymes) were compared to enzymatically dispersed and purified mast cells.[15]

Mast Cell Products

A variety of agents have been shown to induce mast cell degranulation, as defined histologically or by measurement of granule products released in tissue or fluids obtained from stimulated cells. Examples of these agents include antigen (bound to specific IgE antibody on the mast cell surface), anaphylatoxins (C3a, C5a), calcium ionophore, the drugs morphine and curare, endotoxin, synthetic chemotactic peptides, and complex carbohydrates. Activation of mast cells leading to release of their granular products has been best studied in the IgE antibody system. Here, binding of IgE antibody to specific membrane receptors and bridging of these antibody-bound receptors triggers the process. The subsequent release of mast cell products (mediators) occurs as a noncytolytic secretory process. Mast cell mediators released can be classified based on their biological activity and whether they exist preformed in granules or are newly synthesized on activation of the cell. A partial list of these mediators is shown in Table I.

The mast cell and its secreted mediators have been most comprehensively studied, and regarded to play a central role, in the production of immediate hypersensitivity reactions. Since it would be difficult to detail in

TABLE I
MAST CELL MEDIATORS

Preformed mediator		Mediator newly generated as a consequence of mast cell activation
Rapidly eluted from granule	Firmly associated with granule	
Histamine	Heparin	Prostaglandins
Serotonin	Chymotrypsin/trypsin	Leukotrienes
Eosinophil chemotactic factor	Peroxidase	Thromboxanes
Neutrophil chemotactic factor	Superoxide dismutase	Platelet-activating factor
Superoxide dismutase	Arylsulfatase B	
Arylsulfatase	(*N*-acetylgalactosamine-4-sulfatase)	

this chapter methods for measuring all mast cell products, in the discussion that follows methods for the measurement of histamine, leukotrienes, and chemotactic factors will be described. These mediators (among others) are being particularly studied as important participants in immediate and delayed hypersensitivity reactions. In the general methodology for measuring mediator release by isolated mast cells, it is important to remember that calcium is required for most stimulus-specific mediator release. Also, in the measurement of mediators, spontaneous release of mediators and a potential interference of the mast cell activating stimulus or buffer containing the stimulus must be controlled for in all experimental designs.

Assay of Serotonin

Serotonin is synthesized by sequential hydroxylation and decarboxylation of the amino acid tryptophan and is degraded in plasma or tissue by monoamine oxidase. The rodent, but not human, mast cell stores this amine in its granules. Serotonin is vasoconstrictor but is a relatively ineffective agonist on human pulmonary vasculature or bronchial smooth muscle. The classic methods for detection of serotonin are based on fluorescence measurements.[31,32] More recently methods have been described with greater sensitivity. These newer methods include radioenzyme assay,[33] mass spectrometric assay,[34] and high-performance liquid chromatography

[31] D. F. Bogdanski, A. Pletsher, B. Brodie, and S. Udenfriend, *J. Pharmacol. Exp. Ther.* **117,** 82 (1956).

[32] M. Mumtaz, Naransimhacheri, G. Pandey, and J. David, *Res. Commun. Chem. Pathol. Pharmacol.* **36,** 45 (1982).

[33] M. Hussain and M. Sole, *Anal. Biochem.* **111,** 105 (1981).

[34] S. Markey, R. Colburn, and J. Johannessen, *Biomed. Mass Spectrom.* **8,** 301 (1981).

(HPLC) in combination with electrochemical detection.[35] These methods are effective but have proven either time consuming or to have a low specificity. While refinements for these methods are continually being reported,[36,37] routine measurement of serotonin appears to be best performed using the relatively fast, sensitive, and specific radioimmunoassay. Two double-antibody radioimmunoassay methods have been reported for serotonin.[38,39] One method measures serotonin,[38] and the other measures *N*-acetylserotonin.[39] The latter assay has overcome the problems with instability of serotonin in aqueous solution and is more sensitive and specific. This assay procedure is outlined below.

Procedure for Radioimmunoassay. The primary difference in this radioimmunoassay system is that *N*-acetylserotonin instead of serotonin is used throughout the assay. For the preparation of the antiserum, serotonin chemisuccinimide is coupled to BSA as described elsewhere.[40] Rabbits are immunized with an intradermal injection of 0.5 ml of serotonin–BSA (1 mg) emulsified in 0.5 ml of Freund's complete adjuvant, followed by identical injections at 4-week intervals. *N*-Acetylserotonin is prepared as follows: Serum samples containing serotonin are diluted 1 : 4 with water. One hundred microliters of the diluted serum is treated with 2 ml of ethanol, and the mixture is kept for 30 min at 4°. After centrifugation, the clear supernatant is transferred to another tube, and 50 μl of a 1 : 1 mixture of pyridine : acetic anhydride is added. After mixing, the tube is placed in a water bath at 50°, the organic solvent is evaporated under a gentle stream of nitrogen, and the dry sample is taken up in 10 m*M* phosphate buffer, pH 7.4, containing 1.5 m*M* EDTA, 3 m*M* sodium azide, and 10% glycerol. Experiments performed using tritiated serotonin demonstrated 93% recovery of *N*-acetylserotonin from samples.

For the radioimmunoassay, 100 μl of the *N*-acetylserotonin extract or standards, 100 μl of ^{125}I-labeled compound (^{125}I-labeled Bolton–Hunter reagent, New England Nuclear, 20–100 μCi used for labeling), and 100 μl of diluted antiserum reagent (dilution able to bind 30% of isotope under assay conditions) are added to a polystyrene tube and incubated overnight at 4° followed by 1 hr at 37°. After this incubation, 1 ml of goat anti-rabbit γ-globulin (Miles Laboratories) (precipitating antibody) is added to the tube, and a further incubation for 10 min at room temperature is per-

[35] E. Korpi, *Clin. Chem.* **30,** 487 (1984).

[36] H. Scheinin and M. Scheinin, *Eur. J. Pharmacol.* **113,** 345 (1985).

[37] W. Wolf and D. Kuhn, *J. Neurochem,* **46,** 61 (1986).

[38] F. Engbaek and B. Voldby, *Clin. Chem.* **28,** 624 (1982).

[39] B. Manz, H. Kosfeld, G. Harbauer, H. Grill, and K. Pollow, *J. Clin. Chem. Clin. Biochem.* **23,** 657 (1985).

[40] M. A. Delaage and J. J. Puizilliout, *J. Physiol. (Paris)* **77,** 339 (1981).

formed. The tube is then centrifuged (1500 g, 15 min, 4°), the supernatant decanted, and the pelleted radioactivity determined in a gamma counter. Estimation of specificity for this assay has shown less than 0.01% cross-reactivity with serotonin, melatonin, tryptamine, L-dopa, dopamine, and *N*-acetyldopamine among other compounds.[39] There is a 34% cross-reactivity with *N*-propionylserotonin. The sensitivity of this assay has been demonstrated to be 25 fmol/tube.[39]

The use of N-acetylated serotonin in this radioimmunoassay has several advantages over other assay systems used to measure serotonin[39]: (1) Antigen, isotope, and *N*-acetylserotonin are all acyl homologs of serotonin. (2) The method is relatively easy, fast to perform, and more sensitive than previous assays. (3) As can be seen above, acetylation of the serotonin prior to assay increases greatly the specificity of the method. (4) Endogenous *N*-acetylserotonin, a metabolite of serotonin which could interfere with the assay, is detectable by direct comparison of acetylated and non-acetylated extracts.

Assay of Histamine

Histamine was the first mast cell mediator identified and is one of its most comprehensively studied products. It is formed by the action of histidine decarboxylase on the amino acid histidine. Histamine exerts its inflammatory effects by interacting with H1 and H2 cell membrane-associated receptors. Measurement of this mediator was classically performed by bioassay using guinea pig ileum. Bioassay, however, has largely been supplanted by fluorometric[41] and radioisotopic enzyme assays.[42–44]

The fluorometric assay is most often used for histamine measurement and is based on the coupling of histamine with *o*-phthalaldehyde (OPT) at an alkaline pH to form a fluorescent product. The manual fluorometric method is laborious and time consuming and has largely been replaced by the automated fluorometric assay system (available commercially from Technicon or Alpkem) of Siraganian.[41] Using the automated method, 30 samples in 1 hr (as opposed to approximately 50 samples per 3–4 hr using the manual method) can be analyzed for histamine content. A sample volume of 0.5 ml is required for measurement, and the limit of detection is 0.1 ng/ml. Since some buffers or mast cell activating substances may auto-

[41] R. P. Siraganian, *Anal. Biochem,* **57,** 383 (1974).

[42] M. Tharp. R. Suvunrungsi, and T. Sullivan, *J. Immunol.* **130,** 1896 (1983).

[43] J. Brown, M. Frey, B. Reed, A. Leff, R. Shields, and W. Gold, *J. Allergy Clin. Immunol.* **73,** 473 (1984).

[44] J. Dyer, K. Warren, S. Merlin, D. Metcalfe, and M. Kaliner, *J. Allergy Clin. Immunol.* **70,** 82 (1982).

fluoresce or quench fluorescence of the histamine–*o*-phthaladehyde complex, appropriate controls for these processes must be added to the experimental design. Percent histamine content is calculated by subtracting spontaneous histamine release (plus any autofluorescence of stimulating substance or buffer) from the sample and dividing this by the total histamine content (isolated cells plus 3% perchloric acid) and multiplying the quantity by 100.

Procedure for Radioisotopic Enzyme Assay. Radioisotopic enzyme assays have been described and have been shown to be sensitive.[42–44] This method has been utilized to measure the histamine content of plasma, urine, and some tissues. Methods for this procedure are described below.

The radioisotopic enzyme assay is based on the transfer of radiolabeled methyl groups (using the enzyme histamine *N*-methyltransferase, HNMT) from *S*-adenosylmethionine (SAMe) to unlabeled histamine in the sample to be measured. Radiolabeled SAMe (labeled with ^{3}H or ^{14}C) is available commercially (New England Nuclear). HNMT is prepared from rat kidneys.[45] This enzyme is not usually purified to homogeneity, and is, therefore, used as an aliquot of the final extracted material. In the procedure, 0.1 ml of sample or histamine standards (1–60 ng) is added to 0.05 ml of HNMT, 0.05 ml [^{3}H]SAMe (0.1 μCi), and 0.3 ml of distilled water. The mixture is incubated for 60 min at 37°, and the resultant radiolabeled 1-methylhistamine is extracted into chloroform (2 ml) after the addition of 0.2 ml of 0.1 *N* NaOH to stop the reaction. After aspiration of the aqueous layer, 1 ml of the chloroform layer is placed into a scintillation vial, evaporated to dryness, 3 ml of scintillation fluid added, and the radioactivity is determined in a scintillation counter. A standard curve is constructed for histamine concentration from the histamine standards. The sensitivity of this radioisotopic enzyme assay is approximately 1 ng/ml.

Sensitivity has been improved with the use of a double-isotopic label technique.[43] Sensitivity to 0.2 ng/ml by the addition of thin-layer chromatography (TLC) to the single radioisotopic enzyme method has also been described.[44] In the latter method, 0.01 ml of plasma or histamine standards (0–5 ng histamine/ml) in dialyzed plasma is incubated in microcentrifuge tubes with histamine *N*-methyltransferase (0.02 ml), [^{3}H]SAMe (0.5 ml, 236 nCi) and 0.01 ml of 0.1 *M* phosphate buffer, pH 7.4. The incubation lasts 90 min (37°), and is terminated by addition of 0.01 ml 2.5 *M* perchloric acid. Vortexed samples are centrifuged (12,800 *g*, 1 min at room temperature), supernatants are transferred to a new microcentrifuge tube, 0.01 ml of 10 *N* NaOH and 0.3 ml chloroform are added, and the tubes are capped and shaken for 5 min. After centrifugation of these samples

[45] R. Shaff and M. Beaven, *Anal. Biochem.* **94,** 425 (1979).

(12,800 *g*, 1 min at room temperature), the aqueous layer is removed, and 0.1 ml of 3.3 *N* NaOH is added to the chloroform layer. The tubes are capped, vortexed, shaken for 5 min, and centrifuged as before. The aqueous layer is again discarded, and 0.15 ml of the chloroform layer is transferred to a new tube and evaporated to dryness under a stream of nitrogen at 37°. The residue remaining in this is suspended in 0.5 ml of chloroform for TLC.

Channeled TLC plates (Whatman) are washed with acetone, dried, and activated in an oven at 100° for 10 min. Samples and 0.01 ml (1 mg/ml in methanol) of a 1-methylhistamine marker (Sigma) are applied to the bottom of each TLC channel. The plates are developed with acetone:ammonium hydroxide (95:10) in an ascending fashion for approximately 45 min (or until migration is within 0.5 inch of the top of the plate). After allowing the plates to dry at room temperature, they are sprayed with OPT (Pierce Chemical Co.) and the 1-methylhistamine spots visualized under ultraviolet light. Each spot is scraped from the plates into a scintillation counting vial containing 1 ml of 0.1 *N* HCl and allowed to sit for 30 min at room temperature. Ten milliliters of Ultrafluor (National Diagnostics, Somerville, NJ) is added to each vial, the vials are then vortexed, and counted in a scintillation counter. The histamine concentration of unknown samples is calculated from the standard curve constructed for each run.

Assay of Oxidative Products of Arachidonic Acid (see also this series, Vol. 163 [32] and [33])

Arachidonic acid is a prevalent fatty acid present in mammalian membrane phospholipids. It is liberated from intact phospholipids by the action of phospholipase A_2 or from diacylglycerol by the action of diacylglycerol lipase (lipoprotein lipase). Two major enzymes, lipoxygenase and cyclooxygenase, regulate the fate of arachidonic acid. The action of the cyclooxygenase enzyme on arachidonic acid leads to the production of prostaglandin metabolites. The lipoxygenase enzyme releases hydroperoxyeicosatetranoic acid which is further metabolized to a series of monohydroxyeicosatetranoic acids (HETEs) or to leukotrienes (LTs). Slow reacting substance of anaphylaxis (SRS-A), an important mediator of bronchoconstriction in asthmatic syndromes, has recently been shown to be a family of leukotrienes (LTC_4, LTD_4, LTE_4). The spasmogenic activity of LTC_4 and LTD_4 on animal lung parenchymal strips has been observed to be 200 to 20,000-fold greater than that of histamine. LTD_4 has also been reported to produce human bronchial contraction when administered to stable asthmatic subjects.

The classic method of assay for SRS-A has been bioassay in guinea pig illeum. Drawbacks to this bioassay include a detection limit of approximately 1 pmol, lack of speficity, and potential for significant losses in the extraction required before assay. Speculation for an important contributory role for leukotrienes in inflammation has been a major impetus for the development of alternative methods for quantitation of these substances. Reversed-phase HPLC has been used but lacks sensitivity. In the past several years the development and use of radioimmunoassays (RIA), either alone or in combination with HPLC, have become state of the art for the measurement of leukotrienes and other arachidonic acid metabolites. Because of the similarity in structure of the leukotrienes, RIAs with a wide range of selectivity (cross-reactivity) and sensitivity have been reported.[46-48] An antibody with broad cross-reactivity to LTC_4, LTD_4, and LTE_4 can be potentially advantageous, especially if combined with HPLC, since both selectivity and sensitivity can be achieved. In this regard, a sensitive RIA for leukotrienes with a broad cross-reactivity has been recently described.[46] In this method an antibody to LTD_4 was produced (detection to 0.43 pmol) and in RIA had cross-reactivity with LTC_4 (159%), and LTE_4 (57%) and no other arachidonic acid metabolites. Since little evidence for the predominance of any of the leukotrienes in various diseases is published, it seems advantageous to have an RIA that detects all three leukotrienes of importance.

RIAs for a number of arachiodonic acid metabolites (prostaglandins, HETEs, leukotrienes) are commercially available. An RIA kit for LTB_4 is available from Seragen Inc. (Boston, MA). The antibody in this assay kit is relatively selective for LTB_4, having 1% cross-reactivity with other leukotrienes and less than 0.1% cross-reactivity with prostaglandins. An LTC_4 RIA kit is available from New England Nuclear. The antibody in this kit demonstrates a 55% cross-reactivity with LTD_4 and 9% cross-reactivity with LTE_4; minimal cross-reactivity with other leukotrienes and prostaglandins has been observed. Because of the ever-increasing availability of RIA kits for leukotrienes and other arachidonic acid metabolites, detailed methods for RIA will not be discussed. The reader is referred to the specific references cited above for detailed descriptions of antibody production to leukotrienes if they desire to establish their own RIA.

[46] D. Aharony, P. Dobson, P. Bernstein, E. Kusner, R. Krell, and J. B. Smith, *Biochem. Biophys. Res. Commun.* **117,** 574 (1983).

[47] E. Hayes, D. Lombardo, Y. Girard, A. Maycock, J. Rokach, A. Rosenthal, R. Young, R. Egan, and H. Zweernik, *J. Immunol,* **131,** 429 (1983).

[48] R. Young, M. Kakushima, and J. Robach, *Prostaglandins* **23,** 603 (1982).

Assay of Chemotactic Factors (see also this volume [2]–[6])

Mast cell granules contain a number of chemotactic factors that differ considerably in molecular weight but are relatively selective for cells of the granulocytic series. Eosinophil chemotactic factor of anaphylaxis (ECF-A) is released immunologically from mast cell-rich tissues and from isolated rat and human mast cells. A factor of high molecular weight (600,000) chemotactic for neutrophils has been extracted from human lung tissue and rat mast cells and is present in serum of patients after experimental induction of IgE-dependent bronchospasm and cold urticaria. Other mediators found to have chemoattractant properties include histamine, LTB_4, and platelet-activating factor.

The most commonly used methods for measurement of chemotactic activity are the micropore filter method[49] and the under agarose technique.[50] In the micropore filter method, cells (usually pure cell populations are used) and chemoattractant are placed on either side of the filter in a chemotaxis chamber. Cells will migrate into the channels of the filter if a chemotactic stimulus is present. In the under agarose method, cells placed in a central well are allowed to migrate under agarose toward peripheral wells containing a chemoattractant. In this technique chemotactic activity is measured by the number of cells and/or distance of migration in the agarose. The latter method has gained acceptance because of its simplicity and its amenability to special staining procedures which permits determination of the identity of special cell types in mixed cell preparations. The sensitivity of this method, however, has been reported to be 50–100 times lower than the micropore filter method.

Several types of chambers are available for chemotaxis. The microchemotaxis chamber has gained wide popularity because smaller volumes of sample and less cells are required; there are fewer parts; the filter can be easily mounted on glass slides; and the chamber-filling devices are much less expensive. In preparing this chamber for assay, membrane filters of $5\mu m$ pore size (without polyvinylpyrrolidone wetting agent) are used for measurement of chemotactic activity when cells in the granulocytic series are employed. Generally $5-7 \times 10^5$ cells per well are needed for best results. Commercially available chemotaxis chambers come with detailed instructions for filling of wells and staining of membranes, and these methods will not be discussed here. In the chemotaxis assay, however, one must remember that it is necessary to discriminate between chemotaxis

[49] E. G. Maderazo and C. Woronick, *Clin. Immunol. Immunopathol.* **11,** 196 (1978).
[50] R. D. Nelson, P. G. Quie, and R. L. Simmons, *J. Immunol.* **115,** 1650 (1975).

and chemokinesis. The latter property reflects the tendency of a substance to increase random cell migration. This discrimination is best measured by a checkerboard design of chemotaxis. In this design, the substance to be evaluated is incorporated in multiple doses on either or both sides of the membrane. In this manner it is possible to distinguish pure chemotactic activity, pure chemokinetic activity, and mixed activity in the same preparation.[51]

Many methods for assessing the results of the micropore chemotactic assay have been described. Generally, however, these fall into three basic types: (1) counting the number of cells that have arrived at a preselected distance, such as the distal surface of the filter; (2) measurement of the distance migrated by faster migrating cells (leading-front technique); and (3) measurement of the average distance migrated per cell (the locomotion or leukotactic index). The first method described is very sensitive for detecting differences in nonstimulated and chemotactic responses. It can be used effectively for measurement of chemotactic activity released from mast cells. The latter two methods are much more sensitive to defects in cell migration and are used for investigation of clinical defects in this process.

[51] S. H. Zigmond and J. G. Hirsch, *J. Exp. Med.* **137,** 387 (1973).

[43] Basophils in Allergic and Inflammatory Reactions

By FRANK M. GRAZIANO

Introduction

Basophils are granulocytic cells that differentiate in bone marrow, circulate in blood, and migrate into tissues in response to appropriate stimuli. They are the least common of the granulocytic cells and comprise 0.5–1% of the total leukocyte population. Basophil morphology is dominated by intracytoplasmic granules, and the ultrastructure of these granules appears to differ from species to species. In humans, basophil granules vary in shape (round, oval, angular) and size (113–260 Å), are embedded in a less dense matrix, and are surrounded by a membrane.[1] In guinea pigs, basophil granules are large oval structures that, depending on the plane of section, display a parallel array, hexagonal array, finely granular, or rectan-

[1] A. Dvorak and H. Dvorak, *Arch. Pathol. Lab. Med.* **103,** 551 (1979).

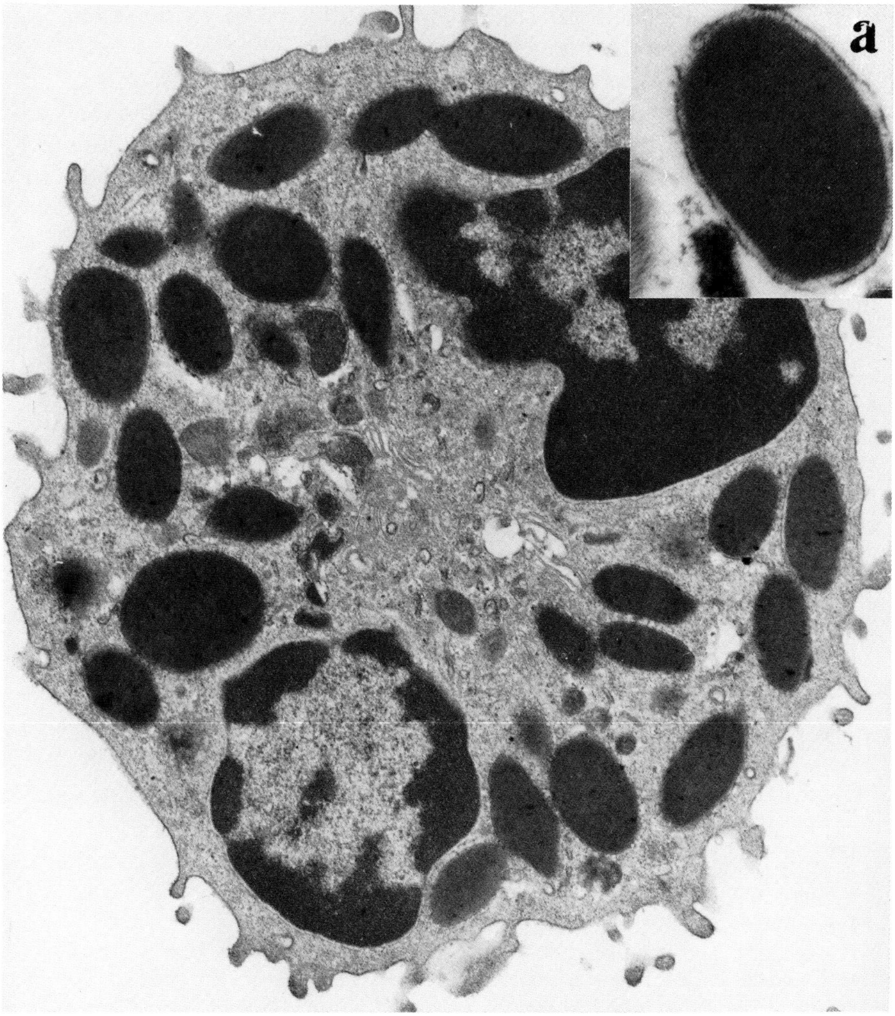

FIG. 1. Electron micrograph of a guinea pig basophil. The cytoplasm demonstrates typical large oval mediator-containing granules. ×14,080. (Insert a) Cytoplasmic granule. The dense granule particle appears in a less dense matrix surrounded by a perigranular membrane. ×27,500.

gular lattice of granular material. A rim of less dense finely granular or fibrillar material surrounds the matrix and separates it from the granule membrane[1] (Fig. 1). A close relationship appears to exist between basophils and the granule-containing mononuclear mast cells, but in both functional and biochemical studies distinct differences between the two cell types have been reported.[2,3]

In humans, basophils contain almost all of the blood histamine.[4] For many years, *in vitro* release of basophil histamine in allergic individuals has served as an assay of immediate hypersensitivity responsiveness.[4] Knowledge of the actual participation of basophils in disease, however, has only come about with the appreciation that significant tissue infiltrates accompany some allergic diseases.[4,5] Attempts in the past to clarify the role of basophils have been greatly limited by the low number of these cells found in peripheral blood.[6,7] Recently, success in achieving high levels of purity and yield of basophils has been accomplished in humans with affinity chromatography,[8-10] Percoll density gradient centrifugation,[11-14] centrifugal elutriation,[15] cell panning procedures,[16,17] and flow microfluorometry.[18] Normal circulating blood basophils have been identified with great difficulty in rodents,[19] appear to be abundant in turtles,[20] and have been isolated in substantial purity from guinea pigs.[21,22] It is the purpose of this chapter to review basophil isolation procedures with special emphasis on human and guinea pig techniques.

[2] D. MacGlashan, R. Schleimer, S. Peters, E. Schulman, G. K. Adams III, A. Sobotka, H. Newball, and L. Lichtenstein, *Fed. Proc., Fed. Am. Soc. Exp. Biol.* **142,** 2504 (1983).

[3] H. Dvorak and A. Dvorak, *Hum. Pathol.* **3,** 454 (1972).

[4] E. B. Mitchell and P. Askenase, *Clin. Rev. Allergy* **1,** 427 (1983).

[5] H. Dvorak, *J. Allergy Clin. Immunol.* **58,** 229 (1976).

[6] R. Day, *Clin. Allergy* **2,** 205 (1972).

[7] T. Ishizaka, R. DeBernardo, H. Tomioka, L. Lichtenstein, and K. Ishizaka, *J. Immunol.* **108,** 1000 (1972).

[8] D. MacGlashan and L. Lichtenstein, *J. Immunol.* **124,** 2519 (1980).

[9] J. Toll, J. Wikberg, and R. Andersson, *Allergy* **36,** 411 (1981).

[10] J. Pruzansky and R. Patterson, *J. Immunol. Methods* **44,** 183 (1981).

[11] P. Raghuprasad, *J. Immunol.* **129,** 2128 (1982).

[12] H. Kaufman, P. Levering, and K. DeVries, *J. Immunol. Methods* **57,** 1 (1983).

[13] E. Leonard, R. Roberts, and A. Skeel, *J. Leuk. Biol.* **35,** 169 (1984).

[14] E. Leonard and A. Skeel, *J. Allergy Clin. Immunol.* **76,** 556 (1985).

[15] M. DeBoer and D. Roos, *J. Immunol.* **136,** 3447 (1986).

[16] F. Landry and S. Findlay, *J. Immunol. Methods* **63,** 329 (1983).

[17] M. Lett-Brown, S. Robinson, H. Juneja, J. Farnam, and J. A. Grant, *Fed. Proc., Fed. Am. Soc. Exp. Biol.* **44,** 7438 (1985).

[18] G. Weil, W. Leiserson, and T. Chused, *J. Immunol. Methods* **58,** 359 (1983).

[19] A. Dvorak, G. Nabel, K. Pyne, H. Cantor, H. Dvorak, and S. Galli, *Blood* **59,** 1279 (1982).

[20] K. Mead, M. Borysenko, and S. Findlay, *J. Immunol.* **130,** 334 (1983).

[21] H. Dvorak, S. Selvaggio, A. Dvorak, R. Colvin, D. Lean, and J. Rypysc, *J. Immunol.* **113,** 1694 (1974).

[22] F. M. Graziano, W. Lipham, N. Swaminathan, J. Brendel, F. M. Graziano, Jr., and T. Warner, *J. Immunol. Methods* **67,** 157 (1984).

Methods for the Isolation of Human Basophils

Studies of human basophil function have progressed much more rapidly than studies of basophil biochemistry, pharmacology, and physiology. Recently, a number of methods have been developed to purify human basophils, and a more intensive study of this cell has proceeded. Since each method has its own advantages and disadvantages and no one method has proved to be clearly superior to all others, in this author's opinion, several important and productive methods for the purification of human basophils are discussed in detail below.

Procedures Using Affinity Chromatography

There have been three major procedures developed which take advantage of a positive selection of basophils using affinity chromatography.[8-10] Two of the methods utilize antibody linked to an insoluble support, and the third utilizes the property of basophil adherence to glass beads.

Procedure 1. Methods utilized to purify human circulating basophils developed by MacGlashan and Lichtenstein[8] were the first major step forward, allowing for the direct biochemical study of IgE-mediated phenomena in normal human basophils. The method of isolation takes advantage of unoccupied IgE receptors on basophils and the high affinity of these receptors for IgE antibody. In essence, the procedure entails passively sensitizing a mixed population of cells with antigen-specific IgE antibody (in this procedure antipenicillin IgE antibody) and passing this sensitized population of cells over a column of Sepharose to which the specific antigen [penicillin, i.e., benzylpenicilloyl (BPO)][22a] has been coupled. IgE-coated basophils adhering to the column are then eluted with hapten antigen (penicillin hapten).

The IgE-specific antipenicillin antibody (antibenzylpenicilloyl, prepared against the major penicillin determinant, BPO) critical to this procedure is isolated from penicillin-allergic individuals by passage of their serum over an affinity column of BPO linked to Sepharose CL-4B (Pharmacia, Piscataway, NJ). This column is prepared by first incubating the

[22a] Abbreviations: BPO, benzylpenicilloyl; CNBr, cyanogen bromide; PA, PIPES plus human serum albumin; PIPES, piperazine-*N,N'*-bis(2-ethanesulfonic acid); BPO-fLys, benzylpenicilloyl-*N*-formyl-L-lysine; EDTA, ethylenediaminetetraacetic acid; REH, RPMI 1640 plus EDTA plus HEPES; Tris–ACM, Tris buffer plus human serum albumin, calcium, and magnesium; HBSS, Hanks' balanced salt solution; HEPES, *N*-2-hydroxyethylpiperazine-*N'*-2-ethanesulfonic acid; PBS, phosphate-buffered saline; HLA, human leukocyte antigen; FCS, fetal calf serum; BSA, bovine serum albumin; FITC, fluorescein isothiocynate; FACS, fluorescence antibody cell sorter; PBS–TC, PBS plus trisodium citrate; HG, HEPES plus gelatin; RBL, rat basophilic leukemia.

cyanogen bromide (CNBr)-activated Sepharose beads with a solution of 0.1 *M* ethylenediamine (Sigma, St. Louis, MO) and 1 *M* NaCl, at pH 8.0 for 2 hr at room temperature. After washing, the beads are incubated for 90 min with a solution of 0.15 *M* sodium penicillin G (Sigma), 1 *M* NaCl, and 0.1 *M* $NaHCO_3$ (maintained at pH 11). The beads are subsequently washed, and the penicillin adsorbent in PA buffer [piperazine-*N,N'*-bis(2-ethanesulfonic acid) (PIPES) 25 m*M*, NaCl 150 m*M*, KCl 5 m*M*, adjusted to pH 7.3 with NaOH and containing 0.003% human serum albumin] is incubated batchwise with penicillin-allergic serum (1 volume of beads with 3 volumes of serum) for 5 hr at 4°. After washing once again, bound IgE antibody is eluted with BPO–fLys (1×10^{-3} *M* benzylpenicilloyl-*N*-formyl-L-lysine), concentrated with no attempt to remove the BPO–fLys hapten, and used for passive sensitization of basophils.

In the cell isolation procedure 50 ml of blood is obtained from nonallergic donors and mixed with saline containing ethylenediaminetetraacetic acid (EDTA) (0.8 ml of 0.1 *M* EDTA and 25 ml of saline/10 ml blood). The diluted blood is carefully layered over Hypaque (Winthrop, New York, NY)–Ficoll (Pharmacia) with a density of 1.086 g/ml (10 ml gradient solution/35 ml diluted blood) and subsequently centrifuged at room temperature for 30 min at 400 *g*. The lymphocyte–monocyte–basophil layer is removed, washed once with saline, twice with PA buffer, and then resuspended in a passive sensitization mix (total volume 0.125 ml) consisting of 1000–2000 ng/ml of purified IgE antibody (IgE–BPO–fLys, described above), 4 m*M* EDTA, 10 μg/ml heparin, and 1×10^{-4} *M* BPO–fLys. Sensitization proceeds for 90 min at 37°. Cells are then pelleted, resuspended in 0.2 ml of PA buffer, layered onto 1 ml of autologous serum containing 4 m*M* EDTA, and centrifuged at 400 *g* for 8 min. After centrifugation, the cells are resuspended in 1 ml REH buffer [RPMI 1640 (Gibco) with L-glutamine, 25 m*M* HEPES, 5% fetal calf serum, 4 m*M* EDTA, 10 μg/ml heparin], pelleted once again, and resuspended in 0.1 ml of REH buffer. Sensitized cells are applied to a 1 ml packed column of BPO–Sepharose 6MB (Pharmacia, prepared with ethylenediamine as described above) in REH buffer and allowed to slowly sink into the column for 2 min. After this time, 15 ml of ice-cold REH buffer is washed over the column at a flow rate of 5 ml/min. Subsequent to this wash, cells bound to the column are desorbed at room temperature with 8 ml of REH buffer containing 2×10^{-3} *M* BPO-fLys. After the first 1 ml of this elution buffer is collected, the column flow is stopped for 5 min (to allow for full desorption of cells) then continued at 5 ml/min. Cells are collected, washed, stained with Wright stain for differential counts and with Alcian blue[23] for basophil counts in a Spier–Levy chamber.

[23] H. Gilbert and L. Ornstein, *Blood* **46**, 279 (1975).

Basophil purities ranging from 65 to 80% and average yields of 60% have been reported with this procedure.[8] The purity of the basophil preparation depends on the concentration of basophils after the Hypaque–Ficoll step (greater than 2% concentration yielded greater purity) and on the concentration of IgE-specific antibody used for passive sensitization (less than 1000 ng/ml of IgE anti-BPO antibody gave poor results). Morphologically, purified basophils appear intact and functionally the cells respond normally by releasing histamine to anti-IgE antibody. Histamine content (1 pg/cell) appears to be unaltered by the purification procedure. Disadvantages to the procedure include binding of the IgE antibody to the IgE Fc receptors (making passive sensitization with other IgE-specific antibody systems difficult to perform) and the low number of basophils that can be obtained.

Procedure 2. The methods of Toll *et al.*[9] take advantage of IgE Fc receptors being occupied by IgE antibody on the surface of the basophil. In this procedure, basophils are passed over a column of Sepharose to which anti-IgE antibody is coupled. Basophils adhering to the column are eluted by differential buffer flow rate or mechanical agitation.

Because of the nonspecific binding properties of CNBr-activated Sepharose 6MB (Pharmacia), a critical portion of this method is the preparation of the anti-IgE affinity column. In this procedure, 1 g of Sepharose 6MB is swollen in a beaker for 15 min with 1 m*M* HCl solution. The beads are subsequently washed on a glass filter with 200 ml of the same HCl solution, then resuspended in 5 ml of a 0.1 *M* carbonate–bicarbonate buffer containing 0.5 *M* NaCl, pH 8.3, and allowed to settle for 3 hr. At the end of this time, 3 mg of anti-IgE (Hoechst) in 6 ml of 0.1 *M* carbonate–bicarbonate buffer (containing 0.5 *M* NaCl, pH 7.0) is added to the Sepharose 6MB and allowed to incubate at 4° in an end-over-end mixer overnight. After this coupling period, the beads are washed and resuspended in 1 *M* ethanolamine (pH 9.0) for a 2-hr incubation (ethanolamine blocks unreacted sites). Subsequently, 3 washing cycles are used to remove noncovalently adsorbed protein. Each cycle consists of a wash at pH 4 (0.1 *M* acetate buffer containing 0.5 *M* NaCl) followed by a wash at pH 8.3 (0.1 *M* carbonate–bicarbonate buffer containing 0.5 *M* NaCl). After the final wash cycle, the beads are resuspended in Tris–ACM buffer (containing Tris 25 m*M*, pH 7.5, NaCl 120 m*M*, KCl 5 m*M*, 0.3 mg/ml human serum albumin, 0.6 m*M* calcium, and 1 m*M* magnesium) and packed into a small plastic column (1 g of Sepharose 6MB yields 3.5 ml of column volume). Using this procedure, a 90% coupling efficiency for anti-IgE antibody has been reported.[9]

In the basophil isolation procedure, 1 part blood is mixed with 3 parts Tris buffer (without calcium and magnesium), and 6 ml of this diluted blood mixture is carefully layered onto 3 ml of a Hypaque–Ficoll gradient solution (10.5% w/v Hypaque, 6.4% w/v Ficoll, density 1.080 g/ml) and centrifuged for 30 min at 400 *g* (at 20°). The mononuclear band containing basophils (at the gradient–buffer interface) is washed in Tris buffer, and resuspended in Tris–ACM buffer at 1×10^7 cells/ml. Five milliliters of this cell suspension is then loaded onto the anti-IgE affinity column at varying flow rates (at 0°). At a flow rate of 8 ml/hr basophils are tightly bound to the column and are eluted only by removing the beads from the column, suspending them in Tris–ACM buffer, stirring for 5 min, and recovering the bound cells in the supernatant fluid. At flow rates of 16 ml/hr fractionation of basophils is observed, and preparations of high purity can be achieved.

Using mechanical agitation, basophil purities of 40–60% may be obtained with a 30–45% recovery from blood.[9] Using selective elution, purities of 80–95% can be obtained, but recovery is only 20–30% of the initial starting basophil concentration.[9] Both basophil preparations are observed to be functional, releasing equivalent amounts of histamine to an anti-IgE stimulus at each step in the purification procedure.[9] Disadvantages to this procedure include the limited lifetime of the affinity column, variability in the coupling efficiency of anti-IgE antibody to Sepharose (leading to variability in basophil purification), and the low number of basophils that can be obtained.

Procedure 3. The methods of Pruzansky and Patterson[10] employ three steps (Hypaque–Ficoll centrifugation, Hypaque centrifugation, glass bead chromatography) for basophil purification. Each step eliminates specific cell types: Hypaque–Ficoll gradients eliminate neutrophils and eosinophils; Hypaque gradient centrifugation, monocytes; glass beads, nonadherent lymphocytes.

In the procedure, blood obtained from allergic donors (0.48 ml 10% EDTA added as anticoagulant) is layered directly onto a Hypaque–Ficoll gradient solution (Hypaque 10.6% w/v, Ficoll 6.6% w/v, density 1.0825 g/ml; 15 ml blood per 10 ml Hypaque–Ficoll gradient) and centrifuged for 5 min at 1400 *g*. The mononuclear cell layer containing basophils (at the gradient–plasma interface) is washed with Tris buffer, pH 7.5 (see above, Procedure 2), and resuspended in a 1-ml volume of this same buffer. The cells are then carefully layered onto a gradient solution of Hypaque (1 ml cushion 12.7% w/v, 1 ml 10.6%, 1 ml 9.5%) and centrifuged at 500 *g* for 5 min (room temperature). Basophils appearing in the cell pellet at the bottom of the tube are washed twice with buffer and are ready for glass bead chromatography. Before use, washed (in water) glass beads (3M, St. Paul, MN) are saturated with whole autologous plasma (sedimented at high speed to remove platelets) supplemented with 4 ml of 0.1 *M* $MgCl_2$ (required for cellular adhesive to beads; calcium is avoided here since plasma clots are found in its presence) and 3 ml of 6% dextran. Saturated beads are added to a height of 8 cm in a glass column (1 × 12 cm), and washed cells (from the Hypaque gradient) resuspended in 1 ml of autologous plasma (supplemented as described above) are added dropwise to this column at room temperature. No incubation of the cell-loaded column is necessary, and nonadherent cells are eluted with autologous plasma (approximately 6 ml). The column is then filled with Tris buffer, pH 7.5, containing 0.1% EDTA, and the glass beads are agitated (gently stirred in the column with a glass rod). Desorbed cells are subsequently eluted with buffer, centrifuged, and differential and basophil counts, using Hansel stain (a modified rapid acting Wright stain, Lide Laboratories, St. Louis, MO), are performed on dried smears.[24]

Basophils of 50% purity and 50% yield from whole blood have been reported with this procedure, and their functional nature is established by histamine release from cells of allergic donors known to be sensitive to mold.[10] The greatest advantage to this procedure is that it can be performed in 1.5 hr. A disadvantage is that the purities obtained are not as great as with other methods.

Procedures Utilizing Percoll Density Gradients

The techniques described above have been successful in purifying basophils. Despite this and the fact that histamine release from these cells has been demonstrated, there is a lingering concern that the multiple manipulations of and positive selection for basophils may cause changes in cell metabolic function which are not readily apparent. Recently, procedures utilizing Percoll density gradient centrifugation have been developed.[12–15] These methods do not involve multiple manipulations or positive selection of basophils and generally are rapidly performed. Each of these reports observed basophils at Percoll densities between 1.072 and 1.082 g/ml.[11–14] The most specific density for basophil banding was observed to be 1.076 g/ml.[11] While each study had slight nuances in forming Percoll gradients, and handling of cells, a general procedure is described below.

Percoll stock solution is prepared by mixing 90 ml of a commercial Percoll solution with 8.96 ml of 10 × Hanks' balanced salt solution (HBSS) (without calcium, magnesium, or bicarbonate), 0.45 ml of 1 *N* HCl, and 1 ml of *N*-2-hydroxyethylpiperazine-*N*′-2-ethanesulfonic acid (HEPES) buffer, pH 7.3 (final density 1.123 g/ml). Specific densities of Percoll are prepared by mixing Percoll stock and HBSS without calcium or magnesium. Final adjustment of the density with either stock Percoll or

[24] J. Pruzansky and R. Patterson, *Int. Arch. Allergy Appl. Immunol.* **38**, 522 (1970).

HBSS is facilitated by density measurement with either a pycnometer or refractometer (Fischer Scientific, Silver Spring, MD) (see also this series, Vol. 108 [9]). The density of Percoll is altered by temperature; therefore, if gradient centrifugation is to be performed at room temperature, all solutions should be made and stored at this temperature. Discontinuous gradients of Percoll are formed by layering (approximately 3 ml) each Percoll solution (greatest density first) in a 15-ml polystyrene conical-bottom centrifuge tube (No. 2095, Falcon, Oxnard, CA) using a peristaltic pump (Miniplus II, Gibson) at a rate of 30 ml/hr. In most reports, blood anticoagulated with EDTA (1 ml 0.1 *M* EDTA/10 ml blood) is layered directly onto the Percoll gradient and centrifuged in a swinging bucket rotor at 600 *g* for 20 min at room temperature. Total basophil recovery is assessed by absolute counts using Alcian blue[23] and differential counts on cytocentrifuged smears using Wright stain.

Basophil purity obtained with Percoll density gradient centrifugation depends on various factors including the following: (1) Number of circulating basophils. A greater number of basophils in donor blood facilitates greater purity and yield. Leukocytosis, on the other hand, dramatically decreases basophil purity without affecting basophil recovery. (2) Technical problems. If an abrupt disturbance of the Percoll gradient is made when layering blood (either by pipet tip or blood dropping into the Percoll) a dramatic decrease in basophil purity can be observed. Similar decreases in basophil purity can also be observed if disturbance of the cell band occurs while withdrawing it from the gradient or if a small quantity of the cell layer is left behind. The results of reports utilizing the general procedure described above are summarized in the following discussion.

In the report of Raghuprasad,[11] a discontinuous Percoll gradient of densities 1.072, 1.074, 1.076, and 1.078 g/ml is used to separate whole anticoagulated blood. Basophil purity is highest at 1.076 and 1.078 g/ml densities, with average purities of approximately 25% obtained (range 6–50% and 6–60%, respectively). Basophils are also found at 1.072 g/ml (average 5% purity) and 1.074 g/ml densities (average 15% purity). Basophil recovery and purity are highest in donors having high basophil counts (approximately 2% of total leukocytes; does not necessarily correspond to an atopic state), and lymphocytes and neutrophils are the most frequent contaminants present in basophil preparations. Centrifugation of cells pooled from the 1.076 and 1.078 g/ml densities and application to a second discontinuous Percoll gradient consisting of two densities (1.072 and 1.080 g/ml) lead to an enhancement of purity (average 40%, range 15–83%) but a decrease in yield (20% of total basophils). The functional nature of the isolated basophils may be verified by histamine release using several secretagogues (anti-IgE, complement component 5a, and alternaria antigen in sensitive individuals).

Kaufman *et al.*[12] also used discontinuous Percoll gradients (densities 1.072, 1.082, 1.100 g/ml) and observed basophils to band between the 1.072 and 1.082 g/ml densities. The range of purity is between 8 and 20% with a basophil recovery of 70% of those applied to the gradient. This method differs from that previously described by the fact that a leukocyte preparation (obtained by unit gravity sedimentation for 30–40 min at 37° in 6% dextran) instead of whole blood is layered onto the Percoll gradient. In another method,[13] whole blood (EDTA anticoagulated) is layered onto a discontinuous Percoll gradient (densities 1.070, 1.079, 1.090 g/ml), and, as in the previous method, basophils are observed to band between densities of 1.070 and 1.079 g/ml. The mean purity of the isolated basophils is 19% (range 5–53%), and lymphocytes are the main contaminating cells.[13] However, on closer scrutiny, basophils from some donors consistently band in two different layers of the discontinuous Percoll gradient (the plasma–1.07 g/ml Percoll interface and the 1.070–1.079 g/ml Percoll interface).[14] While the greatest number of basophils is still found to be in 1.070–1.079 g/ml interface, in some donors the majority of basophils are found in the plasma–1.070 g/ml Percoll interface. From this and other observations, the existence of two populations of basophils has been suggested. These data alone do not prove the existence of multiple populations of basophils, but they do raise the possibility that subpopulations of basophils exist and perhaps have distinct chemical composition and function.

Procedures Utilizing Percoll Density Gradient Centrifugation and Cell Panning Techniques

Landry and Findlay[16] have recently described a basophil purification procedure which combines both Percoll density gradient centrifugation and cell panning techniques, using monoclonal antilymphocyte antibodies. Cell panning techniques negatively select for basophils by the selective removal of unwanted or contaminating cells.

In this procedure, stock Percoll (Pharmacia) is prepared by mixing 9 parts commercial Percoll, and 1 part PIPES (10× buffer) (PIPES 250 m*M*, NaCl 1180 m*M*, KCl 50 m*M*, NaOH 430 m*M*, and 0.1% 0.1 *M* EDTA). Leukocyte-rich plasma is prepared by mixing whole blood (anticoagulated with 0.1 *M* EDTA) with 20% v/v dextran and allowing the mixture to settle at unit gravity for 90 min. The resultant leukocyte preparation is then carefully layered over a discontinuous Percoll gradient prepared by layering 10 ml of a 55% Percoll solution (5.5 ml stock Percoll plus 4.5 ml PIPES 1× buffer) over 10 ml of a 65% Percoll solution (6.5 ml stock Percoll plus 3.5 ml PIPES 1× buffer). This gradient is centrifuged at 400 *g* for 20 min at room temperature. After centrifugation three bands of cells can be seen. The second band of cells found between the 55 and 65% Percoll layers contains an enriched population of basophils plus contaminating mononuclear cells and neutrophils. This layer is carefully withdrawn, washed twice (250 *g* for 10 min at 4°) with 30 ml of PIPES buffer containing 0.03% human serum albumin (PA), and further purification of basophils is obtained with cell panning techniques.

For the preparation of panning plates (see also this series, Vol. 108 [11]), plastic petri dishes (Fisher 8-757-12) are coated with 100 μl (0.1 mg) of goat anti-mouse IgG (Tago No. 4150) in 10 ml of phosphate-buffered saline (PBS) by incubating at 4° for 12 hr. After incubation, the antibody solution is decanted. The plates washed 3 times with 10 ml of PBS after which they are ready for use. Plates may be stored (wrapped in Parafilm) with PBS containing 2% fetal calf serum (FCS) at $-20°$. In the panning procedure, cells from band 2 of the Percoll gradient ($2-4 \times 10^5$) are resuspended in 1 ml of PA buffer. Monoclonal antibodies (anti-Leu-1 and anti-HLA-DR, Becton Dickinson, Sunnyvale, CA) are added to the cells at a final concentration of 1 $\mu g/10^6$ cells, and the mixture is incubated for 1 hr at 4°. After incubation, the cells are centrifuged (240 *g* for 5 min at 4°), resuspended in 3 ml of PBS containing 2% FCS (PBS–FCS, at 4°), added to panning plates, and incubated at 4° for 60 min (swirling of plate is performed every 20 min). After this incubation, the unabsorbed cells are decanted, combined with cells obtained by 3 gentle washings of the plate (with 10 ml of PBS–FCS), centrifuged at 240 *g* for 10 min at 4°, and resuspended in PA buffer. Counting of basophils is performed using the Alcian blue method.[23]

The second cell band in the Percoll gradient obtained from fractionated leukocyte-rich plasma contains basophils of highest purity (mean $14.6 \pm 7.9\%$)/[16] Further purification of these basophils using cell panning techniques increases the purity to a mean of $34 \pm 15\%$ (range 19–54%), with 55% yield of basophils.[16] This is approximately a 30-fold purification. The functional capability of the basophils is verified by histamine release using anti-IgE as a secretagogue. Lett-Brown *et al.*[17] have recently reported a basophil purification technique similar to that described above. The primary difference in their procedure is the Percoll density gradient centrifugation step. In their method, the procedure described by Leonard *et al.*[13,14] (outlined above) is utilized. Basophil-enriched fractions withdrawn from Percoll are subsequently panned as described above (an OKT11 monoclonal antibody is used in place of the anti-Leu-1 monoclonal antibody). In selected donors, functional basophils with purities of greater than 90% and yields from blood of approximately 50% may be achieved.[17] The major drawback is the small number of highly purified basophils ($5-10 \times 10^5$ cells) that can be obtained. While mediator release experiments are feasible with small numbers of basophils, biochemical studies usually require larger numbers of pure cells. The use of leukopheresis products may enhance the productivity of these techniques.

Procedures Utilizing Flow Microfluorometric Techniques (see also this series, Vol. 108 [19])

Flow microfluorometry allows for the isolation of viable and highly purified basophils from blood. This method is the simplest to perform, requiring only commercially available anti-IgE antibody and access to a fluoroescence cell sorter. The technique as described by Weil *et al.*[18] consists of two steps. In the first step, blood collected in EDTA (final concentration 5 m*M*) and diluted with an equal volume of HBSS containing 0.1% bovine serum albumin (BSA) (HBSS–BSA) is layered over a Hypaque–Ficoll solution (specific gravity 1.077 g/ml) and centrifuged at 800 *g* for 15 min, and the mononuclear cell layer containing basophils washed 3 times with HBSS–BSA. Washed cells (5×10^7/ml) are incubated with a 1:50 dilution of fluorescein isothiocyanate (FITC)-conjugated goat anti-human IgE (epsilon-specific IgG fraction, Cappel Laboratories, Cochranville, PA) at 5° for 30 min. Following this incubation, the cells are washed twice with cold HBSS–BSA, and labeled and unlabeled cells are sorted with a FACS II cell sorter (Becton Dickinson) using two sorting rates. The first enrichment sort is performed at a rate of 12×10^3/sec without coincidence constraints. These enriched cells are then sorted at a rate of 4×10^3/sec with coincidence constraints (i.e., only positive cells without cells in adjacent droplets sorted). Basophil purities of 50–65% could be obtained with the first sorting and purities of 97–99% with the second sort.[18] These cells are viable, and a histamine content of 1.7 pg/cell agrees with other reports published for human basophils. Limitations to this procedure include the following: (1) Sorted cells having been exposed to anti-IgE under conditions preventing histamine release are inactivated and cannot be used to study release using antigen-dependent mechanisms. (2) The number of cells able to be prepared by this technique is small and would not lend itself to biochemical studies.

Procedures Utilizing Centrifugal Elutriation (see also this series, Vol. 108 [20])

All of the methods described above have been successful to some extent in isolating and purifying human basophils. The number of basophils isolated using these procedures is usually low (of the order of 10^5–10^6 cells), and generally the techniques are not suitable or capable of being modified for the isolation and purification of large numbers of basophils. The recent report of DeBoer and Roos[15] utilizing Percoll density gradient centrifugation, centrifugal elutriation, and increased quantities of donor blood has, however, demonstrated much greater success in isolating and purifying large numbers of human basophils.

In this method a stock Percoll solution is prepared by mixing 93 ml of Percoll with 7 ml of a solution containing 1.54 *M* NaCl and 0.1 *M* NaH_2PO_4 (all solutions are kept at room temperature). Specific Percoll densities are prepared by mixing stock Percoll with PBS (140 m*M* NaCl, 9.2 m*M* Na_2HPO_4, 1.3 m*M* NaH_2PO_4, pH 7.4). The final Percoll solution also contains human serum albumin (5 μg/ml) and trisodium citrate (13 m*M*). In the isolation procedure, 1 unit of normal donor blood (500 ml) is collected in a plastic bag containing 75 ml of an ACD solution (containing a hydrous disodium citrate, 100 m*M*, and anhydrous glucose, 128 m*M*, pH 5.0). Within 1 hr after collection, the blood is centrifuged in the plastic bag (4000 *g*, 5 min, room temperature), and the plasma is separated from cells. The upper 50–100 ml of the packed cells (buffy coat) contains 70% of the leukocytes. This buffy coat is collected[15] and diluted to 200 ml with PBS containing 13 m*M* trisodium citrate (PBS–TC).

Mononuclear cells and basophils are separated from neutrophils, eosinophils, and red blood cells by centrifugation (1000 *g*, 20 min, room temperature) of the buffy coat over a Percoll solution of 1.077 g/ml density (8 tubes each containing 12 ml of Percoll solution). The supernatant autologous plasma (SP) from this centrifugation is saved, and the cells at the plasma–Percoll interface are diluted to 200 ml with PBS–TC and centrifuged at 400 *g* for 5 min at room temperature. To remove platelets, the resultant cell pellets are resuspended first in 100 ml then in 50 ml of the SP solution. After each resuspension of cells in SP, the cells are centrifuged at 400 *g* for 5 min at room temperature. The final cell pellet is resuspended in 10–20 ml of the SP solution and injected into a JE-6 elutriation rotor powered by a J-21C centrifuge (Beckman). The elutriation medium is PBS–TC plus 5 mg/ml human serum albumin. To separate the various cell populations, the flow rate is maintained at 20 ml/min while the rotor speed is diminished from 4000 to 0 rpm. Fractions of 100–200 ml of the outflow are collected at rotor speeds of 3700, 2600, 2500, and 0 rpm. The last fraction (0 rpm) contains 80% of the basophils present in the starting preparation. This fraction is centrifuged (400 *g*, 5 min, room temperature), resuspended in 4 ml of Percoll 1.067 g/ml), and layered over 2 ml of Percoll of 1.075 g/ml density. A top layer of 0.2 ml of PBS–TC containing 5 mg/ml of human serum albumin is added, and the gradient is centrifuged at 1000 *g* for 10 min at room temperature.

After centrifugation, two interfaces containing cells are observed. The lower Percoll interface (1.067–1.075 g/ml densities) contains the basophils. Basophil purity of 70% (range 45–87%, as measured by Alcian blue) and yield of 60% from the buffy coat have been reported.[15] More importantly, however, $4-19 \times 10^6$ basophils can be obtained using this procedure. Histamine content (1–2 pg/cells) and releasability (30–50% to anti-IgE) have been shown in these preparations.[15] The success and consistency of this procedure has been predicated on the following: (1) the capabilities of centrifugal elutriation; (2) suspension of the cells in Percoll (1.067 g/ml density) before applying to the Percoll solution of higher density (this prevents carryover of cells into a layer with a higher specific gravity than the cells); and (3) prevention of partial degranulation of basophils in the procedure (this is avoided by preventing platelet aggregation), accomplished by platelet removal with a low-speed centrifugation in autologous plasma.

Methods for the Isolation of Basophils from Other Species

Although present, basophils in rodents are much less abundant and, therefore, are less well studied.[19] In dogs, basophils make up less than 1% of the leukocyte population, and their concentration can be enhanced to 1–2% using double-density Hypaque–Ficoll centrifugation.[25] Naturally abundant basophils have been observed in the snapping turtle *Chelydra serpentina.*[20] Normal circulating basophil levels of 50–63% have been reported, and these cells have been shown to release histamine in response to a rabbit anti-turtle immunoglobulin. Studies aimed at purifying and further characterizing these cells have not as yet been reported. To date, basophil function and techniques for isolation from blood have been most thoroughly studied in humans and guinea pigs.[21,22] Procedures for isolation of guinea pig basophils are discussed in detail below.

Procedure

Techniques for purification of guinea pig basophils utilizing Ficoll density gradient centrifugation[21] and negative selection using a rabbit anti-leukocyte antibody and Percoll density gradient centrifugation have been reported.[22] However, the most consistent and reliable technique utilized for guinea pig basophil purification involves both centrifugal elutriation and Percoll density gradient centrifugation.[26] This procedure is described below.

[25] D. Austin, S. Chan, A. Malley, J. Hanifin, and C. Hirshman, *J. Lab. Clin. Med.* **17,** 53 (1985).

[26] B. Undem, J. Brendel, T. Hirth, C. Buckner, and F. M. Graziano, *Am. Rev. Respir. Dis.* **133,** 763 (1986).

Basophils comprise only a small percentage of the total nucleated cells in guinea pig blood. However, the animals can be treated to increase the number of basophils which, in turn, facilitates purification. A striking basophilia can be obtained in guinea pigs sensitized with whole sheep blood for 12 days (daily i.p. injections of 2 ml whole blood diluted 1:2 with PBS).[21,22] On day 13, blood obtained from six animals and containing 0.1 *M* EDTA is thoroughly mixed with 3% gelatin (2 parts blood to 1 part gelatin) in a test tube and placed at an angle of 30° in a 37° water bath. After incubation for 45 min, the leukocyte-rich plasma is withdrawn and centrifuged at 400 *g* (room temperature) for 30 min. Pelleted cells are washed once in HG (HEPES 10 m*M*, NaCl 137 m*M*, KCl 5 m*M*, 0.5 mg/ml gelatin, 5.5 m*M* glucose, pH adjusted to 7.4 with NaOH) containing 4 m*M* EDTA, and residual red cells are removed using hypotonic lysis (see this series, Vol. 108 [6]). Leukocyte preparations (containing an average of 6% basophils)[22,26] are then fractionated using a JE-6 elutriator rotor driven by a J-21B Centrifuge (Beckman Instruments, Palo Alto, CA). Rotor speed is controlled with a 10-turn potentiometer which allows for accurate rpm selection (within 10 rpm). Before each run, the system is washed sequentially with ethanol (200 ml), distilled water (1000 ml), and PBS (200 ml). Blood leukocyte suspensions containing basophils are loaded into the elutriator with a rotor speed of 4150 rpm, and the flow rate is adjusted to and maintained at 30 ml/min. Leukocytes are fractionated by incrementally decreasing the rotor speed to 3600, 2930, 2760, 2100, and 0 rpm. After each adjustment in rotor speed, two to four 45-ml fractions are collected. An additional 90-ml collection is obtained at the end of each run with the centrifuge turned off. Elutriated fractions containing the highest percentage of basophils (fractions collected after reduction in rotor speed to 2760 and 2100 rpm) are pooled, centrifuged, resuspended in 1 ml of HG ($10-50 \times 10^6$ cells/ml), and layered onto a discontinuous Percoll gradient (1.06, 1.07, and 1.08 g/ml).[26] Centrifugation is carried out at 400 *g* for 20 min at room temperature, and cell bands are carefully removed and washed with HG. Total basophil counts are performed at each step during the purification procedure using the Alcian blue method.[23] Differential counts are obtained on cytocentrifuged (Cytospin, Shandon Elliot Corp.) smears stained with Wright stain. Cell viability (90–98%) is monitored at each step during the isolation procedure by Trypan blue dye exclusion.

Guinea pig basophils ($5-10 \times 10^6$ basophils) of greater than 90% purity and 40–50% yield from whole blood are consistently observed at the 1.070–1.080 g/ml Percoll interface using this procedure.[26] Use of centrifugal elutriation alone will consistently give basophil purity of 40–50%.[26] Morphologically, the cells appear to be intact (see Fig. 1), and, functionally, histamine release from these cells can be demonstrated with antigen (sheep γ-globulin), concanavalin A, rabbit anti-IgG_1 antibody, and 12-*O*-tetradecanoylphorbol 13-acetate (TPA).

Rat Basophilic Leukemia Cells

The problems inherent in studying normal basophils (obtaining large quantities of pure cells for biochemical studies, possible heterogeneity of basophil populations) discussed above have focused attention on the study of a rat basophilic leukemia tumor cell line. In 1973 Eccleston *et al.*[27] first described the development of basophilic leukemia in a Wistar rat fed β-chlorethylamine (a potent inducer of myeloid or lymphatic leukemia). Leukemic cells obtained from such animals (RBL cells) can be serially transplanted by intraperitoneal injection into rats, resulting in the development of a solid tumor.[27]

A number of investigators have shown these tumor cells to contain histamine and serotonin and have used them for study of the structure and binding characteristics of IgE surface receptors present on these cells.[27-31] RBL cells have been adapted to suspension cell culture and have been maintained for prolonged periods of time with little morphologic or functional change in cell characteristics compared to original minced tumor cells.[28] The original cell lines, while useful in studying the IgE receptor, failed to be triggered for IgE-mediated histamine release and, therefore, were not useful in studying the mechanisms of this process.[27,28] Subsequently, several cloned histamine-releasing cell lines and nonhistamine-releasing cloned variant sublines defective at different steps in the histamine-releasing process have been reported.[32-34] The defects at different sites of the release pathway have allowed for a dissection of the biochemical events involved in histamine release.[33-38] RBL cell lines have been and will

[27] E. Eccleston, J. Leonard, S. Lowe, and H. Welford, *Nature (New Biol.)* **244,** 73 (1973).

[28] A. Kulczycki, C. Isersky, and H. Metzger, *J. Exp. Med.* **139,** 600 (1974).

[29] A. Froese, R. Helm, D. Conrad, C. Isersky, T. Ishizaka, and A. Kulczycki, *Immunology* **46,** 107 (1981).

[30] H. Metzger, *Immunol. Rev.* **41,** 186 (1979).

[31] H. Metzger, A. Goetze, J. Kanellopoulos, D. Holowka, and C. Fewtrell, *Fed. Proc., Fed. Am. Soc. Exp. Biol.* **41,** 8 (1982).

[32] E. Barsumian, C. Isersky, M. Petrino, and R. Siraganian, *Eur. J. Immunol.* **11,** 317 (1981).

[33] A. McGivney, F. Crews, F. Hirata, J. Axelrod, and R. Siraganian, *Proc. Natl. Acad. Sci. U.S.A.* **78,** 6176 (1981).

[34] R. Siraganian, A. McGivney, E. Barsumian, F. Crews, F. Hirata, and J. Axelrod, *Fed. Proc., Fed. Am. Soc. Exp. Biol.* **41,** 30 (1982).

[35] C. Meyer, L. Wahl, B. Stadler, and R. Siraganian, *J. Immunol.* **131,** 911 (1983).

[36] J. Rivera, J. M. Mullins, K. Furuichi, and C. Isersky, *J. Immunol.* **136,** 623 (1986).

[37] M. Garcia-Gil and R. Siraganian, *J. Immunol.* **136,** 259 (1986).

[38] M. Garcia-Gil and R. Siraganian, *J. Immunol.* **136,** 3825 (1986).

continue to be a convenient model for study of the mechanisms of mediator release. A rat basophilic leukemia cell line (CRL 1378-RBL-1) is available from the American Type Culture Collection, Rockville, MD.

Mediator Release by Basophils

Mediator release (e.g., histamine, arachidonic acid metabolites) from isolated and purified basophils has been documented in humans[8-15,39] and other species.[19-21,25,26] In the general procedure for mediator release, purified cells (generally $5-10 \times 10^4$ basophils per assay tube) in 0.5 – 1 ml of isotonic buffer (e.g., HEPES or PIPES buffer) containing 1 m*M* calcium and magnesium are reacted with varying concentrations of secretagogue, e.g., anti-IgE (0.01 – 1 μg/ml)[8] (Calbiochem-Behring Corp.), antigen (e.g., sheep γ-globulin, 0.1 – 10 μg/ml),[22,26] or concanavalin A (0.1 – 3 μg/ml).[40] Mediators released into the supernatant fluid can be measured with one of the available methods (e.g., this volume [42]).

[39] R. Schleimer, C. Fox, R. Naclerio, M. Plaut, P. Creticos, A. Togias, J. Warner, A Kagey-Sobotka, and L. Lichtenstein, *J. Allergy Clin. Immunol.* **76,** 369 (1985).

[40] P. Siraganian and R. Siraganian, *J. Immunol.* **112,** 2117 (1974).

[44] Human Neutrophil Degranulation

By DANIEL G. WRIGHT

Introduction

Neutrophils are wandering, phagocytic white blood cells that are critical to host defenses against bacterial and fungal infections. They are also the first circulating leukocytes to arrive at a site of acute inflammation. Indeed, the extravascular accumulation of these cells in tissues is a fundamental histologic hallmark of the acute inflammatory response.

Inflammatory neutrophils are recognized to play a key role in mediating the dissolution and remodeling of connective tissues at sites of inflammation. Neutrophils injure or destroy connective tissue cells and digest or alter extracellular matrix proteins by at least two mechanisms, generation of toxic oxygen metabolites (see this series, Vols. 105 and 132) and secretion of proteolytic enzymes.

Unlike the monocyte/macrophage family of phagocytes, neutrophils retain a very limited capacity for *de novo* protein synthesis once they are terminally differentiated and enter the circulation from the bone marrow. Almost all proteolytic enzymes and other biologically active proteins released by neutrophils at sites of inflammation are derived from cytoplasmic storage granules formed during the development of these cells in the marrow.[1] This chapter describes techniques for isolating these subcellular organelles from normal, unstimulated human blood neutrophils and for measuring granule exocytosis by activated neutrophils.

In considering neutrophil granules, granule constituents, and degranulation, it is of utmost importance to bear in mind that these cytoplasmic organelles are functionally heterogeneous, not only in terms of what they contain but in how they are mobilized and used by the cell. There are two principal, functionally distinct types of neutrophil granules: primary and secondary (also called, respectively, "azurophilic" and "specific," older terms derived from the light microscopic appearance of stained neutrophil precursors). Primary granules are formed early in neutrophil development and contain myeloperoxidase. Of the different types of neutrophil granules, primary granules are the most analogous to lysosomes. They contain essentially all the lysosomal acid hydrolases found in neutrophils. Moreover, the normal function of these granules appears to be associated largely with lysosomelike, intracellular digestive processes. The contents of primary granules are relatively inaccessible for extracellular release, unless neutrophils encounter certain types of particulate or surface stimuli, or unless they are damaged to the point of cytolysis.

Secondary granules, on the other hand, are formed at the intermediate, metamyelocyte stage of neutrophil maturation in the marrow, once primary granule formation is largely completed. In mature neutrophils, secondary granules outnumber primary granules by about 2:1. Secondary granules are peroxidase negative and do not contain lysosomal acid hydrolases (Table I). Furthermore, these granules are much more accessible for exocytosis than are primary granules and are similar in various respects to the storage granules of secretory cells. Some degree of secondary granule secretion occurs with most forms of neutrophil activation.

The existence of an even more highly secretory, "tertiary" granule compartment, which is also peroxidase negative and is associated with the enzyme gelatinase, has been described.[2] However, it is not entirely clear that gelatinase is stored in a functionally distinct type of peroxidase-negative neutrophil granule, since recent ultrastructural, immunocytochemical studies have localized gelatinase to the same granule compartment as that

[1] D. G. Wright, *Host Defense Mechanisms* **1,** 75 (1982).

[2] B. Dewald, U. Bretz, and M. Baggiolini, *J. Clin. Invest.* **70,** 518 (1982).

TABLE I
CONSTITUENTS OF HUMAN NEUTROPHIL GRANULES

Constituent type	Primary granules (peroxidase positive)	Secondary granules[a] (peroxidase negative)
Microbicidal enzymes	Lysozyme	Lysozyme
	Myeloperoxidase	
Neutral serine proteases	Elastase	
	Cathepsin G	
	Proteinase 3	
Metalloproteases		Collagenase (Type 1)
		Gelatinase[a]
Acid hydrolases	β-Glycerophosphatase	
	β-Glucuronidase	
	N-Acetyl-β-glucosaminidase	
	α-Mannosidase	
	Arylsulfatase	
	β-Galactosidase	
	α-Fucosidase	
	Cathepsins B and D	
Other	Sialidase	Lactoferrin
		Vitamin B_{12}-binding protein
		Sialidase
		β_2-Microglobulin
		Uncharacterized acidic proteins

[a] There is evidence for a distinct type of peroxidase-negative granule which contains gelatinase (see text).

which contains well-characterized secondary granule proteins, such as lactoferrin.[3]

Methods

Isolation of Human Neutrophils from Venous Blood (see also this series, Vol. 108 [9])

The most widely used method for purifying neutrophils from freshly drawn, anticoagulated blood involves three steps: a Hypaque–Ficoll density gradient centrifugation step to separate neutrophils from mononuclear cells and platelets, a sedimentation step to separate neutrophils from erythrocytes, and a hypotonic lysis step to remove residual erythrocytes. Similar results are obtained whether the density centrifugation step is done first or

[3] M. S. Hibbs, A. H. Kang, and D. F. Bainton, *Clin. Res.* **34,** 459A (1986).

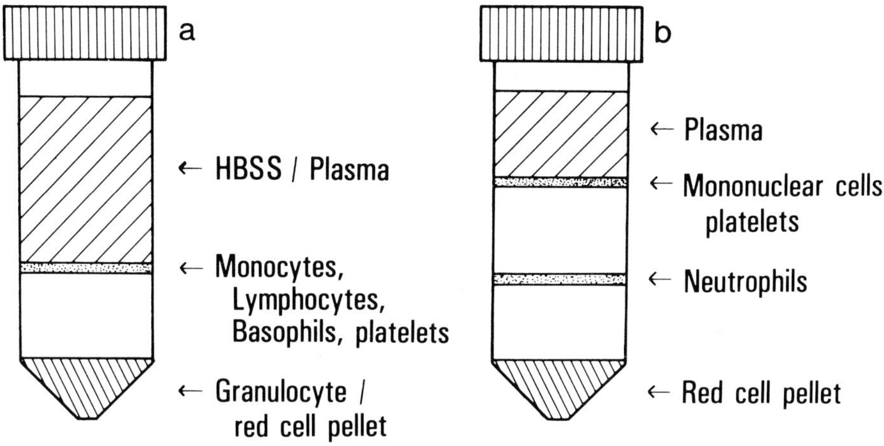

FIG. 1. (a) Blood cell separation by Hypaque–Ficoll gradient centrifugation. (b) Blood cell separation by Percoll gradient centrifugation.

last in sequence. Alternatively, neutrophils may be purified by a single-step density centrifugation procedure using discontinuous Percoll gradients.

Three-Step Procedure. Blood is drawn into 50-ml plastic (polycarbonate) syringes which contain EDTA [1.5% disodium EDTA in Hanks' balanced salt solution (HBSS)[3a] without Ca^{2+} or Mg^{2+} and without phenol red; 1 ml EDTA solution/10 ml blood drawn]. Anticoagulated blood is diluted 1:1 with HBSS without Ca^{2+} or Mg^{2+} and aliquoted into 50-ml conical centrifuge tubes, 35 ml diluted blood per tube. Ten milliliters of Ficoll-paque (Pharmacia) is underlaid beneath the diluted blood by slow injection through a narrow catheter attached to a syringe. The tubes are then centrifuged for 30 min at 500 *g* at room temperature. (Maintenance of HBSS, gradient solutions, and centrifugation conditions at 22° is important for efficient cell separation by density.) Cell separation following centrifugation is depicted diagrammatically in Fig. 1a. Ten milliliters of plasma is aspirated and saved from the top of each gradient tube; the remaining liquid and mononuclear cell/platelet layers are then aspirated and discarded (unless mononuclear cells are to be isolated also) down to the red cell/granulocyte pellet.

[3a] Abbreviations: EDTA, ethylenediaminetetraacetic acid; HBSS, Hanks' balanced salt solution; PMSF; phenylmethylsulfonyl fluoride; PIPES, piperazine-*N,N'*-bis(2-ethanesulfonic acid); EGTA, ethylene glycol bis(β-aminoethyl ether)-*N,N,N',N'*-tetraacetic acid; MPO, myeloperoxidase; BSA, bovine serum albumin; PBS, phosphate-buffered saline; LF, lactoferrin; DMSO, dimethyl sulfoxide.

Ten milliliters of the plasma/HBSS saved from the gradient tubes is added back to each red cell/granulocyte pellet. Then, 3% dextran/saline solution [3% dextran of MW 200,000–500,000 (Pharmacia) in sterile 0.85% NaCl] is added to each tube up to the 50-ml mark. The contents of each tube are mixed well by repeated inversion of the tubes and then allowed to sediment for 20–30 min at room temperature. The supernatants (which contain enough residual red cells to be pink) are aspirated and combined, and the sedimented red cell pellets are discarded. Supernatants are then centrifuged at 250 *g* for 10 min at 4°. The resulting supernatants are aspirated and discarded, with care being taken not to disturb the neutrophil-rich pellets.

Twenty milliliters of cold 0.2% NaCl is added in turn to each cell pellet and mixed by gentle pipetting for a period of 20 sec (measured by stopwatch), followed immediately by the addition of 20 ml cold 1.6% NaCl. Tubes are then capped, mixed by gentle inversion, and centrifuged at 250 *g* for 10 min at 4°. The resulting supernatant is discarded. If necessary, hypotonic lysis of red cells may be repeated once; otherwise, the cell pellets are washed once in HBSS without Ca^{2+} or Mg^{2+} and counted. The cells are then resuspended in media appropriate to subsequent procedures.

One-Step Procedure. Fifteen-milliliter aliquots of 63% Percoll are added to 50-ml conical centifuge tubes; 15 ml of 72% Percoll is then underlaid through a narrow catheter attached to a syringe, with care to avoid the introduction of air bubbles which might disturb the discontinuous gradient [63% Percoll is prepared by diluting 63 parts 100% Percoll, density 1.130 g/ml (Pharmacia), with 7 parts 10× PBS and 30 parts 1× PBS; 72% Percoll is prepared by diluting 72 parts 100% Percoll and 8 parts 10× PBS and 20 parts 1× PBS]. Undiluted, EDTA-anticoagulated blood (15 ml) is then layered on top of each Percoll gradient which is then centrifuged for 30 min at 500 *g* at room temperature. The neutrophil-rich band (see Fig. 1b) is aspirated, and the cells washed 2 times in cold HBSS without Ca^{2+} or Mg^{2+}, counted, and resuspended in appropriate media.

Neutrophils are functionally fragile cells that are sensitive to nonspecific surface activation. These cells should not be stored prior to study, even for brief periods of time. While carefully cleaned, sterile, siliconized glassware can be used successfully in these preparative procedures, sterile, polypropylene tubes and pipets are preferable. Also, after the density gradient and sedimentation steps, the separated cells may be kept on ice. Cell clumping is a sign of neutrophil injury and/or activation during preparation. By either of these procedures one can expect to isolate approximately 2×10^7 neutrophils per 10 ml of normal, freshly drawn blood. Cell preparations should contain over 95% neutrophils with eosinophils as the major cell contaminant. While the one-step Percoll procedure may avoid the hypotonic lysis step, the polyvinylpyrrolidone-coated silica particles which make up the Percoll gradient material cannot be completely removed from the cell preparations and may exert independent effects on these highly sensitive cells; moreover, these particles may complicate ultrastructural studies.

Preparation of Subcellular Fractions Enriched for Primary and Secondary Granules (see also this series, Vol. 163 [28])

Subcellular fractionation of neutrophils is accomplished by disruption of the cells, differential centrifugation to prepare a crude microsomal fraction, and density gradient ultracentrifugation to resolve distinct microsomal components.

Neutrophil Disruption. Nitrogen cavitation is the best method for disrupting human neutrophils, while avoiding lysis of the cytoplasmic granules and nuclei. This procedure may be carried out with a pressure homogenizer of relatively large (950 ml; Parr Instrument Co., Moline, IL) or small capacity (10 ml; Kontes Glass Co.). However, cavitation of relatively small volumes of highly concentrated cell suspensions generally gives the best results. Also, since the release of protease activities is unavoidable during neutrophil disruption, protease inhibitors must be added to the cells for most experimental applications of this procedure.

Isolated neutrophils are suspended in 50 ml HBSS without Ca^{2+} or Mg^{2+} on ice. Phenylmethylsulfonyl fluoride (PMSF, Sigma Chemical Co.) leupeptin (Sigma), and aprotinin (Sigma) are then added at final concentrations of 0.5 mM, 0.1 mM, and 10 μg/ml, respectively. PMSF, like other esterase inhibitors, is very toxic and demands careful handling. The cell suspension is mixed, allowed to stand for 1 hr on ice and then centrifuged at 250 g for 10 min. The cells are then resuspended in 10–20 ml of a PIPES buffer with ATP (pH 7.3) at a cell concentration of 5×10^7/ml [cavitation buffer: 10 mM KCl, 3 mM NaCl, 1 mM ATP (disodium salt, Sigma), 3.5 mM $MgCl_2 \cdot 6H_2O$, and 10 mM PIPES (Sigma)]. The cell suspension is placed in a precooled cavitation device in a cold room (4°), and the valves of the apparatus are closed. The cavitation device is then flushed with nitrogen, filled to 350 psi N_2, and allowed to equilibrate for 20 min at 4°. During equilibration a small magnetic stirring bar at the bottom of the vessel containing the cells is used to keep the cells suspended. Pressure is released very gradually such that the cell suspension is expelled dropwise into a 50-ml centrifuge tube containing 10–20 ml of cavitation buffer with EGTA (Sigma) at a final concentration of 1.25 mM. Some foaming of the cavitate will occur. The cavitate is then centrifuged for 10 min at 1400 g at 4° and the supernatant saved for density gradient ultracentrifugation. The pellet which contains nuclei and unbroken cells may be resuspended and recavitated.

Subcellular Fractionation. Described below are two methods which successfully resolve neutrophil plasma membrane vesicles, secondary granules, and primary granules. Each method has certain advantages and disadvantages. The method using concentrated, hypertonic sucrose gradients may complicate assays for certain granule constituents (e.g., sialidase), while the other method using Percoll is reported to interfere with the detection of β-glucuronidase and, as noted above, Percoll particles cannot be removed completely from the purified fractions.

Continuous sucrose gradient method: Sucrose gradients are prepared in cellulose nitrate ultracentrifuge tubes (1×3.5 inches) (Beckman Instruments) with a gradient mixer. Each gradient is formed with 16 ml heavy (65 g sucrose/100 ml, in distilled H_2O) and 17 ml light (35 g/100 ml) sucrose solutions which are mixed and layered on top of a 1.0-ml cushion of the heavy sucrose solution. These sucrose solutions should be prepared in volumetric flasks. Also, considerable practice is required to generate gradients evenly and reproducibly. Five milliliters of postnuclear cavitate (containing the cellular constituents of $2-3 \times 10^7$ neutrophils per ml of cavitate) is layered on top of each of the gradients, which are then ultracentrifuged in a prechilled swinging bucket rotor at 95,000 *g* for 4 hr at 4° (27,000 rpm, SW-27 rotor; Beckman Instruments ultracentrifuge). Gradients are eluted from the bottom using a tube-piercing apparatus (Buchler Instruments). This method resolves two populations of myeloperoxidase-positive primary granules, as well as the peroxidase-negative secondary granules and plasma membrane vesicles (Fig. 2a). Simple turbiometric measurements of the elution fractions in a spectrophotometer (500 nm) may be used to identify the granule-rich fractions.

Discontinuous Percoll gradient method: Heavy (1.12 g/ml) and light (1.05 g/ml) Percoll solutions are prepared using 10× cavitation buffer with 12.5 m*M* EGTA which is added in amounts that are one-tenth the final volume of each Percoll solution to achieve isotonicity. The heavy Percoll (14 ml) is underlaid beneath 14 ml of the light Percoll solution using a narrow catheter attached to a syringe. Postnuclear cavitate (5–10 ml), prepared as described above, is layered at the top of each gradient, and the gradients are centrifuged at 4° for 15 min at 48,000 *g* in a SS34 rotor (high-speed centrifuge). This centrifugation separates the peroxidase-positive primary granules, secondary granules, and plasma membrane vesicles (see Fig. 2b). Gradients may then be eluted from the bottom as described above or by aspiration from the top using a Pasteur pipet. Most of the Percoll may then be removed from individual or pooled fractions by ultracentrifugation at 4° for 2 hr at 180,000 *g* (SW-41 Beckman rotor).

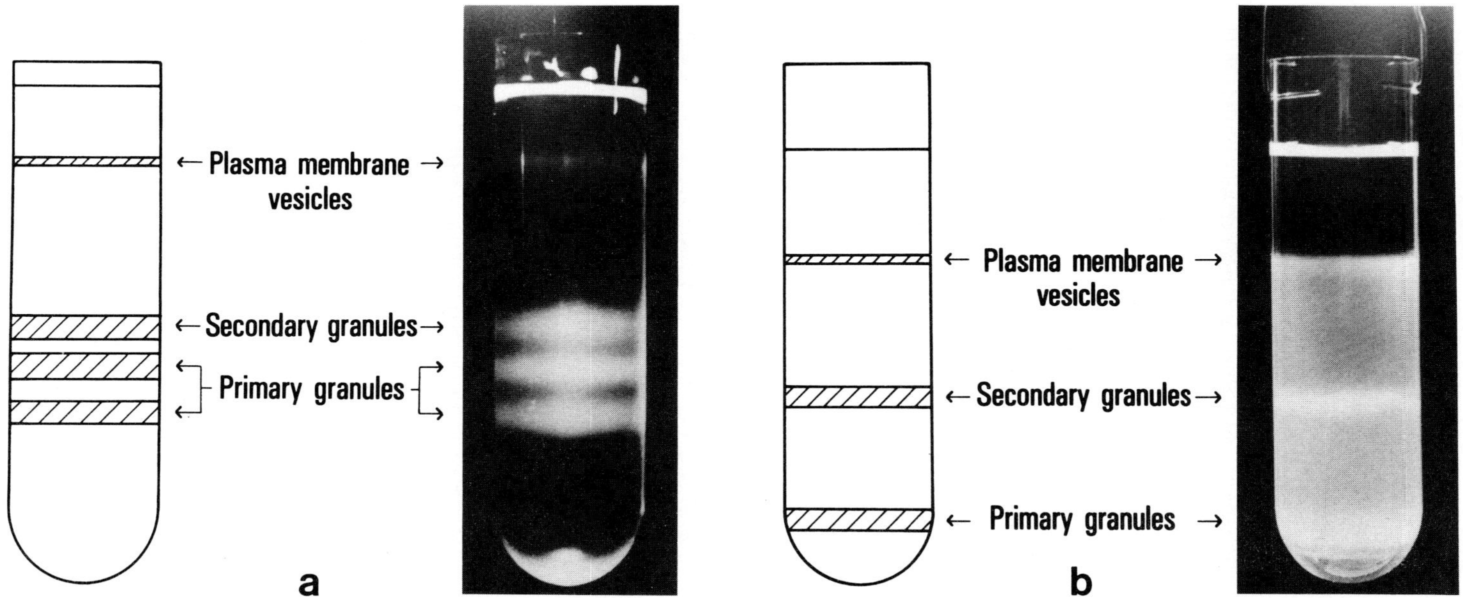

FIG. 2. (a) Subcellular fractionation by continuous sucrose gradient ultracentrifugation. (b) Subcellular fractionation by discontinuous Percoll gradient ultracentrifugation.

High density Percoll solution (1.22 g/ml) made up in cavitation buffer should be used for adjusting the volumes in tubes for balancing. After this high-speed centrifugation, the membrane- or granule-rich material is separated from the Percoll gradient material; the granules or plasma membrane vesicles separate as a layer on top of a hard-packed pellet of Percoll and may be removed by aspiration from above with a Pasteur pipet.

These subcellular fractionation procedures have been shown to be useful for characterizing the granule content of unstimulated and activated neutrophils and for identifying the subcellular localization of neutrophil constituents.[4,5]

Measurement of Neutrophil Granule Constituents (see also this series, Vol. 163 [28])

As is summarized in Table I, a number of enzymes and other proteins have been localized to the primary and/or secondary neutrophil granules, and undoubtedly more will be identified in the future. Elastase (a primary granule constituent) and collagenase Type 1 (a secondary granule constituent), in particular, appear to play important roles in mediating the damage and remodeling of connective tissues which occur during acute inflammation; these enzymes are described in detail elsewhere in this series (Vol. 163 [28] or Vols. 80 or 132). Certain granule constituents (myeloperoxidase, β-glucuronidase, and lactoferrin) have proved to be of particular value as biochemical markers of neutrophil degranulation and secretion, and approaches for measuring these are described below. In studies concerned with human neutrophil degranulation and secretion these granule constituents provide discrete information about the relative mobilization and exocytosis of the two principal neutrophil granule species under the experimental conditions chosen for study.

Primary Granules. As noted above, primary granules are relatively inaccessible for exocytosis. Experimental conditions *in vitro,* which provoke the extracellular release of proteolytic enzymes (such as elastase) contained in these granules, have been closely associated with pathologic conditions *in vivo* that are characterized by connective tissue injury and destruction mediated by inflammatory neutrophils.

β-Glucuronidase (see also this series, Vol. 132 [12]): Measurement of β-glucuronidase in neutrophil lysates and in postsecretory supernatants is carried out by the method of Talalay *et al.*[6] Three reagent solutions are

[4] D. G. Wright and J. I. Gallin, *J. Immunol.* **123,** 285 (1979).

[5] N. Borregaard, J. M. Heiple, E. R. Simons, and R. A. Clark, *J. Cell Biol.* **97,** 52 (1983).

[6] P. Talalay, W. H. Fishman, and C. Huggins, *J. Biol. Chem.* **166,** 757 (1946).

prepared [soln. I, 0.04 *M* acetate buffer (pH 4.5); soln. II, 0.01 *M* phenolphthalein glucuronic acid (Sigma); soln. III, 0.4 *M* glycine buffer (pH 10.5)]. Varying concentrations of phenolphthalein (Sigma) from 2 to 200 μg/ml are prepared by dilution with H_2O for generation of a standard curve. For the assay, 0.1 ml of unknown is mixed with 0.7 ml soln. I and 0.2 ml soln. II in glass tubes (12 × 75 mm) and incubated for 4 to 18 hr in a water bath at 37°. Controls (0.7 ml soln. I plus 0.3 ml 0.85% NaCl, and 0.7 ml soln. I plus 0.2 ml soln. II plus 0.1 ml 0.85% NaCl) and standard phenolphthalein dilutions (0.1-ml aliquots mixed with 0.7 ml soln. I and 0.2 ml soln. II) are incubated in parallel. After incubation 1.0 ml of soln. III is added to each tube, mixed, and the optical density of the mixtures read at 540 nm. The acetate buffer control is used to zero the spectrophotometer. Enzyme activity is measured in terms of the phenolphthalein standard curve and expressed as phenolphthalein released/time of incubation.

Longer incubation times are most useful when assaying samples with low enzyme activity (e.g., cell supernatants); however, the reaction may not be linear after 6–8 hr with samples that have high enzyme activities (e.g., whole cell lysates). When enzyme secretion is expressed as percent total cellular enzyme released, falsely high values may be obtained if total cell enzyme activities are not measured under linear reaction conditions. Normal values for total neutrophil β-glucuronidase are 5–10 μg phenolphthalein released 10^6 neutrophils/4 hr.

Myeloperoxidase (see also this series, Vol. 132 [20]): Myeloperoxidase (MPO) is a heme enzyme that is unique to myeloid cells and their progenitors. This enzyme plays an important role in the microbicidal activity of normal, mature neutrophils, and its activity has been used as the definitive histochemical marker for primary granules.[7] The oxidation of 4-aminoantipyrine by this enzyme in the presence of H_2O_2 is the basis for a functionally specific assay.[8] Reagents for this assay are as follows: 1.7 m*M* H_2O_2 [prepared fresh from 30% H_2O_2 stock solution, a 1:5000 dilution of 30% H_2O_2 in H_2O = 1.7 m*M*] and a 2.5 m*M* solution of 4-aminoantipyrine (Sigma) containing phenol (Sigma) [810 mg phenol is diluted in 40 ml H_2O, followed by the addition of 25 mg 4-aminoantipyrine; this toxic reagent should be stored in the dark at room temperature and handled with care]. Purified MPO (Calbiochem) is used to generate standard curves. To carry out the assay, 1.3 ml of 4-aminoantipyrine/phenol reagent and 1.5 ml of the 1.7 m*M* H_2O_2 are added to a quartz curvette, and the optical

[7] D. F. Bainton and M. G. Farquar, *J. Cell Biol.* **39,** 299 (1968).

[8] S. J. Klebanoff, *in* "The Reticuloendothelial System" (A. J. Sbarra and R. R. Strauss, eds.), Vol. 2, p. 279. Plenum, New York, 1980.

absorbance is recorded continuously in a spectrophotometer at 510 nm. After a basal reaction rate is established, 0.1 ml of the sample to be measured and 0.1 ml of 1.5% (v/v) Triton X-100 are added to the cuvette. The ΔA_{510} is then measured for 4–5 min. Enzyme activity is calculated from the ΔA_{510}/min, taken from the linear portions of reaction curves, in relationship to a standard curve generated with purified MPO. The normal range for total MPO in neutrophils is 0.5–1.0 μg MPO/10^7 cells.

Secondary Granule Constituents. As noted above, mobilization and exocytosis of neutrophil secondary granules is a consequence of most forms of neutrophil activation, and the secretion of secondary (but not primary) granule constituents appears to be an intrinsic component of neutrophil emigration to extravascular spaces.

Lactoferrin: The iron-binding glycoprotein lactoferrin (LF) is specific for neutrophil secondary granules and is a useful marker for secondary granule exocytosis. Its physiologic role is not clear, although antimicrobial functions related to its iron-binding properties have been described. It has also been implicated as a mediator for neutrophil adherence and hydroxyl radical production and as a negative feedback regulator of neutrophil production in the marrow.

Quantitative measurements of lactoferrin may be carried out by ELISA,[9] using microtiter plates (NUNC Immunoplate 1 96F) and rabbit anti-human lactoferrin IgG (Cappel). Stock anti-LF is prepared in sterile PBS at a concentration of 4 mg/ml. A 1:400 dilution of this stock reagent (10 μg/ml) is prepared fresh for each assay in carbonate buffer (pH 9.6; 9.45 g Na_2CO_3 and 16.11 g $NaHCO_3$ in 550 ml H_2O); 10 ml of this reagent is needed for each 96-well immunoplate. Microtiter plates are coated with anti-LF/carbonate buffer reagent, 0.1 ml/well, and then refrigerated (4°) for at least 18 hr before use. Plates are washed 4 times with PBS/Tween [0.5 ml Tween 20 (Sigma) in 1000 ml PBS], 0.3 ml/well, and the plates are inverted and tapped dry. Microtiter wells (inner wells only are used) are inoculated with 0.1-ml aliquots of human lactoferrin standards, 100.0–1.0 ng/ml [prepared from a 10 μg/ml stock solution of purified human lactoferrin (Calbiochem) in PBS, stored at −70°, and diluted in 1% BSA (Sigma) in PBS], 0.1-ml aliquots of 1% BSA/PBS as blanks, and 0.1-ml aliquots of unknown samples which are also diluted at least 1:10 in 1% BSA/PBS. Several dilutions of unknown samples should be assayed (e.g., 1:10, 1:100, 1:500). Plates are incubated for 90 min at 37°, washed 6 times with PBS/Tween reagent, 0.3 ml/well, and tapped dry. Peroxidase-conjugated IgG antilactoferrin (Cappel; stock solution stored at −70° and diluted 1:400 in 1% BSA/PBS fresh for use in assay) is added to the plates, 0.1 ml/well, and the plates are again incubated for 90 min at 37°, followed

[9] S. V. Hetherington, J. K. Spitznagel, and P. G. Quie, *J. Immunol. Methods* **65,** 183 (1983).

by another 6 washes with PBS/Tween, and tapped dry. Peroxidase substrate [0.1-ml aliquots; prepared with 20 mg phenylenediamine (Sigma) and 0.018 ml 30% H_2O_2, 25 ml H_2O, and 25 ml citrate buffer, pH 5.0 (500 ml 0.2 *M* $Na_2HPO_4 \cdot 7H_2O$ added to 472.8 ml 0.1 *M* citric acid)] is added to each column of wells at timed intervals. The plates are then allowed to stand for 15 min. The peroxidase reaction is stopped by adding 0.025 ml 8 *N* H_2SO_4 at the same timed intervals, and the plates read for optical density in a microtiter plate reader at 490 nm, using the buffer controls to blank the instrument. Lactoferrin concentrations in the unknown samples are calculated in relationship to values for the known standards; normal total human neutrophil lactoferrin is 60–80 $\mu g/10^7$ neutrophils.

Neutrophil Granule Exocytosis (see also this series, Vol. 132 [12])

Diverse stimuli, summarized in Table II, have been shown to induce human neutrophil granule exocytosis and secretion of granule constituents *in vitro* (see Ref. 1). Some of these stimuli cause exocytosis of both primary and secondary granules, while others cause secretion of secondary granule constituents primarily or exclusively. Under almost all experimental conditions, other than those which cause neutrophil lysis, the exocytosis of secondary granules is more pronounced than that of primary granules. Indeed, some degree of secondary granule exocytosis occurs spontaneously

TABLE II
STIMULI FOR NEUTROPHIL GRANULE EXOCYTOSIS[a]

Stimulus	Primary granule exocytosis	Secondary granule exocytosis
Phagocytosis of opsonized particles	++	+++
Interaction with surface-bound immune complexes	+++	+++
Streptolysin and other bacterial cytolysins	+++	+++
Nonspecific adherence to highly charged surfaces	±	++
Chemoattractants (e.g., fMLP) together with surface contact	±	++
Monohydroxyeicosatetranoic acids (5- and 12-HETE)	−	++
Ionophore (A23187)	+	+++
Phorbol diesters	−	+++
Lectin (Con A)	−	++
Interleukin-1	−	++
Addition of calcium to CA^{2+}-free incubation media	−	+
Extracellular ATP	−	+

[a] See Ref. 1 for source references.

under most experimental conditions if physiologic calcium concentrations are in the extracellular milieu. While basic mechanisms underlying this difference in the accessibility of primary and secondary granules for fusion with the plasma membrane and exocytosis have not been defined, the extracellular degranulation of these different granule species can be equalized if neutrophils are pretreated briefly (15 min) with cytochalasin B [Sigma, stored at $-20°$ as a stock solution in DMSO, 2.5 mg/ml, and diluted in neutrophil incubation media (e.g., HBSS) for a final concentration of 5 μg/ml], which disrupts cytoplasmic microfilament networks in the neutrophil. Furthermore, cytochalasin B pretreatment converts weak secretagogues, such as *N*-formylated chemotactic peptides [e.g., fMet-Leu-Phe (Sigma), $10^{-8}-10^{-6}$ *M*], into potent secretagogues for both primary and secondary granules.

It is beyond the scope of this chapter to discuss all the experimental conditions that have been shown to induce the mobilization and extracellular release of neutrophil granule contents (for a detailed description of the methods used, see Vol. 132 of this series [12]). However, it is appropriate to make several general comments. First, almost all degranulating stimuli (with the notable exception of phorbol diesters) require the presence of divalent cations, Ca^{2+} and Mg^{2+}, in the extracellular milieu to induce a secretory response. Indeed, the general requirement for divalent cations, (calcium in particular) in neutrophil activation is the underlying reason for using media free of Ca^{2+} and Mg^{2+} in neutrophil isolation procedures, in order to maintain the cells in a resting state before study. Second, neutrophil degranulation in response to an effective stimulus occurs very rapidly; secretory responses can be detected within minutes of exposure of the cells to activating conditions and are usually completed by 15–30 min. Third, the lability of a granule enzyme or protein once released extracellularly, particularly in the presence of reactive oxygen metabolites generated during neutrophil activation, may complicate the detection of secretory responses. For this reason it is appropriate to measure not only total resting cell levels of a granule enzyme or protein, and levels that appear in the extracellular medium, but also total levels recoverable from the cell suspension after activation; substantial losses of total detectable enzyme or protein may occur. A salient example of this problem involves the use of vitamin B_{12}-binding protein as a marker for secondary granule exocytosis. Although this protein is specific for secondary granules, the B_{12}-binding activity of this neutrophil cobalophilin is reduced by exposure to toxic oxygen metabolites.[10]

Another problem relates to the "stickiness" of secreted proteins on the

[10] R. A. Clark and N. Borregaard, *Blood* **65,** 375 (1985).

basis of their charge or other characteristics. Lactoferrin, for example, tends to remain attached to cell surfaces after secondary granule exocytosis. The apparent extracellular release of lactoferrin[11] may be quite different depending on whether or not suspensions of activated cells are diluted with a high molar salt solution (which dissociates surface-bound lactoferrin) before the extracellular media is separated from the cells by centrifugation.

Release of total enzymatic activities from intact neutrophils or granule preparations may be achieved with Triton X-100 (final concentration 0.05–0.1% v/v). However, appropriate controls should be evaluated to rule out independent effects of this detergent on enzyme or protein assays. Finally, in all studies of neutrophil granule exocytosis, the experimental conditions should be evaluated carefully for cell injury or disruption. Dye exclusion methods are not sufficiently sensitive or precise for this purpose. Instead, the extracellular release of cytoplasmic constituents should be investigated. The measurement of lactate dehydrogenase in extracellular incubation media, using standard methods,[12] is an appropriate approach. Under certain experimental conditions, low levels of apparent primary granule exocytosis may be accounted for by damage and lysis of a small proportion of the neutrophils caused by *in vitro* manipulations of the cells under study.

[11] M. S. Klempner, C. Dinarello, and J. I. Gallin, *J. Clin. Invest.* **61,** 1330 (1978).

[12] H. U. Bergmeyer and E. Bernt, *in* "Methods in Enzymatic Analysis" (H. U. Bergmeyer, ed.), p. 737. Academic Press, New York, 1963).

[45] Membrane Attack Complex Proteins C5b-6, C7, C8, and C9 of Human Complement

By ALFRED F. ESSER and JAMES M. SODETZ

Introduction

The cytolytic activity of complement was the first recognized function of the complement system. The five plasma proteins participating in this reaction are C5, C6, C7, C8, and C9, that together form a complex of M_r 1–2 million referred to as the membrane attack complex or MAC.[1–4]

[1] W. P. Kolb, J. A. Haxby, C. M. Arroyave, and H. J. Müller-Eberhard, *J. Exp. Med.* **135,** 549 (1972).

Assembly of the MAC starts with the cleavage of C5 into two fragments, C5a and C5b. Then, without further enzymatic action, a complex is produced with the probable composition of $(C5b, 6, 7, 8)_1(C9)_n$, where n usually varies between 1 and 6 when the complex is formed on natural membranes. However, n can be as high as 15 when C9 is allowed to polymerize and provided that sufficient C9 molecules are available. In the absence of membrane targets, the forming complex may bind to plasma inhibitors. To date three such inhibitors have been identified: lipoproteins, antithrombin III, and the so-called S protein; the latter is the most efficient and is identical with a "spreading protein" called vitronectin.[4-6] In the SC5b-9 complex it appears that n always equals 3. The following scheme summarizes the known reactions:

$$\text{C5b} + \text{C6} \longrightarrow \text{C5b-6} \qquad (1)$$

$$\begin{array}{lllll} & \xrightarrow{+\text{S}} \text{SC5b-7} & \longrightarrow & \text{SC5b-8} \longrightarrow & \text{SC5b-9}_3 \\ \text{C5b-6} + \text{C7} \rightarrow \text{C5b-7*} & & +\text{C8} & +n\,\text{C9} & \\ & \xrightarrow{+\text{Membrane}} \text{MC5b-7} & \longrightarrow & \text{MC5b-8} \longrightarrow & \text{MC5b-9}_n \end{array} \qquad (2)$$

$$n\,\text{C9} \xrightarrow{\text{Zn}^{2+}} \text{poly(C9)}_n \qquad (3)$$

* metastable binding site

Reaction (1) can be easily carried out *in vitro* by activation of acute-phase serum or more efficiently by activation of C7-deficient serum. This allows the isolation of active C5b-6 which then can be used to study Reaction (2) under precise conditions chosen by the experimenter.[7] In addition, it is now also known that C9 can aggregate to form a tubular polymerization

---

[2] Complement proteins are named in accordance with recommendations in *Bull. W.H.O.* **39,** 935 (1986). Conductance of various buffers mentioned in this chapter was measured at room temperature and is stated as specific conductivity ($\text{mS} = \text{m}\Omega^{-1}\ \text{cm}^{-1}$). Abbreviations: C, complement; Ab, antibody; E, erythrocytes; EA, antibody-sensitized erythrocytes; EAC1-7, EAC1-8, and EAC1-9, EA bearing C proteins C1 through C7, or C1 through C8, or C1 through C9, respectively; MAC, membrane attack complex; R-7, R-8, and R-9, serum immunochemically depleted of C7, C8, or C9, respectively; VB, isotonic veronal-buffered saline (145 m*M* NaCl, 5 m*M* veronal, pH 7.4); VB-M, VB plus 0.15 m*M* $CaCl_2$ and 0.5 m*M* $MgCl_2$; GVB-M, VB-M plus 0.1% gelatin; GVB-E, VB plus 10 m*M* EDTA and 0.1% gelatin; PEG, polyethylene glycol; EDTA, ethylenediamine tetraacetic acid; PMSF, phenylmethylsulfonyl fluoride.

[3] S. Bhakdi, P. Ey, and B. Bhakdi-Lehnen, *Biochim. Biophys. Acta* **419,** 445 (1976).

[4] H. J. Müller-Eberhard, *Annu. Rev. Immunol.* **4,** 503 (1986).

[5] E. R. Podack, W. P. Kolb, and H. J. Müller-Eberhard, *J. Immunol.* **120,** 1841 (1978).

[6] D. Jenne and K. K. Stanley, *EMBO J.* **4,** 3153 (1985).

[7] P. J. Lachmann and R. A. Thompson, *J. Exp. Med.* **131,** 643 (1970).

product referred to as poly(C9) [Reaction (3)]; this reaction is catalyzed by metal ions such as Zn^{2+} and favored at low ionic strength.[4] Because of the striking similarity in appearance of poly(C9), the extracted MAC, and the classical complement lesion when viewed in the electron microscope,[8-10] the proposal has been made that the MC5b–8 complex catalyzes the oligomerization of C9 into a tubular structure that is identical with the complement lesion and is responsible for the cytolytic activity of complement.[8,9]

Not all MACs assembled on a target cell are cytolytic,[11] and sensitive bacteria can mutate to become resistant to the action of the MAC.[12] Traditionally, the easiest target cells to use to study the lytic action of complement have been antibody-coated erythrocytes,[13] and these targets still provide the best assay method for MAC function. However, one should be aware of the fact that erythrocytes are generally resistant to *homologous* complement,[14] and also of the growing list of reports indicating that activation of specific cell responses *in the absence of lysis* may be an important function of the membrane attack pathway.[15-18] Because such cell activation processes, e.g., changes in intracellular calcium levels and release of arachidonic acid or reactive oxygen metabolites, play an important role in inflammation, it is timely to provide some information on isolation of the individual proteins or complexes required to assemble this pathway *in vitro* and thus encourage further studies on nonlytic or sublytic functions of MAC and its precursor complexes.

Methods of Assay and Isolation

Our methodology for isolation of the terminal complement proteins is based on plasma as the starting source for complement proteins. However,

[8] E. R. Podack and J. Tschopp, *Proc. Natl. Acad. Sci. U.S.A.* **79,** 574 (1982).
[9] J. Tschopp, H. J. Müller-Eberhard, and E. R. Podack, *Nature (London)* **298,** 534 (1982).
[10] J. Tranum-Jensen, S. Bhakdi, B. Bhakdi-Lehnen, O. J. Bjerrum, and V. Speth, *Scand. J. Immunol.* **7,** 45 (1978).
[11] M. M. Frank, R. R. Dourmashkin, and H. J. Humphrey, *J. Immunol.* **104,** 1502 (1970).
[12] K. A. Joiner, *Curr. Top. Microbiol. Immunol.* **121,** 99 (1985).
[13] M. M. Mayer, *in* "Experimental Immunochemistry" (E. A. Kabat and M. M. Mayer, eds.), 2nd Ed., p. 131. Thomas, Springfield, Illinois, 1961.
[14] J. J. Houle and E. M. Hoffmann, *J. Immunol.* **133,** 1444 (1984).
[15] A. K. Campbell, R. A. Daw, M. B. Hallett, and J. P. Luzio, *Biochem. J.* **194,** 551 (1981).
[16] P. A. Roberts, B. P. Morgan, and A. K. Campbell, *Biochem. Biophys. Res. Commun.* **126,** 692 (1985).
[17] G. M. Hänsch, M. Seitz, G. Martinotti, M. Betz, E. W. Rauterberg, and D. Gemsa, *J. Immunol.* **133,** 2145 (1984).
[18] T. Wiedmer and P. J. Sims, *J. Biol. Chem.* **260,** 8014 (1985).

we have also used Cohn Fraction III as a source for isolation of C8[19] and C9.[20] The bimolecular C5b-6 complex and proteins C7 and C9 can be isolated conveniently from one batch of fresh or unexpired frozen plasma, thus conserving valuable material, but C8 is purified best from Cohn Fraction III. Others have also described a method for the isolation of multiple human complement components from fresh plasma.[21]

Assays

In general, all assays use hemolysis to determine activity of the individual proteins. C5b-6 activity is tested with regular erythrocytes (E), either from chicken or sheep, and C8 and C9 activities are tested with cell intermediates such as EAC1-7 or EAC1-8 and addition of the subsequently acting proteins required to complete the reaction. C7, however, is assayed most easily by measuring its ability to restore the hemolytic function of C7-deficient serum. Because of slightly different methodologies involved in assaying each protein, the individual assays are described more fully in the sections on isolation of each protein. In general, however, all hemolytic activities are expressed by determining from a graph the dilution of a test sample that lyses 50% of the target cells, e.g., if a certain sample causes 50% cell lysis at a 50,000-fold dilution, the original undiluted sample is said to contain 50,000 CH_{50} units.[13] Unfortunately, the hemolytic activity is strongly dependent on the species of origin and the age of the target cells. In order to minimize problems associated with different target structures, it is best to prepare pooled human serum and construct a standard curve of hemolytic activities of individual complement proteins in the pooled serum on each target cell against which the activity of the test sample can then be compared. A complement standard (normal human serum in which all terminal complement proteins have been assayed) is also available from Cytotech (San Diego, CA).

Activities are also frequently expressed in the form of "effective molecules" or Z values, where $Z = -\ln(1 - Y)$ and Y is the fraction of cells lysed. Thus, $Z = 1$ corresponds to lysis of 63% of the target cells and, assuming a Poisson distribution, each lysed cell has received one effective molecule. This measurement has the added advantage that it provides some indication of activities in molecular terms since one can estimate how many molecules are required to lyse one cell. To be valid, however,

[19] E. W. Steckel, R. G. York, J. B. Monahan, and J. M. Sodetz, *J. Biol. Chem.* **255,** 11997 (1980).

[20] J. R. Dankert, J. W. Shiver, and A. F. Esser, *Biochemistry* **24,** 2754 (1985).

[21] C. H. Hammer, G. H. Wirtz, L. Renfer, H. D. Greshman, and B. F. Tack, *J. Biol. Chem.* **256,** 3995 (1981).

both assay forms require that the test be conducted under "single hit" conditions, i.e., doubling or tripling of the number of target cells, while keeping all other conditions constant, should have no effect on number of cells that are lysed, and hemolysis must be measured at end point (see Gee[22] for further discussion).

Standard Reagents

Buffers. The preparation of all common buffers used in complement research has been described before in this series by Gee[22] and that article should be consulted for details. These buffers are abbreviated as indicated below, with Gee's abbreviations given in square brackets.

VB [BBS-M]: isotonic veronal-buffered saline containing 145 m*M* NaCl, 5 m*M* veronal (barbital), pH 7.4

VB-M: VB plus 0.15 m*M* $CaCl_2$ and 0.5 m*M* $MgCl_2$

GVB-M [BBS-GM]: VB-M plus 0.1% gelatin

GVB-E [BBS-G-EDTA]: VB plus 10 m*M* EDTA and 0.1% gelatin

Erythrocytes (E). Sheep and chicken erythrocytes can be purchased from Colorado Serum Co. (Denver, CO). For standardization it is assumed that complete lysis of 1×10^7 cells in 1 ml yields hemoglobin supernatants with light absorption values at 412 nm (A_{412}) of 0.87 for sheep cells and 1.67 for chicken cells (light path 1.0 cm).

EA and cell intermediates. Antibody-coated erythrocytes (EA) are prepared as described by Gee[22]. EAC1–7 or EAC1–8 intermediates are prepared by incubation of EA with C8 or C9-deficient sera, respectively. Briefly, 50 volumes of EA (5×10^8 cells/ml) are incubated with 1 volume of deficient serum (prepared as described below or purchased from Cytotech) and incubated for 30 min at 37°. The cells are washed 5 times in GVB and restandardized to 5×10^8 cells/ml. EA, EAC1–7, and EAC1–8 are also available from Diamedix (formerly Cordis Laboratories, Miami, FL).

Zymosan (Sigma Chemical Co., St. Louis, MO) for complement activation is prepared by boiling 10 g in 100 ml of distilled water for 30 min. The suspension is cooled, centrifuged and the pellet washed 3 times with water. A 5% (w/v) suspension in water is prepared and stored frozen.

Deficient sera (R-sera). Human sera deficient in C7, C8, or C9 are prepared by immunoadsorbtion of serum on solid-phase anti-C7, anti-C8, or anti-C9 IgG.[20] The respective antibodies are commercially available from several companies (Atlantic Antibodies, Scar-

[22] A. P. Gee, this series, Vol. 93, p. 339.

borough, MA; Calbiochem-Behring, La Jolla, CA; Cytotech) or can be raised in rabbits or goats. We have used the following general immunization schedule. *Rabbits*: First injection is a total of 0.15 ml of protein in Freund's complete adjuvant (CFA) at 1 mg/ml into rear footpad; second and third injections at 10-day intervals of 0.15 ml of protein in Freund's incomplete adjuvant (IFA) at 1 mg/ml in rear footpad; subsequent injections of protein in IFA are made at monthly intervals and are intramuscular. *Goats*: First injection is a total of 0.6 ml of protein in CFA (1 mg/ml) at three intramuscular sites (0.2 ml each; front and rear leg muscles); second and third injections of 3×0.2 ml of protein in IFA (1 mg/ml) are made at 3-week intervals; subsequent injections (in IFA) are made monthly. Titers and specificity of the antisera are assessed by standard immunoelectrophoretic techniques. Other immunization schedules, however, have been used without major differences in antiserum titers or specificity. IgG antibodies are isolated as described[20] and coupled at 10 mg/ml to Sepharose CL-4B activated with cyanogen bromide (CNBr) as detailed by Nishikawa and Bailon.[23] Pooled human plasma is clotted at 37° for 1–2 hr by addition of Ca^{2+} to a final concentration of 20 m*M*, centrifuged, and the serum is removed. The immunoadsorbent columns are equilibrated in VB-E containing 0.5 *M* NaCl and the serum is made 0.5 *M* in NaCl to prevent binding of C1q to the column. One volume of gel usually depletes about 4 volumes of plasma. After passage of serum through the respective columns at 4°, the depleted sera are dialyzed against two changes of 10 volumes of VB, concentrated to the original serum concentration, and stored frozen at −70°. Wear gloves when handling all human plasmas and sera.

Removal of Clotting Enzymes from Plasma

Plasma batches used in isolation of proteins are usually depleted first of clotting enzymes. Vitamin K-dependent enzymes are removed by precipitation with $BaCl_2$. Four volumes of a stock solution of 1.0 *M* $BaCl_2$ are added to 100 volumes of plasma. After incubation at 4° with constant stirring, the precipitate is pelleted by centrifugation at 16,000 *g* for 30 min.

Plasminogen is removed from $BaCl_2$-treated plasma by passage through a L-lysine–Sepharose 4B column.[24] This column is prepared by coupling 200 g of L-lysine to 1 liter of packed CNBr-activated Sepharose CL-4B.[23]

[23] A. H. Nishikawa and P. Bailon, *Anal. Biochem.* **64,** 268 (1975).

[24] D. G. Deutsch and E. T. Mertz, *Science* **170,** 1095 (1970).

One volume of beads is sufficient to remove plasminogen from about 3 volumes of plasma. It is convenient to pass the effluent of the lysine–Sepharose column directly over the next column used in the isolation procedure, e.g., the solid-phase anti-C7, the Sephadex CM-50, or the DEAE-Sephacel columns in the C5b-6, C8, or C9 isolation procedures, respectively, by linking the two columns together in series. All buffers used in the second column step are suitable for plasminogen removal except Tris buffers which should be avoided.[24]

C5b–6 Purification

Introduction

The C5b–6 complex was first recovered by precipitation of euglobulin from acute-phase serum by Lachmann and Thompson.[7] Because of the difficulty in obtaining such sera and the presence of C7 in these sera which leads to formation of SC5b–9, it is better to use activated C7-deficient plasma or serum.[5] In sera devoid of C7, complement activation stops after formation of C5b–6, no other terminal complexes are formed, and thus the yield of C5b–6 is greatly increased.

C5b–6 Assay

The reaction mixture consists of 0.1 ml sheep E (5×10^9 cells/ml) in GVB-E, 10 μl guinea pig serum, and 10 μl test samples containing between 0 and 25 μg C5b–6. After incubation at 37° for 30 min, the reaction is stopped by addition of ice-cold 0.15 *M* NaCl, and unlysed cells are pelleted by centrifugation. Absorbance of the free hemoglobin in the supernatant is measured at 541 nm when hemolysis is high or at 412 nm when the final degree of hemolysis is low. Activities are expressed as percent lysis since C5b–6-mediated lysis is dependent on the cell concentration[5] and, therefore, the *Z* value notation should not be used. The efficiency to lyse sheep E is very low in this system but is considerably higher for chicken E. Thus, small amounts of C5b–6 can be assayed better with chicken E.

C7 Depletion

Freshly prepared or unexpired frozen plasma (about 250 ml) is treated with $BaCl_2$, made 10 m*M* in EDTA and 0.5 *M* in NaCl, and then passed consecutively through lysine–Sepharose and anti-C7 Sepharose columns (100 ml each) coupled together and equilibrated with VB-E, 0.5 *M* NaCl. The colored effluent (~0.4 liters) is collected and used as C7-depleted plasma. The columns are washed further with VB-E, 0.5 *M* NaCl until A_{280} remains unchanged. Since the bound C7 is used for C7 purification (see below) it is best to wash the anti-C7 column overnight to remove as many impurities as possible.

Isolation Procedure

A significant advancement in the original isolation procedure for C5b–6[5] was made by Podack and co-workers,[25,26] who noticed that C5b–6 binds to a lysine–Sepharose column and that it can be crystallized to yield a highly pure preparation. However, the crystalline material does not dissolve easily and, therefore, is difficult to use in studies where defined amounts of C5b–6 protein are required. In addition, not all lysine–Sepharose gels bind C5b–6 with sufficient capacity to be useful; sometimes three or four different batches have to be prepared, and it is not clear why some work and some do not. The basis for C5b–6 binding to the lysine–Sepharose column is unknown but is probably related to the weak ion-exchange capacity of this material. For these reasons we have developed an alternative method that does not use this column nor crystallization; however, the yield of pure C5b–6 is lower, and in cases where yields are of major importance the earlier procedures should be considered.

C5b–6 Formation and Plasma Fractionation. The recovered C7-depleted plasma is made 3 m*M* in $CaCl_2$ and 10 m*M* in $MgCl_2$ and incubated with zymosan (10 mg/ml) at 37° for 60 min. Zymosan is then removed by centrifugation, washed once with 50 ml of 2 *M* NaCl, both supernatants are combined, and protease inhibitors are added [10 m*M* EDTA, 1 m*M* phenylmethylsulfonyl fluoride (PMSF), and 5 m*M* benzamidine]. The activated plasma is fractionated by precipitation with polyethylene glycol (PEG) 4000 by slowly adding, with stirring, powdered PEG to a final concentration of 4% (w/v). The suspension is gently stirred for about 1 hr. The precipitate is removed by centrifugation (14,000 *g*, 30 min), and the PEG concentration in the supernatant is brought to 8% (w/v). After stirring for 1 hr the precipitate is collected and resolubilized in buffer (75 m*M* Tris–HCl, 50 m*M* NaCl, 1 m*M* PMSF, pH 8.0). The supernatant can be used to isolate C9 as described later.

Ion-Exchange Chromatography. A glass column (0.1 × 20 cm) (Omni column; purchased from Bodman Chemicals, Doraville, GA) is slurry packed with Synchroprep AX-1000 (Synchrom, Linden, IN) ion-exchange material (30-μm bead size, 100-nm pore size) and equilibrated with 75 m*M*

[25] E. R. Podack and H. J. Müller-Eberhard, *J. Immunol.* **124,** 332 (1980).

[26] E. R. Podack, A. F. Esser, G. Biesecker, and H. J. Müller-Eberhard, *J. Exp. Med.* **151,** 301 (1980).

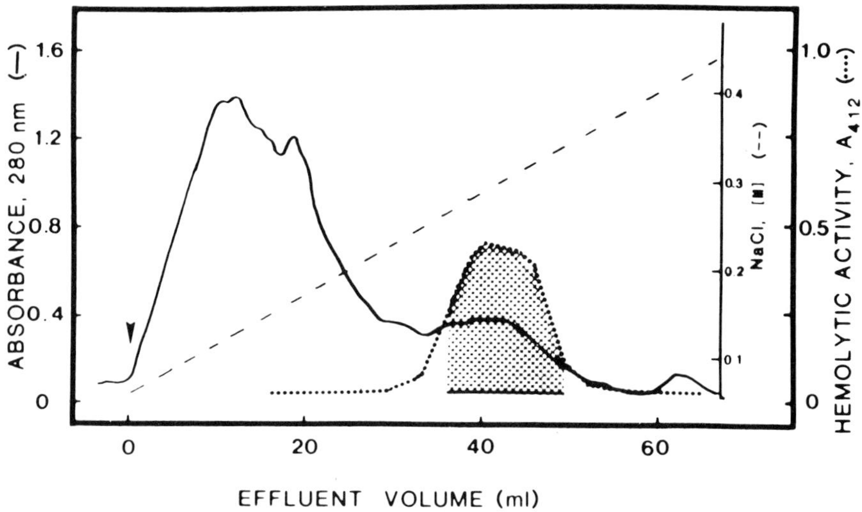

FIG. 1. FPLC anion-exchange chromatography of resolubilized 4–8% PEG 4000 precipitate of activated C7-depleted serum. The C5b–6-containing solution was chromatographed on Synchropak AX-1000 as described in the text. The shaded fractions were pooled for further purification.

Tris–HCl containing 50 m*M* NaCl and 1 m*M* PMSF, pH 8.0. The solubilized 4–8% PEG fraction is pumped onto the column at a flow rate of 1.0 ml/min and then washed with the same buffer until the A_{280} of the effluent reaches baseline values. The column is developed with an 80-ml linear salt gradient. The limit buffer is identical in composition to the starting buffer, except that it contains 1.0 *M* NaCl. The elution is completed by washing the column with 10 ml of 2 *M* NaCl, 75 m*M* Tris, pH 7, followed by 50 ml of 75 m*M* Tris, pH 7. This silica-based column should not be exposed to high pH for long periods of time and should be stored in neutral buffers or water.

The results of this step are shown in Fig. 1. C5b–6 activity is followed by hemolysis of sheep E as described above.

Chromatofocusing. The pooled fractions containing the C5b–6 activity from the AX-1000 chromatography step are diluted with 75 m*M* Tris–HCl, pH 8.0, to a final concentration of about 50 m*M* NaCl. An Omni glass column (0.1 × 20 cm) is slurry-packed with Synchropak AX-300 (10-μm bead size, 30-nm pore size) and equilibrated with start buffer (75 m*M* Tris, pH 8.0). After washing with start buffer until the A_{280} reaches baseline values, the column is eluted with Polybuffer 74 (Pharmacia) diluted 12-fold and titrated to pH 4.0 with HCl. As shown in Fig. 2 the C5b–6 activity elutes between pH 6.2 and 5.5. The eluent is collected in

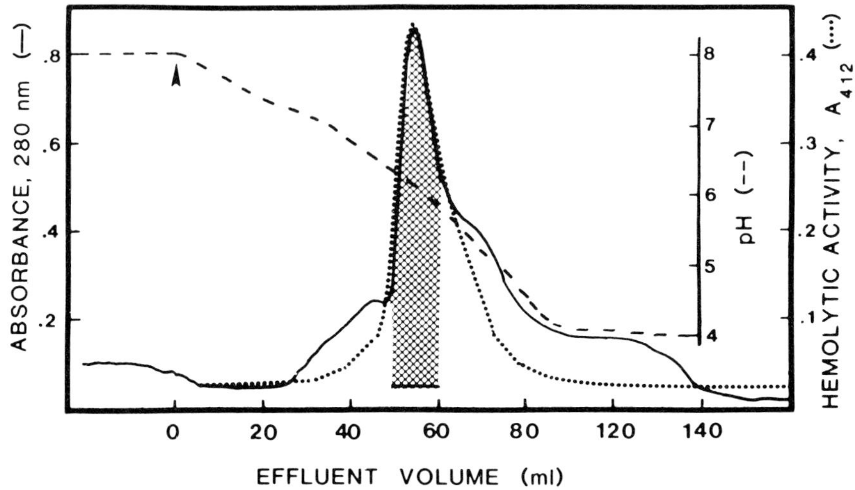

FIG. 2. Chromatofocusing of C5b–6. The pooled fractions shown in Fig. 1 were applied to a Synchropak AX-300 column and eluted with 1 : 12 diluted Polybuffer 74 (pH 4.0). The shaded fractions were pooled.

2-ml fractions in test tubes containing 0.2 ml of 3.3 *M* NaCl. The salt is necessary to prevent precipitation of C5b–6. The active fractions are then pooled and concentrated by vacuum dialysis in a MicroProDiCon apparatus (Bio-Molecular Dynamics, Beaverton, OR).

Gel Filtration. The purity of the C5b–6 preparation is evaluated by sodium dodecyl sulfate–polyacrylamide gel electrophoresis (SDS–PAGE) analysis. If necessary an additional gel filtration step on Superose 12 (Pharmacia) or on BioGel A-1.5m (Bio-Rad) is performed. The resulting protein complex consists of C5b and C6 which upon reduction produces three major bands of approximately M_r 115,000 (C6), 110,000 (C5bα), and 76,000 (C5bβ) and a minor band of M_r 27,000 (C5bα_2) which is presumably derived from C5bα. The C5α_1 fragment ($M_r \sim 76{,}000$) is masked by the C5bβ band.[25,27] C5b–6 is best stored in buffered (10 m*M* HEPES) 0.3 *M* NaCl in 50% glycerol at $-20°$; it is also very stable at 4° in the crystalline form in azide-containing buffers.

Values of $D = 3.22 \times 10^{-7}$ cm²/sec for the diffusion coefficient and of 11.5 S for the sedimentation (*s*) rate have been reported for C5b–6.[5] Since the published extinction coefficients $\epsilon^{1\%}_{1\,\text{cm}}$ at 280 nm are 10.9 for C5b[28] and

[27] K. I. Yamomoto and H. Gewurz, *J. Immunol.* **120,** 2008 (1978).

[28] R. G. DiScipio, C. A. Smith, H. J. Müller-Eberhard, and T. E. Hugli, *J. Biol. Chem.* **258,** 10629 (1983).

10.8 for C6,[29] an extinction coefficient of 10.8 can be used for C5b–6. Another reported $\epsilon_{1\,cm}^{1\%}$ value of 17.1[30] for C6 is most probably in error. The elution of C5b–6 between pH 6.2 and 5.5 from the chromatofocusing column is consistent with reported p*I* values of around 4.9 for C5b[28] and 6.4 for C6.[30]

C7 Purification

Introduction

The seventh component of complement (C7) is thought to be related to the sixth component (C6) because of similar physical and chemical properties and because family studies on genetic polymorphisms of C6 and C7 indicate a close linkage between the loci for both genes.[30,31] This close similarity makes it difficult to separate C6 and C7 using classic purification schemes, and immunoaffinity techniques are, therefore, preferred.[29,30] In studies where C7 is used together with C5b–6 to assemble the MAC, contamination of C7 by C6 is not a problem, and the purification procedure outlined below yields C7 suitable for such experiments. If monospecific anti-C7 affinity adsorbents are used, then C6 contamination can be avoided completely.

C7 Assay

Isotonic C7-depleted serum, reconstituted with calcium and magnesium ions (R-7) is prepared as described above. One-tenth milliliter EA (5×10^8 cells/ml), 0.2 ml GVB-M, 10 μl R-7, and 10- to 40-μl of the appropriately diluted test samples are incubated together at 37° for 60 min, the reaction is stopped by addition of ice-cold GVB-E to a total of 2 ml, and the mixture is centrifuged to pellet unlysed red cells. Hemolysis is measured in either CH_{50} units or expressed in relative A_{412} values.

Isolation Procedure

Immunoaffinity Chromatography. Plasma depleted of vitamin K-dependent enzymes and of plasminogen is passed through a solid-phase anti-C7 column as described above in the C5b–6 purification procedure. The column is washed with VB-E containing 0.5 *M* NaCl and 1 m*M*

[29] W. P. Kolb, L. M. Kolb, and J. R. Savary, *Biochemistry* **21,** 294 (1982).

[30] E. R. Podack, W. P. Kolb, A. F. Esser, and H. J. Müller-Eberhard, *J. Immunol.* **123,** 1071 (1979).

[31] P. J. Lachmann and M. J. Hobart, *J. Immunol.* **120,** 1781 (1978).

PMSF until the A_{280} reaches baseline levels, and the adsorbed C7 is then eluted with a solution of 4 *M* guanidine–HCl containing 2 *M* NaCl at neutral pH. C7 hemolytic activity can be detected in diluted column fractions provided the guanidine concentration is less than 10 m*M*. C7-containing fractions from the anti-C7 column are pooled and dialyzed against 3×10 volumes of a buffer containing 1 m*M* PMSF and 40 m*M* Tris–HCl (pH 7.2; conductance < 5 mS).

The success of this affinity purification step is strongly dependent on the quality of the anti-C7 antibody preparation. To obtain monospecific antibodies we have found it convenient to remove antibodies of unwanted specificity by passing the anti-C7 IgG preparation through a Sepharose 4B column to which C7-depleted serum (R-7) had been coupled. R-7 prepared with the aid of monoclonal anti-C7 is commercially available from Cytotech.

Ion-Exchange Chromatography. Glycerol is added to a final concentration of 25% (v/v) to the C7 pool from the previous step and the solution is applied (60 ml/hr) to a column of DEAE-Sephacel (2.6×30 cm) equilibrated with the dialysis buffer (see above) to which 25% glycerol had been added. The column is washed with the starting buffer until the A_{280} reaches baseline values and then developed with a linear salt gradient. The limit buffer (1 liter) is identical in composition to the starting buffer (1 liter) except that it also contains sufficient NaCl to raise the conductance to 25 mS. Figure 3A shows the chromatogram obtained with this column. Conservative pooling is required at this step to obtain functionally pure C7.

Gel Filtration. An additional gel filtration step on Sephacryl S-200 is necessary to obtain C7 of good purity. The pooled fractions from the DEAE-Sephacel column are either precipitated with 20% (w/v) PEG 4000 and redissolved or concentrated by vacuum dialysis (MicroProDicon, Bio-Molecular Dynamics). The concentrated C7 sample is then chromatographed on Sephacryl S-200 equilibrated with 10 m*M* HEPES, 140 m*M* NaCl, 0.02% sodium azide. The C7-containing fractions (Fig. 4) are pooled, glycerol to a final concentration of 50% (v/v) is added, and the protein is stored at $-20°$.

If smaller quantities of starting material are used then it is better to purify C7 on Synchropak Q-300 (Synchrom) ion-exchange material (10 μm bead size, 30-nm pore size) slurry-packed into an Omni glass column (0.1×20 cm) that is equilibritted with 25 m*M* Tris buffer, pH 7.6. Of course, the pH of the dialysis buffer after the affinity chromatography step (see above) should have been raised to 7.6. The elution profile is shown in Fig. 3B. The pooled and concentrated C7 sample is then sieved on Superose 12 (Pharmacia FPLC system) to yield highly purified C7. For

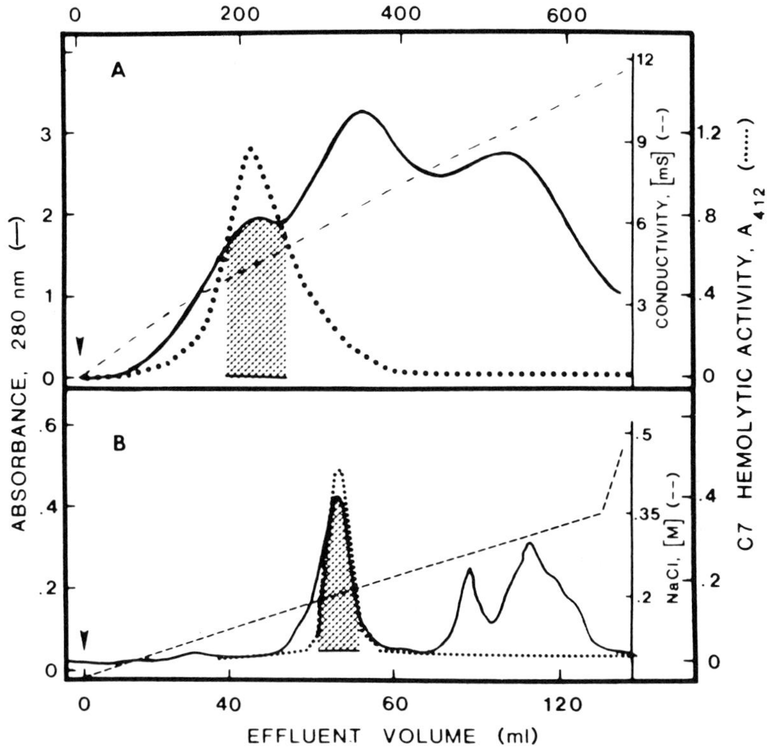

FIG. 3. Ion-exchange chromatography of C7. (A) Recovered fractions from the C7 immunoaffinity chromatography step were applied to a DEAE-Sephacel column and chromatographed as described in the text. (B) Synchropak Q-300 elution profile of C7 recovered from the anti-C7 column. The shaded fractions were pooled for further purification as described in the text.

certain applications, the final product should be tested for esterolytic activity that may be associated with C7[32] and also C6[29] and which may be caused by residual thrombin activity.[4] The protein can be stored in phosphate-buffered saline/50% glycerol at −20° or frozen at −70° for several months without significant loss of hemolytic activity.

According to DiScipio and Gagnon[33] the protein consists of about 795 amino acids and 6.4% carbohydrate with a calculated molecular weight of 92,100; M_r values of 92,400 and 105,000 were obtained by sedimentation equilibrium and by SDS–PAGE analysis, respectively. The same authors

[32] L. J. Floyd, J. R. Savary, and W. P. Kolb, *Fed. Proc., Fed. Am. Soc. Exp. Biol.* **42,** 1948 (1983).

[33] R. G. DiScipio and J. Gagnon, *Mol. Immunol.* **19,** 1425 (1982).

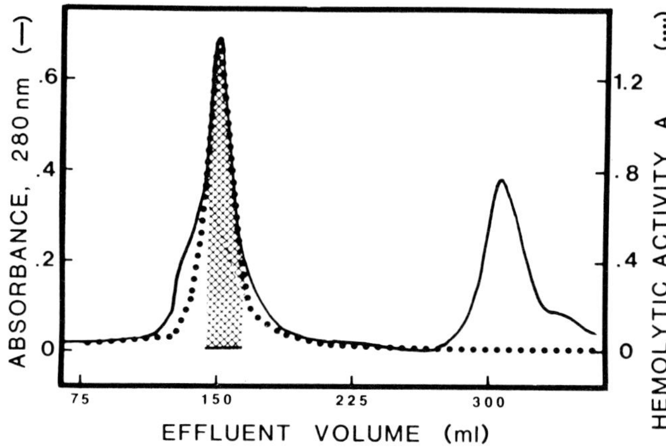

FIG. 4. Gel filtration of C7. The shaded fractions shown in Fig. 3A were pooled, concentrated, and chromatographed on Sephacryl S-200.

published the following physical parameters for C7: $s_{20,w} = 5.6$ S; $d_{20,w} = 5.3 \times 10^{-7}$ cm²/sec; frictional ratio $f/f_0 = 1.36$, and $\epsilon^{1\%}_{1\,cm} = 9.9$ at 280 nm. An extinction coefficient of 19.2 has also been reported[30]; however, since the protein contains only 1.5 tryptophan and 3.2 tyrosine residues per 100 residues it is unlikely that C7 should have such high absorption values.

C8 Purification

Introduction

Human C8 purified from blood has an unusual subunit structure consisting of an α (M_r 64,000), β (M_r 64,000), and γ (M_r 22,000) chain.[19,34] These subunits are present in equimolar amounts and are arranged as a disulfide-linked $\alpha-\gamma$ dimer that is noncovalently associated with β. Studies of C8 deficiencies in humans suggest that separate genes may exist for $\alpha-\gamma$ and β and that noncovalent association occurs postsynthetically. Supporting evidence for this hypothesis includes the rather atypical subunit structure of C8 and the fact that two types of C8 deficiencies have been described, one which is characterized by an abnormality in β[35] and the other by an abnormality in $\alpha-\gamma$.[36] The recent characterization of α[37] and

[34] W. P. Kolb and H. J. Müller-Eberhard, *J. Exp. Med.* **141**, 724 (1975).

[35] J. Tschopp, A. F. Esser, T. J. Spira, and H. J. Müller-Eberhard, *J. Exp. Med.* **154**, 1599 (1980).

[36] F. Tedesco, P. Densen, M. A. Villa, B. H. Petersen, and G. Sirchia, *J. Clin. Invest.* **71**, 183 (1983).

[37] A. G. Rao, O. M. Z. Howard, S. C. Ng, A. S. Whitehead, H. R. Colten, and J. M. Sodetz, *Biochemistry* **26**, 3556 (1987).

β[38] cDNA clones isolated from a human liver cDNA library also supports this hypothesis. Sequencing of these clones has yielded the complete amino acid sequence of α and β.

Human C8 contains a number of distinct functional domains. The β subunit mediates binding of C8 to C5b–7 as evidenced by the fact that it is the only purified subunit that exhibits a high affinity for this complex.[39] The α subunit has at least four distinct functional domains. One is involved in a direct interaction with β and thereby facilitates $\alpha-\gamma$ and β association to form C8.[40] A second domain interacts with γ, regardless of whether the disulfide linkage is intact.[41] A third domain inserts into the target membrane bilayer as C5b–8 is formed and remains so in C5b–9.[42] Such insertion is probably responsible for the low but significant lytic activity of C5b–8 and is likely essential for C5b–9 function. A fourth domain has been identified from studying interactions between C8 subunits and C9 in solution.[43] Results indicate that a single C9 binding site exists on α. This suggests that recognition of C9 and binding to C5b–8 requires direct association with α.

C8 Assays

Assays for C8 use EAC1–7 intermediates prepared as described above or purchased from Diamedix. These cells specifically bind C8 and lyse in proportion to the amount present when excess C9 is added. Assay mixtures are prepared by incubating the following: 0.05 ml of C8 sample diluted with buffer described below; 0.05 ml of human C9 (100 CH_{50} units/ml or 10 ng C9 purified as described below or purchased from Cytotech); 0.1 ml of buffer containing 5 m*M* imidazole, 72.7 m*M* NaCl, 0.15 m*M* $CaCl_2$, 0.5 m*M* $MgCl_2$, 2.5% glucose, and 0.05% gelatin, pH 7.3; 0.05 ml of washed EAC1–7 cellular intermediates diluted to 1×10^8 cells/ml. After 90 min at 37°, 0.5 ml of 0.15 *M* NaCl is added, the solutions are centrifuged, and the extent of lysis is measured by absorbance at 412 nm. Activity is determined from a standard curve prepared by substituting dilutions of pooled normal serum for C8.

Each lot of pooled serum can be standardized by the assay described above as follows. For each dilution of serum, Z values are calculated from $Z = -\ln(1 - Y)$ where Y is the fraction of total cells lysed after correction

[38] O. M. Z. Howard, A. G. Rao, and J. M. Sodetz, *Biochemistry* **26,** 3565 (1987).
[39] J. B. Monahan and J. M. Sodetz, *J. Biol. Chem.* **256,** 3258 (1981).
[40] A. Brickner and J. M. Sodetz, *Biochemistry* **23,** 832 (1984).
[41] A. Brickner and J. M. Sodetz, *Biochemistry* **24,** 4603 (1985).
[42] E. W. Steckel, B. E. Welbaum, and J. M. Sodetz, *J. Biol. Chem.* **258,** 4318 (1983).
[43] J. L. Stewart and J. M. Sodetz, *Biochemistry* **24,** 4603 (1985).

for nonspecific lysis.[22] These values are plotted against serum dilutions and 1 serum CH_{50} unit of C8 activity is defined as the amount which produces $Z = 0.69$ or 50% lysis of 5×10^6 cells in an incubation volume of 250 μl. The concentration of C8 activity in undiluted serum is then expressed as CH_{50} units/ml by correcting for the volume (0.05 ml) added to the assay and the serum dilution. A typical value for C8 activity in pooled serum as measured under these conditions is 1.4×10^6 CH_{50} units/ml. The standard serum is stored at $-70°$ and used to generate curves of Z values versus CH_{50} units for each set of assays. The Z values for unknowns are determined experimentally, and serum CH_{50} units/ml of C8 activity are assigned from the standard curve. Use of standardized serum allows a meaningful and quantitative comparison of activities between unknowns in a given assay and between unknowns assayed on different days. Using purified C8, the lower limit of sensitivity is 0.05 ng.

As an alternative to the above assay, one can quantitate C8 by measuring its ability to restore hemolytic activity to commercially available C8-deficient serum (Cytotech). Varying amounts of C8 are added to C8-deficient serum, and a dilution of reconstituted serum is substituted for C8 and C9 in the above assay. Activities are quantitated by comparing results to normal, standard serum.

Isolation Procedure

The following procedure is applicable to either whole plasma or plasma Fraction III (Cohn Fraction III[44]) as the starting material.[19] Fraction III is commercially available (e.g., United States Biochemical Corp., Cleveland, OH) or can be prepared more economically by fractionating plasma with ethanol.[45]

For Fraction III, 100 g of frozen paste is broken into small pieces and extracted by stirring for 1 hr at 4° in 1 liter of buffer containing 25 m*M* imidazole, 150 m*M* NaCl, 1 m*M* benzamidine, 1 m*M* PMSF, and 0.02% sodium azide, pH 7.2. The mixture is centrifuged at 6000 *g*, and the supernatant is adjusted to pH 6.9 with 0.2 *N* HCl. For plasma, 1 liter of unexpired frozen plasma is thawed and adjusted to 1 m*M* benzamidine and 1 m*M* PMSF. Our experience indicates that removal of plasminogen or vitamin K-dependent factors is not required as long as inhibitors of these proteases are included in all buffers. The plasma is diluted by adding a 40% volume of deionized water and is then adjusted to pH 6.9 at 4° with

[44] E. J. Cohn, L. E. Strong, W. L. Hughes, Jr., D. J. Mulford, J. N. Ashworth, M. Melin, and H. L. Taylor, *J. Am. Chem. Soc.* **68,** 459 (1946).

[45] P. Kistler and H. Friedli, *in* "Methods of Plasma Protein Fractionation" (J. M. Curling, ed.), p. 3. Academic Press, London, 1980.

0.2 *N* HCl. The Fraction III extract or the diluted plasma is then subjected to the following steps.

Ammonium Sulfate Fractionation. Solid ammonium sulfate is added with stirring to 37.5% saturation at 4°. After 20 min, the suspension is centrifuged at 6000 *g*, and the supernatant is adjusted to 50% saturation with ammonium sulfate. After 20 min, the precipitate is collected by centrifugation at 6000 *g* and solubilized in 100 ml of 70 m*M* sodium phosphate, 50 m*M* NaCl, 1 m*M* benzamidine, 1 m*M* PMSF, pH 6.1, at 4°. The sample is then dialyzed against the CM-Sephadex equilibration buffer which is the same buffer but without PMSF.

Ion-Exchange Chromatography. The dialyzate is applied to a CM-Sephadex C-50 column (4 × 50 cm) and washed at 70 ml/hr with equilibration buffer at 4°. The column is developed with a linear gradient of 1200 ml of equilibration buffer and 1200 ml of limit buffer containing 0.5 *M* NaCl. Hemolytic assays are used to detect C8 activity in the effluent. The profile shown in Fig. 5 is typical of that obtained for either Fraction III or plasma. CM-Sepharose can also be used in this step, but a larger volume must be used because of the lower binding capacity per volume of this gel.

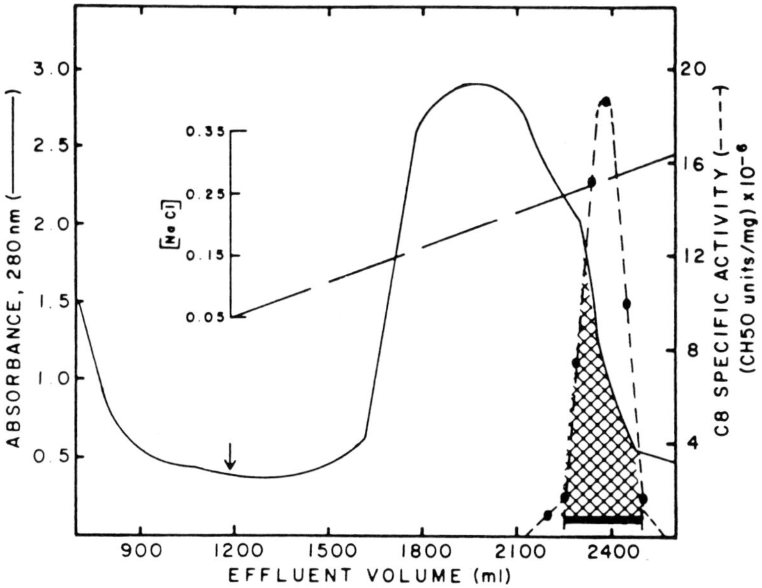

FIG. 5. CM-Sephadex C-50 elution profile of C8. The ammonium sulfate precipitate from the Fraction III extract was applied to the column and chromatographed as described in the text. Shaded fractions were pooled for further purification. Modified from Steckel *et al.*[19]

The pool from Fig. 5 is concentrated by ultrafiltration to 40 ml and dialyzed at 4° into the QAE-Sephadex equilibration buffer which is 25 m*M* Tris-HCl, 70 m*M* NaCl, 1 m*M* benzamidine, pH 8.0. The dialyzate is applied to a QAE-Sephadex A-50 column (2.5 × 25 cm) and eluted at 60 ml/hr with a linear gradient of 500 ml of equilibration buffer and 500 ml

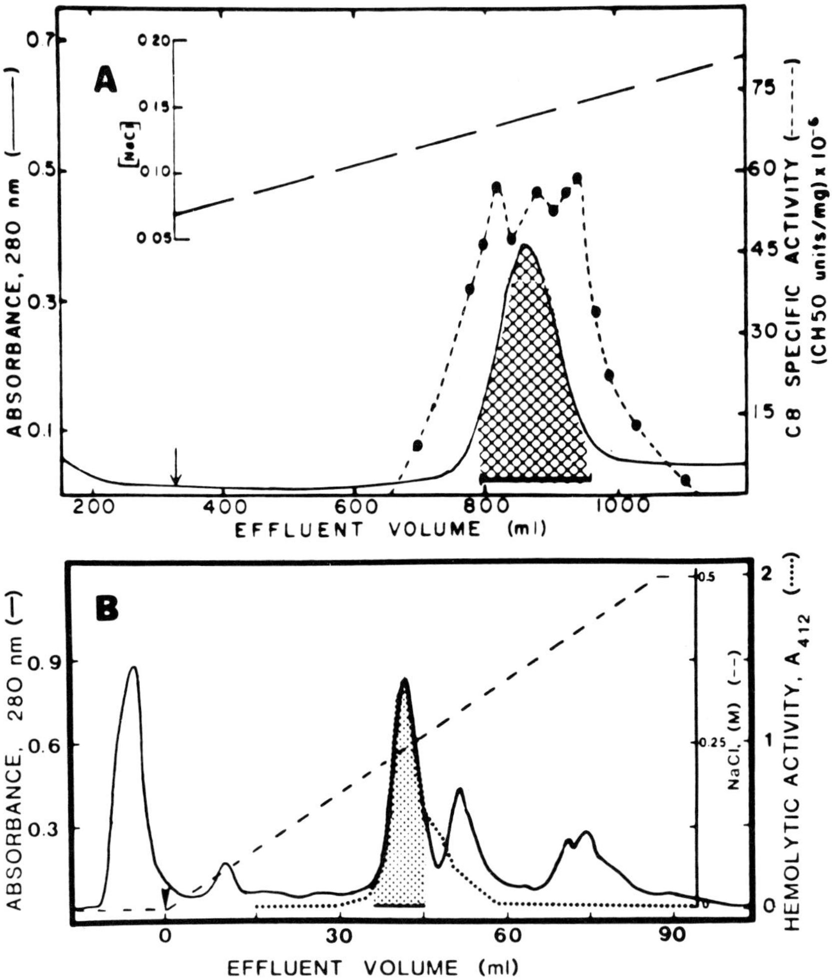

FIG. 6. (A) QAE-Sephadex A-50 elution profile of C8 from Fraction III. Pooled fractions from the CM-Sephadex chromatography step were applied to the column, and the shaded fractions were pooled. Modified from Steckel *et al.*[19] (B) Synchropak Q-300 elution profile of C8 from Fraction III. C8-containing fractions from a CM-Sepharose chromatography step were applied to the column and chromatographed as described in the text. The shaded fractions were pooled for further purification.

of limit buffer containing 220 m*M* NaCl. Specific activities are measured across the effluent protein peak by hemolytic assays.

A typical QAE-Sephadex elution profile for Fraction III is shown in Fig. 6A. It is noted that the specific activity across the single protein peak is nearly constant. This is consistent with results from SDS–PAGE analyses which reveal essentially pure C8 in this peak.[19] If desired, a few minor contaminants may be removed by concentrating the pool and chromatographing it further on a Sephacryl S-200 column (3 × 100 cm) equilibrated in 25 m*M* imidazole, 150 m*M* NaCl, 0.02% sodium azide, pH 7.2, at 4°. The C8 pool is then concentrated to 2 mg/ml by ultrafiltration, adjusted to 1 m*M* diisopropyl fluorophosphate/50% glycerol, and stored at −20°. Purification results using Fraction III are summarized in Table I.

The QAE-Sephadex profile obtained from whole plasma differs significantly, as seen in Fig. 7. C8 specific activity is not constant across the single protein peak; instead, the highest activity is located in the descending region. It is apparent from this behavior and SDS–PAGE analyses that a different population of impurities elutes here relative to those observed for Fraction III.[19] Consequently, conservative pooling is required to obtain C8 of acceptable purity at this stage. It is recommended that activity assays as well as gel analyses be used to assist in pooling. Results from the purification using plasma are summarized in Table II. It is noted that the low yield

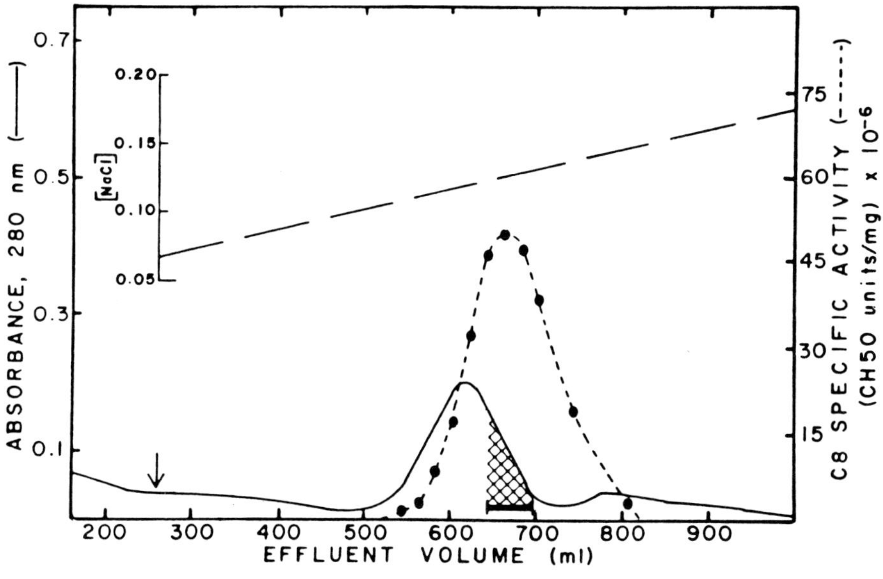

FIG. 7. QAE-Sephadex A-50 elution profile of C8 from plasma. Shaded fractions containing C8 activity were pooled. Modified from Steckel *et al.*[19]

TABLE I

PURIFICATION OF C8 FROM PLASMA FRACTION III[a]

Step	Total volume (ml)	Protein		Activity		Specific activity (CH_{50} units/mg $\times 10^{-6}$)	Purification (-fold)
		Total[b] (mg)	Yield (%)	CH_{50} units ($\times 10^{-6}$)	Yield (%)		
Extraction of Fraction III (100 g)	1,000	14,548	100	2,970	100	0.21	1
$(NH_4)_2SO_4$ fraction (50% saturated)	215	4,945	34	2,460	83	0.51	2.4
CM-Sephadex C-50	257	167	1.15	1,470	51	8.8	43
QAE-Sephadex A-50	181	23.3	0.16	1,320	44	56.4	277
Sephacryl S-200	15.6	16.8 (22.4)[c]	0.12	970	33	58.0 (43.3)[c]	283

[a] From Steckel *et al.*[19]

[b] Protein was measured relative to bovine serum albumin [W. Schaffner and C. Weissman, *Anal. Biochem.* **56,** 506 (1973)].

[c] Based on $\epsilon_{1\,cm}^{1\%} = 14.9$ at 280 nm for purified C8.

TABLE II
PURIFICATION OF C8 FROM HUMAN PLASMA[a]

Step	Total volume (ml)	Protein: Total[b] (mg)	Protein: Yield (%)	Activity: CH_{50} units ($\times 10^{-6}$)	Activity: Yield (%)	Specific activity (CH_{50} units/mg $\times 10^{-6}$)	Purification (-fold)
Plasma	1,000	42,500	100	1,232	100	0.03	1
$(NH_4)_2SO_4$ fraction (50% saturated)	145	4,604	10.8	917	74	0.20	7
CM-Sephadex C-50	337	239	0.56	377	31	1.57	54
QAE-Sephadex A-50	2.5	2.4	0.006	120	10	49.1	1,693
Sephacryl S-200	1.3	2.2 (2.9)[c]	0.005	113	9	53.4 (39.0)[c]	1,841

[a] From E. W. Steckel, Ph.D. dissertation. University of South Carolina, Columbia, South Carolina, 1983.

[b] Protein measured as in Table I.

[c] Based on $\epsilon_{1cm}^{1\%} = 14.9$ at 280 nm for purified C8.

is primarily a consequence of selective pooling required after the QAE-Sephadex column. A better yield can be obtained by using a Synchropak Q-300 (Synchrom) column (1 × 10 cm) in a FPLC system (Pharmacia) in this step. However, because a small column size must be chosen to limit column pressure to no more than 500 psi, the amount that can be processed is smaller. Starting with 100 g of Fraction III, the CM-Sephadex (or CM-Sepharose) pool must be split into four or more aliquots to avoid overloading of the column. But since each chromatogram requires only about 90 min, total workup time is not prohibitive. Figure 6B shows the elution profile of C8 from a Synchropak Q-300 column. No significant differences in activity profiles between plasma or Fraction III-derived C8 are observed with this column.

Properties

Tables I and II indicate that Fraction III is the preferred starting material, primarily because of higher yields. The above purification procedure has also been used successfully with serum, the results being comparable to plasma. In all cases, a highly purified and nondegraded C8 can be obtained, provided the recommended proteolytic inhibitors are included during purification. Once purified, C8 is stable for months in 50% glycerol at −20°. Freezing should be avoided since it results in aggregation and precipitation.

Sedimentation analyses of C8 indicate a molecular weight of 151,000, and $s^{\circ}_{20,w} = 8.2$.[19] Considering this and molecular weights of the individual subunits as measured by SDS–PAGE, it is concluded that α, β, and γ occur in a 1 : 1 : 1 stoichiometry. Although α and β are similar in size, amino acid compositions and amino-terminal sequences confirm that they are different subunits.[19] Using quantitative amino acid analysis, C8 was found to have $\epsilon^{1\%}_{1\,cm} = 14.9$ at 280 nm and a corresponding $\epsilon = 2.25 \times 10^5$ (M^{-1} cm^{-1}) in 10 m*M* sodium phosphate, 150 m*M* NaCl, pH 7.0.[46,47]

Regarding the subunits of C8, the affinity between $\alpha-\gamma$ and β is high. Circular dichroism studies indicate C8 resists complete unfolding by denaturants such a guanidine–HCl and urea.[19] Removal of denaturants results in restoration of C8 conformation and activity. This behavior is attributed in part to the affinity between $\alpha-\gamma$ and β. Procedures for purifying $\alpha-\gamma$ and β use high salt (1.5 *M* NaCl) to disrupt their interaction.[48] Although individually inactive, purified $\alpha-\gamma$ and β retain the ability to associate as evidenced by the fact they fully recombine when equimolar amounts of each are mixed in dilute solution (1 μg/ml) in the presence of 1 mg/ml

[46] J. B. Monahan, J. L. Stewart, and J. M. Sodetz, *J. Biol. Chem.* **258,** 5056 (1983).
[47] J. L. Stewart, J. B. Monahan, A. Brickner and J. M. Sodetz, *Biochemistry* **23,** 4016 (1984).
[48] A. G. Rao and J. M. Sodetz, *Complement* **1,** 182 (1984).

carrier serum albumin.[49] Recombination results in recovery of C8 conformation and activity. Values for $\epsilon_{1\,cm}^{1\%}$ of 11.8 and 13.7 at 280 nm have been determined for purified $\alpha-\gamma$ and β, respectively, in 10 m*M* sodium phosphate, 150 m*M* NaCl, pH 7.0.[46]

C9 Purification

Introduction

Human C9 is a single-chain glycoprotein of M_r 66,000.[50] Its amino acid sequence has been predicted from the cDNA-derived sequence by two groups.[51,52] Although both sequences are almost identical, one discrepancy is noteworthy. According to Stanley *et al.*,[52] C9 has a Cys in position 23, and 24 cysteines altogether, whereas DiScipio *et al.*[51] find an Arg in this position and a total of 23 cysteines. The former result would be compatible with previous observations of Yamamoto and Migita[53] who could not incorporate radioactive iodoacetamide into C9 even in the presence of 6 *M* guanidine–HCl. Thus, irrespective of which amino acid occupies position 23, apparently no sulfhydryl groups are available in C9 which could be used for side chain-specific labeling or be exploited in affinity purification schemes.

C9 Assays

Assays for C9 use EAC1–7 intermediates and excess C8, both prepared as outlined above or purchased from Diamedix and Cytotech, respectively. Fifty microliters EAC1–7 in GVB-M (5×10^8 cells/ml), 15 ng purified C8, and 10–50 μl of appropriately diluted, C9-containing test samples are incubated together in a total volume of 0.2 ml at 37° for 45 min. Then 1.8 ml of ice-cold GVB-E is added to stop the reaction, the samples are centrifuged to pellet unlysed cells, and hemolysis is measured by absorbance at 412 nm. C9 activity can be expressed in CH_{50} units or *Z* values. The sensitivity limit of the assay is about 0.06 ng or 1 fmol of C9. It should be noted that C9 activity should not be assayed with R-9 and EA directly. Because of substantial C8-mediated lysis of EAC1–8 at end point, lysis of

[49] J. B. Monahan and J. M. Sodetz, *J. Biol. Chem.* **255,** 10579 (1980).

[50] R. G. DiScipio and T. E. Hugli, *J. Biol. Chem.* **260,** 14802 (1985).

[51] R. G. DiScipio, M. R. Gehring, E. R. Podack, C. C. Kan, T. E. Hugli, and G. H. Fey, *Proc. Natl. Acad. Sci. U.S.A.* **81,** 7298 (1984).

[52] K. K. Stanley, H.-P. Kocher, J. P. Luzio, P. Jackson, and J. Tschopp, *EMBO J.* **4,** 374 (1985).

[53] K.-I. Yamamoto and S. Migita, *J. Biol. Chem.* **258,** 7887 (1983).

EA by R-9 is indistinguishable from lysis by whole human serum under these conditions.[54]

Isolation Procedure

The following procedure is applicable to plasma Fraction III, whole human plasma or serum, or activated R-7 from the C5b–6 preparation as starting materials. The purification scheme is based on the method of Biesecker and Müller-Eberhard[54] with several modifications.[20,55]

For Fraction III, 100 g of frozen paste is broken into small pieces and extracted by stirring for 1–6 hr at 4° in 1 liter of buffer containing 10 m*M* citrate, 10 m*M* phosphate, 80 m*M* NaCl, and 1 m*M* PMSF, pH 7.4. The mixture is centrifuged at 3000 *g* to remove insoluble material. For human plasma, 1 liter of unexpired frozen plasma is thawed and adjusted to 10 m*M* EDTA, 1 m*M* benzamidine, and 1 m*M* PMSF, diluted by addition of 0.4 volumes of deionized water, and titrated to pH 7.4. The Fraction III extract or human plasma is then subjected to the following steps at 4°.

$BaCl_2$ and PEG Fractionation. To 1 liter of Fraction III solution or plasma, 40 ml of 1 *M* $BaCl_2$ is added with stirring to precipitate vitamin K-dependent enzymes. After 15 min, the suspension is centrifuged at 3000 *g* for 10 min. The supernatant is brought to 7% PEG 4000 by slowly adding 500 ml of a 21% (w/v) aqueous PEG 4000 solution. After stirring for 15 min the suspension is centrifuged for 30 min at 16,000 *g*. (If activated R-7 from the C5b–6 preparation is used as a starting material, the 8% PEG 4000 supernatant can be used from here on.) The supernatant is brought to 20% by addition of solid PEG 4000, and after an additional stirring period of 1 hr the suspension is centrifuged at 2,500 *g* for 15 min. The precipitate is dissolved in solubilizing buffer (10 m*M* phosphate, 90 m*M* NaCl, 1 m*M* PMSF, pH 7.4, conductance <11 mS), and insoluble material is removed by centrifugation for 15 min at 15,000 *g*. Up to this point all preparations are kept in plastic containers.

Plasminogen Removal and Ion-Exchange Chromatography. The C9-containing solution is applied to a lysine–Sepharose CL-4B column (5×16 cm) which is serially linked to a DEAE-Sephacel column (2.6×40 cm). Both columns are equilibrated with solubilizing buffer. The sample is pumped onto the columns (100 ml/hr) followed by elution buffer (=solubilizing buffer without PMSF), and both columns are washed until the A_{280} of the eluent is below 0.05. The DEAE-Sephacel is then developed at a flow rate of 60 ml/hr with a linear NaCl concentration gradient formed from 1

[54] G. Biesecker and H. J. Müller-Eberhard, *J. Immunol.* **124,** 1291 (1980).

[55] J. W. Shiver, J. R. Dankert, J. J. Donovan, and A. F. Esser, *J. Biol. Chem.* **261,** 9629 (1986).

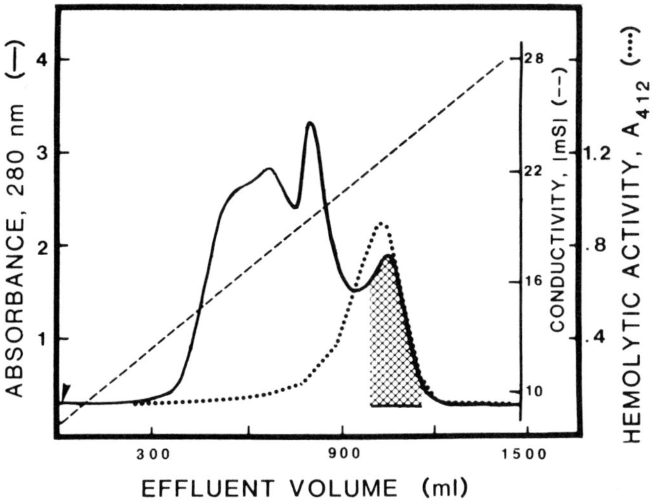

FIG. 8. DEAE-Sephacel anion-exchange chromatography of C9. The resolubilized 7–20% PEG 4000 precipitate was applied to the column and chromatographed as described in the text. The shaded fractions with C9 activity were pooled.

liter elution buffer and 1 liter of buffer adjusted to 30 mS with solid NaCl (~0.3 *M*). Figure 8 shows a typical DEAE-Sephacel elution profile. Fractions containing C9 activities are analyzed by SDS–PAGE and pooled to minimize contaminating proteins.

Hydroxylapatite Chromatography. The C9 pool is diluted with 10 m*M* phosphate buffer (pH 7.0) to a conductance of 10 mS and applied (40 ml/hr) to a hydroxylapatite column (1.6 × 12 cm; LKB HA-Ultrogel) which had been equilibrated in 10 m*M* phosphate, 100 m*M* NaCl, pH 7.0. The column is washed with 100 ml of the same buffer followed by 50 m*M* phosphate buffer at pH 8.1 until the A_{280} reaches baseline levels. The column is then eluted with a linear phosphate gradient formed from 250 ml of 50 m*M* phosphate (pH 8.1) and 250 ml of 350 m*M* phosphate buffer (pH 8.1). (To prolong the lifetime of HA-Ultrogel the column should be equilibrated as soon as possible with neutral buffers and stored according to the manufacturers recommendation.) Fractions containing C9 activity (Fig. 9) are pooled with the aid of SDS–PAGE analysis, concentrated by vacuum dialysis (MicroProDiCon, Bio-Molecular Dynamics), and finally sieved on a Sephacryl S-200 column (2.6 × 75 cm) with 10 m*M* HEPES, 200 m*M* NaCl, 1 m*M* $CaCl_2$, 0.02% sodium azide, pH 7.2, at a flow rate of 15 ml/hr. The protein is pooled and stored frozen at −70° or in soluble form, after addition of 40% glycerol, at −18°. Table III indicates the yields that can be achieved for C9 purification.

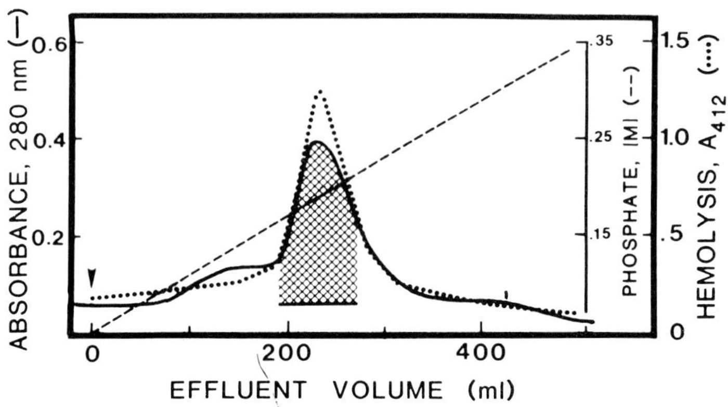

FIG. 9. Elution profile of C9 from a hydroxylapatite HA-Ultrogel column. The active fractions shown in Fig. 8 were applied and eluted with a phosphate gradient at pH 8.1. The shaded fractions were pooled and concentrated.

It should be noted that it is difficult to accurately measure C9 activities in plasma or solubilized Fraction III because of the presence of lysis inhibitors such as lipoproteins. Therefore, no purification values are listed in Table III. Our protocol allows purification of C9 without lipoprotein contamination, but it is advisable to test the final product for high- and low-density lipoproteins using immunodiffusion techniques and commercially available antilipoprotein antisera (Calbiochem-Behring).

Properties

Using quantitative amino acid analysis the extinction coefficient for C9 $\epsilon_{1\,\mathrm{cm}}^{1\%}$ was reported to be 9.6 at 280 nm.[8] Contrary to several earlier reports,[4,8,9] the protein is very stable at the physiological temperature of 37° and in sterile buffers of physiological ionic strength.[20] However, trace contaminants of heavy metal ions, such as Zn^{2+} or Cu^{2+}, inactivate the protein[56,57] especially at elevated temperatures. It is now known that this inactivation results from aggregation of the protein into tubular and string-like assemblies.[3,8-10,20] Oligomerization of C9 into a tubular complex, also referred to as poly(C9),[8] can also be achieved by limited trypsinization and by interaction with small unilamellar lipid vesicles.[20]

The secreted protein has a blocked amino-terminal end (pyroglutamate) and a total of 537 amino acids.[51] It can be split by α-thrombin between residues His(244) and Gly(245) into two fragments, C9a and C9b,

[56] O. Götze, I. Haupt, and H. Fischer, *Nature (London)* **217,** 1165 (1968).

[57] K. Yamamoto and M. Takahashi, *Int. Arch. Allergy Appl. Immunol.* **48,** 653 (1975).

TABLE III
PURIFICATION OF C9 FROM PLASMA FRACTION III

Step	Total volume (ml)	Protein: Total[a] (mg)	Protein: Yield (%)	Activity: Z units[b] (× 10⁻⁶)	Activity: Yield[c] (%)	Specific activity (Z units/mg × 10⁻³)
Extraction of Fraction III (100 g)	1,000	40,000	100	*d*	*d*	*d*
7–21% PEG precipitate	1,000	25,000	62.5	*d*	*d*	*d*
Lysine–Sepharose	1,200	8,600	21.5	112	100	13
DEAE-Sephacel	120	60	1.5	79	70.5	1,317
Hydroxylapatite	95	28.5	0.071	66	58.9	2,316
Sephacryl S-200	12	22.0[e]	0.055	55	49.1	2,500

[a] Protein was measured as A_{280}, using an arbitrary extinction coefficient of $\epsilon_{1cm}^{1\%} = 10.0$.
[b] One *Z* unit = 10^7 effective molecules.
[c] Because of removal of lysis inhibitors these values are apparent yields.
[d] The presence of lysis inhibitors interferes with determination of *Z* values.
[e] Based on an extinction coefficient at 280 nm of $\epsilon_{1cm}^{1\%} = 9.6$ for the purified protein.

that remain associated noncovalently,[58] and the resulting nicked protein ($C9^n$) retains its full hemolytic,[58,59] bacteriolytic,[60] and cytotoxic[61] capacity. However, $C9^n$ has lost the ability to form tubular poly(C9) and the classic complement lesion as seen in the electron microscope.[59] For these reasons, it was postulated that poly(C9) does not constitute the functional transmembrane channel in complement-mediated hemolysis and cytolysis but may be an end product that limits autoimmune damage to the host.[20,59–62] The membranolytic domain of C9 resides most probably in the C9b fragment since the isolated fragment is capable of forming single ion-conducting channels across planar lipid bilayers.[56] Photolabeling studies with membrane-restricted probes revealed that C9 penetrates into the hydrocarbon phase[63] but that only the C9b fragment is accessible to the reactive probe; the C9a fragment resides completely in the aqueous phase.[64] The C9b fragment also contains an interesting sequence homology to melittin,

[58] G. Biesecker, C. Gerard, and T. E. Hugli, *J. Biol. Chem.* **257,** 2584 (1982).
[59] J. R. Dankert and A. F. Esser, *Proc. Natl. Acad. Sci. U.S.A.* **82,** 2128 (1985).
[60] J. R. Dankert and A. F. Esser, *Biochemistry* **25,** 1094 (1986).
[61] B. P. Morgan, J. R. Dankert, and A. F. Esser, *J. Immunol.* **138,** 246 (1987).
[62] A. F. Esser, *UCLA Symp. Mol. Cell. Biol.* **45,** 411 (1987).
[63] V. W. Hu, A. F. Esser, E. R. Podack, and B. J. Wisnieski, *J. Immunol.* **127,** 380 (1981).
[64] B. Ishida, B. J. Wisnieski, C. H. Levine, and A. F. Esser, *J. Biol. Chem.* **257,** 10551 (1982).

another lytic peptide present in bee venom.[62] If an Arg residue occupies sequence position 23 as proposed by DiScipio *et al.*,[51] then this residue is part of a potential phosphorylation site (i.e., Arg-Arg-Met-Ser) in C9a, and indeed C9 is a substrate for the catalytic subunit of cAMP-dependent kinase.[65]

Acknowledgments

Work reported in this article was supported by research Grants RO1 AI 19478 (A.F.E.) and RO1 AI 16856 (J.M.S.) from the National Institutes of Health and an Established Investigatorship to J.M.S. from the American Heart Association (AHA 82-121).

Note Added in Proof

cDNA sequencing of human C7 by DiScipio *et al.*[66] has indicated that the protein consists of 821 amino acids with a calculated M_r of 97,300 including 6.4% carbohydrate. A significant degree of sequence similarity with C8α, C8β, and C9 has been established, indicating a close ancestral relationship among all four proteins.[66] The cDNA for human C8γ has been sequenced by Ng *et al.*,[67] and the results have demonstrated that C8α, C8β, and C8γ, are the products of three different genes. The protein has an amino-terminal pyroglutamyl residue, no carbohydrate, and consists of 182 amino acids corresponding to a M_r 20,329. No apparent sequence similarity to other terminal complement proteins was detected; rather, C8γ, belongs to a family of proteins termed "lipocalins."[68] Furthermore, cloning and sequencing of the gene for human C9 by Stanley and Fey and collaborators[69] have removed the earlier differences between the two published sequences,[51,52] demonstrated an additional Val at position 293, bringing the total number of residues to 538 and confirmed the Cys in position 23 and the Asn at position 394 (formerly position 393 without the additional valine). Finally, human C9 was shown to be a metalloprotein that binds 1 mol of Ca^{2+} per mole of C9 with a dissociation constant of 3 μM. C8α was found to be the only other terminal complement protein that also binds calcium.[70]

[65] A. Dan-Gaspar and A. F. Esser, *Complement* **4,** 145 (1987).

[66] R. G. DiScipio, D. N. Chakravarti, H. J. Müller-Eberhard, and T. E. Hugli, *J. Biol. Chem.* **263,** 549 (1988).

[67] S. C. Ng, A. G. Rao, O. M. Z. Howard, and J. M. Sodetz, *Biochemistry* **26,** 5229 (1987).

[68] J. M. Sodetz, *Curr. Topics Microbiol. Immunol.,* in press (1988).

[69] K. K. Stanley, *Curr. Topics Microbiol. Immunol.,* in press (1988).

[70] N. M. Thielens, K. Lohner, and A. F. Esser, *J. Biol. Chem.* **263,** in press (1988).

[46] C3, C5 Components and C3a, C4a, and C5a Fragments of the Complement System

By JARMILA JANATOVA

Introduction

The proteins C3 and C5 are two components of the complement system,[1,2] whose apparent main function is to help to defend the human body against bacteria, viruses, yeast, and other foreign particles or substances. The proteins C3 and C5 are precursors of several fragments that can serve as ligands for specific receptors present on blood and tissue cells.[3]

Once the complement system is triggered, classical[4] or alternative[5] pathway activation leads to formation of a C3 convertase (C4b,C2a or C3b,Bb), the enzyme which cleaves and activates C3, the central protein of complement (Fig. 1). C3 is an acute-phase protein, and its concentration in plasma rises rapidly in inflammatory conditions. The first cleavage of C3 generates two fragments. The smaller one, the activation peptide C3a (M_r 9,100), is an anaphylatoxin.[6] The large cleavage product, the disulfide-bridged two-chain C3b ($M_r \sim 179{,}000$), has the ability, when in its short-lived nascent state, to bind covalently via transacylation to immune complexes or cell surfaces.[7-13] The bound C3b can exhibit several functions. Covalent binding of C3b to foreign particles or cells results in their opsonization. Subsequently, bound C3b mediates their interaction with phagocytic cells by its binding to CR1 (C3b receptor),[14] and the iC3b molecule

[1] K. B. M. Reid and R. R. Porter, *Annu. Rev. Biochem.* **50,** 433 (1981).
[2] H. J. Müller-Eberhard and R. D. Schreiber, *Adv. Immunol.* **29,** 1 (1980).
[3] D. T. Fearon and W. W. Wong, *Annu. Rev. Immunol.* **1,** 243 (1983).
[4] K. B. M. Reid, *Biochem. Soc. Trans.* **11,** 1 (1983).
[5] M. K. Pangburn and H. J. Müller-Eberhard, *Springer Semin. Immunopathol.* **7,** 163 (1984).
[6] T. E. Hugli, *Crit. Rev. Immunol.* **1,** 321 (1981).
[7] S. K. Law and R. P. Levine, *Proc. Natl. Acad. Sci. U.S.A.* **74,** 2701 (1977).
[8] J. Janatova, B. F. Tack, and J. W. Prahl, *Biochemistry* **19,** 4479 (1980).
[9] B. F. Tack, R. A. Harrison, J. Janatova, M. L. Thomas, and J. W. Prahl, *Proc. Natl. Acad. Sci. U.S.A.* **77,** 5764 (1980).
[10] M. K. Pangburn and H. J. Müller-Eberhard, *J. Exp. Med.* **152,** 1102 (1980).
[11] S. K. Law, N. A. Lichtenberg, and R. P. Levine, *Proc. Natl. Acad. Sci. U.S.A.* **77,** 7194 (1980).
[12] R. B. Sim, T. M. Twose, D. S. Paterson, and E. Sim, *Biochem. J.* **193,** 115 (1981).
[13] K. J. Gadd and K. B. M. Reid, *Biochem. J.* **195,** 471 (1981).
[14] D. T. Fearon, *Immunol. Today* **5,** 105 (1984).

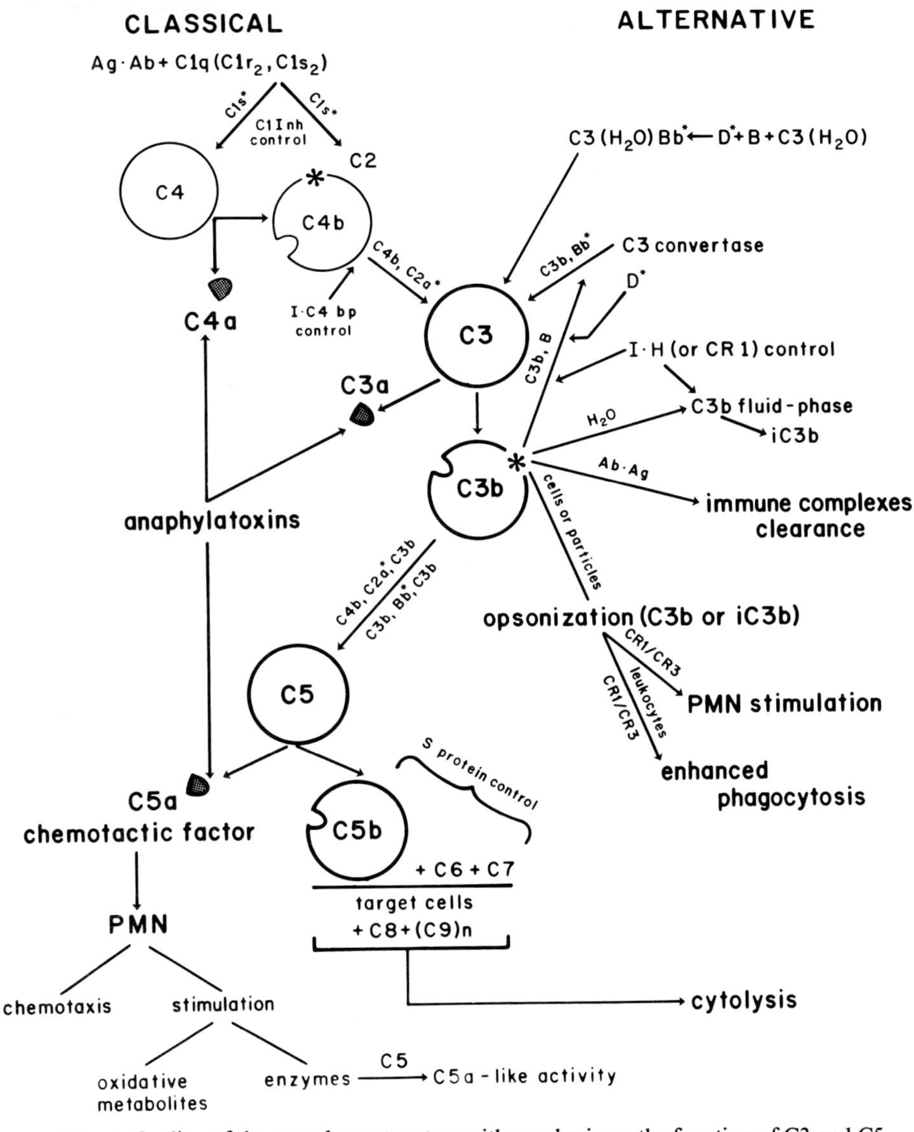

FIG. 1. Outline of the complement system with emphasis on the function of C3 and C5. The asterisk on the C3b or C4b fragments indicates the labile binding site with the reactive thiol ester which allows this protein to bind covalently to immune complexes and/or cell surfaces. For further details, see text.

(C3b cleaved at the carboxy end by factor I) binds to CR3 receptor. These are necessary steps leading to enhanced phagocytosis and/or stimulation of neutrophils to release lysosomal enzymes. The C3b fragment can also function as a noncatalytic subunit of the alternative pathway C3 convertase (C3b,Bb) that in turn activates more C3.

The level of C3 activation is under intrinsic and extrinsic control.[1,2] Unlike native C3, the C3b fragment is susceptible to cleavage by factor I in the presence of a cofactor (factor H or CR1).[15,16] C3b with a fragmented α chain (iC3b) can no longer participate in the formation of C3/C5 convertase,[17] but it can react with complement receptors, particularly CR3, present on a variety of blood or tissue cells.[14,18]

The C3b molecule, as a part of the C5 convertase (C4b,C2a,C3b or C3b,Bb,C3b), can participate in either pathway in cleavage of C5 to C5a and C5b. The large two-chain C5b fragment in the presence of C6 can initiate assembly of the C5b–9 membrane attack complex that mediates complement-dependent cytolysis of target cells.[19]

The smaller C5a fragment[6] that is generated from the amino terminus of the α chain during the activation of C5, in addition to being an anaphylatoxin, is a potent *chemotactic factor* for polymorphonuclear leukocytes (PMNs),[19a] and it also induces PMN to release lysosomal enzymes.[20] Of these, elastase and cathepsin G can cleave C5 at neutral pH to yield biologically active polypeptides. The C5a-like biological activities, namely PMN chemotaxis and release of PMN lysosomal enzymes and of oxidative metabolites have also been observed after treatment of C5 with other physiologically relevant enzymes such as α-thrombin and plasmin.[21] There are several assays available allowing the detection of the presence and the

[15] M. K. Pangburn, R. D. Schreiber, and H. J. Müller-Eberhard, *J. Exp. Med.* **146,** 257 (1977).

[16] M. E. Medof, K. Iida, C. Mold, and V. Nussenzweig, *J. Exp. Med.* **156,** 1739 (1982).

[17] K. Whaley and S. Ruddy, *J. Exp. Med.* **144,** 1147 (1976).

[18] M. A. Arnaut and H. R. Colten, *Mol. Immunol.* **21,** 1191 (1984).

[19] H. J. Müller-Eberhard, *Annu. Rev. Immunol.* **4,** 503 (1986).

[19a] Abbreviations: PMNs, polymorphonuclear leukocytes; IEP, immunoelectrophoresis; SDS–PAGE, sodium dodecyl sulfate–polyacrylamide gel electrophoresis; EDTA, ethylenediaminetetraacetic acid; PBS, phosphate-buffered saline; NHS, normal human serum; IA, immunoadsorbents; $GVBS^{2+}$, veronal-buffered saline (see Ref. 63 for composition); CDS, component-depleted serum; C(n), native forms of complement component; C(i), inactive forms of component; DTNB, 5,5′-dithiobis(2-nitrobenzoic acid); TNB, 5-thio (2-nitrobenzoic acid); PMSF, phenylmethylsulfonyl fluoride; SB, standard buffer (see below for composition); DIC, denaturant-induced cleavage; PEG, polyethylene glycol; SCPN, serum carboxypeptidase N; CVF, cobra venom factor.

[20] I. M. Goldstein and H. D. Perez, *Prog. Hemostas. Thromb.* **5,** 41 (1980).

[21] R. A. Wetsel and W. P. Kolb, *J. Exp. Med.* **157,** 2029 (1983).

concentration of C5a. In some instances, however, the binding of C5a to its receptor on PMNs[22] interferes with its quantitation.

Although considerable homology in size and structure exists between C5 and C3, including the existence of single polypeptide chain pro-forms[23-26] and similar activation mechanisms,[1,2] there are, nevertheless, some substantial differences between the two. Native C5 does not contain an internal thiol ester,[11,27,28] and thus cannot react with activating surfaces to form a covalent bond.[29] However, the activation of C5 is accompanied by conformational changes similar to those observed in C3, and the half-life of the C5b "labile binding site" is also very short.[30] Only "nascent" C5b has an affinity for C6. The C5b fragment does not seem to be susceptible to cleavage by factor I.[31] Upon binding of C7 to the hydrophilic C5b–6 complex, a membrane binding site is generated. The formation of the membrane attack complex can be aborted at a stage of C5b–7 complex after 3 molecules of the control S protein bind to this complex.[19,32]

Unlike C3 or C4, the hemolytic activity of C5 is not affected by treatment with nitrogen nucleophiles,[11,27] C5 is not inactivated upon storage, and it does not undergo denaturant-induced cleavage of the α chain.[27,33] In C3 and C4, the loss of the hemolytic activity due to spontaneous, chemical, or enzymatic inactivation is accompanied by an appearance of an SH group[34] and by conformational changes.[8,35] The appearance of a free SH group can thus serve as a convenient indicator of inactivation. This phenomenon makes possible quantitative determination of hemolytically active C3 or C4 molecules. The hemolytically inactive forms may or may

[22] D. E. Chenoweth and T. E. Hugli, *Proc. Natl. Acad. Sci. U.S.A.* **75,** 3943 (1978).
[23] Y. M. Ooi and H. R. Colten, *J. Immunol.* **123,** 2494 (1979).
[24] Å. B. Lundwall, R. A. Wetsel, T. Kristensen, A. S. Whitehead, D. E. Woods, R. C. Ogden, H. R. Colten, and B. F. Tack, *J. Biol. Chem.* **260,** 2108 (1985).
[25] V. Brade, R. E. Hall, and H. R. Colten, *J. Exp. Med.* **146,** 759 (1977).
[26] K. M. Morris, G. Goldberger, H. R. Colten, D. P. Aden, and B. B. Knowles, *Science* **215,** 399 (1982).
[27] J. Janatova and B. F. Tack, *Biochemistry* **20,** 2394 (1981).
[28] R. G. DiScipio, *Biochem. J.* **199,** 497 (1981).
[29] S. K. Law, N. A. Lichtenberg, F. H. Holcombe, and R. P. Levine, *J. Immunol.* **125,** 634 (1980).
[30] R. G. DiScipio, C. A. Smith, H. J. Müller-Eberhard, and T. E. Hugli, *J. Biol. Chem.* **258,** 10629 (1983).
[31] E. R. Podack and H. J. Müller-Eberhard, *J. Immunol.* **124,** 332 (1980).
[32] E. R. Podack, W. P. Kolb, and H. J. Müller-Eberhard, *J. Immunol.* **120,** 1841 (1978).
[33] R. B. Sim and E. Sim, *Biochem. J.* **193,** 129 (1981).
[34] J. Janatova, P. E. Lorenz, A. N. Schechter, J. W. Prahl, and B. F. Tack, *Biochemistry* **19,** 4471 (1980).
[35] D. E. Isenman and N. R. Cooper, *Mol. Immunol.* **18,** 331 (1981).

not have a reagent-accessible SH. Although they exhibit a $\alpha\beta$ or $\alpha\beta\gamma$ polypeptide chain structure identical to that of native C3 or C4, they share several properties[36–38] with C3b or C4b, the activated forms of C3 or C4. Therefore, it became necessary to develop procedures for effective isolation and separation of native C3 or C4 from their inactive forms. These processes should be fast so that the amount of newly hydrolyzed C3 or C4 acquired during the isolation and separation would be minimal. Some of the assays and the procedures described below fulfill such objectives.

Methods of Antigenic and Functional Assay

Detection and Quantitation of C3, C5, C3a, and C5a

This section deals with assays based on the interaction of the antigen with the respective antibody. Monospecific polyclonal antibodies recognizing individual proteins of the complement system are most often raised in rabbits or goats and can be obtained from several sources (e.g., Cappel/Cooper Biomedical, Malvern, PA; Dako Corp., Santa Barbara, CA; Miles Scientific, Naperville, IL; Genzyme Corp., Boston, MA 02111; Cytotech, San Diego, CA 92121). Addresses and names of additional suppliers of immunochemicals have been compiled.[39] Reports on generation of monoclonal antibodies recognizing the separate determinants on individual complement proteins have been published,[40–43] and some of these monoclonal antibodies are now commercially available (Genzyme, Cytotech). Descriptions of those techniques that are widely used and where detailed recent protocols for determinations of complement components are available[44] are limited here to only brief outlines and comments.

[36] I. von Zabern, R. Nolte, and W. Vogt, *Scand. J. Immunol.* **13,** 413 (1981).

[37] M. K. Pangburn, R. D. Schreiber, and H. J. Müller-Eberhard, *J. Exp. Med.* **154,** 856 (1981).

[38] D. E. Isenman, D. I. C. Kells, N. R. Cooper, H. J. Müller-Eberhard, and M. K. Pangburn, *Biochemistry* **20,** 4458 (1981).

[39] W. D. Linscott, "Linscott's Directory of Immunological and Biological Reagents." 40 Glen Drive, Mill Valley, California 94941 (biannually).

[40] P. J. Lachmann, R. G. Oldroyd, C. Milstein, and B. W. Wright, *Immunology* **41,** 503 (1980).

[41] P. J. Lachmann, M. K. Pangburn, and R. G. Oldroyd, *J. Exp. Med.* **156,** 205 (1982).

[42] J. D. Tamerius, M. K. Pangburn, and H. J. Müller-Eberhard, *J. Immunol.* **135,** 2015 (1985).

[43] W. P. Kolb, B. A. Johnson, L. A. Warczakowska, and J. D. Tamerius, *Fed. Proc., Fed. Am. Soc. Exp. Biol.* **44,** 990 (1985).

[44] K. Whaley, *in* "Methods in Complement for Clinical Immunologists" (K. Whaley, ed.), pp. 50 and 120. Churchill-Livingstone, Edinburgh, Scotland, 1985.

Immunodiffusion and Nephelometry. The presence of the C3 or C5 antigen can be detected qualitatively by the Ouchterlony[45] double-immunodiffusion method; diffusion plates are prepared with phosphate-buffered saline containing 0.02% sodium azide, 2 m*M* disodium EDTA, and 1% agarose. A sample (10 μl) of serum containing about 10 μg C3 or 0.8 μg C5 is applied to a well of 3-mm diameter in a 1.5-mm-thick agarose gel. The same volume of the appropriate antiserum is used in the assay. Only 1 μl of each, the serum with the antigen and the antiserum, is sufficient in the microtechnique, where wells of 1-mm diameter 4 mm apart are used. Determinations of concentration of C3 or C5 antigen can be performed by single radial immunodiffusion according to Mancini *et al.*[46] In this assay, the antibody-containing agarose gel is prepared by using about 10–40 μl of antiserum/ml of gel. Samples (5 or 10 μl) are applied into 2- or 3-mm diameter wells. A lower limit of sensitivity has been reported to be about 5 mg/liter or 0.05 μg/10 μl. The amount of C5 present in serum is at least 10 times higher.

A quantitative evaluation can also be obtained by the rate nephelometry technique[47] that measures the amount of scattered light which is proportional to the concentration of insoluble immune complexes formed during the assay. This method has been automated for use in clinical laboratories for C3, C4, and factor B. Protocols are provided by companies which distribute nephelometers, and procedures vary depending on the type of apparatus. The sensitivity of the assay in clear sera is approximately 20 mg/liter or 20 μg/ml. Consequently, the amount of C3 at 2% (or less) of its original concentration in serum cannot be detected. One-dimensional rocket immunoelectrophoresis[48] is also used for quantitation purposes. In this assay the antibody-containing gel is prepared with 10–15 μl of antiserum/ml of gel. Samples (2 μl) of serum, containing about 2 μg C3 or 0.16 μg C5, are applied to 1.2-mm-diameter wells. These amounts are well above the lower limit of sensitivity, i.e., 1 mg/liter or 0.002 μg/2 μl. Immunotechniques listed above do not differentiate, in general, between native and hemolytically inactive or fragmented C3 or C5.

Immunoelectrophoretic Techniques. Samples are subjected to a zone electrophoresis (usually in 1% agarose at pH 8.6) that is followed, in the second step of the assay, by formation of insoluble antigen–antibody complexes. In ordinary immunoelectrophoresis[49] (IEP) precipitin arcs are

[45] Ö. Ouchterlony, *Prog. Allergy* **5,** 1 (1958).

[46] G. Mancini, A. O. Carbonara, and J. F. Heremans, *Immunochemistry* **2,** 235 (1965).

[47] J. C. Sternberg, *Clin. Chem.* **23,** 1456 (1977).

[48] C. B. Laurell, *Anal. Biochem.* **15,** 45 (1966).

[49] P. Grabar and C. A. Williams, *Biochim. Biophys. Acta* **10,** 193 (1953).

formed by double immunodiffusion, and this assay is used for a qualitative evaluation. Usually 2 μl of undiluted serum, containing 1.6–3 μg C3 and 0.12–0.16 μg C5 are applied to 1.2-mm-diameter wells. In two-dimensional immunoelectrophoresis (antigen–antibody crossed electrophoresis)[50] the antigen–antibody complexes form "rockets" as the antigen migrates in the second dimension in the electrical field through agarose containing the appropriate antibody. This approach can be used for quantitative determination. Samples (5 μl) of serum are electrophoresed. The aliquots (20–30 μl) of antisera are used in preparation of 1 ml of antibody-containing gel. Native as well as the activated and fragmented forms of C3 or C5 can be detected by these procedures since they differ in their electrophoretic mobilities. C3b and C5b, as well as C3c, C3d,g, or C3d fragments, are more acidic than corresponding native proteins and therefore migrate faster toward the anode at pH 8.6. These assays will not differentiate, however, between native and hemolytically inactive uncleaved forms of C3. The purified antigen should be tested at physiological, as well as higher (2- or 3-fold), concentrations with the monospecific anti-human C3 or C5 and also with antibodies raised against whole human serum.

Western Blot Analysis. The presence and the activation of C3 or C5 in body fluids can also be detected by using sodium dodecyl sulfate–polyacrylamide gel electrophoresis (SDS–PAGE) followed by Western blot analysis,[51] utilizing either polyclonal or monoclonal antibodies. When the second antibody is conjugated to peroxidase, quantities as low as 0.1 ng of protein could be detected.[51] This method may also give information about the polypeptide structure of the activation products, provided that all the fragments are recognized by the antibody used.

SDS–PAGE Analysis of C3, C4, or C5 Antigen Bound to Immunoadsorbent. The degree of activation and fragmentation of C3 (and that of C4 or C5) can be evaluated by SDS–PAGE of the antigens that are withdrawn from the untreated or activated plasma or serum using the respective immunoadsorbent (IA) beads.[52]

Immunoaffinity adsorbents: Immunoaffinity adsorbents (IgG–IA) are prepared by reacting 20 ml of the IgG fraction of goat anti-human C3 (Cappel/Cooper Biomedical) with 10 ml of pretreated CNBr-activated Sepharose 4B (Pharmacia Fine Chemicals) as outlined below and described in detail by Janatova and Gobel.[53] The IgG fraction can also be isolated

[50] C. B. Laurell, *Anal. Biochem.* **10,** 358 (1965).

[51] H. Towbin, T. Staehelin, and J. Gordon, *Proc. Natl. Acad. Sci. U.S.A.* **76,** 4350 (1979).

[52] J. Janatova and R. J. Gobel, *J. Immunol. Methods* **85,** 17 (1985).

[53] J. Janatova and R. J. Gobel, *Biochem. J.* **221,** 113 (1984).

from commercial antiserum as a supernatant following the precipitation of the other serum proteins with 6.8% caprylic acid at pH 5.0.[54] Covalent coupling of IgG to the Sepharose matrix[53] is carried out at pH 6.8 in 50 m*M* sodium phosphate, containing 150 m*M* NaCl at 4° for 16–20 hr. Any residual active sites on the IA beads are blocked with 0.2 *M* glycine at pH 8.2. Depending on the concentration of monospecific antibody in IgG fraction, the resulting binding capacity is usually in the range of 2–4 mg C3/ml of anti-C3,IgG–IA beads. After thorough washing with buffer of pH 5.0 (0.1 *M* sodium acetate) alternating with that of pH 8.3 (0.1 *M* sodium bicarbonate), both containing 250 m*M* NaCl, IA beads are subsequently converted to anti-C3,F(ab′)$_2$–IA by treatment with pepsin at pH 4.3 as described below. Purified anti-C3,F(ab′)$_2$ is commercially available (Cappel/Cooper Biomedical) and can be immobilized as described above, but at 1 : 1 ratio.

Conversion of IgG–IA into F(ab′)$_2$–IA[53]: Immunoadsorbent beads with covalently bound IgG (IgG–IA) are equilibrated with, and then suspended (1 : 1) in 0.4 *M* sodium acetate, pH 4.3. The slurry is incubated with 1 part of pepsin per 100 parts (w/w) of immobilized IgG in a shaker bath for 2 hr at 37°. The action of pepsin is stopped by adjusting the reaction mixture to pH 8.0 with solid Tris base (0.22 g/ml). Occasionally, a small amount of H chains remains undigested. IA beads are placed into a small column and washed sequentially with 10 vol of 50% (v/v) ethylene glycol in EQB (equilibrating buffer for IA: 50 m*M* sodium phosphate, 250 m*M* NaCl, 5 m*M* disodium EDTA, pH 7.4), 3 *M* NaCl in EQB, 0.1% (w/v) SDS in EQB, 1 *M* acetic acid, and finally with EQB. Beads are stored as a 1 : 1 suspension in EQB containing 0.2% sodium azide at 4°. A sample of every new batch of F(ab′)$_2$–IA is analyzed by SDS–PAGE as indicated below.

SDS–PAGE of IA beads without or with bound antigen[52,53]: A 25-μl aliquot of F(ab′)$_2$–IA beads suspended in an equal volume of EQB is pipetted into a disposable 1-ml tuberculin syringe with a porous polypropylene disk placed at the bottom and is incubated with 100 μl of plasma or serum containing 5 or 10 m*M* disodium EDTA for 5–15 min. The liquid phase is then drained, and the antigen-saturated IA beads are washed with 1 ml EQB, then with 1 ml 3 *M* NaCl in EQB to remove nonimmunologically bound proteins such as serum albumin, and subsequently with 1 ml of EQB. Five- or 10-μl portions of antigen-saturated IA beads, or unused IA beads as a control, suspended in an equal volume of EQB, are then mixed 2 : 1 (v/v) with the Laemmli sample buffer [0.18 *M* Tris–HCl/glycine, pH 6.8, containing 9% (w/v) SDS and 30% (v/v) glycerol] and analyzed in

[54] M. Steinbuch and R. Audran, *Arch. Biochem. Biophys.* **134,** 279 (1969).

the absence or presence of 5% 2-mercaptoethanol (v/v) by SDS–PAGE,[55] using a 10% (w/v) polyacrylamide resolving gel in the presence of 0.1% SDS. Gels are stained with 0.25% (w/v) Coomassie Brilliant Blue R-250 in methanol/acetic acid/water (5 : 1 : 4, by volume).

When unused $F(ab')_2$ beads are analyzed under nonreducing conditions only small Fc peptides, which migrate close to the dye front, are observed. In the presence of 2-mercaptoethanol, reduced $F(ab')_2$ fragments (i.e., L chains and pepsin-generated Fd′ portions of H chains) are detected as well.[53] The presence of Fc peptides does not seem to interfere with binding or eluting of the antigen, unless the latter or its fragments are in the same M_r range as the Fc peptides. It is difficult to apply exactly the same volumes of IA beads into wells, differences being reflected in the amount of reduced $F(ab')_2$, the internal standard. It should be mentioned at this point that the mobility of C3, C4, or C5 antigens, their constitutive chains, as well as that of immunoglobulins is the same whether they are bound to IA beads or free in solution at the time of application to SDS–PAGE.

Radioimmunoassay to Detect Activation of C3. The activation of C3 in human plasma can be detected by double-antibody radioimmunoassay, in which a monoclonal antibody recognizes a neoantigen on iC3b (three separate neoantigens are known).[40–43] It should be noted that all C3b molecules are immediately cleaved by factor I in the presence of factor H or erythrocyte CR1 receptor to yield iC3b, which consists of disulfide-bonded β-chain, α68, and α43 fragments.[15]

Radioimmunoassay to Detect Activation of C3 or C5: Determination of Anaphylatoxins C3a and C5a. The overall degree of complement activation needs not be high to provoke or propagate acute inflammatory reactions. Physiologically significant activation may not be detectable by hemolytic or immunoelectrophoretic techniques. Determination of the concentration of activation peptides C3a and/or C5a by radioimmunoassay as described in detail by Hugli and Chenoweth[56] is the most sensitive chemical method for the detection of complement activation, with the lower limits in the nanogram range. Radioimmunoassay kits for determination of C3a, C5a, and C4a are prepared and supplied by Upjohn Pharmaceutical Company (Kalamazoo, MI) or Amersham Corporation (Arlington Heights, IL); the detailed protocols for determination of each of the activation peptides are also provided. A more recent and thorough study by Wagner and Hugli[57] compares the results obtained by using the com-

[55] U.K. Laemmli, *Nature (London)* **227,** 680 (1970).

[56] T.E. Hugli and D.E. Chenoweth, *in* "Immunoassays: Clinical Laboratory Techniques for the 1980's" (R.M. Nakamura, W.R. Dito, and E.S. Tucker III, eds.), p. 443. Liss, New York, 1980.

[57] J.L. Wagner and T.E. Hugli, *Anal. Biochem.* **136,** 75 (1984).

mercial radioimmunoassay kits of Upjohn Co. and those obtained by the originally described procedure.[56]

To remove cross-reactive macromolecules (uncleaved C3 or C5), samples are first precipitated with the precipitating agent (a part of the radioimmunoassay kit) or with 1 *M* HCl. The acidification procedure denatures and precipitates both C3 and C5 while leaving the acid-stable anaphylatoxins C3a and C5a in solution. Immunoglobulins are also precipitated. The acid supernatant, containing C3a and C5a, is neutralized with NaOH. Then radiolabeled C3a or C5a (also a part of the kit) is added, followed by anti-C3a or anti-C5a rabbit antibody. After incubation, goat antibody raised against the rabbit immunoglobulin is added to aid precipitation of immunoglobulin-bound ligand [IgG-Sorb or saturated $(NH_4)_2SO_4$ can be used to separate immunoglobulin-bound ligand from the unbound anaphylatoxins]. The amount of radiolabeled antigen bound to antibody is determined. A standard curve is obtained by using a fixed concentration of both the first antibody and the radiolabeled antigen and increasing known amounts of unlabeled antigen. The concentration of C3a or C5a is then determined by comparing the displacement of radioactive antigen by the unlabeled antigen from plasma or serum sample with the standard curve. Despite the extensive homology between the two anaphylatoxins, the antibody specific for each activation peptide does not cross-react with the others.

A similar radioimmunoassay to determine C4a (which can also be measured with the method described above), the least effective of the anaphylatoxins but a reliable indicator of activation of the classical pathway, has been developed by Gorski.[58] A radioimmunoassay kit is also available from Upjohn Co. or Amersham Corp. The major difference in this procedure is in the precipitation step, as it uses 50% saturated $(NH_4)_2SO_4$ rather than 1 *M* HCl as described for C3a and C5a.[56] Commercial radioimmunoassay kits utilize one precipitating agent (of an undisclosed composition) in determination of all three anaphylatoxins C3a, C5a, and C4a.

The sensitivity of the radioimmunoassay for C3a is approximately 50 times greater than that of the other techniques.[56] Conversion of 0.02–0.1% of the plasma C3 can be detected (conventional hemolytic titrations can detect about 5% conversion; radial immunodiffusion and immunoelectrophoretic techniques are even less sensitive). C3 can be activated by either of the two or by both pathways, and the presence and concentration of C3a in EDTA–plasma can serve as a relatively reliable marker of complement activation. When assaying the concentration of C5a in plasma samples,

[58] J. P. Gorski, *J. Immunol. Methods* **47,** 61 (1981).

one should keep in mind that the majority of C5a binds, under physiological conditions, irreversibly to its receptors[22,59,60] on neutrophils and monocytes, and it is not found free in plasma until this capacity is exceeded. Therefore, the concentration of C5a in plasma can not be considered as a true indication of complement activation.

The background levels obtained for EDTA–plasma by using the commercial kits[57] are 152 ± 69 and 5.4 ± 6.6 ng/ml for C3a nd C5a, respectively. It has also been demonstrated in the same study[57] that coagulation during the preparation of serum stimulates a sizable activation of both C3 and C5; the concentration of C3a increased 10-fold (1597 ± 706 ng/ml) and that of C5a was elevated even more (312 ± 155 ng/ml).

Extreme care should be observed during the preparation of samples prior to the assay. First, blood is drawn into EDTA-containing vacutainers (Becton Dickinson CBC tubes). The samples are centrifuged immediately in order to remove all cells. EDTA–plasma samples are then frozen and stored at $-70°$. Other biological fluids should be treated similarly. In some instances, when the samples have to be diluted in order to obtain concentrations of the peptides within the range of the standard curve, it is the supernatants following the precipitation step that are diluted with phosphate-buffered saline (PBS) containing 10 m*M* disodium EDTA. The assay to obtain the standard curve and the concentration of anaphylatoxins in samples must always be run in duplicate and simultaneously. It is also advisable to include positive controls in the assay. If all three activation peptides are to be determined, serum samples incubated with activators of both pathways should be used (aggregated IgG for classical and cobra venom factor for the alternative pathway; the latter can also be activated by endotoxin or by zymosan) (see also this volume [48]).

Detection and Quantitation of Functional C3 or C5

Evaluation of Hemolytic Activity. The ability of individual complement components to mediate hemolysis of EA cells (sheep red blood cells sensitized with the rabbit antibody raised against their membrane antigen) can be estimated by hemolytic titrations. The procedure to be described below is a modification of the one-step method of Gaither *et al.*[61] for determination of C4 in which guinea pig C4-deficient serum is used as a source of all other complement components. The utilization of sera depleted of the respective complement component makes the preparation of cell intermediates involving highly purified functionally active protein components

[59] T. E. Rollins and M. S. Springer, *J. Biol. Chem.* **260,** 7157 (1985).
[60] R. Huey and T. E. Hugli, *J. Immunol.* **135,** 2063 (1985).
[61] T. A. Gaither, D. W. Alling, and M. M. Frank, *J. Immunol.* **113,** 574 (1974).

unnecessary. Sera from patients with hereditary deficiencies of C3 or C5 are relatively uncommon and rarely available. The substitutes can be prepared by depleting fresh normal human serum of C3 or C5 by immunoaffinity chromatography. A simple and fast procedure used in the author's laboratory for preparation of C3- or C5-depleted serum is described below.

Normal human serum: Normal human serum (NHS) is obtained from healthy donors. Blood, freshly drawn into vacutainers with no additives (Becton Dickinson, No. 6432) is allowed to clot for 30 min at room temperature. Serum is then separated by centrifugation at 685 *g* for 15 min at 4°, placed on ice, and immediately applied to the IA column at 4°.

Immunoadsorbents: Immunoadsorbents (IA) are prepared and pre-equilibrated to pH 7.4 with EQB (50 m*M* sodium phosphate, 250 m*M* NaCl, 5 m*M* disodium EDTA) according to earlier described protocols.[53] Note that IgG–IA must be converted to $F(ab')_2$–IA by treatment with pepsin as described above (less than 15% of the original binding capacity of IgG–IA is lost during this process) if IA is to be employed for preparation of depleted serum to avoid C1q binding. The use of immunoadsorbents with high binding capacity (4 or 2 mg C3/ml IA beads to which approximately 20 or 10 mg of monospecific anti-C3 antibody is immobilized) prevents an excessive dilution of NHS and allows preparation of depleted serum, the total protein content of which is as high as 80–85% of the parent NHS.

Preparation of C3-depleted serum (C3DS)[62]*:* Typically, 10 ml of fresh, ice-cold NHS without any additives is passed through a 6-ml column of anti-C3,$F(ab')_2$–IA beads equilibrated to 4° at a flow rate not exceeding 10 ml/cm^2/hr. The eluate (0.5 ml/tube) is monitored for protein content by measuring the absorbance at 280 nm of 1 : 100 diluted aliquots. Five- and 20-μl aliquots from the fractions corresponding to the protein plateau are examined by hemolytic assay. The aliquots are mixed with 600 μl $GVBS^{2+}$ and incubated for 1 hr at 37° after addition of 400 μl of EA cells (5×10^7/ml in $GVBS^{2+}$).[63] The evaluation is performed in the same way as described below for hemolytic titrations. Only those protein fractions that do not exhibit C3 hemolytic activity (a 5-μl aliquot of the starting NHS sample is used as a control yielding 100% lysis) are pooled to a total volume of approximately 8 ml.

[62] B. D. Cheson, H. S. Walker, M. E. Heath, R. J. Gobel, and J. Janatova, *Blood* **63**, 949 (1984).

[63] $GVBS^{2+}$, veronal-buffered saline containing 143 m*M* NaCl, 3 m*M* $NaHCO_3$, 1.45 m*M* sodium barbital, 2.5 m*M* 5,5-barbituric acid, 0.5 m*M* $MgCl_2$, and 0.136 m*M* $CaCl_2$, pH 7.4, is prepared fresh daily from 5-fold concentrated stock and made 0.1% (w/v) in gelatin.

Excess salts and EDTA are removed from 4.0- to 4.5-ml aliquots of postimmunoadsorbent serum by centrifugation (6 min, 1500 *g*, at 4°) through 5 g of coarse Sephadex G-25 (Pharmacia), preequilibrated with PBS. Then 10 ml of the C3DS is supplemented with 50 μl of 0.3 *M* $MgCl_2$ and 100 μl 0.25 *M* $CaCl_2$ to obtain final concentrations of 1.5 m*M* $MgCl_2$ and 2.5 m*M* $CaCl_2$. The individual aliquots (usually 300 μl, in 1.5-ml polypropylene tubes) are quickly frozen in a mixture of solid CO_2 and acetone and stored at −70°. The same procedure is used for the preparation of C4- or C5-depleted sera (C4DS or C5DS)[63] except that the binding capacities of the $F(ab')_2$–IA beads need not be so high (e.g., 0.24 mg C4 or 0.16 mg C5 per ml of the respective IA beads). The hemolytic activity of the CDS is fully restored to the levels of NHS after the addition of a physiological amount of the appropriate purified protein (C3, 0.8–1.5 mg/ml; C4, 0.2–0.4 mg/ml; C5, 60–80 μg/ml).

The specificity of removal of C3 (or C4 or C5) from NHS by affinity chromatography is determined by SDS–PAGE of the respective IA beads before and after chromatography (see above and Ref. 53). No conversion of the respective α chains to their activated forms as a result of immunoaffinity chromatography has been observed. The amount of nonimmunologically adsorbed albumin and IgG is small and not detectable by the criteria of quantitative agarose electrophoresis. Analysis of C1 by double immunodiffusion in the absence of EDTA and the presence of Ca^{2+} according to the protocol by Ziccardi[64] demonstrates that component C1 remains undissociated and unactivated during the whole procedure. Consequently, no reconstitution with C1 or C1q is necessary. All other complement proteins also remain fully functional, as determined by titrations of hemolytic activity. The decrease in their hemolytic titer to 80 or 85% of parent NHS again corresponds to a final dilution of the starting NHS sample.

Regeneration of immunoadsorbent column: Following application of the serum to the column, the flow rate is increased 3-fold, and washing with the equilibrating buffer (EQB) is continued until absorbance values (*A* at 280 nm) of undiluted eluate are below 0.03. The regeneration is carried out in the direction opposite to that of sample application. Elution with EQB, containing 3 *M* NaCl, removes practically all of the nonspecifically bound albumin, whereas the antigen remains attached to the IA beads. Most of the antigen is then eluted with 1 *M* acetic acid (pH 2.5). The pH of the eluate containing antigen is immediately adjusted to 7.4 with 2 *M* Tris base (Sigma). The column is promptly washed with a 10-fold volume of

[64] R. J. Ziccardi, *in* "Immunoassays: Clinical Laboratory Techniques for the 1980's" (R. M. Nakamura, W. R. Dito, and E. S. Tucker III, eds.), p. 433. Liss, New York, 1980.

EQB to decrease the effect of low pH on the antibodies and Sepharose matrix.

The effectiveness of the regeneration is evaluated by SDS–PAGE analysis of the aliquots of IA beads removed from the reequilibrated column. This allows for determination of residual, unstripped antigen. This is an important step, since the capacity of IA column for removing the C3 antigen from NHS decreases with every subsequent cycle.[65] Depending on the batch of antibody, it may take more than 8, or as few as 3, cycles of sample application and subsequent regeneration of the column before IA beads become saturated with the antigen such that it would be impossible to dissociate from IA beads by any other means than by SDS–PAGE. It is noteworthy that the presence of SDS (0.1–3%) does not affect the dissociation of the antigen from IA beads unless an electrical field is applied. The immobilized antibody or its $F(ab')_2$ fragment retain their binding capacity after a thorough removal of SDS.

Hemolytic titrations: Twofold serial dilutions (0.5 ml each) of the samples containing C3 or C5 are made with isotonic $GVBS^{2+}$ (Mg^{2+}-, Ca^{2+}-, and gelatin-containing veronal-buffered saline). Then a 100-μl aliquot of serum, depleted of C3 (or C5) and diluted 1 : 20 with $GVBS^{2+}$, is added to each tube, their contents are mixed, and after the addition of 400 μl of EA cells (5×10^7/ml in $GVBS^{2+}$) the samples are incubated for 1 hr at 37°. Control samples for background lysis contain only EA cells and $GVBS^{2+}$ or $GVBS^{2+}$ and 1 : 20 diluted C3DS (or C5DS). A 5-μl aliquot of undiluted fresh NHS is used as a positive control, yielding 100% lysis. After addition of 1 ml ice-cold PBS the samples are mixed, and unlysed EA cells and EA membranes are pelleted by centrifugation. The degree of lysis is determined by measuring the optical density at 415 nm, which corresponds to the concentration of hemoglobin released by lysed cells into the supernatant. The hemolytic unit, CH_{50}, can be expressed as the reciprocal of the dilution of C3 (or C5) that would lyse 50% of the cells. The hemolytic unit can also be defined as the amount of C3 (or C5) in micrograms (or nanograms) that would mediate lysis of 50% of EA cells in the presence of C3DS (or C5DS) (see also this volume [47]).

An example of the determination of concentration of C3(n) (C3 molecule with the thiol ester) in six different C3 preparations by the means of hemolytic titrations is given in Fig. 2. The same C3 samples were also analyzed by the differential SH assay (described below), and the obtained "absolute" values for the content of C3(n) are given in column (a). The hemolytic titrations give only relative values unless a standard with known concentration of C3(n) is available. In this experiment the C3 sample

[65] J. Janatova and R. J. Gobel, unpublished observations.

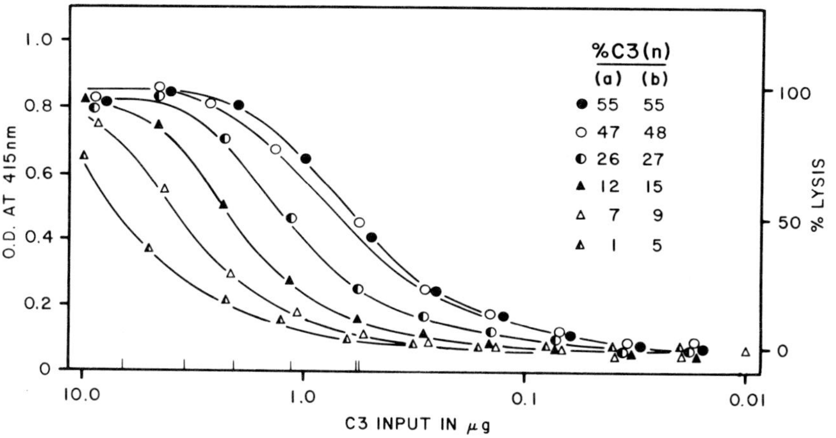

FIG. 2. Determination of C3(n) concentration in six different C3 preparations by hemolytic titrations. This approach gives only relative values, unless a standard of known C3(n) concentration is included in the assay. In this instance the sample containing 55% C3(n) was used as a standard. The concentration of C3(n) in the standard and in the other five samples was determined by the differential sulfhydryl assay [column (a)] according to the protocol described later in this chapter. The values for C3(n) based on CH_{50} hemolytic units obtained in this experiment are given in column (b).

containing 55% C3(n) was used as a standard; the concentrations of C3(n) calculated from CH_{50} values for the other five samples are given in column (b) and agree well with those in column (a).

Quantitative Determination of Native Form in Samples by Differential Sulfhydryl Assay.[66] Both native C3(n) and C4(n) contain the internal thiol ester bond (see Refs. 1, 5, 67, and 68 for details) which is an absolute requirement for their ability to bind covalently to cell surfaces and/or immune complexes, and thus to mediate hemolysis. Whereas no free sulfhydryl (SH) group can be detected in native C3 or C4, one SH group per molecule of C3 or C4 becomes accessible to SH-reactive reagents after cleavage/activation with the respective converting enzyme or treatment with inactivating agents. The thiol ester present in C3(n) or C4(n) undergoes spontaneous hydrolysis during storage and/or freeze–thaw processes. The resulting hemolytically inactive C3(i) or C4(i) forms contain one free SH per molecule. This group, however, is labile so that hemolytically inactive C3(i) or C4(i) may or may not have a reagent-accessible SH.

[66] J. Janatova, *Fed. Proc., Fed. Am. Soc. Exp. Biol.* **41,** 848 (1982).
[67] J. Janatova, *Ann. N.Y. Acad. Sci.* **421,** 218 (1983).
[68] B. F. Tack, *Springer Semin. Immunopathol.* **6,** 259 (1983).

Rationale: Every native C3 or C4 molecule has the potential to bind covalently after activation. In such a molecule, 1 mol of SH per mole of protein should be exposed after the treatment of C3(n) or C4(n) with the activating enzyme or with inactivating agent.[27,34] In this assay the percentage of C3(n) or C4(n) is evaluated by the differential SH content before and after the treatment of sample with converting enzyme. The starting SH values correspond to the amount of C3(i) or C4(i) containing a free SH group. A difference between the final (after treatment) and starting SH values corresponds to the amount of C3(n) or C4(n). The difference between one SH per molecule and the final SH value represents C3(i) or C4(i) without detectable SH. DTNB [5,5′-dithiobis(2-nitrobenzoic acid)] is used as an SH reagent[69] and β-trypsin as an activating agent for C3, while native C4 is activated by C$\overline{1s}$. The reaction proceeds at pH 7.0 at ambient temperature in the presence of 2 m*M* disodium EDTA.

Reagents

- B, 0.1 *M* sodium phosphate buffer (pH 7.0), containing 2 m*M* disodium EDTA
- S, C3 or C4 sample in buffer B at about 1×10^{-5} *M* concentration (1.85 or 2 mg/ml, respectively)
- R, 10 m*M* DTNB in buffer B, purged with nitrogen, and stored in the dark at 4°
- E, converting enzyme β-trypsin (5 mg/ml) made freshly in B or as a frozen stock solution in 1 m*M* HCl; C$\overline{1s}$ (1.6 mg/ml) in PBS (for the preparation of C$\overline{1s}$, see this volume [47])

Procedure: The accuracy of the assay depends on precision of determination of protein concentration and delivery of the *exact* volumes into reaction mixtures. Constriction glass pipets should be used; tips are unacceptable.

Step I. Protein concentration in the C3 or C4 sample is determined from the absorbance of the solution measured at 280 nm in quartz cuvettes of 2-mm path length. An absorption coefficient ($A_{1\,\mathrm{cm}}^{1\%}$ at 280 nm) of 9.7 and a molecular weight of 185,000 are used for C3,[70] and an absorption coefficient of 8.3[71] and 198,000 as molecular weight[72] are used in the evaluation of C4. The absorbance at 412 nm that would correspond to 1 mol SH per mole of C3 or C4 at the given protein concentration in Step II is calculated and taken as equal to 100%.

[69] G. Ellman, *Arch. Biochem. Biophys.* **82,** 70 (1959).

[70] B. F. Tack and J. W. Prahl, *Biochemistry* **15,** 4513 (1976).

[71] S. Nagasawa and R. M. Stroud, *Proc. Natl. Acad. Sci. U.S.A.* **74,** 2998 (1977).

[72] C. Bolotin, S. Morris, B. F. Tack, and J. Prahl, *Biochemistry* **16,** 2008 (1977).

Step II. Aliquots, 400 μl each, of B and S are delivered by glass constriction pipets into separate cuvettes (4-mm wide, 10-mm path length), and the background absorbance of sample S is read at 412 nm (in centrifuged or filtered samples, it can be as low as 0.002) against the absorbance of B which is set at zero. Then 50 μl of R is added to each cuvette, B and S. The contents of both cuvettes are mixed immediately, and the absorbance difference at 412 nm between S and B is recorded. The obtained value is corrected for an original protein absorbance at 412 nm (if any). The content of SH is calculated using the value of 13,600 M^{-1} cm^{-1} as a molar extinction coefficient[69] for the 5-thio(2-nitrobenzoic acid) (TNB) anion at 412 nm, and it corresponds to the content of hemolytically inactive molecules with detectable SH.

Step III. The ability of native C3 or C4 to express a free SH upon cleavage by their converting enzyme is utilized in this step.[27,34] A 5-μl aliquot of E (β-trypsin in place of C3 convertase for C3; $C1\bar{s}$ for C4) is added to 450 μl of a protein–TNB–DTNB mixture from Step II, and the time is recorded. Both cuvettes are immediately covered with parafilm, and their contents are mixed by inverting. The absorbance is followed with time and reaches a plateau when all native C3 or C4 molecules are "activated." The final value is read 30–60 sec after the addition of E. The increase in the SH content due to the activation is calculated from the difference in the absorbance of S between the plateau value and the one before addition of E. This increase corresponds to the amount of hemolytically active C3 or C4 molecules present in the sample.

Step IV. The concentration of hemolytically inactive C3(i) or C4(i) forms without detectable SH is calculated from the difference between one SH (100% calculated in Step I) and the final plateau SH value as obtained in the Step III.

Comment: Unlike any of the current methods, antigenic or functional assays for C3 or C4, determination of the differential SH content before and after treatment with the enzyme serves as a quantitative measure of concentration of native molecules in C3 or C4 samples. It can also differentiate between at least two forms of hemolytically inactive molecules. This assay is also intended for standardization of C3 or C4 samples to be employed as reference points in hemolytic titrations, as demonstrated in Fig. 2. Knowledge of the exact composition of C3 or C4 samples/preparations becomes even more obligatory since many similarities have been identified between C3b or C4b and hemolytically inactive C3(i) or C4(i).[36–38] The ability to participate in the formation of C3 convertases is perhaps the most important.[5,36,73]

[73] C. Parkes, R. G. DiScipio, M. A. Kerr, and R. Prohaska, *Biochem. J.* **193,** 963 (1981).

Purification of the Three Homologous Proteins: C4, C3, C5

Several purification schemes for human C4, C3, or C5 have been published. A method for the isolation of multiple human complement components has been presented by Hammer *et al.*,[74] and the steps relevant to the isolation of C4, C3, and C5, have been described in detail earlier in this series.[75] Two modified protocols for isolation of C4, those after Gigli *et al.*[76] and Bolotin *et al.*,[72] and the protocol for purification of C3 and C5, originally reported by Tack and Prahl,[70] all have been described in detail recently by Whaley.[77] The purification procedures presented below are the ones used in the author's laboratory. The first step, batch adsorbtion to QAE-Sephadex A-50, represents modifications of the protocols published by Lundwall *et al.*[78] for C4 and by Lundwall and Eggertsen[79] for C3 and C5. The choice of Phenyl-Sepharose for the purification of native C3, and its separation from native C5, is based on the observations reported by Al Salihi *et al.*[80]

Procedure. All steps of C4, C3, and C5 isolation are carried out at 4° and in the presence of 0.02% (w/v) sodium azide. In this protocol only PMSF (phenylmethylsulfonyl fluoride) at a final concentration of 0.25 mg/ml plasma is used to prevent activation of the proteins. Unless specified otherwise, a standard buffer (SB), consisting of 20 m*M* sodium phosphate, 2 m*M* disodium EDTA, and 0.02% sodium azide (pH 7.0), is used in all chromatographic steps. To this buffer solid NaCl is added without subsequent adjustment of pH. Concentrations of NaCl vary according to the system used and the protein(s) being adsorbed or eluted. All adsorbents are preequilibrated with the appropriate starting buffer. The pH and mS values of effluents from ion-exchange adsorbents are always checked prior to application of the protein sample. The conductivity of the protein samples is also adjusted prior to its application to an ion-exchange adsorbent. The pH during the adsorbtions is 6.8. The conductivity of all solutions is measured at 4° and reported in units of mS/cm. It is important to keep all glassware at 4° prior to use, especially before the batch adsorbtion.

[74] C. H. Hammer, G. H. Wirtz, L. Renfer, H. D. Gresham, and B. F. Tack, *J. Biol. Chem.* **256,** 3995 (1981).

[75] B. F. Tack, J. Janatova, M. L. Thomas, R. A. Harrison, and C. H. Hammer, this series, Vol. 80, p. 64.

[76] I. Gigli, I. von Zabern, and R. R. Porter, *Biochem. J.* **165,** 439 (1977).

[77] K. Whaley, *in* "Methods in Complement for Clinical Immunologists" (K. Whaley, ed.), pp. 27 and 34. Churchill-Livingstone, Edinburgh, Scotland, 1985.

[78] Å. B. Lundwall, I. Malmheden, G. Stålenheim, and J. Sjöquist, *Eur. J. Biochem.* **117,** 141 (1981).

[79] Å. Lundwall and G. Eggertsen, *J. Immunol. Methods* **81,** 147 (1985).

[80] A. Al Salihi, J. Ripoche, L. Pruvost, and M. Fontaine, *FEBS Lett.* **150,** 238 (1982).

The data presented below for purification of C4, C3, and C5 were obtained from the fractionation of a pool of 5 units of fresh frozen human plasma (approximately 1.1 liter), that were kept for a period of 1–3 months below −20°.

The column eluates are evaluated for the presence of protein by measuring the absorbance at 280 nm. Selected fractions are screened by SDS–PAGE[55] [Laemmli system, 10% (w/v) polyacrylamide resolving gel, Model 360 Mini-Vertical Slab Gel Apparatus by Bio-Rad] to determine the positions of C4, C3, or C5 in the elution diagrams and also to evaluate the degree of contamination by other proteins. The behavior of C4, C3, and C5 on SDS–PAGE has been illustrated elsewhere in this series (Vol. 80 [7]). C3 and C5 consist of two chains each, which yield after a reduction 2 bands on polyacrylamide gels, one corresponding to $M_r \sim 115K$ (α chains) and the other to $M_r \sim 75K$ (β chains). Three chains, α ($M_r \sim 94K$), β ($M_r \sim 75K$), and γ ($M_r \sim 30K$) are present in C4. Fractions containing C4, C3, and/or C5 are further tested for their ability to mediate hemolysis of EA cells in the presence of the appropriate depleted serum. The individual pools containing either C4 or C3 and/or C5 are concentrated by ultrafiltration in 400-ml cells by Amicon (membranes YM100 or PM10). Change of pH is accomplished by ultradialysis with a 5-fold (or larger) volume of a "new" buffer in the same Amicon cell. Ionic strength is adjusted by diluting the protein solution with SB to obtain the required conductivity. Solutions of purified components are transferred by ultradialysis into K buffer (pH 6.5) containing 20 m*M* potassium phosphate and 150 m*M* KCl; 1- or 3-ml aliquots are quickly frozen in 5-ml polypropylene tubes in a mixture of solid CO_2 and acetone and stored at or below −65°.

Treatment of plasma: Each unit of fresh frozen human plasma is obtained from 450 ml of CPDA whole blood. Each transfusion pack with frozen plasma is sealed in a polyethylene bag and placed in a water bath set at 37°. As soon as all plasma is thawed (usually after 1–1.25 hr) the bags are transferred to a container with ice, and then to a cold room (4°). Bags are slit open, and five units of plasma are pooled into a 2-liter heavy duty glass beaker. After removing aliquots for determinations, 13 ml of PMSF (25 mg/ml in 2-propanol) is added to 1160 ml of pooled plasma. When the temperature of the pooled PMSF–plasma reaches 4°, its conductivity (in this instance 7.2 mS) is increased to that of the QAE-I buffer (10.6 mS) by addition of 5.8 g NaCl or as appropriate.

Batch adsorbtion of C4 to QAE-Sephadex A-50 and further purification of C4: QAE-Sephadex A-50 (15 g) is swollen in QAE-I buffer containing 160 m*M* NaCl in SB (10.6 mS, pH 6.8). The preequilibrated QAE-I adsorbent (300 ml) is added to the PMSF–plasma (1173 ml, 10.6 mS), and the mixture is stirred every 10 or 15 min with a glass rod. After 90 min at 4°, the QAE-I plasma supernatant is decanted and used immediately for the isolation of C3 and C5 (see below). The QAE-I adsorbent with bound proteins is filtered under reduced pressure, using a 500-ml glass filter, and the filtrate is added to the QAE-I supernatant. The QAE-I adsorbent is washed once on the filter with the QAE-I starting buffer (160 mM NaCl in SB) and then transferred into a clean 2-liter beaker where it is mixed with DEAE-Sepharose CL-6B (100 ml) preequilibrated with QAE-I. The slurry is poured into a column (5 × 40 cm) with two plungers (Pharmacia) (the height of the adsorbent bed is approximately 20 cm). The outlet is connected to the second column (5 × 30 cm, bed height 10–12.5 cm) containing the same QAE-I/DEAE adsorbent but without proteins. The connected columns are washed overnight with excess (about 3.3 liters) of the QAE-I starting buffer until the absorbance of the eluate at 280 nm decreases to 0.100 or less.

The complement component C4 is separated from other proteins, and it is eluted from the QAE-I/DEAE-adsorbent by using an 8-liter (total volume) linear salt gradient in SB with 160 mM (pH 6.8, 10.6 mS) and 300 mM NaCl (pH 6.6, 18.2 mS) as the starting and limiting concentrations. The gradient is followed with 400 mM NaCl in SB. Fractions (20 ml/tube) are collected at a constant flow rate of 150 ml/hr. An example of the elution profile is given in Fig. 3. SDS–PAGE analysis of selected fractions is shown under nonreducing (inset A) and reducing conditions (inset B). The C4-containing eluate from tubes 210 to 310 (wells 4 through 13) is pooled [total volume of 2200 ml (total 283 A_{280} units)] and, after the addition of 10 ml PMSF (25 mg/ml in 2-propanol), is concentrated by ultrafiltration over a YM100 membrane (Amicon) in three 400-ml cells to the final volume of 53 ml (5.08 A_{280}/ml). The C4 pool is ultradialyzed with an 8-fold volume of K buffer (pH 6.5) and stored in 3-ml aliquots at −65°.

The first step in the isolation of C4 (Fig. 3) allows for copurification of the inter-α-trypsin inhibitor (IαI),[81] a single polypeptide chain glycoprotein with apparent M_r of 190–200K ($A_{1\,\mathrm{cm}}^{1\%}$ at 280 nm = 7.1). This is a major contaminant in the C4 pool. Only partial separation can be obtained by gel filtration on Sephacryl S-300, where IαI elutes before C4.

Purification of native C4 on ω-aminohexylagarose: The second and final step in the purification of C4 involves chromatography on ω-aminohexylagarose. Inter-α-trypsin inhibitor, some unspecified contaminants, and hemolytically inactive C4(i) forms are separated from native C4 as shown in Fig. 4. A portion (28%) of the concentrated post-QAE-I/DEAE C4 pool, containing total 75 A_{280} units in 24 ml, is applied to a ω-amino-

[81] M. Steinbuch, this series, Vol. 45, p. 760.

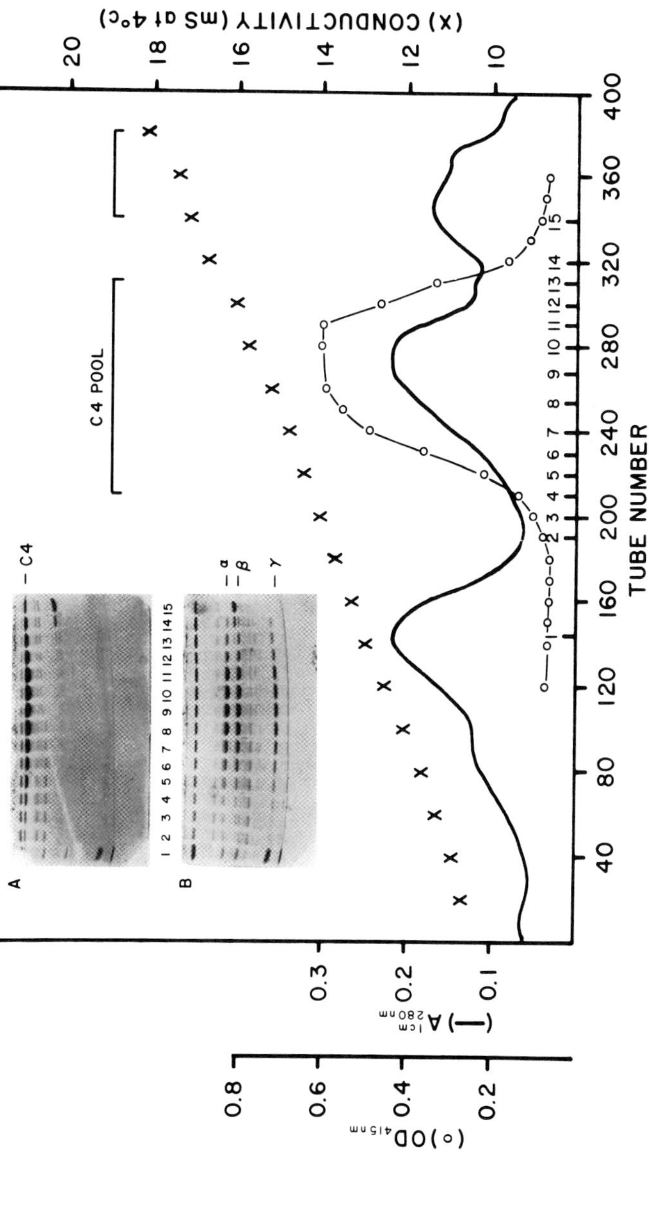

FIG. 3. Isolation of C4 on QAE-I/DEAE adsorbent. For details, see text.

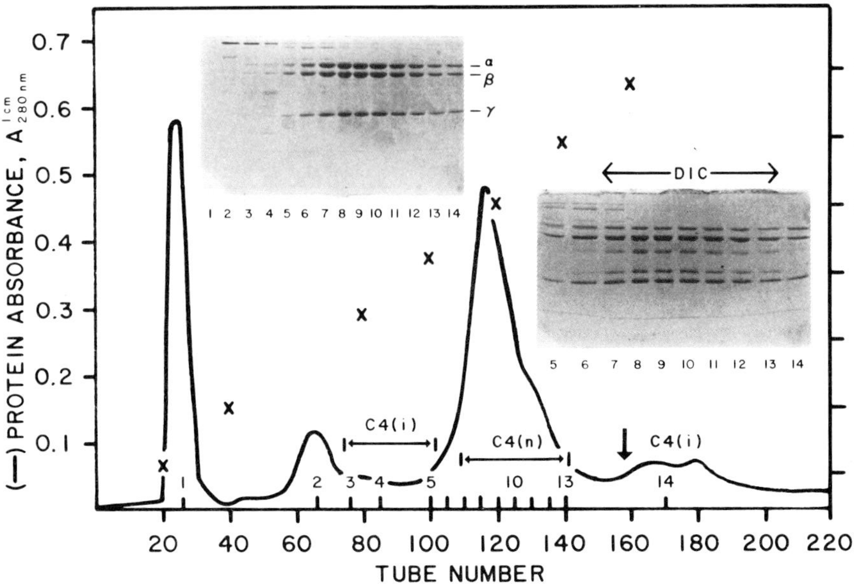

FIG. 4. Isolation of C4 on ω-aminohexylagarose. For details, see text.

hexylagarose column (1.4 × 57 cm) preequilibrated to 13.2 mS and pH 6.8 with 200 m*M* NaCl in SB. The column is washed with 10 ml of the starting buffer, and the adsorbed proteins are eluted with a linear NaCl gradient in SB (total volume 1 liter) to a limiting concentration of 600 m*M* NaCl. The eluate is collected (70 drops or 4.6 ml/tube) at a flow rate of 30 ml/hr. The elution of the proteins is completed with the application of 600 m*M* NaCl in SB (34.0 mS, pH 6.5). The C4-containing fractions are analyzed by hemolytic assay and by SDS–PAGE in the presence of 2-mercaptoethanol, which locates the position of the inter-α-trypsin inhibitor (wells 2 to 4). The first absorbtion peak (well 1) did not yield a band on staining, and it could correspond to the aggregates of IαI. The presence of hemolytically active C4(n) can be conveniently located by using the DIC assay.

The denaturant-induced cleavage (DIC) of the α chain[27,33] has also been used as a criterion for the presence of hemolytically active C4(n) as depicted in the inset on the right (Fig. 4). Aliquots of C4-containing fractions were heated at 95° in a 3% SDS-containing solution in the absence of a reducing agent for 15 min. The α chain of native C4(n) is cleaved to two fragments of M_r 40K and 54K that migrate upon reduction with 2-mercaptoethanol between the β and γ chains (wells 7 to 13). All C4-containing fractions (about 56 mg C4) represent 62% of the total absorbance applied to this column. Approximately 80% of the total C4 eluted from ω-aminohexylagarose (Fig. 4) is present in the C4(n) pool (tubes 110 to 140; wells 7 to 13) and correspond to 45 mg of C4. A total of 135 mg of native C4 is obtained from 1.1 liters of pooled plasma. Additionally, 30 mg of the C4 withdrawn from the PMSF–plasma is converted to hemolytically inactive C4(i) forms during the time of isolation, mainly due to a freezing–thawing process.

Batch adsorbtion of C3 and C5 to QAE-Sephadex A-50: Following the adsorbtion of C4 and other proteins to the QAE-I adsorbent (QAE-Sephadex A-50 in SB with 160 m*M* NaCl), the QAE-I plasma supernatant (1300 ml) is made 5 m*M* in disodium EDTA by the addition of 65 ml of a 100 m*M* EDTA solution. Then the conductivity of the QAE-I plasma supernatant is adjusted to that of the QAE-II buffer (100 m*M* NaCl in SB, 7.7 ± 0.1 mS) by addition of SB (about 870 ml). The preequilibrated QAE-II adsorbent (25 g QAE-Sephadex A-50 swollen to 600 ml in SB containing 100 m*M* NaCl; 7.8 mS, pH 6.8) is added to the adjusted plasma supernatant (2235 ml) in a 4-liter beaker. The mixture is stirred every 10–15 min with a glass rod. After 90 min at 4° the QAE-II plasma supernatant is decanted, and the QAE-II adsorbent with bound proteins is filtered under reduced pressure on a 1-liter glass filter, washed once on the filter with the QAE-II starting buffer (100 m*M* NaCl in SB) and transferred into a clean 2-liter beaker where it is mixed with DEAE-Sepharose (200 ml) preequilibrated with 100 m*M* NaCl in SB. The slurry is then poured into a column (5 × 50 cm) with 2 plungers (Pharmacia) (the height of the adsorbent bed is approximately 30 cm). The outlet is connected to the second column (5 × 30 cm, bed height 10–12.5 cm) containing the same QAE-II/DEAE adsorbent but without proteins. The connected columns are washed overnight with about 4.4 liters of the QAE-II starting buffer (100 m*M* NaCl in SB) until the absorbance of the eluate at 280 nm decreases to 0.200 or less.

Practically all C3 and C5, and a portion of factor H, from the 1.1-liter plasma pool bind to the QAE-II adsorbent. Following the extensive washing with the QAE-II starting buffer, a linear salt gradient is developed by using 3 liters each of 100 m*M* and 200 m*M* NaCl in SB as the starting and limiting concentrations. The elution of proteins is completed by using 400 m*M* NaCl in SB.

An example of separation and elution of C3, C5, and other proteins from the QAE-II/DEAE adsorbent is given in Fig. 5. Only the ascending part of C3/C5 elution curve is overlapped by factor H. The two pools of C3/C5-containing fractions, denoted PS1 (583 A_{280} units in 600 ml) and PS2 (810 A_{280} units in 820 ml), are concentrated by ultrafiltration (YM100, Amicon). As follows from the SDS–PAGE analysis in the

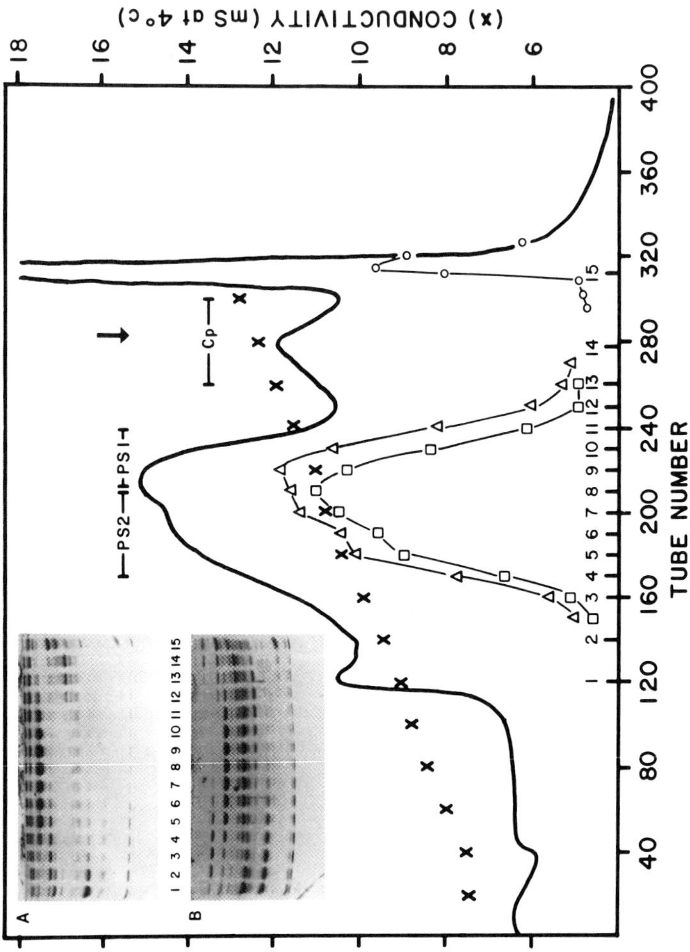

FIG. 5. Isolation of C3 and C5 on QAE-II/DEAE adsorbent. For details, see text.

absence (inset A) or the presence (inset B) of 2-mercaptoethanol, the PS1 pool (wells 8 to 11) contains much less contaminant than the PS2 pool. Factor H[82] is present only in the early eluted PS2 pool (wells 4 to 8). No attempt should be made to recover the small amount of a residual C4 bound by the QAE-II adsorbent. Its presence is detected by the hemolytic assay rather than by SDS–PAGE (well 15). The position of caeruloplasmin is identified by both SDS–PAGE (wells 12 to 14) and by a blue color of the eluate. A single polypeptide chain corresponding by its mobility to C1 esterase inhibitor (M_r 100,000–105,000)[83] is well separated from C3/C5 (wells 12 to 14).

[Chromatography on DEAE-Sephacel[84] as the second step in C3/C5 purification results in an excellent separation of H from both C5 and C3.[85] We have not been successful, however, in obtaining good separation of C5 from C3 for the following reasons: (1) hemolytically inactive C3(i) coelutes with native C5 and (2) in most instances C5 elutes in two peaks, the larger coeluting with C3(i) and the later one with native C3(n). The increase in pH from 6.8 to 7.9 does not improve separation appreciably (the second C5 peak still overlaps substantially that of native C3). A more efficient and faster approach for isolation of native C3 is described below.]

Purification of native C3 by hydrophobic chromatography: In order to separate native C3(n) from hemolytically inactive C3(i) forms and also from C5 and factor H, the concentrated post-QAE-II/DEAE C3/C5-containing pools from the previous step (Fig. 5) are subjected to chromatography on Phenyl-Sepharose (5 × 30 cm, 600-ml bed volume) preequilibrated with K buffer (20 mM potassium phosphate, 150 mM KCl). The two elution profiles depicted in Fig. 6 are obtained after fractionation of the PS1 pool (518 A_{280} units in 87 ml) that is followed at tube 108 by the application of the PS2 pool (741 A_{280} units in 92 ml). Each sample is followed by elution with the K buffer at pH 7.0, and the eluate is collected (20 ml/tube) at a constant flow rate of 150 ml/hr. The majority of the contaminating proteins elute in the first peak as indicated by SDS–PAGE analysis (wells 1 and 2 in Fig. 6). It takes about 13 hr and 2 liters of the K buffer to complete elution of native C3(n) from the preparative column. The C3(n) pools from both runs, PS1 and PS2 (Fig. 6), represent about 37% of the absorbance units applied to Phenyl-Sepharose. The PS1 pool (tubes 54 to 90; total 192 A_{280} units in 740 ml) contains 198 mg, and the PS2 pool (tubes 160 to 195; total 266 A_{280} units in 720 ml) contains

[82] L. G. Crossley, this series, Vol. 80, p. 112.

[83] R. B. Sim and A. Reboul, this series, Vol. 80, p. 43.

[84] J. Ripoche, A. Al Salihi, J. Rousseaux, and M. Fontaine, *Biochem. J.* **221,** 89 (1984).

[85] J. Janatova, unpublished observations.

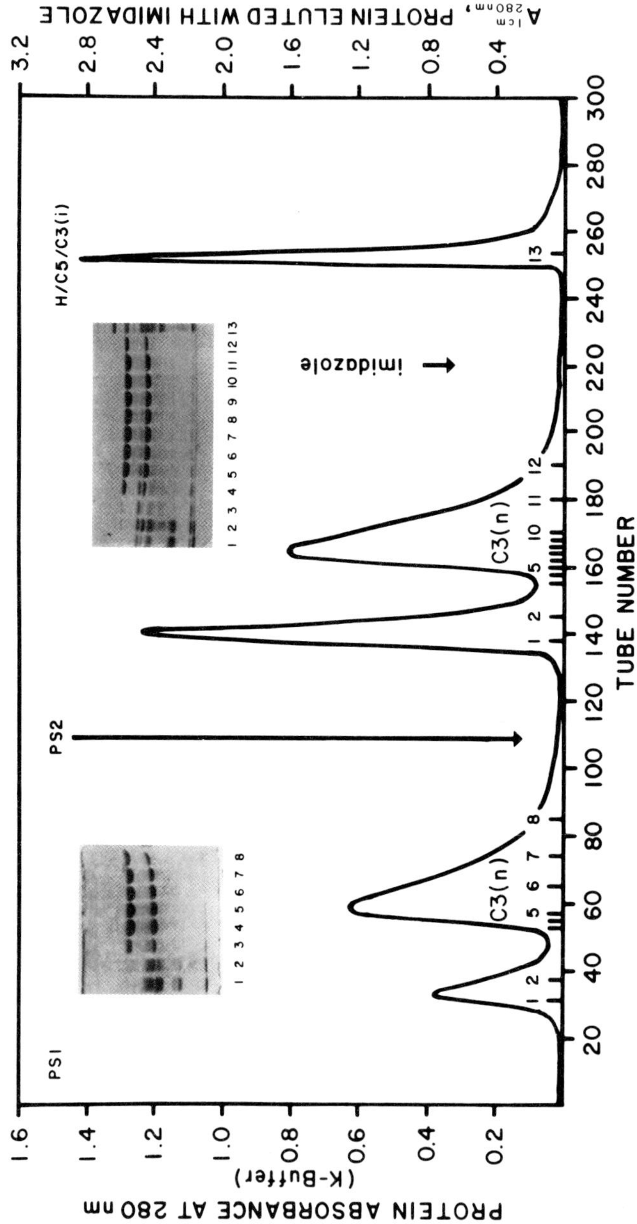

FIG. 6. Separation of C3 from C5 and factor H on Phenyl-Sepharose. For details, see text.

275 mg of the thiol ester-containing C3. After concentration by ultrafiltration to approximately 3.6–4.0 mg C3(n)/ml K buffer, the samples are aliquoted into polypropylene tubes, frozen in a mixture of CO_2 and acetone, and stored below −65°.

Even when all precautions are observed and the sample is stored in the potassium-containing buffer, a certain amount of C3(n) is converted to C3(i) forms by hydrolysis of the thiol ester during the freeze–thaw process. Therefore it is necessary to rechromatograph the C3 preparation immediately prior to its use either on a small Phenyl-Sepharose column (this procedure takes 2.5 hr) or in an overnight procedure using chromatography on ω-aminohexylagarose. Traces of contaminating proteins (if any) are better separated on the latter adsorbent. The detailed protocols for both procedures are described below.

Separation of C3(n) from C3(i) on Phenyl-Sepharose: The hydrolysis of thiol ester results in appearance of free SH,[34] conformational changes,[8,35] and in the case of C3 in increased hydrophobicity.[38,80] When the C3 sample is applied to a Phenyl-Sepharose column, native C3(n) is eluted either with K buffer or with PBS, while all C3(i) forms remain bound to the adsorbent as shown in Fig. 7. The frozen C3 sample (6 or 8 mg in 2 ml) is thawed for 1–2 min at 37° and directly applied in an up-flow direction to a Phenyl-Sepharose column (1.4 × 5 cm). The column is washed with PBS at 40 ml/hr, and 2-ml fractions are collected. The elution of native C3(n) is completed within 2–2.5 hr from the time of sample application. Up to 20 mg of C3 in 4 ml can be fractionated successfully on this small column. All C3(i) forms can be eluted within 1.25 hr with 250 m*M* imidazole–HCl (pH 10), which we have found to be the most efficient solvent for regeneration of Phenyl-Sepharose columns. It takes less than 1 hr to reequilibrate the column. An example of a typical elution profile obtained by chromatography of a C3 preparation stored at 4° for several months is given in Fig. 7 and compared with the efficiency of QAE-Sephadex A-50 and with that of ω-aminohexylagarose in separation of C3(n) from C3(i).

Separation of C3(n) from C3(i) on ω-aminohexylagarose: Elution profiles obtained by fractionation of C3 preparations on ω-aminohexylagarose (Sigma) resemble those obtained on QAE-Sephadex A-50. As demonstrated in Fig. 7, the majority of hemolytically inactive C3(i) forms (peak 1) elute before native C3 (peak 2) in both instances. Unlike the case of QAE-Sephadex A-50, the use of ω-aminohexylagarose results in an almost complete separation of C3(i) forms. The elution parameters depend on the amount of diaminohexane immobilized on 1 ml of agarose. The amount of diaminohexane bound to the gel varies considerably from one batch to another. The higher the diaminohexane content, the higher the salt concentration necessary to elute hemolytically inactive and active forms of C3.

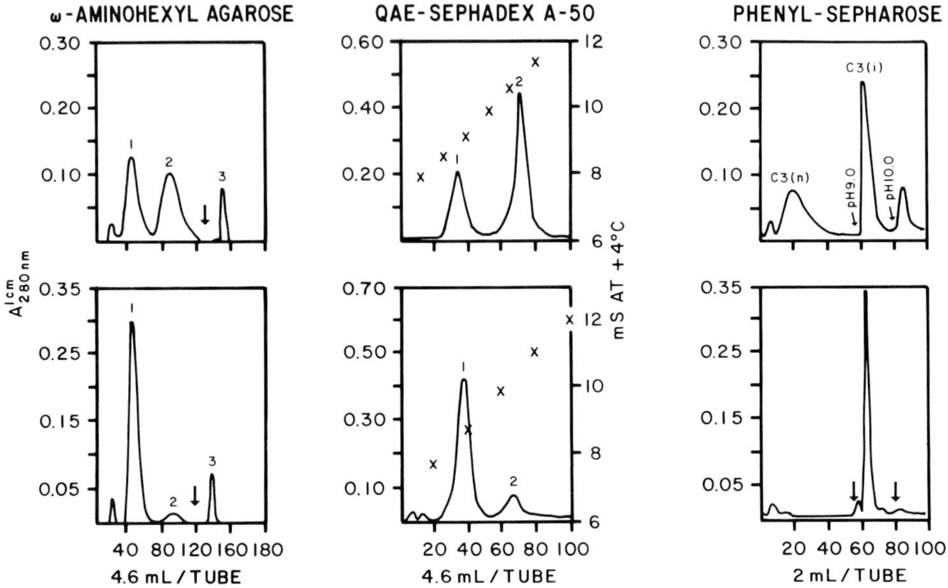

FIG. 7. Two C3 preparations that differed with respect to the content of hemolytically active C3(n) and inactive C3(i) forms were chromatographed on ω-aminohexylagarose, QAE-Sephadex A-50, and Phenyl-Sepharose. The results of fractionation of the C3(n)-containing sample are depicted in the top panels. The elution profiles of the C3(i) sample without any C3(n), which was originally isolated as peak 1 by QAE-Sephadex A-50 chromatography, are presented in the lower panels. Only peak 2, which eluted later from both ω-aminohexylagarose and QAE-Sephadex A-50 (top), contained native C3 and could be subfractionated further by Phenyl-Sepharose to achieve a complete separation of native C3(n) from its inactive C3(i) forms.

For example, the C3 sample (40 mg of freeze–thawed PS2 pool in 10 ml K buffer) is adjusted with SB to the same conductivity as that of the QAE-II buffer (100 m*M* NaCl in SB) and is applied to a column (1.4 × 57 cm) of ω-aminohexylagarose (9.7 μmol diaminohaxane/ml gel). The sample is followed with 10 ml of the starting buffer (100 m*M* NaCl in SB), and the protein is eluted by development of a linear salt gradient (total volume of 1.1 liter) to a limiting concentration of 400 m*M* NaCl. The eluate is collected (90 drops or 6 ml/tube) at 35 ml/hr. When the K buffer at pH 6.5 is used, all conditions are the same as above except that the starting concentration is 150 m*M* KCl in 20 m*M* potassium phosphate. The limiting buffer contains 400 m*M* KCl. The fractionation on ω-aminohexylagarose takes longer (up to 30 hr); it yields, however, a better separation of trace contaminants (they elute at the beginning of the gradient) from C3 than the fast chromatography on Phenyl-Sepharose described in the preceding section.

Purification of C5 by hydrophobic chromatography. Practically all of the C5 protein from the 1.1-liter plasma pool is adsorbed to the QAE-II adsorbent. The conditions for adsorbtion and elution are the same as those for the native C3 component. The C3/C5-containing pools are concentrated by ultrafiltration and applied directly to a column (5 × 30 cm) of Phenyl-Sepharose. The results of fractionation of the two pools PS1 and PS2 are presented in Fig. 6. The majority of contaminating proteins and all native C3(n) are eluted with K buffer, while the C5 protein remains bound to the adsorbent. This protein, as well as factor H, some unspecified contaminants, and hemolytically inactive C3(i) forms, is subsequently eluted in the direction opposite to that of the sample application with 250 m*M* imidazole–HCl at pH 9 (the application of the solvent is marked at tube 220 in Fig. 6). To complete the regeneration of the column, the pH of the solvent is increased to 10. The C5-containing fractions (tubes 251 to 268 in Fig. 6) are pooled, and the C5 pool (272 A_{280} units in 380 ml) is immediately adjusted to pH 7.6 to 8.0 by dropwise addition of 1 *M* HCl, concentrated by ultrafiltration (YM100) to 56 ml, and ultradialyzed with a 5-fold volume of the QAE-II starting buffer (100 m*M* NaCl in SB).

Fractionation of the C5 pool on ω-aminohexylagarose: The post-Phenyl-Sepharose C5 pool (50%) is applied to a column (1.4 × 55 cm) of ω-aminohexylagarose (Fig. 8). The sample (28 ml) is followed by 83 ml of the starting buffer (100 m*M* NaCl in SB). Proteins analyzed (SDS–PAGE) in wells 1 and 2 do not bind to the column. The elution of the proteins present in factor H- (wells 3 and 4) and those in C5- (wells 6 to 12) containing peaks is accomplished at pH 6.8 with SB containing 200 m*M* NaCl. The rest of the proteins bound to the ω-aminohexylagarose are eluted with SB containing 500 m*M* NaCl (wells 13 and 14). The C5 pool (tubes 70 to 100 in Fig. 8) contains cleaved C3(i) and traces of factor H. The C5 pool from the two runs on ω-aminohexylagarose is concentrated by ultrafiltration and is subjected (35 A_{280} units in 32 ml) to immunoaffinity chromatography on anti-C3, and then on anti-H columns (prepared as described above (see also Ref. 53). The final amount of purified C5 recovered after all chromatographic and concentration steps is 21 mg.

Comments on Purification Procedures

1. The addition of only one enzyme inhibitor, PMSF, is sufficient to prevent activation of C3, C5, and particularly of C4 during their isolation and purification.

2. In order to obtain reproducible results, conductivity of the plasma has to be adjusted prior to the adsorption steps. On the other hand, it is not necessary to adjust the pH since small differences in pH do not seem to affect binding of proteins or their elution.

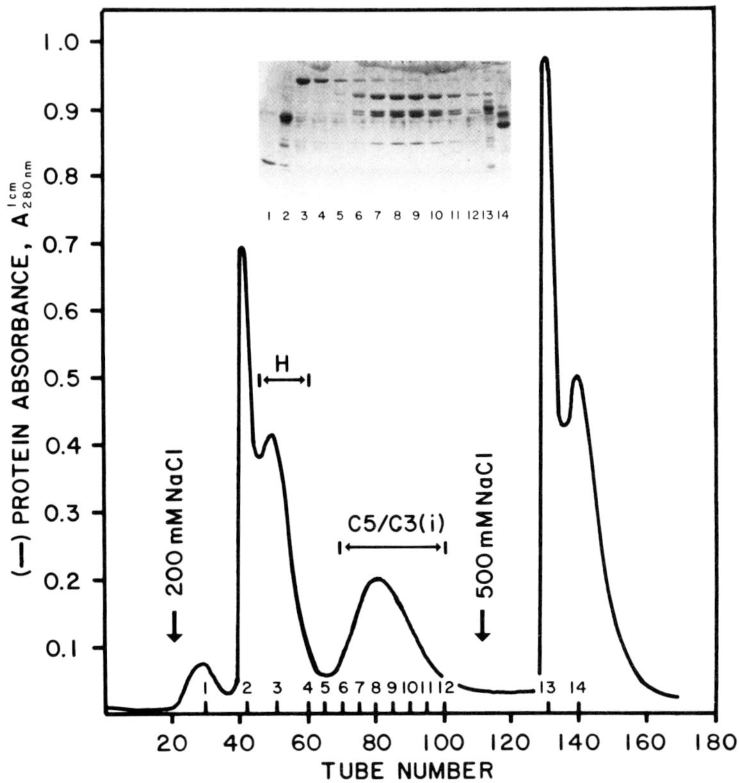

FIG. 8. Fractionation of the post-Phenyl-Sepharose C5 pool by chromatography on ω-aminohexylagarose. The insert shows the results of SDS–PAGE analysis under reducing conditions. For further details, see text.

3. The C1s enzyme which activates C4 binds to QAE-Sephadex A-50 under the same conditions as C4. Moreover the elution profile of C1s overlaps considerably that of native C4. To prevent activation of C4 with C1s, the addition of EDTA is omitted before the first adsorbtion step. Consequently, the C1 component remains undissociated in the QAE-I plasma supernatant. This change in the protocol results in a substantial increase in yields of native C4.

4. The purity of the C4 pool, an eluate from the QAE-I adsorbent (Fig. 3), is not substantially improved when DEAE-Sephacel is used as the second step in C4 purification. Although inter-α-trypsin inhibitor (IαI) elutes at a higher NaCl concentration than native C4, it still overlaps substantially the elution profile of C4.

5. The elution profiles of both C3 and C5 obtained by chromatography of the pooled plasma on the QAE-II column (Fig. 5) are suggestive of the

presence of at least two populations of molecules[86] that may differ in charge. The use of DEAE-Sephacel (100-cm column) does not yield in our hands an effective separation of C5 from C3 (discussed above), under a variety of conditions tested.

6. Detailed analyses of native C3(n) eluted from either Phenyl-Sepharose or ω-aminohexylagarose by hemolytic titrations and differential sulfhydryl assay indicate the existence of at least two populations of C3 molecules (isolated from pooled plasma) that may differ in their efficiency to mediate hemolysis.

7. The presence of native C3 or C4 in the eluate can be conveniently determined by using the DIC (denaturant-induced cleavage) assay[8,27,33] as demonstrated in Fig. 4. An alternative approach is the treatment of C3- or C4-containing eluate with $^{14}CH_3NH_2$.[9-11,27] One mole of the reagent should be incorporated into 1 mol of native C3 or C4.

8. The protocols described above were developed prior to the availability of Q-Sepharose (Pharmacia). Its physicochemical properties should simplify the isolation procedures since the addition of DEAE-Sepharose CL-6B, which was used to improve the physicochemical properties of the chromatographic column[27,75] by increasing the flow rate and preventing excessive shrinkage on application of the NaCl gradient, can be omitted.

9. The procedure can be conveniently scaled down to, e.g., 10 ml of plasma. The use of Q-Sepharose together with the FPLC system (Pharmacia) makes the isolation of C4, C3, and C5 even simpler and faster.

10. Total antigenic levels of C3, C4, and factor B, may be quantified in the starting plasma pool, and in the QAE-I and QAE-II plasma supernatants, by rate nephelometry. While practically no C3 or C5 binds to the QAE-I adsorbent, almost all of C4 binds in the first step of isolation. A total of 473 mg C3(n) obtained after the Phenyl-Sepharose chromatography represents 50% recovery of thiol ester-containing C3 molecules. The yield of C5 (21 mg) is similar to that reported by Hammer *et al.*[74] Assuming that the amount of C5 in the starting material is 84 mg, then the yield would be 25%. One of the main advantages of this protocol is a substantial increase in the recovery of C4. A total of 135 mg of native C4 represents 67% of the starting amount of C4. Additionally 15% has been isolated as hemolytically inactive (but uncleaved) forms of C4. The absence of C1s contamination will also prolong the half-life of the C4 samples isolated by this protocol.

Stability of the Proteins during Isolation and Storage. C5 is a more stable protein than C3 or C4. Its hemolytic activity is not affected by storage at 4° or by freezing and thawing. The pH 6.8 used in the adsorption

[86] N. Behrendt, *Mol. Immunol.* **22,** 1005 (1985).

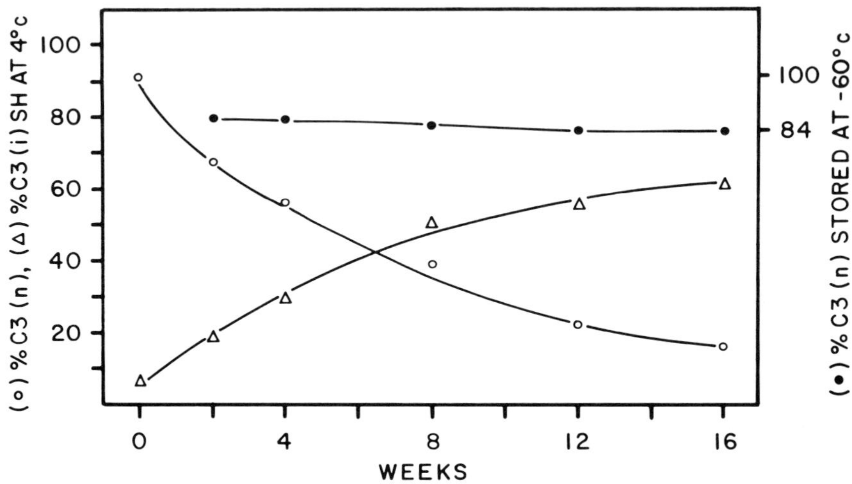

FIG. 9. Stability of native C3(n) during storage. Aliquots of the same C3 sample [91% of C3(n) before storage] were kept in the potassium-containing buffer (pH 6.5) under sterile conditions either at 4° (open circles) or frozen at −60° (filled circles). The degree of inactivation at each time point was determined by using the differential sulfhydryl assay as described above and/or by fractionation on an analytical column of Phenyl-Sepharose.

steps for C4, C3, and C5, and during the elution of C3 and C5, is low enough to decrease the rate of thiol ester hydrolysis in C3 and C4 and yet high enough to prevent C5 "acid activation" and formation of C5–6 complex.[87,88] Hydrolysis of the thiol ester in C3 and C4 is a spontaneous process. The reaction is pH, temperature, and time dependent. The rate of C3(i) formation is between 0.2 and 0.4% per hour,[37] when C3 (1.2 mg/ml) is incubated at 37° in the presence of veronal-buffered saline containing 0.5 m*M* Mg^{2+}. We noted about a 10-fold slower rate when C3 samples are kept at 4° in phosphate/chloride-containing solutions. An example of the effect of temperature and time of the storage on the conversion of C3(n) to C3(i) is given in Fig. 9 (a sterile solution of 3.8 mg C3/ml 20 m*M* potassium phosphate–150 m*M* KCl, pH 6.5). The additional observations[89] concerning the effect of buffer composition, time, temperature, repeated freeze–thaw process, freeze-drying, or precipitation with 16% (w/v) PEG 4000 on the content of native C3 are summarized below.

1. C3 is inactivated quite rapidly when a C3 solution is kept at 4° in the pH range 6.5–7.4. The rate of the thiol ester hydrolysis (0.02–0.04% per

[87] A. Dessauer and U. Rother, *Immunobiology* **164,** 370 (1983).

[88] C. H. Hammer, G. Hänsch, H. D. Gresham, and M. L. Shin, *J. Immunol.* **131,** 892 (1983).

[89] J. Janatova, unpublished observations.

hr) is similar in sodium- or potassium-containing buffers. With Tris-containing buffers a much larger loss of native C3 is observed. Solutions containing ammonium salts should never be used in the isolation of C3 and C4.[36] The usual additives 2 m*M* disodium EDTA and 0.02% sodium azide do not seem to affect the inactivation process at all. Samples stored in 50% glycerol at −20° are hydrolyzed to the same extent as samples stored at 4° (without glycerol).

2. Samples of native C3 (or C4) that are to be frozen should contain only *potassium* salts,[90] since the presence of sodium phosphate during storage, thawing, and freeze-drying substantially enhances the rate of the thiol ester hydrolysis. The following values for C3(n) content were typically obtained when the aliquots of the same C3 sample were treated and stored (at pH 6.8) in the same way except that K samples contained 20 m*M* potassium phosphate, 150 m*M* KCl and corresponding sodium salts were present in N samples: (a) the concentration of C3(n) before treatment was taken as 100%; (b) 90% C3(n) was present in both samples after being frozen for 4.5 hr at −70° and thawed at 37°; (c) after 2 weeks of storage at −60° and thawing at 37°, there were 78% C3(n) in K and 47% C3(n) in N samples; (d) after 2 weeks of storage as in (c), but thawed slowly at 0°, the difference was even more evident, 70% C3(n) in K and only 25% C3(n) in N samples; (e) freeze-drying procedures result in the decrease of C3(n) concentration to 60% in K and 35% in N samples. These observations raise some concern as to the validity of commercially available C3 and C4 standards (Cordis Laboratories). Whether the lower concentrations of salts (5 m*M* sodium phosphate, 70 m*M* NaCl, 0.5 m*M* $MgCl_2$, 0.15 m*M* $CaCl_2$) that are used in the commercial samples in the presence of 2.5% glucose and 0.5% gelatin have a less detrimental effect on C3(n) during the freeze-drying process is not clear.

3. Temperatures used for freezing (−70°), storage (below −60°), and thawing (37°) are very important in protecting C3(n) from conversion to C3(i). As follows from the data presented in the previous paragraph, the C3 or C4 samples should be thawed as quickly as possible and they should never be thawed at 0°.

4. The freeze–thaw process by itself usually results in 8–13% inactivation. Repeating a freeze/thaw step up to 4 times does not change the C3(n) concentration appreciably. This indicates that there is a certain amount of C3(n) molecules which are more susceptible to hydrolysis than others. The length of time (up to 16 weeks) of storage at −60° does not seem to affect the percentage of C3(n) as illustrated in Fig. 9.

5. The composition of several C3 samples that differ with respect to the

[90] R. B. Sim, M.R.C. Immunochemistry Unit, Univ. of Oxford, personal communication.

content of C3(n) and C3(i) forms were determined by the differential sulfhydryl assay before and after precipitation of the C3 protein with 16% (w/v) PEG 4000. The precipitation step does not seem to affect C3(n) adversely. The results also suggest that none of the C3 forms is precipitated preferentially. The C3 precipitation can thus be employed as faster means for the concentration of native C3 following the Phenyl-Sepharose chromatography (Fig. 6).

The active site of the C4 protein seems to be more reactive than the corresponding site in native C3.[27] It may be that the rate of spontaneous hydrolysis in C4(n) is also higher. No attempt has been made in our laboratory to measure the rate of the C4(n) spontaneous inactivation. As in the case of C3, only potassium-containing buffers should be used for C4(n) to be stored in frozen state.

Structure and Function of C3 and C5

C3 and C5 have been subject of several reviews on the complement system,[2,4,5,19] including an article in this series.[75] Therefore, only recent advances in the understanding of the structure and function of C3 and C5 will be discussed below.

Proforms and Mature Proteins

Both C3 and C5 proforms[24,91] are synthesized as large single polypeptide chains (C3 M_r 185,000, C5 M_r 191,000), with the β subunit at the amino terminus. The C3 proform[91] contains a signal peptide of 22 amino acids that is immediately followed by 645 amino acid residues corresponding to the β chain. The junction between the β and α chains (922 residues) is formed by four arginine residues that are removed during the processing of the pro-C3 molecule.[91] The corresponding region in the pro-C5 molecule[24] also consists of four amino acid residues: Arg-Pro-Arg-Arg. The mature native forms of C3 and C5, as present in plasma, consist of the two disulfide-linked chains.[75] The apparent size of the β chains is approximately 75 kDa, and that of the α chains is 115 kDa. However, during SDS–PAGE the β chain of C5 migrates slower and the α chain faster than the corresponding chain of the C3 molecule.

Native and Functional Forms

Neither C3 nor C5 in its native state has affinity for cell surfaces or each other. They have to be activated, and further fragmented, to fulfill the

[91] M. H. L. de Bruijn and G. H. Fey, *Proc. Natl. Acad. Sci. U.S.A.* **82,** 708 (1985).

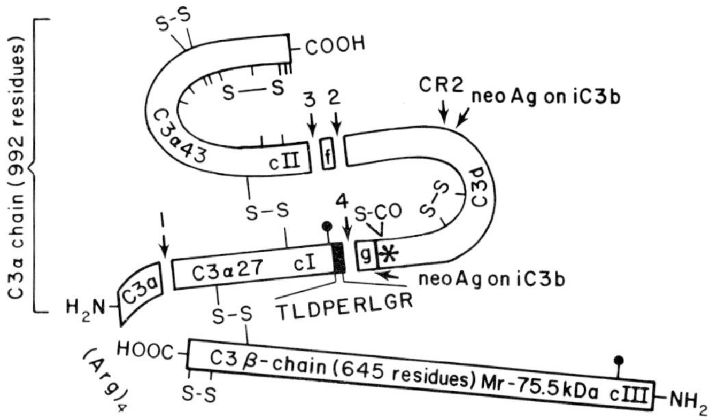

FIG. 10. Schematic representation of the structure and function of the C3 protein. The C3 proform is synthesized as a single polypeptide chain with the β chain at the amino terminus. The junction between the β and α chains is formed by four arginine residues that are removed during the processing of the pro-C3 molecule. Native C3 contains the thiol ester bond (*), and it is a precursor of several fragments with different biological properties and distinct binding sites. All the cleavages occur within the α chain. The first cleavage of C3 by the C3 convertase at site 1 generates the C3a fragment with anaphylactic activity and the C3b fragment with the ability to bind covalently via the thiol ester (*) to immune complexes or cell surfaces or to react with H_2O to form the fluid-phase C3b. The various functions of C3b are outlined in Fig. 1. The C3b molecule can also bind to the CR1 receptor on erythrocytes or leukocytes. It is rapidly cleaved by factor I in the presence of factor H or CR1 at positions 2 and 3, yielding C3f and the iC3b molecule (β, $\alpha68[\alpha27 + \text{C3g} + \text{C3d}]$, $\alpha43$) which is recognized by the receptor CR3. The iC3b molecule is then slowly cleaved at site 4 to form C3d,g and C3c. The C3c molecule consists of two α-chain domains, C3α27 (cI) and C3α43 (cII), and the uncleaved β chain (cIII). The presence and position of carbohydrates is denoted by the symbol ⚲. The one in the C3α27 domain is probably involved in lectin binding. The carboxy terminus of this domain contains the nonapeptide Thr-Leu-Asp-Pro-Glu-Arg-Leu-Gly-Arg, which exhibits the leukocyte mobilizing activity. The interchain (β–α) and interdomain (C3α27–C3α43) disulfide bonds have been identified. It remains to be determined which of the nine yet unassigned cysteinyl residues (indicated by short lines) is involved in the interdomain link. The C3α43 domain is highly cross-linked by six intradomain disulfide linkages; two of these have been determined: residue 1506 is bound to 1511, and 1537 to 1637 (numbering is according to the sequence data of de Bruijn and Fey[91]; the disulfides were assigned by M. S. Strickland, W. R. Gray, and J. Janatova, unpublished observations). The C3α43 domain is apparently involved in the binding of properdin, C5, and factor B. Detailed discussion of all the information presented in this outline with the corresponding references can be found in the text.

variety of their physiological functions. All the cleavages of homologous proteins C3 and C5 are known to occur within their α chain (Fig. 10). This subunit also comprises both the labile binding site (C3b: covalent binding to cell surfaces or immune complexes[7–13] via transacylation; C5b: firm binding to C6 and C7 and a generation of a membrane-binding site[30]) and the stable binding sites for interaction with other complement proteins or the receptors present on the surface of blood or tissue cells. All these binding sites are concealed within the native form of each protein. The role of the β chain in the molecular architecture remains to be determined, whether in native or fragmented C3 or C5. Its participation in the binding of either appropriate complement proteins or receptor molecules is similarly unclear at the present time.

Primary Structure

More recently, additional evidence has been presented in a support of an extensive homology between C3 and C5.[24,91,92] The amino acid sequences of both proteins, deduced from the nucleotide sequences of appropriate cloned complementary DNA, indicate that, apart from the amino terminus of the α chains (C3a and C5a activation peptides), there are several other regions, including the one corresponding to the thiol ester domain in C3, which are homologous in both proteins. However, the cysteine and glutamine residues, which bridge to form the thiol ester bond in C3 (and also in C4 and α_2-macroglobulin), are replaced in C5 by serine and alanine, respectively.[92,93]

An extensive sequence homology between C3 and C5 can thus explain their similar chain structure, ultrastructure, mode of activation, and physical properties. The similarity in size and charge has made separation of C3 and C5 extremely difficult. However, there is a certain difference in the hydrophobicity of both molecules when in their native states. The more hydrophobic C5 can thus be retained on phenyl-Sepharose,[80] and this property allows effective separation from native C3. Resolution of native C3 from C5 can also be obtained by using hydroxylapatite chromatography.[70]

Cysteine Residues[91,92] and Disulfide Bonds[94]

One of the striking features is the similarity in the number and the positions of cysteinyl residues in both proteins.[92] The sequence data show a total of 26 cysteine residues present in C3.[91] An additional cysteine residue is present in the form of a thiol ester in the C3d domain of the molecule.[95] Since no free sulfhydryl group can be detected in native C3, and only one is

[92] R. A. Wetsel, R. T. Ogata, and B. F. Tack, *Biochemistry* **26,** 737 (1987).

[93] R. A. Wetsel, P. D'Eustachio, R. C. Strunk, R. T. Ogata, and B. F. Tack, *Complement* **2,** 86 (1985).

[94] J. Janatova, *Biochem. J.* **233,** 819 (1986).

[95] M. L. Thomas, J. Janatova, W. R. Gray, and B. F. Tack, *Proc. Natl. Acad. Sci. U.S.A.* **79,** 1054 (1982).

present in hemolytically inactive C3 (SH generated from the thiol ester),[34] all other cysteine residues must form disulfide bonds. Three disulfide linkages form a knot in the activation peptide C3a.[6] One intrachain disulfide is located in the C3d fragment. The remaining cysteine residues are present in the C3c molecule that consists of the intact β chain (three cysteine residues at its carboxy terminus) and two α-chain fragments, which can be considered as the additional two domains of the α chain. The smaller C3α27 contains only two cysteine residues; one is bound to the β chain and thus participates in the only interchain bond[94] present in the C3 molecule (Fig. 10). The majority of cysteine residues (13) are present in the carboxy-terminal domain C3α43. This domain seems to be highly cross-linked by six intrachain disulfides and is linked to the C3α27 domain. Thus the carboxy-terminal domain of the α chain is in close proximity to its amino terminus and at the same time close to the carboxy terminus of the β chain. A very similar distribution of cysteinyl residues[96] and disulfide linkages exists in the C4c fragment of the homologous C4 protein.[94]

The results of amino acid analyses[30] and the amino sequence of C5 derived from the cDNA sequence indicate that this protein may contain two additional cysteine residues[92] located in the region of the α chain equivalent to that of the C3α27 domain. There is a possibility of the presence of one free sulfhydryl group[28] located in the region corresponding to the C3d domain.[92] The positions of all the other cysteine residues in the deduced sequence[92] are very similar to those in C3 and C4 proteins.[91,96]

Carbohydrate Content

Although being homologous in many respects, C3 and C5 differ in the content, localization, and composition of their respective carbohydrate moieties.[91,92] More recently, the analysis of carbohydrates was performed by using gas chromatographic and mass spectrometric techniques.[97] The carbohydrates in C3 contribute 1.7% (w/w) of total mass. The ratio of *N*-acetylglucosamine (3.6 residues/C3 molecule) to mannose (15.4 residues/C3 molecule) is approximately 1:4. As follows from the sequence data, there are three possible N-linked carbohydrate attachment sites (β chain, C3α27, and C3α43 domains).[91] However, using either radiolabeled concanavalin A[98] or sequence analysis of glycopeptides from C3,[98a] the presence of carbohydrates is detected only in the β chain and the C3α27

[96] K. T. Belt, M. C. Carroll, and R. R. Porter, *Cell* **36,** 907 (1984).
[97] M. Tomana, M. Niemann, C. Garner, and J. E. Volanakis, *Mol. Immunol.* **22,** 107 (1985).
[98] S. Hirani, J. D. Lambris, and H. J. Müller-Eberhard, *J. Immunol.* **134,** 1105 (1985).
[98a] K. G. Welinder and A. Svendsen, *FEBS Lett.* **202,** 59 (1986).

domain. The latter one has also been implicated in binding of the lectin to the iC3b molecule.[98] The oligosaccharides present in C3 are only of high-mannose type.[99] The compositions for those isolated from the α chain are $(Man)_9(GlcNAc)_2$–Asn and $(Man)_8(GlcNAc)_2$–Asn, and those on the β chain have the compositions $(Man)_6(GlcNAc)_2$–Asn and $(Man)_5(GlcNAc)_2$–Asn. It can be assumed that the steric inaccessibility of the oligosaccharides prevents these carbohydrate chains from further processing. Other roles, except for the binding of the lectin,[98] are not known as yet.

The number of potential carbohydrate binding sites and their locations are apparently different in the C5 protein.[92] Oligosaccharides attached to C5 represent 3.8% of the total mass or 5.2 kDa.[97] While the activation peptide C3a does not contain any carbohydrate, the oligosaccharides in C5a represent a mass of 2.7 kDa.[30,97] The whole C5 molecule contains the following saccharides (number of residues): galactose (6), mannose (11.5), *N*-acetylglucosamine (13.2), and *N*-acetylneuraminate (5.8).

Generation and Properties of Anaphylatoxins[6,100,101]

Both native C3 and C5 serve as substrates for the same enzymes (either C2a or Bb, the catalytic subunits of the complex C3/C5 convertases of the classical or alternative pathway). During activation of C3 or C5, two fragments are generated. The smaller amino-terminal activation peptides (C3a M_r 9,100; C5a M_r 11,200) are known to be effective anaphylatoxins.[6,100] The interaction with their receptors can result in (1) contraction of smooth muscle, (2) release of vasoactive amines from mast cells, or (3) increased vascular permeability. The activity of these fragments is abrogated by carboxypeptidase N, which removes the carboxy-terminal Arg (residue 77 in C3a, 74 in C5a) that is essential for their function. All three activation peptides, C5a, C3a, and also C4a, are thus converted to anaphylactically inactive desArg derivatives [e.g., C5adesArg or C5a(i)].

Primary structures of C3a and C5a have been determined both on protein[100] and cDNA.[24,91] The positions of six cysteinyl residues are highly conserved, and the disulfide linkages are determined based on X-ray crystallography studies of C3a.[102] These humoral factors have remarkable conformational stability. Extensive studies with synthetic peptides indicate

[99] S. Hirani, J. D. Lambris, and H. J. Müller-Eberhard, *Biochem. J.* **233,** 613 (1986).

[100] T. E. Hugli, *Springer Semin. Immunopathol.* **7,** 193 (1984).

[101] T. E. Hugli, C. Gerard, M. Kawahara, M. E. Scheetz II, R. Barton, S. Briggs, G. Koppel, and S. Russell, *Mol. Cell. Biochem.* **41,** 59 (1981).

[102] R. Huber, H. Scholze, E. P. Paques, and J. Deisenhofer, *Hoppe-Seyler's Z. Physiol. Chem.* **361,** 1389 (1980).

that the anaphylactic site resides in a short linear portion of the carboxy terminus (reviewed by Hugli).[6,100] However, to achieve the functional behavior similar on a molar basis to that of native C3a, the carboxy-terminal portion of C3a must be 21 residues long (residues 57 to 77) to include the α-helical portion. The amino-terminal region of C3a comprising the other α helix in C3a (residues 1 to 21) and disulfide-linked core region (residues 22 to 57) are believed to serve primarily to stabilize ordered conformation in the carboxy-terminal region (residues 58 to 77) and thereby orient side chains at the essential active site for optimal receptor interaction.[103] Moreover, the C3a peptide has been suggested to have an immunoregulatory function.[104]

Isolation of Anaphylatoxins[101]

It is practically impossible to isolate C3a or C5a, and also C4a, in the biologically active state, unless exopeptidase serum carboxypeptidase N (SCPN) is properly blocked. The inhibitor 6-aminohexanoic acid has to be used in a 1 *M* concentration in order to efficiently protect C3a and C5a from inactivation. Mercapto derivatives of arginine analogs have also been reported to be effective inhibitors of SCPN. Thus, the biologically active C3a, C5a, and also C4a can be isolated from fresh serum, in which both pathways have been activated, in the presence of the inhibitor 2-mercaptomethyl-5-guanidinopentanoic acid at a 1 m*M* concentration.[101] A similar inhibitor of SCPN, 2-mercaptomethyl-3-guanidinoethylthiopropanoic acid (*d, l*),[105] is commercially available (Calbiochem-Behring Corp., La Jolla, CA), and it can provide protection for 30–60 min at a 2 m*M* concentration.[106] The latter inhibitor also induces or enhances complement activation by the alternative pathway.[106,107] The following is an outline of a procedure for the isolation of all three anaphylatoxins that was developed by Hugli and his collaborators.[101]

Step 1. Generation of C3a, C5a, and C4a. The inhibitor of SCPN[105] is added to fresh normal human serum at 2 m*M* concentration,[106] and both pathways of complement are activated as follows: zymosan is added first at 20 g/liter of serum to generate C3a and C5a; after 10 min at 37°, aggregated human IgG (prepared by heating 10 mg IgG/ml solution in PBS at 64° for 20 min) is added to a final concentration of 0.5 g/liter of serum to

[103] Z.-X. Lu, K.-F. Fok, B. W. Erickson, and T. E. Hugli, *J. Biol. Chem.* **259**, 7367 (1984).

[104] W. O. Weigle, M. G. Goodman, E. L. Morgan, and T. E. Hugli, *Springer Semin. Immunopathol.* **6**, 173 (1983).

[105] T. H. Plummer, Jr., and T. J. Ryan, *Biochem. Biophys. Res. Commun.* **98**, 448 (1981).

[106] T. E. Hugli, Molecular Immunology, Scripps Clinic & Research Foundation, La Jolla, CA, personal communication.

[107] I. von Zabern and R. Nolte, *Complement* **3**, 97 (1986).

generate C4a. Incubation of the activated serum is allowed to proceed at 37° for an additional 30–35 min.

Step 2. Separation of Anaphylatoxins from Parent Protein Molecules. A major portion of the serum proteins, including uncleaved C3, C5, or C4, precipitates after the activated serum is acidified by addition of 10 *M* HCl to a final concentration of 1 *M*. All further steps are carried out at 4°. The precipitate is removed by centrifugation, and the acidic supernatant containing all three anaphylatoxins is dialyzed against 0.1 *M* ammonium formate at pH 4.0, freeze-dried, and then dissolved in water. The final volume is approximately one-fifth of the original volume of the supernatant. This and all subsequent dialysis steps have to be carried out in a small-pore dialysis tubing (cutoff $M_r \sim 3.5$ K).

Step 3. Gel Filtration and Ion-Exchange Chromatography of the Supernatant. The anaphylatoxin-containing sample from Step 2 is first subjected to gel filtration on a column of BioGel P-60, equilibrated with 0.1 *M* ammonium formate, pH 5.0. Fractions containing anaphylatoxins are pooled and applied to a SP-Sephadex column (1.6 × 30 cm), preequilibrated with 0.1 *M* ammonium formate, pH 5.0. A majority of the contaminating serum proteins is eluted with the 0.1 *M* ammonium formate, pH 7.0 (about 170 ml). The anaphylatoxins are eluted from the column at 60 ml/hr with a linear gradient of ammonium formate (from 0.1 to 0.8 *M*; total volume 650 ml) at pH 7.0. The effluent is monitored at 280 nm. To aid the detection of anaphylatoxins in the eluate, ^{125}I-labeled C3a and C5a and ^{131}I-labeled C4a are added to the sample prior to the fractionation on a SP-Sephadex column. Smooth muscle contracting activity can be used to identify the positions of C3a or C5a in the eluate. Anaphylatoxins elute from the SP-Sephadex column in the order C5a, C4a, and C3a, when a linear gradient of ammonium formate is used. Alternatively, the gel filtration step on BioGel P-60 can be replaced by batchwise adsorption of anaphylatoxins from the supernatant to SP-Sephadex preequilibrated in 0.1 *M* ammonium formate. The anaphylatoxins are then desorbed with 0.8 *M* ammonium formate at pH 5.0. After dialysis they are applied to the SP-Sephadex column and fractionated as described above.

Step 4. Purification of C5a, C4a, and C3a to Homogeneity. Following chromatography on SP-Sephadex, further steps in the purification of the individual anaphylatoxins differ from each other, and are dealt with separately as described below.

The C5a-containing fractions are pooled, concentrated, and the sample is applied to a CM-Sephadex C-25 column (0.6 × 14 cm) preequilibrated with 0.1 *M* ammonium formate (pH 7.0). The elution of C5a is accomplished at a flow rate of 2.8 ml/hr with the equilibrating buffer. The presence of protein/peptide materials in the eluate can be determined colorimetrically, by the Folin or biuret assay.

The C4a-containing fractions are pooled and freeze-dried. The sample is dialyzed against 0.15 *M* ammonium formate at pH 5.5 and applied to a column (1.6 × 15 cm) of CM-cellulose (CM 52) equilibrated in the same buffer. The elution of C4adesArg and that of C4a is accomplished at pH 5.5 by using a linear gradient of ammonium formate (from 0.15 to 0.35 *M*; total volume 400 ml) and a flow rate of 30 ml/hr. The effluent is monitored, for example, by the Folin assay at 750 nm. Human C4a elutes approximately with 0.24–0.26 *M* ammonium formate. The final C4a preparation is reported to be free of any contaminations; specifically, no C3a or C5a is present.

The C3a-containing fractions are processed similarly as those containing C4a except that in this case the pH of the dialysis buffer (0.15 *M* ammonium formate) is 7.0. The material is then applied to a column (1.5 × 27 cm) of CM-cellulose (CM 52) preequilibrated with 0.15 *M* ammonium formate, pH 7.0 and eluted with a linear gradient of ammonium formate, pH 7.0 (from 0.15 to 0.45 *M*; total volume 600 ml), at a flow rate of 30 ml/hr. Functional C3a, free of the other two anaphylatoxins, is eluted with 0.31–0.34 *M* ammonium formate.

The procedure described above can lead to the isolation of all three anaphylatoxins, functionally and biochemically pure, that can be used as standards in sensitive radioimmunoassays discussed earlier in this chapter. For further details, the reader is referred to the original article.[101]

Chemotactic and Other Biological Activities of C5a

Unlike C3-derived spasmogen, the C5a activation peptide has a high carbohydrate content (2.7 kDa out of total 11.2 kDa).[30] More importantly, C5a is one of the most potent chemotactic factors.[6,20] The derivative lacking the carboxy-terminal Arg (C5adesArg) is still active, but less efficient. The binding of C5a to its receptor can result in (1) mobilization of CR1 and CR3 receptors (Ca^{2+}-dependent translocation from intracellular pools to the cell surface) that are necessary for the interaction of PMNs with C3b- or iC3b-opsonized particles,[3,108,109] (2) stimulation of PMNs to release lysosomal enzymes,[20] (3) enhancement of cellular adherence,[109] and (4) generation and release of oxidative metabolites.[110]

The generation of C5a and its presence in the medium can be detected by the PMN chemotaxis assay[20] (see this volume [2]–[7]) (i.e., the ability of C5a to cause unilateral migration of phagocytic cells along its gradient;

[108] T. A. Fleisher and M. Berger, *Clin. Immunol. Immunopathol.* **33,** 391 (1984).

[109] H. S. Jacob, P. R. Craddock, D. Hammerschmidt, and C. F. Moldow, *N. Engl. J. Med.* **302,** 789 (1980).

[110] S. J. Klebanoff and R. A. Clark, "The Neutrophil: Function and Clinical Disorders." North-Holland, Amsterdam, 1978.

several assays for leukocyte chemotaxis have been recently compiled and discussed by Whaley[111]) or by the ability of C5a to induce release of lysosomal enzymes from PMNs. For example, *N*-acetyl-β-glucosaminidase released from cytochalasin B-pretreated PMNs is quantified.[112] Cytochalasin B-pretreated PMNs very actively release granule but not cytoplasmic constituents into the extracellular fluid following stimulation.[110] C5a stimulation of the oxidative metabolism of PMN can be measured via superoxide anion generation, which is quantified as the superoxide dismutase-inhibitable reduction of ferricytochrome according to the published protocol.[113]

C5a is chemotactic for PMNs at concentrations in the range of $0.04-1.7 \times 10^{-8}$ *M*, C5a desArg at $0.2-1.2 \times 10^{-6}$ *M*.[114] Also, 85- to 175-fold more C5a desArg than C5a is necessary in order to stimulate cytochalasin B-treated PMNs to release granule constituents such as myeloperoxidase, lysozyme, and β-glucuronidase.[115] It is noteworthy that C5a exhibits its anaphylactic activity at concentrations as low as 10^{-15} *M*. The PMN chemotaxis assay seems to be as sensitive for detection of C5a (3–500 ng/ml plasma) as the radioimmunoassay discussed above. In the case of C5a desArg, however, the sensitivity of PMN chemotaxis is much lower (100–10,000 ng/ml) than that of a radioimmunoassay (10–400 ng C5a desArg/ml plasma). It should be emphasized that the PMN chemotaxis assay is valuable for the detection of C5 activation by noncomplement enzymes, such as elastase, α-thrombin, and plasmin.[21] This is an example where the radioimmunoassay for C5a could not be used, since the cleavage products remain linked by disulfide bonds and therefore cannot be separated in the precipitation step.[21,112]

Metastable Binding Site of the C3b or C5b Fragment

Both C3 and C5 proteins are precursors of several fragments with distinct binding sites and different biological activities.[2,116] The first cleavage of C3 or C5 results in the generation of two fragments, a and b. The smaller, amino-terminal activation peptides C3a and C5a with vasoactive properties were already discussed above. Conformational changes accom-

[111] K. Whaley, *in* "Methods in Complement for Clinical Immunologists" (K. Whaley, ed.), p. 160. Churchill-Livingstone, Edinburgh, Scotland, 1985.

[112] R. A. Wetsel and W. P. Kolb, *J. Immunol.* **128,** 2209 (1982).

[113] I. M. Goldstein, H. B. Kaplan, A. Radin, and M. Frosch, *J. Immunol.* **117,** 1282 (1976).

[114] H. N. Fernandez, P. M. Henson, A. Otani, and T. E. Hugli, *J. Immunol.* **120,** 109 (1978).

[115] R. O. Webster, S. R. Hong, R. B. Johnston, Jr., and P. M. Henson, *Immunopharmacology* **2,** 201 (1980).

[116] H. J. Müller-Eberhard, *Springer Semin. Immunopathol.* **7,** 93 (1984).

pany the activation process of C3 or C5, and the large cleavage product, the disulfide-linked two-chain C3b or C5b fragment, acquires the transient binding site with a half-life of microseconds.[12,30]

In nascent C3b the reactive thiol ester allows this protein to form, by a transacylation reaction, amide or oxygen ester bonds with proteins or polysaccharides present in immune complexes and/or on cell surfaces (reviewed by Janatova[67] and Tack[68]). However, only 10–15% of activated C3b molecules are bound covalently.[117] The majority of C3b reacts with water and is present in the fluid phase. The bound or fluid-phase C3b can exhibit several functions, as already discussed in the Introduction. Its further cleavage by physiological enzymes results in the fragments that may stimulate additional cellular responses.[3,108,118,119]

Nascent C5b lacks a thiol ester bond,[11,27,28] however. Its metastable binding site (apparently due to the transient expression of hydrophobic sites) exhibits a unique specificity and forms a firm complex with C6.[30] In the absence of C6 the C5b fragment forms aggregates. The C5b–6 complex is still hydrophilic. The membrane-binding site is generated upon binding of the additional protein C7. The resulting C5b–7 complex can bind firmly to cell membranes and represents a solid foundation for the formation of the membrane attack complex C5b–9.[19,116] Very little is known about physiological fragmentation of the C5b fragment.

Physiological Fragmentation of C3b

The first two cleavages of the C3 α chain are mediated by factor I in the presence of cofactor (H or CR1) at a distance of 46 and 43 kDa from the carboxy terminus of the α chain, thus yielding a three-chain iC3b (β, α68, α43) and a single chain C3f of M_r 2 or 3K.[120,121] The iC3b form, unlike C3b, seems to be quite stable and is recognized by the CR3 receptor which mediates phagocytosis and degradation of iC3b-opsonized particles by macrophages and granulocytes (reviewed in Refs. 3, 18, and 118).

The further scission of the α chain may take place in the C3α68 fragment, a part of iC3b.[117] The C3d,g fragment (M_r 40K), the part from the C3b or iC3b molecule which remains covalently attached to cell surfaces or immune complexes, can thus be removed from the middle of the C3 α chain.[41] The larger C3c is released into the fluid phase and consists of

[117] G. D. Ross, J. D. Lambris, J. A. Cain, and S. L. Newman, *J. Immunol.* **129,** 2051 (1982).
[118] R. D. Schreiber, *Springer Semin. Immunopathol.* **7,** 221 (1984).
[119] Z. Fishelson, *Immunol. Lett.* **11,** 261 (1985).
[120] A. E. Davis III and R. A. Harrison, *Biochemistry* **21,** 5745 (1982).
[121] E. Sim, A. B. Wood, L.-M. Hsiung, and R. B. Sim, *FEBS Lett.* **132,** 55 (1981).

the β chain (M_r 75K) and two α-chain fragments, amino-terminal C3α27 (M_r 27K) and carboxy-terminal C3α43 (M_r 43K).[122]

C3c Represents the Domains of C3

The disulfide-bonded fragments in C3c can be considered as domains of the C3 molecule. C3c turns out to be ideal for a determination of interchain and interdomain linkages and for a study of the effect of disulfides on the shape/size of the highly cross-linked carboxy-terminal domain of the C3 α chain.[94] Amino-terminal sequence analyses of the isolated constituents of C3c confirm their order within the α chain[122] and allow the localization of two cleavage sites by factor I within the proposed amino acid sequence of human C3.[91] The presence of carbohydrates was localized, by using radiolabeled concanavalin A, to the β chain and the C3α27 domain.[98] In the same study, a lectin-binding site has also been localized in this part of C3 molecule. By using a similar approach, the carboxy-terminal domain C3α43 has been identified as the one with a binding site for properdin.[123] It has also been suggested that this part of the C3 molecule may interact with factor B and the C5 component.[42]

Generation and Properties of C3d-Containing Fragments

Generation and levels of C3d-related fragment(s) can serve as a measure of complement activation in certain disease states, such as rheumatoid arthritis, systemic lupus erythromatosus, glomerulonephritis, or other disorders involving immune complexes, or in patients undergoing hemodialysis.[124–127] Several attempts have been made to detect and quantify levels of C3d-containing fragments. The most frequently used techniques, radial immunodiffusion or rocket immunoelectrophoresis of supernatant obtained after treatment with 22% PEG 6000, cannot differentiate between C3d–K, C3d,g, or C3d. There have also been reports that the C3d-containing fragments may coprecipitate with other C3 forms.[128] Even use of monoclonal antibodies cannot give a straightforward

[122] A. E. Davis III, R. A. Harrison, and P. J. Lachmann, *J. Immunol.* **132,** 1960 (1984).

[123] J. D. Lambris, J. Alsenz, T. F. Schultz, and M. P. Dierich, *Biochem. J.* **217,** 323 (1984).

[124] A. Berkowitz, E. Kappelgaard, J. Petersen, H. Nielsen, J. Ingmann-Hansen, J. Halkjaer-Kristensen, and H. Sorensen, *Acta Pathol. Microbiol. Immunol. Scand., Sect. C* **91,** 397 (1983).

[125] W. J. W. Morrow, D. J. P. Williams, C. Ferec, R. Casburn-Budd, D. A. Isenberg, E. Paice, M. L. Snaith, P. Youinou, and P. Le Goff, *Ann. Rheum. Dis.* **42,** 668 (1983).

[126] L. H. Perrin, P. H. Lambert, and P. A. Miescher, *J. Clin. Invest.* **56,** 165 (1975).

[127] B. Collet, A. Alhaq, N. B. Abdullah, L. Korjtsas, R. J. Ware, N. J. Dodd, E. Alimo, J. Ponte, and D. Vergani, *Br. Med. J.* **289,** 1251 (1984).

[128] S. Bhakdi, M. Roth, and W. Nürnberger, *J. Immunol. Methods* **74,** 79 (1984).

answer, since anti-C3g (amino-terminal portion of C3d,g fragments) will recognize not only C3d,g but also neoantigen on iC3b.[41] Likewise, anti-C3d is not specific enough as it will react not only with C3d or C3d,g but also with iC3b, C3b, or even C3.[41] In the author's laboratory SDS–PAGE analysis of the antigens withdrawn and bound to immunoadsorbents (described above) is used to evaluate the degree of activation and fragmentation of C3 and related proteins.[52,53]

Several C3d-containing fragments have been generated from the iC3b molecule. Limited proteolysis with plasma kallikrein at the site Ala-Val-Arg/Thr-Leu-Asp splits the iC3b into two fragments designated C3c–K and C3d–K.[129] The latter seems to be an effective immunomodulator, as it has been capable of inhibiting T-cell-specific and nonspecific proliferative responses.[104,130] Moreover, it possesses leukocyte mobilizing activity,[130] which has been localized to the nonapeptide Thr-Leu-Asp-Pro-Glu-Arg-Leu-Gly-Arg, present at its amino terminus.[131] However, the fragment, with apparent size of 40 kDa, termed C3d,g, has been isolated from CVF-activated ACD blood or from patient's serum.[122] Its amino-terminal sequence analysis identified the cleavage site as Leu-Gly-Arg/Glu-Gly-Val. The same size fragment with an identical amino-terminal sequence has also been generated in the absence of any blood cells[132] from fresh serum activated with both CVF and aggregated IgG. SDS–PAGE analysis of concomitantly generated C3c reveals the heterogeneity of its amino-terminal C3α27 fragment that has been attributed to the difference at its carboxy end.[122] Since the smaller fragment C3α26 has always been present in excess, and only C3d,g rather than C3d–K has been detected in the activated blood or serum,[52,132] it is reasonable to assume that the leukocyte mobilizing factor, the nonapeptide Thr-Leu-Asp-Pro-Glu-Arg-Leu-Gly-Arg,[131] functions under the physiological conditions as a separate moiety or as a part of C3c. Although there are several enzymes considered as possible candidates for the generation of C3d,g from the iC3b molecule, this question has not been satisfactorily resolved as yet.

The C3d domain is recognized by the CR2 receptor present on B lymphocytes.[118,133] More recently, the existence of another receptor that will interact with C3d has been revealed on the surface of phagocytic cells,

[129] J. L. Meuth, E. L. Morgan, R. G. DiScipio, and T. E. Hugli, *J. Immunol.* **130,** 2605 (1983).

[130] E. L. Morgan, M. L. Thoman, P. D. Hoeprich, and T. E. Hugli, *Immunol. Lett.* **9,** 207 (1985).

[131] P. D. Hoeprich, Jr., C. A. Dahinden, P. J. Lachmann, A. E. Davis III, and T. E. Hugli, *J. Biol. Chem.* **260,** 2597 (1985).

[132] J. Janatova and R. J. Gobel, unpublished observations.

[133] J. J. Weis, T. F. Tedder, and D. T. Fearon, *Proc. Natl. Acad. Sci. U.S.A.* **81,** 881 (1984).

such as neutrophils and macrophages (CR4).[134] The CR2-binding site and the neoantigenic site (which appears on cleavage of C3b to iC3b) has been localized within the amino acid sequence of C3d.[135] The hexapeptide Leu-Tyr-Asn-Val-Glu-Ala (residues 1227–1232 in the published sequence[91] which includes the signal peptide and $\beta-\alpha$ junction) has been identified as being specific for binding of CR2 receptor. This region of the C3d domain, including the preceding 10 amino acid residues (residues 1217–1232), seems to be responsible for the neoantigenic properties detected in the iC3b molecule.[135] Also the C3g region of the C3d,g fragment has been identified as another neoantigenic site that is detected in iC3b.[41] More recently, the presence of the third neoantigen on the iC3b molecule, different from the other two, has been detected.[43]

Concluding Remarks

Preparative procedures and some of the assays described and discussed in this chapter are based on knowledge acquired on the properties of individual complement proteins, specifically those of C3, C4, and C5, and their physiological fragments, during the past decade. During this time period several important advances have also been made that have contributed to our better understanding of the structure and function of C3 and C5. The nature and mechanism of binding of the nascent C3b to immune complexes or cell surfaces, and its homology to C4 and α_2-macroglobulin, have already been reviewed extensively.[67,68,136] Conformations of the thiol ester site have been proposed,[95,137] the specific site of acylation in C3b has been determined,[138] and organic models showing that such a structure could exist have been synthesized.[139] Its position within the α chain has been determined by several different approaches,[67,95] and the importance of the thiol ester for the opsonization of bacteria has been demonstrated.[140] Spontaneously inactivated C3 (C3 molecules in which the thiol ester is

[134] D. P. Vik and D. T. Fearon, *J. Immunol.* **134,** 2571 (1985).

[135] J. D. Lambris, V. S. Ganu, S. Hirani, and H. J. Müller-Eberhard, *Proc. Natl. Acad. Sci. U.S.A.* **82,** 4235 (1985).

[136] L. Sottrup-Jensen, T. M. Stepanik, T. Kristensen, P. B. Longblad, C. M. Jones, D. M. Wierzbicki, S. Magnusson, H. Domdey, R. A. Wetsel, Å. B. Lundwall, B. F. Tack, and G. H. Fey, *Proc. Natl. Acad. Sci. U.S.A.* **82,** 9 (1985).

[137] S. G. Davies and R. B. Sim, *Biosci. Rep.* **1,** 461 (1981).

[138] M. K. Hostetter, M. L. Thomas, F. S. Rosen, and B. F. Tack, *Nature (London)* **298,** 72 (1982).

[139] B. W. Erickson and S. A. Khan, *Ann. N.Y. Acad. Sci.* **421,** 167 (1983).

[140] M. K. Hostetter, R. A. Krueger, and D. J. Schmeling, *J. Infect. Dis.* **150,** 653 (1984).

hydrolyzed with water) acquires structural and functional properties of C3b (reviewed in Ref. 67), and it has been recognized as the important factor in the initiation of the alternative pathway of complement.[5]

Native C3 is the precursor of several fragments, each with various binding sites and distinct physiological functions.[2] The importance of C3 fragments in triggering a variety of cellular responses,[119] and also their possible participation in immune regulation,[104,130] has lately received considerable attention. A summary of the interactions of C3 fragments with their ligands, namely, the complement receptors, has recently been published.[141] The *in vitro* or physiologically generated fragments can be now positioned within the known amino acid sequences (deduced from the sequences of cDNA and supported by the protein sequence data[24,91,92]), and the peptides can be synthesized in order to test various activities or binding sites (e.g., receptor CR2,[135] properdin,[142] C3b receptor[143]).

Acknowledgments

I am obliged to Dr. Charles Parker for helpful discussions and for sharing his technical assistant, and I thank Mirjam McMullan, Nicole Bernshaw, and R. Joseph Gobel for excellent technical assistance during the various stages of development of the protocols described above in this chapter and Don Morse for preparation of the figures. The work by the author was supported by Public Health Service Grants AI 17122, AI 19319, and by funds from the Department of Pathology, University of Utah School of Medicine. Sincere thanks are extended to my colleagues at the M.R.C. Immunochemistry Unit, Department of Biochemistry, University of Oxford, for providing a friendly and stimulating environment during the time of manuscript preparation, and specifically to Drs. K. B. M. Reid, R. B. Sim, and E. Sim for reading the manuscript. The excellent secretarial assistance of Carolyn Brooks is greatly appreciated.

[141] R. B. Sim, V. Malhotra, J. Ripoche, A. J. Day, K. J. Micklem, and E. Sim, *Biochem. Soc. Symp.* **51,** 83 (1986).
[142] M. E. Daoudaki and J. D. Lambris, *Complement* **4,** 146 (1987).
[143] J. D. Becherer and J. D. Lambris, *Complement* **4,** 132 (1987).

[47] C1, C4, and C2 Components of the Classical Pathway of Complement and Regulatory Proteins

By IRMA GIGLI and FRANCISCO A. TAUSK

Introduction

The complement system is a group of plasma proteins that play an important role in host defense[1,2]; it may also have a direct effect in the immune response by facilitating intercellular interactions.[3] A number of these proteins exist as precursor enzymes that are converted to active products during activation and are controlled by inactivators or inhibitors.

Activation of the classical pathway is initiated by the binding of C1[3a] to aggregates of IgG or IgM antibodies produced either specifically by interaction with antigens or nonspecifically by heat or by chemical cross-linking. C1 is a calcium-dependent macromolecular complex composed of three proteins, C1q, C1r, and C1s, which are present in a molar ratio of 1 : 2 : 2.[4] Binding of C1q produces an allosteric change in proenzyme C1r that triggers its autocatalytic activation. The latter acquires a serine protease activity ($C\overline{1}r$) that hydrolyzes a single peptide bond in precursor C1s, resulting in its activation ($C\overline{1}s$).[5]

$C\overline{1}s$ is responsible for the activation of the next two components, C4 and C2.[5] C4 is a 200,000 MW protein composed of three disulfide-linked polypeptide chains, α (93,000 MW), β (75,000 MW), and γ (32,000 MW).[6,7] During complement activation $C\overline{1}s$ cleaves the α chain at the amino-terminal end, releasing a peptide consisting of 74–77 amino acid residues, C4a.[6-8] The larger fragment, C4b, expresses tran-

[1] K. F. Austen, *Bull. W.H.O.* **59,** 489 (1981).

[2] R. R. Porter and K. B. M. Reid, *Adv. Protein Chem.* **33,** 1 (1979).

[3] S. Yachmin and J. M. Ruthenberg, *J. Clin. Invest.* **44,** 518 (1965).

[3a] Abbreviations: C1, C2, C3, C4, first, second, third, and fourth components of complement; $C\overline{1}$, activated C1; $C\overline{1}$INH, C1 inhibitor; C4bp, C4b binding protein; FI, factor I; FH, factor H; VBS, veronal-buffered saline; EDTA, ethylenediaminetetraacetic acid; GVB–EDTA, gelatin–veronal buffer with EDTA; $DGVB^{2+}$, dextrose–gelatin–veronal buffer with calcium and magnesium; C–EDTA, guinea pig complement–EDTA; C4-def gp, C4-deficient guinea pig serum; SDS–PAGE, sodium dodecyl sulfate–polyacrylamide gel electrophoresis; DFP, diisopropyl phosphofluoridate; PMSF, phenylmethylsulfonyl fluoride; EACA, ϵ-aminocaproic acid.

[4] I. Gigli, R. R. Porter, and R. B. Sim, *Biochem. J.* **157,** 541 (1976).

[5] R. B. Sim, R. R. Porter, K. B. M. Reid, and I. Gigli, *Biochem. J.* **163,** 219 (1977).

[6] C. Bolotin, S. Morris, and B. Tack, *Biochemistry* **16,** 2008 (1977).

[7] I. Gigli, I. Von Zabern, and R. R. Porter, *Biochem. J.* **165,** 439 (1977).

[8] K. Moon, J. Gorski, and T. Hugli, *J. Biol. Chem.* **256,** 8685 (1981).

siently an active binding site provided by an internal thioester.[9] The cleavage of C4 exposes an acceptor site in the resulting C4b molecule that permits, in the presence of magnesium, the reversible binding of C2 and the subsequent cleavage by C$\overline{1}$s of an Arg-Lys bond generating two fragments, C2a and C2b.[9] The C2a portion of the molecule (74,000 MW) remains attached to C4b in the formation of the C4b2a complex and contains the active site of the enzyme responsible for C3 activation.[10]

The third component (C3) is a glycoprotein of 180,000 MW composed of two polypeptide chains, α (115,000 MW) and β (75,000 MW). The amino terminus of the α chain is hydrolyzed by the C4b2a enzyme, resulting in the release of a 9,000 M_r fragment (C3a).[11] The C3b fragment is composed of the remaining of the α chain (α') and the intact β chain; like C4b it has a transient capability of forming a stable ester linkage through an internal thioester within the α' chain. C3b molecules that associate with or are in the proximity of its own activating enzyme bind C5, thus facilitating its cleavage by C4b2a. The generation of C5b initiates the nonenzymatic assembly of the membrane attack complex composed of C5b–C6, C7, C8, and C9.[12]

The activation of the classical pathway is modulated by the presence in blood of proteins that function as inhibitors or inactivators. Conversion of C1 to its enzymatically active form C$\overline{1}$ is controlled by the C1 inhibitor (C$\overline{1}$INH). This effect is the result of the inhibition of both the activation of C1s by C$\overline{1}$r and of the activated C$\overline{1}$s.[13]

The function of C4b2a is regulated by the inherited temperature-dependent instability of the enzyme and by the effect of two serum proteins, C4b binding protein (C4bp) and the C4b/C3b inactivator, factor I (FI). C4 binding protein accelerates the decay of C2a from the C4b2a enzyme,[14] prevents the uptake of C2 by C4b, and renders C4b susceptible to the proteolytic activity of FI.[15] Factor I is an endopeptidase that cleaves two peptide bonds from the α' chain of C4b generating three fragments: C4d, which contains the thioester binding site, and $\alpha 3$ and $\alpha 4$ which are covalently bonded to the β and α chains forming C4c. Factor I also hydrolyzes the α' chain of C3b, but this reaction requires the presence of a different cofactor, factor H.[16,17]

[9] B. F. Tack, *Springer Semin. Immunopathol.* **6,** 259 (1983).
[10] H. J. Müller-Eberhard, M. J. Polley, and M. A. Calcott, *J. Exp. Med.* **125,** 359 (1967).
[11] U. R. Nilsson and J. Mapes, *J. Immunol.* **111,** 293 (1973).
[12] H. J. Müller-Eberhard, *Springer Semin. Immunopathol.* **7,** 93 (1984).
[13] G. B. Naff and O. D. Ratnoff, *J. Exp. Med.* **128,** 571 (1968).
[14] I. Gigli, T. Fujita, and V. Nussenzweig, *Proc. Natl. Acad. Sci. U.S.A.* **76,** 6596 (1979).
[15] T. Fujita, I. Gigli, and V. Nussenzweig, *J. Exp. Med.* **148,** 1044 (1978).
[16] J. M. Weiler, M. R. Daha, K. F. Austen, and D. T. Fearon, *Proc. Natl. Acad. Sci. U.S.A.* **73,** 3268 (1976).
[17] K. Whaley and S. Ruddy, *Science* **193,** 1011 (1976).

Assays of Complement Components

Functional and immunochemical methods may be used for the analysis of complement proteins in serum or during purification procedures.

Functional Analysis

The observation that sheep erythrocytes sensitized with specific antibodies are lysed by the classical pathway of complement led to the development of convenient methods for the functional assessment of the proteins of this system. The hemolytic assay of a given complement component is based on the capacity of that component to convert the appropriate intermediate product to the next step in a series of reactions leading to cell lysis.

Reagents. Glass-double-distilled deionized water is used for the preparation of all reagents.

Stock solution of veronal-buffered saline (VBS), 5× concentrated: Dissolve 83 g of NaCl and 10.19 g of sodium 5,5-diethyl barbiturate in distilled H_2O and adjust the pH to 7.5 with 1 N HCl (~34.5 ml) in a final volume of 2 liters.

Stock solution of ethylenediaminetetraacetic acid (EDTA) 0.1 *M* stock solution: Dissolve 37.2 g of EDTA (disodium salt) in 800 ml of H_2O, adjust the pH to 7.5 with 50% w/w NaOH, and bring to 1 liter with H_2O.

Gelatin–veronal buffer containing 0.04 or 0.01 *M* EDTA (GVB–EDTA): Dissolve 1 g of gelatin in boiling H_2O, add 200 ml of stock VBS, 400 ml (0.04 *M*) or 100 ml (0.01 *M*) of stock EDTA, and bring the volume to 1 liter.

Dextrose–gelatin–veronal buffer with Ca^{2+} and Mg^{2+} $(DGVB)^{2+}$: Dissolve 25 g of dextrose in 100 ml H_2O and 1 g of gelatin in 100 ml of boiling H_2O, add 200 ml of stock VBS, and bring to 1 liter with H_2O.

Guinea pig complement–EDTA (C–EDTA): Add 0.1 ml of whole guinea pig serum (Colorado Serum Co., Gibco Life Tech, or Pel-Freeze Biologicals) to 1.9 ml of GVB–EDTA (1 : 20 dilution).

Guinea pig serum deficient in component C4 (C4-def gp) in $DGVB^{2+}$: Add 0.1 ml of whole serum from guinea pigs genetically deficient in C4 (Cordis Laboratories, Miami, FL, or Dimension Laboratories, ON) to 7.5 ml of $DGVB^{2+}$ (1 : 75 dilution).

Human serum: Blood obtained from healthy individuals is clotted at room temperature for 30 min and centrifuged at 3000 rpm for 30 min at 4°. The serum is stored in aliquots at −70° until used.

Sheep erythrocytes sensitized with rabbit anti-sheep hemolysin (EA): Sheep erythrocytes preserved in sterile Alsever's solution are washed twice with 0.15 *M* NaCl (normal saline) by centrifugation for 10 min at 2,000 *g*, followed by two washes with $DGVB^{2+}$. The cells are resuspended in $DGVB^{2+}$ at a concentration of 1×10^9/ml. The IgG fraction of rabbit anti-sheep erythrocyte hemolysin (Cordis Laboratories) at the dilution indicated by the manufacturer is then added to the cells and the mixture incubated with shaking for 30 min at 30°, followed by 15 min on ice. The cells are washed twice in $DGVB^{2+}$, resuspended to 1×10^9/ml in the same buffer containing 500 units/ml of penicillin and 50 μg/ml streptomycin, and stored at 4° for up to 2 weeks.

EAC$\overline{1}$4b and EAC4b intermediates: Although this intermediates may be prepared using purified C1 and C4, an alternative reliable technique takes advantage of the inhibitory effect of methylprednisolone on the deposition of the late-acting complement components on EA.[18] Human serum (0.9 ml) is incubated with 0.1 ml of methylprednisolone (100 mg/ml) in $DGVB^{2+}$ at 37° for 30 min; 5 ml of EA (5×10^8/ml) is then added and incubated in an ice bath for 30 min. The intermediates generated (EAC$\overline{1}$4b cells) in the same buffer are washed by centrifugation in cold $DGVB^{2+}$ and resuspended to the original volume. To prepare EAC4b cells, EAC$\overline{1}$4b are incubated in GVB–0.04 *M* EDTA at 37° for 30 min, washed twice in the same buffer, once in $DGVB^{2+}$, and then resuspended and stored in cold $DGVB^{2+}$.

Hemolytic Titration of Components of the Classical Pathway. In a complement component titration, aliquots of a suspension of optimally sensitized erythrocytes are each mixed with serial dilutions of serum or the sample of the component to be tested, in the presence of an excess amount of all the other components of the classical pathway (Table I).[19,20]

Functional Detection of the Control Proteins: Factor I (FI) and C4b Binding Protein (C4bp). The cleavage of C4b to C4c and C4d provides the basis for the functional enzymatic assays of factor I and the cofactor activity of C4bp. Factor I in the presence of C4bp cleaves the α' chain of

[18] F. Kitamura, K. Shimada, T. Suzuki, and K. Nishioka, *J. Immunol. Methods* **85,** 363 (1985).

[19] R. A. Nelson, Jr., J. Jensen, I. Gigli, and N. Tamura, *Immunochemistry* **3,** 111 (1966).

[20] H. J. Rapp and T. Borsos, "Molecular Basis of Complement Action." Appleton-Century-Crofts, New York, 1970.

TABLE I

PROCEDURE AND REAGENTS FOR THE HEMOLYTIC TITRATIONS OF PROTEINS OF THE CLASSICAL PATHWAY OF COMPLEMENT[a]

C1	Volume (ml)	C4	Volume (ml)	C2	Volume (ml)	C1INH	Volume (ml)
Sample dilutions	0.5	Sample dilutions	0.5	Sample dilutions	0.5	Sample dilutions	0.5
EAC4	0.5	EA	0.5	EACl4	0.5	C1	0.5
(20 min 30°)		C4-def gp	0.5	(t_{max} 30°)[b]		(30 min 30°)	
C2	0.5	(60 min 37°)		GVB–EDTA	0.5	EAC4	0.5
(10 min 30°)				C–EDTA	1	C2	0.5
GVB–EDTA	0.5			(60 min 37°)		(20 min 30°)	
C–EDTA	1					C–EDTA	0.5
(60 min 37°)						(60 min 37°)	

[a] Following the completion of the last incubation, normal saline solution is added to each sample to a final volume of 7.5 ml, and they are centrifuged for 10 min at 2000 rpm. The optical density of the supernatants is then measured in a spectrophotometer adjusting the wavelength to 414 nm (A_{414}). The "reagent lysis" control represents the lysis of the cells produced by the reagents used in the assay in the absence of the component to be tested. The A_{414} of the supernatant of this control sample must be subtracted from the A_{414} of all the samples (usually $<5\%$). The corrected value of each tested dilution is expressed as percent lysis using as 100% a sample of erythrocytes lysed with H_2O. The resulting values from each reaction mixture (y) are plotted on log–log graph paper as the log of $y/(1-y)$ (obtained from a log table) on the x axis. The log of the reciprocal of each dilution is plotted on the y axis. Values of y less than 20% or greater than 80% are discarded. The value where the best-fit line crosses the x axis where $y/(1-y) = 1.0$ (50% lysis) represents the reciprocal of the titer for that complement component.

[b] t_{max} represents the time required for maximal lysis, which must be determined by trial (usually approximately 5 min).

C4 generating three fragments, $\alpha 2$ or C4d (46,000 MW), $\alpha 3$ (25,000 MW), and $\alpha 4$ (16,000 MW). ^{125}I-labeled C4b exposed to FI and C4bp and subjected to sodium dodecyl sulfate–polyacrylamide gel electrophoresis (SDS–PAGE) under reducing conditions shows the presence of the cleaved fragments in addition to the intact β and γ chains.[16,21]

Steps in the procedure are as follows:

1. Purified C4 is radiolabeled with ^{125}I to a specific activity of 10^6 cpm/μg using the Iodogen method[22] according to the instructions of the manufacturer (Pierce Chemicals).
2. Two micrograms of ^{125}I-C4 is incubated for 60 min at 37° with 0.5 μg of purified $\overline{\text{C1}}$s to generate ^{125}I-C4b.
3. Two-tenth microgram of ^{125}I-C4b is incubated with serial dilutions of the FI or C4bp preparation in the presence of 12 μg of C4bp or 0.5 μg of FI, depending on whether factor I or C4bp is to be detected.
4. After 90 min at 37°, the samples are reduced with 200 m*M* dithiothreitol (DTT), alkylated with 400 m*M* iodoacetamide, and subjected to SDS–PAGE in 7.5% polyacrylamide gels, followed by autoradiography.
5. The degree of conversion of the α' chain into $\alpha 2$, $\alpha 3$, and $\alpha 4$ may be quantitated by gel scanning, and expressed as the dilution required to produce 50% cleavage of the α' chain.

Immunochemical Analysis

Rocket immunoelectrophoresis (RIE)[23] is a well-suited procedure to measure the concentration of a particular complement component or control protein in a protein mixture or serum. The principle of this method is based on the migration of an antigen in an electric field and the formation of precipitates with a specific antiserum incorporated in the support media.

Reagents and Technique

1. A solution of barbital buffer is prepared containing 23 m*M* sodium barbital, 4 m*M* barbituric acid, and 2 m*M* EDTA, pH 8.6.
2. Next, 1.2% agarose and 2.5% PEG 6000 are dissolved in hot barbital buffer and maintained at 55° in a heating block. To this mixture 1% goat antiserum to the specific complement component to be tested (available from Cytotech, San Diego, CA) is added and poured onto the hydrophilic

[21] I. von Zabern, E. L. Bloom, V. Chu, and I. Gigli, *J. Immunol.* **128,** 1433 (1982).

[22] M. K. Markwell and C. F. Fox, *Biochemistry* **17,** 4807 (1978).

[23] C. B. Laurell, *Anal. Biochem.* **15,** 45 (1966).

side of a Gel Bond (FMC Corp., Rockland, ME) sheet (150 × 100 mm). After cooling, a row of 3-mm wells is cut and suctioned from the plate.

3. The plate is placed in the immunoelectrophoresis chamber filled with barbital buffer precooled to 80°. Wicks made from strips of blotting paper are soaked in the buffer and placed on the edges of the gel. A current of 100 V is applied immediately prior to the careful addition to each well of 4 μl of serial dilutions of the sample to be tested. The electrophoresis is continued for 18 hr at 8°.

4. The plates are then washed by immersion in 75 m*M* NaCl. After drying, they are stained for 15 min with 0.05% Coomassie blue in a solution of H_2O/acetic acid/methanol (5 : 1 : 5, v/v/v), and destained with the same solution lacking Coomassie blue.

5. The height of protein precipitates is measured at their peak in millimeters and compared to a standard curve derived from dilutions of a sample of known antigen concentration plotted on semilog paper.

Purification of Complement Proteins

Unless indicated otherwise, all complement proteins are prepared from outdated plasma. When serum is required, 20 m*M* $CaCl_2$ is added to plasma and allowed to clot overnight at 4°, the clot is then removed by centrifugation. All procedures are performed at 4° unless otherwise stated.

Preparation of Precursor and Activated Component C1

Five hundred milliliters of serum is made 5 m*M* in diisopropyl phosphofluoridate (DFP) using a stock solution of 2.5 *M* DFP in 2-propanol and centrifuged at 12,500 *g* for 30 min at 4°. The supernatant is diluted with 2 liters of double-distilled water containing 5 m*M* $CaCl_2$, 2.5 m*M* iodoacetamide, and 0.2 m*M* phenantracine, and the pH is adjusted to 7.4 with 1 *M* NaOH. The suspension is stirred at 4° for 2 hr and centrifuged at 12,500 *g* for 30 min. The supernatant is discarded and the euglobulin precipitate in each centrifuge flask is carefully suspended in 30 ml of 0.04 *M* sodium acetate, pH 5.5, containing 5 m*M* $CaCl_2$ and 5 m*M* DFP. After 3 more washes with the same buffer the precipitates are dissolved in 15 ml of 0.2 *M* NaCl in the buffer mentioned above. This solution is centrifuged for 30 min at 75,000 *g*, and the film of fat and the undissolved precipitate are discardrd. The slightly opalescent supernatant containing C1 is applied to a column (5 × 90 cm) of Sepharose 6B equilibrated in the same buffer. Fifteen-milliliter fractions are collected and screened for C1 activity. C1 elutes between fractions 55 and 100. The active fractions are pooled, DFP at the final concentration of 5 m*M* is added, and the pool is concentrated to 15 ml by ultrafiltration. When this material is examined by electrophoresis in a 5.6% polyacrylamide gel containing sodium dodecyl sulfate, C1r, C1s, and the subunits of C1q are seen, as well as a high-molecular-weight contaminant that does not enter the gel. After reduction this material is apparent with a MW of approximately 70,000.[4] Antigenically it is identified as C4bp and can be easily removed by passage through an immunoabsorbent column prepared with the $F(ab)_2$ fragment of IgG anti-C4bp (Cytotech) bound to CNBr-activated Sepharose according to the manufacturer's instructions (Pharmacia).

Based on the hemolytic activity, this technique yields a C1 recovery of 40%. No activation occurs during isolation, as judged by the behavior of the complex on SDS–PAGE. After reduction, the unactivated subcomponents, C1r and C1s, behave as a single peptide chain of apparently higher molecular weight than that observed under nonreducing conditions, while activated subcomponents separate into two polypeptide chains of lower molecular weight. Activated C1 ($C\bar{1}$) can be obtained by using the same procedures but omitting DFP from all steps.

More recently, purification methods have been reported utilizing *o*-nitrophenyl-*p*′-guanidinobenzoate to prevent activation of C1 and anti-IgG-protein A affinity chromatography to remove contaminating IgG.[24]

Separation of Subcomponents C1q, C1r, and C1s

The subcomponents C1q, C1r, and C1s are dissociated from the purified C1 macromolecular complex by adding EDTA to a final concentration of 10 m*M* to the concentrated pool from the Sepharose 6B column.[5] Following dialysis, the sample is applied to a DEAE-cellulose (Whatman DE-32) column (1.2 × 30 cm) and eluted with a salt gradient of 500 ml of a buffer containing 10 m*M* NaH_2PO_4, 47 m*M* Na_2HPO_4, pH 7.4, and 5 m*M* EDTA, used for dialysis of the pool and to equilibrate the column, and 500 ml of a final buffer of the same composition but containing 0.5 *M* NaCl. C1q is eluted at the beginning of the gradient, C1r elutes at approximately 15 mS (0°) and C1s at 20 mS. Subcomponent C1r is purified further by chromatography on a column (2.5 × 90 cm) of Sephadex G-200 in 0.05 *M* sodium acetate buffer, pH 5.5, 0.1 *M* NaCl, and 1 m*M* EDTA. This step is necessary only if C4bp was not removed from the intact C1 preparation. All the subcomponents appear to be pure as judged by SDS–PAGE.

Purification of Component C4

The initial steps for the isolation of human component C4 are those described for the purification of component C1. Following the separation

[24] J. A. Schifferli and G. Steiger, *J. Immunol. Methods* **76,** 283 (1985).

of the euglobulin precipitate used for C1 purification, 1 m*M* DFP in 2-propanol is added to 2500 ml of the supernatant, and the pH of the solution is adjusted to 5 with 1 *M* HCl. After stirring for 15 min, the precipitate formed is removed by centrifugation at 12,500 *g* for 30 min, and the pH of the supernatant (pseudoglobulin) is immediately adjusted to 8.6 with 2 *M* Tris. The ionic strength is raised to 0.185 *M* by adding 2 *M* NaCl, and the solution is applied to a column (7 × 25 cm) of DEAE-Sephadex A-50 equilibrated in 0.01 *M* Tris–HCl, pH 8.6, 0.185 *M* NaCl, and 0.02% sodium azide. The column is washed with the starting buffer until the A_{280} of the eluate has fallen to below 0.05. A linear gradient composed of 2.5 liters of the equilibrating buffer and 2.5 liters of the same buffer with a final NaCl concentration of 0.25 *M* is applied. Following this gradient the column is washed with the latter buffer until no further protein is eluted. A flow rate of 250 ml/hr is used during the initial washing of the column and is decreased to 100 ml/hr during the gradient elution. Hemolytic C4 activity is found in fractions eluted with the gradient at about 11.5 mS/cm (2°) and subsequent washing.

The active fractions are pooled, concentrated to approximately 20 ml by ultrafiltration through an Amicon PM10 membrane, and diluted with ice-cold distilled water to lower the NaCl concentration to 0.195 *M*. Following the addition of DFP to a concentration of 1 m*M* the sample is applied to a column (5 × 35 cm) of DEAE-Sepharose CL-6B equilibrated in 0.01 *M* Tris–HCl, pH 8.6, 0.2 *M* NaCl, and 0.02% sodium azide. The column is eluted with the same buffer (isocratic conditions) using a flow rate of 50 ml/hr. C4 and prealbumin are retarded in the column (eluting between 2.4 and 3.5 bed volumes of buffer) and are separated from other contaminating proteins which elute ahead of them.

The fractions containing C4 hemolytic activity are pooled, concentrated by ultrafiltration, and applied to a Sephacryl S-200 superfine column (5 × 90 cm) equilibrated in 0.02 *M* sodium phosphate buffer, pH 7.0, containing 0.5 *M* NaCl and 0.02% sodium azide at a flow rate of 30 ml/hr, eluting at an apparent MW of 200,000. SDS–PAGE analysis under both reducing and nonreducing conditions of the two protein peaks eluted from the column show that the first peak contains C4 free from contaminants while the second peak contains prealbumin. C4 hemolytic activity is found in samples in the first peak.[7,25]

Purification of Component C2

One liter of serum, containing 2.5 m*M* DFP with the pH adjusted to 5.5 with 1 *M* HCl, is dialyzed overnight against 4 liters of H_2O containing

[25] I. Gigli, *Nature (London)* **272,** 836 (1978).

5 m*M* EDTA and 0.5% benzamidine hydrochloride (w/v). The serum is centrifuged at 25,000 *g* for 30 min, and the euglobulin precipitate is discarded. Fifty milliliters of 0.4 *M* sodium phosphate, pH 6.0, is added to the serum supernatant, the pH is adjusted to 6.0 with NaOH, and the mixture is applied to a CM-Sephadex C-50 column (8 × 10 cm), equilibrated with 0.1 *M* sodium phosphate, 0.5% benzamidine (w/v), pH 6.0. The column is washed with 1 liter of equilibrating buffer, eluted with a linear gradient of 2 liters of the same buffer and 2 liters of 0.25 *M* sodium phosphate (pH 6.0), followed by a wash with 1 liter of 0.4 *M* sodium phosphate, pH 6.0. All buffers are applied at a flow rate of 2 liters/hr.

The fractions containing C2 activity which elute between 0.07 and 0.12 *M* sodium phosphate, are pooled, and solid ammonium sulfate to give a 50% saturation (29 g/liter) is added, as well as 2.5 m*M* DFP. The suspension is stirred for 1 hr at 4°. The precipitate is separated by centrifugation at 23,000 *g* for 30 min and discarded; the supernatant is passed through Whatman #1 filter paper. Additional ammonium sulfate is added to the filtrate to give a final saturation of 75%, the suspension is brought to room temperature (20°) with gentle stirring and centrifuged at the same temperature for 90 min at 23,000 *g*. The precipitate is resuspended in 50 ml of cold 0.4 *M* sodium phosphate buffer, pH 6.0, and the undissolved material is separated by centrifugation at 26,000 *g* for 30 min and discarded.

The supernatant is desalted over a column (4 × 30 cm) of Sephadex G-25 (Pharmacia) equilibrated with 5 m*M* veronal buffer, pH 8.5, containing 0.04 *M* NaCl, 2 m*M* $MgCl_2$, and then applied to a column containing CNBr-activated Sepharose prepared by suspending packed Sepharose 4B (50 ml) washed in water in a minimal volume of 0.1 *M* sodium phosphate, pH 7.8. Seven grams of CNBr dissolved in 7 ml of dimethyl formamide is added. The mixture is stirred, keeping the temperature at 20° by adding crushed ice and the pH at 10–11 with 4 *M* Na OH. When the reaction is complete the Sepharose is extensively washed with 0.1 *M* sodium phosphate, pH 7.8, and stirred for 16 hr in the same buffer. Alternatively, commercially available CNBr-Sepharose 4B is incubated overnight in the last buffer. The column is washed with 5 m*M* VBS, 0.04 *M* NaCl, 2 m*M* $MgCl_2$. The decay of the active groups from CNBr-activated Sepharose 4B results in the formation of sites with affinity for C2. The capacity of this CNBr-activated Sepharose is approximately 1 mg protein/ml packed resin. The pool containing C2 activity is applied to this column and further washed with the equilibrating buffer until the A_{280} is below 0.05. C2 is eluted with 0.4 *M* sodium phosphate buffer, pH 6.0.

The pool of fractions containing C2 activity is equilibrated with 5 m*M* veronal buffer containing 40 m*M* NaCl (VBS) by passing it through a column (5 × 40 cm) of Sepharose G-25 and then applying it to a column

(2.5 × 50 cm) of DEAE-Sepharose equilibrated in the same buffer. The column is washed and eluted with a linear gradient consisting of 500 ml of 5 m*M* VBS (40 m*M* NaCl), and 500 ml of the same buffer containing 80 m*M* NaCl. Fractions with C2 hemolytic activity elute at approximately 27 mS/cm (measured at 2°), a NaCl molarity of 0.05, associated with a protein peak of low A_{280} (<0.06). Because the physicochemical properties of C2 and factor B are very similar, the latter remains as a trace contaminant even after the DEAE-Sepharose chromatography. This can be resolved by an additional step of affinity chromatography utilizing anti-factor B IgG (Dako Corp., Santa Barbara, CA) immobilized on CNBr-activated Sepharose (Pharmacia), according to the manufacturer's instructions. The final pool of purified C2 is concentrated, the pH adjusted to 6, and stored at −70°.[26-28]

Purification of the Inhibitor of Component C1 (C1INH)

Solid PEG 4000 is added to 1 liter of freshly drawn human plasma, precooled at 4°, containing 10 m*M* EDTA and 10 m*M* benzamidine, to a final concentration of 5%. The plasma is stirred for 1 hr, and the precipitate formed is removed by centrifugation for 30 min at 10,000 *g*. The supernatant is applied to a column containing lysine–Sepharose equilibrated with 0.1 *M* phosphate buffer pH 7.0, containing 0.5 *M* KCl, 0.01 *M* EDTA, and 5 m*M* benzamidine, at a ratio of 1 : 5 gel to plasma to remove plasminogen. The column is washed with the same buffer, and the effluent fractions containing unbound proteins are pooled, applied to a 400-ml DEAE-Sephadex A-50 column equilibrated with 0.02 *M* phosphate buffer pH 7.0, containing 0.1 *M* NaCl, 5 m*M* EDTA, and 5 m*M* benzamidine. The column is washed and developed with a liner gradient of 5 column volumes each of equilibration buffer and the same buffer containing 0.4 *M* NaCl. The C1INH elutes approximately between 12 and 15 mS/cm.

Fractions containing $\overline{C1}$INH are concentrated to approximately 2% of the original plasma volume and dialyzed against 0.02 *M* phosphate buffer, pH 7.0, containing 0.1 *M* KCl, 20 m*M* benzamidine and 2 m*M* citrate. The samples are centrifuged at 4° for 10 min at 20,000 *g* and applied to a column (2.5 × 95 cm) of Sephadex G-150 Superfine (Pharmacia) equilibrated with the above buffer. Fractions containing C1INH which elute at an apparent MW of 110,000 are concentrated to approximately 20 ml, dialyzed against 0.02 *M* phosphate buffer, pH 7.0, containing 0.15 *M* KCl,

[26] M. A. Kerr and R. R. Porter, *Biochem. J.* **171,** 99 (1978).
[27] B. Ngan and J. O. Minta, *Mol. Immunol.* **18,** 1035 (1981).
[28] M. A. Kerr, this series, Vol. 80, p. 54.

and applied to a column (2.5 × 50 cm) of hydroxylapatite equilibrated with the same buffer. A width-to-length ratio of 1:20 is important to obtain good separation. C1INH is eluted with 2 bed volumes of dialysis buffer. Occasionally the protein does not elute under these conditions and requires the application of a gradient from the equilibration buffer to one containing 1 *M* KCl; however this preparation appears contaminated with large amounts of albumin. The $C\bar{1}INH$ is identified in the column fractions using rocket immunoelectrophoresis, and the final preparation is quantitated antigenically and functionally. Seventy percent of the C1INH present in plasma is recovered using this procedure.[29]

Purification of C4b Binding Protein (C4bp)

Human C4bp is purified from outdated plasma by a modification of previously published procedures.[30-32] Two liters of citrated plasma containing 5 m*M* DFP, 10 m*M* EDTA, and 50 m*M* ϵ-aminocaproic acid (EACA) are centrifuged at 10,000 *g* for 15 min and passed through a 500-ml gelatin–Sepharose column equilibrated with 50 m*M* Tris–HCl, pH 7.4, containing 50 m*M* EACA, 5 m*M* EDTA, and 0.5 *M* NaCl, to remove cold-insoluble globulins which remain bound to the the gel. The pass-through is dialyzed overnight in 5 m*M* Tris–HCl, pH 7.4, containing 5 m*M* EDTA, 50 m*M* EACA, and 5 m*M* benzamidine–HCl. Polyethylene glycol 4000 is added to the dialysate to a concentration of 5% (w/v), and after stirring for 1 hr the precipitate formed is separated by centrifugation at 10,000 *g* for 30 min. The supernatant is discarded and the pellet is redissolved in 50 m*M* Tris–HCl, pH 7.4, containing 50 m*M* EACA and 5 m*M* EDTA, and applied to a 500-ml column of Bio-Rex 70 equilibrated with the same buffer. The column is washed and eluted with a linear salt gradient of NaCl containing 400 ml of the starting buffer and 400 ml of the same buffer containing 0.5 *M* NaCl. C4bp elutes as a broad heterogeneous peak, at approximately 10 mS (conductivity at 0°).

The C4bp-containing fractions are pooled, dialyzed against 50 m*M* Tris–HCl, pH 7.4, containing 50 m*M* EDTA and 0.15 *M* NaCl, and chromatographed on a column (1.5 × 15 cm) of heparin–Sepharose equilibrated with the same buffer. The cofactor is eluted with this same buffer by a linear salt gradient of increasing NaCl concentration from 0.15 to 0.4 *M* using a total volume of 1 liter of elution buffer. The eluted fractions

[29] R. A. Harrison, *Biochemistry* **22**, 5001 (1983).

[30] J. Scharfstein, A. Ferreira, I. Gigli, and V. Nussenzweig, *J. Exp. Med.* **148**, 1044 (1978).

[31] K. B. M. Reid and J. Gagnon, *FEBS Lett.* **137**, 75 (1982).

[32] L. P. Chung, J. Gagnon, and K. B. M. Reid, *Mol. Immunol.* **22**, 427 (1985).

containing C4bp are concentrated by ultrafiltration and applied to a column (5×90 cm) of Sephacryl S-500 eluted with 5 m*M* Tris–HCl, pH 7.4, containing 50 m*M* EACA, 5 m*M* EDTA, and 0.25 *M* NaCl. In order to remove IgM, which is the only residual detectable contaminating protein after the last step of purification, the samples are pooled, concentrated, and applied to a column (2.5×10 cm) of Sepharose 4B coupled with goat IgG specific for human IgM μ chain (Atlantic Antibodies), at a concentration of 10 mg protein/ml CNBr-activated Sepharose 4B. The column is eluted with 50 m*M* Tris–HCl buffer, pH 7.5, containing 0.15 *M* NaCl.

The yield of purified C4bp is approximately 20 mg/liter of starting plasma. The purified protein appears as two closely spaced bands on a nonreduced SDS–polyacrylamide gel with an apparent molecular weight of approximately 550,000. Upon reduction a single band of 70,000 MW is observed. Using these purification procedures, protein S, previously reported to copurify with C4bp,[33] is not detectable.

Purification of the C4b/C3b Inactivator (Factor I)

One hundred ml of serum containing 0.02 *M* EDTA and 0.5 *M* NaCl is applied to a column (5.6×6.2 cm) prepared with CNBr-activated Sepharose which has been coupled to goat anti-factor I IgG (Cytotech) and equilibrated with 10 m*M* phosphate buffer, pH 7.0, containing 0.5 *M* NaCl and 2 m*M* EDTA. The column is washed with 200 ml of the equilibrating buffer, followed by 200 ml of the same buffer containing 20% sucrose to prevent nonspecific binding of proteins to the Sepharose. The bound proteins are eluted with 200 ml of 2 *M* guanidine–HCl, containing 10 m*M* sodium phosphate, 0.5 *M* NaCl, and 2 m*M* EDTA, pH 7–7.5. The eluate is dialyzed extensively against 0.15 *M* NaCl, 2 m*M* EDTA, pH 7.2, concentrated to 10 ml, and further dialyzed against 10 liters of 8 m*M* sodium phosphate buffer, pH 7.0, containing 5 m*M* NaCl (conductance of 2 mS/cm at 21°). It is then applied to a column (1.4×10 cm of DEAE-cellulose equilibrated with the same buffer, washed with 3 column volumes, and eluted with the equilibrating buffer containing 10 m*M* NaCl, conductance of 4.1 mS/cm at 21°. Although the yield of FI at this step is low (33%), it is 85% pure, and the only contaminant is IgG; the latter is removed by passage through an anti-IgG Sepharose column. The breakthrough is concentrated and stored at −70°.[34]

This method, although simple, yields a relatively low recovery of FI and subjects it to harsh conditions such as exposure to guanidine. A milder

[33] B. Dahlback and J. Stenflo, *Proc. Natl. Acad. Sci. U.S.A.* **78,** 2512 (1981).

[34] M. K. Pangburn, R. D. Schreiber, and H. J. Müller-Eberhard, *J. Exp. Med.* **146,** 257 (1977).

method which produces better yields has been reported, utilizing a series of chromatography columns with lysine–Sepharose, QAE-Sephadex, wheat germ agglutinin–Sepharose, hydroxyapatite, and Sephacryl S-200.[35] The best results can be obtained in a one-step procedure using an affinity column of monoclonal antibodies against FI, now commercially available (Cytotech), as described by Hsiung *et al.*[36] This method is ideal for small-scale purifications but is limited by the cost of the reagents. Briefly, plasma containing 2.5 m*M* DFP and 5 m*M* sodium azide is centrifuged and applied to a preclearance column of CNBr-activated Sepharose (Pharmacia) coupled to bovine IgG (10 mg/ml). Following this step the plasma is applied to a column of monoclonal antifactor I IgG immobilized in the same way as the previous column. After washing with a buffer containing 0.15 *M* NaCl, 25 m*M* Tris, and 20 m*M* EDTA, pH 7.3, the bound antigen is eluted with 0.15 *M* NaCl, 0.05 *M* diethylamine–HCl, pH 11.5, and neutralized with glycine to pH 8.2.

Acknowledgment

This work was supported by National Institutes of Health Grant AI 20067.

[35] L. G. Crossley and R. R. Porter, *Biochem. J.* **191,** 173 (1980).

[36] L.-M. Hsiung, A. N. Barclay, M. R. Brandon, E. Sim, and R. R. Porter, *Biochem. J.* **203,** 293 (1982).

[48] Alternative Pathway of Complement

By MICHAEL K. PANGBURN

Introduction

Complement may be activated in the absence of antibody by direct contact between the activating surface and proteins of the alternative pathway. The alternative pathway thus provides an immediately available line of defense that does not require prior immunization. The six plasma proteins of the alternative pathway (Table I[1-7]) recognize a wide variety of

[1] B. F. Tack and J. W. Prahl, *Biochemistry* **15,** 4513 (1976).

[2] M. K. Pangburn and H. J. Müller-Eberhard, *Biochemistry* **22,** 178 (1983).

[3] B. Curman, L. L. Sandberg-Tragard, and P. A. Peterson, *Biochemistry* **16,** 5368 (1977).

[4] K. Nagaki, K. Iida, M. Okubo, and S. Inai, *Int. Arch. Allergy Appl. Immunol.* **57,** 221 (1978).

[5] K. Whaley and S. Ruddy, *J. Exp. Med.* **144,** 1147 (1976).

[6] K. B. M. Reid, D. M. A. Johnson, J. Gagnon, and R. Prohaska, this series, Vol. 80, p. 143.

[7] R. G. DiScipio. *Biochem J.* **199,** 485 (1981).

TABLE I

PROTEINS OF THE ALTERNATIVE PATHWAY

Symbol	Name	Function	Molecular weight	Serum concentration (μg/ml)	$E^{1\%}_{1cm}$ (280 nm)	s rate	pI
C3	C3	Initiation as C3(H_2O); after enzymatic activation covalent attachment via thioester as C3b, which is a nonenzymatic subunit of C3/C5 convertase	185,000	1300 (1050–1650)	9.7[1] 11.0[2]	9.6	—
B	Factor B	Zymogen of Bb (M_r = 60,000) which is a serine protease that forms C3/C5 convertase with C3(H_2O) or C3b; activated by Factor D	93,000	210 (170–260)	12.7[3]	5.9–6.2	6.6
D	Factor D	Serine protease which activates Factor B when in complex with C3b or C3(H_2O)	24,000	1	—	2.5–3	6.6, 7.0
H	Factor H	Negative regulator which accelerates decay–dissociation of C3/C5 convertase; cofactor of Factor I	150,000	475 (360–550)	9.0[4,5] 12.4[2]	5.6–6.4	5.8–6.2
I	Factor I	Serine protease regulating C3b by conversion to C3bi and C3bi by fragmentation to C3c and C3dg; requires Factor H or CRl as cofactor	88,000	35 (30–40)	14.3[2]	4.5	5.7–6.1
P	Properdin	Positive regulator stabilizing C3/C5 convertase by binding to the complex	220,000	20 (10–30)	18[6] 12[7]	5.2 5.4	>9.5

potential pathogens within minutes after these organisms come in contact with plasma. Sensitive organisms include bacteria, fungi, a number of viruses, virus-infected cells, certain tumor cell lines, parasites such as trypanosomes, and erythrocytes from patients with paroxysmal nocturnal hemoglobinuria. Because of a unique amplification process which occurs during activation, the alternative pathway generates large quantities of the anaphylatoxins C3a and C5a and deposits C3b on the surface of the activating particle.[8] C3b and its breakdown products stimulate phagocytosis via receptors on neutrophils and mononuclear phagocytes.

Assays for Alternative Pathway Activity

Assays for alternative pathway activation will be described here. Assays for the individual components will be described in the appropriate sections below. It should be noted that the alternative pathway is very sensitive to dilution and should be assayed with as little dilution as possible. Activity can rarely be detected at 20-fold or higher dilutions. The alternative pathway does not require Ca^{2+} ions for activation, but it does require Mg^{2+}. Thus, alternative pathway activation proceeds in the presence of MgEGTA, which chelates Ca^{2+}. The classical pathway is blocked by this reagent. Both pathways are inhibited by EDTA. Activation of the pathway results in two easily measured events: cell lysis and C3b deposition.

Procedure 1

Perhaps the most convenient assay for the activity of the human alternative pathway is lysis of rabbit erythrocytes[9] (Fig. 1). These cells directly activate the alternative pathway, and the release of hemoglobin is monitored to assess complement activity.

Reagents: Cells. Long-term storage of rabbit erythrocytes at $-80°$ is possible and provides a ready source of cells. Rabbit blood should be drawn into sufficient EDTA or citrate to make the final concentration 10 m*M*. Cells washed free of plasma proteins with GVB[10] may be brought to 1×10^9/ml and used immediately or frozen.

[8] M. K. Pangburn and H. J. Müller-Eberhard, *Springer Semin. Immunopathol.* **7,** 163 (1984).

[9] T. A. E. Platts-Mills and K. Ishizaka, *J. Immunol.* **113,** 348 (1974).

[10] Abbreviations: VBS, 5 m*M* sodium barbital, 150 m*M* NaCl, pH 7.4; GVB, VBS containing 0.1% gelatin; GVBE, GVB containing 10 m*M* EDTA; VBE, VBS containing 10 m*M* EDTA; MgEGTA, 50 m*M* $MgCl_2$ containing 50 m*M* EGTA, [ethylene bis(oxyethylenenitrile]tetraacetic acid, pH 7.4; PBS, 10 m*M* sodium phosphate, 150 m*M* NaCl, pH 7.4.

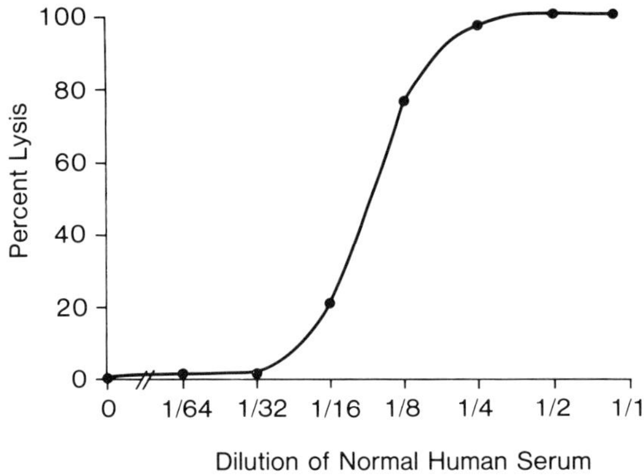

FIG. 1. Effect of dilution on activation of the alternative pathway. Lysis of rabbit erythrocytes by normal human serum. Assays contain 15 μl rabbit erythrocytes (10^9/ml), 5 μl 50 m*M* MgEGTA, various volumes of normal human serum, and GVB in a final volume of 100 μl. After 30 min at 37°, 900 μl cold GVB is added, and the released hemoglobin is determined after centrifugation (1000 *g* for 2 min).

Freezing Cells. Cells are conditioned for freezing by a slow infusion of glycerol. This is accomplished by additions of glycerolizing solution (GS: 114 ml glycerol, 4.75 g sodium acetate trihydrate, 95 mg potassium chloride, 173 mg sodium monobasic phosphate monohydrate brought to 250 ml with water, pH 6.8). To each 1 ml of packed cell volume at 37° add 125 μl of GS, mix immediately, and incubate for 5 min at 37°. After each of the following additions of GS a 5-min incubation at 37° is allowed: 125, 250, 250, 250, 500, 500 μl. The final total volume is 3 ml. Aliquots of 600 μl are made in plastic tubes (12 × 75 mm) so that thawing may be accomplished without transfer and will yield approximately 10^9 cells/tube. The GS-treated cells are cooled on ice and placed at −80°.

Thawing Cells. Slow egress of glycerol permits recovery of approximately 90% of cells. A tube of 600 μl frozen cells is thawed at 37°, and the following additions are made at 5-min intervals: 100 μl 12% NaCl in 25 m*M* sodium phosphate, pH 7.4, 100 μl same, then the following additions of 1.6% NaCl in 25 m*M* sodium phosphate, pH 7.4: 200, 250, 300, 350, 400, 500, 500, and 500 μl. The cells are centrifuged, the supernatant is discarded, and the cells are resuspended in 2 ml of 1.6% NaCl in 25 m*M* sodium phosphate, pH 7.4. After 5 min cells are centrifuged again, resuspended in 1 ml of the 1.6% NaCl–phosphate buffer plus 1 ml of GVB. After 5 min the cells are centrifuged and washed twice in GVB and resuspended at 10^9/ml in this buffer.

Serum. Blood is drawn without anticoagulants, allowed to clot for 2 hr, and the serum decanted, centrifuged, aliquoted, and frozen at −80°. It should not be refrozen after thawing.

MgEGTA and EDTA. MgEGTA is prepared from $MgCl_2$ and EGTA, [ethylene bis(oxyethylenenitrile)]tetraacetic acid (Sigma), as a 50 m*M* solution of each at pH 7.4. A stock EDTA solution is needed at 200 m*M* adjusted to pH 7.4.

Procedure. Assays are set up on ice and transferred to a 37° water bath after all additions are made. Appropriate dilutions of NHS are made in GVB, and 80 μl is added to tubes followed by 5 μl of MgEGTA and 15 μl of rabbit erythrocytes at 1×10^9/ml. After 30 min at 37°, 900 μl of cold GVBE is added, the tubes are centrifuged to remove unlysed cells, and the amount of released hemoglobin in the supernatant is determined spectrophotometrically at 414 nm. Transfer of 200 μl of the supernatant to a microtiter plate allows determination in an automated plate scanner with a 414-nm filter. Percent specific lysis may be determined by subtracting the lysis in a sample containing EDTA (instead of MgEGTA) and dividing by the maximum lysis induced by adding 985 μl H_2O to 15 μl of erythrocytes.

Procedure 2

Large numbers of C3b molecules are covalently attached to the surface of particles during activation. Deposition of C3b during alternative pathway activation may be followed by monitoring uptake of radiolabeled C3, of fluoresceinated C3, or of similarly labeled monoclonal or polyclonal antibodies. Use of radiolabeled C3[11] provides the most convenient and quantitative assay (Fig. 2), and only this will be described here.

Reagents: Serum. Fresh or frozen normal human serum (NHS) or, to prevent lysis, serum depleted of C5, C6, C7, or C8 may be used. This reagent may be prepared by passage of NHS containing 10 m*M* EDTA over an immunoadsorbent column bearing antisera to one of the proteins to be depleted. The column is run in VBE,[10] all fractions above 0.1 A_{280} are pooled, the pool is concentrated by ultrafiltration with a membrane which retains molecules of MW 10,000, and the concentrate is dialyzed versus VBS containing 0.01 m*M* EDTA. The absence of metal ions and handling at 0–4° prevents activation during these manipulations.

Radiolabeled C3. Purified C3 may be safely labeled with iodine-125 to a specific activity between 0.5 and 2 μCi/μg using any of the common melthods (Iodobeads or Iodogen, Pierce Chemical Co.) that do not employ strong reducing agents.

[11] M. K. Pangburn, R. D. Schreiber, and H. J. Müller-Eberhard, *J. Immunol.* **131,** 1930 (1983).

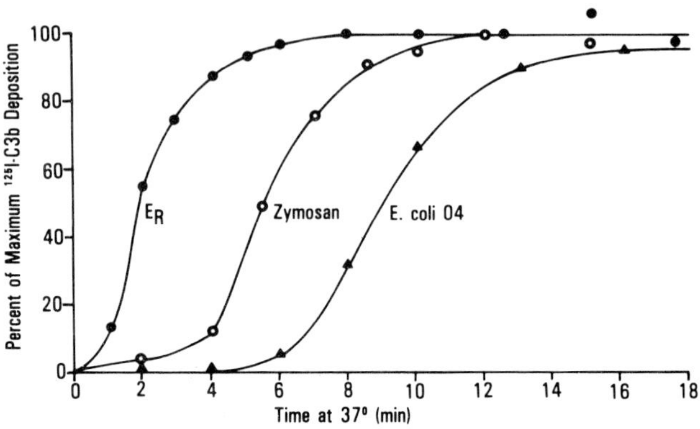

FIG. 2. Time course of C3b deposition on alternative pathway activators. Deposition on rabbit erythrocytes (E_R) is measured by incubating a mixture of 300 μl C7-depleted human serum (prepared by immunoadsorption) containing 2 μg ^{125}I-labeled C3, 50 μl GVB, 50 μl 50 m*M* MgEGTA, and 125 μl E_R (4×10^9/ml in GVB). Samples are removed at the indicated times and layered on 300 μl of 20% sucrose in a Beckman microfuge tube. The cells are sedimented by centrifugation at 8,000 *g* for 30 sec. The pellet and supernatant are separated by cutting the tube just above the pellet, and the percent bound radioactivity is determined. Zymosan and *Escherichia coli* 04 (5×10^9 particles/ml) are treated similarly. The C3 bound by E_R, zymosan, and *E. coli* at the plateau is 7.0, 4.1, and 2.3% of input, respectively.[11]

Activating Particles. Zymosan (Sigma) which is made from yeast cell walls is a convenient and powerful activator of the alternative pathway. Dry zymosan should be boiled before use but may then be stored in VBS at $-80°$ for immediate use. Rabbit erythrocytes, many bacteria (live or fixed, such as Staph A), and particulate inulin may also be used to activate the alternative pathway. Stock suspensions of 1×10^9/ml or higher are needed.

Procedure. Radiolabeled C3 is added to the source of complement (5–10 μCi/ml), and the mixture is centrifuged for 5 min at 8000 *g* at 4° to remove aggregates. Deposition on zymosan is measured by incubating mixtures containing 60 μl serum, 5 μl MgEGTA, and up to 35 μl of zymosan ($1-5 \times 10^9$/ml in GVB) for various times at 37°. Separation of bound and free radiolabel is carried out by washing. This may be conveniently and rapidly accomplished by layering the sample on a 300-μl cushion of 20% sucrose in GVB in a 400-μl polyethylene microfuge tube. The particles are sedimented by centrifugation at 8000 *g* for 30 sec. The pellet and supernatant are separated by cutting the tube just above the pellet, and the two fractions are counted to determine the percentage bound. Maximum uptakes between 1 and 7% can be expected. Washing by multiple dilutions and centrifugations will reduce maximum uptake 50% or more due to slow release of bound C3b during washes. Lysed erythrocyte ghosts bearing C3b will go through 20% sucrose if spun for 3 min.

C3

A detailed description of the purification of C3 from human plasma was presented in an earlier volume of this series and elsewhere.[12,13] Only a modification of the last step will be presented here along with two new assay methods.

Assays

Crude C3 and fractions from chromatography columns may be conveniently assayed using rabbit erythrocytes (prepared as described above) and normal human serum depleted of C3 activity by treatment with methylamine.[14] To 10 ml NHS 0.5 ml of 0.2 *M* EDTA,[10] pH 7.4, and 1.17 ml 1 *M* methylamine, pH 7.8, are added, and the mixture is incubated for 2.5 hr at 30°. It is then dialyzed against VBS and frozen at −80°. The assay is performed by mixing 25 μl of methylamine-treated NHS, 5μl MgEGTA, 15 μl rabbit erythrocytes at 10^9/ml in GVB, and up to 55 μl C3 sample or GVB in a final volume of 100 μl. After incubation at 37° for 20 min, the reaction is stopped by adding 900 μl cold GVBE, the tubes are centrifuged at 1000 *g* for 2 min, and the absorbance of the supernatant is read at 414 nm. NHS may be used as a source of C3 for a positive control. This assay is not as quantitative as the EAC42 cell assay,[15] but it is very convenient for assaying column fractions.

A fluorimetric assay has recently been described[16] for the accurate determination of the content of native C3 in purified preparations of C3. The assay may be performed in the presence of C3b or C3(H_2O), which are hemolytically inactive, but contaminating proteins, such as albumin, or detergents may interfere significantly.[17] The fluorimetric assay is performed at room temperature by mixing in a cuvette approximately 200 μg purified C3 in VBS or PBS (protein concentration does not affect the assay between 20 and 600 μg/ml), 80 μl of 0.4 m*M* ANS (8-anilino-1-naphthalene sulfonate, Sigma) in PBS, and sufficient PBS to give a final volume of 1.6 ml. The cuvette is placed in the spectrofluorometer, and the initial fluorescence intensity is measured using an excitation wavelength of 386 nm and an emission wavelength of 472 nm. Trypsin (10 μl of a

[12] B. F. Tack, J. Janatova, M. L. Thomas, R. A. Harrison, and C. H. Hammer, this series, Vol. 80, p. 64.

[13] C. H. Hammer, G. H. Wirtz, L. Renfer, H. D. Gresham, and B. F. Tack, *J. Biol. Chem.* **256,** 3995 (1981).

[14] T. E. Jessen, V. Barkholt, and K. G. Welinder, *J. Immunol. Methods,* **60,** 89 (1983).

[15] A. P. Dalmasso and H. J. Müller-Eberhard, *J. Immunol.* **97,** 680 (1966).

[16] M. K. Pangburn *J. Immunol. Methods* **102,** 7 (1987).

[17] M. K. Pangburn, unpublished observations.

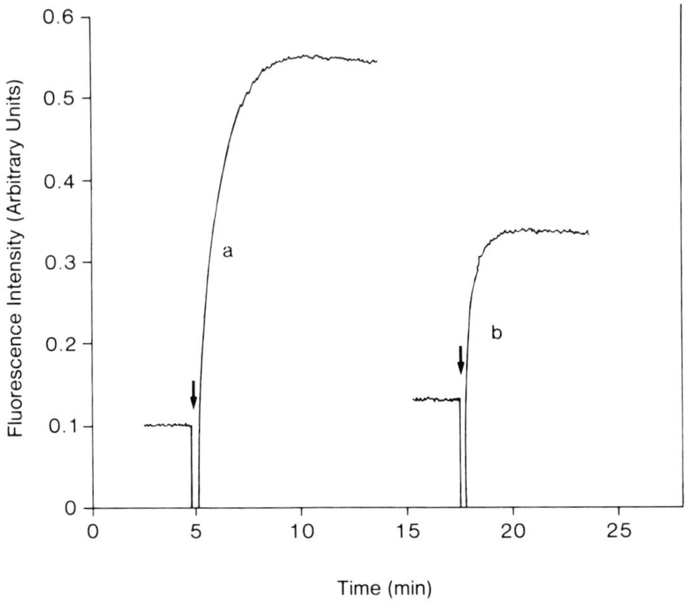

FIG. 3. Fluorimetric assay for native C3. Recorder tracings from two assays showing the fluorescence intensity measured before and after addition of 10 μg trypsin (arrows). (a) The cuvette initially contains 1280 μl PBS, 20 μl 2 mM ANS, and 300 μl of 1 mg/ml C3 purified by FPLC on Mono S (see text). The fluorescence intensity is measured at 22°, 10 μl 1 mg/ml trypsin is added (arrow), and the fluorescence intensity monitored until a maximum is reached. The background fluorescence is 0.055 (1585 μl PBS, 20 μl ANS). The initial fluorescence ($F_i = 0.098 - 0.055$) and the final fluorescence ($F_f = 0.547 - 0.055$) are used to calculate the % native C3 (100%) as described in the text. (b) A C3 preparation frozen at −80° thawed and incubated 2 hr at 37° to convert the unstable C3(H_2O)* formed at −80° to stable C3(H_2O). Assay is as in example a except that the mixture contains 150 μg C3. $F_i = 0.133 - 0.055$, $F_f = 0.338 - 0.055$, and % native C3 = 75%.

1 mg/ml stock solution in 0.1 mM HCl) is added, and the reaction is followed until a maximum in fluorescence intensity is reached 2–4 min after the addition of trypsin (Fig. 3). The background fluorescence intensity of a sample without protein is subtracted from the initial and final fluorescence to give F_i and F_f. The percent native C3 present is calculated from the expression[16]

$$\% \text{ native C3} = 100[11.5(F_f - F_i)/10.5F_f]$$

If the C3 has been stored at −80° a preincubation of the C3 for 2 hr at 37° is necessary for reasons described below to provide an accurate determination of the % native C3 (Table II).

TABLE II

PERCENT NATIVE C3 DETERMINED BY FLUORIMETRIC ASSAY[a]

F_f/F_i[b]	0.0	0.1	0.2	0.3	0.4	0.5	0.6	0.7	0.8	0.9
1.0	0.0	10.0	18.3	25.3	31.3	36.5	41.1	45.1	48.7	51.9
2.0	54.8	57.4	59.7	61.9	63.9	65.7	67.4	69.0	70.4	71.8
3.0	73.0	74.2	75.3	76.3	77.3	78.2	79.1	79.9	80.7	81.4
4.0	82.1	82.8	83.4	84.1	84.6	85.2	85.7	86.2	86.7	87.2
5.0	87.6	88.0	88.5	88.9	89.2	89.6	90.0	90.3	90.6	91.0
6.0	91.3	91.6	91.9	92.1	92.4	92.7	92.9	93.2	93.4	93.7
7.0	93.9	94.1	94.3	94.5	94.7	94.9	95.1	95.3	95.5	95.7
8.0	95.8	96.0	96.2	96.3	96.5	96.6	96.8	96.9	97.1	97.2
9.0	97.4	97.5	97.6	97.7	97.9	98.0	98.1	98.2	98.3	98.5
10.0	98.6	98.7	98.8	98.9	99.0	99.1	99.2	99.3	99.4	99.5
11.0	99.6	99.7	99.7	99.8	99.9	100.0				

[a] Values for percent native C3 were derived from the equation described in the text and the assay shown in Fig. 3.[16] For example, if an initial fluorescence (F_i) of 2.0 and a final fluorescence of 11.0 were found, then $F_f/F_i = 5.5$ and the % native C3 would be read at row 5.0 column 0.5 as 89.6%.

[b] The fluorescence intensity after trypsinization minus background fluorescence (F_f) divided by the initial fluorescence intensity minus background (F_i).

Purification

Methods for purification of C3 have been described in an earlier volume of this series and elsewhere.[11,12,13] These procedures remove most contaminating proteins yet yield C3 that is usually less than 100% active. In addition, freezing and thawing cause partial inactivation, and storage at $-80°$ reduces the specific activity of C3 by approximately 5% per month. These two phenomena appear to be the result of hydrolysis of the critical thioester in C3.

Several methods for removal of inactive forms of C3 have been described.[18-21] Removal of accumulated C3(H_2O) and any proteolytically generated C3 fragments is best achieved by ion-exchange chromatography on Mono S (Pharmacia) at pH 6.0 (Fig. 4). Several forms of C3(H_2O) are resolved. The form that elutes immediately after C3, C3(H_2O*), appears to be a conformational intermediate trapped at low temperature ($-80°$). The subunit structures of C3, C3(H_2O)*, and C3(H_2O) on reduced sodium dodecyl sulfate gels are identical. C3(H_2O)* converts to the stable form (last peak, Fig. 4) with a half-life of approximately 40 min at 37°,[16] suggesting that it corresponds to the intermediate estimated by kinetic studies to have a 35-min half-life.[22] The concentration of C3(H_2O)* is constantly changing as it decays to the stable form of C3(H_2O). Its F_f/F_i after trypsinization is 3.6.[16] Therefore, prior to assaying native C3 by the fluorescence method it must be removed by incubation at 37° or by separation, as shown in Fig. 4. Incubation at 37° is most convenient, since C3(H_2O)* rarely constitutes more than 20% of the protein, a 2-hr incubation at 37° will reduce its concentration to less than 3%. Native C3 is minimally affected by incubation at 37°.[23]

Factor B

Methods for purifying Factor B from human serum have been described in a previous volume of this series and elsewhere and will not be repeated here.[24,25] A convenient assay will be described which is useful in

[18] J. Janatova, B. F. Tack, and J. W. Prahl, *Biochemistry* **19,** 4479 (1980).

[19] C. Parkes, R. G. DiScipio, M. A. Kerr, and R. Prohaska, *Biochem. J.* **193,** 963 (1981).

[20] J. Janatova and B. F. Tack, *Biochemistry* **20,** 2394 (1981).

[21] M. K. Pangburn, R. D. Schreiber, and H. J. Müller-Eberhard, *J. Exp. Med.* **154,** 856 (1981).

[22] D. E. Isenman, D. I. C. Kells, N. R. Cooper, H. J. Müller-Eberhard, and M. K. Pangburn, *Biochemistry* **20,** 4458 (1981).

[23] H. J. Müller-Eberhard and R. D. Schreiber, *Adv. Immunol.* **29,** I (1980).

[24] M. A. Kerr, this series, Vol. 80, p. 102.

[25] O. Gotze and H. J. Müller-Eberhard, *J. Exp. Med.* **134,** 90s (1971).

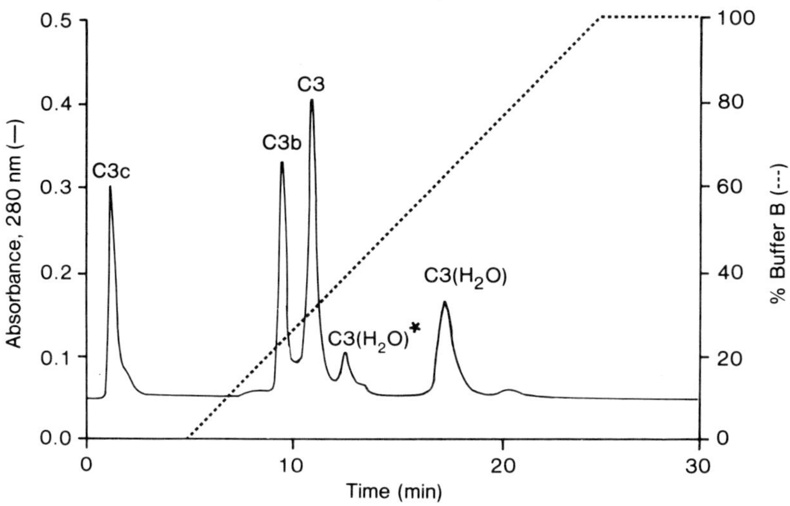

FIG. 4. Separation of native C3 from common contaminants.[16] C3 is separated from a mixture of C3c, C3b, $C3(H_2O)^*$, and $C3(H_2O)$ by FLPC on Mono S 5/5 (Pharmacia). Chromatography is performed at 22° using a flow rate of 1 ml/min. The sample proteins are mixed (2 mg total protein in VBS) with an equal volume of buffer A (50 m*M* sodium phosphate, pH 6.0) and applied to the column. The column is washed with 5 ml of buffer A followed by a 20-ml linear gradient to 100% buffer B (50 m*M* sodium phosphate, 500 m*M* NaCl, pH 6.0). Peaks are identified by hemolytic assay (C3) and by electrophoresis in polyacrylamide gels in the presence of sodium dodecyl sulfate and a reducing agent [C3c, C3b, $C3(H_2O)^*$, and $C3(H_2O)$].

monitoring the purification. It is not as sensitive nor as quantitative as assays utilizing EAC43 cells, but the reagents required are simple to prepare and are stable. Rabbit erythrocytes may be prepared and frozen as described for C3 assays or they may be used fresh. They are used at 10^9/ml in GVB. NHS is depleted of Factor B by heating at 50° for 20 min. It should be tested for inability to lyse rabbit erythrocytes and heated longer if necessary. This Factor B-depleted reagent may be frozen at −80°. Assays are performed by mixing on ice 30 μl B-depleted NHS, 0–50 μl sample, 50–0 μl GVB, 5 μl MgEGTA, and 15 μl rabbit erythrocytes. After 10 min at 37°, 0.9 ml cold GVBE is added, the tube is centrifuged at 1000 *g* for 2 min, and the supernatant is examined at 414 nm for released hemoglobin. Samples between 10 and 200 μg/ml Factor B may be assayed.

Factor D

Factor D is present in plasma at only 1–2 μg/ml as an active serine protease. The two best methods of purification employ initial steps which provide 200- to 400-fold purification.[26,27] A modification of the method of Lesavre[27] will be presented here. This procedure does not require preparation of pseudoglobulin, which necessitates dialysis of very large volumes, and it is considerably faster due to the elimination of one chromatographic step and the utilization of HPLC.

Assay

An excellent assay, utilizing rabbit erythrocytes and NHS depleted of Factor D by passage over BioRex 70 (Bio-Rad Laboratories, Inc.), has been described.[26,27]

Purification

The procedure follows that of Lesavre[27] except that the QAE-Sephadex step is replaced by HPLC on Mono Q (Pharmacia). Five liters of fresh human ACD plasma containing 10 m*M* EDTA is adjusted to pH 7.0 with 1 *N* HCl. The plasma is applied directly to a column (6 × 35 cm) of Bio-Rex 70 equilibrated with VBE. The column is washed successively with 3 liters VBS containing 0.2% Triton X-100 and 6 liters VBS. The bound material is eluted with VBS adjusted to 30 mS (at 4°) with NaCl. The flow rate is 600 ml/hr during washing and 200 ml/hr during elution of Factor D. Fractions containing Factor D activity are pooled (500 ml) and dialyzed overnight against sufficient distilled water to obtain a final conductivity of 2.6 mS (at 4°). The dialyzed pool from Bio-Rex 70 chromatography is centrifuged for 2 hr at 23,000, and the supernatant is decanted and concentrated to 10 ml. This material is applied to a column (4 × 75 cm) of Sephadex G-75 and eluted with VBS at a flow rate of 15 ml/hr and a fraction volume of 6 ml. The glass column and tubes must be siliconized to prevent large losses. Fractions containing Factor D hemolytic activity are pooled, and the pool is concentrated, diluted 3-fold with water in the concentrator, and reconcentrated to 10 ml. This pool is applied directly to the Mono Q column. The column is equilibrated in 20 m*M* Tris–HCl, pH 7.6, the sample loaded, and Factor D eluted with a 20-ml linear gradient to 20 m*M* Tris–HCl, 400 m*M* NaCl, pH 7.6. Factor D elutes at 40% of the gradient. The Factor D pool is dialyzed into VBS and may be frozen indefinitely. The yield of this procedure is in excess of 2 mg Factor D.

[26] K. B. M. Reid, D. M. A. Johnson, J. Gagnon, and R. Prohaska, this series, Vol. 80, p. 134.

[27] P. H. Lesavre, T. E. Hugli, A. F. Esser, and H. J. Müller-Eberhard, *J. Immunol.* **123**, 529 (1979).

Factor H

Factor H is difficult to assay since it has no enzymatic activity. The standard assay[28] employs cells bearing C3b,Bb,P and measures inhibition of lysis due to destruction of this enzyme by Factor H. A modification of this assay will be described here. Purified Factor H may also be assayed by immunodiffusion, by measuring binding of the radiolabeled protein to C3b-bearing cells,[29] or by a fluorescence assay[2] that measures the ability of Factor H to function as a cofactor for Factor I in the cleavage of C3b. Purification of Factor H has been described in a previous volume of this series.[30] The Factor H-containing fractions from the DEAE-Sephacel column used in the C3 purification is also an excellent starting material and eliminates the need for the QAE column.

Factor H Assay

The assay utilizes sheep erythrocytes (E_s) bearing the Factor H-susceptible enzyme C3b,Bb stabilized with Ni^{2+}. These cells lyse in guinea pig serum if the enzyme has not been inactivated by Factor H. Washed E_s (10^9) are centrifuged for 2 min at 1000 *g*, the pellet is resuspended in 200 μl of C3 at 5 mg/ml, 5 μl of 1 mg/ml trypsin is added, and the mixture is incubated 5 min at 37°. The cells now bearing C3b are washed thoroughly, resuspended in 200 μl GVB containing 20 μg Factor B, 1 μg Factor D, and 1 m*M* $NiCl_2$, and incubated for 5 min at 37°. Enzyme formation is stopped by addition of 10 μl 0.2 *M* EDTA, and additional C3b is deposited by adding 50 μl of 5 mg/ml C3 and incubating for 2 hr at 37°. After thorough washing with GVB the E_sC3b must be titered to determine the amount of Factor B needed to obtain 70% lysis. E_sC3b (10^7 cells in 20 μl) is added to 80-μl mixtures containing 1 μg/ml Factor D, 1 m*M* $NiCl_2$, and varying amounts of Factor B. After 5 min at 37°, 100 μl of GVBE (without Factor H) is added, and the cells are incubated for 10 min. Lysis of the cells is determined after addition of 800 μl of 1/50 diluted guinea pig serum in GVBE, incubation for 10 min at 37°, centrifugation, and determination of released hemoglobin in the supernatant at 414 nm. Samples containing Factor H are assayed by making various dilutions in GVBE and adding 100 μl to the E_sC3b,Bb cells for 10 min prior to addition of guinea pig serum. Normal human serum diluted approximately 1/000 may be used as a crude standard.

[28] J. M. Weiler, M. R. Daha, K. F.Austen, and D. T. Fearon, *Proc. Natl. Acad. Sci. U.S.A.* **73,** 3268 (1976).

[29] M. K. Pangburn and H. J. Müller-Eberhard, *Proc. Natl. Acad. Sci. U.S.A.* **75,** 2416 (1978).

[30] L. G. Crossley, this series, Vol. 80, p. 112.

Factor I

Purification of Factor I is probably best initiated by affinity chromatography using antisera (Cytotech, 11045 Roselle St., San Diego, Ca 92121) to Factor I.[31] A conventional purification is described by Crossley in this series.[30] A variety of assays have also been described.[2,30,32]

Factor P (Properdin)

Properdin may be assayed by measuring the enhancement of lysis of rabbit erythrocytes by properdin-depleted NHS in the presence of properdin. This assay is described in detail in an earlier volume of this series.[6]

Purification

The procedure given here combines steps from several published procedures.[7,33,34] These include precipitation with 5% polyethylene glycol, removal of most contaminants by passage over QAE-Sephadex, and removal of IgG by passage over anti-IgG–Sepharose. Properdin does not bind to either column, making the overall procedure very rapid.

Fresh or outdated human citrated plasma (5 liters) from fasting individuals is clotted by adding 1 *M* $CaCl_2$ to a final concentration of 21 m*M*. The recovered serum at 4° is adjusted to pH 7.5 and made 15 m*M* in EDTA. Polyethyleneglycol 4000 is added to make a 5% (w/v) final concentration, and the precipitate is collected by centrifugation (4500 *g* for 30 min). The thick precipitate is dissolved by stirring overnight with 250 ml 20 m*M* Tris, 300 m*M* NaCl, pH 8.5, at 4°, centrifuged to remove insoluble material, then dialyzed against 20 m*M* Tris–HCl, 30 m*M* NaCl, pH 8.5 (3.9 mS at 4°). Any precipitate is again removed by centrifugation, and the sample is applied to a column of QAE-Sephadex A-50 (6 × 72) cm) equilibrated with 20 m*M* Tris, 30 m*M* NaCl, pH 8.5. The column is developed at 100 ml/hr. Properdin elutes between 500 and 1100 ml. The fractions exhibiting properdin activity are pooled and passed over an anti-human IgG–Sepharose column with the capacity of binding 20 mg IgG. The eluate from this column is concentrated by ultrafiltration and dilayzed against

[31] M. K. Pangburn, R. D. Schreiber, and H. J. Müller-Eberhard, *J. Exp. Med.* **146,** 257 (1977).

[32] T. A. Gaither, C. H. Hammer, and M. M. Frank, *J. Immunol.* **123,** 1195 (1979).

[33] D. T. Fearon and K. F. Austen, *Proc. Natl. Acad, Sci. U.S.A.* **74,** 1683 (1977).

[34] O. Gotze, R. G. Medicus, and H. J. Müller-Eberhard, *J. Immunol.* **118,** 525 (1977).

VBS. The procedure described yields approximately 10 mg. Freezing properdin converts a portion of the material to an aggregated form which activates complement when added to NHS. Storage at 4° for several weeks also causes this conversion. Native properdin must therefore be prepared fresh when needed.

[49] Human C5a Anaphylatoxin: Gene Synthesis, Expression, and Recovery of Biologically Active Material from *Escherichia coli*

By ARTHUR E. FRANKE, GLENN C. ANDREWS, NORMA P. STIMLER-GERARD, CRAIG J. GERARD, and HENRY J. SHOWELL

Introduction

Anaphylatoxin C5a is a 74-amino acid residue single-chain glycoprotein derived from the fifth component of complement during activation of the complement cascade.[1] In addition to being the most potent humoral neutrophil chemotactic substance so far identified, its biological properties extend across a multitude of cells and tissues involved in both acute and chronic inflammatory processes (see this volume [46]).[2] As quantities of C5a which can be purified from serum, especially human, are extremely limited, many extensive biological and biochemical/biophysical investigations are made difficult. The production of biologically active C5a by recombinant DNA techniques would provide adequate quantities of anaphylatoxin for such studies and also provide the opportunity to examine its protein structure–activity relationships by site-specific mutagenesis. In this chapter we demonstrate the feasibility of this methodology.

Synthetic Gene for Human C5a

The plan devised to achieve bacterial production of human anaphylatoxin C5a is analogous to previous methods used for synthesis and expression of several small human hormones. It employs the construction of a gene in which the coding sequence for the activated complement component C5a is made by chemical synthesis. The oligodeoxynucleotides used in the C5a structural gene synthesis are chemically prepared by a modified

[1] H. J. Müller-Eberhard, *Annu. Rev. Biochem.* **44,** 697 (1975).

[2] T. E. Hugli, *CRC Crit. Rev. Immunol.* **1,** 321 (1981).

```
                1                                              10                                                  20
EcoRI         thr leu gln lys lys ile glu glu ile ala ala lys tyr lys his ser val val lys lys cys cys tyr asp gly ala cys
                                   (A)                                                                           (B)

AATTCT  ATG ACT CTG CAA AAG AAG ATC GAA GAA ATC GCT GCT AAG TAC AAG CAC TCC GTC GTT AAG AAG TGT TGT TAC GAT GGT GCA TGC
    GA  TAC TGA GAC GTT TTC TTC TAG CTT CTT TAG CGA CGA TTC ATG TTC GTG AGG CAG CAA TTC TTC ACA ACA ATG CTA CCA CGT ACG

                                   (J)                                                                           (I)

      30                                        40                                                50
val asn asn asp glu thr cys glu gln arg ala ala arg ile ser leu gly pro arg cys ile lys ala phe thr glu cys cys val val
                                   (C)                                                                           (D)

GTC AAC AAC GAC GAA ACC TGT GAA CAA CGA GCT GCT CGT ATT TCT CTG GGC CCT CGC TGT ATC AAG GCT TTC ACT GAA TGT TGT GTT GTC
CAG TTG TTG CTG CTT TGG ACA CTT GTT GCT CGA CGA GCA TAA AGA GAC CCG GGA GCG ACA TAG TTC CGA AAG TGA CTT ACA ACA CAA CAG

                                   (H)                                                                           (G)

        60                                     70                74
ala ser gln leu arg ala asn ile ser his lys asp met gln leu gly arg stop
                                   (E)                                      HindIII

GCT TCC CAA CTG CGC GCT AAC ATT TCT CAC AAG GAC ATG CAA CTC GGC CGC TAA A
CGA AGG GTT GAC GCG CGA TTG TAA AGA GTG TTC CTG TAC GTT GAG CCG GCG ATT TTCGA

                                   (F)
```

FIG. 1. Nucleotide sequence of the synthesized DNA fragment containing the C5a gene. The 235-bp synthetic DNA sequence of the human C5a coding region is determined from the subclone plasmid pBR322–C5a (see Fig. 2). The corresponding protein sequence is also shown. The 10 oligodeoxyribonucleotides, labeled A through J, are represented by arrows. The artificial initiation codon ATG (underlined) is immediately preceded by an *Eco*RI site, and the artificial termination codon TAA is followed by a *Hin*dIII site at the 3′ terminus.

phosphoramidate procedure[3] using a controlled glass pore support[4] on a Milligen 6500 (New Bedford, MA) automated DNA synthesizer. The protocol utilizes 3% (w/v) dichloroacetic acid in dichloromethane for detritylation, in-line activation of the phosphoramidates with saturated tetrazole in acetonitrile, capping with diethoxyphosphine tetrazolide,[5] and oxidation with aqueous iodine/THF.[5a] The total time per addition cycle is 14 min. The ten 47-mers are obtained in 98.8% average yield/step (by trityl analysis), deblocked by the procedure of Caruthers,[5] ethanol precipitated from 0.3 *M* sodium acetate, and isolated by preparative gel electrophoresis in 10% polyacrylamide – 6 *M* urea denaturing gels prior to annealing.

Given the known amino acid sequence of human C5a,[2] one can design a DNA fragment encoding the information for its 74 amino acids. Figure 1 shows the amino acid sequence of the desired protein and the arrangement of synthetic oligonucleotides needed to make a gene coding for human C5a. The synthetic gene is chosen to maximize utilization of codons preferred by *E. coli* and designed to allow for several restriction endonuclease sites to facilitate characterization and possibly mutagenesis of the gene sequence. This approach allows for direct expression of anaphylatoxin by introducing an ATG initiation codon for protein synthesis in front of the triplet coding for the first amino acid of the C5a polypeptide. To facilitate its insertion in a desired orientation into plasmid pBR322, the synthetic C5a gene is designed to contain *Eco*RI and *Hin*dIII restriction endonuclease recognition sites at its termini.

The assembly of the gene involves two primary annealing/ligation reactions (Fig. 2) containing oligomers A, B, I, and J; oligomers C, D, E, F, G, and H. All oligomers (except A and F) are phosphorylated at their 5′ ends with T4 polynucleotide kinase.[6] The resulting 94-bp and 141-bp double-stranded DNA fragments are isolated after electrophoresis on a 10% polyacrylamide gel, ligated together, and their 235-bp product isolated by gel electrophoresis. The ligation reaction is loaded onto a preparative

[3] N. D. Sinha, J. Biernat, and H. Koster, *Tetrahedron Lett.* **24,** 5843 (1983).

[4] S. P. Adams, K. S. Kavka, E. J. Wykes, S. B. Holder, and G. R. Galluppi, *J. Am. Chem. Soc.* **105,** 661 (1983).

[5] M. D. Matteucci and M. H. Caruthers, *J. Am. Chem. Soc.* **103,** 3183 (1981).

[5a] Abbreviations: Ap, ampicillin; BAP, bacterial alkaline phosphatase; EACA ϵ-amino-*n*-caproic acid; *Eco*RI, restriction endonuclease from *E. coli* RY13 (G˅AATTC); EDTA, ethylenediaminetetraacetic acid; *Hin*dIII, restriction endonuclease from *Haemophilus influenzae* Rd (A˅AGCTT); HPLC, high-pressure liquid chromatography; LB, Luria broth; M_r, relative molecular weight; PAGE, polyacrylamide gel electrophoresis; PMNs, polymorphonuclear leukocytes; RC5a, recombinant C5a; SDS, sodium dodecyl sulfate; Tc, tetracycline; TFA, trifluoroacetic acid; THF, tetrahydrofuran; *trp, E. coli* tryptophan promoter.

[6] T. Maniatis, E. F. Fritch, and J. Sambrook, "Molecular Cloning: A Laboratory Manual." Cold Spring Harbor Lab., Cold Spring Harbor, New York, 1982.

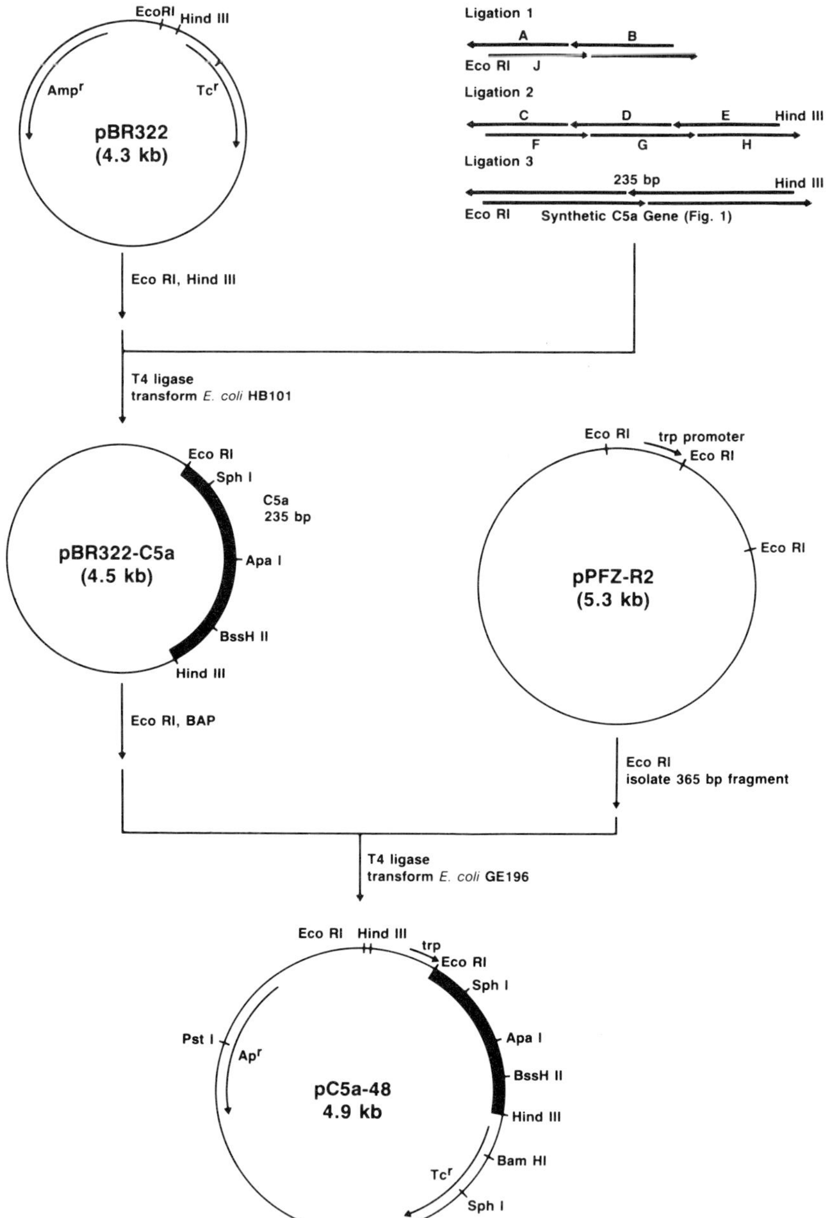

FIG. 2. Schematic presentation of the steps involved in the construction of pC5a–48 used for direct expression of human anaphylatoxin in *E. coli.* The expression plasmid pC5a–48 is an approximately 4.9-kb plasmid containing the *trp* promoter sequence and the C5a gene (enlarged shaded box), made by chemical synthesis, inserted into vector pBR322. The ampicillin (*Ap^r*) and tetracycline (*Tc^r*) resistance genes of pBR322 are also shown; *trp* indicates the *E. coli* tryptophan promoter/operator and ribosome binding; site; *E. coli* GE196 is a derivative of SA1500. Arrows show the 5′ to 3′ direction of transcription.

5% polyacrylamide gel (20 × 15 × 0.15 cm) run in TBE buffer (89 m*M* Tris–borate, 89 m*M* boric acid, 2 m*M* EDTA), and the products are resolved by electrophoresis at 200 V for 1.5 hr. After electrophoresis, the gel is stained (1 μg/ml ethidium bromide in TBE), and the DNA fragments are visualized under illumination from 266-nm ultraviolet light. The C5a gene fragment band is cut out of the gel and purified from the gel slice by electroelution (2 hr, 100 V, 4°) in dialysis tubing (Bethesda Research Laboratories) containing 0.1× TBE as described by Smith.[7] The 235-bp DNA fragment containing a structural gene encoding C5a is inserted between the *Eco*Ri and *Hin*dIII sites of pBR322 vector DNA and transformed into competent cells of *E. coli* K12 strain HB101.[8] Plasmid DNA is prepared in either large-scale[9] or miniprep scale[10] quantities and cleaved with restriction endonucleases (New England Biolabs, Bethesda Research Laboratories) as directed by the supplier. Restriction analysis of plasmid DNA isolated from six transformants shows that five of the six clones contain a *Eco*RI/*Hin*dIII fragment of the correct size. The nucleotide sequence of the C5a gene region of each of these plasmids is determined by the method of Maxam and Gilbert[11] to identify a clone with the correct sequence.

Construction and Characterization of C5a Expression Plasmid

In order to produce anaphylatoxin in *E. coli* it is necessary to place a bacterial promoter and ribosome binding site sequence upstream of the C5a gene. The construction of the C5a expression plasmid is initiated by cleavage of the C5a subclone pBR322–C5a (Fig. 2) with the restriction endonuclease *Eco*RI (obtainable from Bethesda Research Laboratories), followed by dephosphorylation using bacterial alkaline phosphatase (BAP). In the case of BAP treatment for removal of terminal 5′-phosphates from the vector DNA, the plasmid DNA is cleaved with *Eco*Ri and ethanol precipitated from 0.3 *M* sodium acetate, pH 4.8. The DNA pellet is resuspended in buffer (50 m*M* NaCl, 10 m*M* $MgCl_2$, 10 m*M* Tris, 10 m*M* dithiothreitol) at a final concentration of 1 μg per 10 μl of buffer. The BAP (Bethesda Research Laboratories) is added to a final concentration of 50 units of enzyme per 1 μg of DNA fragment. The reaction mixture is incubated at 65° for 2 hr followed by phenol extraction and ethanol precipitation of the DNA fragments.

The *E. coli trp* promoter–operator was originally obtained from

[7] H. O. Smith, this series, Vol. 65, p. 371.

[8] H. W. Boyer and D. Roulland-Dussoix, *J. Mol. Biol.* **41,** 459 (1969).

[9] D. B. Clewell and D. R. Helinski, *J. Bacteriol.* **110,** 667 (1972).

[10] H. C. Birnboim, this series, Vol. 100, p. 243.

[11] A. M. Maxam and W. Gilbert, this series, Vol. 65, p. 499.

ptrpL1.[12] The 360-bp *Eco*RI fragment containing the *trp* promoter–operator and a ribosome binding site sequence used in the C5a expression plasmid (pC5a–48) is isolated from the prorennin expression plasmid pPFZ-R2.[13] This promoter fragment is isolated by 5% polyacrylamide gel electrophoresis and ligated to the *Eco*RI-cleaved pBR322 DNA in 20 μl of 50 mM Tris–HCl, pH 7.5, containing 10 mM $MgCl_2$, 20 mM dithiothreitol, 1 mM rATP, and 200 units of T4 ligase (New England Biolabs). The ratio of fragment insert to pBR322 vector used in the ligation reaction is approximately equilmolar, with a total DNA concentration of about 10–12 μg/ml. The ligation reaction is carried out for 16 hr at 4°. Cells of *E. coli* strain HB101 are rendered compentent for uptake of DNA by the $CaCl_2$ procedure[14] and transformed with 10 μl of the ligation reaction. (Competent cells of HB101 can be obtained from Bethesda Research Laboratories). Tetracycline-resistant recombinants are obtained and screened by size and restriction analysis for the presence and orientation of the *trp* promotor insert fragment. This is done by preparing rapid-plasmid DNA isolates from individual clones, restricting with *Hin*dIII (obtainable from Bethesda Research Laboratories), and examining the DNA fragments on 5% polyacrylamide gels using appropriate size markers. Since this *Eco*RI fragment could insert into pBR322 in either orientation, it is important to select an appropriate restriction analysis (*Hin*dIII) to distinguish the different orientations. The orientation of the promoter fragment which would result in transcription of the C5a gene generates a 565-bp *Hin*dIII DNA fragment, whereas the the undesired orientation generates a 264-bp *Hin*dIII DNA fragment. Multiple isolates from this ligation reaction are identified with plasmids containing the anaphylatoxin gene adjacent to the bacterial promoter sequence in the configuration required for direct expression of C5a. A restriction map of the C5a expression plasmid pC5a–48 is illustrated in Fig. 2.

One method used to increase C5a expression is to decrease the possibility of degradation of the small protein in *E. coli* by cellular proteases. Mutant strains of *E. coli*, designated as *lon*, have been documented to lack one of several proteases present in wild-type cells.[15] Approximately 100 ng of C5a expression plasmid DNA (pC5a–48) is used to transform cells of the protease-deficient strain (GE196) (available from the *E. coli* Stock Center, Yale University, New Haven, CT 06510; strain CC, SC 6436),

[12] J. C. Edman, R. A. Hallewell, P. Valenzuela, H. M. Goodman, and W. J. Rutter, *Nature (London)* **291,** 503 (1981).

[13] A. E. Franke, European Patent No. 0147178 (1985).

[14] R. W. Davis, D. Botstein, and J. R. Roth, *in* "Advanced Bacterial Genetics." Cold Spring Harbor Lab., Cold Spring Harbor, New York, 1981.

[15] S. Gottesman and D. Zipser, *J. Bacteriol.* **133,** 844 (1978).

which are made competent for DNA uptake by $CaCl_2$ treatment.[14] Tetracycline-resistant GE196 cells are obtained and screened for the presence of the C5a expression plasmid pC5a–48. A clone designated GE196(pC5a–48) was chosen for anaphylatoxin production.

Identification of Bacterially Produced Human C5a

Translation of the DNA sequence of the synthetic C5a gene predicts a 75-amino acid polypeptide with a calculated molecular weight of about 8400 (Fig. 1). Total cell protein extracts from control and expression cultures are resolved by SDS–PAGE. Immunodetection of bacterially produced C5a (RC5a) is accomplished by the Western blot procedure[16] employing authentic human C5a, which has a single asparagine-64-linked oligosaccharide,[17] human C5a deglycosylated with endoglycosidase H, and antisera to authentic glycosylated C5a. The goat anti-human C5a antibody is prepared as previously described.[18] Briefly, C5a desArg (50 μg) is emulsified in Freund's complete adjuvant, and a goat is injected intramuscularly at multiple sites followed by booster injections at four monthly intervals. Animals are bled 10 days after each C5a desArg injection, and the antiserum is collected and stored at $-70°$. An IgG fraction (3 mg/ml) obtained from pooled antisera is used at a final dilution of 1/1000.

Bacterial cultures containing plasmid pC5a–48 or pBR322 (control) are grown overnight in LB broth [10 g Bacto-tryptone (Difco), 5 g Bacto-yeast extract (Difco), 10 g sodium chloride; adjusted to pH 7.5 with sodium hydroxide]. M9 media contains (per liter) the following: 0.5 g NaCl, 6 g Na_2HPO_4, 3 g KH_2PO_4, 1 g NH_4Cl, 2 g casamino acids (Difco), adjusted to pH 7.4. The meterial is autoclaved, cooled, and then to it the following are added: 10 ml of 20% glucose, 2 ml of $MgSO_4$, and 0.1 ml of $CaCl_2$, which are sterilized separately. Aliquots of overnight cultures are diluted 1:100 into M9 medium plus 5 μg/ml tetracycline and grown in shaker flasks to densities of $A_{550} = 0.5$. Cell pellets from 2-ml aliquots of these cultures are resuspended in TE (50 m*M* Tris–HCl, pH 7.5, 0.1 m*M* NaEDTA); an aliquot is added to an equal volume of 2× gel sample buffer [4% SDS, 0.125 *M* Tris–HCl, pH 6.8, 5% 2-mercaptoethanol, 20% glycerol, 10 mg bromophenol blue, 10 mg xylene cyanol (Sigma Chemical Co.)] and run on a 10–20% gradient SDS–polyacrylamide gel.[19]

Gel-resolved proteins are electrophoretically transferred to a nitrocellu-

[16] R. Hawkes, E. Niday, and J. Gordon, *Anal. Biochem.* **119,** 142 (1982).

[17] H. N. Fernandez and T. E. Hugli, *J. Immunol.* **117,** 1688 (1976).

[18] G. L. Manderino, A. F. Suarez, S. J. Kunkel, P. A. Ward, A. A. Hirata, and H. J. Showell, *J. Immunol. Methods* **53,** 41 (1982).

[19] U. K. Laemmli, *Nature (London)* **227,** 680 (1970).

lose filter (0.22 μm), producing a replica of the original gel pattern. The filter is overlayed with goat antibodies to authentic C5a, followed by incubation with peroxidase-conjugated rabbit anti-goat IgG antibody (Miles Scientific). The bound antibodies are detected by staining with 4-chloro-1-naphthol (Sigma) and hydrogen peroxide. The C5a produced by the culture containing the pC5a–48 expression plasmid reacts specifically with the antibody prepared against authentic human C5a (Fig. 3, lane 2). Authentic C5a is prepared from human serum activated for complement in the presence of 1 *M* EACA with yeast cell walls. The peptide is then purified to homogeneity by sequential chromatography on BioGel P-60 (Bio-Rad), SP-Sephadex (Pharmacia), and CM-Sephadex (Pharmacia) as previously described.[17,20] There is comigration of RC5a and human deglycosylated C5a (Fig. 3, lane 3). Human C5a is deglycosylated with endoglycosidase H as described.[21] The authentic C5a has a molecular weight of about 11,000 on SDS–PAGE due to the carbohydrate at Asn-64 (Fig. 3, lane 4). The control culture, containing the vector pBR322, is essentially free of immunoreactive protein in that region of the gel (Fig. 3, lane 1). From the intensity of C5a protein bands visible in Coomassie blue-stained gels and comparison of immunoreactive bands to known amounts of human C5a standards on Western analysis, C5a is produced at a level of 1–3% of total *E. coli* proteins with this plasmid construction in the protease-deficient host strain.

In addition, Western blot analysis of soluble and insoluble cellular fractions demonstrates that the majority of the C5a produced by the expression culture (GE196/pC5a–48) is insoluble. For preparation of soluble and insoluble protein fractions from bacteria, cells are suspended in TE at a final concentration of approximately 10^8 cells/ml then placed on ice, and lysozyme to 1 mg/ml, Nonidet P-40 (Sigma) to 0.2%, and NaCl to 0.35 *M* are added. The lysate is adjusted to 10 m*M* $MgCl_2$ and incubated with 50 μg/ml deoxyribonuclease I (Sigma) on ice for 30 min.[22] Insoluble material is removed by centrifugation (15,000 *g*, 10 min) and washed with TE buffer. The supernatant, pellet, and total cellular protein fractions are examined on Western blots. As in the case of some other eukaryotic proteins expressed in *E. coli*, most of the C5a produced by the expression culture is in the insoluble pellet fraction.

[20] C. Gerard and T. E. Hugli, *J. Biol. Chem.* **254,** 6346 (1979).

[21] C. Gerard and T. E. Hugli, *Proc. Natl. Acad, Sci. U.S.A.* **78,** 1833 (1981).

[22] R. M. Lawn, J. Adelman, A. E. Franke, C. M. Houck, M. Gross, R. Najarian, and D. V. Goeddel, *Nucleic Acids Res.* **9,** 1045 (1981).

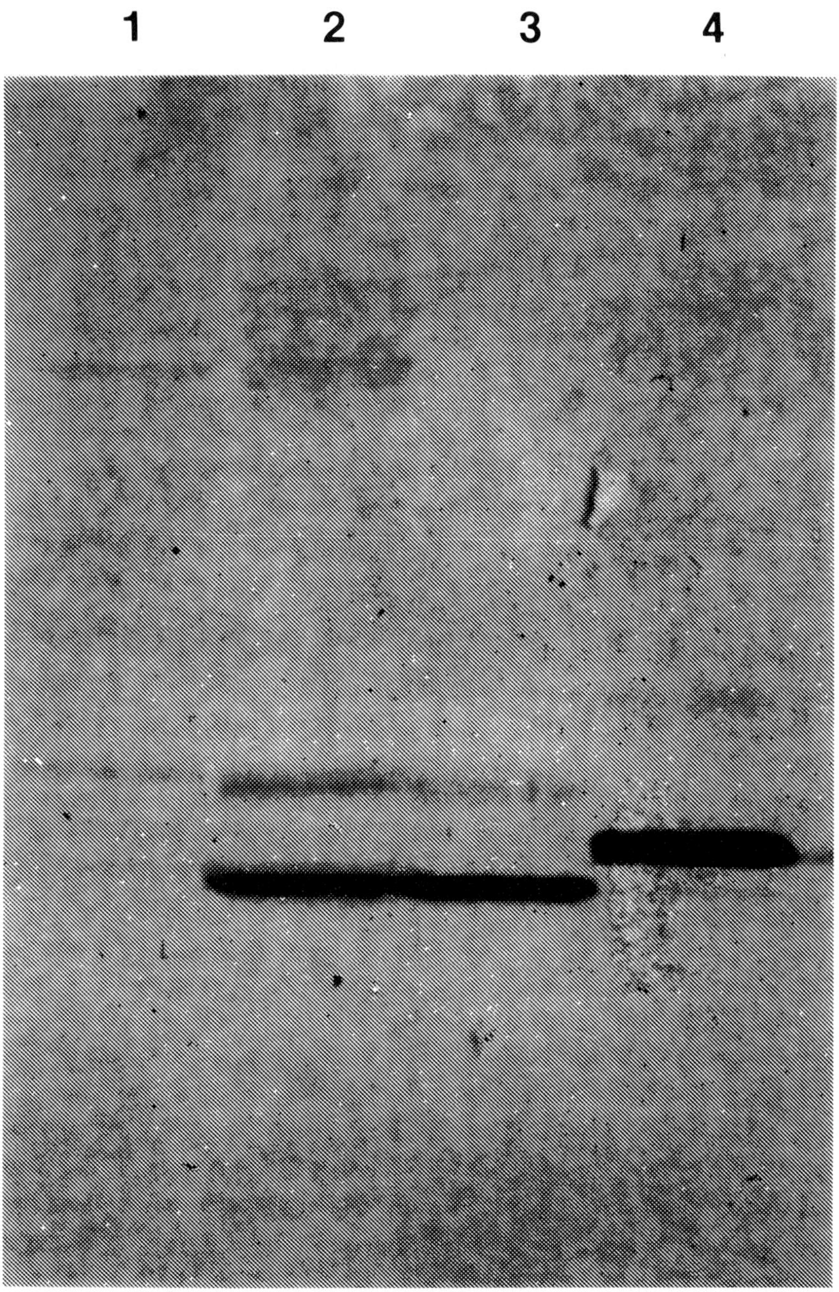

FIG. 3. Western blot of native and bacterial produced C5a. Total cellular proteins from control and expressed cultures are prepared for electrophoresis on a 10–20% gradient SDS–polyacrylamide gel. A nitrocellulose blot is reacted with goat anti-human C5a antibodies followed by rabbit anti-goat IgG which is previously conjugated to horseradish peroxidase. The bound antibodies are detected by staining with 4-chloro-1-naphthol and hydrogen peroxide. Protein samples were loaded as follows: 1, GE196/pBR322; 2, GE196/pC5a–48; 3, 500 ng deglycosylated human C5a; 4, 500 ng purified human C5a.

Reconstitution of RC5a from Bacterial Extracts

Structure–function studies with purified human and porcine C5a anaphylatoxins have previously demonstrated that the biological activity of these molecules requires the presence of three interchain disulfide bonds and that these molecules can undergo reversible denaturation.[23] It was therefore reasonable to process the insoluble pellet recovered from the bacterial lysate through a cycle of reduction and denaturation, followed by dialysis to remove the conformational perturbants, in order to allow the molecules to refold to their presumably unique biologically active conformation. Indeed, we observe that such treatment generated C5a-specific smooth muscle contractile activity in the supernatant recovered from the renaturation step.

Escherichia coli GE196/pC5a–48 is grown to A_{550} of 0.5 in 2 liters of M9 medium containing 10 mg of tetracycline per liter. The cells obtained are processed by the same procedure used to obtain soluble and insoluble protein fractions. The insoluble pellet obtained from lysed cells is suspended in 10 ml of 0.1 *M* potassium phosphate buffer, pH 7.4, containing 0.1 *M* 2-mercaptoethanol and 6 *M* guanidinium hydrochloride. A tracer quantity of ^{131}I-labeled deglycosylated authentic human C5a is added to facilitate subsequent purification steps. Deglycosylated human C5a is labeled with ^{131}I using the solid-phase lactoperoxidase–glucose oxidase method (Enzymobeads, Bio-Rad). After 8 hr at room temperature, the viscous solution is dialyzed (Spectrapor, 3500 M_r) for 12 hr at room temperature against 20 m*M* potassium phosphate buffer, pH 7.6. A second dialysis for 4 hr at room temperature against 20 m*M* potassium buffer, pH 6.6, yields a opalescent supernatant and a flocculent white precipitate which is removed by centrifugation (1800 *g*, 15 min). Bioassay of the supernatant on strips of guinea pig ileum or lung parenchyma reveals C5a-sepcific smooth muscle activity[24,25] (see below).

Purification of RC5a

The cationic nature of the anaphylatoxins has been previously exploited for the purification of these peptides.[21] Therefore, the initial purification step involves adsorption of the renatured RC5a to sulfopropyl-Sephadex. The supernatant containing renatured RC5a is applied to a column (1 × 3 cm) of sulfopropyl-Sephadex C-25 (Pharmacia) equilibrated in 0.1 *M* ammonium formate, pH 6.5. Adsorbed anaphylatoxin is

[23] C. Gerard, D. E. Chenoweth, and T. E. Hugli, *J. Reticuloendothel. Soc.* **26,** 711 (1979).
[24] C. G. Cochrane and H. J. Müller-Eberhard, *J. Exp. Med.* **127,** 371 (1986).
[25] N. P. Stimler, M. K. Bach, C. M. Bloor, and T. E. Hugli, *J. Immunol.* **128,** 2247 (1981).

eluted with 1 *M* ammonium formate, pH 6.5. While virtually all of the ultraviolet-absorbing material in the renatured supernatant passes through the column unretained, approximately 80% of the C5a-specific smooth muscle contractile activity is adsorbed. This fraction is next gel filtered over a column (1.4 × 77 cm) of BioGel P-10 (200–400 mesh) equilibrated in 0.1 *M* ammonium formate, pH 5.5. The column is monitored for protein by following the absorbance at 280 nm, and fractions are assessed for ^{131}I-labeled C5a with a Packard gamma counter.

Chemical Characterization of RC5a

Gel filtration chromatography of the material recovered from the cation-exchange resin over BioGel P-10 yields homogeneous RC5a (Fig. 4A). The purified RC5a is subjected to cellulose acetate electrophoresis as recommended by the manufacturer (Microzone, Beckman) (Fig. 5). The electrophoretic mobility of the purified RC5a is assessed on strips of cellulose acetate at pH 8.6 in barbital buffer as previously described.[20] A single cationic species is detected which has identical electrophoretic mobility to enzymatically deglycosylated human C5a,[21] suggesting that the *N*-formylmethionyl initiating residue has been removed and that the carboxy-terminal arginine-74 is present.

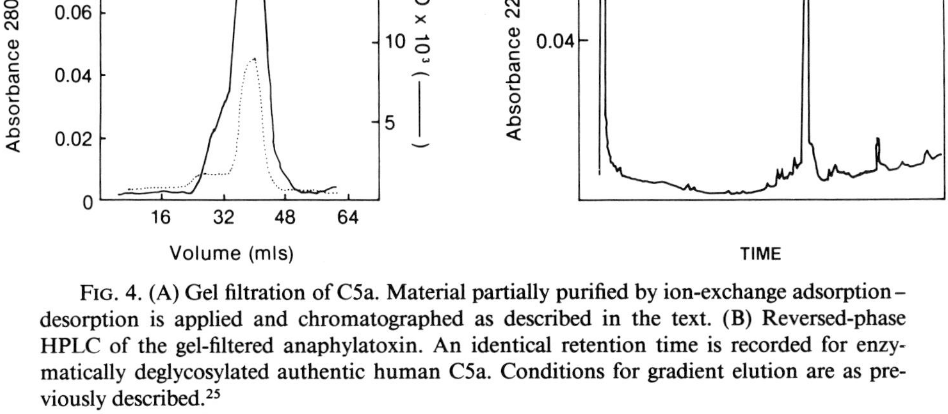

FIG. 4. (A) Gel filtration of C5a. Material partially purified by ion-exchange adsorption–desorption is applied and chromatographed as described in the text. (B) Reversed-phase HPLC of the gel-filtered anaphylatoxin. An identical retention time is recorded for enzymatically deglycosylated authentic human C5a. Conditions for gradient elution are as previously described.[25]

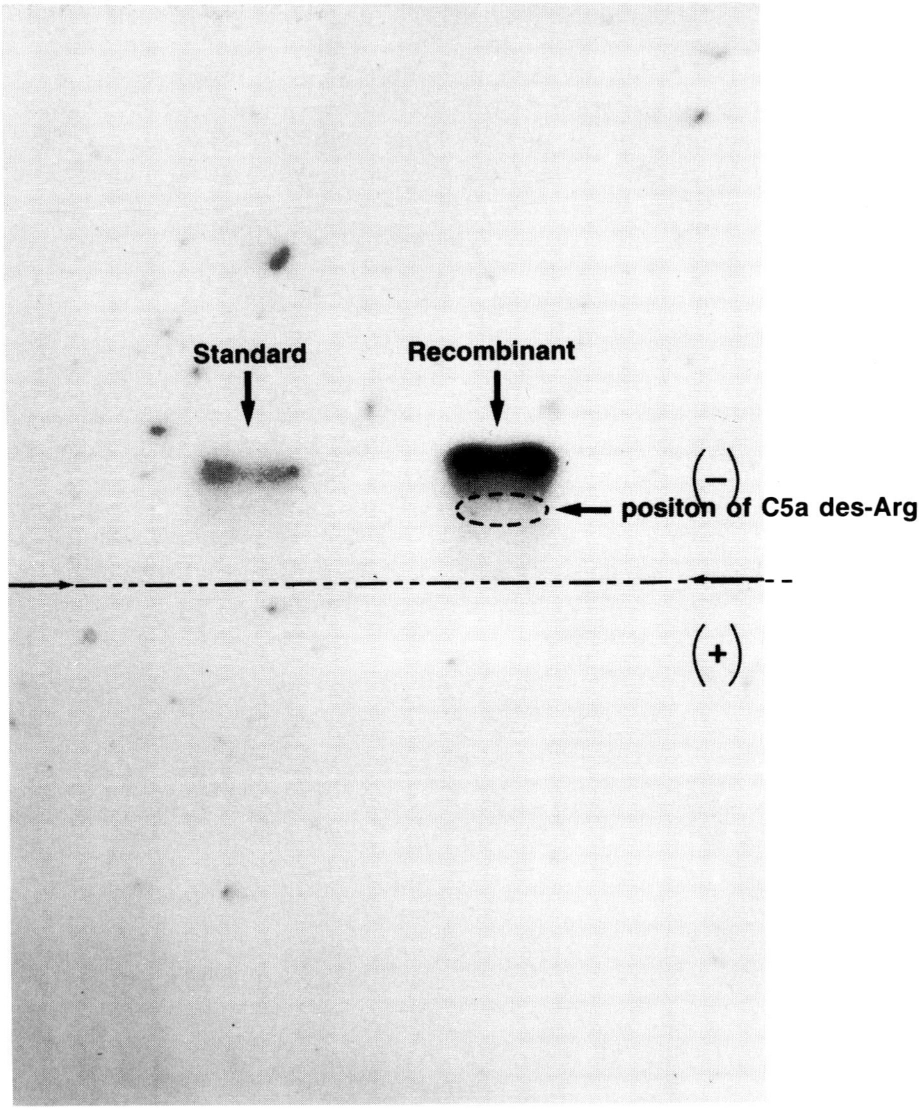

FIG. 5. Cellulose acetate electrophoresis of the protein recovered from gel filtration. Left lane, enzymatically deglycosylated human C5a; right lane, human C5a purified from bacterial extracts.

Human C5a produced by *E. coli* and recovered from the gel filtration is subjected to analytical reversed-phase HPLC using the system previously described by Gerard *et al.*[26] (Fig. 4B). Analytical reversed-phase HPLC is performed on a C_{18} column (Synchropak, RP-P, 100×4.1 mm) in 0.1% TFA in water containing 6% CH_3CN with gradient elution to 30% CH_3CN. The amino acid composition of the RC5a is indistinguishable from that of authentic human C5a and allows calculation of the overall yield of 20–30% for the purification. Protein samples (5 μg) are hydrolyzed in constant boiling HCl *in vacuo* overnight at 10°, derivatized with phenylisothiocyanate, and quantitated using Pico-Tag reversed-phase HPLC as described by the manufacturer (Waters Associates). When 2 nmol of the RC5a preparation recovered from the gel filtration step is subjected to five cycles of automated amino-terminal analysis using an Applied Biosystems Gas Phase System, the unique sequence NH_2-Thr-Leu-Gln-Lys is obtained, matching the NH_2-terminal sequence of authentic human C5a.[2] Phenylthiohydantoin amino acid derivatives are quantitated by reversed-phase HPLC. Therefore, the initiating methionine is quantitatively and selectively removed, and no internal proteolysis has occurred in the isolated preparation.

Biological Characterization of the Purified RC5a

The spasmogenic activity of the RC5a expressed by *E. coli* is assessed on strips of guinea pig ileum as described by Cochrane and Müller-Eberhard.[24] Briefly, the tissue is anchored in an organ bath under 1 g tension and attached to an isotonic muscle transducer (Harvard Apparatus, South Natick, MA 01760). The organ bath is filled with Tyrode's buffer, oxygenated with 95% O_2, 5% CO_2, and maintained at 37°. Responses elicited on addition of C5a are recorded, and the minimum concentration required for contraction by RC5a is compared with that for the authentic peptide. In addition, identity of the peptides is confirmed by testing for cross-tachyphylaxis. A segment of ileal tissue, or lung parenchyma as described below, which responds to C5a will not respond to a second challenge of the same peptide at similar concentration; however, it responds normally to all unrelated agonists (i.e., C5a is tachyphylactic for the tissue). Therefore, C5a-specific activity in RC5a is shown by an inability of the preparation to contract tissues previously exposed to authentic C5a, but with activity comparable to authentic C5a on naive tissues. Similarly, the activity of the preparations is tested using isolated parenchymal strips of guinea pig lung as described by Stimler *et al.*[25] Tissues are mounted in organ baths filled

[26] C. Gerard, H. J. Showell, P. D. Hoeprich, T. E. Hugli, and N. P. Stimler, *J. Biol. Chem.* **260,** 2613 (1985).

with Tyrode's buffer at 37°, oxygenated with 95% O_2, 5% CO_2, and affixed to force-displacement transducers under a resting tension of 300 mg (FTO3C, Grass Instrument Co., Quincy, MA). Contractile responses to C5a as a function of concentration are normalized to the maximal acetylcholine response, using a fresh lung strip for each response. In this way dose–response curves are generated for both authentic and recombinant preparations. In addition, cross-tachyphylaxis is tested on these tissues as described above for ileal tissues.

Human C5a possesses a complex oligosaccharide attached to asparagine-64,[17] while the porcine anaphylatoxin is devoid of carbohydrate. We have previously shown that the biological activity of human C5a is unaffected by enzymatic deglycosylation. In agreement with our previous findings, the RC5a, which is devoid of carbohydrate, is indistinguishable from glycosylated human C5a when assayed on strips of guinea pig ileum or lung parenchyma (Fig. 6A). Specific C5a activity in the bacterial preparation is demonstrated in both assays by the degree of desensitization of the strips to subsequent challenges with authentic human C5a. As expected, authentic human C5a and the RC5a are capable of cross-tachyphylaxis.

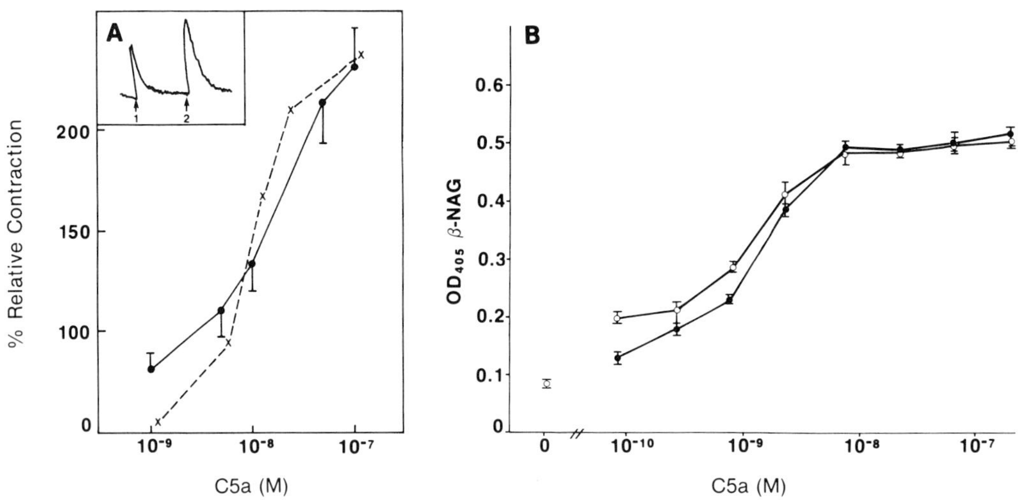

FIG. 6. (A) Dose–response profile for recombinant (●) and authentic human (x) C5a contractile activity on guinea pig lung parenchyman strips. (Inset) Contractile activity of C5a obtained from bacterial extracts on guinea pig ileal strips. Arrow 1, 2 ng/ml histamine; arrow 2, 5×10^{-10} *M* recombinant C5a. (B) Release of *N*-acetyl-β-D-glucosaminidase from rabbit peritoneal PMNs induced by authentic human C5a and RC5a. The lysosomal enzyme release stimulated by human C5a (○) and recombinant C5a (●) is determined as described in the text. The ED_{50} values for authentic and recombinant C5a are 9×10^{-10} and 1.2×10^{-9}, respectively.

In addition, as shown in Fig. 6B, authentic human C5a and RC5a induce the secretion of the lysosomal enzyme *N*-acetyl-β-D-glucosaminidase from rabbit PMNs in the presence of cytochalasin B. C5a-induced lysosomal enzyme release from rabbit and human neutrophils is performed as described.[27] Briefly, rabbit peritoneal exudate cells (98% neutrophils) are obtained by injecting approximately 250 ml of 0.1% oyster glycogen (Type II, Sigma) in 0.9% saline into the peritoneal cavity of female New Zealand White rabbits (3 kg). After 4 hr the cells are harvested, washed, and resuspended in Hanks' balanced salt solution (124 m*M* NaCl, 4.9 m*M* KCl, 0.64 m*M* Na_2HPO_4, 0.52 m*M* KH_2PO_4, 15.1 m*M* $NaHCO_3$, 1.6 m*M* $CaCl_2$ supplemented with 10 m*M* *N*-2-hydroxyethylpiperazine-*N*′-2-ethanesulfonic acid and 1 mg/ml glucose, adjusted to pH 7.2 with 6 *N* HCl) at $0.5-1 \times 10^7$/ml and equilibrated in a shaking water bath at 37° for around 20 min. C5a dilutions (5 μl) and cytochalasin B (25 μl of 100 μg/ml) (Sigma) are combined in glass test tubes (12 × 75 mm) and then incubated at 37° before addition of 0.47 ml neutrophils. After 3 min the test tubes are removed from the water bath, immersed in ice-cold water, and then centrifuged at 4° for 3 min at 400 *g*. The supernatants containing the released lysosomal enzymes are then aspirated and assayed for *N*-acetyl-β-D-glucosaminase. Two tenths milliliter of supernatant is incubated with 0.2 ml of 8 m*M* *p*-nitrophenyl-N-acetyl-β-glucosaminide (Sigma) in 42.5 m*M* sodium acetate, pH 4.5, for 30 min at 37°. The reaction is terminated by the addition of 0.4 ml of 0.4 *M* glycine–NaOH, pH 10.4, and the absorbance is measured spectrophotometrically at 405 m*M*.

Human peripheral leukocytes are isolated using dextran sedimentation of blood. Briefly, 25 ml of heparinized blood is mixed with 20 ml 6% dextran (MW 151,000) (Sigma) in 0.9% saline in a 50-ml plastic syringe and then allowed to stand at room temperature for 30–40 min. The clear, red blood cell-free layer is then removed and diluted 1:4 with Hanks' buffer and centrifuged for 5 min at 800 *g*. Contaminating red cells are lysed with a small volume (5 ml) of 0.15 *M* NH_4Cl and then diluted with Hanks', washed (2 times), and the cells are finally resuspended to 5×10^6/ml. For human neutrophil degranulation experiments the same protocol applies as for rabbit peritoneal neutrophil degranulation experiments. Clearly, there is equal potency between the natural and bacterial source in evoking this response. Table I summarizes the data of replicate runs using rabbit PMNs with human C5a and RC5a and, in addition, includes data generated using deglycosylated human C5a. As discussed above, authentic human C5a with or without the carbohydrate is equally potent in inducing

[27] H. J. Showell, P. H. Naccache, R. W. Walenga, M. Dalecki, M. B. Feinstein. R. I. Sha'afi, and E. L. Becker, *J. Reticuloendothel. Soc.* **30,** 167 (1981).

TABLE I
BIOLOGICAL ACTIVITIES OF AUTHENTIC HUMAN C5a, DEGLYCOSYLATED HUMAN C5a, AND RC5a ON RABBIT AND HUMAN NEUTROPHILS

C5a/Source	Exocytosis, ED_{50}(β-NAG)[a] Rabbit PMNs	Human PMNs
C5a/human	$1.2 \pm 0.5 \times 10^{-9}$ *M* ($n = 4$)	2.6×10^{-9} *M*
C5a-CHO/human	1.6×10^{-9} *M*	2.2×10^{-9} *M*
C5a/*E. coli*	$1.4 \pm 0.2 \times 10^{-9}$ *M* ($n = 4$)	4.4×10^{-9} *M*

[a] Mean ± SD.

degranulation from human PMNs. This is clearly the case also with the C5a prepared from bacterial extracts, but in this instance using rabbit PMNs. Also included in Table I are results from an experiment with human PMNs, which in essence confirm the results obtained from rabbit PMNs.

Acknowledgments

The authors thank Tom C. Crawford for suggesting this collaboration, Alan R. Proctor and L. Fred Wright for advice and valuable discussions, Karen Chan, Maryrose J. Conklyn, Leonard Contillo, Michael E. Eisenhard, and Frank S. Kaczmarek for technical assistance, and Mark O. Lively, Department of Biochemistry, Bowman Gray School of Medicine, in whose lab the protein characterization was performed. This research was supported by Pfizer, Inc. N.S.G. is supported by Grants HL RO-1 36162 and HL KP-4 01777 from the National Institutes of Health.

Author Index

Numbers in parentheses are footnote reference numbers and indicate that an author's work is referred to although the name is not cited in the text.

A

B

C

D

E

F

G

H

I

J

L

M

N

O

P

S

T

U

V

Y

Subject Index

D

E

F

G

H

I

M

Q

R

S

T

U

V

W

X

Z